SIGNALS AND COMMUNICATION TECHNOLOGY

For further volumes:
http://www.springer.com/series/4748

Signals and Communication Technology

Multimodal User Interfaces
From Signals to Interaction
D. Tzovaras ISBN 978-3-540-78344-2

Human Factors and Voice Interactive Systems
D. Gardner-Bonneau, H.E. Blanchard (Eds.)
ISBN 978-0-387-25482-1

Wireless Communications
2007 CNIT Thyrrenian Symposium
S. Pupolin (Ed.) ISBN 978-0-387-73824-6

Satellite Communications and Navigation Systems
E. Del Re, M. Ruggieri (Eds.)
ISBN 978-0-387-47522-6

Digital Signal Processing
An Experimental Approach
S. Engelberg ISBN 978-1-84800-118-3

Digital Video and Audio Broadcasting Technology
A Practical Engineering Guide
W. Fischer ISBN 978-3-540-76357-4

Three-Dimensional Television
Capture, Transmission, Display
H.M. Ozaktas, L. Onural (Eds.)
ISBN 978-3-540-72531-2

Foundations and Applications of Sensor Management
A.O. Hero, D. Castañón, D. Cochran, K. Kastella (Eds.) ISBN 978-0-387-27892-6

Digital Signal Processing with Field Programmable Gate Arrays
U. Meyer-Baese ISBN 978-3-540-72612-8

Adaptive Nonlinear System Identification
The Volterra andWiener Model Approaches
T. Ogunfunmi ISBN 978-0-387-26328-1

Continuous-Time Systems
Y.S. Shmaliy ISBN 978-1-4020-6271-1

Blind Speech Separation
S. Makino, T.-W. Lee, H. Sawada (Eds.)
ISBN 978-1-4020-6478-4

Cognitive Radio, Software Defined Radio, and AdaptiveWireless Systems
H. Arslan (Ed.) ISBN 978-1-4020-5541-6

Wireless Network Security
Y. Xiao, D.-Z.Du, X. Shen
ISBN 978-0-387-28040-0

Terrestrial Trunked Radio – TETRA
A Global Security Tool
P. Stavroulakis ISBN 978-3-540-71190-2

Multirate Statistical Signal Processing
O.S. Jahromi ISBN 978-1-4020-5316-0

Wireless Ad Hoc and Sensor Networks
A Cross-Layer Design Perspective
R. Jurdak ISBN 978-0-387-39022-2

Positive Trigonometric Polynomials and Signal Processing Applications
B. Dumitrescu ISBN 978-1-4020-5124-1

Face Biometrics for Personal Identification
Multi-Sensory Multi-Modal Systems
R.I. Hammoud, B.R. Abidi, M.A. Abidi (Eds.)
ISBN 978-3-540-49344-0

Cryptographic Algorithms on Reconfigurable Hardware
F. Rodríguez-Henríquez ISBN 978-0-387-33883-5

Ad-Hoc Networking Towards Seamless Communications
L. Gavrilovska ISBN 978-0-4020-5065-7

Multimedia Database Retrieval
A Human-Centered Approach
P. Muneesawang, L. Guan
ISBN 978-0-387-25627-6

Broadband Fixed Wireless Access
A System Perspective
M. Engels; F. Petre ISBN 978-0-387-33956-6

Acoustic MIMO Signal Processing
Y. Huang, J. Benesty, J. Chen
ISBN 978-3-540-37630-9

Algorithmic Information Theory
Mathematics of Digital Information Processing
P. Seibt ISBN 978-3-540-33218-3

Continuous-Time Signals
Y.S. Shmaliy ISBN 978-1-4020-4817-3

Interactive Video
Algorithms and Technologies
R.I. Hammoud (Ed.) ISBN 978-3-540-33214-5

Handover in DVB-H
Investigation and Analysis
X. Yang ISBN 978-3-540-78629-0

Building Automation:
Communication Systems with EIB/KNX, LON and BACnet
H. Merz, T. Hansemann, C. Hübner
ISBN 978-3-540-88828-4

LDPC Coded Modulations
M. Franceschini, G. Ferrari, R. Raheli
ISBN 978-3-540-69455-7

Walter Fischer

Digital Video and Audio Broadcasting Technology

A Practical Engineering Guide

Third Edition

Springer

Dipl. Ing. (FH) Walter Fischer
Rohde & Schwarz GmbH & Co. KG
Geschäftsbereich Meßtechnik
Mühldorfstr. 15
81671 München
Germany
Walter.Fischer@Rohde-Schwarz.com

Translator:
Horst von Renouard
36 Chester Road
DA 15 8S6 Sidcup, Kent
United Kingdom

ISSN 1860-4862
ISBN 978-3-642-11611-7 e-ISBN 978-3-642-11612-4
DOI 10.1007/978-3-642-11612-4
Springer Heidelberg Dordrecht London New York

Library of Congress Control Number: 2010923775

Originally published with the title Digital Television

Cover design: WMXDesign GmbH, Heidelberg

Printed on acid-free paper

Springer is part of Springer Science+Business Media (www.springer.com)

Preface

It is not so long ago that the second English edition of this book appeared. In many countries, the switch-over from analog to digital television has now been completed, especially in terrestrial television. Everyone is talking about high-definition TV - HDTV - which is supported by virtually every TV display on the market today. The reason that the HDTV supply chain is not as yet gapless is not to be found in the technology but only in the lack of available HD source material. However, this is expected to change from 2010 onward. The number of studios which are being re-equipped with the new technology is ever increasing. Suitable compression standards for HDTV have been around for years and there is now also sufficient bandwidth available with the second generation of DVB transmission standards.

In comparison with digital television, digital audio broadcasting - DAB - still has its problems. Although DAB is on the air in many countries, it is still largely unknown to the general public, Good old FM radio is still the Number One audio transmission medium. It will be interesting to find out what will happen in this respect in the next few years. Will DVB also "gobble up" DAB?

What is now the third edition of the English version has been updated further to match the facts of present conditions. This book contains all the modern source encoding standards for digital television and digital audio broadcasting. Nevertheless, one or the other areas will still seem to be slightly inadequate to the liking of some readers. As the author, may aim is not simply to copy standards - which I may even have misunderstood in one point or the other - as the standards themselves are publicly available and more or less comprehensible in their form although their interpretation sometimes presents problems. Rather, I am mainly concerned with passing on the knowledge I have actually acquired myself.

The fact that this book has now been published in four languages and thus has found unexpectedly wide circulation all over the world is my greatest compensation for all the work done since 2001. The depth of technical knowledge presented in many chapters would have been impossible to achieve without the numerous discussions with my colleagues from the Broadcasting Division at Rohde&Schwarz. Many more impulses also came from many readers and participants in seminars, and my special thanks are due to the team of my publishers, Springer Verlag and to my translator, collaboration with whom had been outstanding.

Moosburg an der Isar, near Munich, August 2009

Walter Fischer

Preface to the Second English Edition

A few years have passed since the first English edition of this book appeared in the bookshops. Digital television has become a fact of life in many countries, conveyed to the viewer either by satellite, by cable or terrestrially through the rooftop antenna, and there are now also first indications of a fourth distribution path through IPTV, television by Internet, not to forget Mobile TV which is being mentioned more and more frequently in advertising. All these are reasons why it became necessary to update and expand much of the book. But there are also some new chapters such as DAB, Data Broadcasting, Mobile TV in the form of DVB-H, and T-DMB, DRM, etc. Sections on modern source encoding methods such as MPEG-4 have also been amended, incorporating many suggestions from readers and participants in seminars.

My previous publications "Digital Television - A Practical Guide for Engineers" and "Digitale Fernsehtechnik in Theorie und Praxis" have found warm acceptance by a wide circle of readers and both works have also been used as welcome support material in numerous seminars.

My lectureship in the subject of "Television Engineering" at the Munich University of Applied Science, which I am carrying on in the spirit of Prof. Mäusl's lectures on the subject, is also providing me with new impulses in the ways in which knowledge of the subject can be imparted and in the selection of contents, whilst at the same time enriching my own experience.

Since the last edition, many new findings and experiences have been gathered by myself in many seminars throughout the world but also by me personally participating when the DVB-T networks were being switched on in Bavaria. Some of these findings and experiences will be found again in this book.

Many thanks to my publishers, Springer Verlag, especially to Dr. Merkle and Mrs. Jantzen, and to Horst von Renouard, the translator of this book, and to my colleagues from Rohde&Schwarz, for their excellent collaboration in producing the finished book.

Moosburg an der Isar, near Munich, August 2007

Walter Fischer

Preface to the Second English Edition

It is now three years since the first English edition of this book appeared on the bookshelves. [illegible] has become a fact [illegible] [illegible], [illegible] in the [illegible] either by [illegible] [illegible] [illegible] [illegible] [illegible] and there are [illegible] and [illegible]

[illegible] [illegible]

[illegible]

Mönchengladbach [illegible] [illegible] Munich, August 200[illegible]

[illegible]

Preface to the First English Edition

The world of television engineering has long fascinated me and from the day I wrote my diploma paper on "The Generation of Test Lines" at the Fachhochschule München (Munich University or Applied Sciences) under Prof. Rudolf Mäusl in 1983, it has never released its grip on me. My research for this paper led to contacts with Rohde&Schwarz who where subsequently to become my employers. I worked there as development engineer until 1999, always in video test engineering but within various fields of products and activities. For many years, this activity involved analog video testing and there mainly video insertion test signals (VITS), but from the mid-nineties onward, the focus shifted more and more to MPEG-2 and digital video broadcasting (DVB) and then quite generally to the field of digital television. Naturally, as a consequence of my work as a development engineer I also became intensively engaged in the field of firmware and software development and my involvement with the programming language C and C++ led me into the domain of software training where I was increasingly active in-house from the early nineties onward. I have lost count of the number of seminars and of the participants in these seminars who succeeded in implanting in me a joy in this type of "work". Whatever the cause, it was in the course of these, perhaps forty seminars that I discovered my love for instructing and in 1999 I chose this to be my main occupation. Since March 1999, I have been active as instructor in the field of television engineering, main subject "digital television", in the Rohde&Schwarz Training Center. Since then, I have travelled more than 500,000 km by air all over the world, from Stockholm to Sydney, to provide instruction about the new field of digital television and especially about test engineering and transmitter technology.

A key event in my professional life has been a seminar request from Australia in July 1999 which resulted in, thus far, 7 trips to Australia with a total stay of about half a year, more than 50 seminar days and almost 400 participants. From this has sprung a love for this far-distant, wonderful continent which, I am sure, will be apparent between the lines throughout this book. One of the main suggestions to write this book as a résumé of my seminars came from the circle of participants in Australia. These trips gave rise to significant impulses and I have gained a large amount of practical experience during my seminars "Down Under" and during the construction of their DVB-T network, which proved to be invaluable in the creation of this book. I owe special thanks to my colleague, Simon Haynes from Rohde&Schwarz Australia, who provided me with the closest support for the seminars and with helpful suggestions for this book. We often talked about publishing the contents of the seminars but I had underestimated the effort involved. The original documentation for the seminars did not easily lend itself to being directly for the book. Virtually all the texts had be completely revised, but now I had plenty to occupy me during the 100 days or so of travelling a year, even at night, an important factor with all the boredom of being absent from home.

My readers will be people who have a practical interest in the new subject of "Digital Television", engineers and technicians who want to or have to familiarize themselves with this new field and the book, therefore, contains only a minimum ballast of mathematics although, by the nature of thins, there have to be some.

In the meantime, I have been able to extend my seminar travels to other countries as, for example, Greenland, and to gather numerous impressions there, too. However, although it is very nice to see the world as a result of one's professional activities, it is not easy for one's family or for oneself, for that matter. For this reason, I would like to take this opportunity to express special thanks to those who had to stay at home for whom I was then not available. To some extent this also applies to the time when this book was written. In particular, I thank my daughter Christine for her help in writing the manuscript.

I would like to thank Horst von Renouard from London for his successful translation. As chance would have it, he, too, had spent many years in Australia and also comes from the field of television engineering. He thus was able to empathize with what I was trying to express and to convey this in his translation. And while I am on the subject of translation, my gratitude is due also to the Rohde&Schwarz Translation Department who also contributed some chapters which were required in advance for seminar purposes.

To my former patron, Prof. Rudolf Mäusl, who initiated me into the world of television engineering as no-one else could have done, my heartfelt thanks four our many conversations and for all his helpful suggestions. His lectures at the Fachhochschule and his way of imparting knowledge have always been of guiding influence on me and, I hope, have also been a positive influence on how this book has turned out. His many publications and books are models in their field and can only be recommended.

Many thanks also to my publishers, Springer Verlag, to Dr. Merkle, Mrs. Jantzen and Mrs. Maas for their active support, and for the opportunity to have this book published by this renowned publishing house.

And many thanks for the many discussions and suggestions by the participants in my seminars throughout the world, in Australia, Austria, Canada, the Czech Republic, France, Germany, Greenland, Latvia, Mexico, the Netherlands, Portugal, Spain, Sweden, Switzerland, Turkey, the United States and all the other countries in which I have been or from which participants have come to Munich or elsewhere in order to join me in finding out about the complex subject of digital television.

To the present day, there have been worldwide seminars on the subject of analog and digital television for just on 300 days, with about 2000 participants from all corners of the world. These international seminars present a rich personal experience and I am filled with gratitude at the many contacts made, some are still ongoing via email.

Moosburg an der Isar, near Munich, June 2003

Walter Fischer

Foreword

Without a doubt, this book can definitely be called a reference work and is a true "Engineering Guide to Digital Television". Walter Fischer is an outstandingly knowledgeable author and an expert in his chosen field. I have known him since the beginning of the eighties when he attended my lectures at the Fachhochschule München (Munich University of Applied Sciences). He attracted attention even then with his excellent knowledge and with the way he tackled new and complex problems. After he had concluded his studies, continuing contacts with my erstwhile employer Rohde & Schwarz then provided him with the opportunity to give free rein to his talent in their Department of Television Test Engineering.

In 1988 the Fernseh- und Kinotechnische Gesellschaft (Television and Cinematographic Association) awarded him their Rudolf Urtel Price for independently developing a test method for determining the parameters of a video channel by means of the Fast Fourier Transform (FFT).

After a long period of developing test instruments for analog and digital television signals and equipped with the extensive knowledge in digital television practice gained from this, he finally realized his long-standing ambition to change over into the field of teaching. For some years now he has been active for the Rohde&Schwarz Training Center and is passing on this knowledge in seminars all over the world. I may add that I, too, have been able to benefit from Walter Fischer's expertise in my own relatively brief volume on digital television.

I wish Walter Fischer continuing success, particularly with regard to a good acceptance of this reference work throughout the world.

Aschheim near Munich, February 2003

Professor Rudolf Mäusl

Note from the Translator

When I was first asked to translate Walter Fischer's book on digital television a few years ago, digital television was a relatively new concept for me who had 'grown up' in the age of valves and analog black-and-white, and later color, television. I had been a television technician with the Australian Broadcasting Corporation, the ABC, and later, in 1969, operated the TV scan converter at the Honeysuckle Creek tracking station near Canberra in Australia that brought live television of the Apollo 11 moon landing beamed down from the moon over a distance of more than 350,000 kilometers to a world audience ("TELE-vision", seeing at a distance, indeed!).

That was over 40 years ago, narrow-band (to save transmission power) analog black-and-white slow scan television (SSTV) at 10 frames per second that was recorded on a large-diameter magnetic disk from where each frame was read out 5 times to produce the 60 frames-a-second NTSC-standard broadcast picture which was relayed to the world (not "done in a shed", either, as some people want you to believe today - that SSTV signal came into our receivers from the moon as verified by our ranging pulse which indicated that the source of the signal was, indeed, some 250,000 miles distant, and by the vidicon camera mounted on the receiving antenna dish which showed us that the antenna was pointing at the moon).

After the Apollo and Skylab days I left engineering to pander to my second love - languages, combining the two to become a technical translator. But television has long been in my blood, and when Walter Fischer asked me to translate his supremely accomplished reference work on digital television, I felt singularly honoured and, I admit, flattered. In the end, its success has justified our combined efforts and, I feel sure, will continue to do so.

London, December 2009

Horst E. von Renouard

Table of Contents

1 Introduction

For many decades, television and data transmission have followed parallel paths which, however, were completely independent of one another. Although television sets were used as first home computer monitors back in the eighties of the century now past, this was the only interaction between the two fields. Today, however, it is becoming more and more difficult to distinguish between the two media of TV and computers which are converging increasingly in this age of multimedia. There are now excellent TV cards for PCs so that the PC can easily become another TV set. On the other side, teletext was introduced back in the eighties to provide an early medium for supplementary digital information in analog TV. For young people, this type of information is such a natural part of viewing, e.g. as electronic program guide, as if there had been teletext from the beginnings of television.

And now we are living in the age of digital TV, since 1995 in fact, and the distinction between data and television has virtually disappeared. When one is able to follow the developments in this field throughout the world like the author has on numerous seminar trips, one will encounter more and more applications where either both television and data services are found jointly in one data signal, or the services are even just pure data services, e.g. fast Internet access via channels which were actually provided for digital TV. The common factor leading to this fusion is the high data rate. Today's generation is hungry for information and used to getting it in large quantity and variety. Talking to telecommunication specialists about data rates, one hears time and again how envious they are of the data rates used in digital TV. Thus GSM, for example, works with data rates of 9600 bit/sec and UMTS uses a maximum of 2 Mbit/sec under optimum conditions, e.g. for Internet accesses. An ISDN basic access telephone channel has two times 64 kbit/sec. By comparison, the data rate of an uncompressed digital Standard Definition TV signal is already 270 Mbit/sec and High Definition TV begins at about 1.5 Gbit/s and extends into the 3 Gigabit range. One would be fully justified to call television a broadband technology, not only from the point of view of digital TV but even in analog TV where the channels have always been very wide. An analog or digital terrestrial TV channel has a width of 6, 7 or 8 MHz and the chan-

W. Fischer, *Digital Video and Audio Broadcasting Technology*, Signals and Communication Technology, 3rd ed., DOI: 10.1007/978-3-642-11612-4_1, © Springer-Verlag Berlin Heidelberg 2010

nels broadcast via satellite are even 36 MHz wide. It is not surprising that a new boom is being experienced especially in broadband TV cable, which is being used as a medium for high-speed home Internet access in the Mbit/sec range, uplinking via cable modems.

The foundation stone for analog television was laid by Paul Nipkow back in 1883 when he developed what is now known as the Nipkow disc. He had the idea of transmitting a picture by disecting it into lines. The first real analog TV transmissions per se took place in the thirties but, held back by World War II, analog television didn't have its proper start until the fifties, in black and white at first. The television set acquired color towards the end of the sixties and from then on, this technology has been basically only refined, both in the studio and in the home. There have been no further changes in the principles of the technology. Analog TV transmissions are often so perfect, at least in quality if not in content, that it is difficult to interest many people in buying a receiver for digital TV.

In the eighties, an attempt was made to depart from traditional analog TV by way of D2MAC. For various reasons, this did not succeed and D2MAC vanished from view again. In Europe, the PAL system was given a slight boost by the introduction of PALplus but this, too, did not achieve much success in the TV set market, either. At the same time, various approaches were tried, mainly in Japan and in the US, to achieve success with the transmission of HDTV, but these also failed to gain the universal popular appeal hoped for.

In the studio, digital television signals have been used since the beginning of the nineties as uncompressed digital TV signals conforming to "CCIR 601". These data signals have a date rate of 270 Mbit/sec and are highly suitable for distribution and processing in the studio, and are very popular today. But they are not at all suitable for broadcasting and transmission to the end user. The channel capacities available via cable, terrestrial channels and satellite would not be even nearly adequate enough for these signals. In the case of HDTV signals, the data rate is about 1.5 Gbit/sec uncompressed. Without compression, these signals could not be broadcast.

The key event in the field of digital television can be considered to be the establishment of the JPEG standard. JPEG stands for Joint Photographic Experts Group, a group of experts specializing in still frame compression. It was here that the discrete cosine transform (DCT) was used for the first time for compressing still frames towards the end of the eighties. Today, JPEG is a commonly used standard in the data field and is being used very successfully in the field of digital photography. Digital cameras are experiencing quite a boom and are becoming better and better so that this medium has replaced traditional photography in many areas.

The DCT also became the basic algorithm for MPEG, the Motion Picture Experts Group, which developed the MPEG-1 standard by 1993 and the MPEG-2 standard by 1995. The aim of MPEG-1 was to achieve the reproduction of full-motion pictures at data rates of up to 1.44 Mbit/sec, using the CD as a data medium. The aim for MPEG-2 was higher and MPEG-2, finally, was to become the baseband signal for digital television world-wide. Initially, only Standard Definition Television (SDTV) was provided for in MPEG-2, but High Definition Television (HDTV) was also implemented which was apparently originally intended for MPEG-3. However, there is no MPEG-3 (nor does it have anything to do with MP3 files, either). In MPEG-2, both the MPEG data structure was described (ISO/IEC 13818-1) and a method for full-motion picture compression (ISO/IEC 13818-2) and for audio compression (ISO/IEC 13818-3) defined. These methods are now used throughout the world. MPEG-2 allows the digital TV signals of originally 270 Mbit/sec to be compressed to about 2 to 7 Mbit/sec. The uncompressed data rate of a stereo audio signal of about 1.5 Mbit/sec, too, can be reduced to about 100 to 400 kbit/sec, typically to 192 kbit/s. As a result of these high compression factors it is now possible even to combine a number of programs to form one data signal which can then be accommodated in what was originally an e.g. 8-MHz-wide analog TV channel.

In the meantime, there is MPEG-4, MPEG-7 and MPEG-21.

At the beginning of the nineties, Digital Video Broadcasting (DVB) was then created as a European project. In the course of this project, several transmission methods were developed: DVB-S, DVB-C and DVB-T. The satellite transmission method DVB-S has been in use since about 1995. Using the QPSK method of modulation and with channel bandwidths of about 33…36 MHz, a gross data rate of 38 Mbit/sec is possible with satellite transmission. With approximately 6 Mbit/sec per program, up to 6, 8 or even 10 programs can now be transmitted in one channel depending on data rate and content and when mainly audio programs are broadcast, more than 20 programs are often found in one channel. In the case of DVB-C, transmitted via coaxial cable, the 64QAM modulation also provides a data rate of 38 Mbit/sec at a bandwidth of only 8 MHz. Current HFC (hybrid fibre coax) networks now allow data rates of more than 50 Mbit/s per channel. DVB-C, too, has been in use since about 1995. The digital terrestrial TV system DVB-T started in 1998 in Great Britain in 2K mode and is now available nationwide. This terrestrial path to broadcasting digital TV signals is being used more and more, spreading from the UK, Scandinavia and Spain all the way to Australia. DVB-T provides for data rates of between 5 to 31 Mbit/sec and the data rate actually used is normally about 22 to 22 Mbit/sec if a DVB-T network has been designed for roof antenna reception, or about 13 to 15 Mbit/sec for portable indoor use. Germany was changing, region by region, from analog terrestrial TV to DVB-T. This change-over was completed in Germany at the end of 2008.

In North America, other methods are in use. Instead of DVB-C, a very similar system which conforms to ITU-J83B is used for cable transmission. Terrestrial transmission makes use of the ATSC method where ATSC stands for Advanced Television System Committee. In Japan, too, other transmission methods are used, such as ITU-J83C for cable transmission, again very similar to DVB-C (which corresponds to ITU-J83A), and the ISDB-T standard for terrestrial transmission. Yet another terrestrial transmission system is being developed in China. The common factor for all these methods is the MPEG-2 baseband signal.

In 1999, another application was given the green light, namely the digital versatile disc, or DVD. The video DVD also uses an MPEG-2 data stream with MPEG video and MPEG or Dolby Digital audio.

In the meantime, the range of digital television has been extended to mobile reception with the development of standards for use with mobile telephones, designated as DVB-H (Digital Video Broadcasting for Handhelds) and T-DMB (Terrestrial Digital Multimedia Broadcasting) and CMMB.

This book deals with all present-day TV compression and transmission methods, i.e. MPEG, DVB, ATSC, ISDB-T and DTMB. The video DVD is also discussed to some extent. The discussion is focused on dealing with these subjects in as practical a way as possible. Although mathematical formulations are used, they are in most cases only utilized to supplement the text. The mathematical ballast will be kept to a minimum for the practical field engineer. This has nothing to do with any possible aversion the author may have against mathematics. Quite on the contrary. In the course of many seminars involving thousands of participants throughout the world, forms of presentation were developed which have contributed to a better and easier understanding of these in some cases highly complex subjects. The book also contains chapters dealing with basic concepts such as digital modulation or transformations into the frequency domain, some of which can be skipped by a reader if he so desires. Experience has shown, however, that it is better to read these chapters, too, before starting with the actual subject of digital television. A major emphasis is placed on the measuring techniques used on these various digital TV signals. Necessary and appropriate measuring techniques are discussed in detail and practical examples and hints are provided.

Table 1.1. Digital video and audio broadcasting standards and methods

Method/Standards	Application
JPEG	Still picture compression, photography
Motion JPEG	MiniDV, digital home video cameras
MPEG-1	Video on CD
MPEG-2	Baseband signal for digital television, DVD Video
MPEG-4	New codecs for video and audio compression
DVB	Digital Video Broadcasting
DVB-S	Digital Video Broadcasting via satellite
DVB-C	Digital Video Broadcasting via broadband cable
DVB-T	Digital Terrestrial Video Broadcasting
DVB-H	Digital Video Broadcasting for Handhelds; mobile TV standard
MMDS	Multipoint Microwave Distribution System, local terrestrial multipoint transmission of digital television to supplement broadband cable
ITU-T J83A	ITU equivalent of DVB-C
ITU-T J83B	US cable standard
ITU-T J83C	Japanese cable standard
ATSC	Standard for digital terrestrial television (US, Canada)
ISDB-T	Japanese standard for digital terrestrial television
DTMB	Chinese standard for digital terrestrial television
DAB	Digital Audio Broadcasting; standard for digital terrestrial audio broadcasting
DRM	Digital Radio Mondiale; standard for digital terrestrial audio broadcasting
T-DMB	Terrestrial Digital Multimedia Broadcasting; mobile TV standard
DVB-SH	DVB for handheld mobile terminals, satellite and terrestrial, hybrid standard for satellite and terrestrial broadcasting
DVB-S2	Second Generation DVB for Satellite
DVB-T2	Second Generation DVB Terrestrial
DVB-C2	Second Generation DVB for Cable
CMMB	Chinese Mobile Multimedia Broadcasting

Note: Many of the terms listed in the table are protected by copyright ©

As far as possible, practical findings and experiences have been incorporated time and again in the individual chapters. In some cases it will be possible to recognize one or the other experience of the author on his travels. Particularly extensive practical insights were gained especially far away from Europe in Australia during the introductory phase of DVB-T and were written down in this book. But it is not intended to be a travel guide to Australia or the world even though it would be very interesting to tell about these areas and many beautiful locations where digital television is being newly introduced. The content of this book is structured in such a way that it starts with the analog TV baseband signal and then continues with a discussion of the MPEG-2 data stream, digital video, digital audio and the compression methods. After an excursion into the digital modulation methods, all the transmission methods like DVB-S, DVB-C, ITU-J83ABC, DVB-T, ATSC and ISDB-T are discussed in detail. Interspersed between these are found the chapters on the relevant measuring technique. As well, transmission methods based on Digital Audio Broadcasting (DAB) are being discussed. The book deals more intensively with the subject of "digital audio broadcasting" with the DRM (Digital Radio Mondiale) audio transmission standard and the possibilities of transmitting digital audio also by DVB. Since it is no longer possible to separate the subjects of television, audio broadcasting, in any case, the title of the book has been changed to "Digital Video and Audio Broadcasting Technology, A Practical Engineering Guide" from its second edition. Television still claims the greater part, however. Part of the reason for this is that digital audio broadcasting is still having problems with its practical implementation. It will be interesting to see what will be the end result of the competition between DAB and DVB-T2.

The methods and standards relating to the subject of “digital television and digital audio broadcasting” and discussed in this book are listed in Table 1.1. In the meantime, new standards such as DVB-SH, DVB-T2 and DVB-C2 have appeared which are also described as far as possible even if there is only little practical experience available as yet. It should also not be forgotten that digital television and digital audio can also be transmitted via the Internet and IPTV is also mentioned briefly. Digital SDTV is now a reality when it was only being introduced at the time the first English edition of this book appeared. It is now only necessary for HDTV to become reality, and that not only by the complete range of technology now available but also by widely provided contents, i.e. available programs.

Bibliography: [ISO13818-1], [ISO13818-2], [ISO13818-3], [ETS300421], [ETS300429], [ETS300744], [A53], [ITU205], [ETS300401], [ETS101980]

2 Analog Television

Throughout the world, there are only two major analog television standards, the 625-line system with a 50 Hz frame rate and the 525-line system with a 60 Hz frame rate. The composite color video-and-blanking signal (CVBS, CCVS) of these systems is transmitted in the following color transmission standards:

- PAL (Phase Alternating Line)
- NTSC (National Television System Committee)
- SECAM (Séquentiel Couleur a Mémoire)

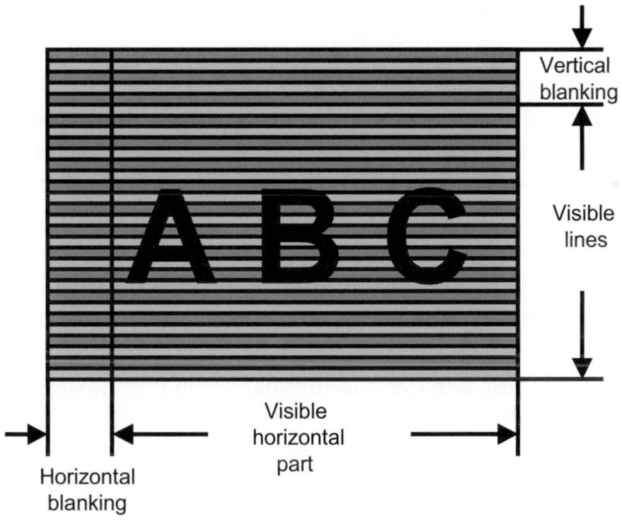

Fig. 2.1. Dividing a frame into lines

PAL, NTSC and SECAM color transmission is possible in 625-line systems and in 525-line systems. However, not all the possible combinations have actually been implemented. The video signal with its composite coding is then modulated onto a carrier, the vision carrier, mostly with negative-going amplitude modulation. It is only in Std. L (France) that positive-

W. Fischer, *Digital Video and Audio Broadcasting Technology*, Signals and Communication Technology, 3rd ed., DOI: 10.1007/978-3-642-11612-4_2, © Springer-Verlag Berlin Heidelberg 2010

going modulation (sync inside) is used. The first and second sound subcarrier is usually an FM-modulated subcarrier but an amplitude-modulated sound subcarrier is also used (Standard L, France). In Northern Europe, the second sound subcarrier is a digitally modulated NICAM subcarrier. Although the differences between the methods applied in the various countries are only minor, together they result in a multiplicity of standards which are mutually incompatible. The analog television standards are numbered through alphabetically from A to Z and essentially describe the channel frequencies and bandwidths in VHF bands I and III (47 ... 68 MHz, 174 ... 230 MHz) and UHF bands IV and V (470 ... 862MHz); An example is Standard B, G Germany: B =7 MHz VHF, G = 8 MHz UHF.

In the television camera, each field is dissected into a line structure of 625 or 525 lines. Because of the finite beam flyback time in the television receiver, however, a vertical and horizontal blanking interval became necessary and as a result, not all lines are visible but form part of the vertical blanking interval. In a line, too, only a certain part is actually visible. In the 625-line system, 50 lines are blanked out and the number of visible lines is 575. In the 525-line system, between 38 and 42 lines fall into the area of the vertical blanking interval.

To reduce the flickering effect, each frame is divided into two fields combining the even-numbered lines and odd-numbered lines in each case. The fields are transmitted alternately and together they result in a field repetition rate of twice the frame rate. The beginning of a line is marked by the horizontal sync pulse, a pulse which is below the zero volt level in the video signal and has a magnitude of -300 mV.. All the timing in the video signal is referred to the front edge of the sync pulse and there exactly to the 50% point. 10 μs after the sync pulse falling edge, the active image area in the line begins in the 625-line system. The active image area itself has a length of 52 μs.

In the matrix in the television camera, the luminance (luminous density) signal (Y signal or black/white signal) is first obtained and converted into a signal having a voltage range from 0 Volt (corresponding to black level) to 700 mV (100% white). The matrix in the television camera also produces the color difference signals from the Red, Green and Blue outputs. It was decided to use color difference signals because, on the one hand, the luminance has to be transmitted separately for reasons of compatibility with black/white television and, on the other hand, color transmission had to conserve bandwidth as effectively as possible. Due to the reduced color resolution of the human eye, it was possible to reduce the bandwidth of the color information. In fact, the color bandwidth is reduced quite significantly compared with the luminance bandwidth: The luminance bandwidth is between 4.2 MHz (PAL M), 5 MHz (PAL B/G) and 6 MHz (PAL D/K, L) whereas the chrominance bandwidth is only 1.3 MHz in most cases.

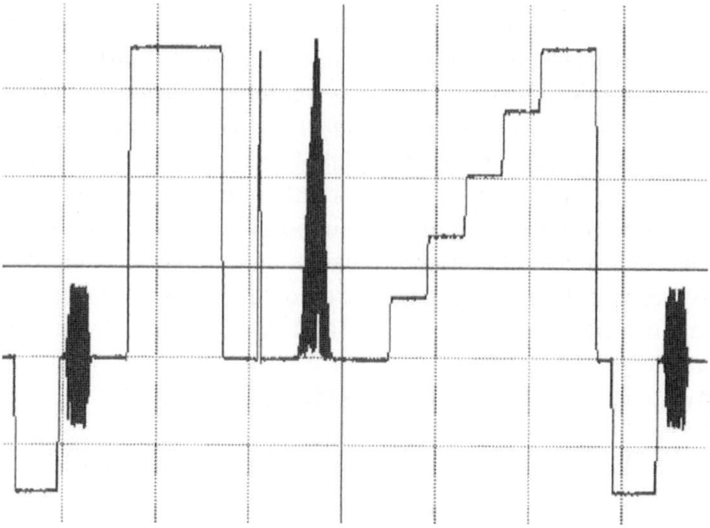

Fig. 2.2. Analog composite video signal (PAL)

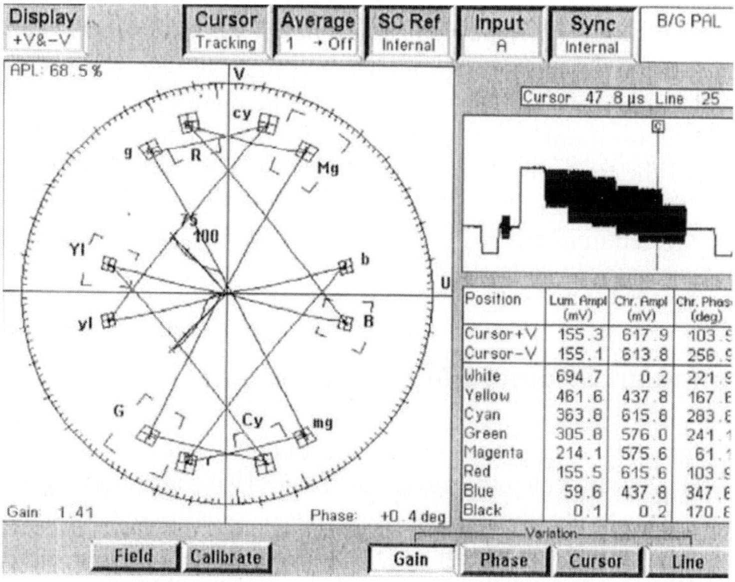

Fig. 2.3. Vector diagram of a composite PAL video signal

In the studio, the color difference signals U=B-Y and V=R-Y are still used directly. For transmission purposes, however, the color difference signals U and V are vector modulated (IQ modulated) onto a color subcarrier in PAL and NTSC. In SECAM, the color information is transmitted frequency-modulated. The common feature of PAL, SECAM and NTSC is that the color information is modulated onto a color subcarrier of a higher frequency which is placed at the upper end of the video frequency band and is simply added to the luminance signal. The frequency of the color subcarrier was selected such that it causes as little interference to the luminance channel as possible. It is frequently impossible, however, to avoid crosstalk between luminance and chrominance and conversely, e.g. if a newsreader is wearing a pinstriped suit. The colored effects which are then visible on the pinstriped pattern are the result of this crosstalk (cross-color or cross-luminance effects).

Vision terminals can have the following video interfaces:

- CVBS, CCVS 75 Ohms 1 V_{PP} (video signal with composite
- coding)
- RGB components (SCART, Peritel)
- Y/C (separate luminance and chrominance to avoid cross color or cross luminance effects)

In the case of digital television, it is advisable to use an RGB (SCART) connection or a Y/C connection for the cabling between the receiver and the TV monitor in order to achieve optimum picture quality.

In digital television only frames are transmitted, no fields. It is only at the very end of the transmission link that fields are regenerated in the set top box or in the decoder of the IDTV receiver. The original source material, too, is provided in interlaced format which must be taken into account in the compression (field coding).

2.1 Scanning an Original Black/White Picture

At the beginning of the age of television, the pictures were only in "black and white". The circuit technology available in the 1950s consisted of tube circuits which were relatively large and susceptible to faults and consumed a lot of power. The television technician was still a real repairman and, in the case of a fault, visited his customers carrying his box of vacuum tubes.

Let us look at how such a black/white signal, the "luminance signal", is produced. Using the letter "A" as an example, its image is filmed by a TV camera which scans it line by line (see Fig. 2.4.). In the early days, this was done by a tube camera in which a light-sensitive layer, onto which the image was projected by optics, was scanned line by line by an electron beam deflected by horizontal and vertical magnetic fields.

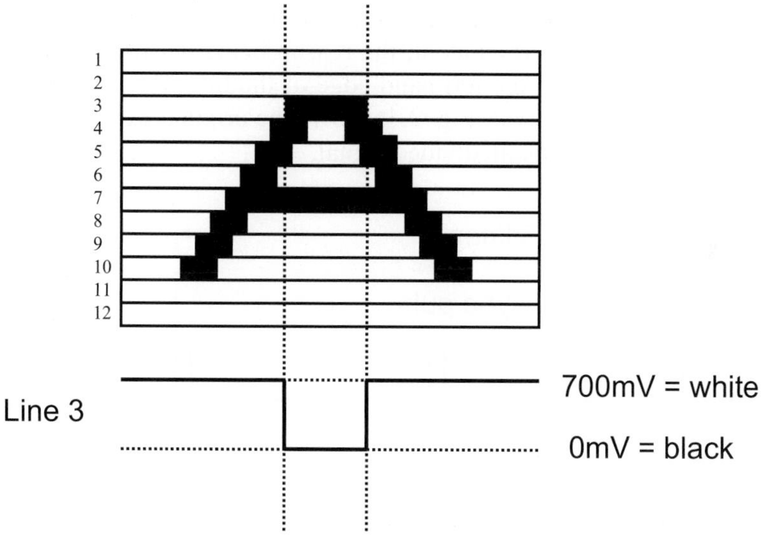

Fig. 2.4. Scanning an original black/white picture

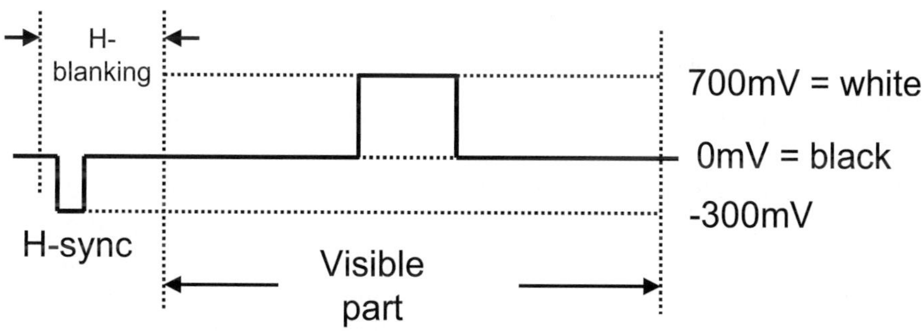

Fig. 2.5. Inserting the horizontal sync pulse

Today, CCD (charge coupled device) chips are universally used in the cameras and the principle of the deflected electron beam is now only preserved in TV receivers; and even there the technology is changing to LCD and plasma screens.

The result of scanning the original is the luminance signal where 0 mV corresponds to 100% black and 700 mV is 100% white. The original picture is scanned line by line from top to bottom, resulting in 625 or 525 active lines depending on the TV standard used. However, not all lines are visible. Because of the finite beam flyback time, a vertical blanking interval of up to 50 lines had to be inserted. In the line itself, too, only a certain part represents visible picture content, the reason being the finite flyback time from the right-hand to the left-hand edge of the line which results in the horizontal blanking interval. Fig. 2.4. shows the original to be scanned and Fig. 2.5. shows the associated video signal.

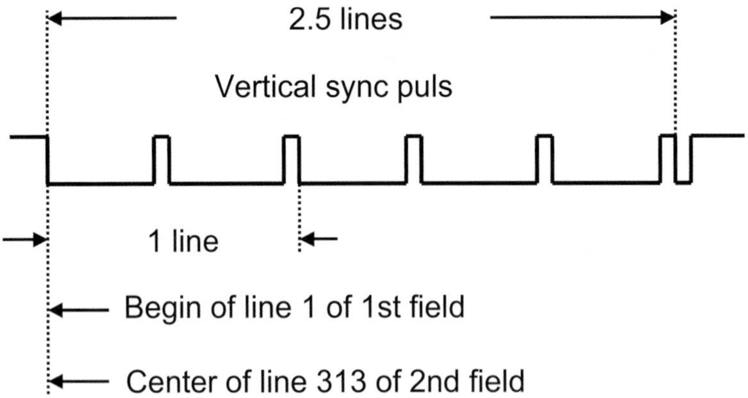

Fig. 2.6. Vertical synchronization pulse

2.2 Horizontal and Vertical Synchronization Pulses

However, it is also necessary to mark the top edge and the bottom edge of the image in some way, in addition to the left-hand and right-hand edges. This is done by means of the horizontal and vertical synchronization pulses. Both types of pulses were created at the beginning of the television age so as to be easily recognizable and distinguishable by the receiver and are located in the blacker than black region below zero volts.

The horizontal sync pulse (Fig. 2.5.) marks the beginning of a line. The beginning is considered to be the 50% value of the front edge of the sync pulse (nominally -150 mV). All the timing within a line is referred to this time. By definition, the active line, which has a length of 52 µs, begins

10 µs after the sync pulse front edge. The sync pulse itself is 4.7 µs long and stays at -300 mV during this time.

At the beginning of television, the capabilities of the restricted processing techniques of the time which, nevertheless, were quite remarkable, had to be sufficient. This is also reflected in the nature of the sync pulses. The horizontal sync pulse (H sync) was designed as a relatively short pulse (appr. 5 µs) whereas the vertical sync pulse (V sync) has a length of 2.5 lines (appr. 160 µs). In a 625-line system, the length of a line including H sync is 64 µs. The V sync pulse can, therefore, be easily distinguished from H sync. The V sync pulse (Fig. 2.6.) is also in the blacker than black region below zero volts and marks the beginning of a frame or field, respectively.

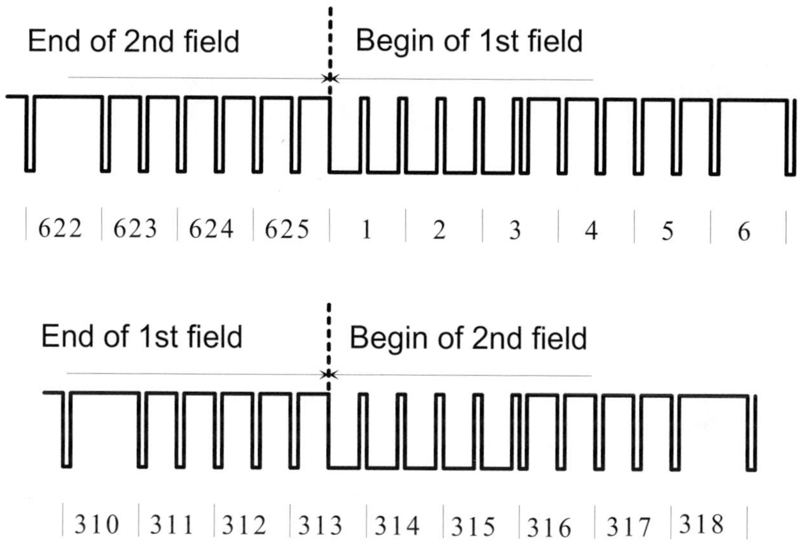

Fig 2.7. Vertical synchronization pulses with pre- and post-equalizing pulses in the 625 line-system

As already mentioned, a frame, which has a frame rate of 25 Hz = 25 frames per second in a 625-line system, is subdivided into 2 fields. This makes it possible to cheat the eye, rendering flickering effects largely invisible. One field is made up of the odd-numbered lines and the other one is made up of the even-numbered lines. They are transmitted alternatingly, resulting in a field rate of 50 Hz in a 625-line system. A frame (beginning of the first field) begins when the V sync pulse goes to the -300 mV level for 2.5 lines at the precise beginning of a line. The second field begins when the, V sync pulse drops to the -300 mV level for 2.5 lines at the center of line 313.

The first and second field are transmitted interlaced with one another, thus reducing the flickering effect. Because of the limitations of the pulse technology at the beginnings of television, a 2.5-line-long V sync pulse would have caused the line oscillator to lose lock. For this reason, additional pre- and post-equalizing pulses were gated in which contribute to the current appearance of the V sync pulse (Fig. 2.7.). Today's signal processing technology renders these unnecessary.

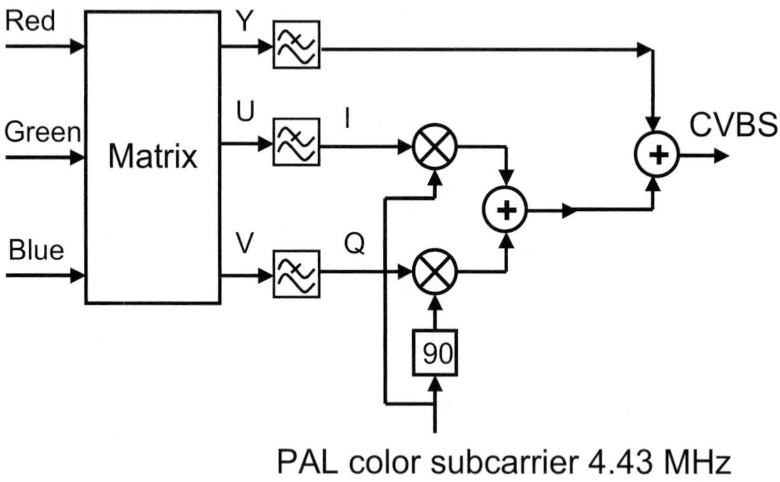

Fig. 2.8. Block diagram of a PAL modulator

2.3 Adding the Color Information

At the beginning of the television age, black/white rendition was adequate because the human eye has its highest resolution and sensitivity in the area of brightness differences and the brain receives its most important information from these. There are many more black/white receptors than color receptors in the retina. But just as in the cinema, television managed the transition from black/white to color because its viewers desired it. Today this is called innovation. When color was added in the sixties, knowledge about the anatomy of the human eye was taken into consideration. With only about 1.3 MHz, color (chrominance) was allowed much less resolution, i.e. bandwidth, than brightness (luminance) which is transmitted with about 5 MHz. At the same time, chrominance is embedded compatibly into the luminance signal so that a black/white receiver was undisturbed but a color receiver was able to reproduce both color and black/white correctly. If a receiver falls short of these ideals, so-called cross-luminance and cross-color effects are produced.

In all three systems, PAL, SECAM and NTSC, the Red, Green and Blue color components are first acquired in three separate pickup systems (initially tube cameras, now CCD chips) and then supplied to a matrix where the luminance signal is formed as the sum of R + G + B, and the chrominance signal. The chrominance signal consists of two signals, the color difference signals Blue minus luminance and Red minus luminance. However, the luminance signal and the chrominance signal formed must be matrixed, i.e. calculated, provided correctly with the appropriate weighting factors according to the eye's sensitivity, using the following formula

$$Y = 0.3 \bullet R + 0.59 \bullet G + 0.11 \bullet B;$$
$$U = 0.49 \bullet (B-Y);$$
$$V = 0.88 \bullet (R-Y);$$

The luminance signal Y can be used directly for reproduction by a black/white receiver. The two chrominance signals are also transmitted and are used by the color receiver. From Y, U and V it is possible to recover R, G and B. The color information is then available in correspondingly reduced bandwidth, and the luminance information in greater bandwidth ("paintbox principle").

To embed the color information into a CVBS (composite video, blanking and sync) signal intended initially for black/white receivers, a method had to be found which has the fewest possible adverse effects on a black/white receiver, i.e. keeps it free of color information, and at the same time contains all that is necessary for a color receiver.

Two basic methods were chosen, namely embedding the information either by analog amplitude/phase modulation (IQ modulation) as in PAL or NTSC, or by frequency modulation as in SECAM. In PAL and NTSC, the color difference signals are supplied to an IQ modulator with a reduced bandwidth compared to the luminance signal (Fig. 2.8.) The IQ modulator generates a chrominance signal as amplitude/phase modulated color subcarrier, the amplitude of which carries the color saturation and the phase of which carries the hue. An oscilloscope would only show, therefore, if there is color, and how much, but would not identify the hue. This would require a vectorscope which supplies information on both.

In PAL and in NTSC, the color information is modulated onto a color subcarrier which lies within the frequency band of the luminance signal but is spectrally intermeshed with the latter in such a way that it is not visible in the luminance channel. This is achieved by the appropriate choice of color subcarrier frequency. In PAL (Europe), the color subcarrier frequency was chosen by using the following formula

$$f_{SC} = 283.75 \bullet f_H + 25\ \text{Hz} = 4.43351875\ \text{MHz};$$

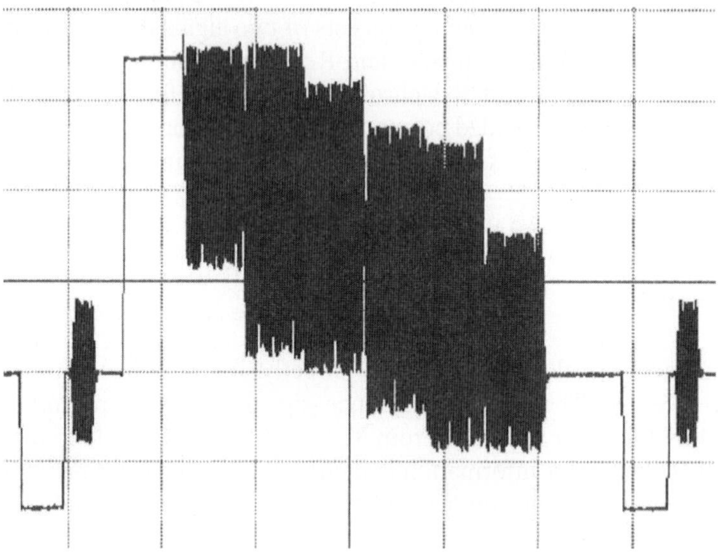

Fig. 2.9. Oscillogram of a CVBS, CCVS (composite color video and sync) signal

In SECAM, the frequency modulated color difference signals are alternately modulated onto two different color subcarriers from line to line. The SECAM process is currently only used in France and in French-speaking countries in North Africa, and also in Greece. Countries of the previous Eastern Block changed from SECAM to PAL in the nineties.

Compared with NTSC, PAL has a great advantage due to its insensitivity to phase distortion because its phase changes from line to line. The color cannot be changed by phase distortion on the transmission path, therefore. NTSC is used in analog television, mainly in North America, where it is sometimes ridiculed as "Never Twice the Same Color" because of the color distortions.

The composite PAL, NTSC or SECAM video signal (Fig. 2.9.) is generated by mixing the black/white signal, the sync information and the chrominance signal and is now called a CCVS (Composite Color, Video and

Sync) signal. Fig. 2.9. shows the CCVS signal of a color bar signal. The color burst can be seen clearly. It is used for conveying the reference phase of the color subcarrier to the receiver so that its color oscillator can lock to it.

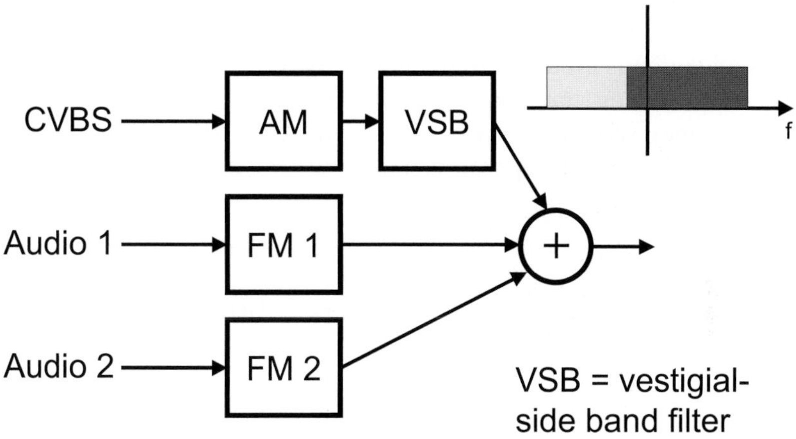

Fig. 2.10. Principle of a TV modulator for analog terrestrial TV and analog TV broadband cable

2.4 Transmission Methods

Analog television is disseminated over three transmission paths which are: terrestrial transmission paths, via satellite and by broadband cable. The priority given to any particular transmission path depends greatly on the countries and regions concerned. In Germany, the traditional analog "antenna TV" has currently only a minor status with fewer than 10%, this term being used mainly by the viewers themselves whereas the actual technical term is "terrestrial TV". The reason for this is the good coverage by satellite and cable, and more programs. This will change when DVB-T is introduced as has already become apparent in some regions.

Transmission of analog television via terrestrial and satellite paths will shrivel away into insignificance within a few years. Whether this will also be true of broadband cable cannot yet be predicted.

In the terrestrial transmission of analog TV signals, and that by cable, the modulation method used is amplitude modulation, in most cases with negative modulation. Positive modulation is only used in the French Standard L.

The sound subcarriers are frequency modulated in most cases. To save bandwidth, the vision carrier is VSB-AM (vestigial sideband amplitude modulation) modulated, i.e. a part of the spectrum is suppressed by bandpass filtering. The principle is shown in Fig. 2.10. and 2.11. Because of the nonlinearities and the low signal/noise ratio on the transmission link, frequency modulation is used in satellite transmission.

Since these analog transmission paths are losing more and more in significance, they will not be discussed in greater detail in this book and the reader is referred to the appropriate literature, instead.

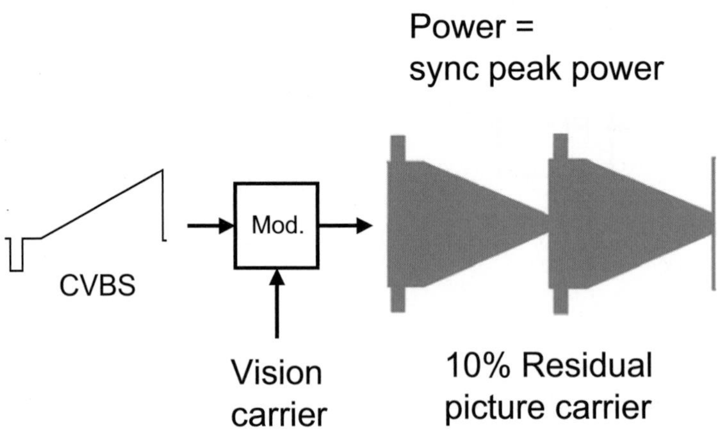

Fig. 2.11. Vision modulator

2.5 Distortion and Interference

Over the entire transmission link, an analog video signal is subjected to influences which have a direct effect on its quality and are immediately visible in most cases. These distortions and interferences can be roughly grouped in the following categories:

- Linear distortion (amplitude and phase distortion)
- Non-linear distortion
- Noise

- Interference
- Intermodulation

Linear distortion is caused by passive electronic components. The amplitude or group delay is no longer constant over a certain frequency range which is 0 ... 5 MHz in the case of video. Parts of the relevant frequency range are distorted to a greater or lesser extent, depending on the characteristic of the transmission link involved. As a result, certain signal components of the video signal are rounded. The worst effect is rounding of the sync pulses which leads to synchronization problems in the TV receiver such as, e.g. horizontal "pulling" or "rolling" of the picture from top to bottom. These terms have been known since the early days of television.

Changing of heads from field to field produces similar effects at the top edge of the picture with some older videorecorders, the picture is "pulling".

These effects have become relatively rare thanks to modern receiver technology and relatively good transmission techniques. In the active picture area, linear distortion manifests itself either as lack of definition, ringing, optical distortion or displacement of the color picture with respect to the luminance picture.

Nonlinear distortion can be grouped into

- Static nonlinearity
- Differential gain
- Differential phase

With non-linear distortion, neither the gray steps nor the color subcarrier are reproduced correctly in amplitude and phase. Non-linear distortion is caused by active components (transmitter tubes, transistors) in the transmission link. However, they become only visible ultimately when many processes are added together since the human eye is very tolerant in this respect. Putting it another way: "Although this isn't the right gray step, who is to know?". And in color television this effect is less prominent, in any case, because of the way in which color is transmitted, particularly with PAL.

One of the most visible effects is the influence of noise-like disturbances. These are simply produced by superimposition of the ever-present gaussian noise, the level of which is only a question of its separation from the useful signal level. I.e., if the signal level is too low, noise becomes visible. The level of thermal noise can be determined in a simple way via the Boltzmann constant, the bandwidth of the useful channel and the normal ambient temperature and is thus almost a fixed constant. Noise is immediately visible in the analog video signal which is the great difference compared with digital television.

Intermodulation products and interference are also very obvious in the video signal and have a very disturbing effect, forming moiré patterns in the picture. These effects are the result of heterodyning of the video signal with an interfering product either from an adjacent channel or interferers entering the useful spectrum directly from the environment. This type of interference is the one most visible and thus also causes the greatest disturbance in the overall impression of the picture. It is also most apparent in cable television because of its multichannel nature.

2.6 Signals in the Vertical Blanking Interval

Since the middle of the seventies, the vertical blanking interval, which was originally used for the vertical flyback, is no longer only "empty" or "black". At first, so-called VITS (vertical insertion test signals), or test lines, were inserted there which could be used for assessing the quality of the analog video signal. In addition, teletext and the data line can be found there. Test lines were and are used for monitoring the transmission quality of a TV transmission link or section virtually on-line without having to isolate the link. These test lines contain test signals which can be used for identifying the causes of faults.

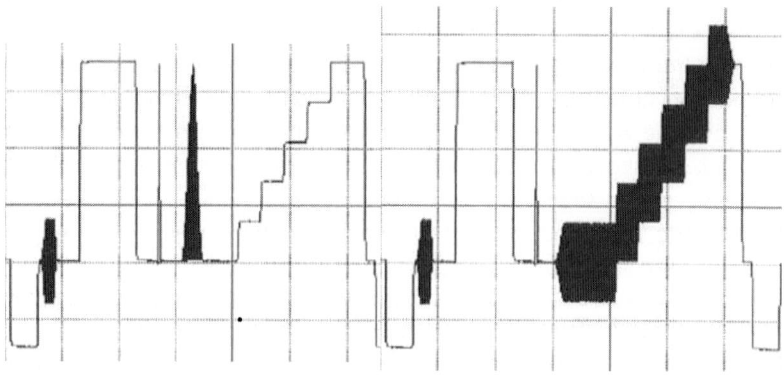

Fig. 2.12. CCIR 17 and 330 test lines

Test line "CCIR 17" (now ITU 17, on the left in Fig. 2.12.) begins with the so-called white pulse (bar) and is used as technical voltage reference for 100% white. Its nominal amplitude is 700 mV. The "roof" of the white pulse is 10 µs long and should be flat and without overshoots. This is followed by the 2T pulse which is a so-called $\cos^2$ pulse with a half-amplitude period of 2T 2 • 100 ns = 200 ns. The main components of its spectrum extend to the end of the luminance channel of 5 MHz. It reacts very sensitively to amplitude response and group delay distortion from 0 ... 5 MHz and can thus be used for assessing linear distortion both visually and by measurement. The next pulse is a 20T pulse, a $\cos^2$ pulse with superimposed color subcarrier and with a half-amplitude period of 20T = 20 • 100 ns = 2 µs. It clearly shows linear distortion of the color channel with respect to the luminance channel.

Linear distortion of the color channel with respect to the luminance channel is

- Differential gain of the color channel with respect to the luminance channel
- Luminance-chrominance delay caused by group delay

Non-linear distortion can be easily identified by means of the 5-step gray scale. All five steps must have identical height. If they do not have equal height due to nonlinearities, this is called static nonlinearity (luminance nonlinearity). In test line 330, the gray scale is replaced by a staircase on which the color subcarrier is superimposed. This can be used for identifying non-linear effects on the color cubcarrier such as differential amplitude and phase. The color bursts superimposed on the staircase should all be ideally of the same amplitude and must not have a phase discontinuity at the transition points of the steps.

Teletext is well known by now (Fig. 2.13. and 2.14.). It is a data service offered in analog television. The data rate is about 6.9 Mbit/s, but only in the area of the lines really used in the vertical blanking interval. In actual fact, the data rate is much lower. In each teletext line, 40 useful characters are transmitted. A teletext page consists of 40 characters times 24 lines. If the entire vertical blanking interval were to be used, just short of one teletext page could be transmitted per field. Teletext is transmitted in NRZ (non-return-to-zero) code. A teletext line begins with the 16-bit-long run-in, a sequence of 10101010... for synchronizing the phase of the teletext decoder in the receiver. This is followed by the framing code. This hexadecimal number 0xE4 marks the beginning of the active teletext. After the magazine and line number, the 40 characters of a line of the teletext are transmitted. One teletext page consists of 24 text lines.

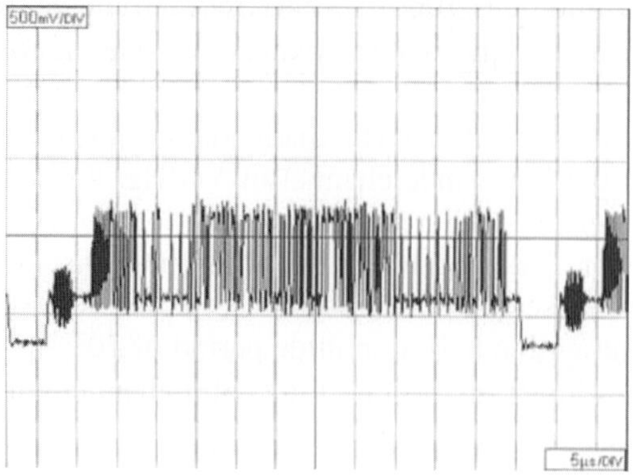

Fig. 2.13. Teletext line

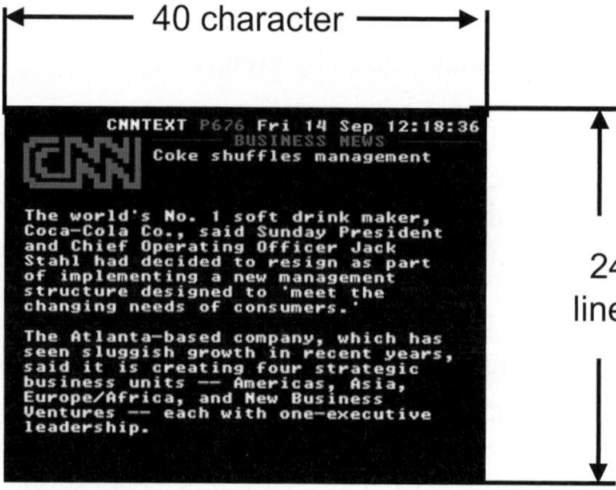

Fig. 2.14. Teletext page

The most important teletext parameters are as follows:

- Non-return-to-zero code
- Data rate: 444 • 15625 kbit/s = 6.9375 Mbit/s
- Error protection: Even parity
- Characters per line: 40
- Lines per teletext page: 24

The data line (e.g. line 16 and corresponding line in the second field, Fig. 2.15.) is used for transmitting control information, signaling and, among other things, the VPS (video program system) data for controlling video recorders. In detail, the data line is used for transmitting the following data:

- Byte 1: Run-in 10101010
- Byte 2: Start code 01011101
- Byte 3: Source ID
- Byte 4: Serial ASCII text transmission (source)
- Byte 5: Mono/stereo/dual sound
- Byte 6: Video content ID
- Byte 7: Serial ASCII text transmission
- Byte 8: Remote control (routing)
- Byte 9: Remote control (routing)
- Byte 10: Remote control
- Byte 11 to 14: Video program system (VPS)
- Byte 15: Reserved

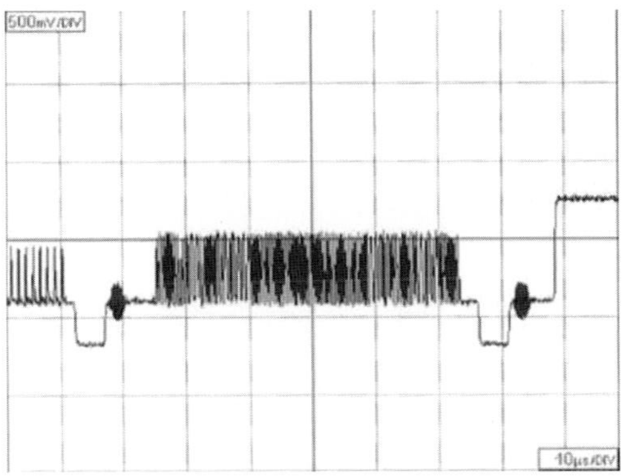

Fig. 2.15. Data line (mostly line 16 in the vertical blanking interval)

The VPS bytes contain the following information:

- Day (5 bits)
- Month (4 bits)
- Hour (5 bits)
- Minute (6 bits) = virtual starting time of the program
- Country ID (4 bits)
- Program source 11) (6 bits)

The transmission parameters of the data line are:

- Line: 16/329
- Code: Return-to-zero code
- Data rate: 2.5 Mbit/s
- Level: 500 mV
- Data: 15 bytes per line

According to DVB, these signals from the vertical blanking interval are partially regenerated in the receiver to retain compatibility with analog television. The test line signals, however, are no longer provided.

2.7 Measurements on Analog Video Signals

Analog video signals have been measured since the beginning of the TV age, initially with simple oscilloscopes and vectorscopes and later with e-ver more elaborate video analyzers, the latest models of which were digital (Fig. 2.22.). These video measurements are intended to identify the distortions in the analog video signal. The following test parameters are determined with the aid of test lines:

- White bar amplitude
- Sync amplitude
- Burst amplitude
- Tilt on the white bar
- 2T pulse amplitude
- 2T K factor
- Luminance-chrominance amplitude on the 20T pulse
- Luminance-chrominance delay on the 20T pulse
- Static nonlinearity on the grayscale
- Differential gain on the grayscale with color subcarrier
- Differential phase on the grayscale with color subcarrier
- Weighted and unweighted luminance signal/noise ratio
- Hum

In addition, an analog TV test receiver also provides information on:

- Vision carrier level

- Sound carrier level
- Deviation of the sound carriers
- Frequencies of vision and sound carriers
- Residual picture carrier
- ICPM (Incidential phase modulation)

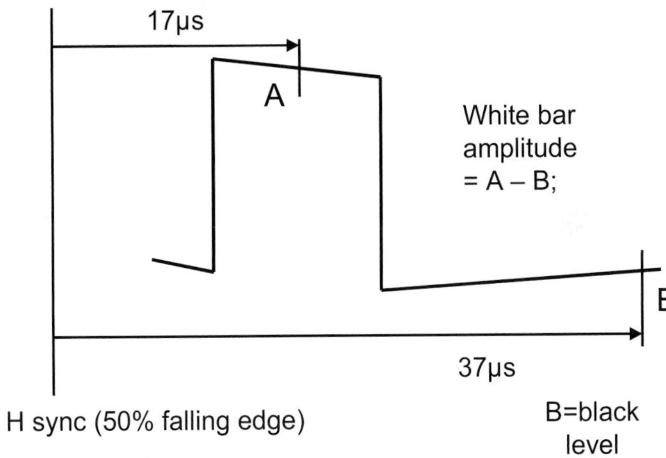

Fig. 2.16. Measuring the white bar amplitude

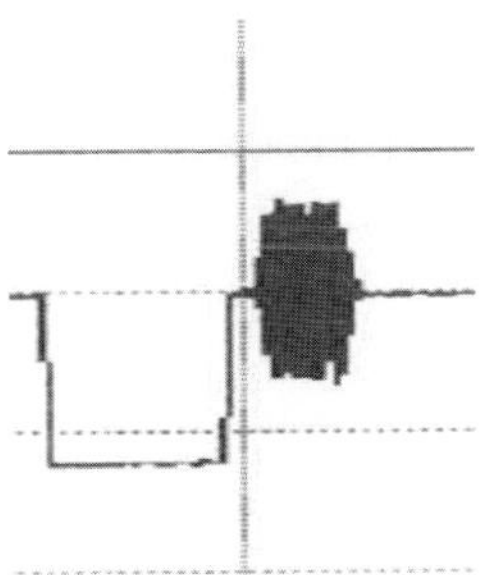

Fig. 2.17. Sync pulse and burst

The most important parameter to be measured on an analog TV signal is the white bar amplitude which is measured as shown in Fig. 2.16. In the worst case, the white bar can also be quite rounded due to linear distortions, as indicated in the figure. The sync amplitude (s. Fig. 2.17.) is used as voltage reference in the terminals and is of special importance for this reason. The sync amplitude is nominally 300 mV below black. The 50% value of the falling edge of the sync pulse is considered to be the timing reference in the analog video signal. The burst (s. Fig. 2.17.) is used as voltage and phase reference for the color subcarrier. Its amplitude is 300 mV_{PP}. In practice, amplitude distortions of the burst have little influence on the picture quality.

Linear distortion leads to tilt on the white bar (Fig. 2.18.). This is also an important test parameter. To measure it, the white bar is sampled at the beginning and at the end and the difference is calculated which is then related to the white pulse amplitude.

The 2T pulse reacts sensitively to linear distortion in the entire transmission channel of relevance. Fig. 2.19. shows the undistorted 2T pulse on the left. It has been used as test signal for identifying linear distortion since the seventies. A 2T pulse altered by linear distortion is also shown on the right in Fig. 2.19. If the distortion of the 2T pulse is symmetric, it is caused by amplitude response errors. If the 2T pulse appears to be unsymmetric, then group delay errors are involved (non-linear phase response).

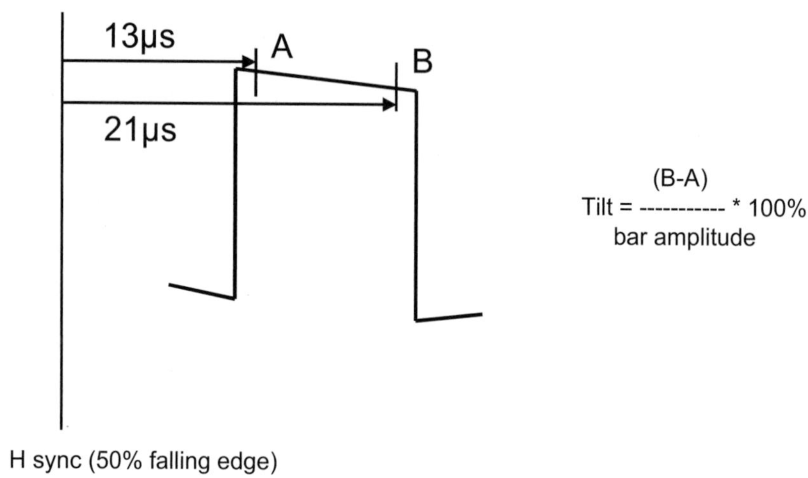

Fig. 2.18. Tilt on the white bar

The 20T pulse (Fig. 2.20., center) was created especially for measurements in the color channel. It reacts immediately to differences between luminance and chrominance. Special attention must be paid to the bottom of the 20T pulse. It should be straight, without any type of indentation. In the ideal case, the 20T pulse, like the 2T pulse, should have the same magnitude as the white pulse (700 mV nominal).

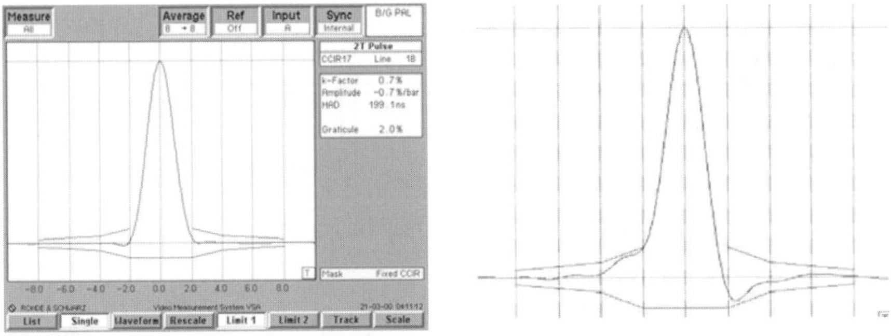

Fig. 2.19. Undistorted (left) and distorted (right) 2T pulse

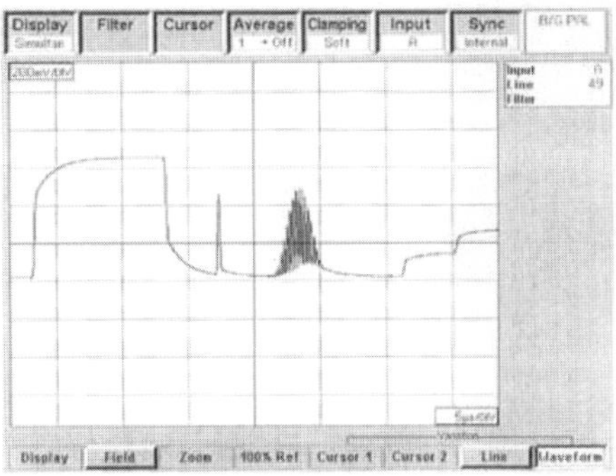

Fig. 2.20. Linearly distorted white pulse, 2T pulse and 20T pulse

Nonlinearities distort the video signal in dependence on modulation. This can be shown best on staircase signals. To this end, the gray scale and the staircase with color subcarrier were introduced as test signal, the steps simply being of different size in the presence of nonlinearitics. Noise and intermodulation can be verified best in a black line (Fig. 2.21.). In most cases, line 22 was kept free of information for this purpose but this is not necessarily so any longer, either, since it carries teletext in most cases. To measure these effects, it is only necessary to look for an empty line suitable for this purpose among the 625 or 525 lines and this differs from program to program.

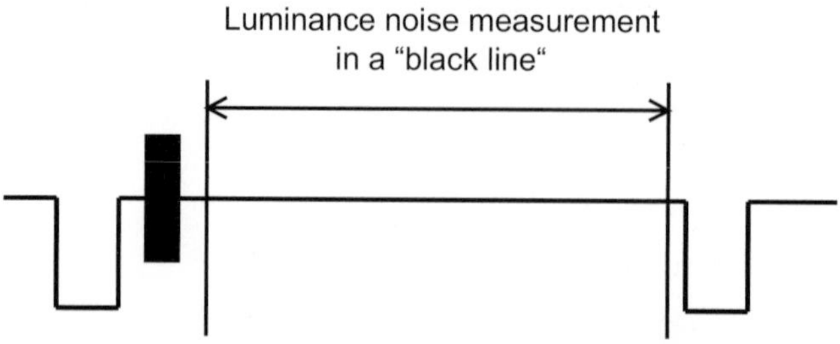

Fig. 2.21. Luminance noise measurement in a ”black line“

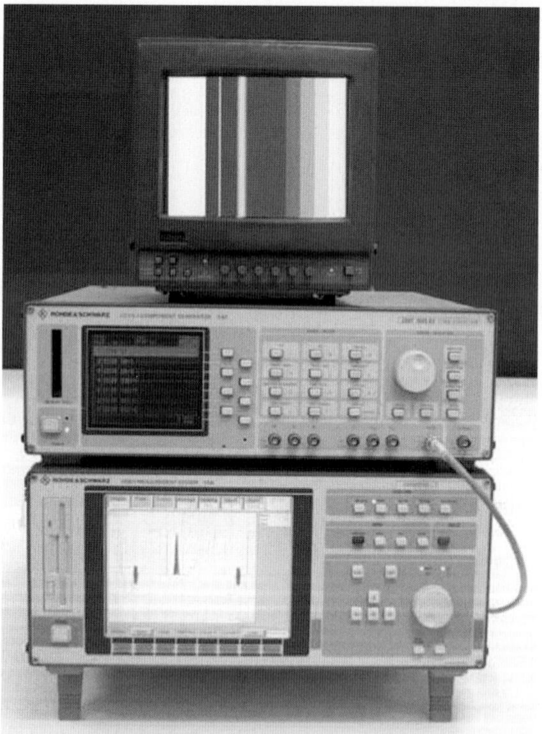

Fig. 2.22. Analog video test and measurement equipment: video test signal generator and video analyzer (Rohde&Schwarz SAF and VSA)

In digital television, test line testing now only makes sense for assessing the beginning (studio equipment) and the end (receiver) of the transmission link. In between - on the actual transmission link - nothing happens that can be verified by this means. The corresponding measurements on the digital transmission links will be described in detail in the respective chapters.

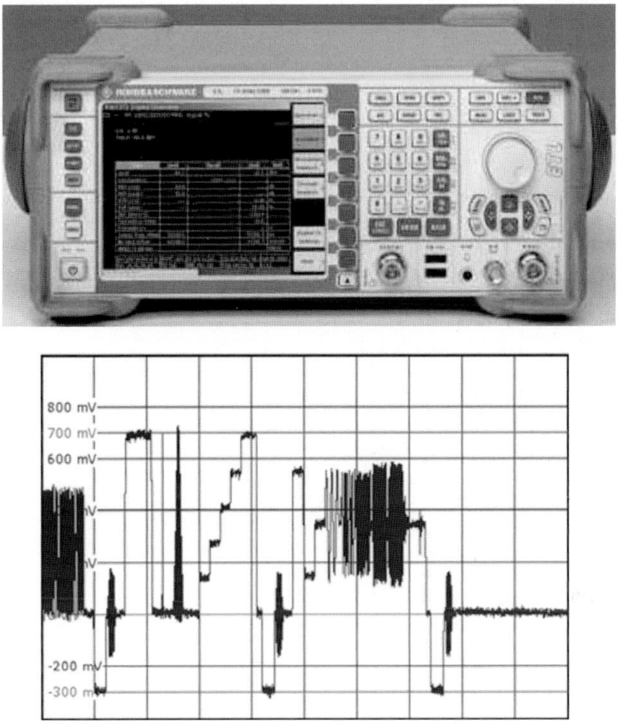

Fig. 2.23. TV Analyzer Rohde&Schwarz ETL for analog and digital TV measurements; ETL offers spectrum analyzer functionality and analog and digital TV measurements

2.8 Analog and Digital TV in a Broadband Cable Network

There is still both analog and digital TV and FM radio in a broadband cable network. That means analog TV and analog TV measurements are still a topic today.

Fig. 2.23. and 2.24. shows a current example of a mix of analog FM audio and analog and digital TV channels in a broadband cable network (year 2008, Germany and Austria).

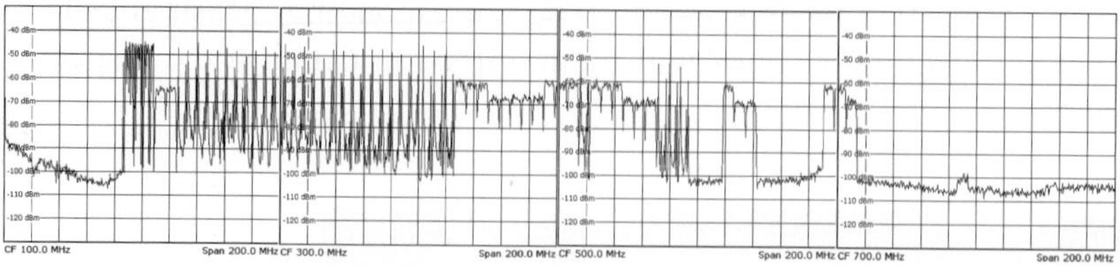

Fig. 2.24. Analog and digital broadband cable channels (example Munich, Germany, 0 … 800 MHz)

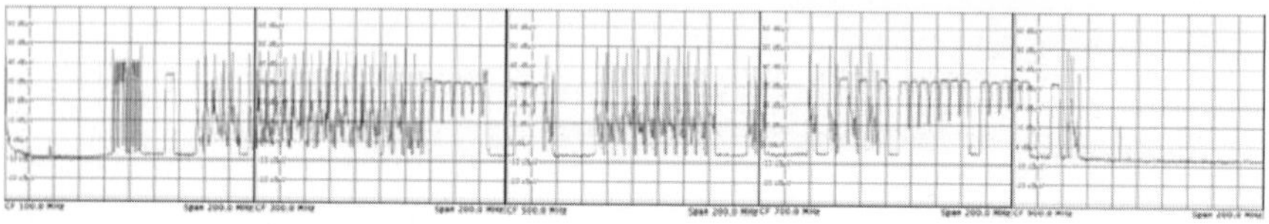

Fig. 2.25. Analog and digital broadband cable channels (example Klagenfurt, Austria, 0 … 1000 MHz)

Bibliography: [MÄUSL3], [MÄUSL5], [VSA], [FISCHER6], [ETL]

3 The MPEG Data Stream

The abbreviation MPEG, first of all, stands for Moving Pictures Experts Group, that is to say MPEG deals mainly with the digital transmission of moving pictures. However, the data signal defined in the MPEG-2 Standard can also generally carry data which have nothing at all to do with video and audio and could be Internet data, for example. And indeed, throughout the world there are MPEG applications in which it would be futile to look for video and audio signals. Thus, in Wollongong, about 70 km south of Sydney in Australia, an Australian pay TV provider is operating a pure data broadcasting service using MPEG-2 data signals via MMDS (Microwave Multipoint Distribution System). "Austar" are here providing their customers with fast Internet links in the Mbit/s range.

MPEG = Moving Pictures Expert Group

MPEG-1	**MPEG-2**	**MPEG-4**	**MPEG-7**	**MPEG-21**
Part1: systems ISO/IEC11172-1 "PES layer"	Part1: systems ISO/IEC13818-1 "Transportation"	Part1: systems ISO/IEC14496	Metadata, XML based ISO/IEC15938 "Multimedia Content Description Interface"	additional "tools" ISO/IEC21000
Part2: video ISO/IEC11172-2	Part2: video ISO/IEC13818-2	Part2: video ISO/IEC14496-2		
Part3: audio ISO/IEC11172-3	Part3: audio ISO/IEC13818-3	Part3: audio (AAC) ISO/IEC14496-3		
	Part6: DSM-CC ISO/IEC13818-6	Part10: video (AVC, H.264) ISO/14496-10		
	Part7: AAC ISO/IEC13818-7			

Fig. 3.1. MPEG standards

As in the MPEG Standard itself, first the general structure of the MPEG data signal will be described in complete isolation from video and audio. An understanding of the data signal structure is also of greater importance

W. Fischer, *Digital Video and Audio Broadcasting Technology*, Signals and Communication Technology, 3rd ed., DOI: 10.1007/978-3-642-11612-4_3, © Springer-Verlag Berlin Heidelberg 2010

in practice than a detailed understanding of the video and audio coding which will be discussed later.

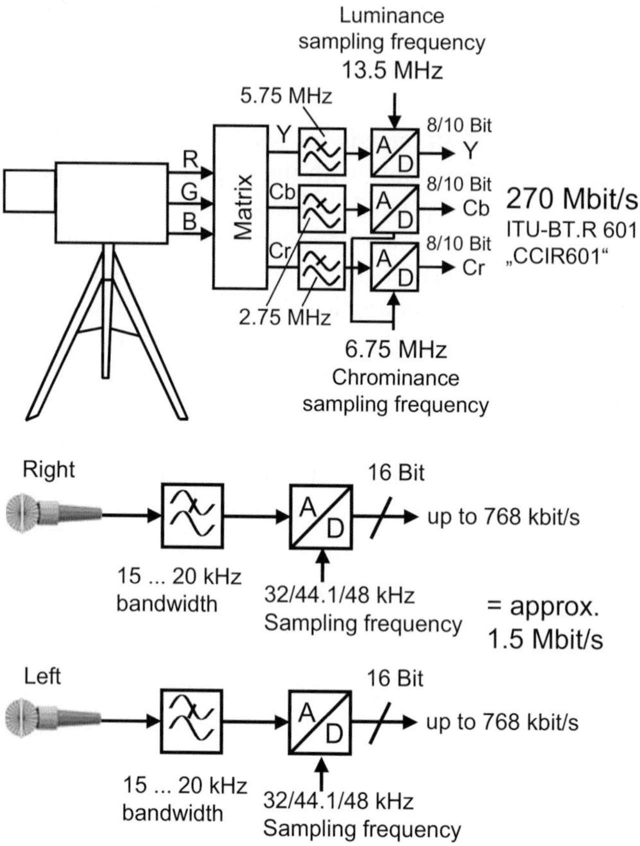

Fig. 3.2. Video and audio data signals

All the same, the description of the data signal structure will begin with the uncompressed video and audio signals. An SDTV (Standard Definition Television) signal without data reduction has a data rate of 270 Mbit/s and a digital stereo audio signal in CD quality has a data rate of about 1.5 Mbit/s (Fig. 3.2.).

The video signals are compressed to about 1 Mbit/s in MPEG-1 and to about 2 - 7 Mbit/s in MPEG-2. The video data rate can be constant or variable (statistical multiplex). The audio signals have a data rate of about 100 - 400 kbit/s (mostly 192 kbit/s) after compression (to be discussed in a separate chapter) but the audio data rate is always constant and a multiple of 8 kbit/s. The compression itself will be dealt with in a separate chapter. The compressed video and audio signals in MPEG are called "elementary streams", ES in brief. There are thus video streams, audio streams and, quite generally, data streams, the latter containing any type of compressed or uncompressed data. Immediately after having been compressed (i.e. encoded), all the elementary streams are divided into variable-length packets, both in MPEG-1 and in MPEG-2 (Fig. 3.3.).

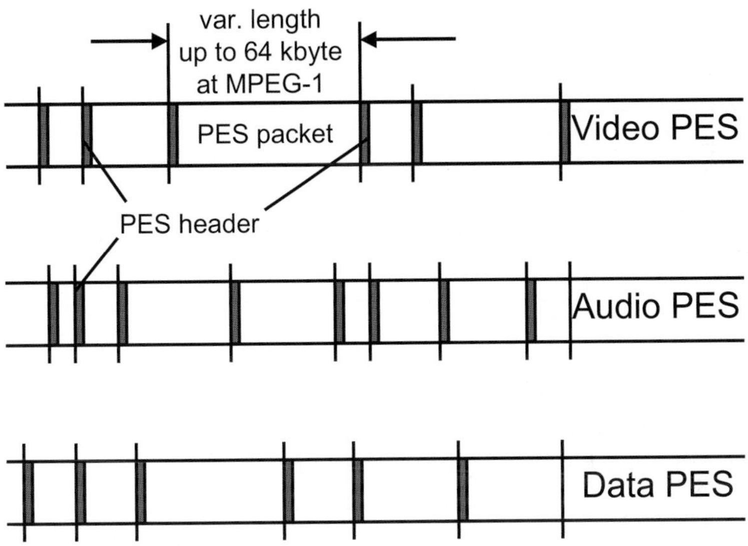

Fig. 3.3. MPEG Elementary Streams

Since it is possible to have sometimes more and sometimes less compression depending on the instantaneous video and audio content, variable-length containers are needed in the data signal. These containers carry one or more compressed frames in the case of the video signal and one or more compressed audio signal segments in the case of the audio signal. These elementary streams (Fig. 3.3.) thus divided into packets are called "packetized elementary streams", or simply PES for short. Each PES packet usually has a size of up to 64 kbytes. It consists of a relatively short header and of a payload. The header contains inter alia a 16-bit-long length indicator for the maximum packet length of 64 kbytes. The payload part contains either the compressed video and audio streams or a pure data stream. According to the MPEG Standard, however, the video packets can also be longer than 64 kbytes in some cases. The length indicator is then set to zero and the MPEG decoder has to use other mechanisms for finding the end of the packet.

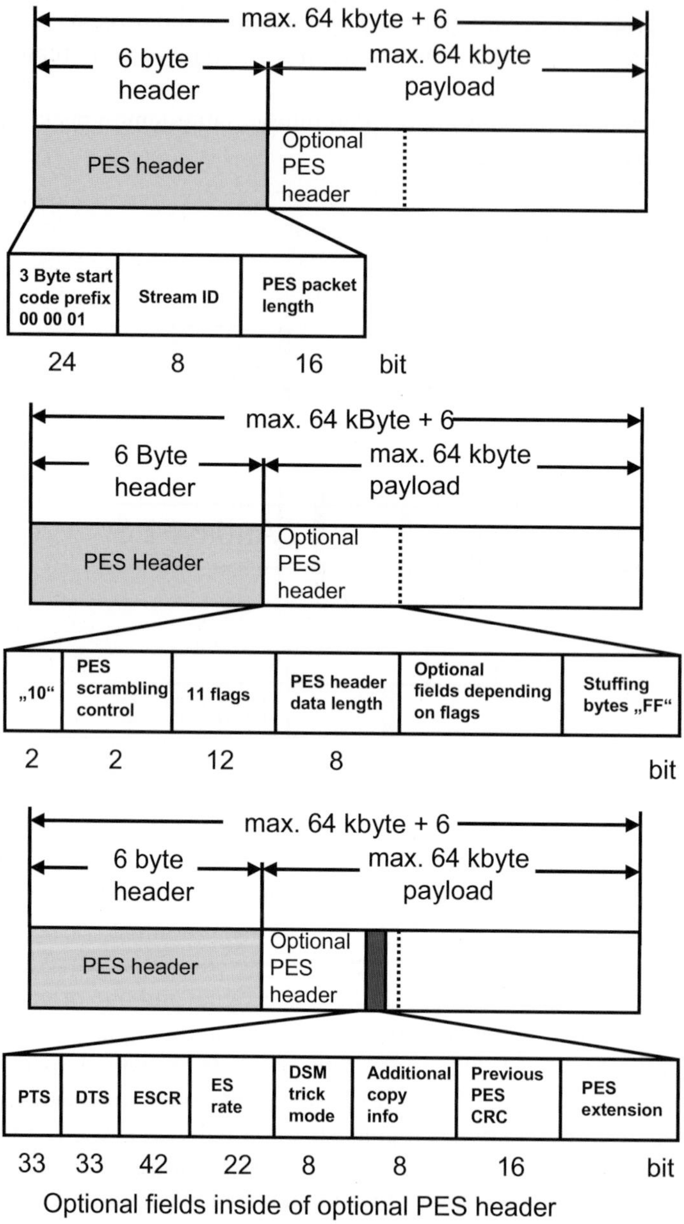

Fig. 3.4. The PES packet

3.1 The Packetized Elementary Stream (PES)

All elementary streams in MPEG are first packetized in variable-length packets called PES packets. The packets, which primarily have a length of 64 kbytes, begin with a PES header of 6 bytes minimum length. The first 3 bytes of this header represent the "start code prefix", the content of which is always 00 00 01 and which is used for identifying the start of a PES packet. The byte following the start code is the "stream ID" which describes the type of elementary stream following in the payload. It indicates whether it is, e.g. a video stream, an audio stream or a data stream which follows. After that there are two "packet length" bytes which are used to address the up to 64 kbytes of payload. If both of these bytes are set to zero, a PES packet having a length which may exceed these 64 kbytes can be expected. The MPEG decoder then has to use other arrangements to find the PES packet limits, e.g. the start code.

After these 6 bytes of PES header, an "optional PES header" is transmitted which is an optional extension of the PES header and is adapted to the requirements of the elementary stream currently being transmitted. It is controlled by 11 flags in a total of 12 bits in this optional PES header. These flags show which components are actually present in the "optional fields" in the optional PES header and which are not. The total length of the PES header is shown in the "PES header data length" field. The optional fields in the optional header contain, among other things, the "Presentation Time Stamps" (PTS) and the "decoding time stamps" (DTS) which are important for synchronizing video and audio. At the end of the optional PES header there may also be stuffing bytes. Following the complete PES header, the actual payload of the elementary stream is transmitted which can usually be up to 64 kbytes long or even longer in special cases, plus the optional header.

In MPEG-1, video PES packets are simply multiplexed with PES packets and stored on a data medium (Fig. 3.5.). The maximum data rate is about 1.5 Mbit/s for video and audio and the data stream only includes a video stream and an audio stream.

This "Packetized Elementary Stream" (PES) with its relatively long packet structures is not, however, suitable for transmission and especially not for broadcasting a number of programs in one multiplexed data signal.

In MPEG-2, on the other hand, the objective has been to assemble up to 6, 10 or even 20 independent TV or radio programs to form one common multiplexed MPEG-2 data signal. This data signal is then transmitted via satellite, cable or terrestrial transmission links. To this end, the long PES packets are additionally divided into smaller packets of constant-length.

From the PES packets, 184-byte-long pieces are taken and to these another 4-byte-long header is added (Fig. 3.6.), making up 188-byte-long packets called "transport stream packets" which are then multiplexed.

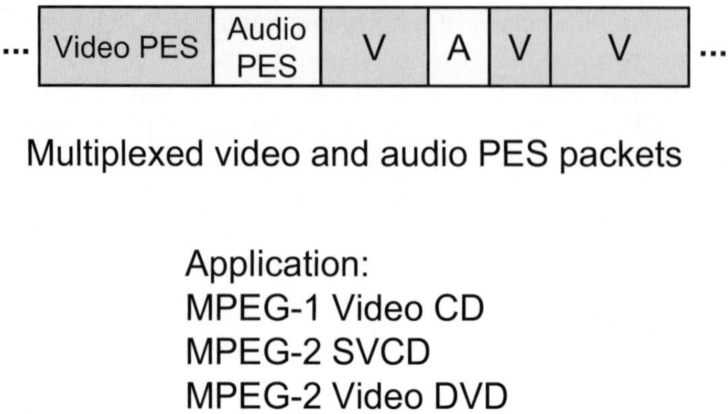

Fig. 3.5. Multiplexed PES packets

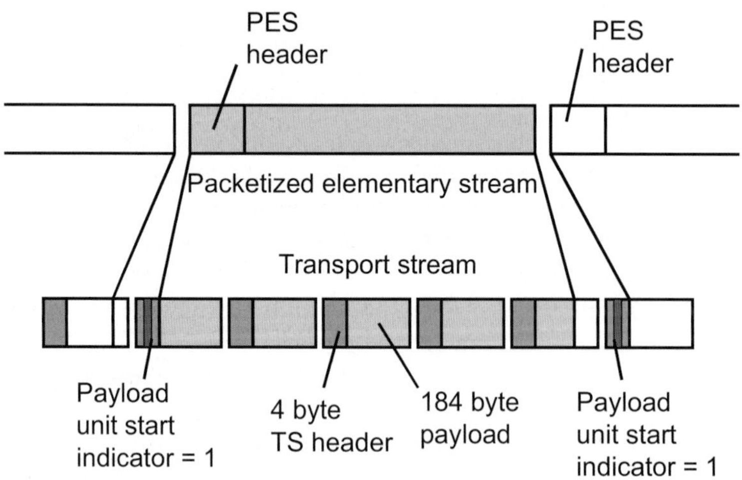

Fig. 3.6. Forming MPEG-2 transport stream packets

To do this, first the transport stream packets of one program are multiplexed together. A program can consist of one or more video and audio signals and an extreme example of this is a Formula 1 transmission with a number of camera angles (track, spectators, car, helicopter) and presented in different languages. All the multiplexed data streams of all the programs are then multiplexed again and combined to form a complete data stream which is called an "MPEG-2 transport stream" (TS for short).

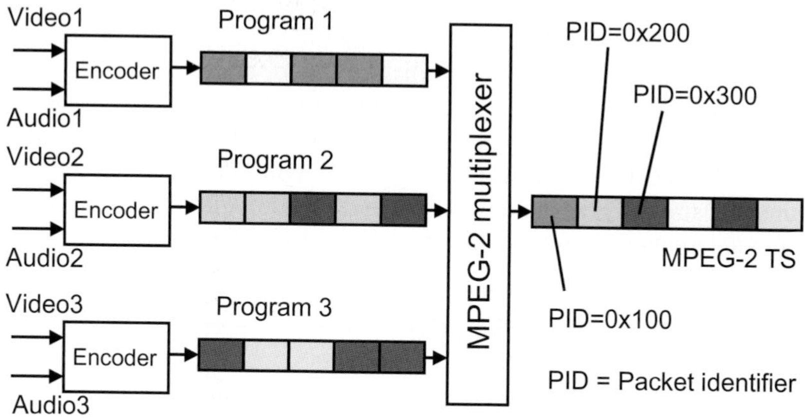

Fig. 3.7. Multiplexed MPEG-2 transport stream packets

An MPEG-2 transport stream contains the 188-byte-long transport stream packets of all programs with all their video, audio and data signals. Depending on the data rates, packets of one or the other elementary streams will occur more or less frequently in the MPEG-2 transport stream. For each program there is one MPEG encoder which encodes all elementary streams, generates a PES structure and then packetizes these PES packets into transport stream packets. The data rate for each program is usually approx. 2 - 7 Mbit/s but the aggregate data rate for video, audio and data can be constant or vary in accordance with the program content at the time. This is then called "statistical multiplex". The transport streams of all the programs are then combined in a multiplexed MPEG-2 data stream to form one overall transport stream (Fig. 3.7.) which can then have a data rate of up to about 40 Mbit/s. There are often up to 6, 8 or 10 or even 20 programs in one transport stream. The data rates can vary during the transmission but the overall data rate has to remain constant. A program can contain video and audio, only audio (audio broadcast) or only data, and the structure is thus flexible and can also change during the transmission. To be able to determine the current structure of the transport stream during the decoding, the transport stream also carries lists describing the structure, so-called "tables".

3.2 The MPEG-2 Transport Stream Packet

The MPEG-2 transport stream consists of packets having a constant length (Fig. 3.8.). This length is always 188 bytes, with 4 bytes of header and 184 bytes of payload. The payload contains the video, audio or general data. The header contains numerous items of importance to the transmission of the packets. The first header byte is the "sync byte". It always has a value of 47_{hex} (0x47 in C/C++ syntax) and is spaced a constant 188 bytes apart in the transport stream. It is quite possible, and certainly not illegal, for there to be a byte having the value 0x47 somewhere else in the packet.

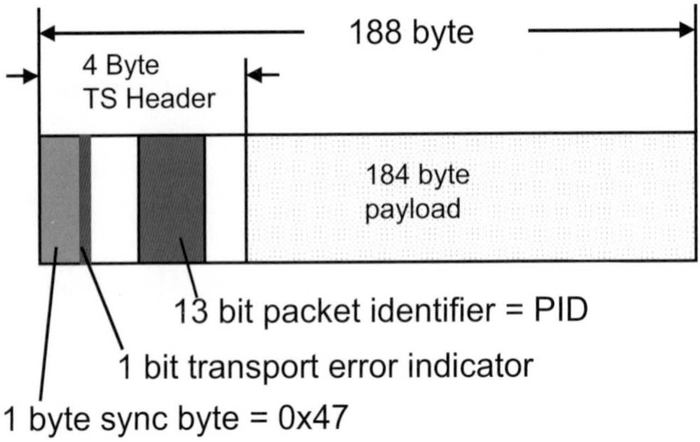

Fig. 3.8. MPEG-2 transport stream packet

The sync byte is used for synchronizing the packet to the transport stream and it is its value plus the constant spacing which is being used for synchronization. According to MPEG, synchronization at the decoder occurs after five transport stream packets have been received. Another important component of the transport stream is the 13 bit-long "packet identifier" or PID for short. The PID describes the current content of the payload part of this packet. The hexadecimal 13 bit number in combination with tables also included in the transport stream show which elementary stream or content this is.

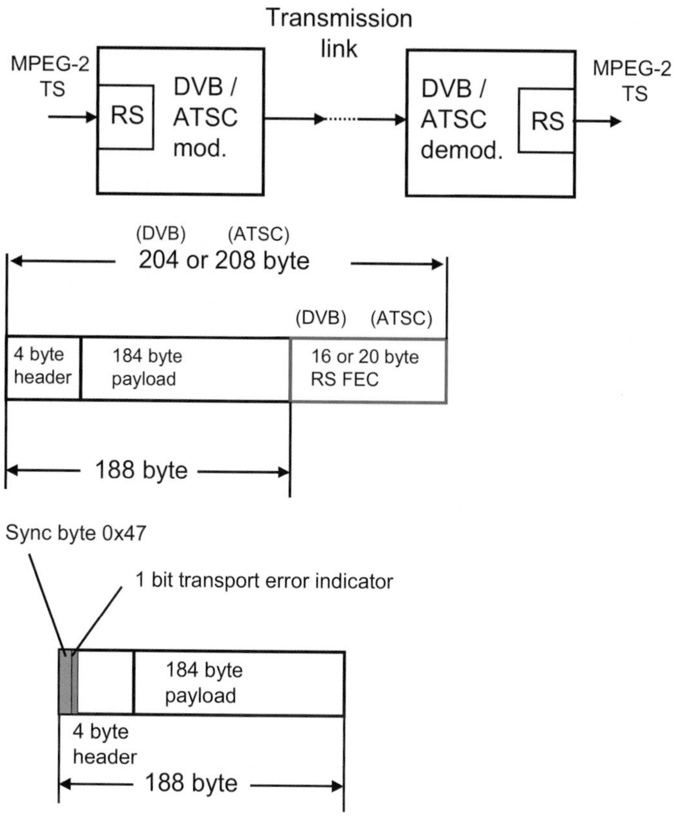

Fig. 3.9. Reed-Solomon FEC

The bit immediately following the sync bit is the "transport error indicator" bit (Fig. 3.8.). With this bit, transport stream packets are flagged as errored after their transmission. It is set by demodulators at the end of the transmission link if e.g. too many errors have occurred and there had been no further possibility to correct these by means of error correction mechanisms used during the transmission. In DVB (Digital Video Broadcasting), e.g., the primary error protection used is always the Reed Solomon error correction code (Fig. 3.9.). In one of the first stages of the (DVB-S, DVB-C or DVB-T) modulator, 16 bytes of error protection are added to the initially 188 bytes of the packet. These 16 bytes of error protection are a special checksum which can be used for repairing up to 8 errors per packet at the receiving end. If, however, there are more than 8 errors in a packet, there is no further possibility for correcting the errors, the error protection has failed and the packet is flagged as errored by the transport error indicator. This packet must now no longer be decoded by the MPEG decoder which, instead, has to mask the error which, in most cases, can be seen as a type of "blocking" in the picture.

It may be necessary occasionally to transmit more than 4 bytes of header per transport stream packet. The header is extended into the payload field in this case. The payload part becomes correspondingly shorter but the total packet length remains a constant 188 bytes. This extended header is called an "adaptation field" (Fig. 3.10.). The other contents of the header and of the adaptation field will be discussed later. "Adaptation control bits" in the 4 byte-long header show if there is an adaptation field or not.

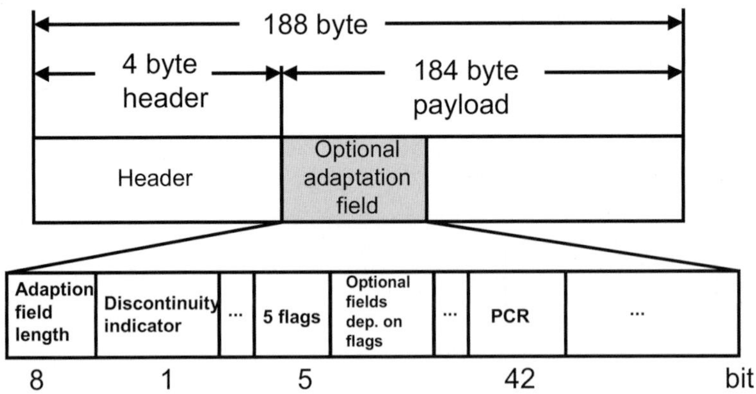

Fig. 3.10. Adaptation field

The structure and especially the length of a transport stream packet are very similar to a type of data transmission known from telephony and LAN technology, namely the "asynchronous transfer mode" or ATM in short. Today, ATM is used both in long-haul networks for telephony and Internet calls and for interconnecting computers in a LAN network in buildings. ATM also has a packet structure. The length of one ATM cell is 53 bytes containing 5 bytes of header and 48 bytes of payload. Right at the beginning of MPEG-2 it was considered to transmit MPEG-2 data signals via ATM links. Hence the length of an MPEG-2 transport stream packet. Taking into consideration one special byte in the payload part of an ATM cell, this leaves 47 bytes of payload data. It is then possible to transmit 188 bytes of useful information by means of 4 ATM cells, corresponding exactly to the length of one MPEG-2 transport stream packet. And indeed, MPEG-2 transmissions over ATM links are nowadays a fact of life. Examples of this are found, e.g. in Austria where all national studios of the Austrian broadcasting institution ORF (Österreichischer Rundfunk) are linked via an ATM network (called LNET). In Germany, too, MPEG streams are exchanged over ATM links.

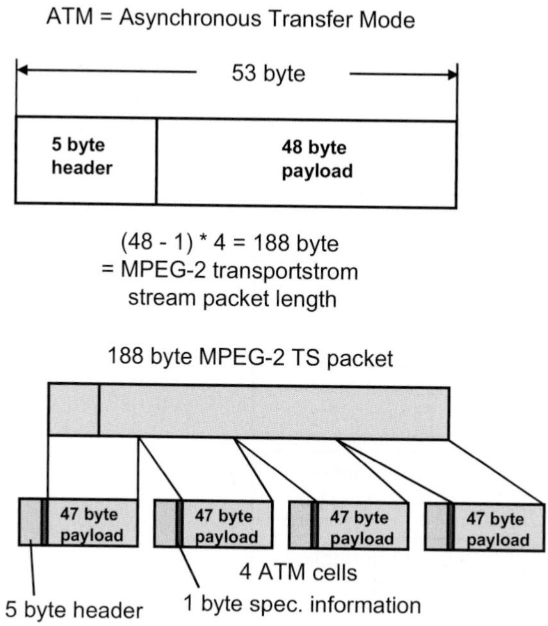

Fig. 3.11. ATM cell

When MPEG signals are transmitted via ATM links, various transmission modes called ATM Adaptation Layers can be applied at the ATM level. The mode shown in Fig. 3.11. corresponds to ATM Adaptation Layer 1 without FEC (i.e. AAL1 without FEC (forward error correction)). ATM Adaptation Layer 1 with FEC (AAL1 with FEC) or ATM Adaptation Layer 5 (AAL5) are also possible. The most suitable layer appears to be AAL1 with FEC since the contents are error-protected during the ATM transmission in this case.

The fact that the MPEG-2 transport stream is a completely asynchronous data signal is of particularly decisive significance. There is no way of knowing what information will follow in the next time slot (= transport stream packet). This can only be determined by means of the PID of the transport stream packet. The actual payload data rates in the payload can fluctuate; there may be stuffing to supplement the missing 184 bytes. This asynchronism has great advantages with regard to future flexibility, making it possible to implement any new method without much adaptation. But there are also disadvantages: the receiver must always be monitoring and thus uses more power; unequal error protection as, e.g., in DAB (digital audio broadcasting) cannot be applied and different contents can not be protected to a greater or lesser degree as required.

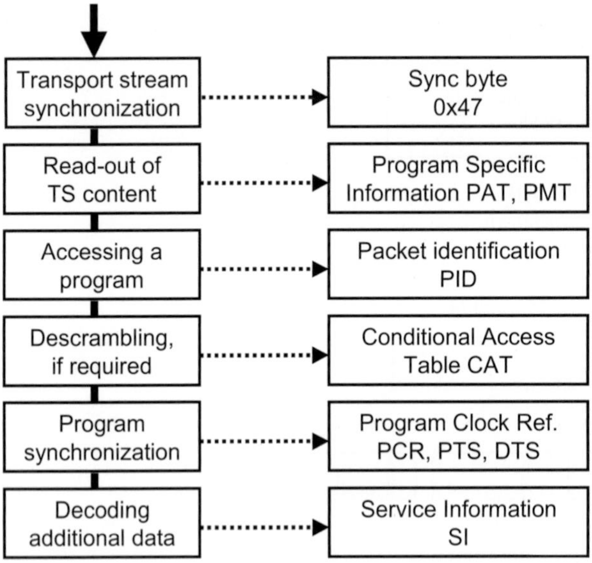

Fig. 3.12. Information for the receiver

3.3 Information for the Receiver

In the following paragraphs, the components of the transport stream which are necessary for the receiver will be considered. Necessary components means in this case: What does the receiver, i.e. the MPEG decoder, need for extracting from the large number of transport stream packets with the most varied contents exactly those which are needed for decoding the desired program? In addition, the decoder must be able to synchronize correctly to this program. The MPEG-2 transport stream is a completely asynchronous signal and its contents occur in a purely random fashion or on demand in the individual time slots. There is no absolute rule which can be used for determining what information will be contained in the next transport stream packet. The decoder and every element on the transmission link must lock to the packet structure. The PID (packet identifier) can be used for finding out what is actually being transmitted in the respective element. On the one hand, this asynchronism has advantages because of the total flexibility provided but there are also disadvantages with regard to power saving. Every single transport stream packet must first be analysed in the receiver.

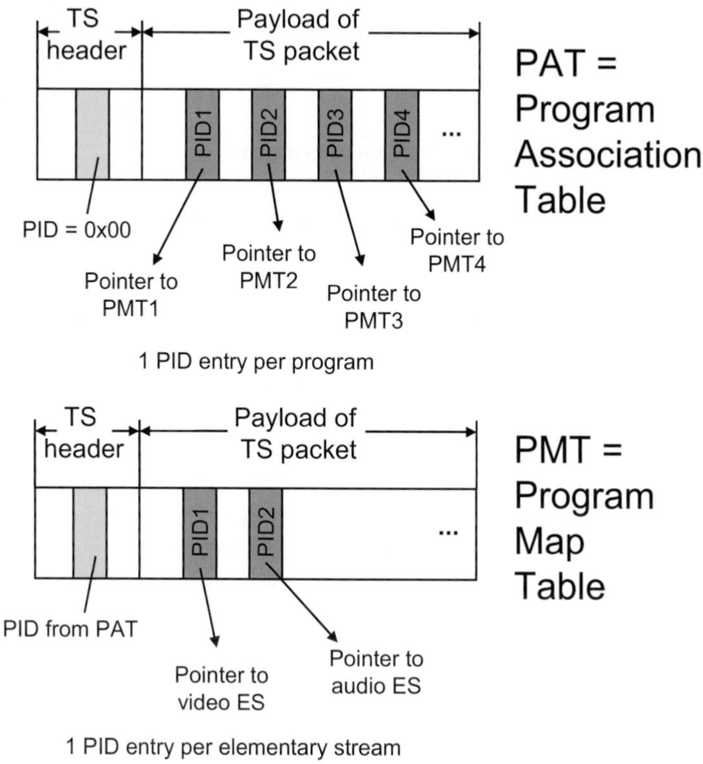

Fig. 3.13. PAT and PMT

3.3.1 Synchronizing to the Transport Stream

When the MPEG-2 decoder input is connected to an MPEG-2 transport stream, it must first lock to the transport stream, i.e. to the packet structure. The decoder, therefore, looks for the sync bytes in the transport stream. These always have the value of 0x47 and always appear at the beginning of a transport stream packet. They are thus present at constant intervals of 188 bytes. These two factors together, the constant value of 0x47 and the constant spacing of 188 bytes, are used for the synchronization. If a byte having a value of 0x47 appears, the decoder will examine the positions of n times 188 bytes before and after this byte in the transport stream for the presence of another sync byte. If there is, then this is a sync byte. If not, then this is simply some code word which has accidentally assumed this value. It is inevitable that the code word of 0x47 will also occur in the continuous transport stream. Synchronization will occur after 5 transport stream packets and the decoder will lose lock after a loss of 3 packets (as quoted in the MPEG-2 Standard).

3.3.2 Reading out the Current Program Structure

The number and the structure of the programs transmitted in the transport stream is flexible and open. The transport stream can contain one program with one video and audio elementary stream, or there can be 20 programs or more, some with only audio, some with video and audio and some with video and a number of audio signals which are being broadcast. It is, therefore, necessary to include certain lists in the transport stream which describe the instantaneous structure of the transport stream.

These lists provide the so-called "program specific information", or PSI in short (Fig. 3.13.). They are tables which are occasionally transmitted in the payload part. The first table is the "Program Association Table" (PAT). This table occurs precisely once per transport stream but is repeated every 0.5 sec.. This table shows how many programs there are in this transport stream. Transport stream packets containing this table have the value zero as packet identifier (PID) and can thus be easily identified. In the payload part of the program association table, a list of special PIDs is transmitted. There is exactly one PID per program in the program association table (Fig. 3.13.).

These PIDs are pointers, as it were, to other information describing each individual program in more detail. They point to other tables, the so-called "Program Map Tables" (PMT). The program map tables, in turn, are special transport stream packets with a special payload part and special PID. The PIDs of the PMTs are transmitted in the PAT. If it is intended to receive, e.g. program No.3, PID no. 3 is selected in the list of all PIDs in the payload part in the program association table (PAT). If this is, e.g. 0x1FF3, the decoder looks for transport stream packets having PID = 0x1FF3 in their header. These packets are then the program map table for program no. 3 in the transport stream. The program map table, in turn, contains PIDs which are the PIDs for all elementary streams contained in this program (video, audio, data).

Since there can be a number of video and audio streams - as for instance in a Formula-1 broadcast in various languages - the viewer must select the elementary streams to be decoded. Ultimately he will select exactly 2 PIDs - one for the video stream and one for the audio stream, resulting e.g. in the two hexadecimal numbers PID1 = 0x100 and PID2 = 0x110. PID1 is then e.g. the PID for the video stream to be decoded and PID2 is the PID for the audio stream to be decoded. From now on, the MPEG-2 decoder will only be interested in these transport stream packets, collect them, i.e. demultiplex them and assemble them again to form the PES packets. It is precisely these PES packets which are supplied to the video and audio decoder in order to generate another video-and-audio signal.

The composition of the transport stream can change during the transmission, e.g. local programs can only be transmitted within certain windows. A set-top box decoder, e.g. for DVB-S signals must, therefore, continuously monitor in the background the instantaneous structure of the transport stream, read out the PAT and PMTs and adapt to new situations. The header of a table contains a so-called version management for this purpose which signals to the receiver whether something has changed in the structure. It is regrettable that this does still not hold true for all DVB receivers. A receiver often recognizes a change in the program structure only after a new program search has been started. In many regions in Germany, so-called "regional window programs" are inserted into the public service broadcast programs at certain times of the day. These are implemented by a so-called "dynamic PMT", i.e. the contents of the PMT are altered and signal changes in the PIDs of the elementary streams.

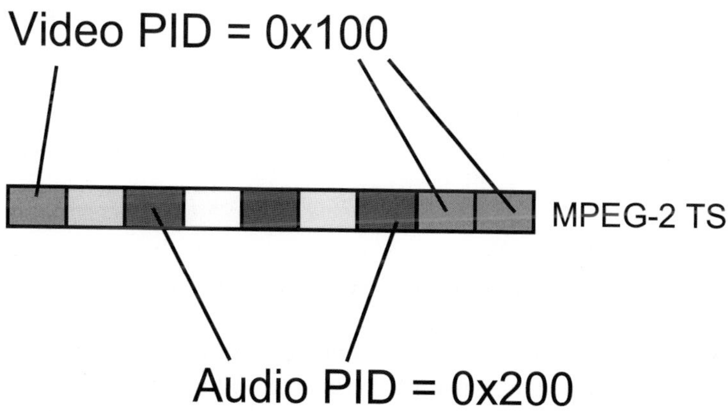

Fig. 3.14. Accessing a program via video and audio PIDs

3.3.3 Accessing a Program

After the PIDs of all elementary streams contained in the transport stream have become known from the information contained in the PAT and the PMTs and the user has committed himself to a program, a video and audio stream, precisely two PIDs are now defined (Fig. 3.14.): the PID for the video signal to be decoded and the PID for the audio signal to be decoded. The MPEG-2 decoder, on instruction by the user of the set-top box, will now only be interested in these packets. Assuming then that the video PID is 0x100 and the audio PID is 0x110: in the following demultiplexing process all TS packets with 0x100 will be assembled into video PES packets and supplied to the video decoder. The same applies to the 0x110 audio packets which are collected together and reassembled to form PES packets which are supplied to the audio decoder. If the elementary streams are not scrambled, they can now also be decoded directly.

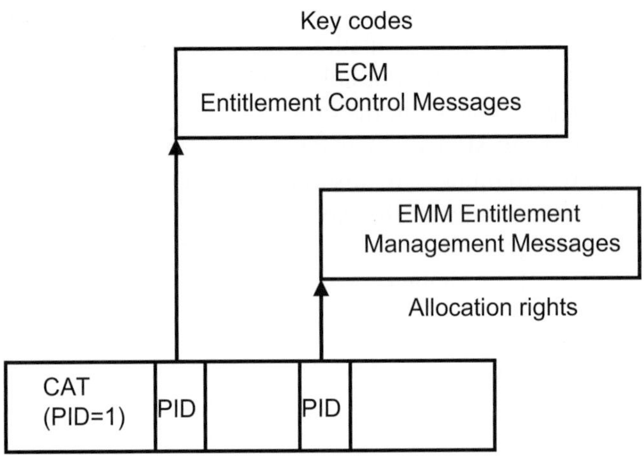

Fig. 3.15. The Conditional Access Table

3.3.4 Accessing Scrambled Programs

However, the elementary streams are transmitted scrambled. All or some of the elementary streams are transmitted protected by an electronic code in the case of pay TV or for licencing reasons involving local restrictions on reception. The elementary streams are scrambled (Fig. 3.17.) by various methods (Viaccess, Betacrypt, Irdeto, Conax, Nagravision etc.) and cannot be received without additional hardware and authorization. This additional hardware must be supplied with the appropriate descrambling and authorization data from the transport stream. For this purpose, a special table is transmitted in the transport stream, the "conditional access table" (CAT) (Fig. 3.15.).

The CAT supplies the PIDs for other data packets in the transport stream in which this descrambling information is transmitted. This additional descrambling information is called ECM (entitlement control message) and EMM (entitlement management message). The ECMs are used for transmitting the scrambling codes and the EMMs are used for user administration. The important factor is that only the elementary streams themselves may be scrambled, and no transport stream headers (or tables, either). Neither is it permitted to scramble the transport stream header or the adaptation field.

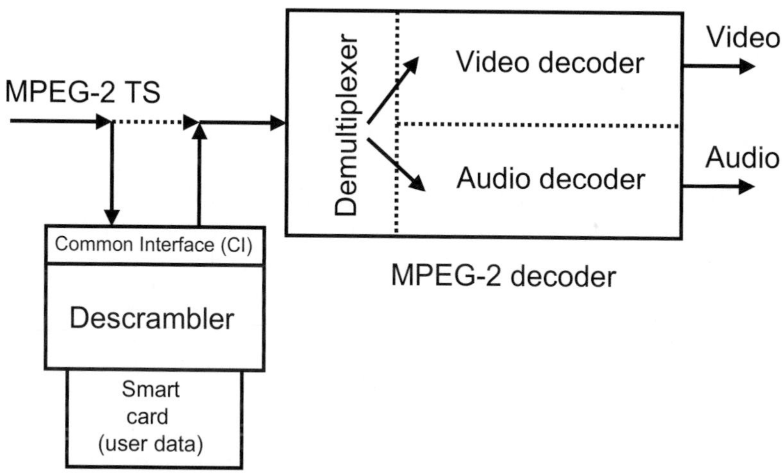

Fig. 3.16. Descrambling in the DVB receiver

The descrambling itself is done outside the MPEG decoder in additional hardware related to the descrambling method, which can be plugged into a so-called "common interface" (CI) in the set-top box. The transport stream is looped through this hardware before being processed further in the MPEG decoder. The information from the ECMs and EMMs and the user's personal code from the smart card then enable the streams to be descrambled.

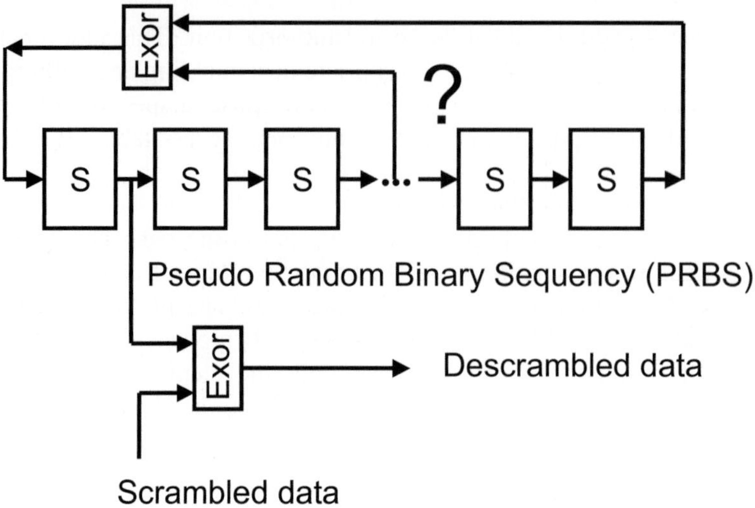

Fig. 3.17. Scrambling and descrambling by PRBS generator in the CA system and the receiver

3.3.5 Program Synchronization (PCR, DTS, PTS)

Once the PIDs for video and audio have been determined and any scrambled programs have been descrambled and the streams have been demultiplexed, video and audio PES packets are generated again. These are then supplied to the video and audio decoder. The actual decoding, however, requires a few more synchronization steps. The first step consists of linking the receiver clock to the transmitter clock. As indicated initially, the luminance signal is sampled at 13.5 MHz and the two chrominance signals are sampled at 6.75 MHz. 27 MHz is a multiple of these sampling frequencies, which is why this frequency is used as reference, or basic, frequency for all processing steps in the MPEG encoding at the transmitter end. A 27 MHz oscillator in the MPEG encoder feeds the "system time clock" (STC). The STC is essentially a 42 bit counter which is clocked by this same 27 MHz clock and starts again at zero after an overflow. The LSB positions do not go up to FFF but only to 300. Approximately every 26.5 hours the counter restarts at zero. At the receiving end, another system time clock (STC) must be provided, i.e. another 27 MHz oscillator connected to a 42 bit counter is needed. However, the frequency of this 27 MHz oscillator must be in complete synchronism with the transmitting end, and the 42 bit counter must also count in complete synchronism.

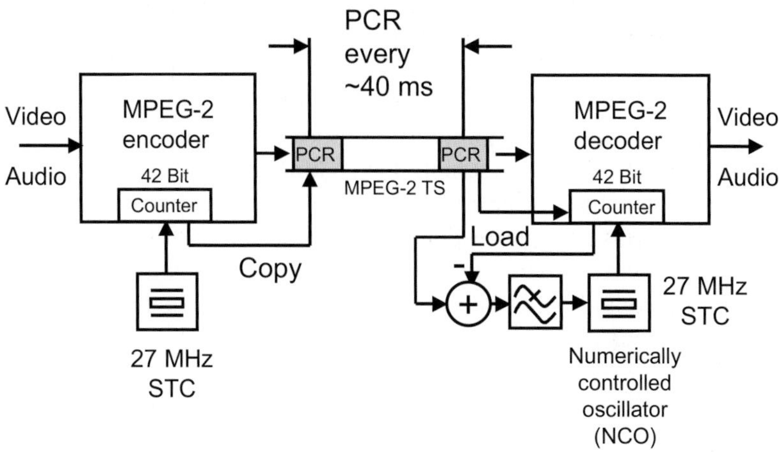

Fig. 3.18. Program Clock Reference

To accomplish this, reference information is transmitted in the MPEG data stream (Fig. 3.18.). In MPEG-2, these are the "Program Clock Reference" (PCR) values which are nothing else than an up-to-date copy of the STC counter fed into the transport stream at a certain time. The data stream thus carries an accurate internal "clock time". All coding and decoding processes are controlled by this clock time. To do this, the receiver, i.e. the MPEG decoder, must read out the "clock time", namely the PCR values, and compare them with its own internal system clock, that is to say its own 42 bit counter.

If the received PCR values are locked to the system clock in the decoder, the 27 MHz clock at the receiving end matches the transmitting end. If there is a deviation, a controlled variable for a PLL can be generated from the magnitude of the deviation, i.e. the oscillator at the receiving end can be corrected. In parallel, the 42 bit count is always reset to the received PCR value, a basic requirement for system initialization and in the event of a program change.

The PCR values must be present in sufficient numbers, that is to say with a maximum spacing, and relatively accurately, that is to say free of jitter. According to MPEG, the maximum spacing per program is 40 ms between individual PCR values. The PCR jitter must be less than ± 500 ns. PCR problems manifest themselves in the first instance in that instead of a color picture, a black/white picture is displayed. PCR jitter problems can occur during the remultiplexing of a transport stream, among other things. The reason is that e.g., the order of the transport stream packets is changed without the PCR information continued in them also being changed. There is frequently a PCR jitter of up to ± 30 μs even though only ± 500 ns is allowed. This can be handled by many set-top boxes but not by all. The PCR information is transmitted in the adaptation field of a transport stream packet belonging to the corresponding program. The precise information about the type of TS packets in which this is done can be found in the corresponding program map table (PMT). The PMT contains the so-called PCR_PID which, however, corresponds to the video PID of the respective program in most cases. After program clock synchronization has been achieved, the video and audio coding steps are then executed in lock with the system time clock (STC).

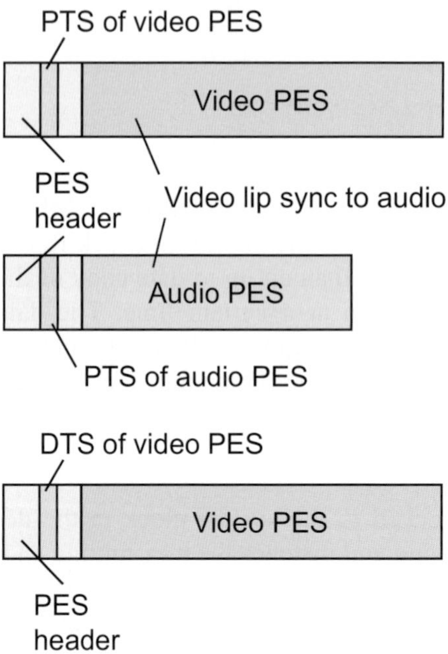

Fig. 3.19. PTS and DTS

However, another problem now presents itself. Video and audio must be decoded and reproduced with lip synchronization. In order to be able to achieve "lip sync", i.e. synchronization between video and audio, additional timing information is keyed into the headers of the video and audio PESs. This timing information is derived from the system time clock (STC, 42 bits). Using the 33 most significant bits (MSB) of the STC, these values are entered into the video and audio PES headers at maximum intervals of 700 ms and are called "presentation time stamps" (PTS)

As will be seen later in the section on video coding, the order in which the compressed picture information is transmitted will differ from the order in which it is recorded. The frame sequence is now scrambled in conformity with certain coding rules, a necessary measure in order to save memory space in the decoder. To recover the original sequence, additional time stamps must be keyed into the video stream. These are called "decoding time stamps" (DTS) and are also transmitted in the PES header.

An MPEG-2 decoder in a set-top box is then able to decode the video and audio streams of a program, resulting again in video and audio signals, either in analog form or in digital form.

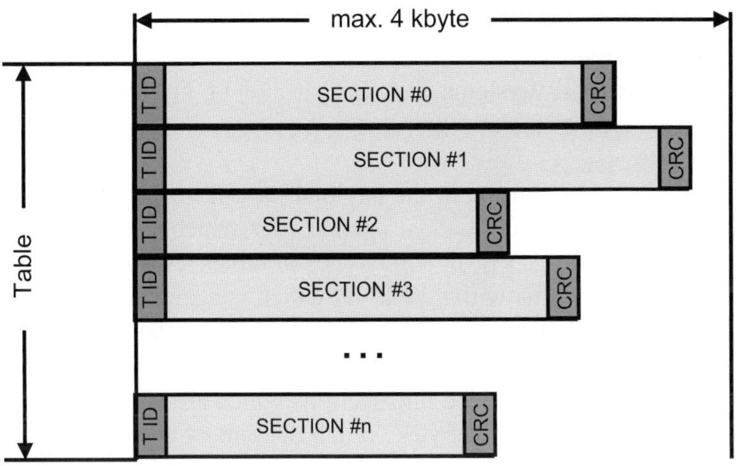

Fig. 3.20. Sections and tables

3.3.6 Additional Information in the Transport Stream (SI/PSI/PSIP)

According to MPEG, the information transmitted in the transport stream is fairly hardware-oriented, only relating to the absolute minimum requirements, as it were. However, this does not make the operation of a set-top box particularly user-friendly. For example, it makes sense, and is necessary, to transmit program names for identification purposes. It is also desirable to simplify the search for adjacent physical transmission channels. It is also necessary to transmit electronic program guides (EPG) and time and date information. In this respect, both the European DVB Project group and the US ATSC Project group have defined additional information for the transmission of digital video and audio programs which is intended to simplify the operation of set-top boxes and make it much more user-friendly.

3.3.7 Non-Private and Private Sections and Tables

To cope with any extensions, the MPEG Group has incorporated an "open door" in the MPEG-2 Standard. In addition to the "program specific information" (PSI), the "program map table" (PMT) and the "conditional access table" (CAT), it created the possibility to incorporate so-called "private sections and private tables" (Fig. 3.20.) in the transport stream. The group has defined mechanisms which specify what a section or table has to look like, what its structure has to be and by what rules it is to be linked into the transport stream.

According to MPEG-2 Systems (ISO/IEC 13818-1), the following was specified for each type of table:

- A table is transmitted in the payload part of one or more transport stream packets with a special PID which is reserved for only this table (DVB) or some types of tables (ATSC).
- Each table begins with a table ID which is a special byte which identifies only this table alone. The table ID is the first payload byte of a table.
- Each table is subdivided into sections which are allowed to have a maximum size of 4 bytes. Each section of a table is terminated with a 32-bit-long CRC checksum over the entire section.

The "Program Specific Information" (PSI) has exactly the same structure. The PAT has a PID of zero and begins with a table ID of zero. The PMT has the PIDs defined in the PAT as PID and has a table ID of 2. The CAT has a PID and a table ID of one in each case. The PSI can be composed of one or more transport stream packets for PAT, PMT and CAT depending on content.

Apart from the PSI tables PAT, PMT and CAT mentioned above, another table, the so-called "network information table" (NIT) was provided in principle but not standardized in detail. It was actually implemented as part of the DVB (Digital Video Broadcasting) project.

All tables are implemented through the mechanism of sections. There are non-private and private sections (Fig. 3.21.). Non-private sections are defined in the original MPEG-2 Systems Standard. All others are correspondingly private. The non-private sections include the PSI tables and the private ones include the SI sections of DVB and the MPEG-2 DSM-CC (Digital Storage Media Command and Control) sections which are used for data broadcasting. The header of a table contains administration of the version number of a table and information about the number of sections of which a table is made up. A receiver must first of all scan through the header of these sections before it can evaluate the rest of the sections and tables. Naturally, all sections must be broken down from an original maximum length of 4 kbytes to maximally 148 bytes payload length of an MPEG-2 transport stream packet before they are transmitted.

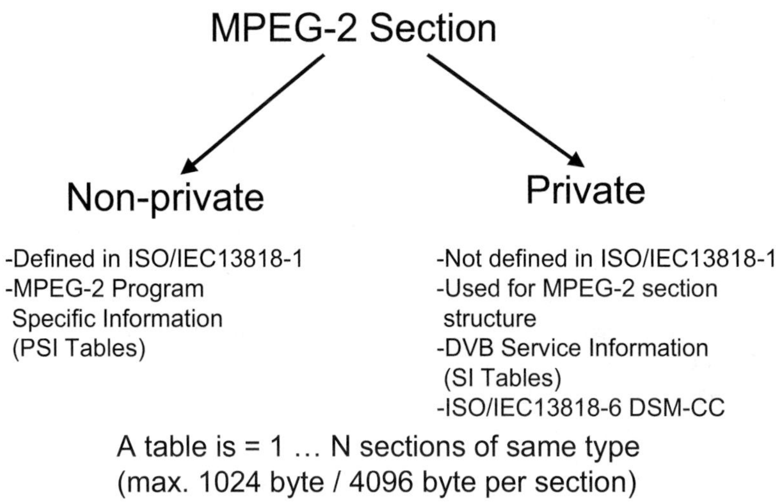

Fig. 3.21. Sections and tables according to MPEG-2

In the case of PSI/SI, the limit of the section length has been lowered to 1 kbyte in almost all tables, the only exception being the EIT (Event Information Table) which is used for transmitting the electronic program guide (EPG). The sections of the EIT can assume the maximum length of 4 kbytes because they carry a large amount of information as in the case of a week-long EPG.

If a section begins in a transport stream packet (Fig. 3.22.), the payload unit start indicator of its header is set to "1". The TS header is then followed immediately by the pointer which points (in number of bytes) to the actual beginning of the section. In most cases (and always in the case of

PSI/SI), this pointer is set to zero which means that the section begins immediately after the pointer.

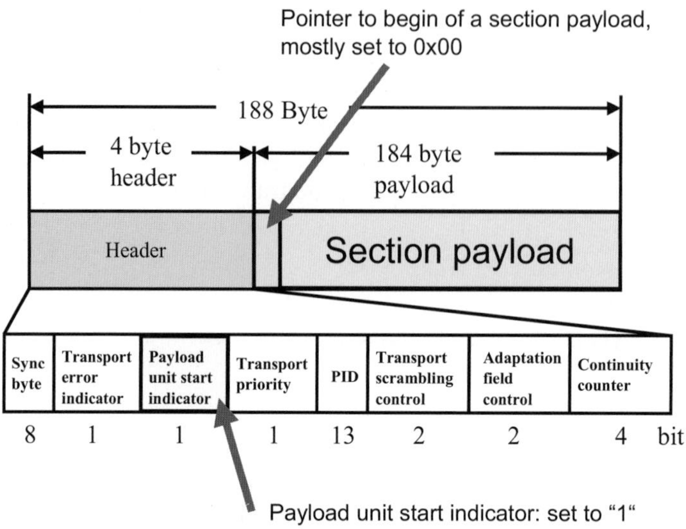

Fig. 3.22. Beginning of a section in an MPEG-2 transport stream packet

```
table_id                                   8 Bit
section_syntax_indicator                   1
private_indicator                          1
reserved                                   2
section_length                             12
if (section_syntax_indicator == 0)
  table_body1() /* short table */
else
  table_body2() /* long table */
if (section_syntax_indicator == 1)
  CRC                                      32 Bit
```

Fig. 3.23. Structure of a section

If the pointer has a value which differs from zero, remainders of the preceding section can still be found in this transport stream packet. This is utilized for saving TS packets, an example being MPE (multi-protocol encapsulation) over DSM-CC sections in the case of IP over MPEG-2 (see DVB-H).

The structure of sections always follow the same plan (Fig. 3.23., Fig. 3.24.). A section begins with the table_ID, a byte which signals the type of table. The section_syntax_indicator bit indicates whether this is a short type of section (bit = 0) or a long one (bit = 1). If it is a long section, this is then followed by an extended header which contains, among other things, the version management of the section and its length and the number of the last section. The version number indicates if the content of the section has changed (e.g. in case of a dynamic PMT or if the program structure has changed). A long section is always concluded with a 32-bit-long CRC checksum over the entire section.

```
table_body1()
{
  for (i=0;i<N;i++)
    data_byte                              8 Bit
}

table_body2()
{
  table_id_extension                       16 Bit
  reserved                                 2
  version_number                           5
  current_next_indicator                   1
  section_number                           8
  last_section_number                      8

  for (i=0;i<N;i++)
    data_byte                              8 Bit
}
```

Fig. 3.24. Structure of the section payload

The detailed structure of a PAT and PMT can now also be understood more easily. A PAT (Fig. 3.25, Fig. 3.26) begins with the table_ID = 0x00. Its type is that of a non-private long table, i.e. the version management follows in the header. Since the information about the program structure to be transmitted is very short, a single section is virtually always sufficient (last_section_no = 0) and it also fits inside a transport packet. In the program loop, the program number and the associated program map ID are listed for each program. Program No. Zero is a special exception, it informs about the PID of the later NIT (program information table). The PAT is then concluded with the CRC checksum. There is one PAT per transport stream but it is repeated every 0.5 sec. In the header of the table, an unambiguous number, the transport stream_ID, is allocated to the transport stream via which it can be addressed in a network (e.g. a satellite network with many transport streams). The PAT does not contain any text information.

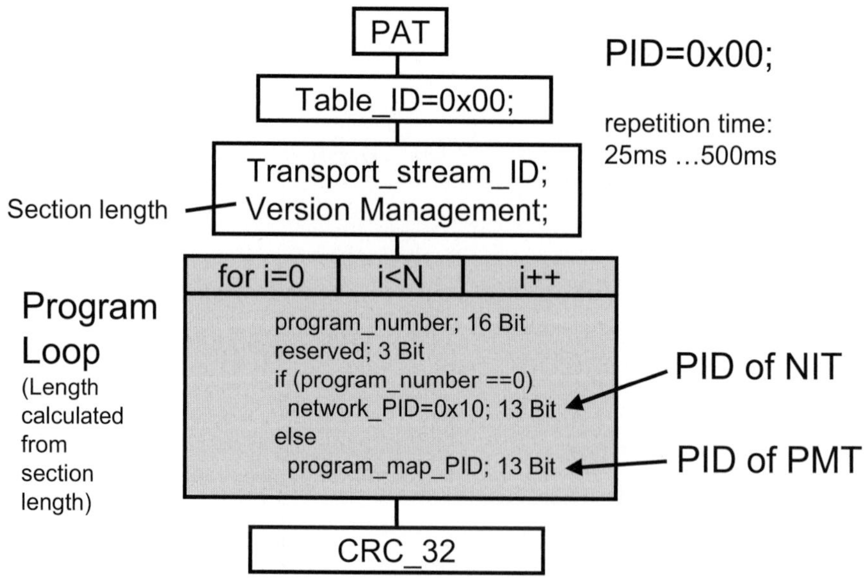

Fig. 3.25. Detailed structure of the PAT

The program map table (PMT) begins with the table_ID = 0x02. The PID is signalled via the PAT and is in the range of 0x20 ... 0x1FFE. The PMT is also a so-called non-private tale with version management and concluding CRC checksum. The header of the PMT carries the program_no, already familiar from the PAT. The program_no in PAT and PMT must match, i.e. be equal.

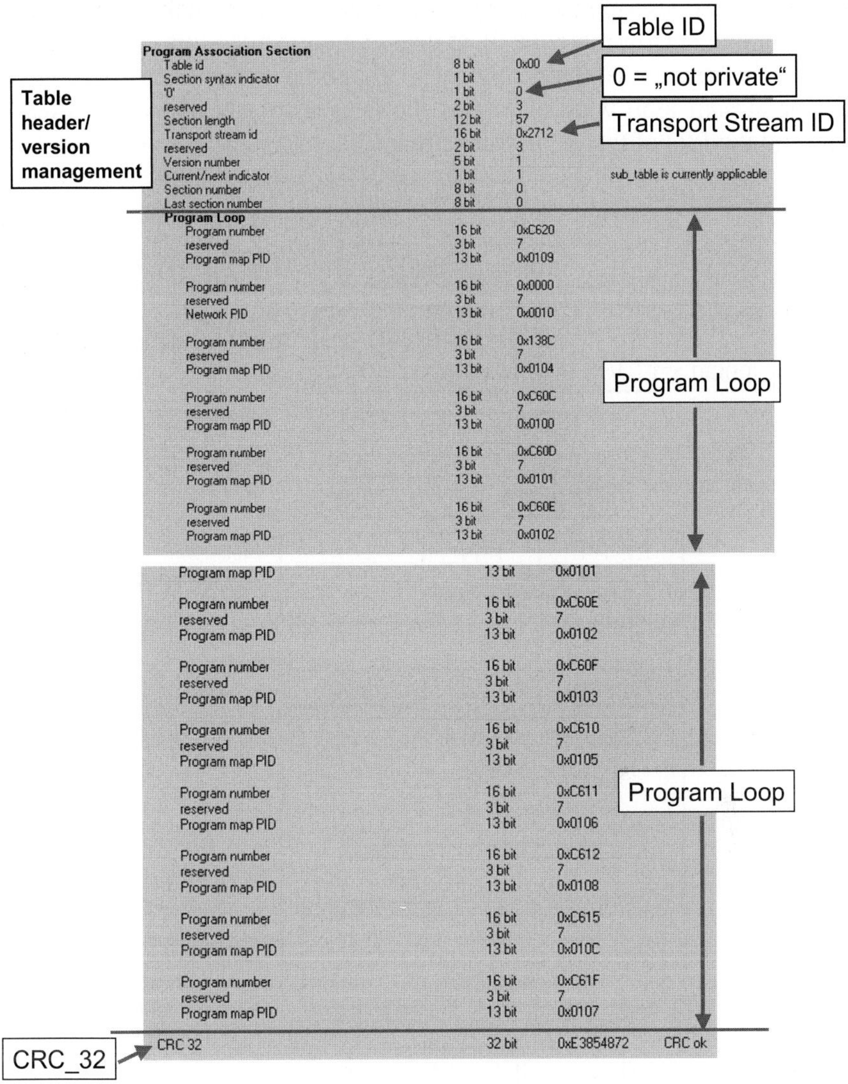

Fig. 3.26. Details of the Program Association Table (practical example)

The header of the PMT is followed by the program_info_loop into which various descriptors can be inserted as required which describe program components in more detail. It does not have to be utilized, however. The actual program components like video, audio or teletext are identified via the stream loop which contains the entries for the respective stream type and the PID of the elementary stream.

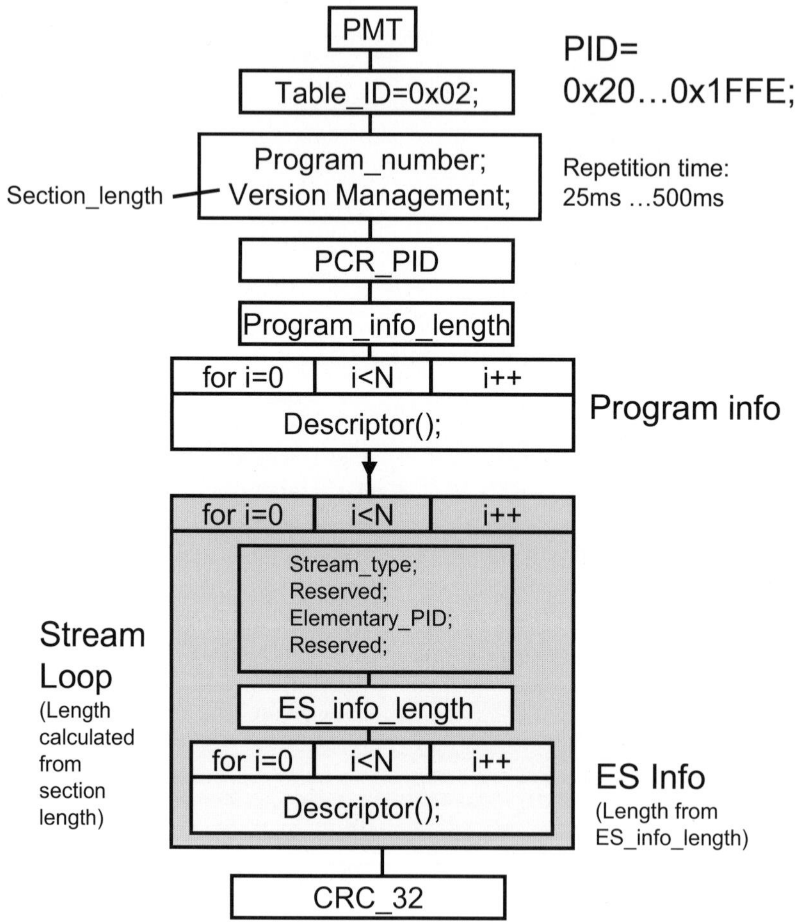

Fig. 3.27. Detailed structure of the Program Map Table

It is possible to include a number of descriptors for each program component in the ES_info_loop. There is one PMT for each program and it is sent out every 0.5 sec. There is no text information in the PMT, either.

Fig. 3.28. shows an actual example of the structure of a Program Map Table, which is quite short in this case. It will be discussed in more detail as representative of many other tables following. The example, recorded with an MPEG-2 analyzer, shows that the PMT begins with the table ID 0x02, a byte which clearly identifies it as such.

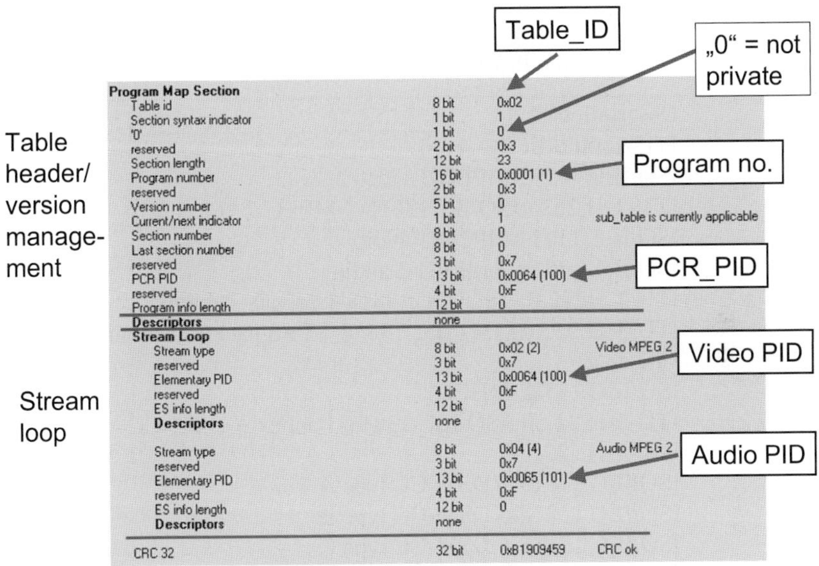

Fig. 3.28. Details of the Program Map Table (practical example)

The section syntax indicator bit is set to “1” and tells one that this is a long table with version management. The subsequent bit is set to “0” and identifies this table as a so/called non/private MPEG table. The section length says how long this current section of this table happens to be, namely 23 bytes long in this case. The field of the table_ID extension contains the program number; there must also be a corresponding entry in the PAT. The version number and the current/next indicator signal a change in the program map table. This information must be continuously checked by a receiver which must respond to a change in the program structure (dynamic PMT) if necessary. The section number tells what section this happens to be and the Last Section No informs about the number of the last section of a table. It is set to zero in this case, i.e. the table consists of only one section.

The PCR_PID (program check reference – packet identifier) provides the PID on which the PCR value is broadcast. This is the video PID in most cases.

There should now be a program_info loop but there is none in this example, a fact which is signalled by the length indicator "program_info_length = 0.

However, the stream loop provides information about the video and audio PID. The stream type (see Table 3.1.) shows the type of payload, namely MPEG-2 video and MPEG-2 audio in this case.

Table 3.1. Stream types of the Program Map Table

Value	Description
0x00	ITU-T/ISO/IEC reserved
0x01	ISO/IEC 11172 MPEG-1 video
0x02	ITU-T H.262 / ISO/IEC13818-2 MPEG-2 video
0x03	ISO/IEC 11172 MPEG-1 audio
0x04	ISO/IEC 13818-3 MPEG-2 audio
0x05	ITU-T H222.0 / ISO/IEC 13818-1 private sections
0x06	ITU-T H.222.0 / ISO/IEC 13818-1 PES packets containing private data
0x07	ISO/IEC 13522 MHEG
0x08	ITU-T H.222.0 /ISO/IEC 13818-1 annex A DSM-CC
0x09	ITU-T H.222.1
0x0A	ISO/IEC 13818-6 DSM-CC type A
0x0B	ISO/IEC 13818-6 DSM-CC type B
0x0C	ISO/IEC 13818-6 DSM-CC type C
0x0D	ISO/IEC 13818-6 DSM-CC type D
0x0E	ISO/IEC 13818-1 auxiliary
0x0F-0x7F	ITU-T H.222.0 / ISO/IEC 13818-1 reserved
0x80-0xFF	User private

3.3.8 The Service Information according to DVB (SI)

Taking advantage of the "private section" and "private table" features, the European DVB Group has introduced numerous additional tables intended to simplify the operation of the set-top boxes or quite generally of the DVB receivers. Called "service information" (SI), they are defined in ETSI Standard ETS300468.

They are the following tables (Fig. 3.29.): the "network information table" (NIT), the "service descriptor table" (SDT), the "bouquet association table" (BAT), the "event information table" (EIT), the "running status table" (RST), the "time&date table" (TDT), the "time offset table" (TOT) and, finally, the "stuffing table" (ST). These eight tables will now be described in more detail.

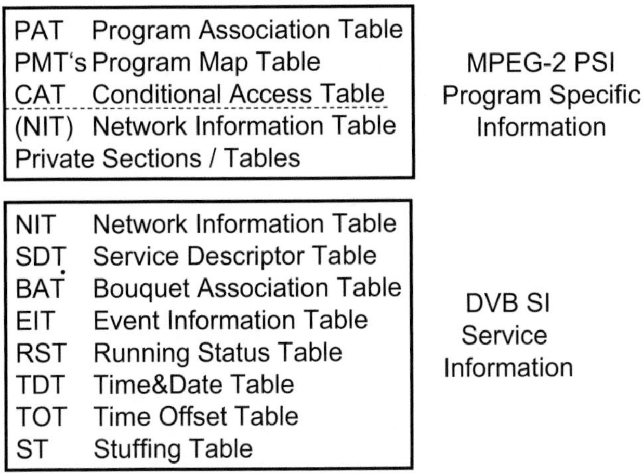

PAT Program Association Table
PMT's Program Map Table
CAT Conditional Access Table
(NIT) Network Information Table
Private Sections / Tables
MPEG-2 PSI
Program Specific
Information
NIT Network Information Table
SDT Service Descriptor Table
BAT Bouquet Association Table
EIT Event Information Table
RST Running Status Table
TDT Time&Date Table
TOT Time Offset Table
ST Stuffing Table
DVB SI
Service
Information

Fig. 3.29. MPEG-2 PSI and DVB SI

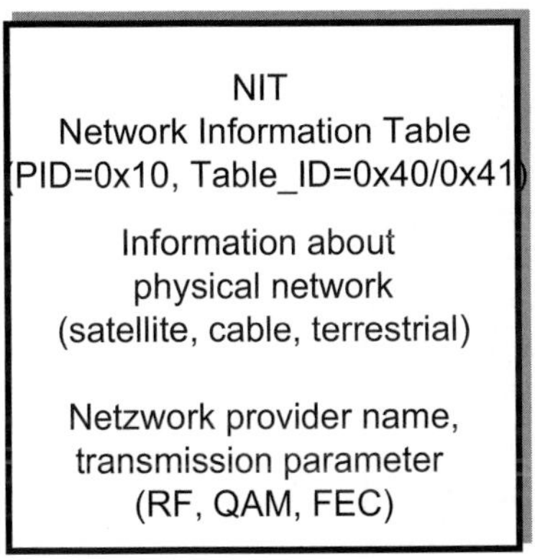

NIT
Network Information Table
(PID=0x10, Table_ID=0x40/0x41)
Information about
physical network
(satellite, cable, terrestrial)
Netzwork provider name,
transmission parameter
(RF, QAM, FEC)

Fig. 3.30. Network Information Table (NIT)

The "network information table" (NIT) (Fig. 3.30., Fig. 3.31., Fig. 3.32.) describes all physical parameters of a DVB transmission channel. It contains, e.g. the received frequency and the type of transmission (satellite, cable, terrestrial) and also all the technical data of the transmission, i.e. error protection, type of modulation etc.. This table has the purpose of optimizing the channel scan as much as possible. A set-top box is able to store all the parameters of a physical channel when scanning during setup, and it is possible, e.g. to broadcast information about all available physical channels within a network (e.g. satellite, cable), making it possible to do away with the actual physical search for channels.

The NIT contains the following information:

- Transmission path (satellite, cable, terrestrial)
- Received frequency
- Type of modulation
- Error protection
- Transmission parameters

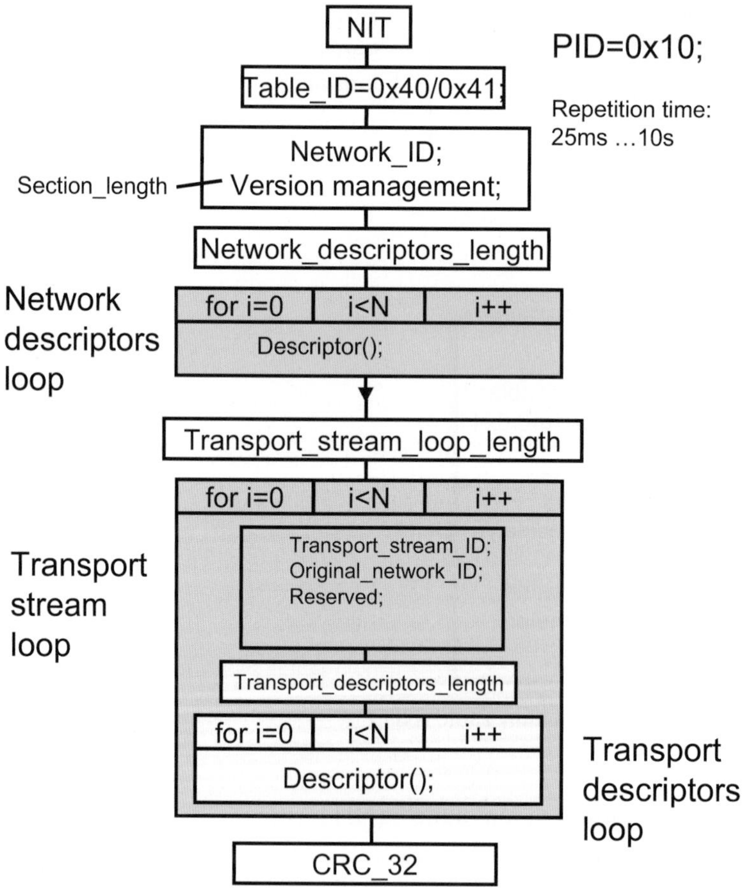

Fig. 3.31. Structure of the Network Information Table (NIT)

The important factor in relation to the NIT is that many receivers, i.e. set-top boxes, may behave in a "peculiar" manner if the transmission parameters in the NIT do not match the actual transmission. If, e.g. the transmit frequency given in the NIT does not correspond to the actual received frequency, many receivers, without any indication of reasons, may simply refuse to reproduce any picture or sound.

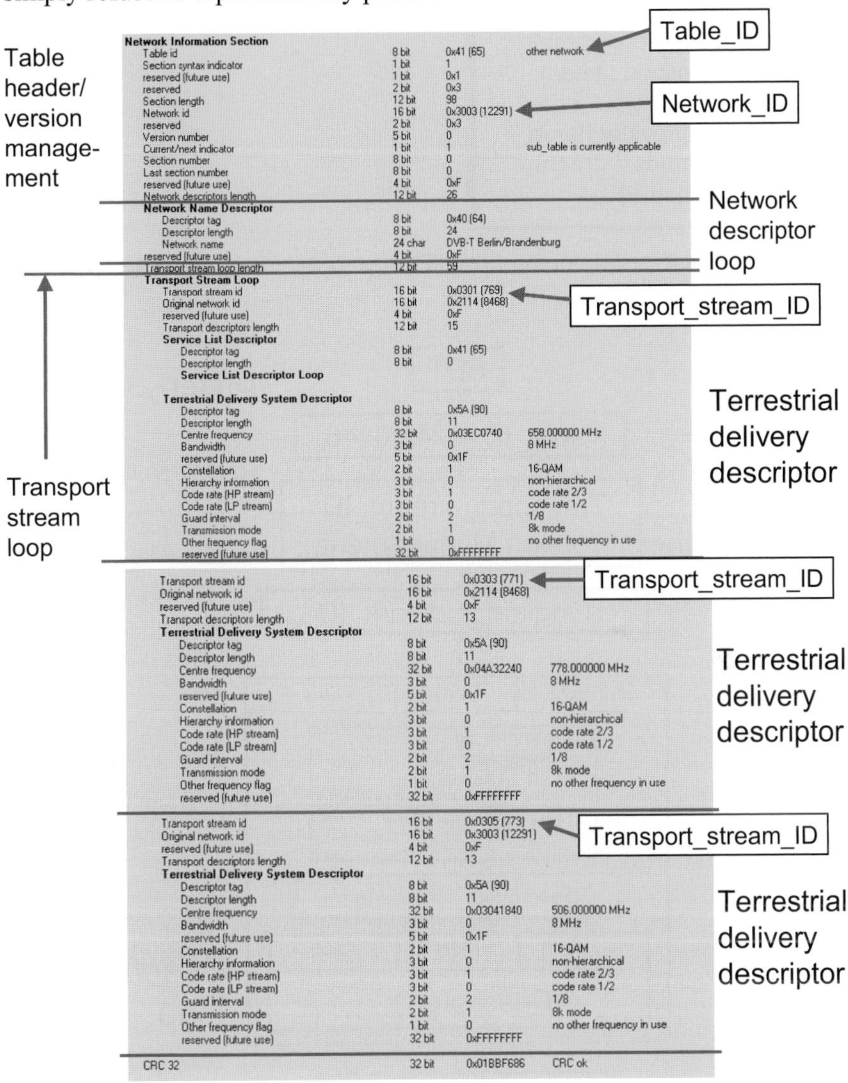

Network Information Section			
Table id	8 bit	0x41 (65)	other network
Section syntax indicator	1 bit	1	
reserved (future use)	1 bit	0x1	
reserved	2 bit	0x3	
Section length	12 bit	98	
Network id	16 bit	0x3003 (12291)	
reserved	2 bit	0x3	
Version number	5 bit	0	
Current/next indicator	1 bit	1	sub_table is currently applicable
Section number	8 bit	0	
Last section number	8 bit	0	
reserved (future use)	4 bit	0xF	
Network descriptors length	12 bit	26	
Network Name Descriptor			
Descriptor tag	8 bit	0x40 (64)	
Descriptor length	8 bit	24	
Network name	24 char	DVB-T Berlin/Brandenburg	
reserved (future use)	4 bit	0xF	
Transport stream loop length	12 bit	59	
Transport Stream Loop			
Transport stream id	16 bit	0x0301 (769)	
Original network id	16 bit	0x2114 (8468)	
reserved (future use)	4 bit	0xF	
Transport descriptors length	12 bit	15	
Service List Descriptor			
Descriptor tag	8 bit	0x41 (65)	
Descriptor length	8 bit	0	
Service List Descriptor Loop			
Terrestrial Delivery System Descriptor			
Descriptor tag	8 bit	0x5A (90)	
Descriptor length	8 bit	11	
Centre frequency	32 bit	0x03EC0740	658.000000 MHz
Bandwidth	3 bit	0	8 MHz
reserved (future use)	5 bit	0x1F	
Constellation	2 bit	1	16-QAM
Hierarchy information	3 bit	0	non-hierarchical
Code rate (HP stream)	3 bit	1	code rate 2/3
Code rate (LP stream)	3 bit	0	code rate 1/2
Guard interval	2 bit	2	1/8
Transmission mode	2 bit	1	8k mode
Other frequency flag	1 bit	0	no other frequency in use
reserved (future use)	32 bit	0xFFFFFFFF	
Transport stream id	16 bit	0x0303 (771)	
Original network id	16 bit	0x2114 (8468)	
reserved (future use)	4 bit	0xF	
Transport descriptors length	12 bit	13	
Terrestrial Delivery System Descriptor			
Descriptor tag	8 bit	0x5A (90)	
Descriptor length	8 bit	11	
Centre frequency	32 bit	0x04A32240	778.000000 MHz
Bandwidth	3 bit	0	8 MHz
reserved (future use)	5 bit	0x1F	
Constellation	2 bit	1	16-QAM
Hierarchy information	3 bit	0	non-hierarchical
Code rate (HP stream)	3 bit	1	code rate 2/3
Code rate (LP stream)	3 bit	0	code rate 1/2
Guard interval	2 bit	2	1/8
Transmission mode	2 bit	1	8k mode
Other frequency flag	1 bit	0	no other frequency in use
reserved (future use)	32 bit	0xFFFFFFFF	
Transport stream id	16 bit	0x0305 (773)	
Original network id	16 bit	0x3003 (12291)	
reserved (future use)	4 bit	0xF	
Transport descriptors length	12 bit	13	
Terrestrial Delivery System Descriptor			
Descriptor tag	8 bit	0x5A (90)	
Descriptor length	8 bit	11	
Centre frequency	32 bit	0x03041840	506.000000 MHz
Bandwidth	3 bit	0	8 MHz
reserved (future use)	5 bit	0x1F	
Constellation	2 bit	1	16-QAM
Hierarchy information	3 bit	0	non-hierarchical
Code rate (HP stream)	3 bit	1	code rate 2/3
Code rate (LP stream)	3 bit	0	code rate 1/2
Guard interval	2 bit	2	1/8
Transmission mode	2 bit	1	8k mode
Other frequency flag	1 bit	0	no other frequency in use
reserved (future use)	32 bit	0xFFFFFFFF	
CRC 32	32 bit	0x01BBF686	CRC ok

Fig. 3.32. Practical example of a Network Information Table (NIT)

SDT
Service Descriptor Table
(PID=0x11, Table_ID=0x42/0x46)
Information about all
services (= programs)
in a transport stream
Service provider name
service names = program
names

Fig. 3.33. Service Descriptor Table (SDT)

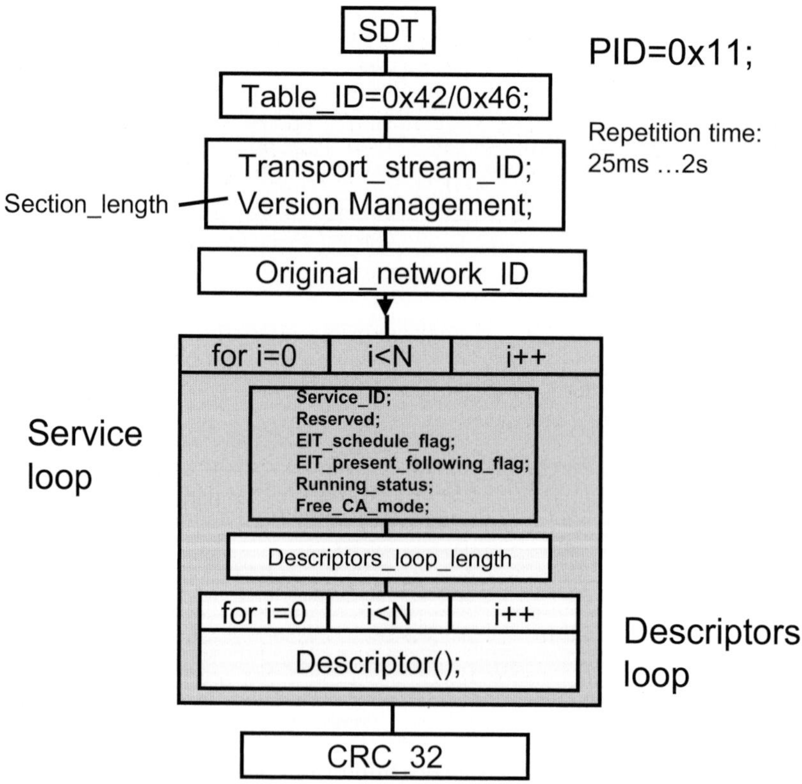

Fig. 3.34. Structure of the Service Descriptor Table (SDT)

The "service descriptor table" (SDT) contains more detailed descriptions of the programs carried in the transport stream, the "services". Among other things, these are the program titles such as, e.g. "CNN", "CBS", "Eurosport", "ARD", "ZDF", "BBC", "ITN" etc.. That is to say, in parallel with the program PIDs entered in the PAT, the SDT now contains textual information for the user. This is intended to facilitate the operation of the receiving device by providing lists of text.

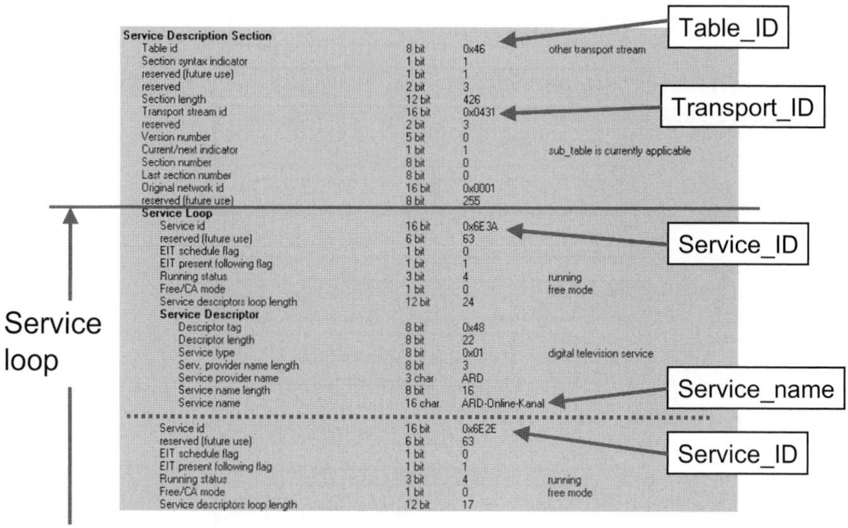

Fig. 3.35. Practical Example of an SDT

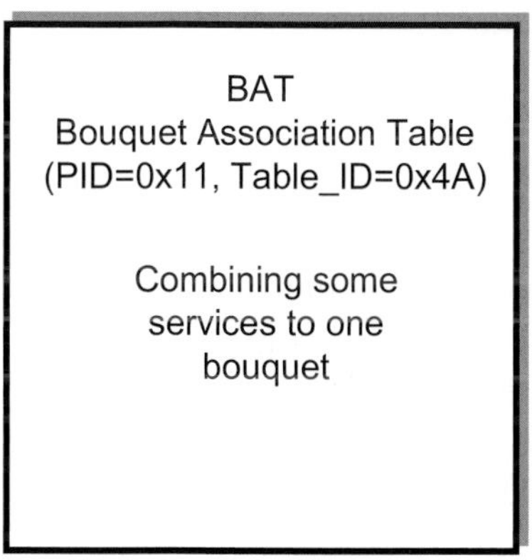

Fig. 3.36. Bouquet Association Table

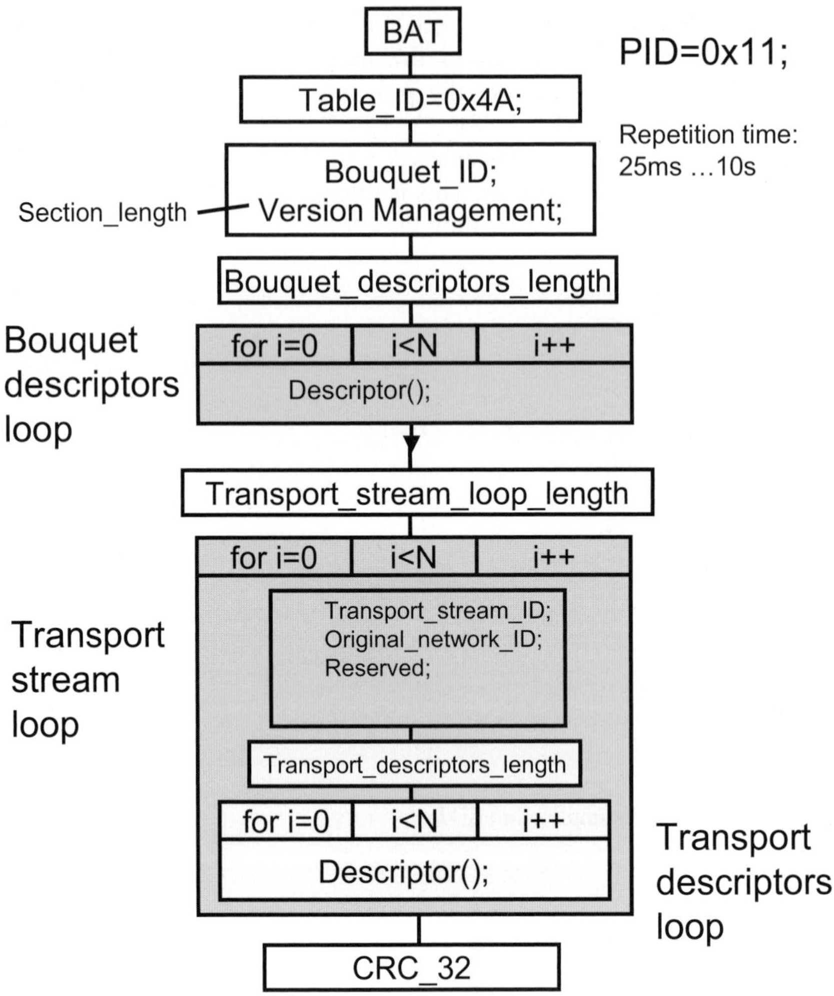

Fig. 3.37. Structure of a Bouquet Association Table (BAT)

A close relative of the service descriptor table is the "bouquet association table" (BAT). SDT and BAT have the same PID and differ only in the table ID. Whereas the SDT describes the program structure of one physical channel, a BAT describes the program structure of several physical channels or of a large number of physical channels.

The BAT is thus nothing else than a multi-channel program table. It provides an overview of all services contained in a group. Program providers can make use of e.g. an entire bouquet of physical channels if a single channel is insufficient for transmitting the complete range of programs provided. An example of this is the pay TV provider “Premiere”. A handful of satellite or cable DVB channels are combined here to form a bouquet of this provider’s channels. The associated BAT is transmitted in all individual channels and links this bouquet together.

In fact, however, a bouquet association table is found very rarely in a transport stream. The broadcasters ARD and ZDF in Germany, and Premiere, are broadcasting a BAT for their respective bouquet and sometimes a BAT can be found in networks of cable network providers.

But frequently, the BAT doesn’t exist at all, as already mentioned. When it does exist, it tells by way of so-called linkage descriptors which service of a particular service ID can be found in which transport streams.

Many providers are also transmitting an “electronic program guide” (EPG) which has its own table in DVB, the so-called “event information table”, or EIT for short (Fig. 3.38. and 3.39.). It contains the planned starting and stopping times of all broadcasts of, e.g. one day or one week. The structure which is possible here is very flexible and also allows any amount of additional information to be transmitted. Unfortunately it is true that this feature is not supported by all set-top boxes, or only inadequately so.

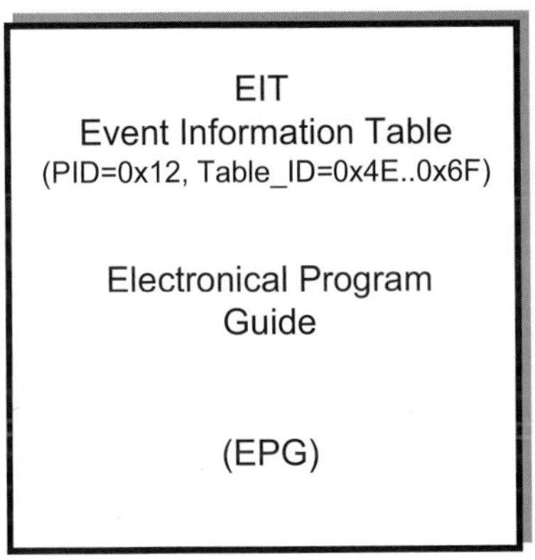

Fig. 3.38. EIT

Frequently, however, there are variations and delays in the planned starting and stopping times of broadcasts. To be able to start and stop, e.g. a video recorder at a given time, the relevant control information is transmitted in the “Running Status Table” (RST). The RST can thus be compared to the VPS (video program system) signal in the data line of an analog TV signal. The RST is currently not being used in practice, or, at least, has not been found by the author in a transport stream anywhere in the world, excepting "synthetic" transport streams. Instead, the data line containing the VPS has been adapted within DVB for controlling video recorders and similar recording media.

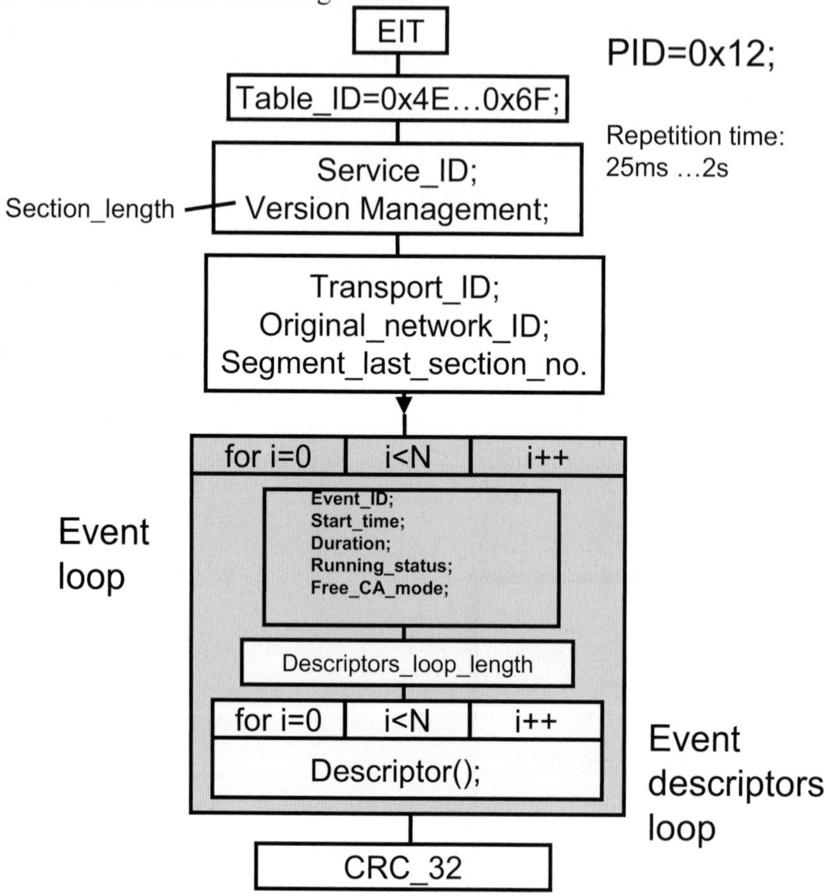

Fig. 3.39. Structure of the Event Information Table (EIT)

The operation of the set-top box also requires the transmission of the current clock time and the current date. This is done in two stages. In the "Time&Date Table" (TDT) (Fig. 3.42. and 3.43.), Greenwich Mean Time (GMT or UTC), i.e. the current clock time on the Zero-Degree meridian without any daylight saving time shift is transmitted. The respective applicable time offsets can then be broadcast in a "time offset table" (TOT) (Fig. 3.42. and 3.43.) for the various time zones. It depends on the software of the set-top box how the information contained in the TDT and TOT is evaluated, and to what extent. Complete support for this broadcast time information would require the set-top box to be informed of its current location and in a country having a number of time zones such as Australia, especially, more attention should be paid to this point.

It may sometimes be necessary to cancel certain information, especially tables in the transport stream. After a DVB-S signal has been received in a CATV head station, it can quite easily happen that, e.g. the NIT must be exchanged or overwritten or that individual programs must be rendered unusable for relaying. This can be done by means of the "stuffing table" (ST) (Fig. 3.44.) which enables information in the transport stream to be overwritten.

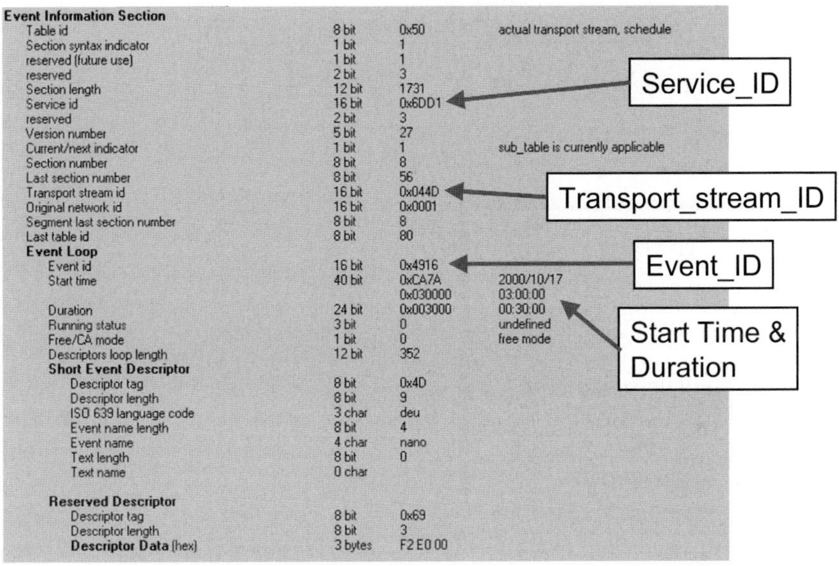

Event Information Section			
Table id	8 bit	0x50	actual transport stream, schedule
Section syntax indicator	1 bit	1	
reserved (future use)	1 bit	1	
reserved	2 bit	3	
Section length	12 bit	1731	
Service id	16 bit	0x6DD1	
reserved	2 bit	3	
Version number	5 bit	27	
Current/next indicator	1 bit	1	sub_table is currently applicable
Section number	8 bit	8	
Last section number	8 bit	56	
Transport stream id	16 bit	0x044D	
Original network id	16 bit	0x0001	
Segment last section number	8 bit	8	
Last table id	8 bit	80	
Event Loop			
Event id	16 bit	0x4916	
Start time	40 bit	0xCA7A	2000/10/17
		0x030000	03:00:00
Duration	24 bit	0x003000	00:30:00
Running status	3 bit	0	undefined
Free/CA mode	1 bit	0	free mode
Descriptors loop length	12 bit	352	
Short Event Descriptor			
Descriptor tag	8 bit	0x4D	
Descriptor length	8 bit	9	
ISO 639 language code	3 char	deu	
Event name length	8 bit	4	
Event name	4 char	nano	
Text length	8 bit	0	
Text name	0 char		
Reserved Descriptor			
Descriptor tag	8 bit	0x69	
Descriptor length	8 bit	3	
Descriptor Data (hex)	3 bytes	F2 E0 00	

Fig. 3.40. Event Information Table (practical example)

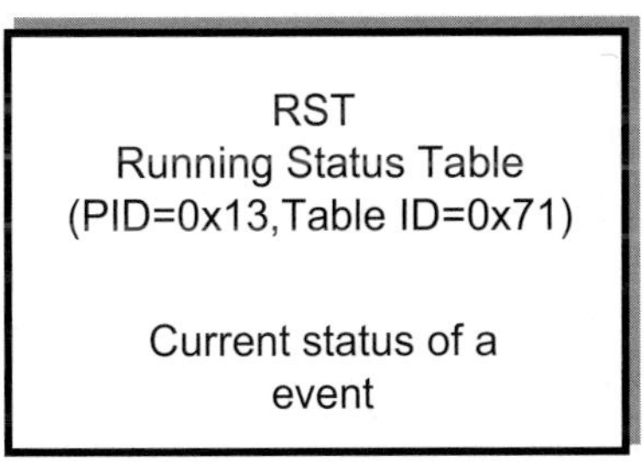

Fig. 3.41. Running Status Table (RST)

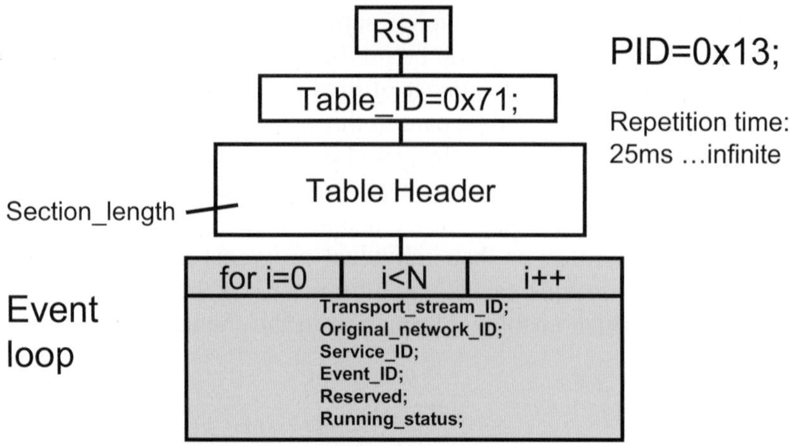

Fig. 3.42. Structure of the Running Status Table (RST)

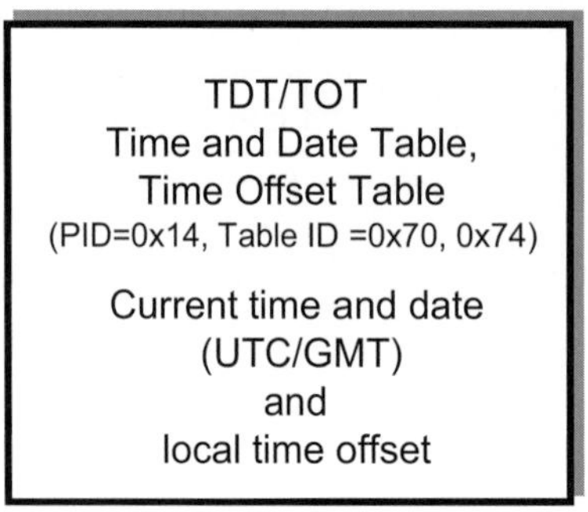

Fig. 3.43. Time and Date Table (TDT) and Time Offset Table (TOT)

The PIDs and the table IDs for the service information have been permanently allocated within DVB in Table 3.2.

The PSI/SI tables are linked to one another via the most varied identifiers (Fig. 3.45.). These are both PIDs and special, table-dependent identifiers. In the PAT, the PMT_PIDs are chained together by way of the prog_no. To each prog_no, a PMT_PID is allocated which refers to a transport stream packet with the corresponding PMT of this associated program. The prog_no can then also be found in the header of the respective PMT. Prog_no = 0 is allocated to the NIT where the PID of the NIT can be found.

Time and Date Section			
Table id	8 bit	0x70	
Section syntax indicator	1 bit	0	
reserved (future use)	1 bit	1	
reserved	2 bit	3	
Section length	12 bit	5	
UTC time	40 bit	0xCA79 0x105827	2000/10/16 10:58:27
Time Offset Section			
Table id	8 bit	0x73 (115)	
Section syntax indicator	1 bit	0	
reserved (future use)	1 bit	0x1	
reserved	2 bit	0x3	
Section length	12 bit	26	
UTC time	40 bit	0xCA79 0x112519	2000/10/16 11:25:19
reserved	4 bit	0xF	
Descriptors loop length	12 bit	15	
Local Time Offset Descriptor			
Descriptor tag	8 bit	0x58 (88)	
Descriptor length	8 bit	13	
Country Loop			
Country code	3 char	DEU	
Country region id	6 bit	0	no time zone extension used
reserved	1 bit	0x1	
Local time offset polarity	1 bit	0	local time is advanced to UTC
Local time offset	16 bit	0x0200	02:00
Time of change	40 bit	0xCA86 0x030000	2000/10/29 03:00:00
Next time offset	16 bit	0x0100	01:00
CRC 32	32 bit	0xD49C603D	CRC ok

Fig. 3.44. Example of a Time and Date Table (TDT and Time Offset Table (TOT)

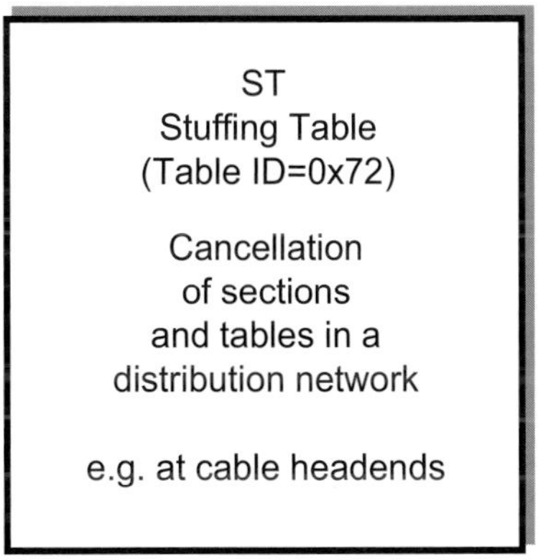

ST
Stuffing Table
(Table ID=0x72)
Cancellation
of sections
and tables in a
distribution network
e.g. at cable headends

Fig. 3.45. Stuffing Table (ST)

Table 3.2. PIDs and table IDs of the PSI/SI tables

Table	PID	Table_ID
PAT	0x0000	0x00
PMT	0x0020...0x1FFE	0x02
CAT	0x0001	0x01
NIT	0x0010	0x40...0x41
BAT	0x0011	0x4A
SDT	0x0011	0x42, 0x46
EIT	0x0012	0x4E...0x6F
RST	0x0013	0x71
TDT	0x0014	0x70
TOT	0x0014	0x73
ST	0x0010...0x0014	0x72

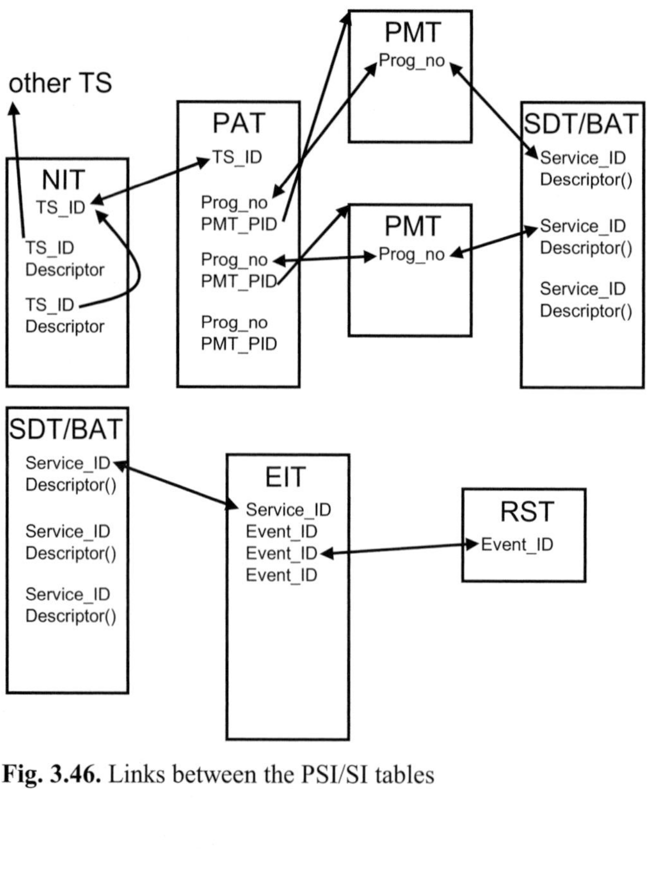

Fig. 3.46. Links between the PSI/SI tables

In the NIT, the physical parameters of all transport streams of a network are described via their TS_IDs. A TS_ID corresponds to the current transport stream; precisely this TS_ID can be found in the header of the PAT at the position of the Table ID extension.

The services (= programs) contained in this transport stream are listed in the service descriptor table via the service IDs. The service IDs must correspond to the prog_no in the PAT and in the PMTs.

This is continued in the EIT: there is an EIT for every service. In the header of the EIT, the table_ID_extension corresponds to the service_ID of the associated program. In the EIT, the events are associated with these by way of event_IDs. If there are associated RSTs, then these are chained to the respective RST via these event_IDs.

Table 3.3. Repetition rates of the PSI/SI tables according to MPEG/DVB

PSI/SI table	Max. interval (complete table)	Min. interval (single sections)
PAT	0.5 s	25 ms
CAT	0.5 s	25 ms
PMT	0.5 s	25 ms
NIT	10 s	25 ms
SDT	2 s	25 ms
BAT	10 s	25 ms
EIT	2 s	25 ms
RST	-	25 ms
TDT	30 s	25 ms
TOT	30 s	25 ms

The repetition rates of the PSI/SI tables are regulated through MPEG-2 Systems [ISO&IEC 13818/1] and DVB/SI [ETS 300468] (Table 3.3)

3.4 The PSIP according to the ATSC

In the US, a separate standard was specified for digital terrestrial and cable TV. This is the ATSC standard, where ATSC stands for Advanced Television System Committee. During the work on the ATSC standard, the decision was made to use the MPEG-2 transport stream with MPEG-2 video and AC-3 Dolby Digital audio as the baseband signal. The type of modulation used is 8 or 16VSB. In addition, it was recognized that other tables going beyond PSI are needed. Like the SI tables in DVB, ATSC, therefore, has the PSIP tables, listed below and described in more detail in the text which follows.

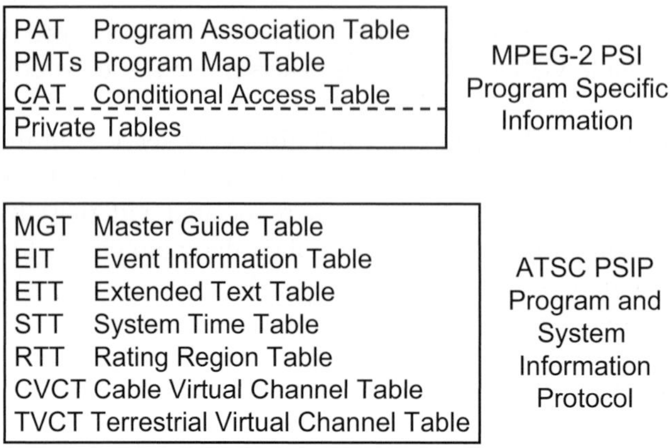

Fig. 3.47. ATSC PSIP tables

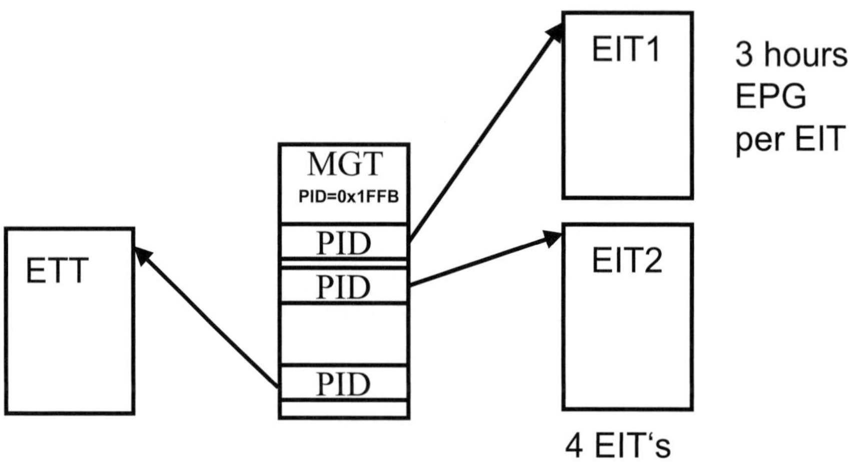

Fig. 3.48. Referencing the PSIP in the MGT

PSIP stands for "program and system information protocol" and is nothing else than another way of representing similar information to that given in the previous section on DVB SI. In ATSC, the following tables are used: the Master Guide Table (MGT) (Fig. 3.47.), the Event Information Table (EIT), the Extended Text Table (ETT), the System Time Table

(STT), the Rating Region Table (RRT), and the Cable Virtual Channel Table (CVCT) or the Terrestrial Virtual Channel Table (TVCT).

According to ATSC, the PSI tables defined in MPEG-2 and provided in the MPEG Standard are used for accessing the video and audio streams, i.e. the transport stream carries one PAT and several PMTs. The conditional access information is also referenced via a CAT.

The actual ATSC tables are implemented as "private tables". The Master Guide Table, the main table, so to say, contains the PIDs for some of these ATSC tables. The Master Guide Table can be recognized by the packet ID = 0x1FFB and the table ID = 0xC7. The transport stream must contain at least four Event Information Tables (EIT-0, EIT-1, EIT-2, EIT-3) and the PIDs for these EITs are found in the Master Guide Table. Up to 128 further Event Information Tables are possible but are optional. An EIT contains a 3-hour section of an electronic program guide (EPG). Together with the 4 mandatory EITs, it is thus possible to cover a period of 12 hours. Furthermore, Extended Text Tables can be optionally accessed through the MGT. Each existing Extended Text Table (ETT) is allocated to one EIT. Thus, e.g. ETT-0 contains extended text information for EIT-0. It is possible to have up to a total of 128 ETTs.

Table 3.4. PSIP tables

Table	PID	Table ID
Program Association Table (PAT)	0x0	0x0
Program Map Table (PMT)	über PAT	0x2
Conditional Access Table (CAT)	0x1	0x1
Master Guide Table (MGT)	0x1FFB	0xC7
Terrestrial Virtual Channel Table (TVCT)	0x1FFB	0xC8
Cable Virtual Channel Table (CVCT)	0x1FFB	0xC9
Rating Region Table (RRT)	0x1FFB	0xCA
Event Information Table (EIT)	über PAT	0xCB
Extended Text Table (ETT)	über PAT	0xCC
System Time Table (STT)	0x1FFB	0xCD

In the Virtual Channel Table, which can be present either as Terrestrial Virtual Channel Table (TVCT) or as Cable Virtual Channel Table (CVCT) depending on the transmission path, identification information for the virtual channels, i.e. programs, contained in a multiplexed transport stream are transmitted. The VCT contains, among other things, the program names. The VCT is thus comparable to the SDT table in DVB:

In the System Time Table (STT), all the necessary time information is transmitted. The STT can be recognized by the packet ID = 0x1FFB and the table ID = 0xCD. In the STT, the GPS (Global Positioning System) time and the time difference between GPS time and UTC (Universal Time Coordinated (= GMT)) is transmitted. The Rating Region Table (RRT) can be used for restricting the size of the audience in terms of age or region. In addition to the information about region (e.g. a Federal State in the US), information relating to the minimum age set for the program currently being broadcast is also included. Using the RRT, a type of parental lock can thus be implemented in the set-top box. The RRT is recognized by the packet ID = 0x1FFB and the table ID = 0xCA.

The PIDs and Table IDs of the PSIP tables are listed in Table 3.4.

3.5 ARIB Tables according to ISDB-T

Like DVB (Digital Video Broadcasting) and ATSC (Advanced Television Systems Committee), Japan, too, has defined its own tables in its ISDB-T (Integrated Services Digital Broadcasting – Terrestrial) standard. These are called ARIB (Association of Radio Industries and Business) tables according to ARIB Std. B.10.

According to the ARIB standard, the following tables are proposed:

Table 3.5. ARIB tables

Type	Name	Note
PAT	Program Association Table	ISO/IEC 13818-1 MPEG-2
PMT	Program Map Table	ISO/IEC 13818-1 MPEG-2
CAT	Conditional Access Table	ISO/IEC 13818-1 MPEG-2
NIT	Network Information Table	like DVB-SI, ETS 300468
SDT	Service Description Table	like DVB-SI, ETS 300468
BAT	Bouquet Association Table	like DVB-SI, ETS 300468
EIT	Event Information Table	like DVB-SI, ETS 300468
RST	Running Status Table	like DVB-SI, ETS 300468
TDT	Time&Date Table	like DVB-SI, ETS 300468
TOT	Time Offset Table	like DVB-SI, ETS 300468
LIT	Local Event Information Table	
ERT	Event Relation Table	
ITT	Index Transmission Table	
PCAT	Partial Content Announcement Table	
ST	Stuffing Table	like DVB-SI, ETS 300468
BIT	Broadcaster Information Table	
NBIT	Network Board Information Table	

LDT	Linked Description Table
and others	
ECM	Entitlement Control Message
EMM	Entitlement Management Message
DCT	Download Control Table
DLT	Download Table
SIT	Selection Information Table
SDTT	Software Download Trigger Table
DSM-CC	Digital Storage Media Command & Control

Table 3.6. PID's and table IDs of the ARIB tables

Table	PID	Table ID
PAT	0x0000	0x00
CAT	0x0001	0x01
PMT	über PAT	0x02
DSM-CC	über PMT	0x3A...0x3E
NIT	0x0010	0x40, 0x41
SDT	0x0011	0x42, 0x46
BAT	0x0011	0x4A
EIT	0x0012	0x4E...0x6F
TDT	0x0014	0x70
RST	0x0013	0x71
ST	all except 0x0000, 0x0001, 0x0014	0x72
TOT	0x0014	0x73
DIT	0x001E	0x7E
SIT	0x001F	0x7F
ECM	via PMT	0x82...0x83
EMM	via CAT	0x84...0x85
DCT	0x0017	0xC0
DLT	via DCT	0xC1
PCAT	0x0022	0xC2
SDTT	0x0023	0xC3
BIT	0x0024	0xC4
NBIT	0x0025	0xC5, 0xC6
LDT	0x0025	0xC7
LIT	via PMT or 0x0020	0xD0

The BAT, PMT and CAT tables fully correspond to the MPEG-2 PSI. Similarly, the NIT, SDT, BAT, EIT, RST, TDT. TOT and ST tables have exactly the same structure as in DVB SI and also have the same functionality. The ARIB Standard thus also makes reference to ETSI 300468.

3.6 DTMB (China) Tables

China, too, have their own digital terrestrial television standard named DTMB – Digital Terrestrial Multimedia Broadcasting. It can be assumed that there is also an independent or modified or copied table of comparable significance to DVB-SI but there have been no publications regarding what modifications, if any, were made.

3.7 Other Important Details of the MPEG-2 Transport Stream

In the section below, other details of the MPEG-2 transport stream will be discussed in more detail.

Apart from the sync bytes (synchronization to the transport stream) already mentioned, the transport stream error indicator and the packet identifier (PID), the transport stream header also contains:

- the Payload Unit Start Indicator,
- the Transport Priority,
- the Transport Scrambling Control,
- the Adaptation Field Control, and
- the Continuity Counter.

The Payload Unit Start Indicator is a bit which marks the start of a payload. If this bit is set, it means that a new payload is starting in this transport stream packet: this transport stream packet contains either the start of a video or audio PES packet plus PES header, or the beginning of a table plus table ID as the first byte in the payload part of the transport stream packet.

3.7.1 The Transport Priority

This bit indicates that this transport stream packet has a higher priority than other TS packets with the same PID.

3.7.2 The Transport Scrambling Control Bits

The two Transport Scrambling Control Bits show whether the payload part of a TS packet is scrambled or not. If both bits are set to zero, this means that the payload section is transmitted unscrambled. If one of the two bits is not zero, the payload is transmitted scrambled. A Conditional Access Table is then needed to descramble the payload.

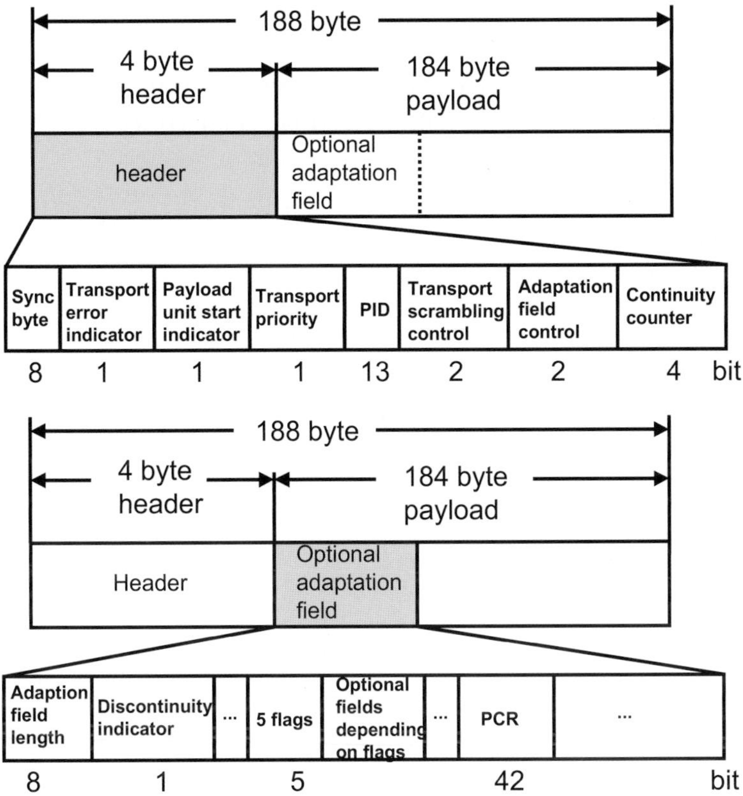

Fig. 3.49. Other details in the MPEG-2 transport stream

3.7.3 The Adaptation Field Control Bits

These two bits indicate whether there is an extended header, i.e. an adaptation field, or not. If both bits are set to zero, there is no adaptation field. If there is an adaptation field, the payload part is shortened and the header becomes longer but the total packet length remains a constant 188 bytes.

3.7.4 The Continuity Counter

Each transport stream packet with the same PID carries its own 4-bit counter. This is the continuity counter which continuously counts from 0 - 15 from TS packet to TS packet and then begins again from 0. The continuity counter makes it possible to recognize missing TS packets and to identify an errored data stream (counter discontinuity). It is possible, and permissible, to have a discontinuity with a program change which is then indicated by the Discontinuity Indicator in the adaptation field.

Bibliography: [ISO 13818/1], [ETS 300468], [A53], [REIMERS], [SIGMUNT], [DVG], [DVDM], [GRUNWALD], [FISCHER], [FISCHER4], [DVM], [ARIB]

4 Digital Video Signal According to ITU-BT.R.601 (CCIR 601)

Uncompressed digital video signals have been used for some time in television studios. Based on the original CCIR Standard CCIR 601, designated as IBU-BT.R601 today, this data signal is obtained as follows:

To start with, the video camera supplies the analog Red, Green and Blue (R, G, B) signals. These signals are matrixed in the camera to form luminance (Y) and chrominance (color difference C_B and C_R) signals.

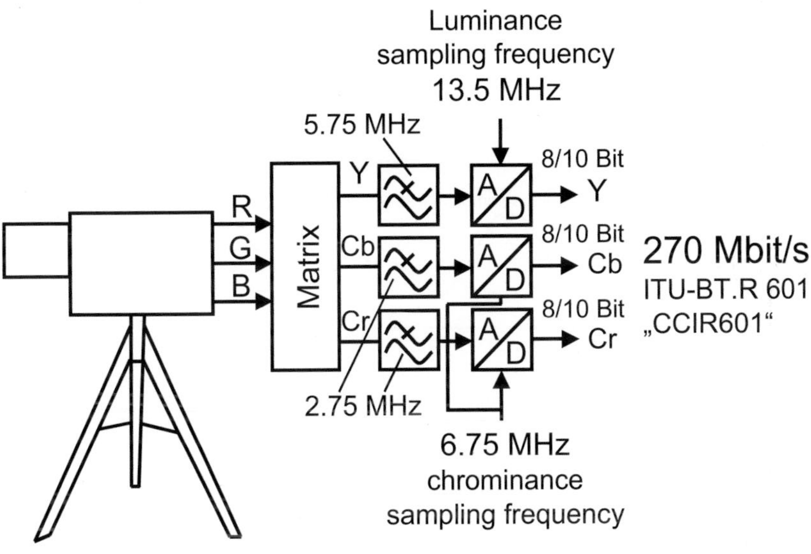

Fig. 4.1. Digitization of luminance and chrominance

These signals are produced by simple addition or subtraction of R = Red, G = Green, B = Blue:

$$Y = (0.30 \bullet R) + (0.59 \bullet G) + (0.11 \bullet B);$$

W. Fischer, *Digital Video and Audio Broadcasting Technology*, Signals and Communication Technology, 3rd ed., DOI: 10.1007/978-3-642-11612-4_4, © Springer-Verlag Berlin Heidelberg 2010

$C_B = 0.56 \bullet (B\text{-}Y);$

$C_R = 0.71 \bullet (R\text{-}Y)\ ;$

The luminance bandwidth is then limited to 5.75 MHz using a low-pass filter. The two color difference signals are limited to 2.75 MHz, i.e. the color resolution is clearly reduced compared with the brightness resolution. This principle is familiar from children's books where the impression of sharpness is simply conveyed by printed black lines. In analog television (NTSC, PAL, SECAM), too, the color resolution is reduced to about 1.3 MHz. The low-pass filtered Y, C_B and C_R signals are then sampled and digitized by means of analog/digital converters. The A/D converter in the luminance branch operates at a sampling frequency of 13.5 MHz and the two C_B and C_R color difference signals are sampled at 6.75 MHz each.

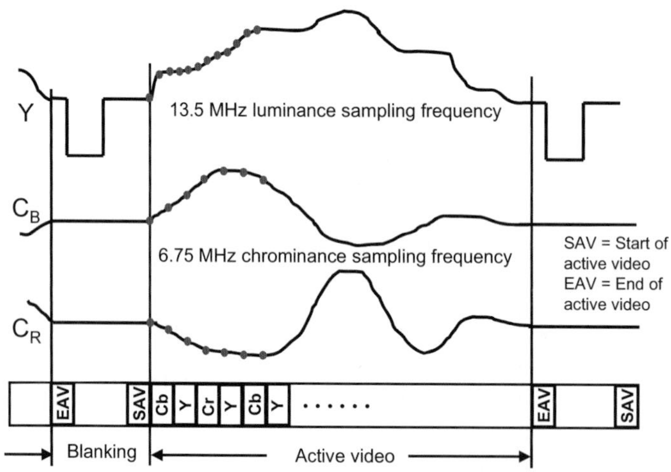

Fig. 4.2. Sampling of the components in accordance with ITU-BT.R601

This meets the requirements of the sampling theorem: There are no more signal components above half the sampling frequency. The three A/D converters can all have a resolution of 8 or 10 bits. With a resolution of 10 bits, this will result in a gross data rate of 270 Mbit/s which is suitable for distribution in the studio but much too high for TV transmission via existing channels (terrestrial, satellite or cable). The samples of all three A/D converters are multiplexed in the following order: C_B Y C_R Y C_B Y ... In this digital video signal (Fig. 4.1.), the luminance value thus alternates with a C_B value or a C_R value and there are twice as many Y values as there are C_B or C_R values. This is called a 4:2:2 resolution, compared with the resolution immediately after the matrixing, which was the same for all components, namely 4:4:4.

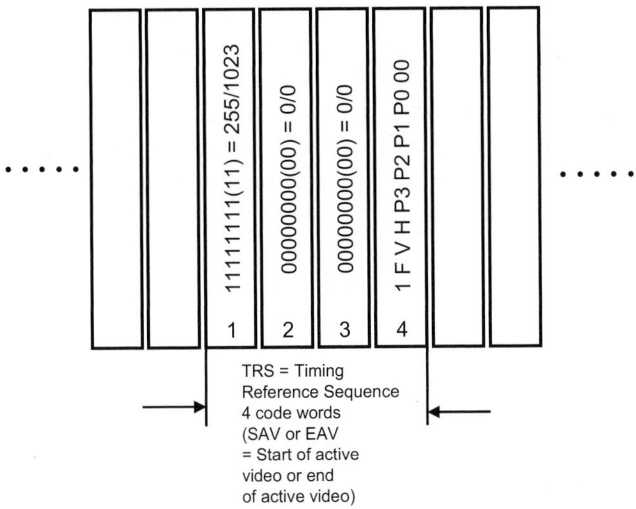

Fig. 4.3. SAV and EAV code words in the ITU-BT.R601 signal

This digital signal can be present in parallel form at a 25 pin sub-D connector or serially at a 75 Ohm BNC socket. The serial interface is called SDI which stands for serial digital interface and has become the most widely used interface because a conventional 75-Ohm BNC cable can be used.

Within the data stream, the start and the end of the active video signal is marked by special code words called SAV (start of active video) and EAV (end of active video), naturally enough (Fig. 4.2.). Between EAV and SAV, there is the horizontal blanking interval which does not contain any information related to the video signal, i.e. the digital signal does not contain the sync pulse. In the horizontal blanking interval, supplementary information can be transmitted such as, e.g. audio signals (embedded audio).

The SAV and EAV code words (Fig. 4.3.) consist of four 8 or 10 bit code words each. SAV and EAV begins with one code word in which all bits are set to one, followed by two words in which all bits are set to zero. The fourth code word contains information about the respective field or the vertical blanking interval, respectively. This fourth code word is used for detecting the start of a frame, field and active picture area in the vertical direction. The most significant bit of the fourth code word is always 1. The next bit (bit 8 in a 10 bit transmission or bit 6 in an 8 bit transmission) flags the field; if this bit is set to zero, it is a line of the first field and if it is set to one, it is a line of the second field. The next bit (bit 7 in a 10 bit transmission or bit 5 in an 8 bit transmission) flags the active video area in the vertical direction. If this bit is set to zero, then this is the visible active video area and if not, it is the vertical blanking interval. Bit 6 (10 bit) or bit 4 (8 bit) provides information about whether the present code word is an SAV or an EAV. It is SAV if this bit is set to zero and EAV if it is not. Bits 5...2 (10 bit) or 3...0 (8 bit) are used for error protection of the SAV and EAV code words. Code word 4 of the timing reference sequence (TRS) contains the following information:

- F = Field (0 = 1st field, 1 = 2nd field)
- V = Vertical blanking (1 = vertical blanking interval active)
- H = SAV/EAV identification (0 = SAV, 1 = EAV)
- P0, P1, P2, P3 = Protection bits (Hamming code)

Neither the luminance signal (Y) nor the color difference signals (C_B, C_R) use the full dynamic range available for them. There is a prohibited range which is reserved as headroom, on the one hand, and, on the other hand, allows SAV and EAV to be easily identified. A Y signal ranges between 16 and 64 decimal (8 bits) or 240 and 960 decimal (10 bits).

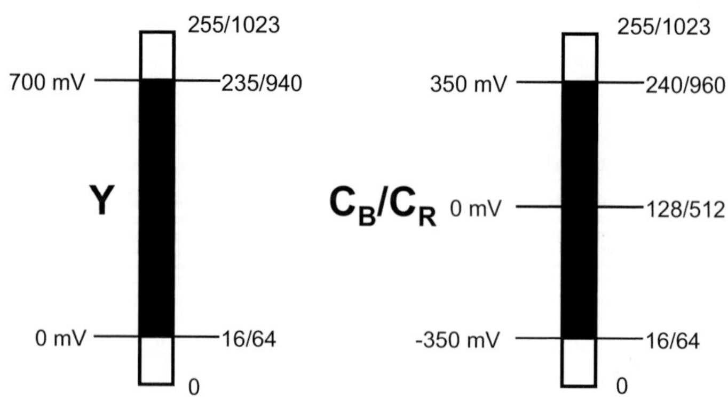

Fig. 4.4. Level diagram

The dynamic range of C_B and C_R is 16 to 240 decimal (8 bits) or 64 to 960 decimal (10 bits). The area outside this range is used as headroom and for sync identification purposes.

This video signal conforming to ITU-BT.R601, which is normally available as an SDI (Serial Digital Interface) signal, forms the input signal to an MPEG encoder.

Bibliography: [ITU601], [MÄUSL4], [GRUNWALD]

5 High Definition Television – HDTV

The Standard Definition Television – SDTV – introduced in the 50s is still virtually the main standard for analog and digital television in all countries throughout the world. However, as in the field of computers, modern TV cameras and terminal devices such as plasma screens and LCD receivers provide for much higher pixel resolution.

In computer monitors, the resolutions are:

- VGA 640 x 480 (4:3)
- SVGA 800 x 600 (4:3)
- XGA 1024 x 768 (4:3)
- SXGA 1280 x 1024 (5:4)
- UXGA 1600 x 1200 (4:3)
- HDTV 1920 x 1080 (16:9)
- QXGA 2048 x 1536 (4:3)

pixels, together with the respective aspect ratios (width x height).

Since the 1990s, there have been efforts in some countries to switch from the standard resolution SDTV to high resolution HDTV (High Definition Television). The first attempts were made in Japan with MUSE (Multiple Sub-Nyquist Sampling Encoding), developed by the broadcaster NHK (Nippon Hoso Kyokai). In Europe, too, HDTV was on the agenda at the beginning of the 1990s as HD-MAC (High-Definition Multiplexed Analog Components) but never entered the market. In the US, it was decided in the mid 90s to introduce HDTV as part of the ATSC (Advanced Television System Committee) effort, and in Australia it was decided to transmit HDTV as part of digital terrestrial television when the DVB-T standard was adopted. Europe, too, is now beginning to introduce HDTV. HDTV is currently implemented by MPEG-2 coding both in Japan, in the US and in Australia.

Europe, too, is now beginning to introduce HDTV. Since 2006, the channels Premiere (now Sky), Pro7 and Sat1 had also been on the air in HD. Pro7 and Sat1 are currently suspending their HD transmissions until

W. Fischer, *Digital Video and Audio Broadcasting Technology*, Signals and Communication Technology, 3rd ed., DOI: 10.1007/978-3-642-11612-4_5, © Springer-Verlag Berlin Heidelberg 2010

2010 when the public broadcasters in Germany will begin to transmit in HD.

The usual field rate in a 625 line TV system is 50 Hz and in a 525 line system it is 60 Hz. This is related to the power line frequency used in the original countries. The aspect ratio for HDTV will normally be 16:9 which is also becoming the norm for SDTV.

Initially, HDTV was to be based on twice the number of lines and twice the number of pixels per line. This would result

- in 1250 lines total with 1152 active lines and 1440 active pixels in a 625 line system
- and 1050 lines total with 960 active lines and 1440 active pixels in a 525 line system.

However, the resolution used for ATSC and HDTV in the US is 1280 x 720 pixels at 60 Hz. In Australia it is usually 1920/1440 x 1080 pixels at 50 Hz. The resolution of the European HDTV satellite channel EURO1080 is 1080 active lines x 1920 pixels at a field rate of 50 Hz.

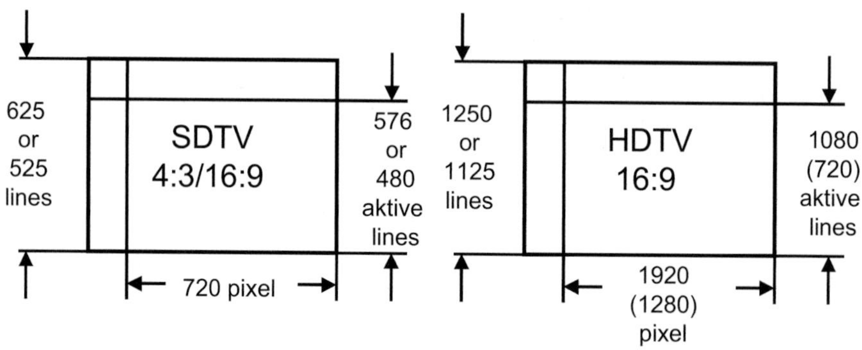

Fig. 5.1. SDTV and HDTV resolution

Once HDTV is introduced throughout Europe, MPEG-2 coding will be replaced by MPEG-4 Part 10, H.264, which will be more effective by a factor of 2 to 3. In this chapter, however, only the uncompressed digital baseband signal for HDTV as defined by the ITU-R BT.709 and ITU-R BT.1120 standards will be described.

The ITU has generally decided on a total number of 1125 lines in the 50 Hz and 60 Hz system, with 1080 active lines (Fig. 5.4.) and 1920 pixels per line both in the 50 Hz and the 60 Hz system. An active image of 1080 lines x 1920 pixels is called the Common Image Format (CIF). The sampling rate of the luminance signal is 74.25 MHz (Fig. 5.2.). The Y:C_B:C_R format is 4:2:2. The sampling rate of the color difference signals is 0.5 x 74.25 MHz = 37.125 MHz. ITU-R BT.709 provided a sampling rate of 72 MHz for the luminance and 36 MHz for the chrominance. To avoid aliasing, the luminance signal bandwidth is limited to 30 MHz and that of the chrominance signals to 15 MHz by low-pass filtering them before they are sampled.

In the 1125/60 system (Fig. 5.2.), and with a 10 bit resolution, this results in a gross physical data rate of:

Y: 74,25 x 10 Mbit/s = 742.5 Mbit/s
C_B: 0.5 x 74,25 x 10 Mbit/s = 371.25 Mbit/s
C_R: 0.5 x 74,25 x 10 Mbit/s = 371.25 Mbit/s

1.485 Gbit/s
gross data rate (1125/60)

Because of the slightly lower sampling rates in the 1250/50 system (Fig. 5.2.), the gross data rate, with 10 bit resolution, is then:

Y: 72 x 10 Mbit/s = 720 Mbit/s
C_B: 0.5 x 72 x 10 Mbit/s = 360 Mbit/s
C_R: 0.5 x 7,2 x 10 Mbit/s = 360 Mbit/s

1.44 Gbit/s
gross data rate (1250/50)

Both interlaced and progressive scanning are provided for. Plasma and LCD screens support progressive scanning due to the technology used, and interlaced scanning can lead to unattractive artefacts. With 50/60 progressively scanned frames the sampling rates are doubled to 148.5 and 144 MHz, respectively, for the luminance signal and 74.25 and 72 MHz, respectively, for the chrominance signals. The gross data rates are then doubled to 2.97 Gbit/s and 2.88 Gbit/s, respectively.

The structure of the uncompressed digital HDTV data signal is similar to ITU-R BT.601. A parallel and a serial interface (HD-SDI) are defined.

Since, apart from a few exceptions, the large-scale introduction of HDTV is still pending, however, the actual technical parameters are still subject to modification.

The European receiver manufacturers have defined the logos "HD Ready" and "Full HD" in order to describe characteristics of a display or projector. "HD Ready" defines a display or projector which offers the following features:

- at least minimum 720 lines physical resolution
- aspect ratio of 16:9
- supports a resolution of 1280 x 720 at 50 Hz or 60 Hz frame rate, progressive
- supports a resolution of 1980 x 1080 at 50 Hz or 60 Hz frame rate, interlaced
- analog Y Pb Pr interface
- digital DVI or HDMI interface
- HDCP encryption in the digital interfaces

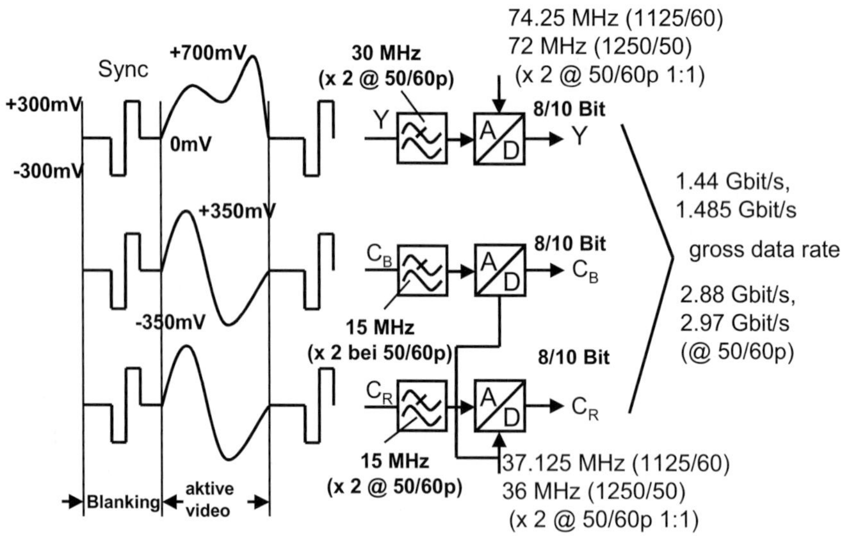

Fig. 5.2. Sampling of an HDTV signal according to ITU-R.BT709

DVI stands for Digital Visual Interface and is already known from PCs where it will replace the usual VGA interface. DVI allows a data rate of 1.65 Gbit/s. HDMI stands for High Definition Multimedia Interface and supports a data rate of up to 5 Gbit/s, transporting both picture and sound. HDCP means High Bandwidth Digital Content Protection and protects digital HD material in the DVI and HDMI interface against illegal recording, a requirement set by the film industry.

In contrast to "HD Ready", "Full HD" provides the full physical resolution of 1920 x 1080 pixels.

Bibliography: [MÄUSL6], [ITU709], [ITU1120]

6 Transforms to and from the Frequency Domain

In this chapter, principles of transforms to and from the frequency domain are discussed. Although it describes methods which are used quite generally throughout the field of electrical communication, a thorough knowledge of these principles is of great importance to understanding the subsequent chapters on video encoding, audio encoding and Orthogonal Frequency Division Multiplex (OFDM), i.e. DVB-T and DAB. Experts, of course, can simply skip this chapter.

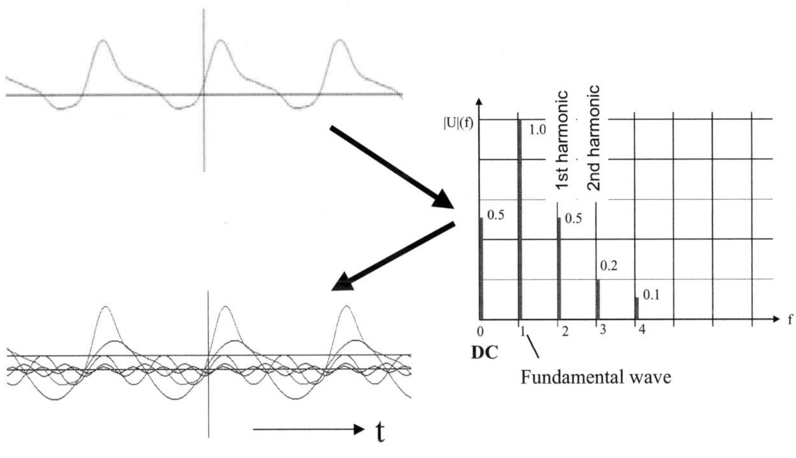

u(t) = 0.5 + 1.0sin(t+0.2)+0.5sin(2t)+0.2sin(3t-1)+0.1sin(4t-1.5);

Fig. 6.1. Fourier Analysis of a periodic time domain signal

Signals are normally represented as signal variation with time. An oscilloscope, for example, shows an electrical signal, a voltage, in the time domain. Voltmeters provide only a few parameters of these electrical signals, e.g. the DC component and the RMS value. These two parameters can also be calculated from the voltage variation by using a modern digital oscilloscope. A spectrum analyzer shows the signal in the frequency domain. It is possible to think of any time domain signal as being composed of an infi-

W. Fischer, *Digital Video and Audio Broadcasting Technology*, Signals and Communication Technology, 3rd ed., DOI: 10.1007/978-3-642-11612-4_6, © Springer-Verlag Berlin Heidelberg 2010

nite number of sinusoidal signals of a certain amplitude, phase and frequency.

The time domain signal is obtained by adding together all the sinusoidal signals at every point in time, i.e. the original signal is obtained from the superposition. A spectrum analyzer, however, only shows us the information about the amplitude or power of these sinusoidal part-signals, the harmonics.

A periodic time domain signal can be resolved into its harmonics mathematically by means of Fourier Analysis (Fig. 6.1.). This signal, which can have any shape, can be thought of as being composed of the fundamental wave which has the same period length as the signal itself, and of the harmonics which are simply multiples of the fundamental. In addition, each time domain signal also has a certain DC component. This direct voltage corresponds to a zero frequency. Non-periodic signals can also be represented in the frequency domain. but non-periodic signals do not have a line spectrum but a continuous spectrum. Thus, the spectral band contains spectral lines not only at certain points but at any number of points.

$$H(f) = \int_{-\infty}^{+\infty} h(t) e^{-j2\pi ft} dt; \text{ Fourier Transform (FT)}$$

$$h(t) = \int_{-\infty}^{\infty} H(f) e^{j2\pi ft} df; \text{ Inverse Fourier Transform (IFT)}$$

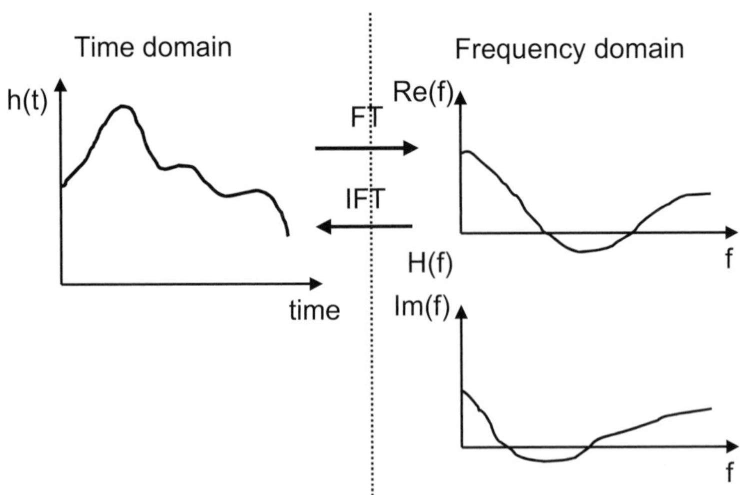

Fig. 6.2. Fourier Transform

6.1 The Fourier Transform

The spectrum of any time domain signal can be obtained mathematically by means of the so-called Fourier Transform (Fig. 6.2.). This is an integral transform in which the time domain signal has to be observed from minus infinity to plus infinity. Such a Fourier Transform can thus only be solved correctly if the time domain signal can be described in unambiguous terms mathematically. The Fourier Transform then calculates the variation of the real components and the variation of the imaginary components versus frequency from the time domain signal. It is possible to assemble any sinusoidal signal of any amplitude, phase and frequency from a cosinusoidal signal component of this frequency with a special amplitude and from a sinusoidal signal component of this frequency and special amplitude. The real component accurately describes the amplitude of the cosinusoidal component and the imaginary component accurately describes the amplitude of the sinusoidal component.

In the vector diagram (Fig. 6.3.), the vector of a sinusoidal quantity is obtained by the vectorial addition of the real and imaginary parts, i.e. of the sine and cosine components. The Fourier Transform thus provides the information about the real part, i.e. the cosine component, and the imaginary part, i.e. the sine component, at any point in the spectrum in infinitely fine resolution. The Fourier Transform is possible forwards and backwards and is referred to as Fourier Transform (FT) and Inverse Fourier Transform (IFT), respectively.

The Fourier transform turns a real time domain signal into a complex spectrum which is composed of real parts and imaginary parts as described. The spectrum consists of positive and negative frequencies and the negative frequency range does not provide any additional information about the time domain signal in question. The real part is mirror-symmetrical with respect to the zero frequency and Re(-f) = Re(f) holds true whereas the imaginary part is point-to-point-symmetrical and Im(-f) = -Im(f) holds true. The Inverse Fourier Transform supplies a single real time domain signal again from the complex spectrum. The Fourier Analysis, i.e. the analysis of the harmonics, is nothing else than a special case of a Fourier Transform where the Fourier Transform is simply applied to a periodic signal and the integral can then be replaced by a summation formula. The signal can be unambiguously described since it is periodic. The information over one period is sufficient.

By applying Pythagoras's theorem or the arc tangent, respectively, amplitude and phase information can be obtained from the real and imaginary parts if required (Fig. 6.4.). The group delay characteristic is obtained by differentiating the phase variation with frequency.

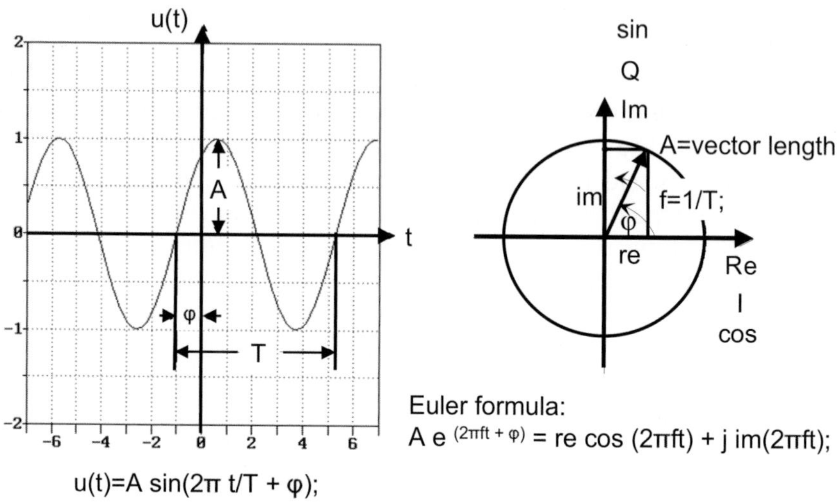

Fig. 6.3. Vector diagram of a sinusoidal signal

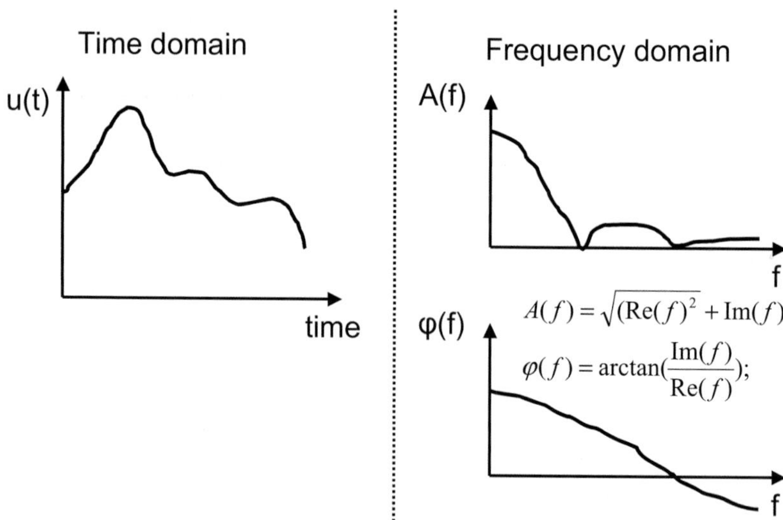

Fig. 6.4. Amplitude and phase characteristic

6.2 The Discrete Fourier Transform (DFT)

Signals of a quite general format cannot be described mathematically; there are no periodicities and they would have to be observed for an infinite period of time which is impossible in practice. There is thus no possible mathematical or numerical approach for calculating its spectrum. One solution which approximately supplies the frequency band is the Discrete Fourier Transform (DFT). Using, e.g. an analog/digital converter, the signal is sampled at discrete points in the time domain at intervals Δt and observed only within a limited time window at N points (Fig. 6.5.).

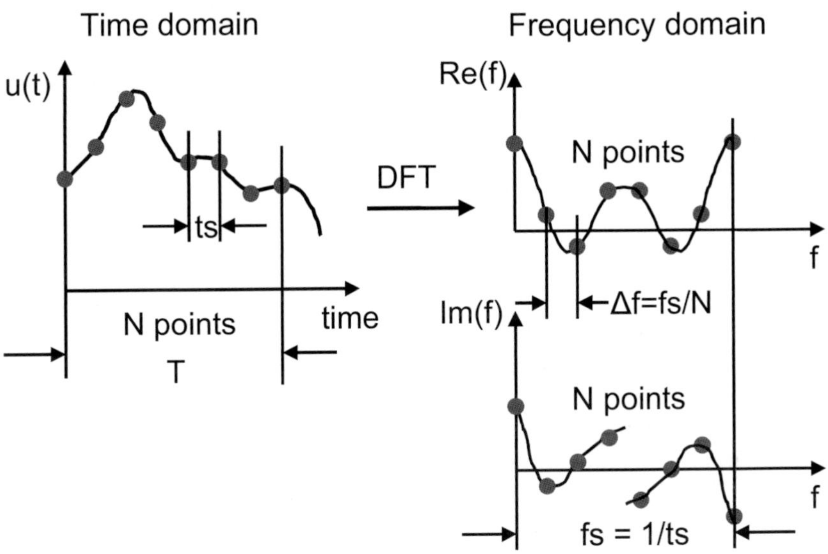

Fig. 6.5. Discrete Fourier Transform (DFT)

Instead of an integral from minus infinity to plus infinity, only a summation formula has to be solved then and this can even be done purely numerically by means of digital signal processing. The discrete Fourier transform results in N points for the real part(f) and N points for the imaginary part(f) in the spectral band.

The Discrete Fourier Transform (DFT) and the Inverse Discrete Fourier Transform (IDFT) are obtained through the following mathematical relations:

$$H_n = \sum_{k=0}^{N-1} h_k e^{-j2\pi k\frac{n}{N}} = -\sum_{k=0}^{N-1} h_k \cos(2\pi k\frac{n}{N}) - j\sum_{k=0}^{N-1} h_k \sin(2\pi k\frac{n}{N});$$

$$h_k = \frac{1}{N}\sum_{n=0}^{N-1} H_n e^{j2\pi k\frac{n}{N}};$$

The frequency band thus no longer has an infinitely fine resolution and is described only at discrete frequency interpolation points. The band extends from DC to half the sampling frequency and then continues symmetrically or point-to-point-symmetrically up to the sampling frequency. The real-time graph is symmetrical up to half the sampling frequency and the imaginary part is point-to-point-symmetrical. The frequency resolution is a function of the number of points in the window of observation and on the sampling frequency.

The following applies: $\Delta f = \frac{f_s}{N}; \Delta t = \frac{1}{f_s};$

The Discrete Fourier Transform (DFT), in reality, actually corresponds to a Fourier analysis within the observed time window of the band-limited signal. It is thus assumed that the signal in the observed time window continues periodically. This assumption results in “uncertainties” in the analysis so that the Discrete Fourier Transform can only supply approximate information about the actual frequency band. ‘Approximate’ in as much as the areas preceding and following the time window are not taken into consideration and the signal window is sharply truncated. However, the DFT can be solved by simple mathematical and numerical means and it functions both forwards and in the reverse direction in the time domain (Inverse Discrete Fourier Transform – IDFT, Fig. 6.6.). The result of performing a DFT on a real time domain signal interval is a discrete complex spectrum (real and imaginary parts). The IDFT transforms the complex spectrum back into a real time domain signal again. In reality, however, the section of time domain signal cut out and transformed into the frequency domain has been converted into a periodic signal.

Once a rectangular time domain signal segment has been windowed, the spectrum corresponds to a convolution of a sin(x)/x function with the original spectrum of the signal. This produces different effects which in a spectrum analysis done by means of the DFT disturb and affect the measurement result to a greater or lesser extent. In test applications, therefore, the choice would be not to select a rectangular window function but, e.g. $\cos^2$ function which would cut out a smoother window and lead to fewer disturbances in the frequency domain. Various types of window function are used, e.g. rectangular windows, Hanning windows, Hamming windows, Blackman windows etc.. Windowing means that the signal segment is first cut out to a rectangular shape and then multiplied by the window function.

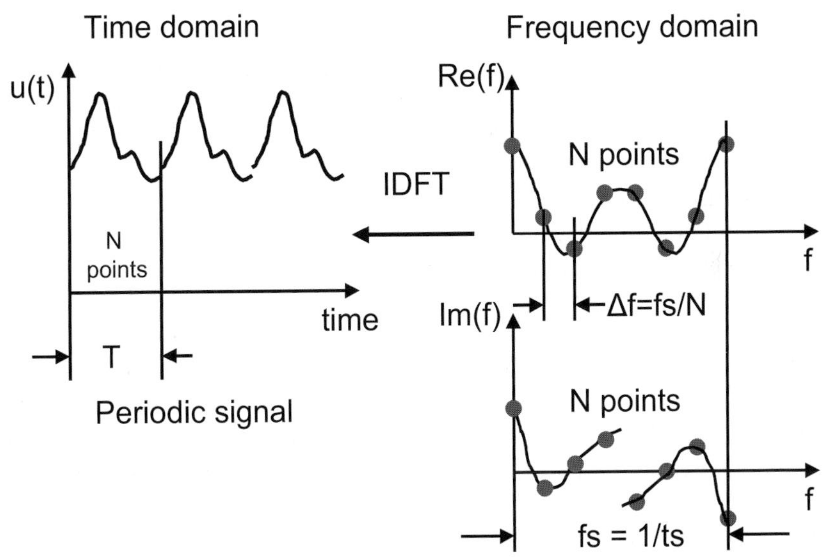

Fig. 6.6. IDFT

6.3 The Fast Fourier Transform (FFT)

The Discrete Fourier Transform is a simple but fairly time-consuming algorithm. However, if the number of points N within the window of observation is restricted to $N=2^x$, i.e. a power of two (Cooley, Tukey, 1965), a more complex, but less time-consuming algorithm, the Fast Fourier Transform (FFT), can be used. This algorithm itself provides exactly the same result as a DFT but is much faster and is restricted to $N=2^x$ points (2, 4, 8, 16, 32, 64, ...,256, …,1024, 2048, …,8192, ...). The Fast Fourier Transform can also be inverted (Inverse Fast Fourier Transform - IFFT).

The FFT algorithm makes use of methods of linear algebra. The samples are presorted in co-called bit reversal and then processed by means of butterfly operations. These operations are implemented as machine codes in signal processors and special FFT chips.

The number of multiplications given below shows the time gained by the FFT compared with the DFT:

Number of multiplications needed:

DFT: N • N
FFT; N • log(2N)

The FFT has long been used in the field of acoustics (surveying concert halls and churches) and in geology (searching for minerals, ores and oil). However, the analyses were performed off-line with fast computers, using a Dirac impulse to excite the medium to be examined (hall, rocks) and then recording the impulse response of the medium under investigation. A Dirac impulse is a very short and very strong impulse, an example of an acoustical Dirac impulse being a pistol shot and a geological Dirac impulse being the explosion of a blasting charge.

Back in 1988, a 256 point FFT still consumed minutes of PC time. Today, an 8192 point FFT (8k FFT) takes less than one millisecond of computing time! This opens the door for new and interesting applications such as video and audio compression or Orthogonal Frequency Division Multiplex (OFDM). FFT has also been used increasingly for spectrum analysis in analog video testing and for detecting the amplitude and group delay response of video transmission links since the late 1980s. In modern storage oscilloscopes, too, this interesting test function is frequently found today and makes it possible to perform a low-cost spectrum analysis, especially also in audio test engineering.

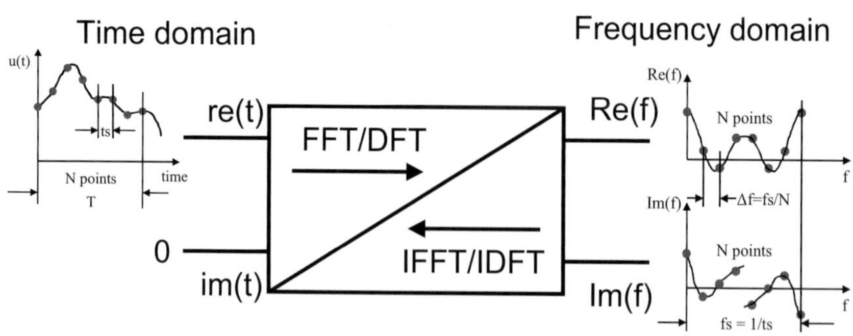

Fig. 6.7. Implementation and practical applications of DFT and FFT

6.4 Implementation and Practical Applications of DFT and FFT

The Fourier Transform, the Discrete Fourier Transform and the Fast Fourier Transform are all defined through the field of complex numbers. This means that both the time domain signal and the frequency domain signal have real and imaginary parts. Typical time domain signals are, however, always purely real, i.e. the imaginary part is zero at every point in time. The imaginary part must, therefore, be set to zero before the Fourier transform or its numerical variations DFT and FFT are performed.

When DFT or FFT and IDFT or IFFT are performed in practice two input signals are required (Fig. 6.7.). The input signals are implemented as real-part and imaginary-part tables and correspond to the sampled time or frequency domain. As the N samples of a typical time domain signal are always real, the corresponding imaginary part must be set to zero for each of the N points. This means that the imaginary-part table for the time domain must be filled with zeros. When the inverse transform is performed, the imaginary part of the time domain signal must again be zero assuming that the frequency range for the real part is about half the sampling frequency and the frequency range for the imaginary part is point-to-point symmetric about half the sampling frequency. If these symmetries are not present in the frequency domain, a complex time domain signal is output, i.e. the signal also has imaginary components in the time domain.

6.5 The Discrete Cosine Transform (DCT)

The Discrete Cosine Transform (DCT), and thus also the fast Fourier Transform which is a special case of the DCT, is a cosine-sine transform as can be seen from its formula; it is an attempt to assemble a time-domain signal segment by the superposition of many different cosine and sine signals of different frequency and amplitude. A similar result can also be achieved by using only cosine signals or only sine signals.

They are then called Discrete Cosine Transform (DCT) (Fig. 6.8.) or Discrete Sine Transform (DST) (Fig. 6.9.). Compared with the DFT, the sum of single signals required remains the same but twice as many cosine or sine signals are required. In addition, half-integral multiples of the fundamental are needed as well as integral multiples. The Discrete Cosine Transform (Fig. 6.8.), especially, has become quite important for audio and video compression.

The formulas of the Discrete Cosine Transform (DCT) and the Discrete Sine Transform (DST) are:

$$F_k = \sum_{z=0}^{N-1} f_z \cos\left(\frac{\pi k(z+\frac{1}{2})}{N}\right); \; F_k = \sum_{z=0}^{N-1} f_z \sin\left(\frac{\pi z k}{N}\right);$$

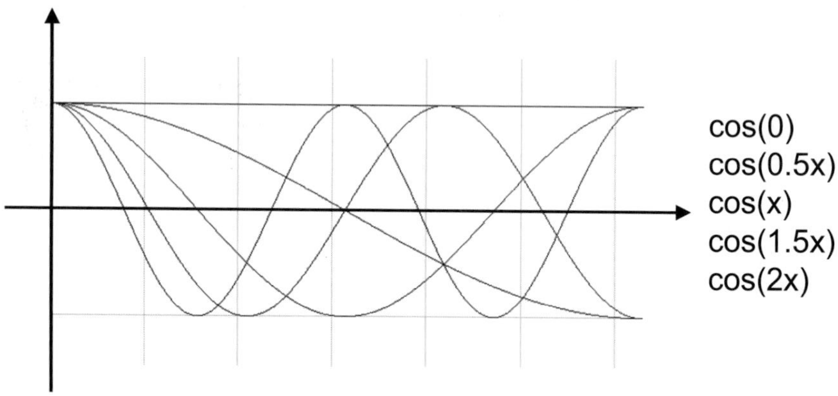

Fig. 6.8. Discrete Cosine Transform (DCT)

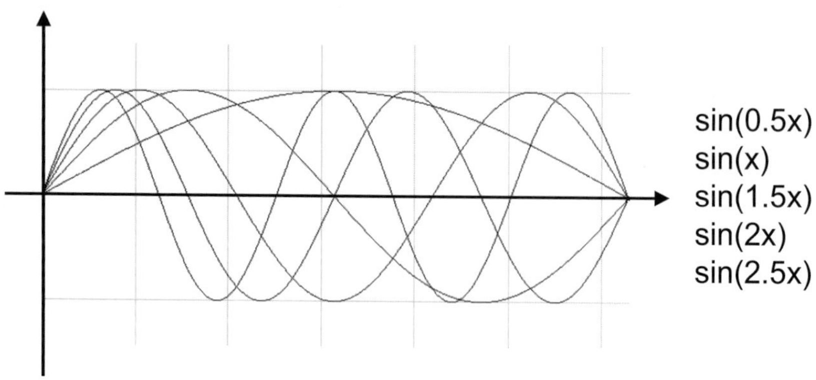

Fig. 6.9. Discrete Sine Transform (DST)

The DCT supplies, in the time domain, the amplitudes of the cosine signals from which the time interval analyzed can be assembled. The zero coefficient corresponds to the DC component of the signal segment. All the other coefficients first describe the low-frequency components, then the medium-frequency and then the higher-frequency components of the signal or, respectively, the amplitudes of the cosine functions from which the time-domain signal segment can be generated by adding them together. The response of the DCT is relatively gentle at the edges of the signal segment cut out and will lead to lesser discontinuities if a signal is transformed and retransformed segment by segment. This may well be the reason why the DCT has attained such great importance in the field of compression.

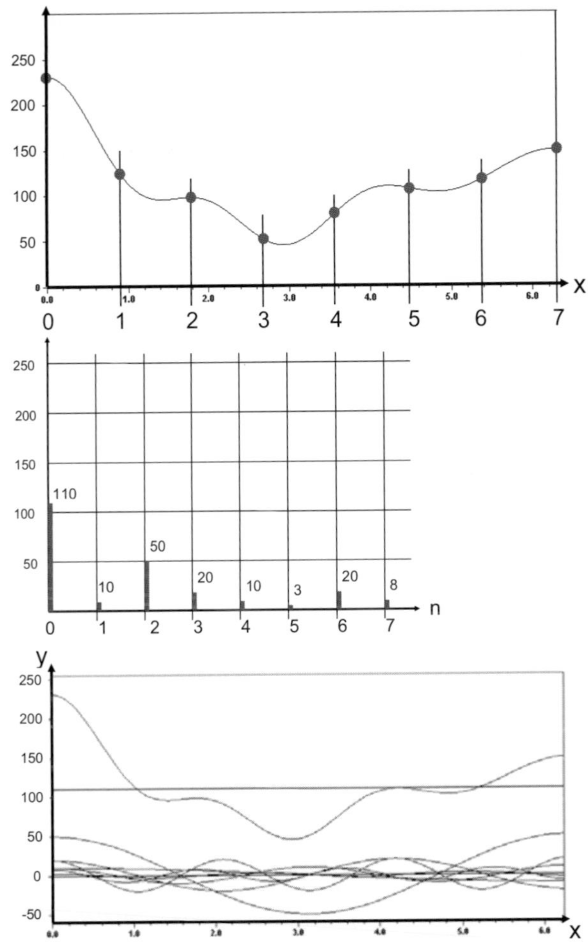

y(x)=110+10cos(0.5x)+50cos(x)+20cos(1.5x)
+10cos(2x)+3cos(2.5x)+20cos(3x)+8cos(3.5x);

Fig. 6.10. DCT and IDCT

The DCT is the algorithm at the core of the JPEG and MPEG image compression (digital photography and video) in which an image is transformed two-dimensionally block by block into the frequency domain and compressed block by block. It is of particular importance that the block edges cannot be recognized in the image after its decompression (no discontinuities at the edges).

The discrete cosine transform does not supply the coefficients in the frequency domain in pairs, i.e. separated according to real and imaginary parts and does not provide any information about the phase, only about the amplitude. Neither does the amplitude characteristic correspond directly to the result of the DFT. But this type of frequency transform is adequate for many applications and is also possible in both directions (Inverse Discrete Cosine Transform - IDCT) (Fig. 6.10.).

In principle, of course, there is also a Discrete Sine Transform (Fig. 6.9) where it is attempted to duplicate a time domain signal by the superposition of pure sinusoidal signals.

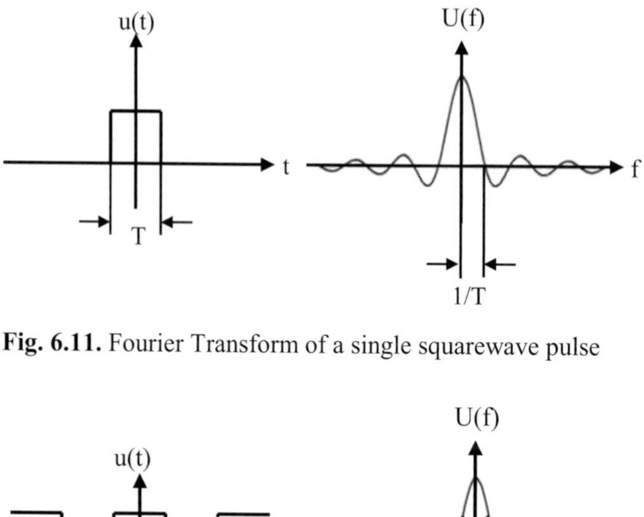

Fig. 6.11. Fourier Transform of a single squarewave pulse

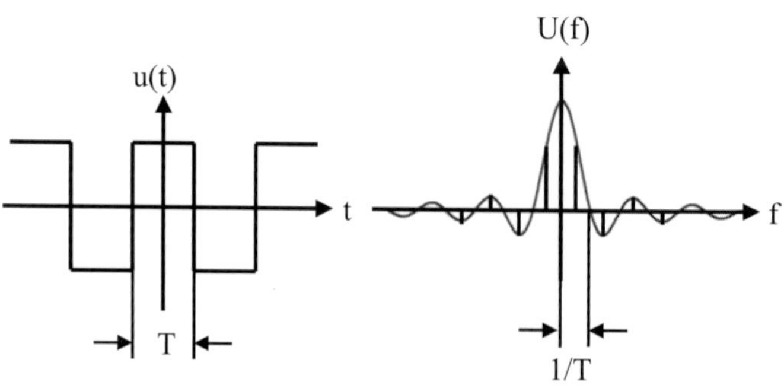

Fig. 6.12. Fourier Transform of a periodic squarewave pulse

6.6 Time Domain Signals and their Transforms in the Frequency Domain

In the following paragraphs, some important time-domain signals and their transforms in the frequency domain will be discussed. The purpose of these observations is to get some feel for the results of the Fast Fourier Transform.

Let us begin with a periodic squarewave signal (Fig. 6.12.): since it is a periodic signal, it has discrete lines in the frequency spectrum; all discrete spectral lines of the squarewave signal are located at integral multiples of the fundamental frequency of the squarewave signal. Most of the energy will be found in the fundamental wave itself. If there is a DC component, it will result in a spectral line at zero frequency (Fig. 6.14.). The envelope of the spectral lines of the fundamental and the harmonics is the sin(x)/x function.

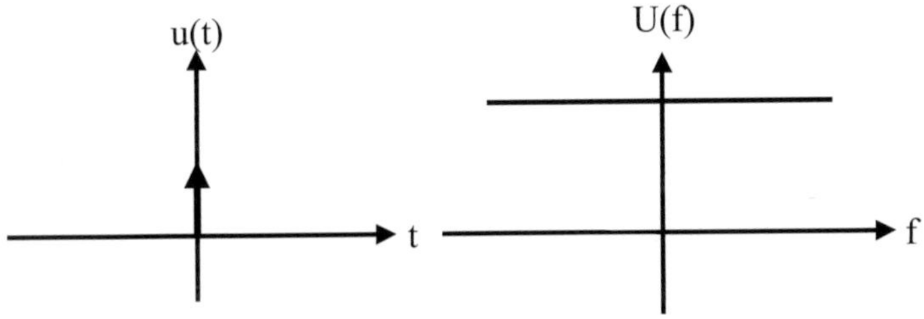

Fig. 6.13. Fourier Transform of a Dirac impulse

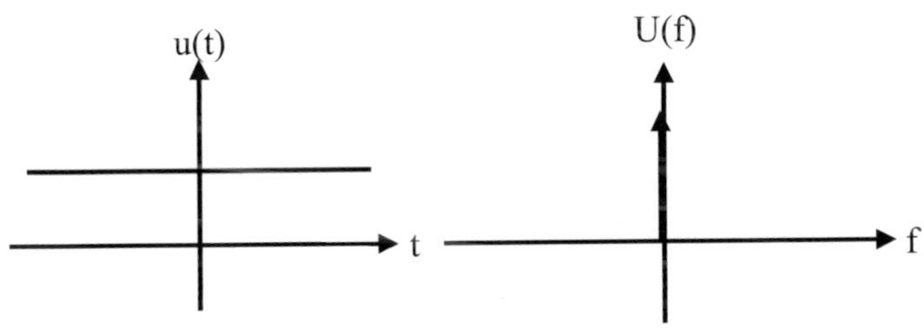

Fig. 6.14. Fourier Transform of a pure direct voltage (DC)

If then the duration of the period T of the squarewave signal is allowed to tend towards infinity, the discrete spectral lines move closer and closer together until a continuous spectrum of a single pulse is obtained (Fig. 6.11.).

The spectrum of a single squarewave pulse is a sin(x)/x function. If then the pulse width T is allowed to become narrower and narrower and to tend towards zero, all zero points of the sin(x)/x function will tend towards infinity. In the time domain, this provides an infinitely short pulse, a so-called Dirac impulse, the Fourier Transform of which is a straight line; i.e. the energy is distributed uniformly from zero frequency to infinity (Fig. 6.13.). Conversely, a single Dirac needle at f=0 in the frequency domain corresponds to a direct voltage (DC) in the time domain.

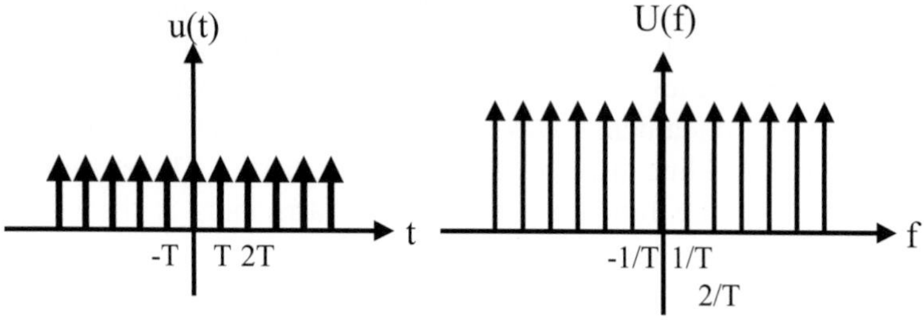

Fig. 6.15. Fourier Transform of a sequence of Dirac impulses

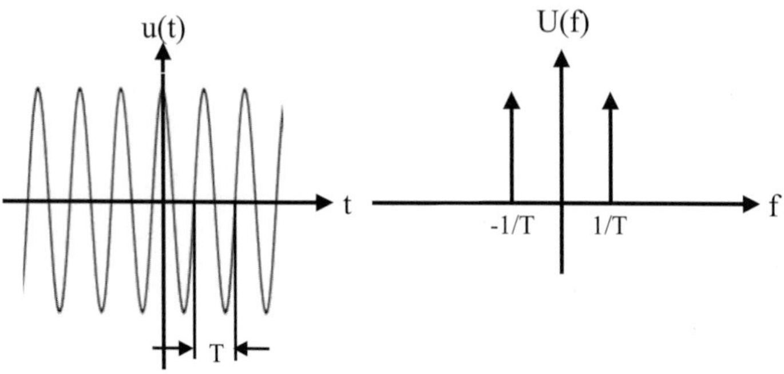

Fig. 6.16. Fourier Transform of a sinusoidal signal

A sequence of Dirac impulses spaced apart at intervals T from one another again results in a discrete spectrum of Dirac needles spaced apart by 1/T (Fig. 6.15.). The Dirac impulse train is of importance when considering a sampled signal. Sampling an analog signal has the consequence that this signal is convoluted with a sequence of Dirac impulses.

To conclude, a purely sinusoidal signal will be considered. Its Fourier transform is a Dirac needle at the frequency of the sinewave fs and –fs (Fig. 6.16.).

6.7 Systematic Errors in DFT or FFT, and How to Prevent them

To obtain the precise result of the Fourier Transform, a time-domain signal would have to be observed for an infinitely long period of time. In the case of the Discrete Fourier Transform, however, a signal segment is only observed for a finite period of time and transformed. The result of the DFT or FFT, respectively, will thus always differ from that of the Fourier Transform. It has been seen that, in principle, this analyzed time segment is converted into periodic signals in the DFT, i.e., the result of the DFT must be considered to be the Fourier Transform of this converted time segment.

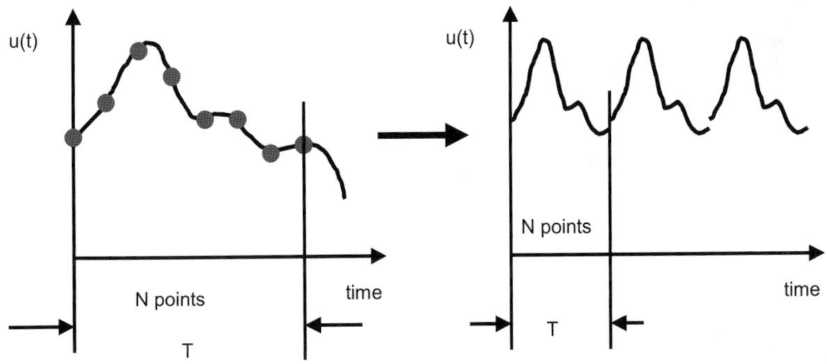

Fig. 6.17. Conversion of a signal segment into periodic signals by the DFT or FFT, resp.

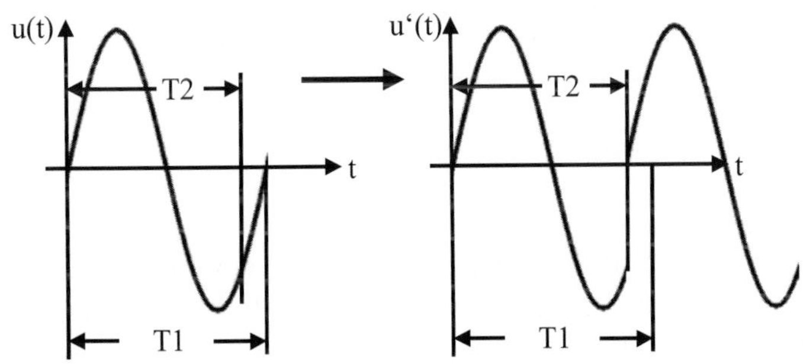

Fig. 6.18. Windowing (T1, T2) a sinusoidal signal

It is clear that, naturally, the result of the transform depends greatly on the type and position of the “cutting-out” process, the so-called windowing. This can be visualized best by performing the DFT on a sinusoidal signal. If exactly one sample is taken from the sinusoidal signal so that it has a length of a multiple n=1, 2, 3 etc of the period, the result of the DFT will exactly match that of the Fourier transform because converting this time segment into periodic signals will again produce a signal which is exactly sinusoidal.

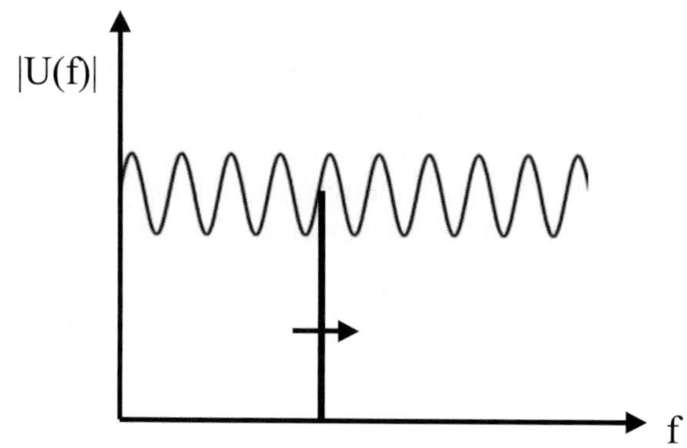

Fig. 6.19. Picket fence effect

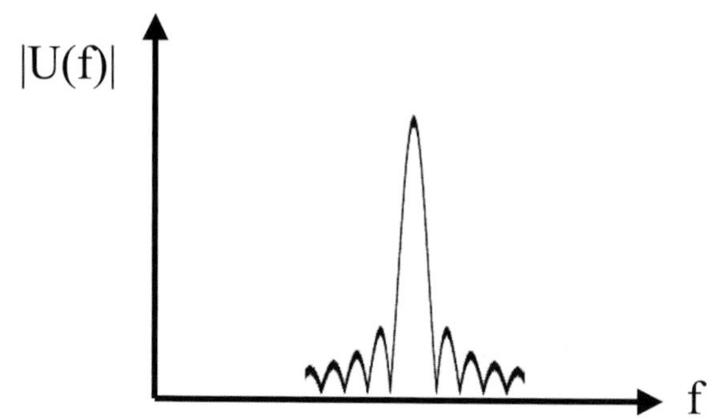

Fig. 6.20. Dispersal of the energy to main and side lobes

If, however, the length of the window (Fig. 6.18.) cut out differs from the length of the period, the result of the transform will differ more or less from the expected value depending on the number of cycles included. A sample of less than one fundamental wave will have the worst effect. A Dirac needle will become a wider “lobe”, in some cases with “sidelobes”. The amplitude of the main lobe will correspond more or less to the expected value. Leaving the period of observation constant and varying the frequency of the signal, the amplitude of the spectral line will fluctuate and will correspond to the expected value whenever there is exactly one multiple of the period within the window of observation; in between that it will become smaller and assume the exact value time and again. This is called the "picket fence" effect (Fig. 6.19.).

The fluctuation in the amplitude of the spectral line is caused by a dispersal of the energy due to a widening of the main lobe and by the appearance of sidelobes (Fig. 6.20.).

In addition, aliasing products may appear if the measurement signal is not properly band-limited; moreover quantization noise becomes visible and will limit the dynamic range.

These systematic errors can be prevented or suppressed by programming an observation time of corresponding length, by good suppression of aliasing products and by using A/D converters having a correspondingly high resolution. In the next section, "windowing" will be discussed as a further aid in suppressing DFT system errors.

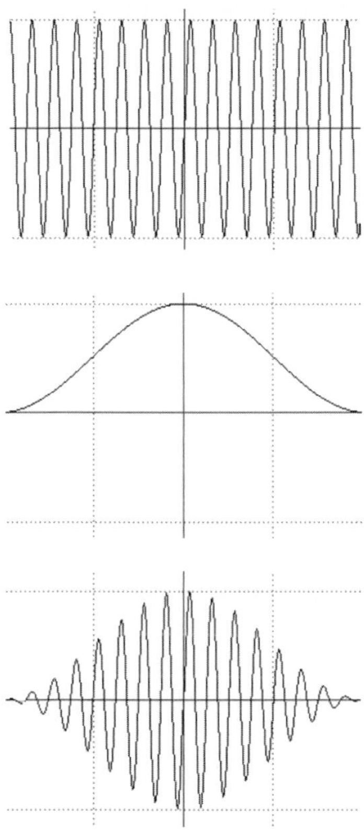

Fig. 6.21. Multiplying a signal by a window function

6.8 Window Functions

In the last section it was shown that windows with abrupt or “hard” edge transitions produced spurious effects, so-called leakage, as picket fence effect and sidelobes. The main lobe is dispersed depending on whether an integral multiple of the period has been sampled or not.

These leakage effects can be reduced by using soft windowing, i.e. a window function with soft edges, instead of a rectangular window with hard rectangular edges.

Fig. 6.21. shows that in windowing, the original signal is weighted, i.e. multiplied by the window function k(t). The signal is cut out softly towards the edge. The window function shown is the Hanning window function - a simple cosine squared window which is the most commonly used window. The sidelobes are attenuated more and the picket fence effect is reduced.

There are a number of windows used in practice, examples of which are:

- Rectangular window
- Hanning window
- Hamming window
- Triangular window
- Tukey window
- Kaiser Bessel window
- Gaussian window
- Blackman window

Depending on the window selected, the main lobes are widened to a greater or lesser extent, the sidelobes are attenuated more or less, and the picket fence effect is reduced to a greater or lesser extent. Rectangular windowing means maximum or no cutting out, the Hanning cosine squared window was shown in the Figure. Regarding the other windows, reference is also made to the relevant literature references and to the article by [HARRIS].

Bibliography: [COOLEY], [PRESS], [BRIGHAM], [HARRIS], [FISCHER], [GIROD], [KUEPF], [BRONSTEIN]

7 MPEG-2 Video Coding

Digital SDTV (Standard Definition Television) video signals (uncompressed) have a data rate of 270 Mbit/s. This data rate is much too high for broadcasting purposes, which is why they are subjected to a compression process before being processed for transmission. The 270 Mbit/s must be compressed to about 2...7 Mbit/s - a very high compression factor which, however, is possible due to the use of a variety of redundancy and irrelevance reduction mechanisms. The data rate of an uncompressed HDTV signal is even higher than 1 Gbit/s and MPEG-2 coded HDTV signals have a data rate of about 15 … 20 Mbit/s.

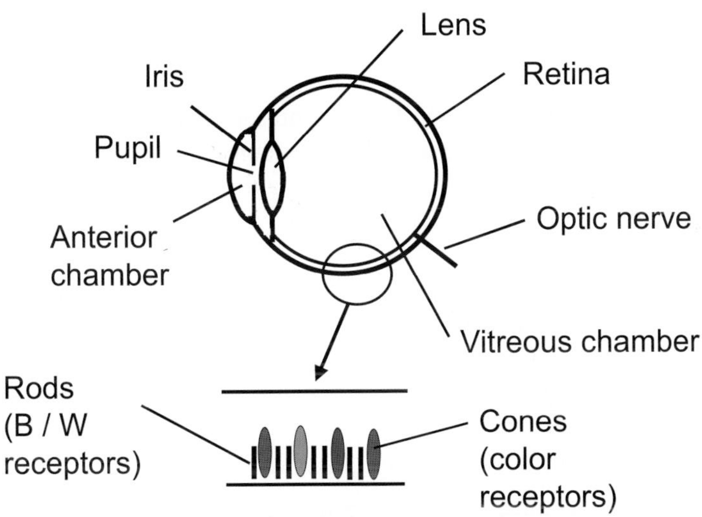

Fig. 7.1. Anatomics of the human eye

7.1 Video Compression

To compress data, it is possible to remove redundant or irrelevant information from the data stream. Redundant means superfluous, irrelevant means

W. Fischer, *Digital Video and Audio Broadcasting Technology*, Signals and Communication Technology, 3rd ed., DOI: 10.1007/978-3-642-11612-4_7, © Springer-Verlag Berlin Heidelberg 2010

unnecessary. Superfluous information is information which exists several times in the data stream, or information which has no information content, or simply information which can be easily and losslessly recovered by mathematical processes at the receiving end. Redundancy reduction can be achieved, e.g. by variable-length coding. Instead of transmitting ten zeroes, the information 'ten times zero' can be sent by means of a special code which is much shorter.

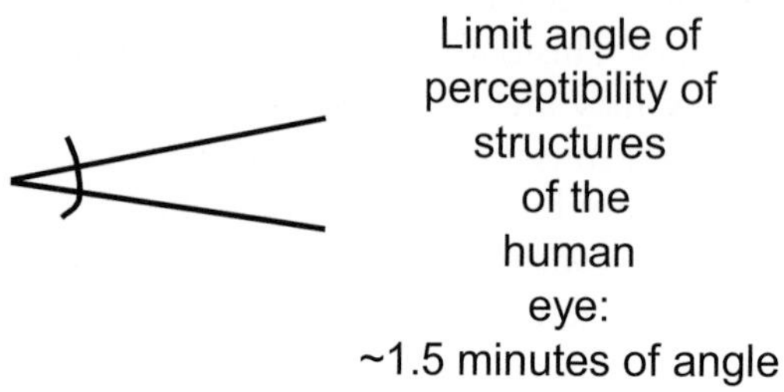

Fig. 7.2. Limit angle of perceptibility of structures of the human eye

The alphabet of the Morse code, too, uses a type of redundancy reduction. Letters which are used frequently are represented by short code sequences whereas letters which are used less frequently are represented by longer code sequences. In information technology, this type of coding is called Huffman coding or variable length coding.

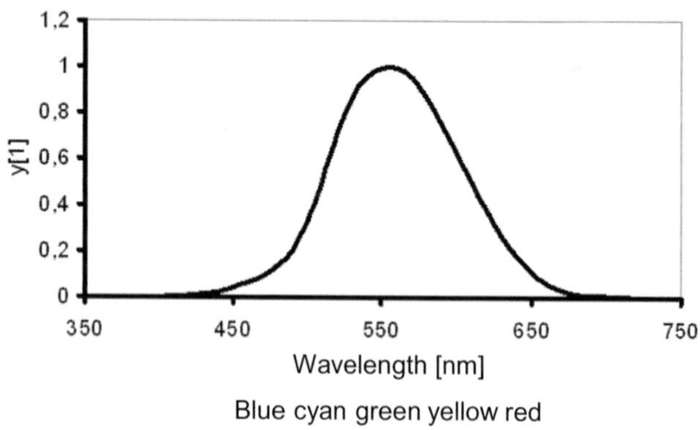

Fig. 7.3. Luminance sensitivity of the human eye

Irrelevant information is the type which cannot be perceived by the human senses. In case of the video signal, they are the components which the eye does not register due to its anatomy. The human eye (Fig. 7.1.) has far fewer color receptors than detection cells for brightness information. For this reason, the "sharpness in the color" can be reduced which means a reduction in the bandwidth of the color information. The receptors for black/white are called rods and the color receptors are cones, both of which are located on the retina of the eye, behind the lens. The lens focuses the image sharply onto the retina. The rods have their main function in night vision and are much more sensitive and present in much greater numbers. The limit angle of the perceptibility of structures is a function of the number of rods in the human eye and is about 1.5 minutes of angle (Fig. 7.2.). There are red-, green- and blue-sensitive cones, the sensitivity to green being much more pronounced than that for blue and red and that for red, in turn, being greater than that for blue (Fig. 7.3.). This also finds its expression in the matrixing formula for forming the luminance signal:

$$Y = 0.30 \cdot R + 0.59 \cdot G + 0.11 \cdot B$$

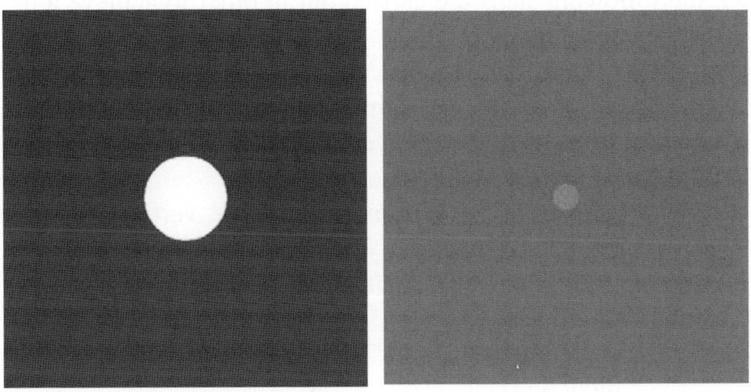

Fig. 7.4. Perception test for coarse and fine image structures

It is also known that we cannot discern fine structures in a picture, e.g. thin lines, as well as coarse structures. This can be illustrated well by perception tests (Fig. 7.4). If, e.g., one varies the size of a spot and its brightness against a background, the color of which can also be varied, it can be demonstrated that at some point the human eye can no longer see a small spot which differs only slightly from the background. This is precisely the main point of attack for data reduction methods like JPEG and MPEG, where coarse structures are transmitted with much greater accuracy, i.e. with many more bits, than fine structures, performing, in fact, an irrelevance reduction, a so-called perception coding. However, irrelevance reduction is always associated with an irretrievable loss of information which is why the only method considered in data processing is redundancy reduction as, e.g. in the well known ZIP files.

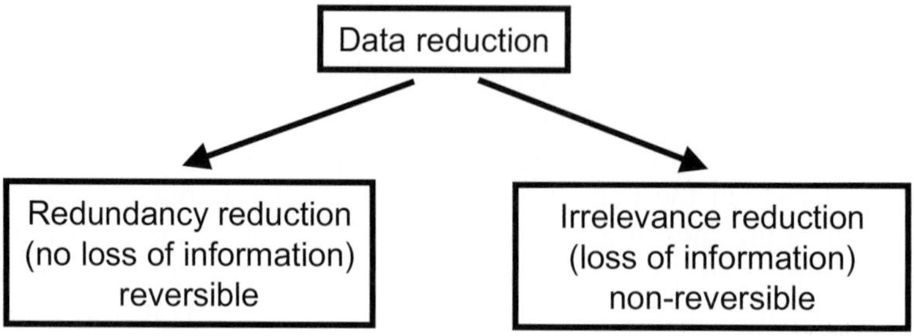

Fig. 7.5. Data reduction

In MPEG, the following steps are carried out in order to achieve a data reduction factor of up to 130:

- 8 bits resolution instead of 10 bits (irrelevance reduction)
- Omitting the horizontal and vertical blanking interval (redundancy reduction)
- Reducing the color resolution also in the vertical direction (4:2:0) (irrelevance reduction)
- Differential pulse code modulation (DPCM) of moving pictures (redundancy reduction)
- Discrete cosine transform (DCT) followed by quantization (irrelevance reduction)
- Zig-zag scanning with variable-length coding (redundancy reduction)
- Huffman coding (redundancy reduction)

Let us begin again with the analog video signal from a television camera. The red, green and blue (RGB) output signals are matrixed to become Y, C_B and C_R signals. After that, the bandwidth of these signals is limited and they are analog/digital converted. According to ITU-BT.R601, this provides a data signal with a data rate of 270 Mbit/s. The color resolution is reduced in comparison with the brightness resolution, making the number of brightness samples twice that of the C_B and C_R values and resulting in a 4:2:2 signal; there is thus already an irrelevance reduction in ITU-

BT.R601. It is this 270 Mbit/s signal which must be compressed to about 2...7 (15) Mbit/s in the MPEG video coding process.

7.1.1 Reducing the Quantization from 10 Bits to 8

In analog television, the rule of thumb was that when a video signal has a signal/noise ratio, referred to white level and weighted, of more than 48 dB, the noise component is just below the threshold of perception of the human eye. Given the appropriate drive to the A/D converter, the quantization noise from the 8 bit resolution is already well below this threshold so that a 10 bit resolution in Y, C_B and C_R is unnecessary outside the studio. In the studio, 10 bit resolution is better because post-processing is easier and gives better results. Reducing the data rate from 10 bits to 8 bits compared with ITU BT.R601 means a reduction in the data rate of 20 % ((10-8)/10 = 2/10 = 20 %), but this is an irrelevance reduction and the original signal cannot be recovered in the decoding at the receiving end. According to the rule of thumb that S/N [dB] = 6•N, the quantization noise level has now risen by 12 dB.

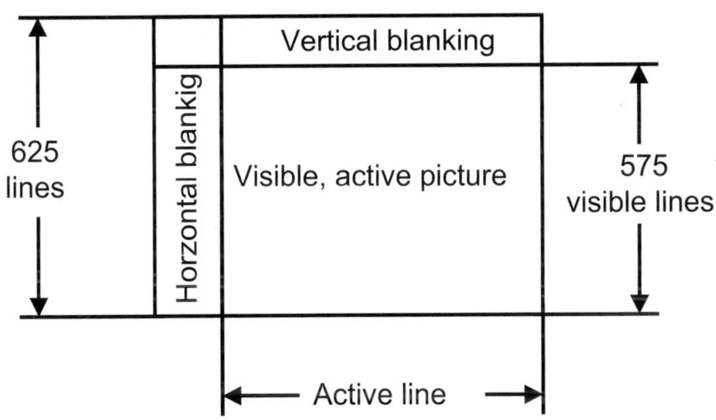

Fig. 7.6. Horizontal and vertical blanking

7.1.2 Omitting the Horizontal and Vertical Blanking Intervals

The horizontal and vertical blanking intervals of a digital video signal according to ITU BT.R601 (Fig. 7.6.) do not contain any relevant information, not even teletext. These areas can contain supplementary data such as sound signals but these must be transmitted coded separately according to MPEG. The horizontal and vertical blanking intervals are, therefore, left out completely in MPEG. The horizontal and vertical blanking intervals and all signals in them can be regenerated again without problems at the receiving end.

A PAL signal has 625 lines, only 575 of which are visible. The difference of 50 lines, divided by 625, is 8 % which is the saving in data rate achieved when the vertical blanking is omitted. The length of one line is 64 μs but the active video area is only 52 μs which, divided by 64, amounts to a further saving of 19 % in the data rate. Since there is some overlap in the two savings, the total result of this redundancy reduction is about 25 %.

7.1.3 Reduction in Vertical Color Resolution (4:2:0)

The two color difference signals C_B and C_R are sampled at half the data rate compared with the luminance signal Y. In addition, the bandwidth of C_B and C_R is also reduced to 2.75 MHz in comparison with the luminance bandwidth of 5.75 MHz - a 4:2:2 signal (Fig. 7.7.). However, the color resolution of this 4:2:2 signal is only reduced in the horizontal direction. The vertical color resolution corresponds to the full resolution resulting from the number of lines in a television frame.

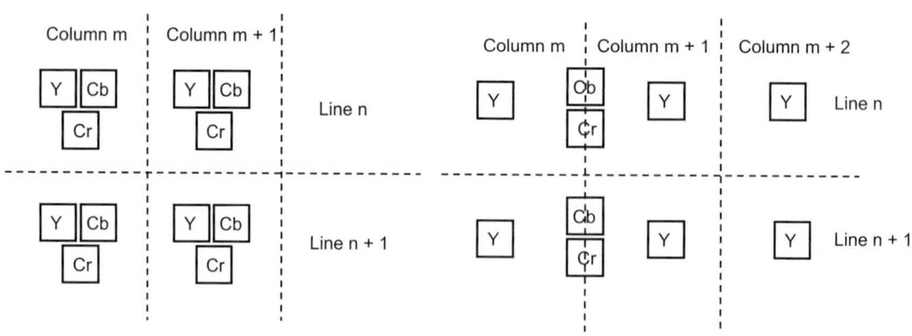

Fig. 7.7. 4:4:4 and 4:2:2 resolution

However, the human eye cannot distinguish between horizontal and vertical as far as color resolution is concerned. It is possible, therefore, to also reduce the color resolution to one half in the vertical direction without perceptible effect. MPEG-2 does this usually in one of the first steps and the signal then becomes a 4:2:0 signal (Fig.7.8.). Four Y pixels are now in each case associated with only one C_B value and one C_R value each. This type of irrelevance reduction results in another saving of exactly 25 % data rate.

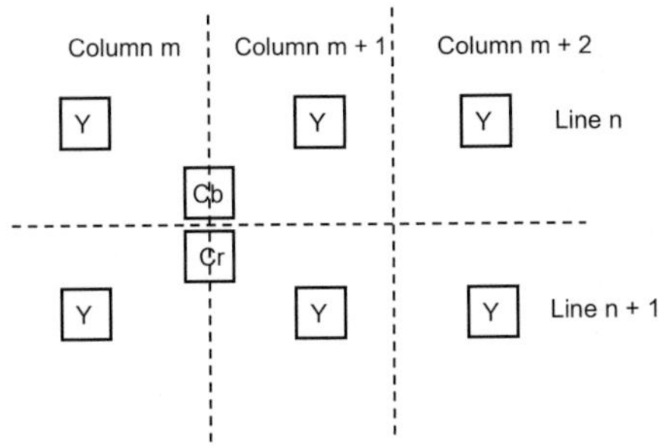

Fig. 7.8. 4:2:0 resolution

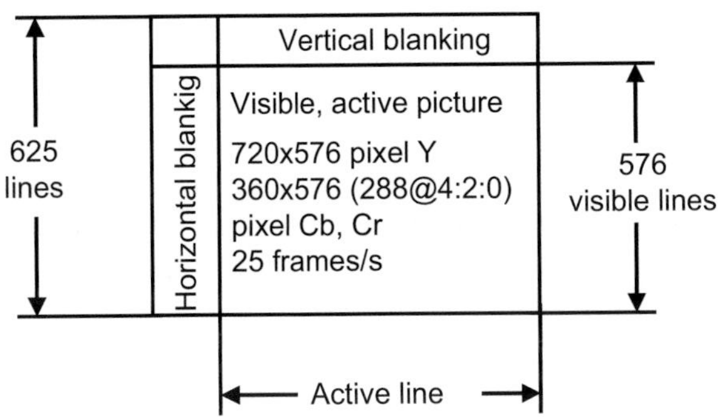

Fig. 7.9. Physical parameters of a SDTV signal

7.1.4 Further Data Reduction Steps

The data reduction carried out up to now has produced the following result: Beginning with an original data rate of 270 Mbit/s, this ITU BT.R601 signal has now been compressed to 124.5 Mbit/s, i.e. to less than half its original rate, by applying the following steps:

- ITU BT.R601 = 270 Mbit/s
- 8 bits instead of 10 (-20%) = 216 Mbit/s
- Hor. and vert. blanking (appr. -25%) = 166 Mbit/s
- 4:2:0 (-25%) = 124.5 Mbit/s

There is, however, still a large gap between the 124.5 Mbit/s now achieved and the required 2...6 Mbit/s with its upper limit of 15 Mbit/s, and this gap needs to be closed by means of further steps which are much more complex.

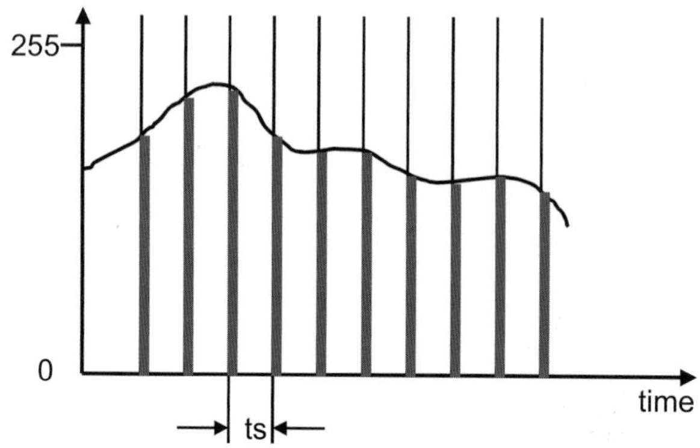

Fig. 7.10. Pulse Code Modulation

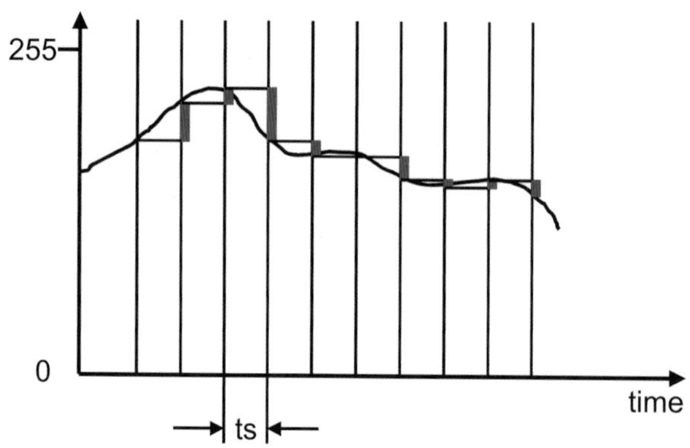

Fig. 7.11. Differential Pulse Code Modulation

7.1.5 Differential Pulse Code Modulation of Moving Pictures

Adjoining moving pictures differ only very slightly from each other. They contain stationary areas which won't change at all from frame to frame; there are areas which only change their position and there are objects which are newly added. If each frame were to be transmitted completely every time, some of the information transmitted would always be the same, resulting in a very high data rate. The obvious conclusion is to differentiate between these types of picture areas and to transmit only the difference, i.e. the delta value, from one frame to the next. This particular method of redundancy reduction, which is based on a method which has been known for a long time, is called differential pulse code modulation (DPCM).

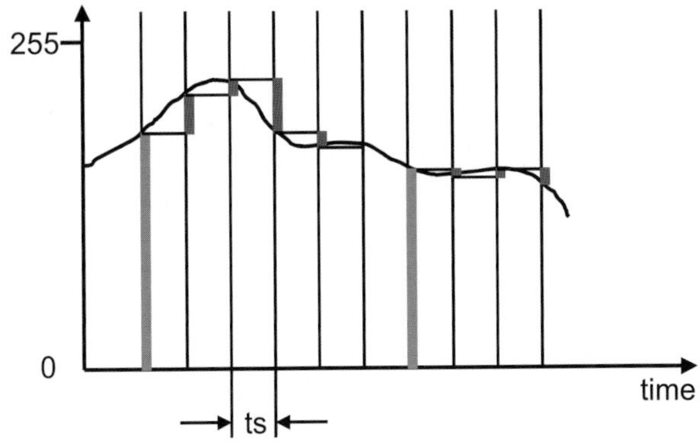

Fig. 7.12. Differential Pulse Code Modulation with reference values

Fig. 7.13. Dividing a picture into blocks and macroblocks

What then is differential pulse code modulation? If a continuous analog signal is sampled and digitized, discrete values, i.e. values which are no longer continuous, are obtained at equidistant time intervals (Fig.7.10.). These values can be represented as pulses spaced apart at equidistant intervals, which corresponds to a pulse code modulation. The height of each pulse carries information in discrete, non-continuous form about the current state of the sampled signal at precisely this point in time.

In reality, the differences between adjacent samples, i.e. the PCM values, are not very large because of the previous band-limiting. If only the difference between adjacent samples is transmitted, transmission capacity can be saved and the required data rate is reduced. This type of pulse code modulation is a relatively old idea and is now called differential pulse code modulation (Fig. 7.11.).

The problem with the usual DPCM is, however, that after a switch-on or after transmission errors it takes a very long time until the demodulated time domain signal again matches the original signal to some extent. This problem can be eliminated though by employing the small trick of transmitting at regular intervals firstly complete samples, then a few differences followed again by a complete sample etc. (Fig.7.12.) This very closely approaches the differential pulse code modulation method used in the MPEG-1/-2 image compression.

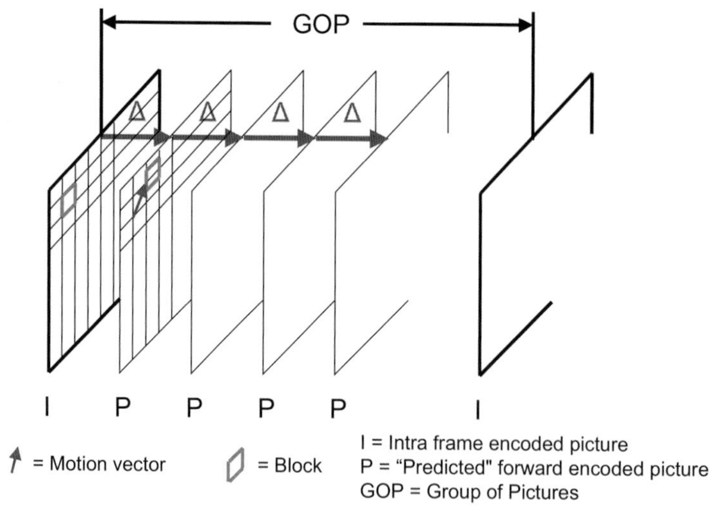

Fig. 7.14. Forward predicted delta frames

Before a frame is examined for stationary and moving components, it is first divided into numerous square blocks of 16x16 luminance pixels and 8x8 C_B and C_R pixels each (Fig.7.13.). Due to the 4:2:0 pattern, 8x8 C_B pixels and 8x8 C_R pixels are in each case overlaid on one layer of 16x16 luminance pixels each. This arrangement is now called a macroblock (Fig.7.25.). One single frame is composed of a large number of macroblocks and the horizontal and vertical number of pixels is selected to be such that it is divisible by 16 and also by 8 (Y: 720 x 576 pixels). At certain intervals complete reference frames, so-called I (intracoded) frames, formed without forming the difference, are then repeatedly transmitted and interspersed between them the delta frames (interframes).

Forming the difference is done at macroblock level, i.e. the respective macroblock of a following frame is always compared with the macroblock of the preceding frame. Put more precisely, this macroblock is first examined to see whether it has shifted in any direction due to movement in the picture, has not shifted at all or whether the picture information in this macroblock is completely new. If there is a simple displacement, only a so-called motion vector is transmitted. In addition to the motion vector, it is also possible to transmit the difference, if any, with respect to the preceding macroblock. If the macroblock has neither shifted nor changed in any way, nothing needs to be transmitted at all. If no correlation with an adjoining preceding macroblock can be found, the macroblock is completely recoded. Such pictures produced by simple forward prediction are called P (predicted) pictures (Fig.7.14.).

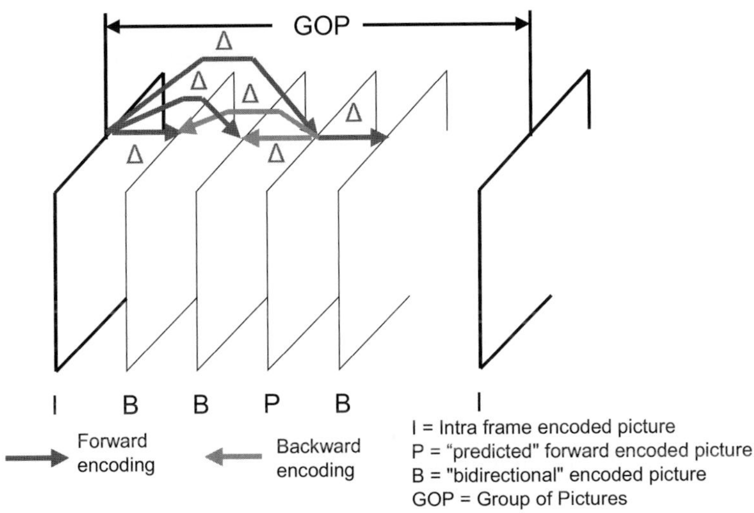

Fig. 7.15. Bidirectionally predicted delta frames

Apart from unidirectionally forward predicted frames there are also bidirectionally, i.e. forward and backward, predicted delta frames, so-called B pictures. The reason for this is the much lower data rate in the B pictures compared with the P pictures or even I pictures, which becomes possible as a result of this. The arrangement of frames occurring between two I pictures, i.e. complete pictures, is called a group of pictures (GOP) (Fig.7.14.).

The motion estimation for obtaining the motion vectors proceeds as follows: Starting with a delta frame to be encoded, the system looks in the preceding frame (forward prediction P) and possibly also in the subsequent frame (bidirectional prediction B) for suitable macroblock information in the environment of the macroblock to be encoded. This is done by using the principle of block matching within a certain search area around the macroblock.

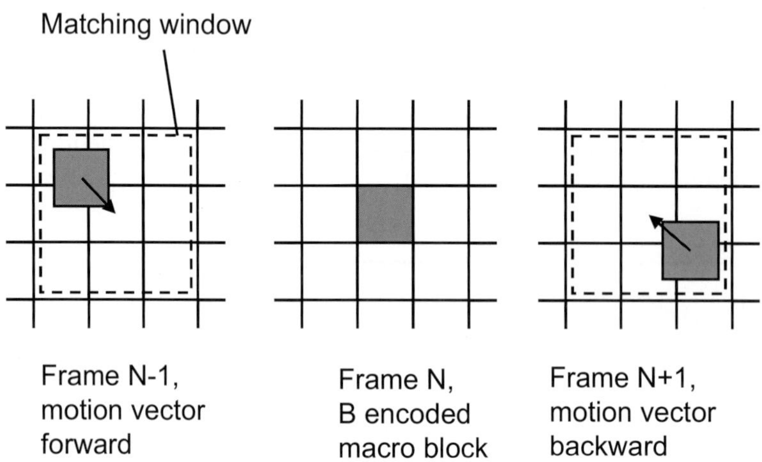

Fig. 7.16. Motion vectors

If a matching block is found in front, and also behind in the case of bidirectional coding, the motion vectors are determined forward and backward and transmitted. In addition, any additional block delta which may be necessary can also be transmitted, both forward and backward. However, the block delta is coded separately by DCT with quantization, described in the next chapter, a method which saves a particularly large amount of storage space.

A group of pictures (GOP) then consists of a particular number and a particular structure of B pictures and P pictures arranged between two I pictures. A GOP usually has a length of about 12 frames and corresponds to the order of I, B, B, P, B, B, P, ... The B pictures are thus embedded between I and P pictures. Before it is possible to decode a B picture at the receiving end, however, it is absolutely necessary to have the information of the preceding I and P pictures and that of the following I or P picture in each case. But according to MPEG, the GOP structure can be variable. So that not too much storage space needs to be reserved at the receiving end, the GOP structure must be altered during the transmission so that the respective backward prediction information is already available before the actual B pictures. For this reason, the frames are transmitted in an order which no longer corresponds to the original order.

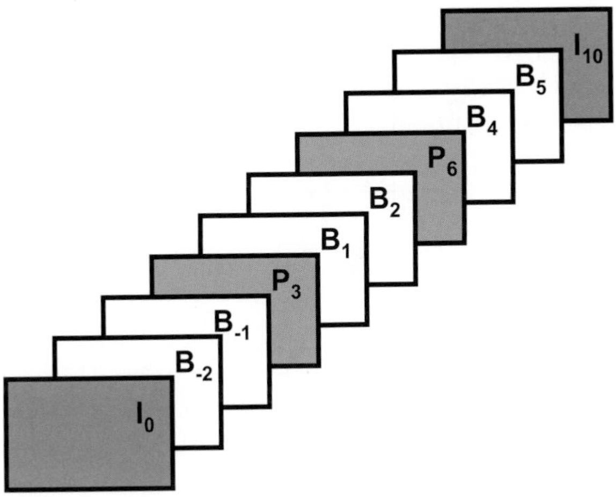

Fig. 7.17. Order of picture transmission

Instead of the order I_0, B_1, B_2, P_3, B_4, B_5, P_6, B_7, B_8, P_9, the pictures are now transmitted in the following order: I_0, B_{-2}, B_{-1}, P_3, B_1, B_2, P_6, B_4, B_5, P_9, etc. (Fig. 7.17.). That is to say, the P or I pictures following the B pictures are now available at the receiving end before the corresponding B pictures are received and must be decoded. The storage space to be reserved at the receiving end is now calculable and limited. To be able to restore the original order, the frame numbers must be transmitted coded in some way. For this purpose, the DTS (decoding time stamp) values contained in the PES header are used, among other things (see Section 3, The MPEG Data Stream).

7.1.6 Discrete Cosine Transform Followed by Quantization

A very successful method for still-frame compression has been in use since the end of the eighties: the JPEG method, which today is also being used for digital cameras and produces excellent picture quality. JPEG stands for Joint Photographic Experts Group, i.e. a group of experts in the coding of still frames. The basic algorithm used in JPEG is the Discrete Cosine Transform (Fig.7.18.), or DCT in brief. This DCT also forms the central algorithm for the MPEG video coding method.

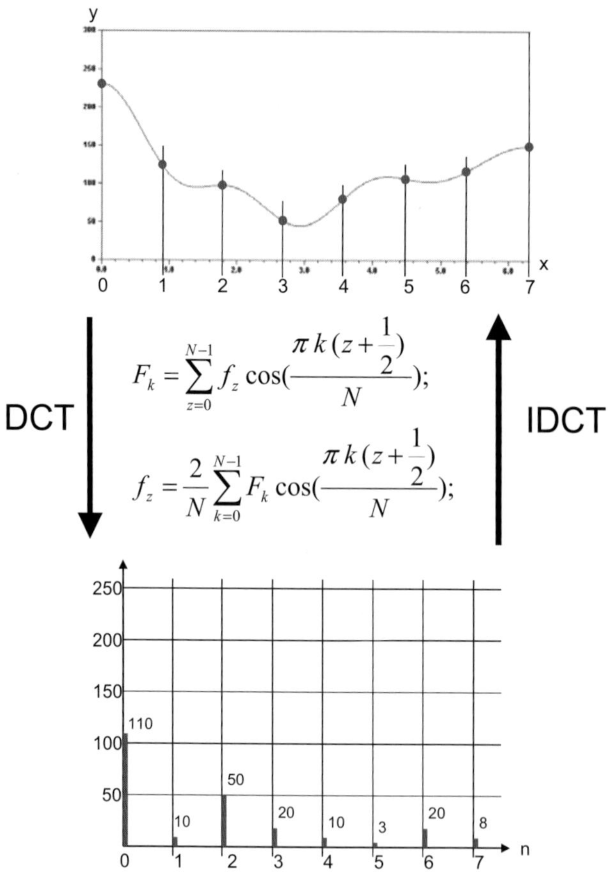

Fig. 7.18. One-dimensional Discrete Cosine Transform

The human eye perceives fine structures in a picture differently from coarse structures. In analog video test engineering it was already known that low-frequency picture disturbances, i.e. picture disturbances which correspond to coarse image structures or interfere with these are perceived more readily than high-frequency disturbances, i.e. those corresponding to fine image structures or interfering with these.

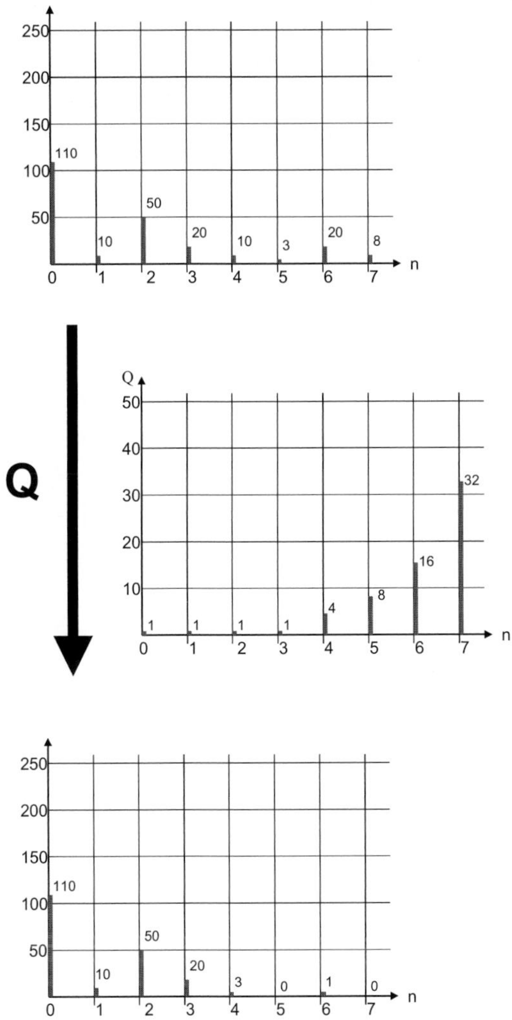

Fig. 7.19. Quantization of the DCT coefficients

For this reason, the signal/noise ratio had been measured weighted, i.e. referred to the sensitivity of the eye, even at the beginning of video testing. It was possible to allow for much more noise in the direction of higher-frequency image structures than with coarse, low-frequency image components. This knowledge is utilized in JPEG and in MPEG. Low-frequency, coarse image components are coded with finer quantization and fine image components are coded with coarser quantization in order to save data rate.

But how to separate coarse components from medium and fine image components? It is done by means of transform coding (Fig.7.18.).

Firstly, a transition is made from the time domain of the video signal into the frequency domain. The Discrete Cosine Transform is a special case of the Discrete Fourier Transform or the Fast Fourier Transform, respectively.

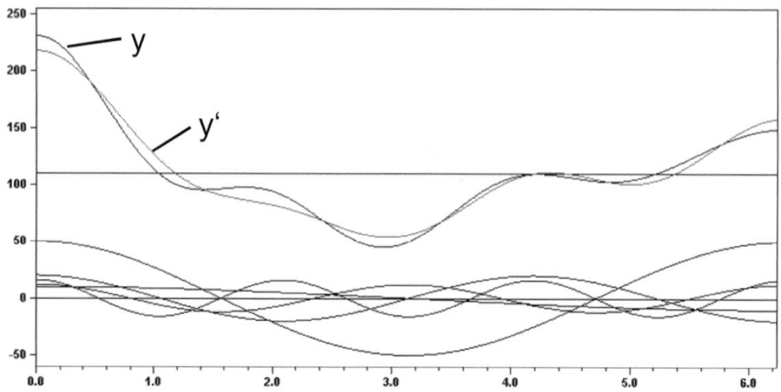

$$y(x)=110+10\cos(0.5x)+50\cos(x)+20\cos(1.5x)+10\cos(2x)+3\cos(2.5x)+20\cos(3x)+8\cos(3.5x);$$

$$y'(x)=110+10\cos(0.5x)+50\cos(x)+20\cos(1.5x)+12\cos(2x)+0\cos(2.5x)+16\cos(3x)+0\cos(3.5x);$$

Fig. 7.20. Original curve (y) and quantized Curve (y')

These transforms are dealt with in a separate Section (5) of this book. To begin with a simple example: Using the DCT, 8 samples in a video line are transformed into the frequency domain (Fig.7.18.). This again provides 8 values which, however, no longer correspond to video voltage values in the time domain but to 8 power values in the frequency domain, graded into DC, low- and medium- to high-frequency components within these 8 transformed video voltage values. The first value (DC coefficient) in the frequency domain corresponds to the energy of the video component with the lowest frequency in this section up to medium- or higher-frequency signal components. The information in one video signal section has now been processed in such a way that an irrelevance reduction can be performed which corresponds to the sensitivity characteristic of the human eye.

In a first step in this process, these coefficients are quantized in the frequency domain, i.e. each coefficient is divided by a certain quantization factor (Fig.7.19.). The higher the value of the quantization factor, the coarser the quantization. In the case of coarse image structures, the quantization must be changed only a little or not at all and in the case of fine image structures, the quantization is reduced more, meaning that the quantization factors increase in the direction of finer image structures. Due to the quantization, many values which have become zero are obtained as the fineness of the image structure increases, that is to say in the direction of higher frequency coefficients.

183	198	220	239	244	236	222	211
198	209	222	231	229	215	198	186
144	154	170	184	190	190	185	180
162	164	166	167	165	161	157	154
195	191	185	180	178	178	179	181
174	168	161	156	160	170	183	192
174	160	138	119	112	115	125	133
152	138	119	105	104	115	133	146

f(x,y)

Fig. 7.21. 8 x 8 pixel block

55	70	92	111	116	108	94	83
70	81	94	103	101	87	70	58
16	81	42	56	62	62	57	52
34	36	38	39	37	33	29	26
67	63	57	52	50	50	51	53
46	40	33	28	32	42	55	64
46	32	10	-9	-16	-13	-3	5
24	10	-9	-23	-24	-13	5	18

Fig. 7.22. Subtracting 128

These values can then be coded in a special space-saving way. However, the characteristic recovered by decoding at the receiving end after the quantization then no longer corresponds perfectly to the original curve (Fig.7.20.) and exhibits quantization errors.

In practice, however, the coding in JPEG and MPEG is two-dimensional transform coding. For this, the picture is divided into 8 x 8 pixel blocks (Fig.7.13.). Each 8 x 8 pixel block (Fig.7.23.) is then transformed into the frequency domain by means of the two-dimensional Discrete Cosine Transform. Before that is done, the value 128 is first subtracted from all pixel values in order to obtain signed values (Fig.7.22.).

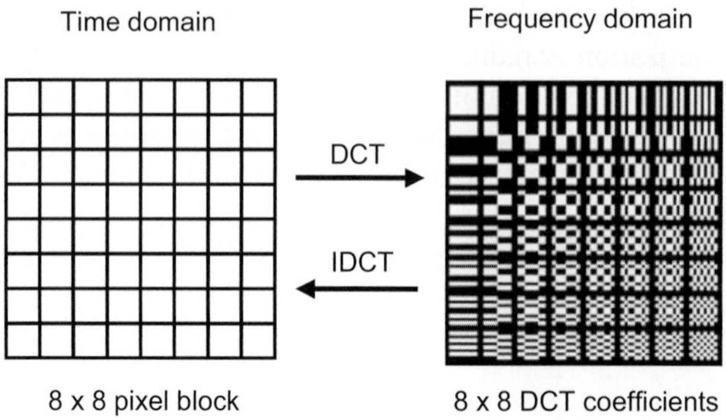

DC coefficient

1384	0	0	0	0	0	0	0
216	-36	-99	0	0	0	0	0
0	0	0	0	0	0	0	0
99	0	0	0	0	0	0	0
0	0	0	0	0	0	0	0
-58	0	0	0	0	0	0	0
0	-60	0	0	0	0	0	0
0	0	0	0	0	0	0	1

F(v,u) = DCT(f(x,u);

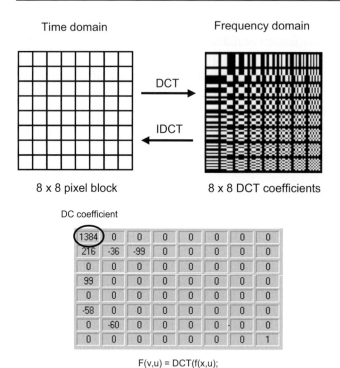

Fig. 7.23. Two-dimensional DCT

The result (Fig.7.23.) of the two-dimensional Discrete Cosine Transform of an 8 x 8 pixel array is another 8 x 8 pixel array, but now in the frequency domain. The first coefficient of the first row is the DC coefficient which corresponds to the DC component of the entire block. The second coefficient corresponds to the energy of the coarsest image structures in the horizontal direction and the last coefficient of the first row corresponds to the energy of the finest image structures in the horizontal direction. The first column of the 8 x 8 pixel block contains from top to bottom the energies of the coarsest image structures down to the finest image structures in the vertical direction. The coefficients of the coarse to fine image structures in the diagonal direction can be found diagonally.

The next step is the quantization (Fig.7.24.). All coefficients are divided by suitable quantization factors. The MPEG standard defines quantization tables but these can be exchanged by any encoder which can replace them with its own tables. These are then made known to the decoder by being transmitted to it. The quantization usually results in a great number of values which have now become zero. After the quantization, the matrix is also relatively symmetric to the diagonal axis from top left to bottom right. The matrix is, therefore, read out in a zig-zag scanning process which then provides a large number of adjacent zeroes. These can then be variable-length coded in the next step, resulting in a very large data reduction. The quantization is the only 'adjusting screw' for controlling the data rate of the video elementary stream.

8	16	19	22	26	27	29	34
16	16	22	24	27	29	34	37
19	22	26	27	29	34	34	38
22	22	26	27	29	34	37	40
22	26	27	29	32	35	40	48
26	27	29	32	35	40	48	58
26	27	29	34	38	46	56	69
27	29	35	38	46	56	69	83

Q(v,u)

scale_factor = 2 ;

173	0	0	0	0	0	0	0
6	-1	-2	0	0	0	0	0
0	0	0	0	0	0	0	0
2	0	0	0	0	0	0	0
0	0	0	0	0	0	0	0
-1	0	0	0	0	0	0	0
0	-1	0	0	0	0	0	0
0	0	0	0	0	0	0	0

QF(v,u) = F(v,u) / Q(v,u) / scale_factor ;

Fig. 7.24. Quantization after the DCT

With 4:2:0, four 8 x 8 Y pixel blocks and one 8 x 8 C_B and 8 x 8 C_R pixel block each are combined to form one macroblock (Fig.7.25.). The quantization for Y, C_B and C_R can be changed by means of a special quantizer scale factor from macroblock to macroblock. This factor alters all quantization factors either of the standard MPEG tables or of the quantization tables provided by the encoder, by a simple multiplication by a certain factor. The complete quantization table can only be exchanged at sequence level at certain times, as will be seen later.

This transform coding followed by quantization must be performed for the Y pixel plane and for the C_B and C_R planes.

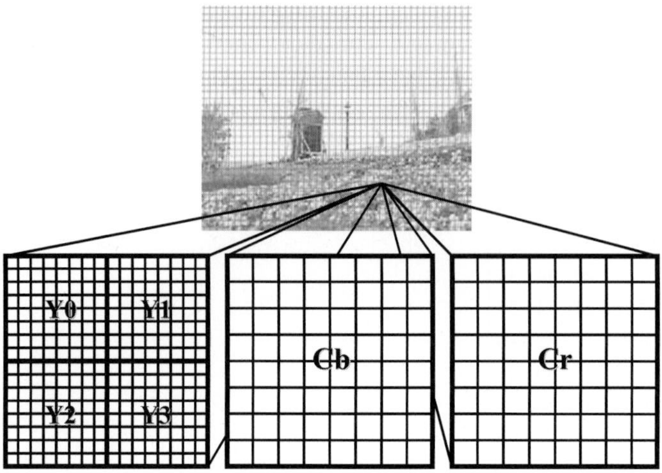

Fig. 7.25. Macroblock Structure with 4:2:0

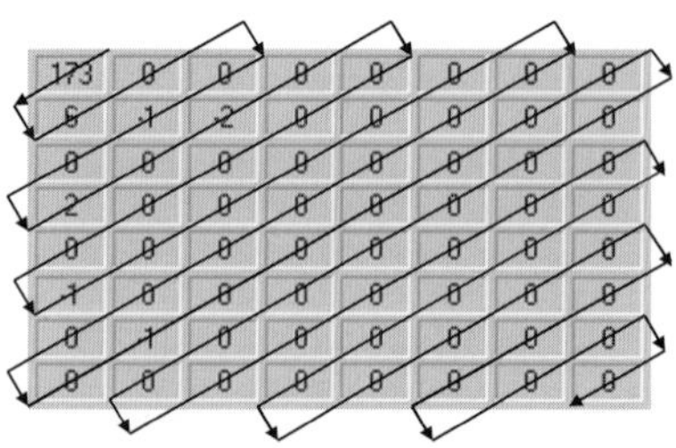

173	6	0	0	-1	0	2	0
-2	0	0	0	0	0	0	-1
0	0	0	0	0	0	0	0
0	0	0	0	0	-1	0	0
0	0	0	0	0	0	0	0
0	0	0	0	0	0	0	0
0	0	0	0	0	0	0	0
0	0	0	0	0	0	0	0

QFS(v,u)

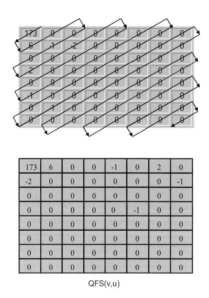

Fig. 7.26. Zig-zag scanning

In the case of I frames, all macroblocks are coded in the manner described above. In the case of P and B frames, however, the pixel differences are transform coded from macroblock in one frame to macroblock in another frame. I.e., first the macroblock of the preceding frame may have to be shifted to a suitable position with the aid of the motion vector of the macroblock and then the difference with respect to the macroblock at this position is calculated. Using the DCT, these 8 x 8 difference values are then transformed into the frequency domain and then quantized. The same also applies to the backward prediction of B pictures.

7.1.7 Zig-Zag Scanning with Run-Length Coding of Zero Sequences

After the zig-zag scanning (Fig.7.26.) of the quantized DCT coefficients, a large number of adjacent zeroes is obtained. Instead of these many zeroes, only their number is then simply transmitted by using run-length coding (RLC) (Fig.7.27.), transmitting, e.g. the information 10 times 0 instead of 0, 0, 0, ...0. This type of redundancy reduction, in conjunction with DCT and quantization, provides the main gain in the data compression.

173	6	0	0	-1	0	2	0
-2	0	0	0	0	0	0	-1
0	0	0	0	0	0	0	0
0	0	0	0	0	-1	0	0
0	0	0	0	0	0	0	0
0	0	0	0	0	0	0	0
0	0	0	0	0	0	0	0
0	0	0	0	0	0	0	0

↓ RLC

173, 6, 2*0, -1, 1*0, 2, 1*0, -2, 6*0, -1, 13*0, -1, 34*0

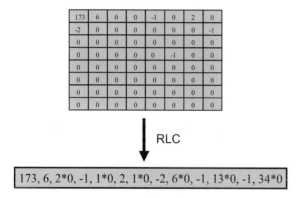

Fig. 7.27. Run-length coding (RLC)

7.1.8 Huffman Coding

Codes occurring frequently in the RLC-coded data stream are also subjected to Huffman coding (Fig.7.28.), i.e., the code words are suitably recoded, resulting in further redundancy reduction. In this type of coding, the codes used frequently are recoded into particularly short codes as in Morse code.

7.2 Summary

By using a few methods of redundancy reduction and irrelevance reduction, it has been possible to reduce the data rate of a standard definition television signal with an initial data rate of 270 Mbit/s in the 4:2:2 format according to ITU BT.R601 to about 2...6 Mbit/s with an upper limit of 15 Mbit/s. The heart of this compression method can be considered to be a differential pulse code modulation with motion compensation in combination with DCT transform coding. MPEG-2 signals intended for distribution to homes have their color resolution reduced both in the horizontal and in the vertical direction. This is then the 4:2:0 format. For studio-studio contribution, MPEG also provides the 4:2:2 format and the data rate is naturally somewhat higher.

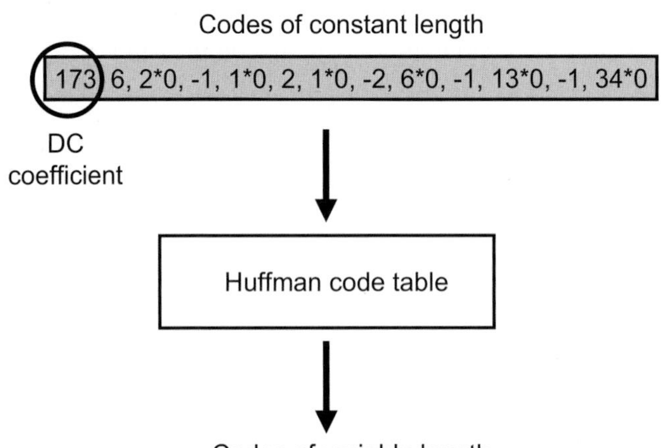

Fig. 7.28. Huffman coding (variable length coding - VLC)

Standard Definition (4:2:0) is called Main Profile@Main Level (Fig.7.29.) and Standard Definition (4:2:2) is called High Profile@Main Level. However, the MPEG standard also implements High Definition Television, both as a 4:2:0 signal (Main Profile@High Level) and as a 4:2:2 signal (High Profile@High Level). At over 800 Mbit/s, the initial data rate of an HDTV signal is clearly higher than that of an SDTV signal but the compression processes in HDTV, SDTV and 4:2:2 and 4:2:0 are the same as described before. The resultant signals only differ in their different quality and, naturally, their data rate.

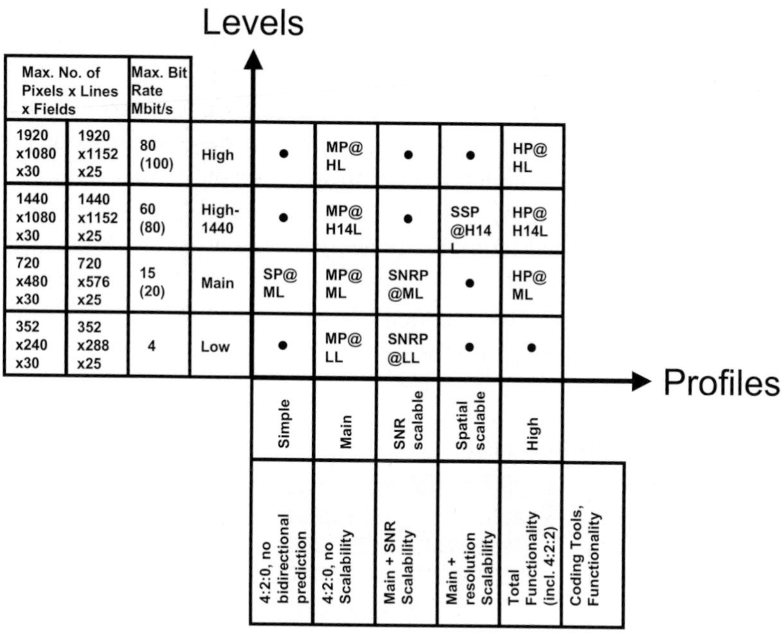

Fig. 7.29. MPEG-2 profiles and levels

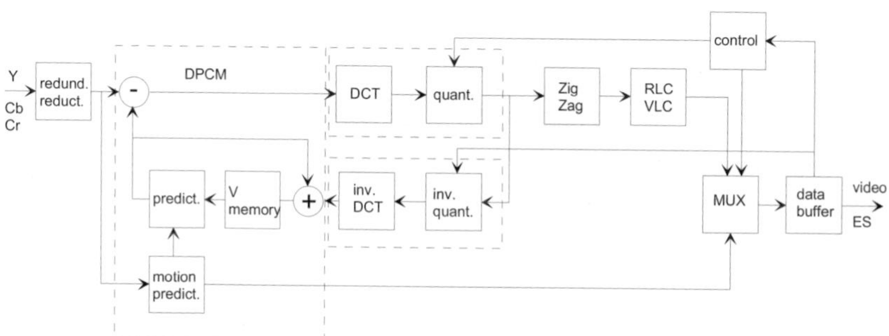

Fig. 7.30. MPEG-2 encoder

The quality of a 6 Mbit/s SDTV signal in 4:2:0 format approximately corresponds to the quality of a conventional analog TV signal. In practice, however, there are data rates ranging from 2 to 7 Mbit/s which, naturally, determines the picture quality. Correspondingly high data rates are needed, especially for sports broadcasts.

The data rate of the elementary video stream can be constant or can vary depending on the current picture content. The data rate is controlled by changing the quantization factors in dependence on the level of the output buffer of the MPEG encoder (Fig.7.30.).

The macroblocks of an I, P or B picture can be coded in various ways. The most numerous variants occur especially in the case of the B picture where a macroblock can be coded in the following ways:

- Intraframe coded (completely new)
- Forward coded
- Forward and backward coded
- Skipped (not coded at all)

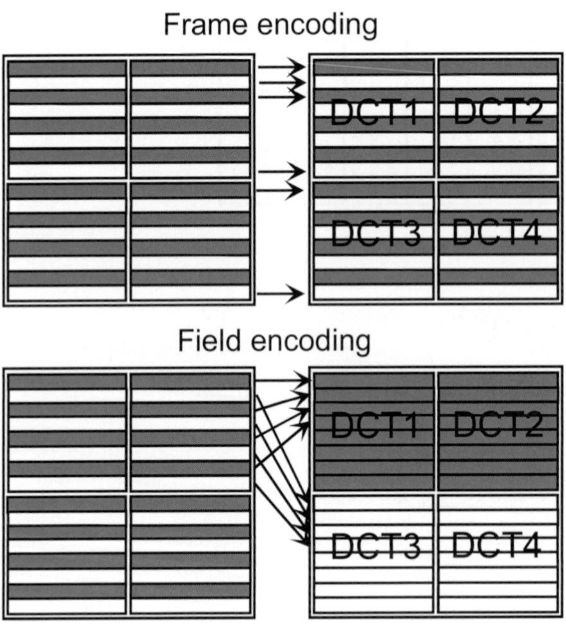

Fig. 7.31. Frame/field coding of macroblocks

The type of coding is decided by the encoder (Fig.7.30.) with reference to the current picture content and the available channel capacity (data rate).

In contrast to analog television, no fields are transmitted but only frames. The fields are then recreated at the receiving end by reading out the frame buffer in a particular way.

There is, however, a special type of DCT coding which results in better image quality for the interlaced scanning method. This involves frame and field coding of macroblocks (Fig.7.31.). In this method, macroblocks are first recoded line by line before being subjected to DCT coding.

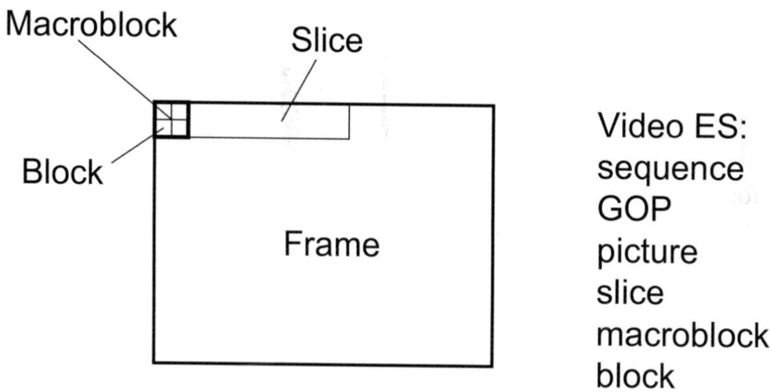

Fig. 7.32. Block, macroblock, slice and frame

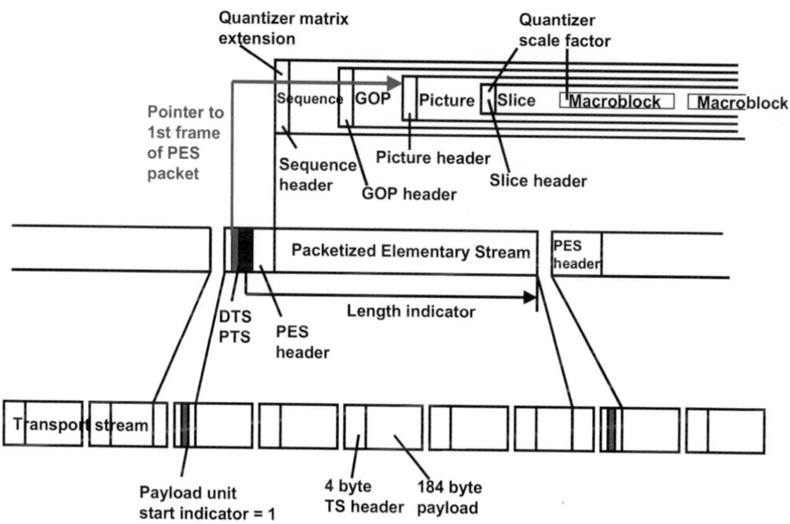

Fig. 7.33. Structure of MPEG-2 video elementary stream

7.3 Structure of the Video Elementary Stream

The smallest unit of the video stream is a block consisting of 8 x 8 pixels. Each block is subjected to a separate Discrete Cosine Transform (DCT) during the encoding. In the case of a 4:2:0 profile, four luminance blocks and one C_B block and one C_R block in each case together form one macroblock. Each macroblock can exhibit a different amount of quantization, i.e. be compressed to a greater or lesser extent. To this end, the video encoder can select different scaling factors by which each DCT coefficient is additionally divided. These quantizer scaling factors are the actual "set screws" for the data rate of the video PES stream. The quantization table itself cannot be exchanged from macroblock to macroblock. Each macroblock can be either frame encoded or field encoded. This is decided by the encoder on the basis of necessity and opportunity. One necessity for field encoding arises from the existence of motion components between the first and second field and an opportunity is presented by the available data rate.

Together, a certain number of macroblocks in a row form a slice (Fig.7.32.). Each slice starts with a header which is used for resynchronization, e.g. in the case of bit errors. At the level of the video stream, error concealment mainly takes place at slice level, i.e. in the case of bit errors, the MPEG decoders copy the slice of the preceding frame into the current frame. The MPEG decoder can resynchronize itself again with the beginning of a new slice. The shorter the slices, the lower the interference caused by bit errors.

Many slices together will then form a frame (picture). A frame, too, starts with a header, the picture header. There are different types of frames, called I (intraframe) frame, P (predicted) frame and B (bidirectionally predicted) frame. Because of the bidirectional differential coding, the order of the frames does not correspond to the original order and the headers and especially the PES headers, therefore, carry a time stamp so that the original order can be restored (DTS).

Together, a certain number of frames corresponding to a coding pattern of the I, P and B frame coding predetermined by the encoder, form a group of pictures (GOP). Each GOP has a GOP header. In broadcasting, relatively short GOPs are used which, as a rule, have a length of about 12 frames, i.e about half a second. The MPEG decoder can only lock to the signal and begin to reproduce pictures when it receives the start of a GOP, i.e. the first I frame. Longer GOPs can be chosen for mass storage devices such as the DVD since it is easy to position their read head on the first I frame.

One or more GOPs produce a sequence, each of which also starts with a header. At the sequence header level, it is possible to change essential video parameters such as the quantization table. If an MPEG encoder uses its own table which differs from the standard, this is where it will be found or, respectively, where the decoder is informed of this.

The structure of the video PES stream (Fig. 7.33.) described above is embedded wholly or partially in the video PES packets. The manner of this embedment and the length of a PES packet are determined by the video encoder. On mass storage devices such as the DVD, the PES packets are additionally inserted in so-called packs. PES packets and packs also start with a header.

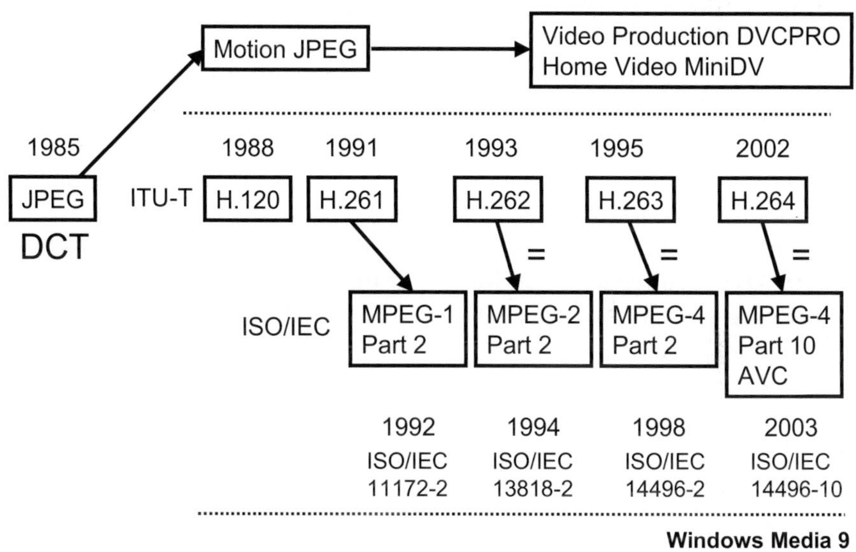

Fig. 7.34. History of the development of video coding

7.4 More Recent Video Compression Methods

Time has not stood still. Today, more modern, more advanced compression methods such as MPEG-4 Part 10 Advanced Video Coding (AVC) (H.264) or Windows Media 9 (= VC-1) are already available. With data rates which are lower by a factor of 2 to 3, a better image quality than with MPEG-2 can be achieved in many cases. Although the basic principle of video coding has not changed, the difference lies in the details. Thus, variable transform block sizes are used e.g. in H.264. Fig.7.34. shows the history of the development of video coding. As has already been mentioned several times, establishment of the JPEG standard was also a milestone of sorts for motion picture coding. DCT was used for the first time in JPEG and was only replaced by a similar transform, an integer transform, in MPEG-4 Part 10 (= H.264). Video coding was developed as part of the ITU-T H.xxx standards and then incorporated in the series of MPEG video coding methods as MPEG-1, MPEG-2 and MPEG-4. MPEG-2 Part 2 Video corresponds to H-262, MPEG-4 Part 2 Video to H.263 and MPEG-4 Part 10 AVC (Advanced Video Coding), finally, to ITU-T H.264.

7.5 MPEG-4 Advanced Video Coding

Compared with MPEG-2, the much improved MPEG-4 Part 10 AVC (H.264) video codec enables the data rates to be decreased by 30 to 50%. This means that an SDTV signal can now be compressed to approx. 1.5 ... 3 Mbit/s compared with a data rate of 2 ... 7 Mbit/s, the original uncompressed data rate having been 270 Mbit/s. Using MPEG-4, an HDTV signal can be shrunk to about 10 Mbit/s from its original 1.5 Gbit/s. MPEG-2 would have required about 20 Mbit/s for this.

MPEG-4 Part 10 Advanced Video Coding (H.264) is distinguished by the following features:

- Formats 4:2:0, 4:2:2 and 4:4:4 are supported
- Up to 16 reference frames maximum
- Improved motion compensation (1/4 pixels accuracy)
- Switching P (SP) and Switching I (SI) frames
- Higher accuracy due to 16 bit implementation
- Flexible macroblock structure (16x16, 16x8, 8x16, 8x4, 4x8, 4x4)
- 52 selectable sets of quantization tables
- Integer or Hadamard transform instead of a DCT (block size 4x4 or 2x2 pixels, resp.)
- In-loop deblocking filter (eliminates blocking artefacts)
- Flexible slice structure (better bit error performance)
- Entropy encoding; variable length coding (VLC) and context adaptive binary arithmetic coding (CABAC)

The details are as follows:

In MPEG-2 video coding, a 4:2:0 format macroblock consists of 4 luminance blocks of 8x8 pixels and one C_B and C_R block each of 8x8 pixels. MPEG-4 provides much more flexibility in this respect. Here, a macroblock has a size of either 16x16, 16x8, 8x16, 8x4, 4x8 or 4x4 pixels in the luminance layer. The block itself comprises either 4x4 or 2x2 pixels whereas it was always fixed at 8x8 pixels in MPEG-2 and MPEG-1.

The accuracy of the motion compensation is now 1/4 pixel instead of 1/2 pixel in MPEG-2. In the MPEG-2 interframe coding it was only possible to use one reference in each direction. In MPEG-4, it is possible to form several reference frames which enables the data rate to be reduced considerably.

In MPEG-2, a slice was always a multiple of macroblocks in the horizontal direction whereas MPEG-4 provides for a flexible macroblock allocation in a slice.

But it is mainly in the field of transform coding that MPEG-4 shows great changes.

In principle, MPEG-2 transform coding by means of the DCT is actually performed by a matrix multiplication in the encoder which is then inverted in the decoder. For this purpose, a lookup table is stored in the hardware. The formula for the two-dimensional DCT is:

$$F(u,v) = \frac{2}{N} C(u)C(v) \sum_{x=0}^{N-1} \sum_{y=0}^{N-1} f(x,y) \cos\frac{(2x+1)u\pi}{2N} \cos\frac{(2y+1)v\pi}{2N}$$

$$C(u), C(v) = \begin{cases} \frac{1}{\sqrt{2}} & \text{for } u,v = 0 \\ 1 & \text{otherwise} \end{cases}$$

Fig. 7.35. Definition of Discrete Cosine Transform (DCT)

It can be split into matrix multiplications based on a matrix of cosine values:

$M_{ab}[]$ → a $M_{ab}[\ ] = \cos(\frac{(2a+1)b\pi}{16});$

↓ b

cos(0) =1	cos(0) =1	cos(0) =1	cos(0) =1	cos(0) =1	cos(0) =1	cos(0) =1	cos(0) =1
cos(π/16) =0.9808	cos(3π/16) =0.8315	cos(5π/16) =0.5556	cos(7π/16) =0.1951	cos(9π/16) =-0.1951	cos(11π/16) =-0.5556	cos(13π/16) =-0.8315	cos(15π/16) =-0.9808
cos(π/8) =0.9239	cos(3π/8) =0.3827	cos(5π/8) =-0.3827	cos(7π/8) =-0.9239	cos(9π/8) =-0.9239	cos(11π/8) =-0.3827	cos(13π/8) =0.3827	cos(15π/8) =0.9239
cos(3π/16) =0.8315	cos(9π/16) =-0.1950	cos(15π/16) =-0.9808	cos(21π/16) =-0.5556	cos(27π/16) =0.5556	cos(33π/16) =0.9808	cos(39π/16) =0.1951	cos(45π/16) =-0.8315
cos(π/4) =0.7071	cos(3π/4) =-0.7071	cos(5π/4) =-0.7071	cos(7π/4) =0.7071	cos(9π/4) =0.7071	cos(11π/4) =-0.7071	cos(13π/4) =-0.7071	cos(15π/4) =0.7071
cos(5π/16) =0.5556	cos(15π/16) =-0.9807	cos(25π/16) =0.1951	cos(35π/16) =0.8315	cos(45π/16) =-0.8315	cos(55π/16) =-0.1951	cos(65π/16) =0.9808	cos(75π/16) =-0.5556
cos(3π/8) =0.3827	cos(9π/8) =-0.9239	cos(15π/8) =0.9239	cos(21π/8) -0.3827	cos(27π/8) =-0.3827	cos(33π/8) =0.9239	cos(39π/8) =-0.9239	cos(45π/8) =0.3827
cos(7π/16) =0.1951	cos(21π/16) =-0.5556	cos(35π/16) =0.8315	cos(49π/16) =-0.9808	cos(63π/16) =0.9808	cos(77π/16) =-0.8315	cos(91π/16) =0.5556	cos(105π/16) =-0.1951

Fig. 7.36. Cosine matrix lookup table

The Discrete Cosine Transform can be represented and executed as a matrix multiplication in both directions:

$$F[] = C \bullet f[] \bullet M_{ab}[] \bullet M_{ab}[]^T;$$

where $M_{ab}[]^T$ is the transposed matrix of $M_{ab}[]$, i.e. columns and rows have been swapped. This makes it possible to simultaneously perform both a horizontal and a vertical transformation, i.e. two-dimensional transformation. Linking it to matrix C makes the matrix M_{ab} into a so-called orthonormal matrix which is of great practical significance for implementing the transformation process. An orthogonal matrix is a matrix in which the inverted matrix corresponds to the transposed matrix. The following thus applies in the case of an orthogonal matrix:

$$M^T = M^{-1};$$

An orthogonal matrix has the additional property that the vectors of the matrix all have the same length. The matrix of cosine values becomes an orthonormal matrix if the first row is multiplied by $1/\sqrt{2}$ which is achieved by multiplying it by matrix C. Reversing the transformation process requires an inverted matrix.

Naturally, inverting the multiplication

$M_1 = M_2 \bullet M_3$

is not

$M_2 = M_1/M_3^{-1}$

but is defined by

$M_2 = M_1 \bullet M_3^{-1}$;

i.e. by the multiplication by the transposed matrix.
In principle, a matrix multiplication is defined as follows:

$$A \cdot B = \left[\sum_{j=1}^{n} a_{ij} \cdot b_{jk} \right];$$

$$\begin{bmatrix} a_{11}, a_{12} \\ a_{21}, a_{22} \end{bmatrix} \cdot \begin{bmatrix} b_{11}, b_{12} \\ b_{21}, b_{22} \end{bmatrix} \bullet = \begin{bmatrix} a_{11}b_{11} + a_{12}b_{21}, a_{11}b_{21} + a_{12}b_{22} \\ a_{21}b_{11} + a_{22}b_{21}, a_{21}b_{12} + a_{22}b_{22} \end{bmatrix};$$

Fig. 7.37. Definition of a matrix multiplication

Apart from the Discrete Cosine Transform (DCT), other transformation processes are also conceivable for compressing frames and can be represented as matrix multiplications, these being the

- Karhunen Loeve Transform (1948/1960)
- Haar's Transform (1910)
- Walsh-Hadamard Transform (1923)
- Slant Transform (Enomoto, Shibata, 1971)
- Discrete Cosine Transform (DCT, Ahmet, Natarajan, Rao, 1974)
- Short Wavelet Transform

A great advantage of the DCT is the great energy concentration (Fig. 7.38.) to a very few values in the spectral domain, and the avoidance of Gibbs' phenomenon which would lead to overshoots in the inverse transformation and thus to clearly visible blocking. Gibbs' phenomenon (Fig. 7.39.), known from DCT, is based on the sinusoidal component of this transformation process.

183	198	220	239	244	236	222	211
198	209	222	231	229	215	198	186
144	154	170	184	190	190	185	180
162	164	166	167	165	161	157	154
195	191	185	180	178	178	179	181
174	168	161	156	160	170	183	192
174	160	138	119	112	115	125	133
152	138	119	105	104	115	133	146

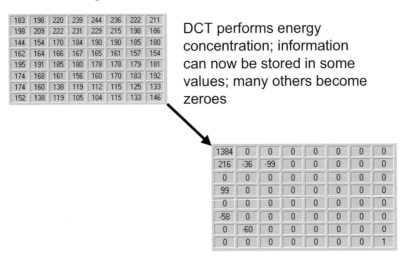

DCT performs energy concentration; information can now be stored in some values; many others become zeroes

1384	0	0	0	0	0	0	0
216	-36	-99	0	0	0	0	0
0	0	0	0	0	0	0	0
99	0	0	0	0	0	0	0
0	0	0	0	0	0	0	0
-58	0	0	0	0	0	0	0
0	-60	0	0	0	0	0	0
0	0	0	0	0	0	0	1

Fig. 7.38. Energy concentration of the DCT

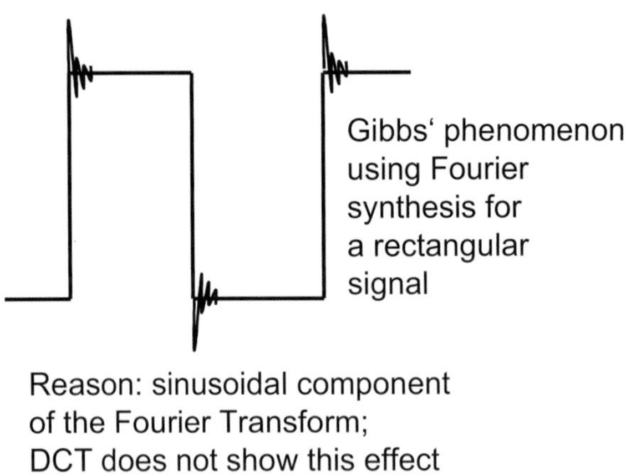

Gibbs‘ phenomenon using Fourier synthesis for a rectangular signal
Reason: sinusoidal component of the Fourier Transform; DCT does not show this effect

Fig. 7.39. Gibbs’ phenomenon

Since the cosine matrix of the DCT has now been converted into $1/\sqrt{2}$ by the conversion of the first row which consisted of all ones, and has thus become orthonormal, implementation of the transform and its inverse is quite simple.

The transform and its inverse can now be represented as follows:

$F = f \bullet M_{ab} \bullet M_{ab}^{T};$

$f = F \bullet M_{ab}^{T} \bullet M_{ab};$

During the quantization, the results of the transform and its inverse were additionally influenced by a scalar multiplication

$F = f \bullet M_{ab} \bullet M_{ab}^{T} \bullet Q;$

$f = F \bullet M_{ab}^{T} \bullet M_{ab} \bullet Q`;$

If only ones are entered in Q and Q', nothing changes. However, the quantization of the DCT coefficients is reduced towards higher frequencies via Q.

In various transformation methods, only other matrices M_{ab} are used, in principle, i.e. "basic functions" from which it is attempted to represent the original functions are others. In the case of the DCT, these are cosine patterns.

In MPEG-4, these basic patterns, or the coefficients of the matrix M_{ab}, respectively, are replaced by others. In the case of MPEG-4, this is called an integer matrix multiplication or also Hadamard transformation. The transformation matrices used in MPEG-4 AVC are the following:

$$T=\begin{bmatrix} 1 & 1 & 1 & 1 \\ 2 & 1 & -1 & -2 \\ 1 & -1 & -1 & 1 \\ 1 & -2 & 2 & -1 \end{bmatrix} \quad H=\begin{bmatrix} 1 & 1 & 1 & 1 \\ 1 & 1 & -1 & -1 \\ 1 & -1 & -1 & 1 \\ 1 & -1 & 1 & -1 \end{bmatrix} \quad C=\begin{bmatrix} 1 & 1 \\ 1 & -1 \end{bmatrix}$$

T = integer transform for luma and chroma samples
H = Hadamard transform for luma DC coefficients
C = Hadamard transform for chroma DC coeffients

Fig. 7.40. Transformation matrices in MPEG-4 AVC

The matrices used in MPEG-4 AVC have a size of only 4x4 or 2x2 pixels, respectively. In the case of luminance, the transformation is performed in two steps. In the first step, the original 4x4 pixel blocks are transformed into the spectral domain by means of the matrix T. Following this, the DCT coefficients of 16 blocks are again transformed by means of the Hadamard matrix H so that they can be compressed further (Fig. 7.41.)

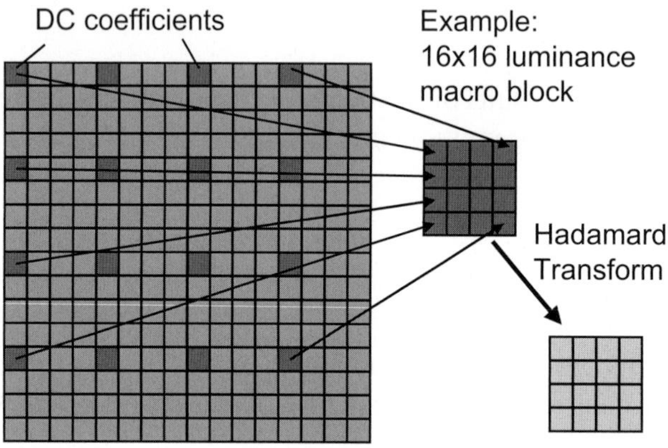

Fig. 7.41. Hadamard transform of the DC coefficients in MPEG-4 AVC

In MPEG-2, it is either the matrices specified in the Standard which are used, or they are specified by the encoder and modified and in each case transmitted to the receiver in the sequence header at the beginning of a sequence. In addition, in MPEG-2, each coefficient is divided by the quantizer scale factor which ultimately determines the actual data rate. MPEG-4 uses a set of 52 quantization matrices.

MPEG-4 also uses a deblocking filter (Fig. 7.42.) which is intended to additionally suppress the visibility of blocking artefacts. This is also aided by the smaller block size and the variable macroblock and slice size.

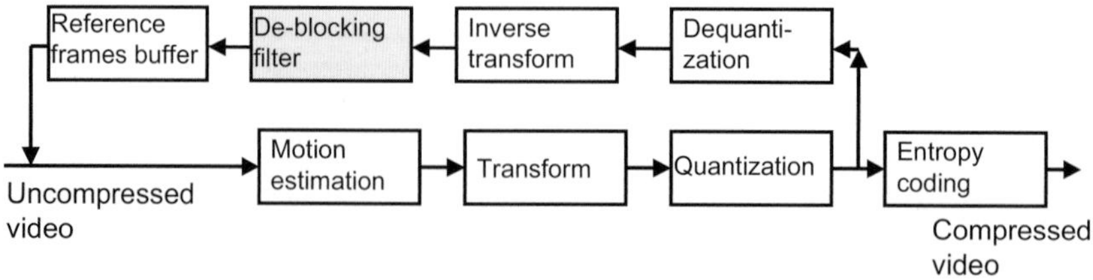

Fig. 7.42. Deblocking filter in MPEG-4

Like MPEG-2, MPEG-4 also has profiles and levels. SDTV (standard definition TV) largely corresponds to Main Profile @ Level 3 (MP@L3). HDTV (high definition TV) is then Main Profile @ Level 4 (MP@L3).

Table 7.1. MPEG-4 AVC profiles

Coding tools	Baseline profile	Extended profile	Main profile
I, P slices	x	x	x
CAVLC	x	x	x
Error resilience	x	x	
SP and SI slices		x	
B slices		x	x
Interlaced coding		x	x
CABAC			x

Table 7.2. MPEG-4 levels

Level number	Typical picture size	Max. frame rate for typ. picture size	Max. compressed bit rate	Max. number of reference frames of typ. picture size
1	QCIF	15	64 kbit/s	4
1.1	320 x 240	10	192 kbit/s	3
	QCIF	30		9
1.2	CIF	15	384 kbit/s	6
1.3	CIF	30	768 kbit/s	6
2	CIF	30	2 Mbit/s	6
2.1	HHR	30/25	4 Mbit/s	6
2.2	SD	15	4 Mbit/s	5
3	SD	30/25	10 Mbit/s	5
3.1	1280 x 720p	30	14 Mbit/s	5
3.2	1280 x 720p	60	20 Mbit/s	4
4	HD 720p, 1080i	60p/30i	20 Mbit/s	4
4.1	HD 720p, 1080i	60p/30i	50 Mbit/s	4
4.2	1920 x 1080p	60p	50 Mbit/s	4
5	2k x 1k	72	135 Mbit/s	5
5.1	2k x 1k	120	240 Mbit/s	5
	4k x 2k	30		

MPEG-4 Part 10 AVC allows an image compression which is more effective by at least 30% up to 50%, with better image quality. The SDTV data rate after compression is now less than 3 Mbit/s and the HDTV data rate is less than 10 Mbit/s. MPEG-4 AVC also makes it possible to use clearly less than 1 Mbit/s for mobile TV with SDTV quality.

MPEG-4 AVC is used today for HDTV in DVB-S2 and mobile TV as part of DVB-H and T-DMB. MPEG-4 AVC can be incorporated without problems in the MPEG-2 transport stream. On the contrary, there have been no attempts at changing anything in the transport layer. The lip synch mechanisms are the same, too, and have their origin in the MPEG-1 PES layer.

Bibliography: [ISO13818-2], [TEICHNER], [GRUNWALD], [NELSON], [MAEUSL4], [REIMERS], [H.264], [ISO14496-10]

8 Compression of Audio Signals to MPEG and Dolby Digital

8.1 Digital Audio Source Signal

The human ear has a dynamic range of about 140 dB and a hearing bandwidth of up to 20 kHz. High-quality audio signals must, therefore, match these characteristics. Before the analog audio signals are sampled and digitized, they have to be band-limited by means of a low-pass filter. Then analog-to-digital conversion is performed at a sampling rate of 32 kHz, 44.1 kHz or 48 kHz (and now also at 96 kHz), and with a resolution of at least 16 bits. The 44.1 kHz sampling rate corresponds to that of audio CDs, 48/96 kHz are studio quality. While the 32 kHz sampling frequency is still provided for in the MPEG standard, it is in fact obsolete. A sampling rate of 48 kHz at 16 bit resolution yields a data rate of 786 kbit/s per channel, which means approx. 1.5 Mbit/s for a stereo signal (Fig. 8.1.).

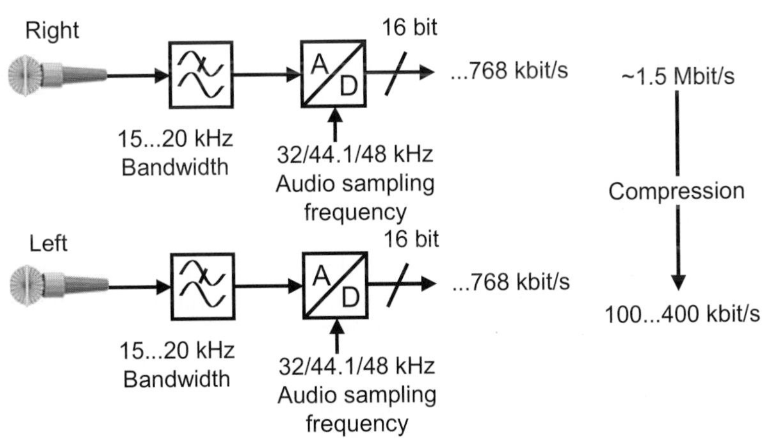

Fig. 8.1. Digital audio source signal

W. Fischer, *Digital Video and Audio Broadcasting Technology*, Signals and Communication Technology, 3rd ed., DOI: 10.1007/978-3-642-11612-4_8, © Springer-Verlag Berlin Heidelberg 2010

The objective of audio compression is to reduce the 1.5 Mbit/s data rate to between about 100 kbit/s and 400 kbit/s. MP3 audio files, which are very widely used today, often have a data rate as low as 32 kbit/s. Similarly as with video compression, this is achieved by way of redundancy reduction and irrelevance reduction. In redundancy reduction, superfluous information is simply omitted; there is no loss of information. By contrast, in irrelevance reduction information is eliminated that cannot be perceived at the receiving end, in this case the human ear. All audio compression methods are based on a psychoacoustic model, i.e. they make use of the "imperfection" of the human ear to remove irrelevant information from the audio signal. The human ear is not capable of perceiving sound events close to strong sound pulses in frequency or in time. This means that, to the ear, certain sound events will mask other sound events of lower amplitude.

8.2 History of Audio Coding

In the year 1988, the MASCAM method was developed at the Institut für Rundfunktechnik (IRT) in Munich in preparation for the digital audio broadcasting (DAB) system. From MASCAM, the MUSICAM (masking pattern universal subband integrated coding and multiplexing) method was developed in 1989 in cooperation with CCETT, Philips and Matsushita.

MUSICAM-coded audio signals are used in DAB. MASCAM and MUSICAM are both based on subband coding. The audio signal is split into a large number of subbands, each of which is subjected to irrelevance reduction to a greater or lesser degree.

At the same time as the subband coding method was developed, the Fraunhofer Gesellschaft together with Thomson devised the ASPEC (Adaptive Spectral Perceptual Entropy Coding) method, which is based on transform coding. The audio signal is transformed from the time to the frequency domain using DCT (Discrete Cosine Transform), and then irrelevant signal components are removed.

Both the subband-coding MUSICAM and the transform-coding ASPEC method were included in the MPEG-1 audio compression method, which was established in 1991 (ISO/IEC 11172-3 standard). MPEG-1 audio comprises three possible layers: II essentially use MUSICAM coding, and layer III principally uses ASPEC coding. MP3 audio files are coded to MPEG-1 layer III. MP3 is often mistaken for MPEG-3. MPEG-3 was originally aimed at implementing HDTV (high definition television), but HDTV was already integrated in the MPEG-2 standard, so MPEG-3 was skipped and abandoned altogether. Therefore the MPEG-3 standard does not exist.

In MPEG-2 audio, the three layers of MPEG-1 audio were taken over, and layer II was extended to form layer II MC (multichannel). The ISO/IEC 13818-3 MPEG-2 audio standard was adopted in 1994.

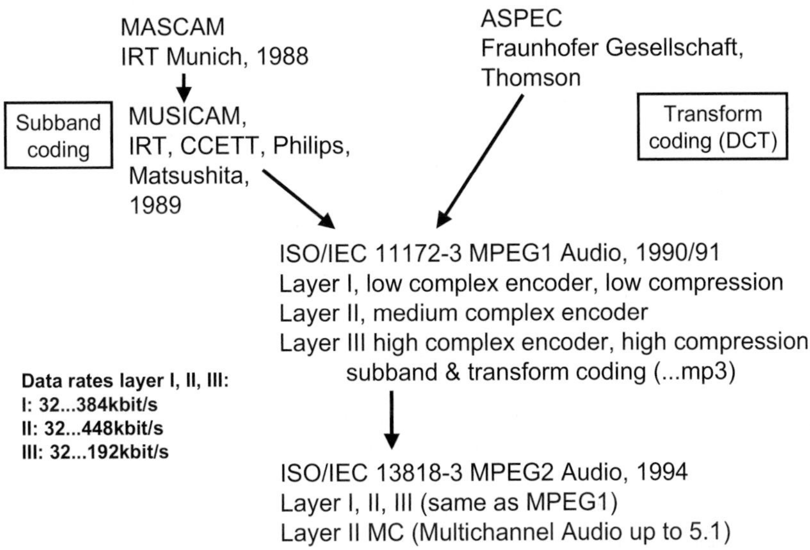

Fig. 8.2. Development of MPEG audio [DAMBACHER]

Simultaneously with MPEG audio, the Dolby digital audio standard (also known as AC-3 audio) was developed by Dolby Labs in the USA. This standard was laid down in 1990 and first presented to the public in the movie "Star Trek VI" shown in December 1991. Nowadays, many movies employ the Dolby digital technique. In the USA, digital terrestrial TV broadcasts to ATSC use AC-3 audio coding exclusively. Some other countries too (e.g. Australia) will introduce AC-3 audio in addition to MPEG audio. The use of both AC-3 audio and MPEG audio is meaningful, if only because of the fact that this does away with the recoding of movies. As from the point of quality, there is practically no difference between MPEG audio and Dolby digital. Modern MPEG decoder chips, therefore, support both methods. DVD video discs too may use Dolby digital AC-3 audio in addition to PCM audio and MPEG audio. Below is a short overview of the development of Dolby digital:

- 1990 Dolby digital AC-3 audio
- 1991 First AC-3 audio coded movie show
- Dec. 1991 "Star Trek VI" coded in AC-3 audio

Today:

- AC-3 audio is used as standard in many movies, in ATSC and, in addition to MPEG audio, in MPEG-2 transport streams all over the world, and on DVDs.
- Dolby AC-3 audio transform coding based on Modified Discrete Cosine Transform (MDCT); 5.1 audio channels (left, center, right, left surround, right surround, subwoofer), 128 kbit/s per channel.

MPEG, too, has come up with new audio coding methods:

- MPEG-2 AAC ISO/IEC 13818-7
  AAC = Advanced Audio Coding
- MPEG-4 ISO/IEC 14496-3:
  AAC and AAC Plus

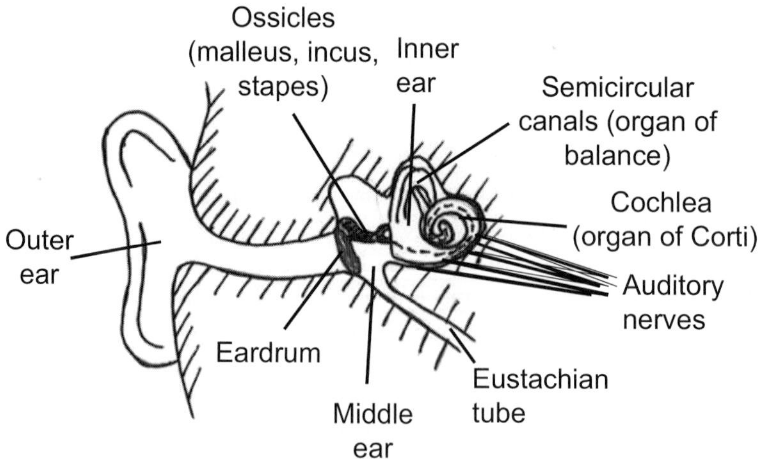

Fig. 8.3. Anatomy of the human ear

8.3 Psychoacoustic Model of the Human Ear

In the following section, the process of audio compression will be discussed. Redundancy reduction (lossless) and irrelevance reduction (lossy) lower the data rate of the original audio signal by about 90 %. Irrelevance reduction relies on the psychoacoustic model of the human ear, which essentially goes back to Professor Zwicker, former holder of a professorship for electroacoustics at the Technical University of Munich. This type of reduction is based on what is referred to as perceptual coding. This means that audio components which are not perceived by the human ear are not transmitted.

Let us first have a look at the anatomy of the human ear (Fig. 8.3., 8.4.). The ear consists of three main parts: the outer ear, the middle ear, and the inner ear. The outer ear performs the functions of impedance matching, sound transmission over air, and acts as a filter with a slight resonance step-up in the region of 3 kHz. It is in the same region, i.e. from 3 kHz to 4 kHz, that the human ear exhibits its maximum sensitivity. The eardrum or tympanic membrane converts sound waves to mechanical vibrations, which are transmitted via the malleus, incus and stapes to a membranous window leading to the sensory inner ear. The air pressure must be the same, ahead of and behind the eardrum. This is ensured by a tube connecting the region behind the eardrum with the pharynx; the tube is called the Eustachian tube. Everyone knows the problem of pressure building up in the ear when climbing large heights. By swallowing, the mucous membrane in the Eustachian tube provides for pressure compensation.

In the inner ear we find the organ of balance, which is made up of several liquid-filled arches, and the cochlea. The cochlea is the actual hearing organ (organ of Corti) by which sound is directly perceived. If the cochlea were to be uncoiled, the sensors for the high frequencies would be found at its entrance, then the sensors for the medium frequencies, and at the end of the cochlea would be the sensors for the low frequencies.

The cochlea consists of a spiral canal in which lies a smaller membranous spiral passage that becomes wider from the front to the rear. On the inner membrane rest the frequency-selective sound-collecting sensors from which the auditory nerves extend to the brain. The auditory nerves transport electrical signals with an amplitude of approx. 100 mV_{pp}. The repetition rate of the electrical pulses is in the order of 1 kHz. The information contained in this rate is the volume of a tone at a given frequency. The louder the tone, the higher the repetition rate. Each frequency sensor communicates with the brain via a separate neural line. The frequency selectivity of the sensors is highest at low frequencies and decreases towards higher frequencies.

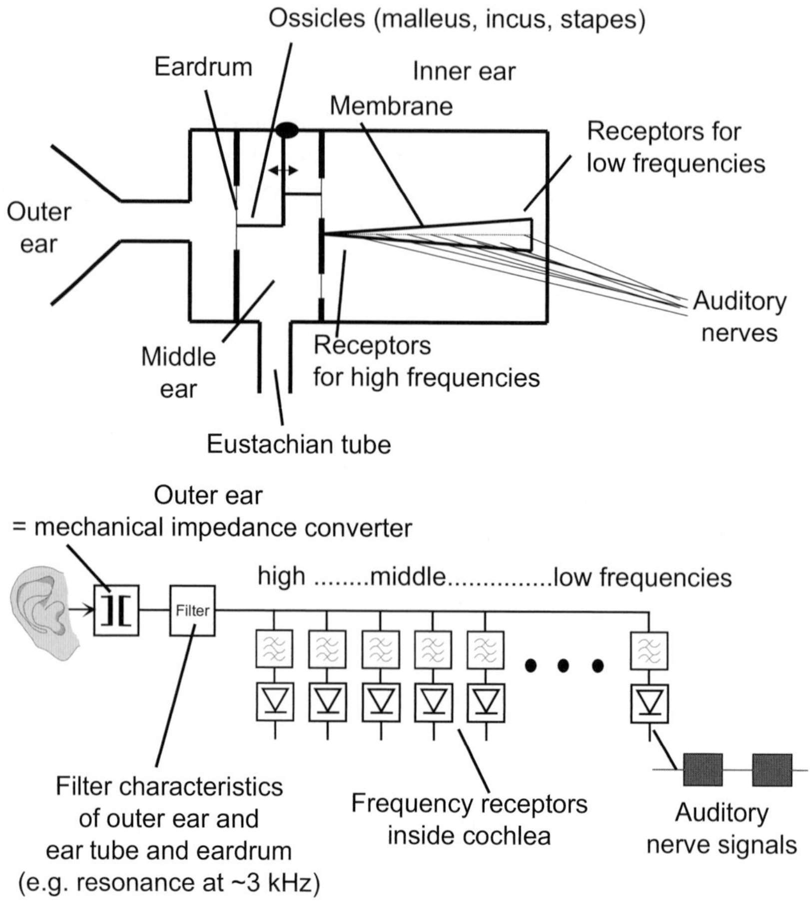

Fig. 8.4. Technical model of the human ear

In the following section, we want to investigate those characteristics of the human ear that are of interest for audio coding. To begin with, the sensitivity of the ear is to a great extent dependent on frequency. Sound signals below 20 Hz and above 20 kHz are practically not audible. The maximum sensitivity of the ear is in the range around 3 kHz to 4 kHz; outside this range the sensitivity decreases towards higher or lower frequencies. Sounds with a level below a certain threshold (referred to as threshold of audibility) are not perceived by the human ear. The threshold of audibility is frequency-dependent. Any components of audio signals whose level is below the audibility threshold need not be transmitted; they are irrelevant for the human ear. Fig. 8.5. illustrates the general relationship of audibility threshold versus frequency.

The next characteristic of the human ear that is of significance for audio coding is a characteristic known as masking. For example, a sinusoidal carrier at 1 kHz with constant amplitude is applied to the ear of a test person, and the region around 1 kHz is investigated by applying other sinusoidal carriers, the frequency and amplitude of which is varied. It will be found that the other test signals are not audible below a certain frequency-dependent level threshold around 1 kHz. This is known as the masking threshold (Fig. 8.6.). The shape of the masking threshold depends on the frequency of the masking signal. The higher the frequency of the masking signal, the wider the masked range.

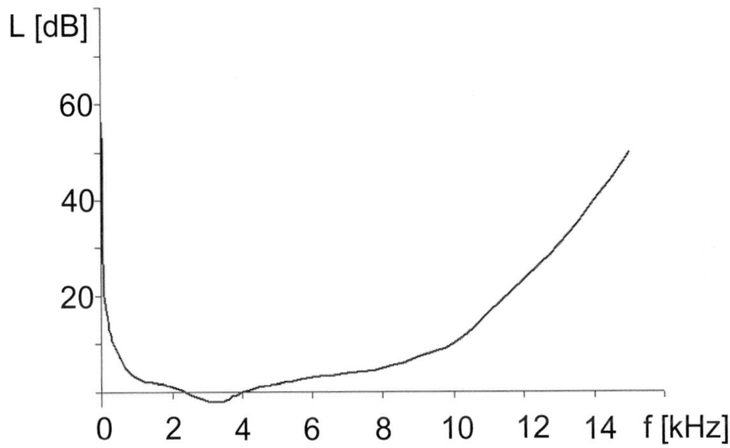

Fig. 8.5. Threshold of audibility

This characteristic of the ear is known as masking in the frequency domain (Fig. 8.6.). The relevant factor for audio coding is the fact that audio components below a defined masking threshold need not be transmitted.

However, masking not only occurs in the frequency domain but also in the time domain (Fig. 8.7.). A strong pulse in the time domain masks sound signals before and after the pulse, provided the levels of these signals are below a certain threshold. This effect, and in particular premasking, is difficult to imagine but very well explicable. It is due to the finite time resolution of the human ear in conjunction with the way signals are transported to the brain via the auditory nerves.

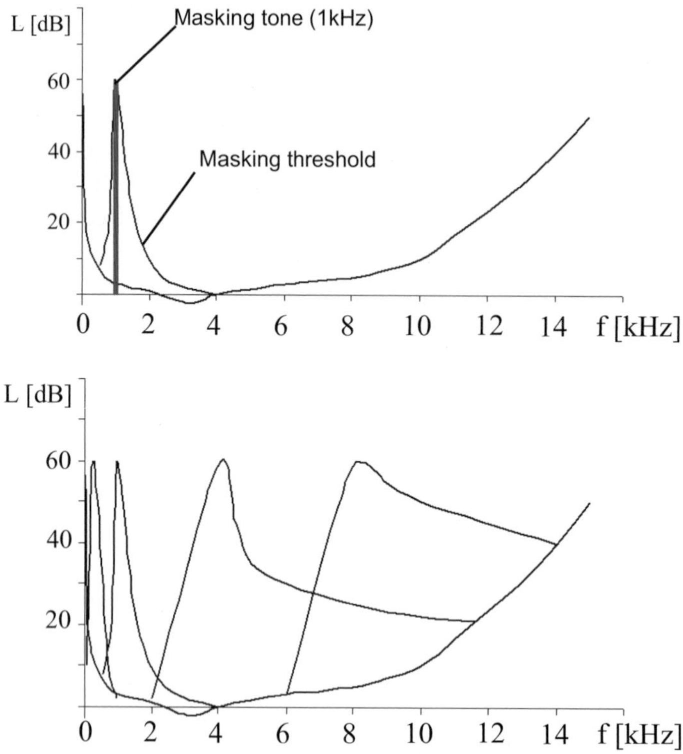

Fig. 8.6. Masking in the frequency domain

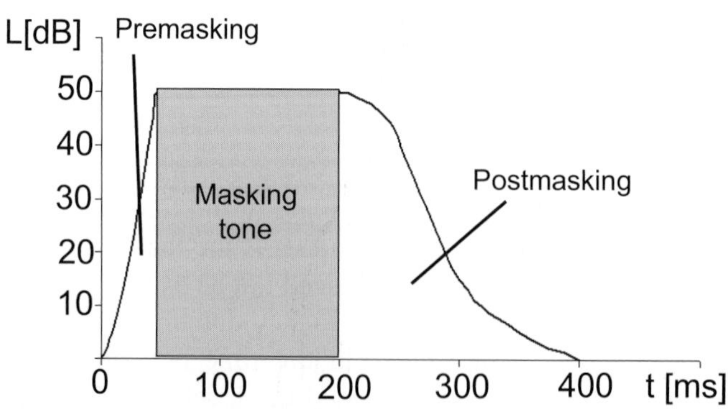

Fig. 8.7. Masking in the time domain

The audio compression methods known so far use masking only in the frequency domain, the techniques employed being very similar in all cases.

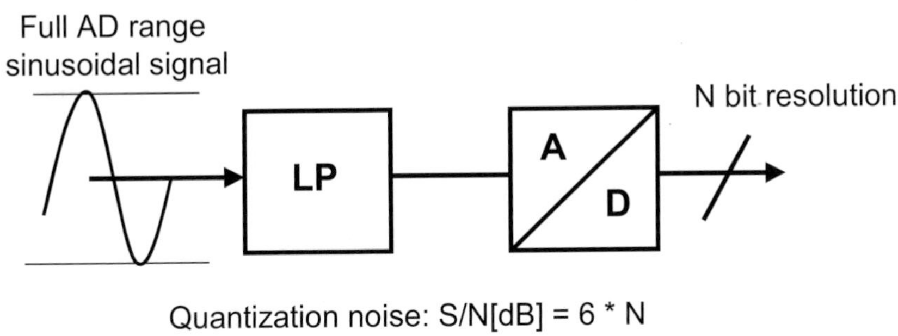

Fig. 8.8. Quantization noise

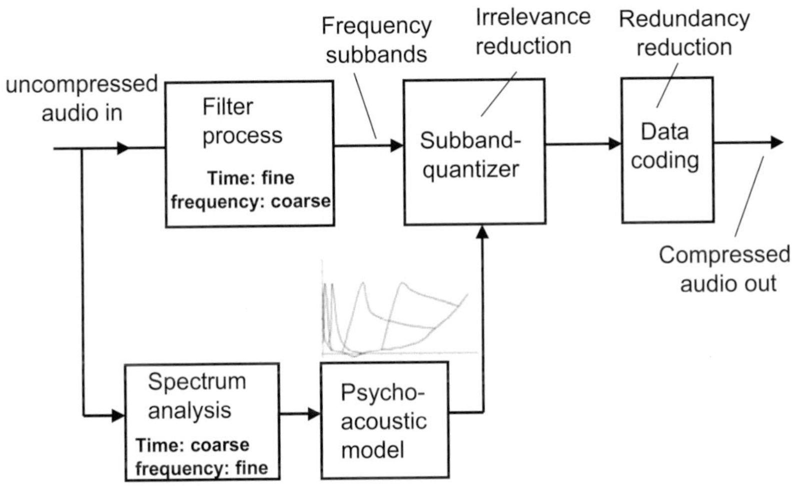

Fig. 8.9. Principle of audio coding based on perceptual coding

8.4 Basic Principles of Audio Coding

Prior to discussing the principle of irrelevance reduction for audio signals, quantization noise will be examined briefly. If an analog-to-digital converter is driven to full modulation with a sinusoidal signal, an S/N ratio of approx. 6 • N dB (rule of thumb) is obtained for a resolution of N bits due to quantization noise (Fig. 8.8.). This means that approx. 48 dB are obtained for 8 bit resolution and 96 dB for 16 bit resolution. Audio signals are usually sampled with 16 bits or more. 16 bit resolution, however, still does not match the dynamic range of the human ear, which is about 140 dB.

Let us now discuss the basic principle of audio coding (Fig. 8.9.). The digital audio source signal is split into two branches in the coder, filtered and taken to a frequency analyzer. The frequency analyzer performs spectrum analysis by means of a Fast Fourier transform (FFT) and determines the components of the audio signal with low time resolution and high frequency resolution.

Based on the knowledge of the psychoacoustic model (masking effect), irrelevant frequency components of the current signal can be identified.

Simultaneously with spectrum analysis, the audio signal undergoes filtering by which it is split into many subbands. It may happen that a complete subband is masked by signals of other subbands, i.e. the signal level in this subband is below the masking threshold. If this is the case, the subband in question need not be transmitted; the information carried in this band is completely irrelevant to the human ear. The filtering process by which the audio signal is spread to subbands must use very high time resolution so that no information in the time domain will be lost. In contrast for the frequency domain, coarse resolution will do. As far as irrelevance reduction is concerned, there is another possibility. Sometimes, signals in a subband are above the masking threshold, but only by a slight margin. In such cases, quantization in the subband concerned is reduced to the extent that quantization noise in this band is below the masking threshold and is therefore not audible.

Likewise, signals below the threshold of audibility need not be transmitted. Here, too, coarser or finer quantization can be selected depending on the different audibility thresholds of the subbands so that the resulting quantization noise always remains below the threshold. Lower bit resolution is possible especially at higher frequencies.

The decision of whether a subband is to be suppressed completely, or if coarser or finer quantization is to be applied is made in the "psychoacoustic model" block, which is fed with the information from the spectrum analysis block. Quantization is suppressed or controlled by means of the subband quantizer. It may be followed by redundancy reduction, which is effected by a special data coding. After these processes are completed, the compressed audio signal is available.

Perceptual coding may be implemented in various ways. There is pure subband coding and transform coding, and there are mixed forms which are referred to as hybrid coding.

8.5 Subband Coding in Accordance with MPEG Layer I, II

First the method of subband coding will be discussed. In accordance with MPEG layer I and II (Fig. 8.10.), the audio signal is passed through a filter bank of 32 filters that split the signal into frequency subbands of 750 Hz. For each subband there is a separate quantizer controlled by an FFT block and a psychoacoustic model. The quantizer either completely suppresses the subband in question or reduces the number of quantization steps. In the case of layer II coding, FFT is carried out every 24 milliseconds on 1024 samples. This means that the information fed to the psychoacoustic model changes every 24 milliseconds. During the 24 ms intervals, the subbands are subjected to irrelevance reduction in accordance with the information received from the psychoaoustic model block. In other words, the signal is treated as if its composition had not altered for 24 ms.

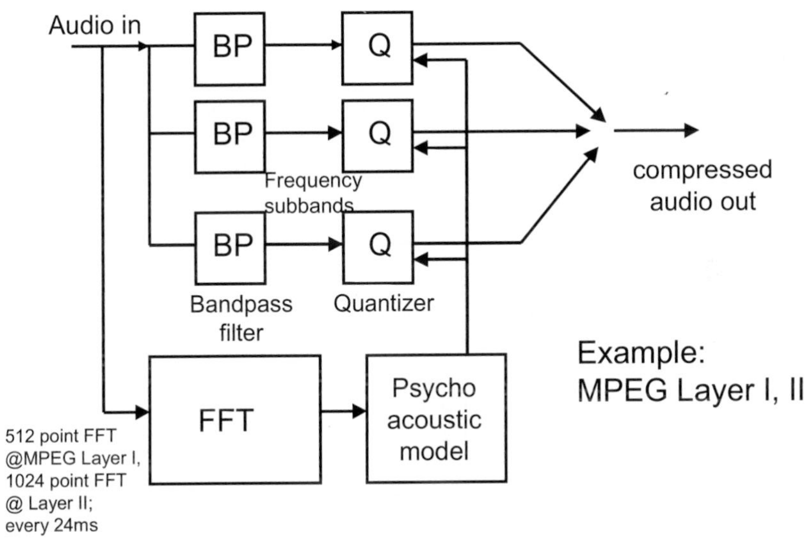

Fig. 8.10. Subband coding using 32 bandpass filters in MPEG-1 and MPEG-2 layer I, II

Because of the different audibility thresholds, bit allocation and thus quantization is different for the different subbands. Quantization must be finest at low frequencies; it may be reduced towards higher frequencies.

Fig. 8.11. illustrates the principle of irrelevance reduction in audio transmission by means of two examples. In one subband, there is a signal at about 5 kHz with a level above the masking threshold. In the case of this subband, only the number of quantization steps can be reduced. In another subband, we find a signal at about 10 kHz with a level below the masking threshold. This means that this subband is fully masked by signals of neighbouring subbands and can therefore be suppressed completely.

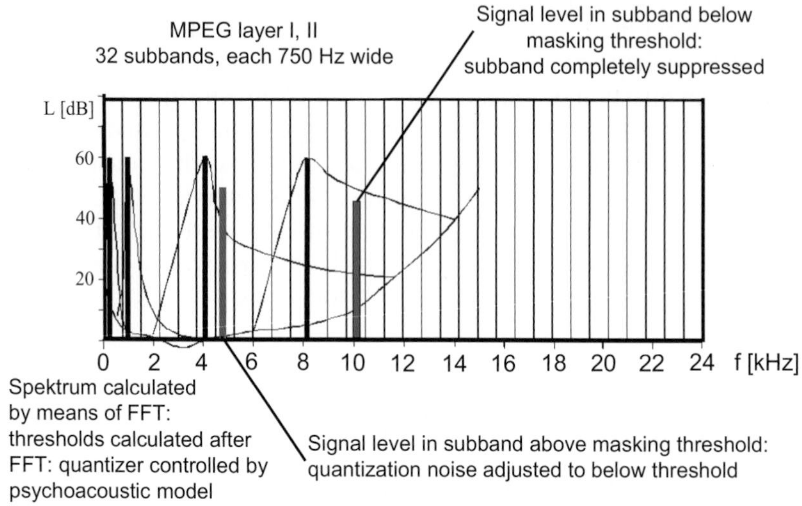

Fig. 8.11. Irrelevance reduction utilizing masking effects

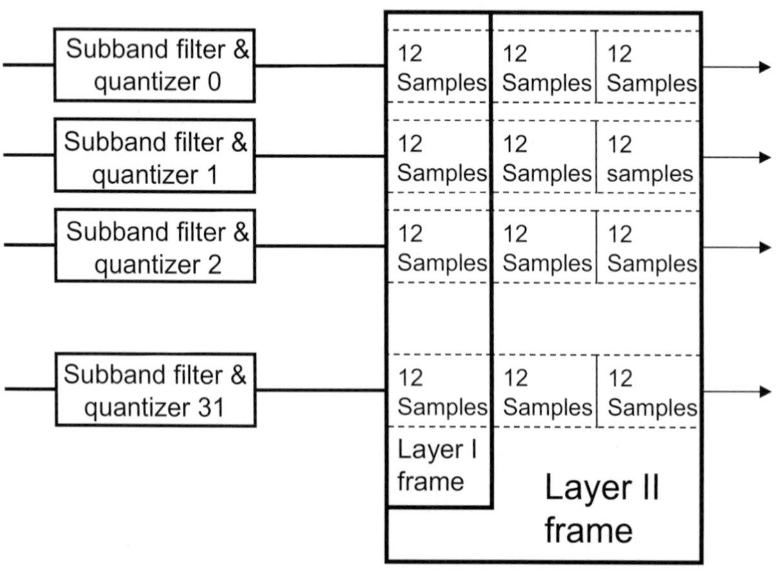

Fig. 8.12. MPEG-2 layer I, II data structure

In irrelevance reduction, subbands are also evaluated as to whether they contain harmonics of signals belonging to a lower subband, i.e. whether the masked signals are tonal (harmonic) or non-tonal components. Only non-tonal, masked signals may be completely suppressed.

In MPEG coding, a certain number of samples are always combined into frames. A layer I frame is formed with 12 samples for each subband. A layer II frame is formed with 3 x 12 samples for each subband (Fig. 8.12.).

For each 12-sample block, the highest sample is determined. This sample is used as a scaling factor which is applied to all 12 samples of the block to provide for redundancy reduction (Fig. 8.13.).

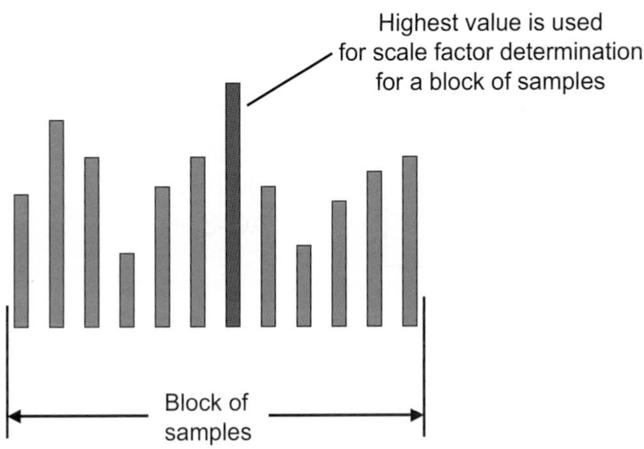

Fig. 8.13. Redundancy reduction to MPEG-2 layer I, II

8.6 Transform Coding for MPEG Layer III and Dolby Digital

Transform coding, in contrast to subband coding, uses no filter bank for subband filtering; the splitting of audio information in the frequency domain is effected by a variation of the Discrete Fourier Transform. Using a Discrete Cosine Transform (DFT) or Modified Discrete Cosine Transform (MDFT), the audio signal is processed to give 256 or 512 spectral power values. At the same time, in the same way as with subband coding, A Fast Fourier Transform (FFT) is carried out with relatively high resolution in the frequency domain. Controlled by the psychoacoustic model created from the FFT output data, the power values of the audio signal obtained through MDFT are subjected to coarser or finer quantization or are suppressed completely. The advantage of this method over subband coding is that it offers higher frequency resolution for the process of irrelevance reduction. This type of coding is used, for example, in Dolby Digital AC-3 Audio (Fig. 8.14.) (AC-3 stands for audio coding 3).

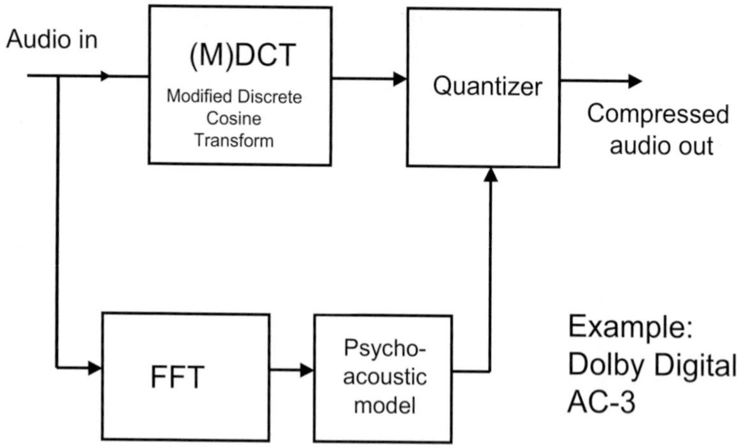

Fig. 8.14. Transform coding

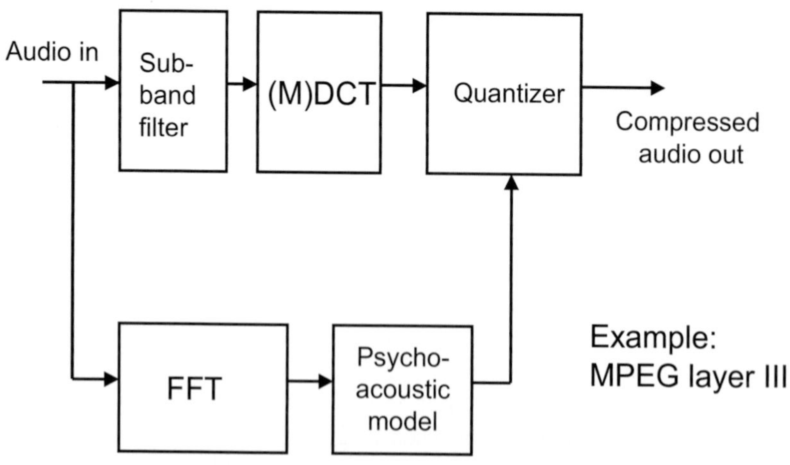

Fig. 8.15. Hybrid subband and transform coding

There is also mixed subband coding and transform coding. which is known as hybrid coding. For example, in MPEG layer III coding, subband filtering is performed prior to (M)DCT (Fig. 8.15.). This means that first a coarse splitting into subbands takes place, then (M)DCT is applied to each subband to obtain a finer resolution. After (M)DCT, data are subjected to irrelevance reduction controlled by the psychoacoustic model, which in turn is fed with information from the Fast Fourier Transform. Audio data coded to MPEG layer III are commonly referred to as MP3 audio files and are used all over the world today.

8.7 Multichannel Sound

In multichannel audio coding, irrelevances between the channels can be determined and omitted in transmission. This means that the channels are investigated for correlated components that do not contribute to the spatial hearing impression. This procedure is employed, for example, in MPEG layer II MC and Dolby digital 5.1 surround. In 5.1 audio, the following channels are transmitted: left, center, right, left surround, right surround and a low-frequency enhancement (LFE) channel for a subwoofer.

Fig. 8.16. shows the loudspeaker configuration for 5.1 multichannel audio.

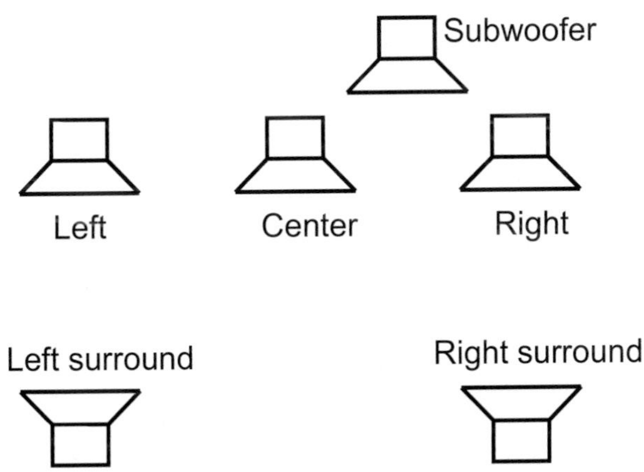

Fig. 8.16. Multichannel audio

The more detailed structure of these audio coding methods is not relevant in terms of practical applications and will not be further discussed here. For more information, consult the relevant literature and the standards.

8.8 New Developments - MPEG-4

Time has not stood still in the field of audio compression, either. MPEG-4 Advanced Audio Coding - MPEG-4 AAC - is a standard which contains a number of newly developed audio codecs. Fig. 8.17. shows once again the complete development history of audio compression including MPEG-4 AAC. MPEG-4 AAC = ISO/IEC 14493-3, i.e. MPEG-4 Part 3 includes both the previous MPEG-2 AAC codec and various new MPEG-4 audio codecs up to the MPEG-4 HE (High Efficiency) AAC, which is equivalent to the AAC+ developed by Coding Technologies in Nuremberg. AAC+ allows broadcasting-type audio quality at data rates of 64 kbit/s, i.e. 1/3 of the data rate in comparison with MPEG-1 Layer II.

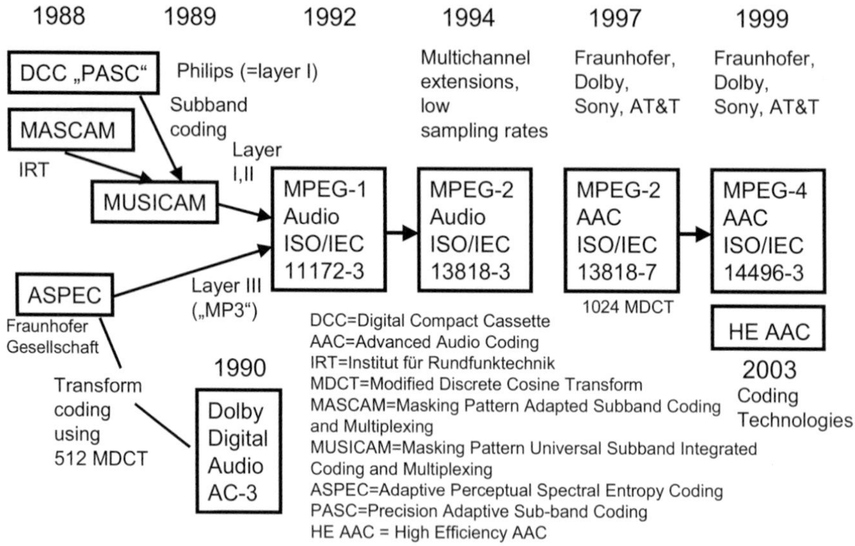

Fig. 8.17. History of the development of audio coding

The magic word of the latest audio coding method is "Spectral Band Replication" (SBR), i.e. the effective transmission and recovery of higher-frequency audio components. This audio compression method is used by all mobile TV standards. DAB+ and DRM also use the latest MPEG-4

AAC algorithms. Fig. 8.18. once again shows the audio structure of all MPEG standards. The audio coding method is always described in Part 3 of the respective MPEG standard. DVB uses currently MPEG-1 Layer II Audio with a data rate of 192 kbit/s in most cases. The MPEG-2 extensions do not provide any advantages in DVB. Multichannel capability is implemented via Dolby Digital Audio transmitted in parallel. And DVB does not need the lower sampling rates provided by MPEG-2 Layer II Audio. Audio players mostly supported MPEG-1 Layer III Audio, MPEG-2 Layer III Audio or MPEG-4 AAC. These devices, called MP3 players in most cases, are known to everyone by now and are also replacing almost every other audio recording and replaying device.

MPEG-1 Part 3	**MPEG-2 Part 3**	**MPEG-4 Part 3**
Layer I (Philips, DCC, PASC)	Layer I	(includes MPEG-2 AAC) AAC LC
Layer II (DAB, MUSICAM)	Layer II	AAC LTP AAC scalable
Layer III (ASPEC, Fraunhofer, MP3)	Layer III (all layers: low sampling rates and multichannel 3/2 +LFE)	Twin VQ CELP HVXC TTSI BSAC HE AAC = AAC+
	MPEG-2 Part 7 AAC	

Fig. 8.18. Audio codec structure in MPEG

Bibliography: [ISO13818-1], [DAMBACHER], [DAVIDSON], [THIELE], [TODD], [ZWICKER]

9 Teletext, Subtitles and VPS for DVB

In analog television, teletext, subtitles and VPS (Video Program System for VCR control) have been much used supplementary services for many years. Apart from being able to create completely new, comparable services in DVB, standards were set up for enabling these familiar services to be incorporated compatibly in MPEG-2 data streams conforming to DVB. The approach is for the DVB receiver to insert these services back into the vertical blanking interval at the composite CVBS video output. This does not affect any parallel DVB data services such as EPG (Electronic Program Guide) or MHP (Multimedia Home Platform).

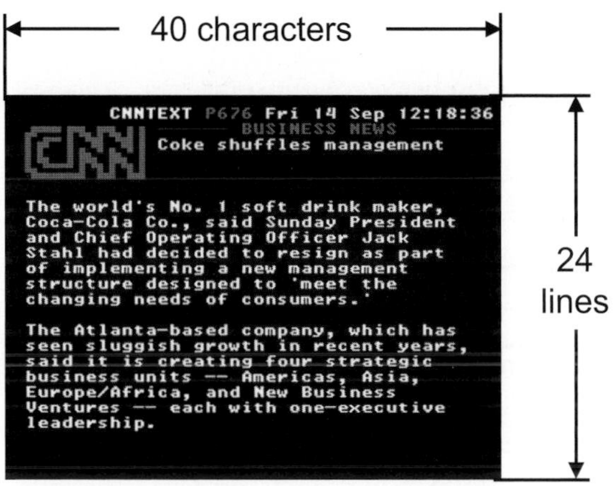

Fig. 9.1. Teletext page

9.1 Teletext and Subtitles

In analog TV, teletext (Fig. 9.1.) is inserted with roll-off filtering as an NRZ (non-return-to-zero) coded supplementary signal into the vertical blanking interval. In DVB, by contrast, a teletext elementary stream is simply multiplexed directly into the MPEG-2 transport stream. The

W. Fischer, *Digital Video and Audio Broadcasting Technology*, Signals and Communication Technology, 3rd ed., DOI: 10.1007/978-3-642-11612-4_9, © Springer-Verlag Berlin Heidelberg 2010

teletext data are processed to give magazines and lines, i.e. the same structure as in British teletext, and combined to form a packetized elementary stream. A teletext page according to the British or EBU standard is composed of 24 lines of 40 characters each. The data of each line are transmitted in a teletext line in the vertical blanking interval.

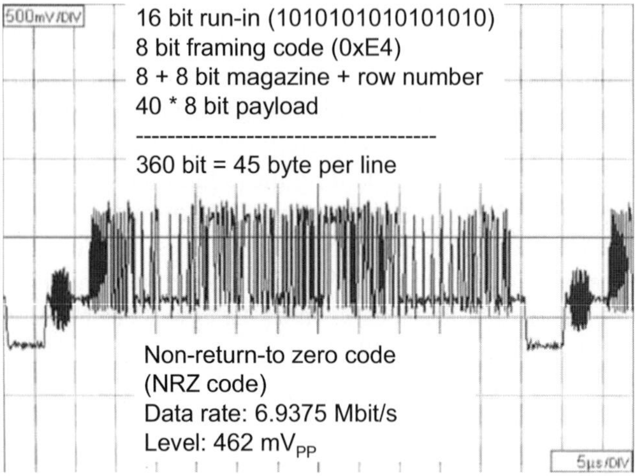

Fig. 9.2. Analog TV teletext in the vertical blanking interval

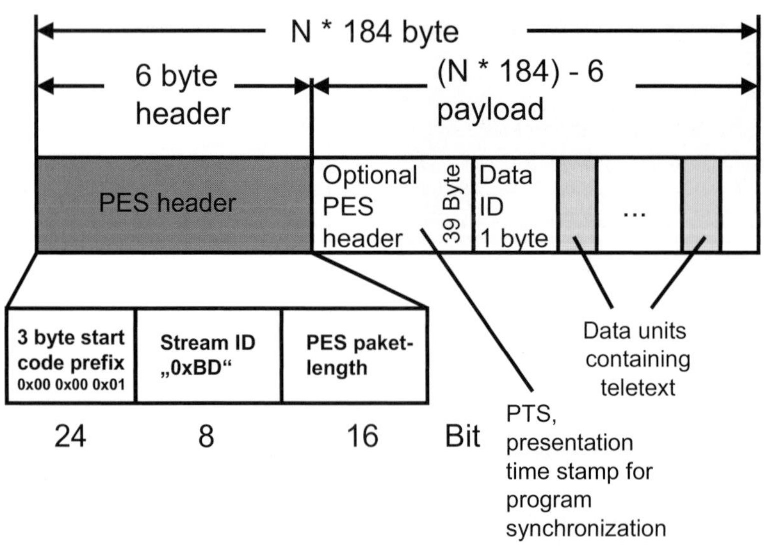

Fig. 9.3. PES Packet with teletext

The analog TV teletext line shown in Fig. 9.2. begins with the 16-bit-long run-in (1010 sequence) followed by the 1-byte-long framing code with a value of 0xE4. This marks the beginning of the active teletext. It is followed by the magazine and line number of 1 byte each. After that, 40 payload characters consisting of 7 bits payload and 1 (even) parity bit are transmitted. The total amount of data per line is 360 bits (= 45 bytes) and the data rate is 6.9375 Mbit/s. In DVB teletext (ETS 300472), the teletext data are inserted into the PES packets after the framing code (Fig. 9.3.). The 6-byte PES header starts with a 3-byte start code (0x00 0x00 0x01) which is followed by the stream ID 0xBD which corresponds to a "Private_Stream_1". Next comes a 16-bit (= 2-byte) length indicator which in the case of teletext is always set so that the total PES length corresponds to an integral multiple of 184 bytes.

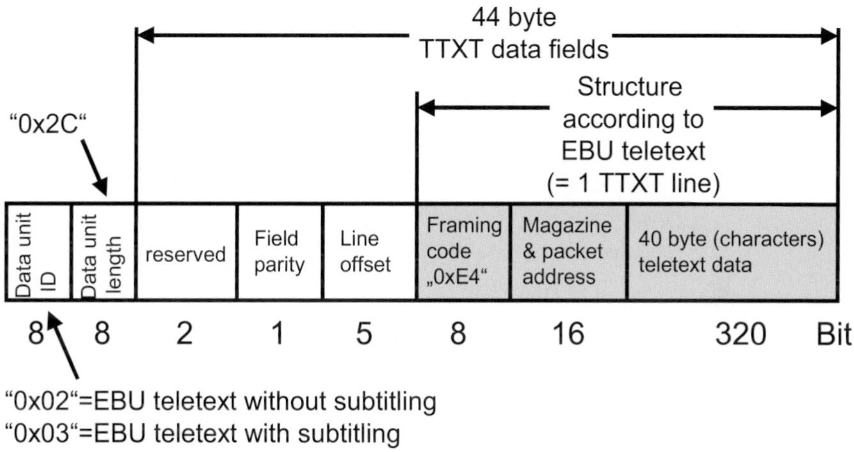

Fig. 9.4. Teletext data block in a PES packet

Then comes a 39 byte optional PES header so that the overall PES header length for teletext is 45 bytes. The actual teletext information is divided into blocks of 44 bytes. The last 43 bytes are identical to the structure of a teletext line of an EBU TTXT after the run-in-code. These bytes include the magazine and line information as well as the actual 40 bytes of teletext characters per line. A teletext page consists of 24 lines of 40 characters and the coding is identical to that of EBU or British teletext.

The teletext, processed to form long PES packets, is divided into short transport stream packets comprising the 184 byte payload and a 4 byte transport stream header, and multiplexed into the transport stream for transmission the same as video and audio data.

The packet identifiers (PIDs) of transport stream packets containing teletext are included as PIDs for private streams in the program map table (PMT) of the respective program (Fig. 9.5.).

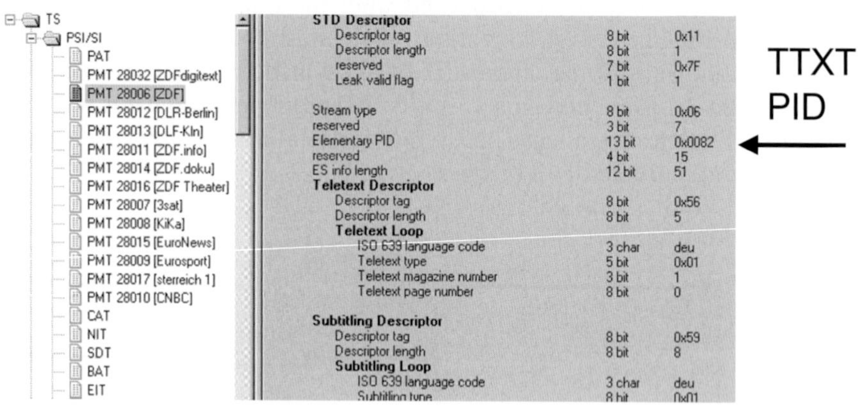

Fig. 9.5. Entry of a teletext service in a Program Map Table (PMT)

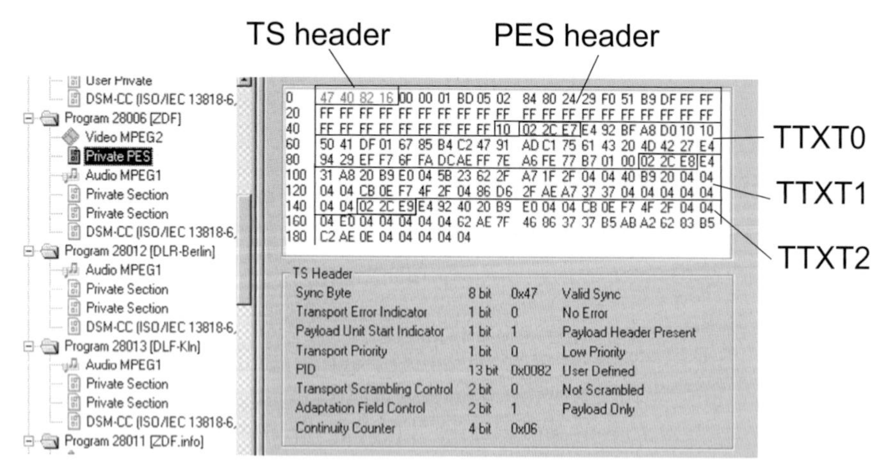

Fig. 9.6. Transport stream packet with teletext content

These PIDs can then be used for accessing transport stream packets containing teletext. A transport stream packet containing a PES header can be recognized by its payload start indicator bit being set. The payload unit of this packet contains the 45 byte PES header and the first teletext packets. The further teletext packets follow in the next transport stream packets having the same PID. The length of a teletext PES packet is adjusted so that an integral number of many transport stream packets will yield one complete PES packet. After a teletext PES packet has been completely transmitted, it will be retransmitted or a new packet sent if there are any changes to the teletext. Immediately preceding the teletext data in the PES packet, the field parity and the line offset indicate the field and line in which the teletext data are to be inserted back into the composite video signal by the DVB receiver.

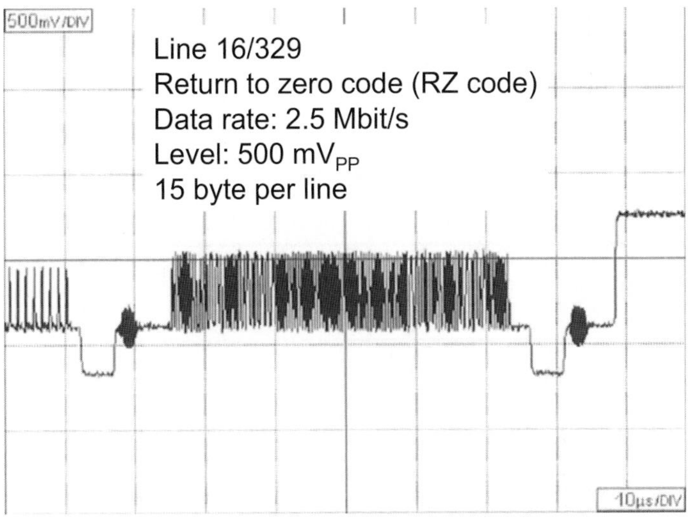

Fig. 9.7. Analog TV data line in the vertical blanking interval

9.2 Video Program System

VPS, the video program system for controlling video recorders, has long been known and used in public-service TV broadcasting, especially in Europe. It can be used for controlling recording in video recorders via the data line, mostly line 16 in the first field. In the data line (Fig. 9.7.), 15 bytes are transmitted in RZ (return-to-zero) coding, including the VPS information. According to ETSI ETS 301775, bytes 3 to 15 of the data line are simply inserted into the payload part of a PES packet, similarly to DVB teletext (Fig. 9.8. and 9.9.). The data unit ID is set to 0xC3 in this case, corresponding to the VBI (vertical blanking interval) according to DVB. As in DVB teletext, this is followed by the Data Unit Length, the field ID and the line number in the field.

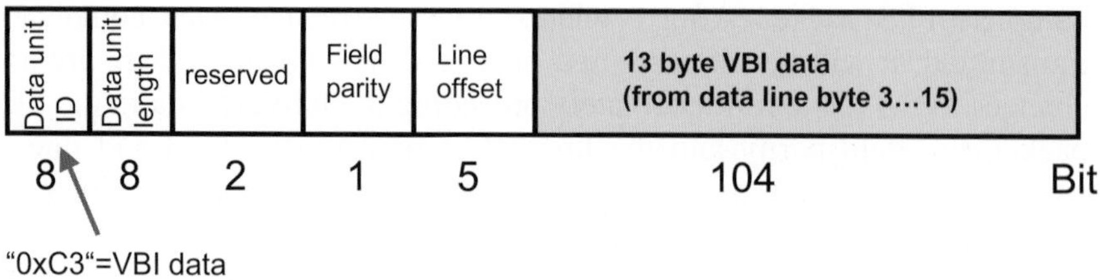

Fig. 9.8. PES packet with VBI data

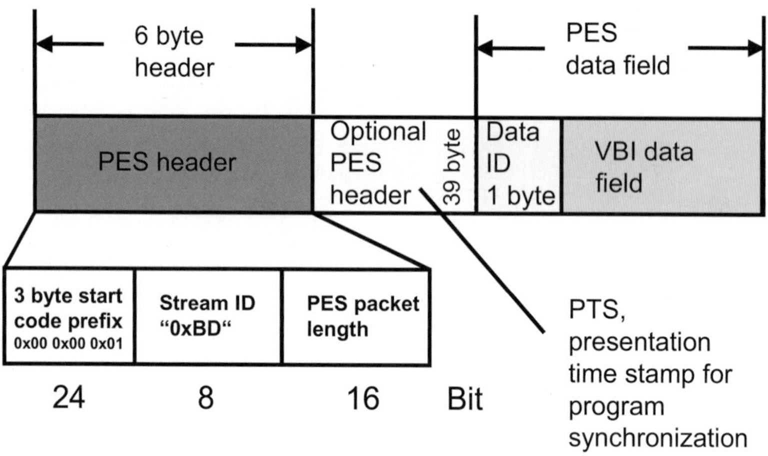

Fig. 9.9. VBI data field

The data line (Fig. 9.7.) contains the following information:

- Byte 1: Run-in 10101010
- Byte 2: Start code 01011101
- Byte 3: Source ID
- Byte 4: Serial ASCII text transmission (source)
- Byte 5: Monaural/stereo/binaural
- Byte 6: Video content ID

- Byte 7: Serial ASCII text transmission
- Byte 8: Remote control (routing)
- Byte 9: Remote control (routing)
- Byte 10: Remote control
- Byte 11 to 14: Video Program System
- Byte 15: Spare

4 bytes of VPS data (bytes 11 to 14):

- Day (5 bits)
- Month (4 bits)
- Hour (5 bits)
- Minute (6 bits)
- Country (4 bits)
- Program source ID (6 bits)

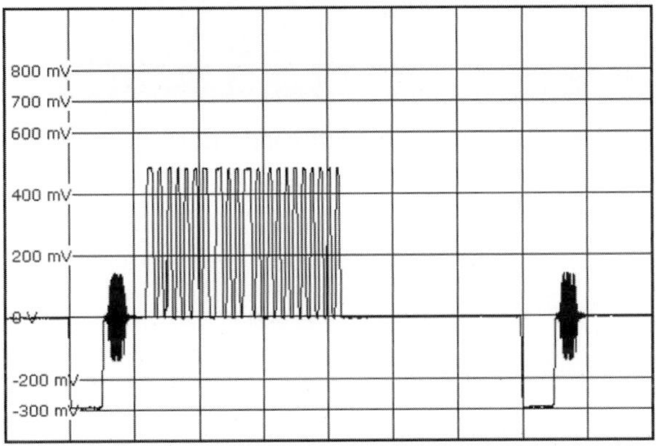

Fig. 9.10. WSS signal in line 23 of an analog TV signal

9.3 WSS – Wide Screen Signalling

Since PALplus so-called wide screen signalling (WSS) can be provided in video line 23. WSS informs the monitor or display about the current display format 4:3 or 16:9. For this purpose a digital control signal can be inserted into the first part of line 23 (Fig. 9.10.). A WSS signal can also be tunnelled via DVB in private PES packets. The principle is the same as for teletext and VPS. For technical reasons, this signal is visible on some TV screens at the top the left-hand side of the picture (Fig. 9.11.).

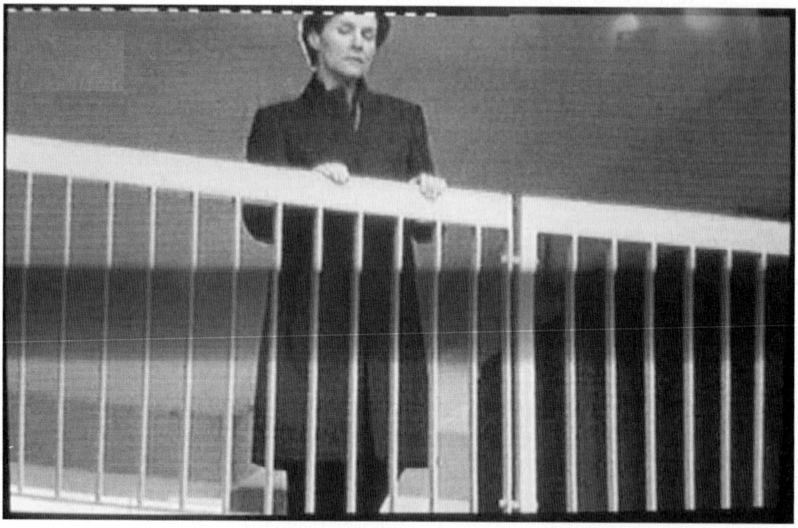

Fig. 9.11. Visible WSS signal on a TV screen (top, left-hand side)

5th Stream			
Stream type	8 bit	0x06 (6)	Private PES
reserved	3 bit	0x7	
Elementary PID	13 bit	0x0C8F (3215)	
reserved	4 bit	0xF	
ES info length	12 bit	10	
Teletext Descriptor			
Descriptor tag	8 bit	0x56 (86)	
Descriptor length	8 bit	5	
Teletext Loop			
ISO 639 language code	3 char	ger	
Teletext type	5 bit	0x01	initial teletext page
Teletext magazine number	3 bit	1	
Teletext page number	8 bit	0	
Stream Identifier Descriptor			
Descriptor tag	8 bit	0x52 (82)	
Descriptor length	8 bit	1	
Component tag	8 bit	0x04 (4)	

Fig. 9.12. Declaration of teletext in a PMT

9.4 Practical examples

In this chapter, the screen shots from a MPEG analyzer show some practical examples of teletext, VBI and WSS tunnelling in a DVB compliant MPEG-2 transport stream (Fig. 9.12., 9.13., 9.14., 9.15.).

```
4th Stream
Stream type                              8 bit  0x06 (6)     Private PES
reserved                                 3 bit  0x7
Elementary PID                          13 bit  0x0C8E (3214)
reserved                                 4 bit  0xF
ES info length                          12 bit  19
VBI Data Descriptor
 Descriptor tag                          8 bit  0x45 (69)
 Descriptor length                       8 bit  6
 Data service id                         8 bit  0x04 (4)     VPS
 Data service descriptor length          8 bit  1
 reserved                                2 bit  0x3
 Field parity                            1 bit  1
 Line offset                             5 bit  16
 Data service id                         8 bit  0x05 (5)     WSS
 Data service descriptor length          8 bit  1
 reserved                                2 bit  0x3
 Field parity                            1 bit  1
 Line offset                             5 bit  23
User Defined Descriptor
 Descriptor tag                          8 bit  0xC3 (195)
 Descriptor length                       8 bit  6
 Descriptor Data (hex)                  6 byte  04 01 F0 05 01 F7
Stream Identifier Descriptor
 Descriptor tag                          8 bit  0x52 (82)
 Descriptor length                       8 bit  1
 Component tag                           8 bit  0x06 (6)
```

Fig. 9.13. Declaration of VBI and WSS in a PMT [DVM]

```
Packetized Elementary Stream (PES)
  Packet Start Code Prefix                 24 bit  0x000001
  Stream Id                                 8 bit  0xBD          privat
  PES packet length                        16 bit  1282
  PES header data
     ...
  PES data
    Data identifier                         8 bit  0x10          EBU da
    data_unit_id                            8 bit  0x02          EBU Te
    data_unit_length                        8 bit  44
    txt_data_field (includes one line of text)
      reserved_future_use                   2 bit  3
      field_parity                          1 bit  1             first
      line_offset                           5 bit  7             videol
      framing_code                          8 bit  0xE4          bit re
      magazine_and_packet_address          16 bit  0xD91C        magazi
      data_block                                   0xE0234F75E337F746
      data block in bit reverse                    0x07 D r . G l o
    data_unit_id                            8 bit  0x02          EBU Te
    data_unit_length                        8 bit  44
    txt_data_field (includes one line of text)
```

Fig. 9.14. Teletext transmission in a private PES packet [DVM]

```
txt_data_field (includes one line of text)
  reserved_future_use                  2 bit  3
  field_parity                         1 bit  1            first videofie
  line_offset                          5 bit  8            videoline 8
  framing_code                         8 bit  0xE4         bit reverse 0x
  magazine_and_packet_address         16 bit  0x7AF4       magazine: 3
  data_block                                  0xE0234F75B3864F2F977604CB
  data block in bit reverse                   0x07 Dr. Martin
data_unit_id                           8 bit  0x02         EBU Teletext n
data_unit_length                       8 bit  44
txt_data_field (includes one line of text)
  reserved_future_use                  2 bit  3
  field_parity                         1 bit  1            first videofie
  line_offset                          5 bit  9            videoline 9
  framing_code                         8 bit  0xE4         bit reverse 0x
  magazine_and_packet_address         16 bit  0xD9F4       magazine: 3
  data_block                                  0xE00B4FF7677504CB97B6F776
  data block in bit reverse                   0x07 Prof.  Simc
data_unit_id                           8 bit  0x02         EBU Teletext n
data_unit_length                       8 bit  44
vps_data_field (includes one line of vertical blanking)
  reserved_future_use                  2 bit  3
  field_parity                         1 bit  1            first videofie:
  line_offset                          5 bit  16           videoline 16
  not_relevant                         8 bit  0xB1
  not_relevant                         8 bit  0x62
  PCS                                  8 bit  0xA0
  not_relevant                         8 bit  0x00
  not_relevant                         8 bit  0x00
  not_relevant                         8 bit  0x00
  not_relevant                         8 bit  0x00
  not_relevant                         8 bit  0x00
  Net or provider bin                  2 bit  0x3
  Day                                  5 bit  24
  Month                                4 bit  8
  Hour                                 5 bit  10
  Minute                               6 bit  30
  Country                              4 bit  0xD
  Net or provider bin                  6 bit  0x01
  Programmetype bin                    8 bit  0xFF
stuffing_bytes all                     8 bit  0xFF
wss_data_field
  reserved_future_use                  2 bit  3
  field_parity                         1 bit  1
  line_offset                          5 bit  23
  Aspect ratio                         4 bit  0xE
  Film bit                             1 bit  0x0
  Colour coding bit                    1 bit  0x1
  Helper bit                           1 bit  0x0
  reserved bit                         1 bit  0x0
  Subtitles within Teletext bit        1 bit  0x0
  Subtitling mode                      2 bit  0x0
  Suround sound bit                    1 bit  0x0
  Copyright information                2 bit  0x0
  reserved_future_use                  2 bit  3
stuffing_bytes all                     8 bit  0xFF
data_unit_id                           8 bit  0x02
data_unit_length                       8 bit  44
```

Fig. 9.15. Teletext, VBI and WSS data blocks in a PES packet [DVM]

Bibliography: [ETS 300472], [ETS 301775], [DVM]

10 A Comparison of Digital Video Standards

10.1 MPEG-1 and MPEG-2, VCD and DVD, M-JPEG and MiniDV/DV

In 1992, MPEG-1 was created as the first standard for encoding moving pictures accompanied by sound. The aim was to achieve a picture quality close to that of VHS at CD data rates (< 1.5 Mbit/s). MPEG-1 was provided only for applications on storage media (CD, hard disk) and not for transmission (broadcasting) and its data structures correspond to this objective. The audio and video coding of MPEG-1 is quite close to that of MPEG-2 and all the fundamental algorithms and methods are already in place. There are both I, P and B frames, i.e. forward and backward prediction, and naturally there are the DCT-based irrelevance reduction methods already found in JPEG. The picture resolution, however, is limited to about half the VGA resolution (352 x 288). Neither is there any necessity for field encoding (interlaced scanning method). In MPEG-1, there is only the so-called Program Stream (PS) which is composed of multiplexed packetized elementary stream (PES) packets of audio and video. The variable-length (64 kbytes max) audio and video PES packets are simply assembled interleaved in accordance with the present data rate to form a data stream. This data stream is not processed any further since it is only intended to be stored on storage media and not used for transmission. A certain number of audio and video PES packets are combined to form a so-called pack which consists of a header and the payload just like the PES packets themselves. A pack is often based on the size of a physical data sector of the storage medium.

In MPEG-2, the coding methods were developed further in the direction of higher resolution and better quality. In addition, transmission was also considered, in addition to the storage of such data. The MPEG-2 transport stream is the transportation layer, providing much smaller packet structures and more extensive multiplexing mechanisms. In MPEG-1, there is

W. Fischer, *Digital Video and Audio Broadcasting Technology*, Signals and Communication Technology, 3rd ed., DOI: 10.1007/978-3-642-11612-4_10, © Springer-Verlag Berlin Heidelberg 2010

only one program (only one movie), whereas MPEG-2 can accommodate a multiplexed data stream with up to 20 programs and more.

In addition to Standard Definition TV (SDTV), MPEG-2 also supports High Definition TV (HDTV). MPEG-2 is used throughout the world as digital baseband signal in broadcasting.

A Video CD (VCD) contains an MPEG-1 coded data signal as a program stream, i.e. there is one program consisting of multiplexed PES packets. The total data rate is about 1.5 Mbit/s. Many pirate copies of movies are available as Video CD and can be downloaded from the Internet or bought on the Asian market.

A Super Video CD (SVCD) carries an MPEG-2 data signal coded with 2.4 Mbit/s, also as a program stream with multiplexed PES packets. A Super Video CD approximately corresponds to VHS type quality, sometimes even better.

On a DVD (Digital Versatile Disk - NOT 'Digital Video Disk'), the data material is MPEG-2 coded with data rates of up to 10.5 Mbit/s and exhibits a much better picture quality than that recorded on VHS tape. A DVD also carries a multiplexed PES data stream. Subtitles and much else besides are also possible.

The DVD is intended for a variety of applications including video, audio and data. In contrast to the CD (approx. 700 Mbytes), the data volume on a DVD is up to 17 Gbytes and it is possible to have 1, 2 or 4 layers with 4.7 Gbytes each per layer (see table below).

Table 10.1. DVD types

Type	Sides	Layers / side	Data [Gbytes]	x CD-ROM
DVD 5	1	1	4.7	7
DVD 9	1	2	8.5	13
DVD 10	2	1	9.4	14
DVD 18	2	2	17.1	25

Technical data of the Video DVD:

- Storage capacity: 4.7 to 17.1 Gbytes
- MPEG-2 Video with variable data rate, 9.8 Mbit/s video max.

Audio:

- Linear PCM (LPCM) with 48 kHz or 96 kHz sampling frequency at 16, 20 or 24 bits resolution

- MPEG Audio (MUSICAM) mono, stereo, 6-channel sound (5.1), 8-channel sound (7.1)
- Dolby Digital (AC3) mono, stereo, 6-channel sound (5.1)

Table 10.2. Digital video standards

Standard	Video coding	Resolution	Video data rate [Mbit/s]	Total data rate [Mbit/s]
MPEG-1	MPEG-1	352 x 288 192 x 144 384 x 288	0.150 - (1.150) - 3.0	max. approx. 3.5 (1.4112)
MPEG-2	MPEG-2	720 x 576 (SDTV, 25 frames per second) different resolutions up to HDTV	up to 15	basically open, from the interfaces up to 270
MPEG-4	MPEG-4 Part 2 and Part 10 (H.264)			
Video CD	MPEG-1	352 x 288	1.150	1.4112
Super VCD	MPEG-2	480 x 576	2.4	2.624
Video DVD	MPEG-2	720 x 576	up to 9.8, variable	10.5
MiniDV	MJPEG variant	720 x 576	25	approx. 30
DVPro	MJPEG variant	720 x 576	25/50	approx. 30/55

Apart from MPEG, there are also proprietary methods based on JPEG all of which have that in common that the video material is only DCT coded and not interframe coded. Both DV and MiniDV are such methods. MiniDV has become widely used in the home video camera field and has revolutionized this field with respect to the picture quality. The data rate is 3.6 Mbyte/s total or 25 Mbit/s video data rate. The picture size is 720 x 576 pixels, the same as in MPEG-2, with 25 frames per second. MiniDV can be edited at any point since it virtually only consists of frames comparable to I frames. DVCPro is the big brother to MiniDV. DVCPro is a studio standard and supports video data rates of 25 and 50 Mbit/s. The 25

Mbit/s data rates corresponds to the MiniDV format. DVCPro and MiniDV are special variants of Motion JPEG. In contrast to MPEG, no quantizer tables are transmitted, neither are quantizer scale factors varied from macroblock to macroblock. Instead, a set of quantizing tables is provided locally, from which the coder selects the most suitable one from macroblock to macroblock. MiniDV and DVPro exhibit a very good picture quality at relatively high data rates and lend themselves easily to postprocessing. Home editing software for the PC is now available at a cost of around 100 Euros and provides functions available only to professionals a few years ago. Apart from the actual editing, which is now free of losses and is easy to handle, the software also allows video material to be coded in MPEG-1, MPEG-2, VCD, SVCD and video DVD.

Table 10.2. shows the most important technical data of the methods discussed.

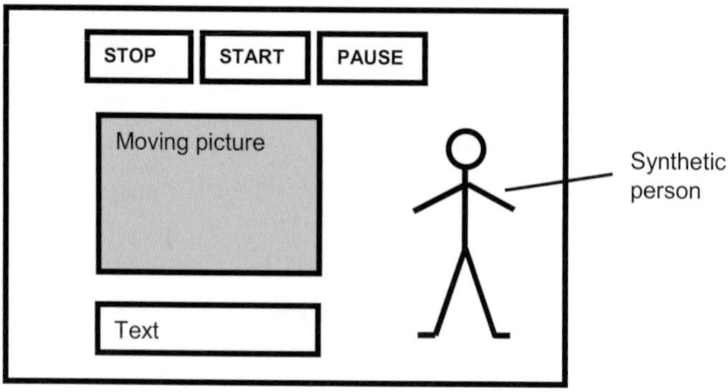

Fig. 10.1. MPEG-4 example

10.2 MPEG-3, MPEG-4, MPEG-7 and MPEG-21

In the previous chapters, MPEG-2 and MPEG-1 have been discussed in detail. However, the Moving Picture Experts Group also considered, and is still working on, other standards such as MPEG-4, MPEG-7 and MPEG-21. There was also MPEG-3 but this only had a temporary existence as an approach to HDTV and has now been completely absorbed into the MPEG-2 standard. MPEG-4 is a standard for multimedia applications with interactive components and has been in existence since the end of 1999. This involves not only video and audio but also applications which can be composed of a number of different objects. Its structure is object oriented similar to the programming language C++ and an MPEG-4 application can thus be composed of, e.g., the following audiovisual objects (Fig. 10.1.):

- A fixed colored background which may be patterned
- An MPEG-4 coded moving picture in a fixed frame
- A synthetic figure which moves three-dimensionally in synchronism with the video, e.g. a synthetic person which "mimes" the sound in gestures (deaf-and-dumb alphabet)
- Stop, start, pause, forward and rewind buttons (interactive elements)
- Accompanying text
- MPEG-coded audio signal

The development of MPEG-4 was continued with respect to the video and audio coding, adopting and refining the methods known from MPEG-1 and MPEG-2 instead of looking for a wholly new approach. The only new thing is that MPEG-4 can also cope with synthetic visual and audiovisual elements such as synthetic sound. MPEG-4 objects can be present as PES stream both within an MPEG-2 transport stream and as an MPEG-4 file. MPEG-4 can also be transmitted as a program stream in IP packets.

MPEG-4 applications can be typically employed in

- The Internet
- Interactive multimedia applications on the PC
- New video compression applications with greater compression factors as required, e.g. for HDTV

MPEG-4 was made into a standard in 1999. At the beginning of the new millenium, a further new video compression standard H.264 was developed and standardized. Compared with MPEG-2, this method is more effective by a factor of 2 to 3 and thus allows data rates which are lower by a factor of 2 to 3, often even with improved picture quality. The relevant standard is ITU-T H.264. H.264 has also been incorporated in the group of MPEG-4 standards as MPEG-4 Part 10.

The most important standard documents covered by the heading MPEG-4 are:

- MPEG-4 Part 1 – Systems, ISO/IEC 14496-1
- MPEG-4 Part 2 – Video Encoding, ISO/IEC 14496-2
- MPEG-4 Part 3 – Audio Encoding, ISO/IEC 14996-3
- MPEG-4 Part 10 – H.264 Advanced Video Coding. ISO/IEC 14496-10

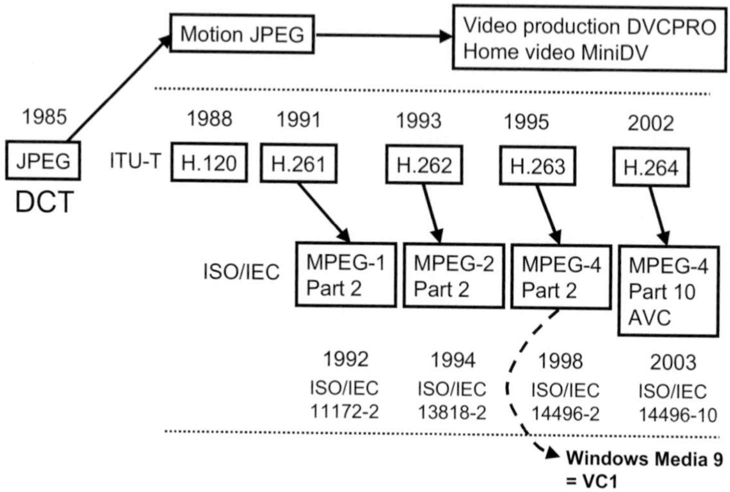

Fig. 10.2. History of the development of video encoding

MPEG-4 Part 10 – Advanced Video Coding (AVC) is scheduled for HDTV applications in Europe as part of the DVB project. Whereas HDTV requires data rates of about 15 Mbit/s for the video signal with MPEG-2, these are about 9 Mbit/s or even lower when they are encoded as MPEG-4 AVC signals. In H.264/MPEG-4 Part 10 AVC, the block size is not a constant 8 x 8 pixels but variable within certain limits. Up to 9 motion vectors are possible and measures for masking blocking have been implemented.

This would be a suitable point at which to pause and to explore some of the history of the development of video encoding (Fig. 10.2.). A key event is considered to be the establishment of the JPEG (Joint Photographic Experts Group) Standard in 1985. This was the first time the DCT (Discrete Cosine Transform) was used for compressing still pictures. Today, JPEG is a common standard used mainly in digital photography. From JPEG, Motion JPEG applications such as DVCPro for studio applications and MiniDV for home video applications evolved in one line of development. The advantage of Motion JPEG lies mainly in the fact that the video material can be edited unrestrictedly at every frame and the picture quality is extremely good. A further line of development formed through video telephony and video conferencing via the ITU-T standards H.120, H.261 etc. ITU-T H.261 merged into the MPEG-1 video standard ISO/IEC 11172.2 and H.262 became the MPEG-2 video standard ISO/IEC 13818-2. H-263 formed the basis for MPEG-4 Part 2 video encoding ISO/IEC 14496-2.

And finally, H.264 was developed, also known as MPEG-4 Part 10 AVC (Advanced Video Coding), or as ISO/IEC 14496-10. In parallel to this, there is also Microsoft Windows Media 9 (now also called VC-1) which probably arose as a result of Microsoft's collaboration in MPEG-4 Part 2 and Part 10. Fig. 10.2. shows a rough overview of the history of the development of moving picture encoding.

MPEG-7, in contrast and as a supplement to MPEG-2 and -4, deals exclusively with program-associated data, the so-called meta-data, as a complement to MPEG-2 and MPEG-4. The aim is to transmit background information for a program on air, a type of electronic program guide, with the aid of XML- and HTML-based data structures together with the program, e.g. in an MPEG-2 transport stream. MPEG-7 has been a standard since 2001 but has yet to make its debut in practice, at least for the end user.

MPEG-21 was to be transformed into a full standard by 2003 and was intended to contain tools and methods to supplement all other MPEG standards (including end-user-to-end-user applications, e.g. via the Internet). It is not clear what has become of it.

Broadcasting, multimedia and the Internet are converging more and more. In broadcasting, however, much higher point-to-multipoint data rates are possible in the downstream than will ever be possible over the Internet.

Table 10.3. MPEG standards

Standard	Description	Status
MPEG-1	Moving pictures and sound, approx. in VHS quality with CD data rate (< 1.5 Mbit/s)	Standard since 1992
MPEG-2	Digital television (SDTV+HDTV)	Standard since 1993
MPEG-3	Existed only temporarily (no relation to MP3)	not applicable
MPEG-4	Multimedia, interactive	Standard since 1999
MPEG-7	Program-associated supplementary data (Meta-data)	Standard since 2001
MPEG-21	Supplementary tools and methods	Current status ???

10.3 Physical Interfaces for Digital Video Signals

Analog SDTV (Standard Definition Television) signals have a bandwidth of appr. 4.2 to 6 MHz and are transmitted over 75 Ohm coaxial lines. These cables, which in most cases are green-jacketed, are fitted with BNC connectors both in professional applications and in high-end consumer applications. If they are terminated in exactly 75 Ohms, analog video signals have an amplitude of 1 V_{PP}. The first interfaces for digital TV signals were designed as parallel interfaces, using the 25-pin Cannon connector known from the PC printer interface. Because of its noise immunity, transmission was conducted as low-voltage differential signaling via twisted pair lines. Today, however, 75 Ohm technology is again being used in most cases.

Digital video signals are transmitted as a serial data signal with a data rate of 270 Mbit/s via 75 Ohm coax cable fitted with the familiar, rugged BNC connectors, making no distinction between uncompressed video signals according to the ITU 601 standard and MPEG-2 transport streams. The distribution paths in the studio are the same, the cables are the same, amplifiers and cable equalizers are also the same. Engineers often talk of SDI or of TS-ASI. The physical interface is the same in both cases, only the content differs. SDI stands for Serial Digital Interface, meaning the serial digital uncompressed video signal of the 601 standard with a data rate of 270 Mbit/s. TS-ASI stands for Transport Stream Asynchronous Serial Interface, meaning the MPEG-2 transport stream on a serial interface, the transport stream having a data rate which is distinctly lower than the data rate on this serial transmission link. The data rate of the transport stream is asynchronous to the constant data rate of 270 Mbit/s on the TS-ASI interface. If, e.g., the transport stream has a data rate of 38 Mbit/s, stuffing information is used to fill up the data rate of 270 Mbit/s. The reason for working with a constant 270 Mbit/s is clear: In the studio, it is desirable to have uniform distribution paths for 601 signals and the MPEG-2 transport streams.

10.3.1 Parallel and Serial CCIR 601

Uncompressed SDTV video signals have a data rate of 270 Mbit/s. They are distributed either as parallel signals via twisted pair lines or serially via 75 Ohm coax cables. The parallel interface is the familiar 25 pin Cannon socket also known as a PC printer interface. The signals are LVDS (low voltage differential signaling) signals which means that ECL levels and not TTL levels are used as voltage levels (800 mV_{PP}). In addition, for each data bit, the inverted data bit is also transmitted in order to keep the noise level as low as possible over twisted lines. Table 10.4 shows the pin allocation at the 25 pin parallel interface. The compatible allocation of the parallel MPEG-2 transport stream interface is also entered. In most cases, however, only the serial ITU 601 interface is used today. It is also called the Serial Digital Interface (SDI) and uses a 75 Ohm BNC socket with a voltage level of 800 mV_{PP}. In contrast to the parallel interface, the signals can be distributed over relatively long distances if cable equalizers are used.

Table 10.4. Parallel CCIR601 and TS interface

Pin	Signal	Pin	Signal
1	Clock	14	Inverted clock
2	System ground	15	System ground
3	601 data bit 9 (MSB) TS data bit 7 (MSB)	16	601 inverted data bit 9 (MSB) inverted TS data bit 7 (MSB)
4	601 data bit 8 TS data bit 6	17	inverted 601 data bit 8 inverted TS data bit 6
5	601 data bit 7 TS data bit 5	18	inverted 601 data bit 7 inverted TS data bit 5
6	601 data bit 6 TS data bit 4	19	inverted 601 data bit 6 inverted TS data bit 4
7	601 data bit 5 TS data bit 3	20	inverted 601 data bit 5 inverted TS data bit 3
8	601 data bit 4 TS data bit 2	21	inverted 601 data bit 4 inverted TS data bit 2
9	601 data bit 3 TS data bit 1	22	inverted 601 data bit 3 inverted TS data bit 1
10	601 data bit 2 TS data bit 0	23	inverted 601 data bit 2 inverted TS data bit 0
11	601 data bit 1 TS data valid	24	inverted 601 daten bit 1 inverted TS data valid
12	601 data bit 0 TS packet sync	25	inverted 601 data bit 0 inverted TS packet sync
13	Case ground		

10.3.2 Synchronous Parallel Transport Stream Interface (TS Parallel)

The parallel MPEG-2 transport stream interface is designed to be fully compatible with the parallel ITU 601 interface. The signals are also LVDS signals, i.e. signals with ECL level which are transmitted with balanced levels on twisted pairs. The connector is also a 25 pin Cannon connector with pin allocations which are compatible with the ITU 601 interface. The pin allocations of the data signal, which is only 8 bits wide in contrast to the ITU 601 signal, can be found in the table in the previous section.

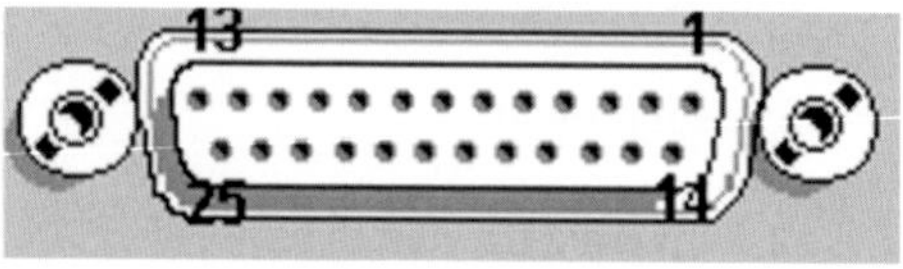

Fig. 10.3. Parallel TS connector

The data stream transmitted via the transport stream interface (Fig. 10.3., 10.4. and 10.5.) is always synchronous with the MPEG-2 transport stream to be transmitted, i.e. if the transport stream has a data rate of, e.g., 38 Mbit/s, the data rate will be 38 Mbit/s here, too. The transport stream remains unchanged.

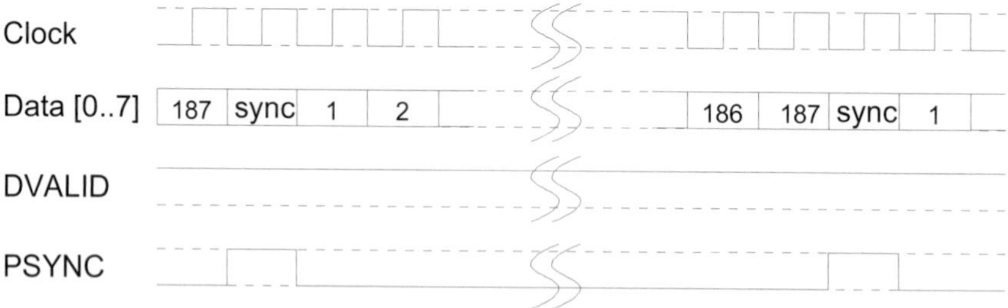

Fig. 10.4. Transmission Format with 188-Byte Packets [DVG]

However, the transport stream interface can be operated with 188 byte-long packets or 204 or 208 byte-long MPEG-2 transport stream packets. The 204 or 208 byte-long packets are due to the Reed Solomon error protection of DVB or ATSC signals on the transmission link. At the transport stream interface, however, any data going beyond 188 bytes are only dummy bytes and their content can be ignored. Many devices can be configured to these various packet lengths or have the capability of handling all formats.

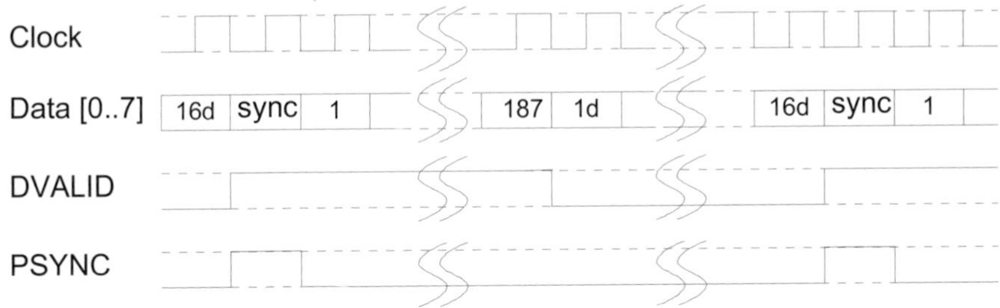

Fig. 10.5. Transmission format with 188 byte packets and 16 dummy bytes (=204 Bytes) [DVG]

Fig. 10.6. TS-ASI

10.3.3 Asynchronous Serial Transport Stream Interface (TS ASI)

The asynchronous serial transport stream interface (Fig. 10.6.) is an interface with a constant data rate of 270 Mbit/s. Data bytes (8 bits each) will be transmitted via this interface at a maximum rate of 270 Mbit/s, i.e. the data rate on this interface is not synchronous with the actual MPEG-2 transport stream but always a constant 270 Mbit/s. The advantage of this is, however, that the same distribution system system can be used as with the SDI. Each byte is supplemented by 2 additional bits in accordance with a standardized table. On the one hand, this identifies the data bytes (dummy bytes) which are irrelevant but necessary for filling up the data rate of 270 Mbit/s and, on the other hand, it prevents the occurrence of a DC component in the serial signal.

The connector uses a BNC socket with an impedance of 75 Ohms. The level is 800 mV_{PP} (+/-10%).

The TS ASI interface can be operated in two modes: Burst mode, in which the TS packets remain unchanged in themselves and dummy packets are inserted to attain the data rate of 270 Mbit/s, and single byte mode, in which dummy bytes are inserted to provide "stuffing" to the output data rate of 270 Mbit/s.

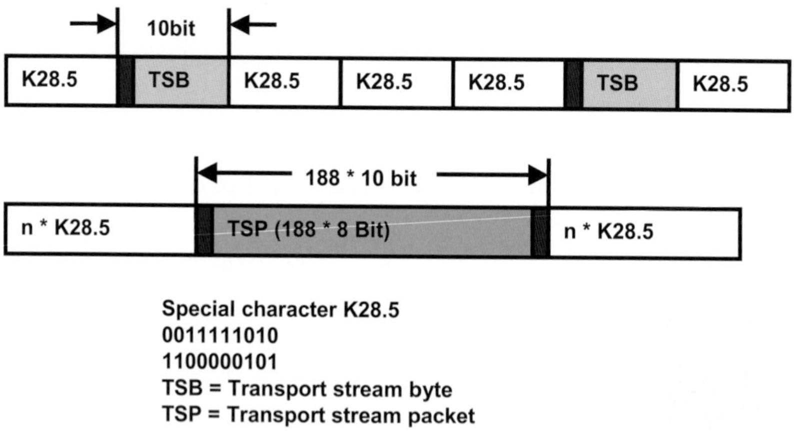

Fig. 10.7. TS-ASI in Single-Byte Mode (top) and in Burst Mode (bottom)

10.3.4 SMPTE 310 Interface

The SMPTE 310 interface defined for ATSC is a special version of the TS ASI interface. This interface is fixed at the data rate of precisely 19.39 Mbit/s of the ATSC standard. The connector used is a BNC connector.

10.3.5 DVI Interface

DVI stands for Digital Visual Interface (Fig. 10.8.) and is an interface which will replace the VGA interface as display interface in the PC domain. It exists as integrated model (DVI-I) and in digital form (DVI-D). The DVI-I interface additionally contains the analog VGA components, i.e. using a passive adapter plug it can also be converted into a VGA interface. The DVI-D interface only has the digital monitor signals. The data rate of the DVI interface is 1.65 Gbit/s. DVI is a possible interface of an HD Ready monitor or beamer (data projector).

Table 10.5. The DVI interface

Pin	Signal
01	TMDS data 2-
02	TMDS data 2+
03	Ground for TDMS data 2, 4
04	TMDS data 4-
05	TMDS data 4+
06	DDC clock
07	DDC data
08	Analog V sync
09	TMDS data 1-
10	TMDS data 1+
11	Ground for TMDS data 1, 3
12	TMDS data 3-
13	TMDS data 3+
14	+5V
15	Ground for +5V
16	Hotplug detect
17	TMDS data 0-
18	TMDS data 0+
19	Ground for TMDS data 0, 5
20	TMDS data 5-
21	TMDS data 5+
22	Ground for TMDS clock
23	TMDS clock+
24	TMDS clock-
C1	Analog red
C2	Analog green
C3	Analog blue
C4	Analog ground

TDMS - Transmission Minimized Digital Signalling
DDC - Display Data Channel

10.3.6 HDMI Interface

HDMI (High Definition Multimedia Interface) is also an interface (Fig. 10.9.) which can be found on an HD Ready or Full HD monitor. Apart from the video data, it also contains the audio signals and supports a data rate of up to 5 Gbit/s. If the data rates are suitable, DVI video can be converted to HDMI video and conversely. This can be done purely passively by using adapter plugs.

Fig. 10.8. DVI interface (DVI-D on the left, DVI-I on the right)

Fig. 10.9. HDMI interface

10.3.7 HD-SDI Interface

HD-SDI (High Definition Serial Digital Interface) is the big brother of the SDI interface. It is used for distributing uncompressed HD video with 10 bits resolution at a data rate of 1.485 Gbit/s or supplying it to an HD MPEG encoder. This interface can be provided as a BNC interface with an impedance of 75 ohms and a level of 800 mV or as an optical interface.

10.3.8 Gigabit Ethernet Interface as Transport Stream Distributor

There are more and more applications in which TS ASI is replaced by a Gigabit Ethernet interface as MPEG-2 transport stream interface. It is assumed that this interface will replace the TS ASI interface completely in the medium term. In this arrangement, several transport streams can also be distributed over one interface. The addressing is then carried out via a "socket", known from the PC world, which is composed of the port number and the IP address.

Bibliography: [GRUNWALD], [DVG], DVMD], [DVQ], [FISCHER4], [ITU601], [REIMERS], [TAYLOR], [MPEG4]

11 Measurements on the MPEG-2 Transport Stream

With the introduction of digital television, neither the hopes of the users nor the fears of the test equipment makers were confirmed: there is still a large need for test instruments for digital television, but of a different type. Where it was mainly video analyzers for evaluating the test lines of an analog baseband signal in analog television, it is mainly MPEG-2 test decoders which are being used in digital TV. Throughout the world, taking measurements directly at the transport stream has become the most important digital TV test technology with regard to turnover and demand. Thus, e.g., every MPEG-2 transport stream to be broadcast is analyzed and monitored by means of a MPEG-2 test decoder at almost every transmitter site of DVB-T networks in some countries.

The input interface of an MPEG-2 test decoder is either a parallel 25-pin MPEG-2 interface or a serial TS-ASI BNC connector, or both at the same time.

The MPEG-2 analyzer consists of the essential circuit blocks of MPEG-2 decoder, MPEG-2 analyzer - usually a signal processor - and a control computer which acquires all the results, displays them on the display and performs and manages all operating and control operations. A test decoder is capable of decoding all the video and audio signals contained in the transport stream and of performing numerous analyses and measurements on the data structure. The MPEG-2 transport stream analysis is a special type of logic analysis.

The Measurement Group in the DVB Project has defined numerous measurements on the MPEG-2 transport stream within its Measurement Guidelines ETR 290. These measurements will be described in more detail in the following chapters. According to ETR 290, the errors to be detected by means of these measurements were graded into three levels of priority: Priority 1, 2, 3.

MPEG-2 transport stream errors:

- Priority 1 - no decodability

W. Fischer, *Digital Video and Audio Broadcasting Technology*, Signals and Communication Technology, 3rd ed., DOI: 10.1007/978-3-642-11612-4_11, © Springer-Verlag Berlin Heidelberg 2010

- Priority 2 - partially no decodability
- Priority 3 - errors in the supplementary information/SI

If there is a Priority 1 error, there is often no chance to lock to the transport stream or even to decode a program. Priority 2, in contrast, means that there is partially no possibility of reproducing a program faultlessly. The presence of a category 3 error, on the other hand, only indicates errors in the broadcasting of the DVB service information. The effects are then dependent on how the set-top box used reacts.

Apart from the category 3 errors, all measurements can also be applied with the American ATSC standard where comparable analyses can be made on the PSIP tables.

The following measurements on the MPEG-2 transport stream are defined in the DVB Measurement Guidelines ETR 290:

Table 11.1. MPEG-2 measurement to ETR290 / TR101290 [ETR290]

Measurement	Priority
TS_sync_loss	1
Sync_byte_error	1
PAT_error	1
PMT_error	1
Continuity_count_error	1
PID_error	1
Transport_error	2
CRC_error	2
PCR_error	2
PCR_accuracy_error	2
PTS_error	2
CAT_error	2
SI_repetition_error	3
NIT_error	3
SDT_error	3
EIT_error	3
RST_error	3
TDT_error	3
Undefined_PID	3

11.1 Loss of Synchronisation (TS_sync_loss)

The MPEG-2 transport stream consists of 188-byte-long data packets composed of 4 bytes of header and 184 bytes of payload. The first byte of the header is the synchronisation or sync byte which always has the value 0x47 and occurs at constant intervals of 188 bytes. In special cases a spacing of 204 or 208 bytes is also possible, namely when the data frame with Reed Solomon error protection according to DVB or ATSC is similar. The additional 16 or 20 bytes are then dummy bytes and can be simply ignored. At any rate, there is no useful information present since it is not the Reed Solomon coder and decoder which represent the first or, respectively, last element of the transmission link but the energy dispersal unit, and thus any Reed Solomon error protection bytes present would not fit in with the actual transport stream packet. According to DVB, synchronism is achieved after 5 successive sync bytes have been received at correct intervals and with the correct content. When 3 successive sync bytes or transport stream packets have been lost, the MPEG-2 decoder or the corresponding transmission device drops lock again.

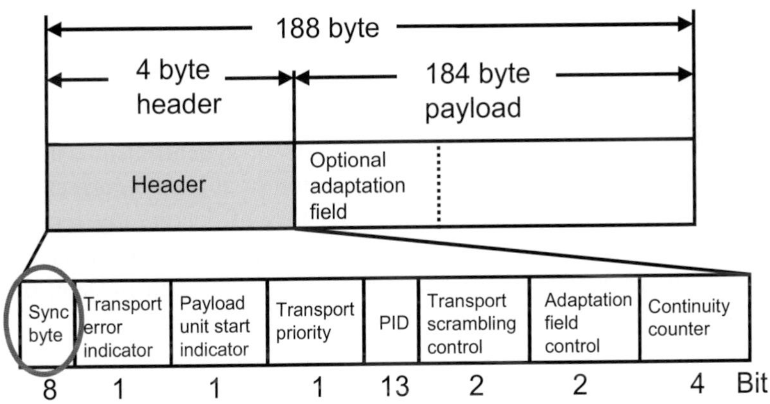

Fig. 11.1. TS_sync_loss

The state of loss of transport stream synchronization, which may occur either because of severe interference or simply because of a break in a line, is called "TS_sync_loss" (Fig. 11.1.).

"TS_sync_loss" occurs when

- the content of the sync bytes of at least 3 successful transport stream packets is not equal to 0x47.

The conditions of synchronization (acquisition of lock, loss of lock) can be adjusted in test decoders.

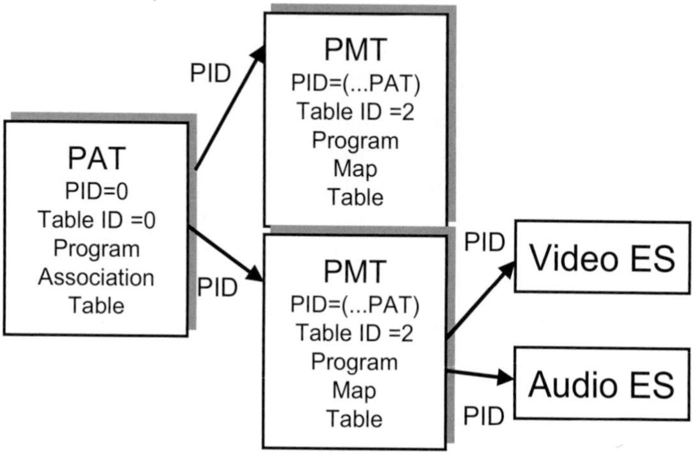

Fig. 11.2. PAT and PMT errors

11.2 Errored Sync Bytes (sync_byte_error)

As explained in the previous chapter, the state of synchronism with the transport stream is considered to be the reception of at least 5 correct sync bytes. Loss of synchronism occurs after the loss of 3 correctly received sync bytes. However, incorrect sync bytes may occur occasionally here and there in the transport stream due to problems in the transmission link. This state, caused in most cases by too many bit errors, is called "sync_byte_error" (Fig. 11.2.).

A "sync_byte_error" occurs when

- the content of a sync byte in the transport stream header is not equal to 0x47.

11.3 Missing or Errored Program Association Table (PAT) (PAT_error)

The program structure, i.e. the composition of the MPEG-2 transport stream is variable or, in other words, open. For this reason, lists for describing the current transport stream composition are transmitted in special TS packets in the transport stream. The most important one of these is the Program Association Table (PAT) which is always transmitted in transport stream packets with PID=0 and Table ID=0. If this table is missing or is errored, identification, and thus decoding, of the programs becomes impossible. In the PAD, the PIDs of all Program Map Tables (PMTs) of all programs are transmitted. The PAT contains pointer information to many PMTs. A set top box will find all necessary basic information in the PAT.

A PAT which is missing, is transmitted scrambled, is errorred or is transmitted not frequently enough will lead to an error message "PAT error". The PAT should be transmitted free of errors and unscrambled every 500 ms at a maximum.

A PAT error occurs when

- the PAT is missing,
- the repetition rate is greater than 500 ms,
- the PAT is scrambled,
- the table ID is not zero.

Details in the PAT are not checked at that time.

11.4 Missing or Errored Program Map Table (PMT) (PMT_error)

For each program, a Program Map Table (PMT) is transmitted at maximum intervals of 500 ms. The PIDs of the MAPs are listed in the PAT. The PMT contains the respective PIDs of all elementary streams belonging to this program. If a PMT referred to in the PAT is missing, there is no way for the set top box or decoder to find the elementary streams and to demultiplex and decode them. A PMT listed in the PAT and is missing, errored or scrambled will lead to the error message "PMT_error".

A "PMT error" occurs when

- a PMT listed in the PAT is missing,
- a section of the PMT is not repeated after 500 ms at the latest,
- a PMT is scrambled,
- the table ID is not 2.

Details in the PMT are not checked.

Like any other table, the PMTs can also be divided into sections. Each section begins with the table_ID=2 and with a PID, specified in the PAT, of between 0x0010 and 0x1FFE according to MPEG-2 and between 0x0020 and 0x1FFE according to DVB. PID 0x1FFF is intended for the zero packets.

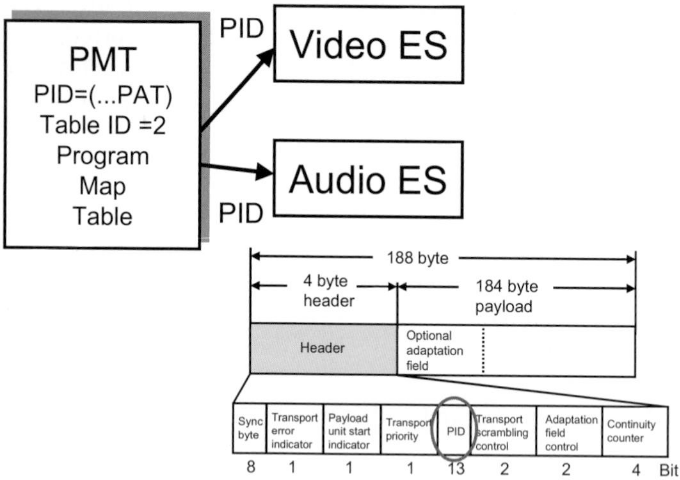

Fig. 11.3. PID_error

11.5 The PID_Error

The PIDs of all elementary streams of a program are contained in the associated program map table (PMT). The PIDs are pointers to the elementary streams: they are used for the addressed access to the corresponding packets of the elementary stream to be decoded. If a PID is listed in some PMT but this is not contained in any packet in the transport stream, there is no way for the MPEG-2 decoder to access the corresponding elementary stream since this is now not contained in the transport stream or has been multiplexed with the wrong PID information. This is what one might call a "classical PID_error". The time limit for the expected repetition rate of transport packets having a particular PID must be set in dependence on application during the measuring. This is usually of the order of magnitude of half a second but is a user-definable quantity, in any case.

A “PID_error” (Fig. 11.3.) occurs when

- transport stream packets with a PID referred to in a PMT are not contained in the transport stream or
- if their repetition rate exceeds a user-definable limit which is usually of the order of magnitude of 500 ms.

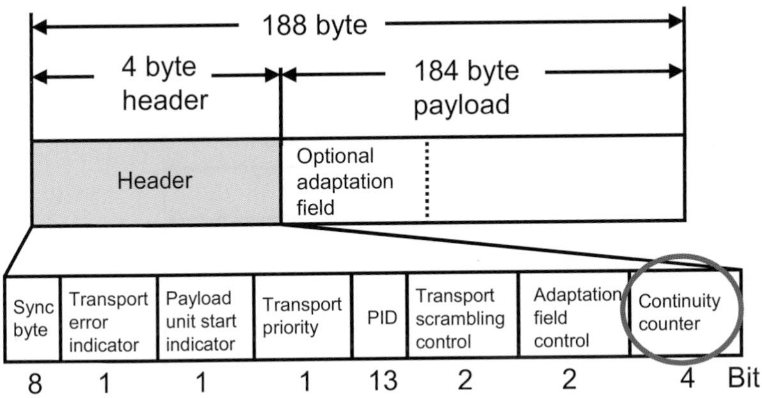

Fig. 11.4. Continuity_count_error

11.6 The Continuity_Count_Error

Each MPEG-2 transport stream packet contains in the 4-byte-long header a 4-bit counter which continuously counts from 0 to 15 and then begins at zero again after an overflow (modulo 16 counter). However, each transport stream packet of each PID has its own continuity counter, i.e. packets with a PID=100, e.g., have a different counter, as do packets with a PID=200. It is the purpose of this counter to enable one to recognize missing or repeated transport stream packets of the same PID in order to draw attention to any multiplexer problems.

Such problems can also arise as a result of errored remultiplexing or sporadically due to bit errors on the transmission link. Although MPEG-2 allows discontinuities in the transport stream, they must be indicated in the adaptation field, e.g. after a switch-over (discontinuity indicator=1). In the case of zero packets (PID=0x1FF), on the other hand, discontinuities are allowed and is not checked, therefore.

A continuity_error (Fig. 11.4.) occurs when

- the same TS packet is transmitted twice without a discontinuity being indicated, or
- if a packet is missing (count incremented by 2) without a discontinuity being indicated, or
- the sequence of packets is completely wrong.

Note: The way in which an MPEG-2 decoder reacts to a continuity counter error when the packet sequence is, in fact, correct depends on the decoder and the decoder chip used in it.

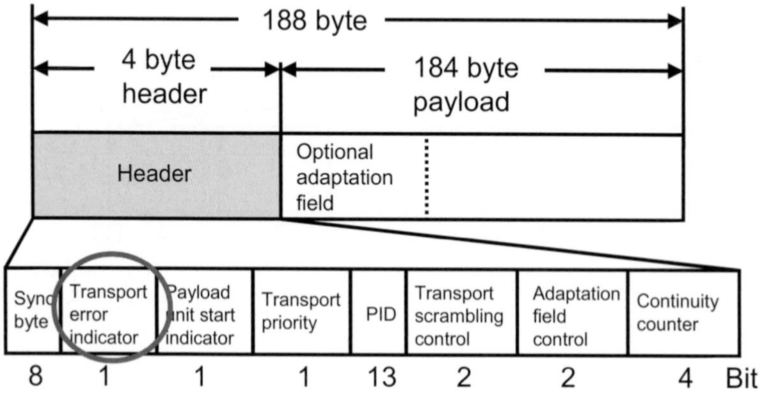

Fig. 11.5. Transport_error

11.7 The Transport_Error (Priority 2)

Every MPEG-2 transport stream packet contains a bit called Transport Error Indicator which follows directly after the sync byte. This bit flags any errored transport stream packets at the receiving end. During the transmission, bit errors may occur due to various types of influences. If error protection (at least Reed Solomon in DVB and ATSC) is no longer able to repair all errors in a packet, this bit is set. This packet can no longer be utilized by the MPEG-2 decoder and must be discarded.

A transport_error (Fig. 11.5.) occurs when

- the transport error indicator bit in the TS header is set to 1.

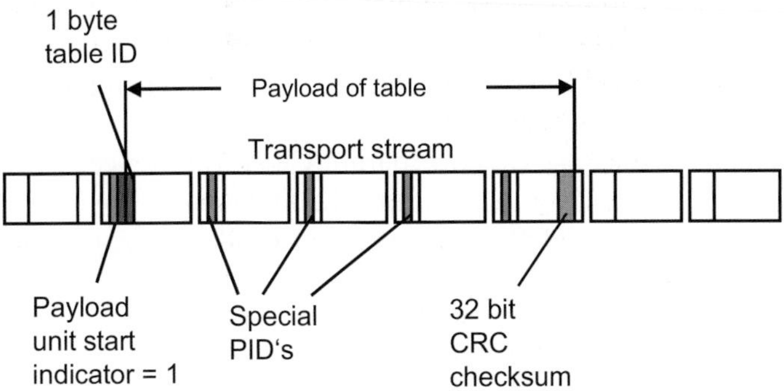

Fig. 11.6. CRC_error

11.8 The Cyclic Redundancy Check Error

During the transmission, all tables in the MPEG-2 transport stream, whether they are PSI tables or other private tables according to DVB (SI tables) or according to ATSC (PSIP tables), are protected by a CRC checksum. It is 32 bits long and is transmitted at the end of each sector. Each sector, which can be composed of many transport stream packets, is thus additionally protected. A CRC error has occurred if these checksums do not match the content of the actual section of the respective table. The MPEG-2 decoder must then discard this table content and wait for this section to be repeated. The cause of a CRC error is in most cases interference on the transmission link. If a set top box or decoder were to evaluate such errored table sections it could become "confused".

A CRC_error (Fig. 11.6.) occurs when

- a table (PAT, PMT, CAT, NIT,...) in a section has a wrong checksum which doesn't match its content.

11.9 The Program Clock Reference Error (PCR_Error, PCR_Accuracy)

All coding processes at the MPEG-2 encoder end are derived from a 27 MHz clock reference. This 27 MHz clock oscillator is coupled to a

42 bit-long counter which provides the System Time Clock (STC). For each program, a separate system time clock (STC) is used. To be able to link the MPEG-2 decoder to this clock, copies of the current program system time are transmitted about every 40 ms per program in the adaptation field. The PMT of the respective program carries information about the TS packets in which this clock time can be found.

The STC reference values are called Program Clock Reference (PCR). They are nothing else than a 42 bit copy of the 42 bit counter. The MPEG-2 decoder links itself to these PCR values via a PLL and derives its own system clock from them.

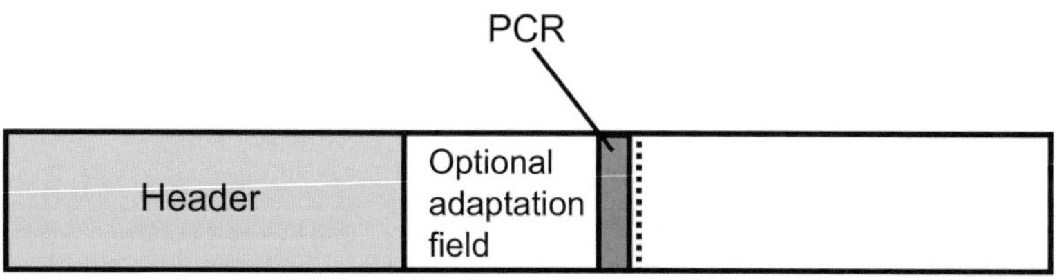

Fig. 11.7. PCR value

If the repetition rate of the PCR values is too slow, it may be due to the fact that the PLL of the receiver has problems in locking to it. MPEG-2 specifies that the maximum interval between two PCR values must not exceed a period of 40 ms. According to the DVB Measurement Guidelines, a PCR_error has occurred if this time is exceeded.

The timing of the PCR values with respect to one another should also be relatively accurate, i.e. there should not be any jitter. Jitter may occur, for example, if the PCR values are not corrected, or are corrected inaccurately, during remultiplexing.

If the PCR jitter exceeds ±500 ns, a PCR_accuracy_error has occurred. PCR jitter frequently extends into the ±30 µs range which can be handled by many set top boxes, but not by all. The first indication that the PCR jitter is too great is a black/white picture instead of a colour picture. The actual effect, however, depends on how the set top box is wired to the TV receiver. An RGB connection (e.g. via a SCART A/V cable) is certainly less critical than a composite video cable connection.

A PCR_error occurs when

- the difference between two successive PCR values of a program is greater than 100 ms and no discontinuity is indicated in the adaptation field, or

- the time interval between two packets with PCR values of a program is more than 40 ms.

A PCR_accuracy_error occurs when

- the deviation between the PCR values is greater than ±500 ns (PCR jitter).

11.10 The Presentation Time Stamp Error (PTS_Error)

The Presentation Time Stamps (PTS) transmitted in the PES headers contain a 33 bit-long timing information item about the precise presentation time. These values are transmitted both in the elementary video streams and in the elementary audio streams and are used, e.g. for lip synchronisation between video and audio. The PTS values are derived from the system time clock (STC) which has a total width of 42 bits but only the 33 MSBs are used in this case. The spacing between two PTS values must not be greater than 700 ms to avoid a PTS error.

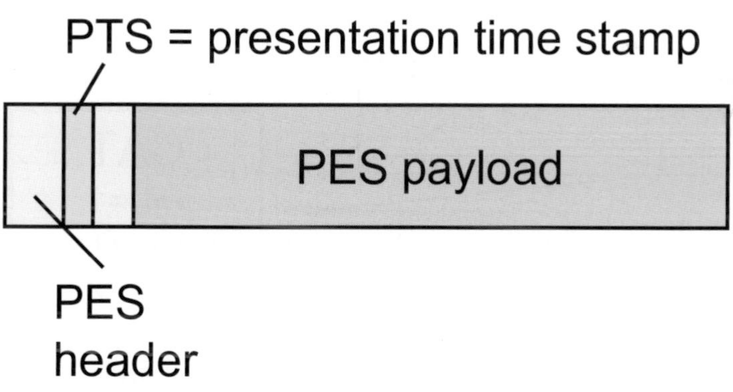

Fig. 11.8. PTS value in the PES header

A PTS_error occurs when

- the spacing between two PTS values of a program is greater than 700 ms.

Although real PTS errors occur only rarely, a perceptible lack of lip sync between video and audio happens quite frequently. In practice, the causes of this are difficult to detect and identify during a broadcast and can be attributable both to older MPEG-2 chips and to faulty MPEG-2 decoders. The direct measurement of lip synchronism would be an important test parameter.

11.11 The Conditional Access Table Error (CAT_Error)

An MPEG-2 transport stream packet can contain scrambled data but only the payload part must be scrambled and never the header or the adaptation field. A scrambled payload part is flagged by two special bits in the TS header, the Transport Scrambling Control bits. If both bits are set to zero, there is no scrambling. If one of the two is not zero, the payload part is scrambled and a Conditional Access Table (CAT) is needed to descramble it. If this is missing or only rarely there, a CAT_error occurs. The CAT has a 1 as PID and also a 1 as table ID. Apart from the EIT in the case of the transmission of a program guide, all DVB tables must be unscrambled.

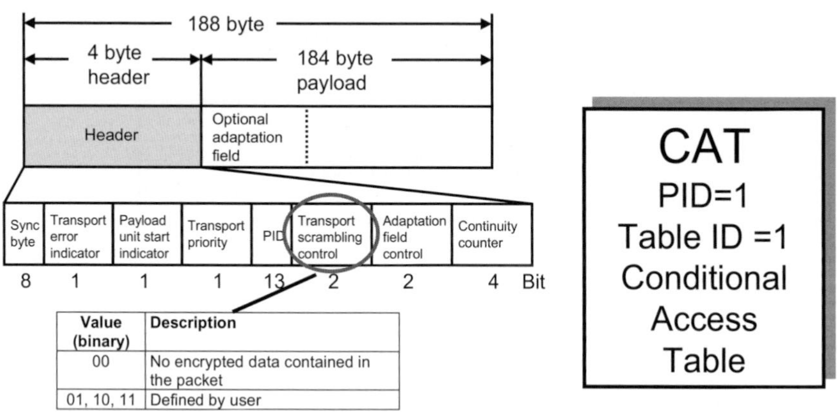

Fig. 11.9. CAT_error

A CAT_error (Fig. 11.9.) occurs when

- a scrambled TS packet has been found but no CAT is being transmitted,

- a CAT has been found by means of PID=1 but the table ID is not equal to 1.

11.12 Service Information Repetition Rate Error (SI_Repetition_Error)

All the MPEG-2 and DVB tables (PSI/SI) must be regularly repeated at minimum and maximum intervals. The repetition rates depend on the respective type of table.

The minimum time interval of the table repetition rate (s. Table 11.2.) is normally about 25 ms and the maximum is between 500 ms and 30 s or even infinity.

Table 11.2. PSI/SI table repetition time

Service information	Max. interval (complete table)	Min. intervall (single sections)
PAT	0.5 s	25 ms
CAT	0.5 s	25 ms
PMT	0.5 s	25 ms
NIT	10 s	25 ms
SDT	2 s	25 ms
BAT	10 s	25 ms
EIT	2 s	25 ms
RST	-	25 ms
TDT	30 s	25 ms
TOT	30 s	25 ms

A SI_repetition_error occurs when

- the time interval between SI tables is too great,
- the time interval between SI tables is too small.

The limit values depend on the tables.

Since not every transport stream contains all the types of tables, the test decoder must be capable of activating or deactivating the limit values.

11.13 Monitoring the NIT, SDT, EIT, RST and TDT/TOT Tables

In addition to the PSI tables of the MPEG-2 standard, the DVB Group has specified the NIT, SDT/BAT, EIT, RST and TDT/TOT SI tables.

The DVB Measurement Group recognized that these tables needed to be monitored for presence, repetition rate and correct identifiability. This does not include checking the consistency, i.e. the content of the tables. A SI table is identified by means of the PID and its table ID. This is because there are some tables which have the same PID and can thus only be recognized from the table ID (SDT/BAT and TDT/TOT).

Table 11.3. SI tables

Service Information	PID [hex]	Table_id [hex]	Max. Interval [sec]
NIT	0x0010	0x40, 0x41, 0x42	10
SDT	0x0011	0x42, 0x46	2
BAT	0x0011	0x4A	10
EIT	0x0012	0x4E to 0x4F, 0x50 to 0x6F	2
RST	0x0013	0x71	-
TDT	0x0014	0x70	30
TOT	0x0014	0x73	30
ST	0x0010 to 0x0013	0x72	-

A NIT_error, SDT_error, EIT_error, RST_error or TDT_error occurs when

- a corresponding packet is contained in the TS but has the wrong table index,
- the time interval between two sections of these SI tables is too great or too small.

11.14 Undefined PIDs (Unreferenced_PID)

All PIDs contained in the transport steam are conveyed to the MPEG-2 decoder via the PAT and the PMTs. There are also the PSI/SI tables. However, it is perfectly possible that the transport stream contains TS packets whose PID is not indicated by this mechanism, the so-called unreferenced PIDs. According to DVB, an unreferenced PID may be contained there only for half a second during a program change.

An unreferenced_PID (Fig. 11.10.) occurs when

- a packet having an unknown PID is contained in the transport stream and is not referenced within a PMT after half a second at the latest.

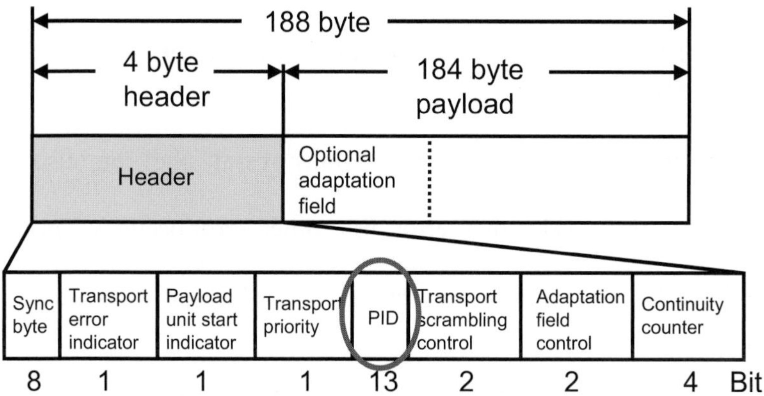

Fig. 11.10. Unreferenced PID

11.15 Errors in the Transmission of Additional Service Information

Apart from the usual information, additional service information (SI_other_error) can be transmitted for other channels, according to DVB. These are the NIT_other, SDT_other and EIT_other tables.

The SI_other tables can be recognized from the PIDs and table_IDs in Table 10.4. also lists the time limits.

An SI_other_error occurs when

- the time interval between SI_other tables is too great,
- the time interval between SI_other tables is too small.

Table 11.4. SI Other

Service information	Table_ID	Max. interval (complete table)	Min. interval (single sections)
NIT_OTHER	0x41	10 s	25 ms
SDT_OTHER	0x46	2 s	25 ms
EIT_OTHER	0x4F, 0x60 to 0x6F	2 s	25 ms

11.16 Faulty tables NIT_other_error, SDT_other_error, EIT_other_error

In addition to monitoring the 3 SI_other tables overall, they can also be monitored individually:

A NIT_other_error, SDT_other_error, EIT_other_error occurs when

- the time interval between sections of these tables is too great.

11.17 Monitoring an ATSC-Compliant MPEG-2 Transport Stream

According to the DVB Measurement Guidelines, the following measurements can be made unchanged on an ATSC-compliant MPEG-2 transport stream:

- TS_sync_error
- Sync_byte_error
- PAT_error
- Continuity_count_error
- PMT_error
- PID_error
- Transport_error
- CRC_error
- PCR_error
- PCR_accuracy_error
- PTS_error
- CAT_error

It is only necessary to adapt all Priority 3 measurements to the PSIP tables.

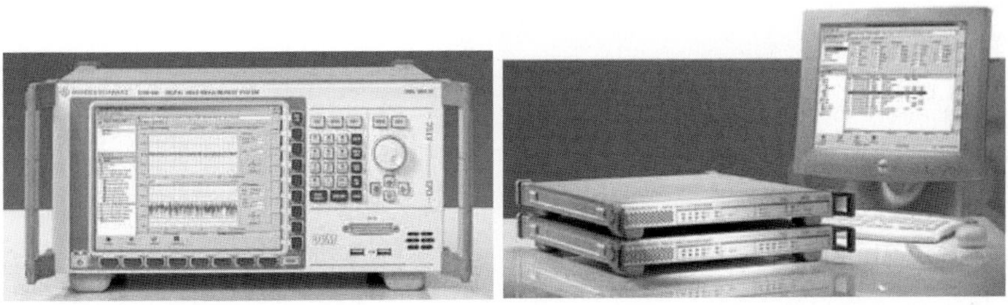

Fig. 11.11. MPEG-2 analyzer, Rohde&Schwarz, on the left: DVM400, on the right: DVM100

Bibliography: [TR100290], [DVMD], [DVM]

12 Picture Quality Analysis of Digital TV Signals

The picture quality of digital TV signals is subject to quite different effects and influences than that of analog TV signals. Whereas noise effects in analog TV signals manifest themselves directly as 'snow' in the picture, they initially only produce an increase in the channel bit error rate in digital television. Due to the error protection included in the signal, however, most of the bit errors can be repaired up to a certain limit and are thus not noticeable in the picture or the sound. If the transmission path for digital television is too noisy, the transmission breaks down abruptly ('brick wall' effect, also called 'fall off the cliff'). Neither does linear or nonlinear distortion have any direct effect on the picture and sound quality in digital television but in the extreme case it, too, leads to a total transmission breakdown. Digital TV does not require VITS (vertical insertion test signal) lines for detecting linear and nonlinear distortion or black-level lines for measuring noise, neither are they provided there and would not produce any test results concerning the transmission link if they were. Nevertheless, the picture quality can still be good, bad or indifferent but it now needs to be classified differently and detected by different means. There are mainly two sources which can disturb the video transmission and which can cause interference effects of quite a different type:

- the MPEG-2 encoder or sometimes also the multiplexer, and
- the transmission link from the modulator to the receiver.

The MPEG-2 encoder has a direct effect on the picture quality due to the more or less severe compression imposed by it. The transmission link introduces interference effects resulting in channel bit errors which manifest themselves as large-area blocking effects, as frozen picture areas or frames or as a total loss of transmission. If the compression of the MPEG-2 encoder is too great, it causes blocks of unsharp picture areas. All these effects are simply called blocking. This section explains how the effects caused by the MPEG-2 video coding are produced and analyzed.

All video compression algorithms work in blocks, i.e. the image is in most cases initially divided into blocks of 8 x 8 pixels. Each of these

W. Fischer, *Digital Video and Audio Broadcasting Technology*, Signals and Communication Technology, 3rd ed., DOI: 10.1007/978-3-642-11612-4_12, © Springer-Verlag Berlin Heidelberg 2010

blocks is individually compressed to a greater or lesser extent, independently of the other blocks. In the case of MPEG-2, the image is additionally divided into 16 x 16 pixels called macroblocks which form the basis for the interframe coding. If the compression is excessive, the block boundaries become visible and blocking occurs. There are discontinuities between blocks in the luminance and chrominance signals and these are perceptible. With a predetermined compression, the amount of blocking in an image also depends on the picture material, among other things. Some source images can be compressed without problems and almost without errors at a low data rate whereas other material produces strong blocking effects when compressed. Simple moving picture sources for moving-picture compression are, for example, scenes with little movement and little detail. Animated cartoons, but also classical celluloid films, can be compressed without loss of quality with relatively few problems. The reason for this is, among other things, that there is no movement between the first and second fields. In addition, the image structures are relatively coarse in animated cartoons. The most critical sources are sports programs, and this, in turn, depends on the type of sport. By their nature, Formula I programs will be more difficult to compress without interference than programs involving the thinker's sport of chess. In addition, however, the actual picture quality depends on the MPEG-2 encoder and the algorithms used there. In recent years, the picture quality has clearly improved in this department. Fig. 12.1. shows an example of blocking.

Apart from the blocking, the excessively compressed image also shows the DCT structures, i.e. patterned interference suddenly occurs in the picture.

The decisive factor is that it is always the MPEG-2 encoder which is responsible for such interference effects. Although it is difficult to measure good or bad picture quality caused by compression processes, it can be done. Of course, picture quality will never be 100% measurement - there is always some subjectivity involved. Even so-called objective video quality analyzers are calibrated by test persons using subjective tests. At least, this applies to analyzers which do not use a reference signal for quality assessment, but in practice, there is no reference signal with which the compressed video signal could be compared. The requirement that it should be possible to use reference signals is unrealistic, at least with regard to transmission testing.

The basis for all video quality analyzers throughout the world - and there are not many - is the ITU-R BT.500 standard. This standard describes methods for subjective video quality analysis where a group of test persons analyses video sequences for their picture quality.

Fig. 12.1. Blocking effects with excessive compression

12.1 Methods for Measuring Video Quality

The Video Quality Experts Group (VQEG) in the ITU has defined methods for assessing picture quality which have then been incorporated in the ITU-R BT.500 standard.

In principle, these are two subjective methods for picture quality assessment by test persons, namely:

- the DSCQS (Double Stimulus Continual Quality Scale) method and
- the SSCQE (Single Stimulus Continual Quality Evaluation) method.

The two methods basically differ only in that one method makes use of a reference video signal and the other one does not have a reference signal. The basis is always a subjective picture quality analysis by a group of test persons who assess an image sequence in accordance with a particular procedure. It is then attempted to reproduce these subjective methods by means of objective methods in a test instrument by performing picture analyses on the macroblocks and using adaptation algorithms.

12.1.1 Subjective Picture Quality Analysis

In subjective picture quality analysis, a group of test persons assesses an image sequence (SSCQE - Single Stimulus Continual Quality Scale Method, Fig. 12.2.) or compares an image sequence after compression with the original (DSCQS - Double Stimulus Continual Quality Scale Method) and issues marks on a quality scale from 0 (Bad) to 100 (Excellent) by means of a sliding control. The positions of the sliding controls are detected by a computer connected to them which continually determines (e.g. every 0.5 sec) a mean value of all the marks issued by the test persons.

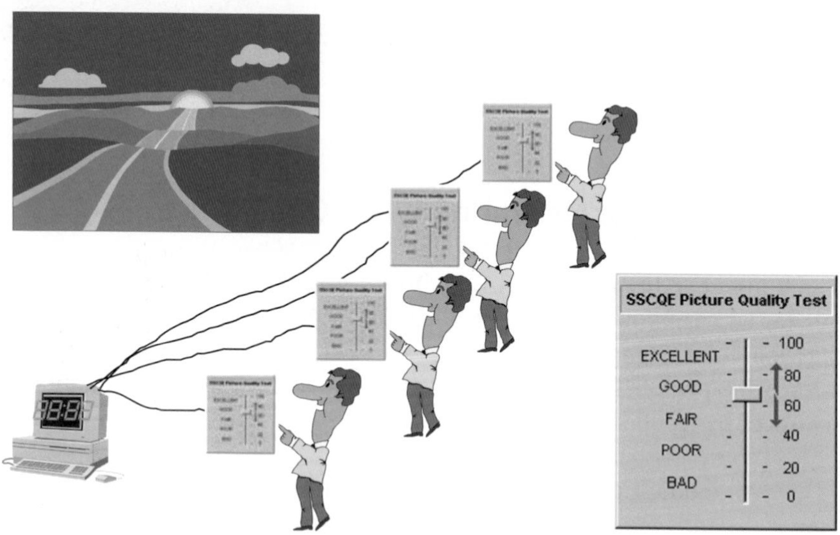

Fig. 12.2. Subjective Picture Quality Analysis

A video sequence will then provide a picture quality value versus time, i.e. a quality profile of the video sequence.

12.1.2 Double Stimulus Continual Quality Scale Method DSCQS

In the Double Stimulus Continual Quality Scale method according to ITU-R BT.500, a group of test persons compares an edited or processed video sequence with the original video sequence. The result obtained is a comparative quality profile of the edited or processed video sequence, i.e. a picture quality value from 0 (Bad) to 100 (Excellent) versus time.

The DSCQS method always requires a reference signal, on the one hand, but, on the other hand, the purely objective analysis can then be performed very simply by forming the difference. In practice, however, a reference signal is frequently no longer provided. Transmission link measurements cannot be performed using this method. There are test instruments on the market which reproduce this method (Tektronix).

12.1.3 Single Stimulus Continual Quality Evaluation Method SSCQE

Since the Single Stimulus Continual Quality Evaluation method SSCQE deliberately dispenses with a reference signal, this method can be used much more widely in practice. In this method, a group of test persons only assesses the processed video sequence and issues marks from 0 (Bad) to 100 (Excellent) which also provides a video quality profile versus time.

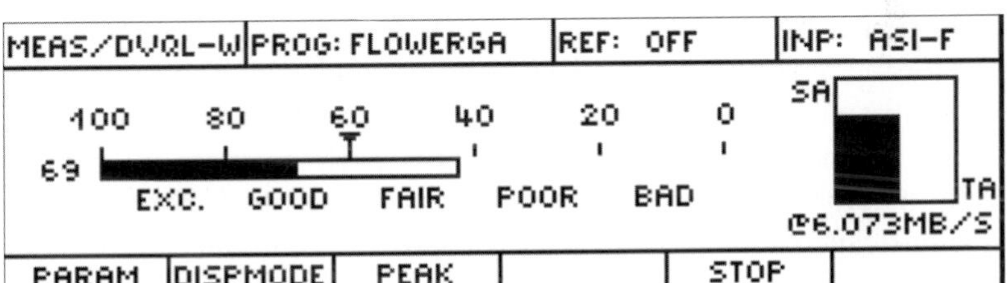

Fig. 12.3. Objective picture quality analysis using a test instrument [DVQ]

12.2 Objective Picture Quality Analysis

In the following sections, an objective test method for assessing the picture quality analysis in accordance with the Single Stimulus Continual Quality Scale Evaluation method is described. A digital picture analyzer operating in accordance with this method may provide by Fig. 12.3.

Since the DCT-related artefacts of a compressed video signal are always associated with blocking, an SSCQE type digital picture analyzer will attempt to verify the existence of this blocking in the picture. To be able to do this, the macroblocks and blocks must be analyzed in detail.

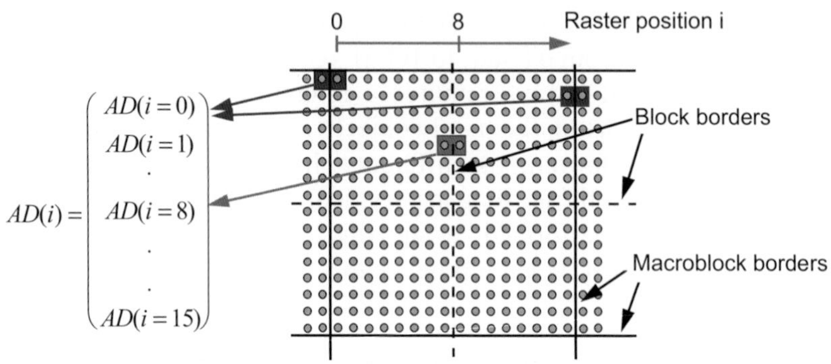

Fig. 12.4. Pixel difference at the block and macroblock boundaries

In a test procedure developed by the Technical University of Braunschweig (Germany) and Rohde&Schwarz, the differences between adjoining pixels within a macroblock are formed. Pixel difference means that simply the amplitude values of adjacent pixels of the Y signal within a macroblock, and also separately those of the Cb and Cr signals are subtracted. For each macroblock, 16 pixel differences are then obtained per line, e.g. for the Y signal. Then all 16 lines are analyzed. The same is also done vertically which also provides 16 pixel differences per column for the macroblock of the Y signal. This analysis is performed for all columns within the macroblock. The pixel differences at the block borders are of special significance here and will be particularly large in the case of blocking.

The pixel differences of all macroblocks within a line are then combined by adding them together in such a way that 16 individual values are obtained per line (Fig. 12.4.). The 16 pixel difference values of the individual lines are then also added together within a frame, resulting in 16 values per frame as pixel difference values. This, finally. provides information about the mean pixel difference 0 ... 15 in the horizontal and vertical direction within all macroblocks. The same process is repeated for Cb and Cr, i.e. the color difference signals.

Considering then the pixel differences of a video sequence with good picture quality and one with poor picture quality, it can be seen quite clearly what this objective test method for assessing the picture quality amounts to:

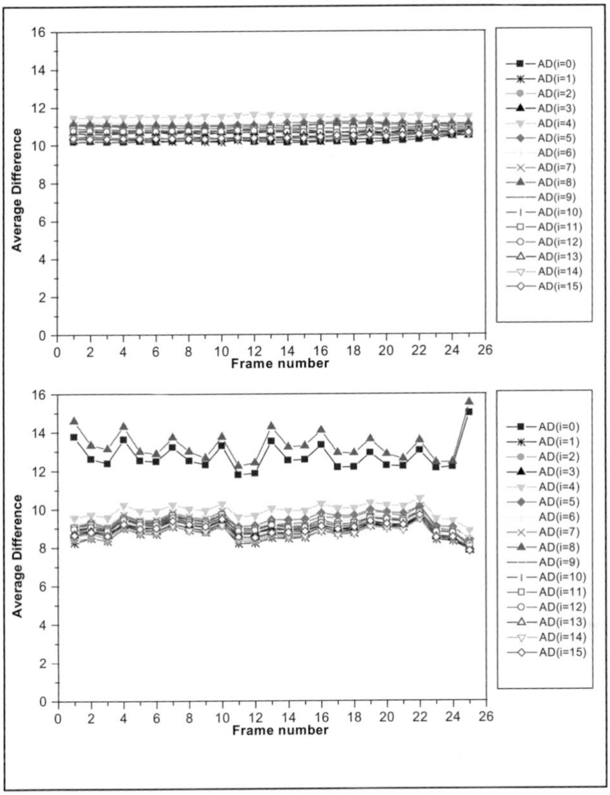

Fig. 12.5. Averaged macroblock pixel differences in a video sequence with good picture quality (top, flower garden/original, 6 Mbit/s) and with poorer picture quality (bottom, flower garden/MPEG-2, 2 Mbit/s)

Fig. 12.5. shows clearly that the pixel amplitude differences in the "good" video sequence are very close to each other for all 16 pixel differences within the macroblocks. In the present example, they are all at about 10 ... 12.

In a video sequence with "poor" quality (bottom display) with blocking, it can be seen that the macroblock borders exhibit greater jumps, i.e the pixel differences are greater there.

It can be seen clearly that pixel differences No. 0 and No. 8, in the bottom display, are obviously greater than the remaining difference values.

No. 0 corresponds to the macroblock border and No. 8 corresponds to the block boundary within a macroblock.

Clearly, this simple analysis of the pixel amplitude differences makes it possible to verify the existence of blocking (Fig. 12.6.).

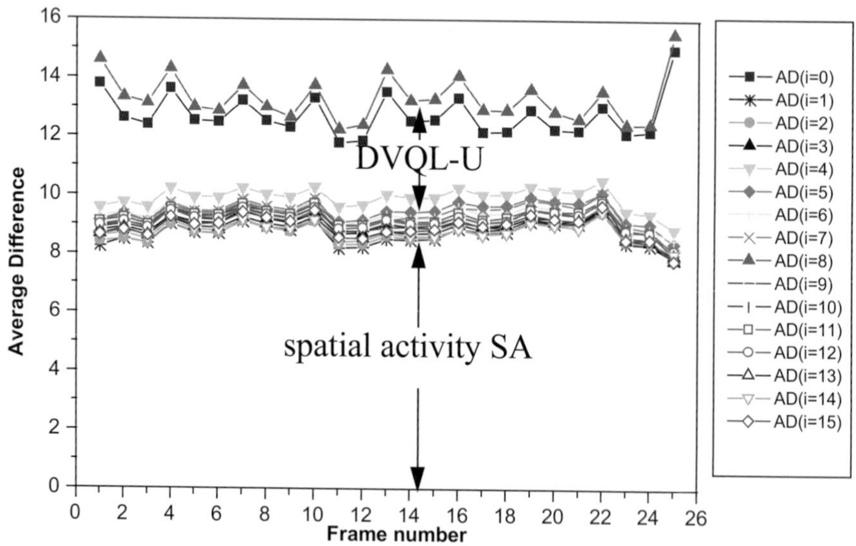

Fig. 12.6. Determining digital video quality level unweighted (DVQL-U) and spatial activity (SA) from the macroblock pixel differences

The basic test value of a Digital Video Quality Analyzer by Rohde&Schwarz for calculating the picture quality of DCT coded video sequences is the picture quality test value DVQL-U (digital video quality level - unweighted). DVQL-U is used as absolute value for the existence of blocking type interference patterns within an original frame. In contrast to DVQL-W (digital video quality level - weighted), DVQL-U is a direct measure of these blocking types of interference. Depending on the original frame, however, the test value is not always correlated with the impression of quality of a subjective observation.

To bring the objective picture quality test value closer to the subjectively perceived picture quality, other quantities in the moving picture must also be taken into consideration. These are

- the spatial activity (SA), and
- the temporal activity (TA).

This is because both spatial and temporal activity can render blocking structures invisible, i.e. they can mask them. These artifacts in the picture are then simply not seen by the human eye.

The spatial activity is a measure of the existence of fine structures in the picture. A picture rich in detail, i.e. one with many fine structures, exhibits high spatial activity. An unstructured monochrome picture, on the other hand, would correspond to a spatial activity of zero. The maximum theoretically achievable spatial activity would occur if a white pixel always alternates with a black pixel both horizontally and vertically in a frame (fine grid pattern).

Fig. 12.7. Low (left) and high (right) spatial activity

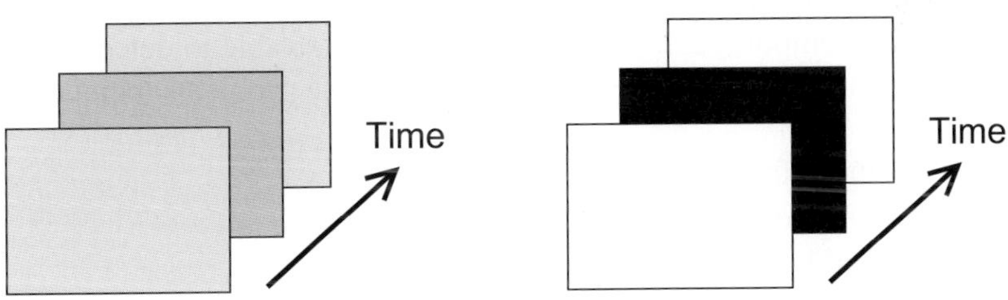

Fig. 12.8. Low (left) and high (right) temporal activity

In addition to the spatial activity in the picture (Fig. 12.7.), the temporal activity (TA) must be taken into consideration (Fig. 12.8.). The temporal activity is an aggregate measure of the change (movement) in successive frames. The maximum temporal activity which could be achieved in theory would be if all pixels were to change from black to white or conversely in successive frames. Accordingly, a temporal activity of 0 corresponds to a sequence of frames without movement.

The two parameters SA and TA must be included when calculating the weighted video quality from the unweighted DVQL (blocking level).

In a first process, the unweighted digital quality level DVQL-U for all Y, Cb and Cr signals and the spatial activity SA and temporal activity TA are determined in the above-mentioned digital video quality analyzer.

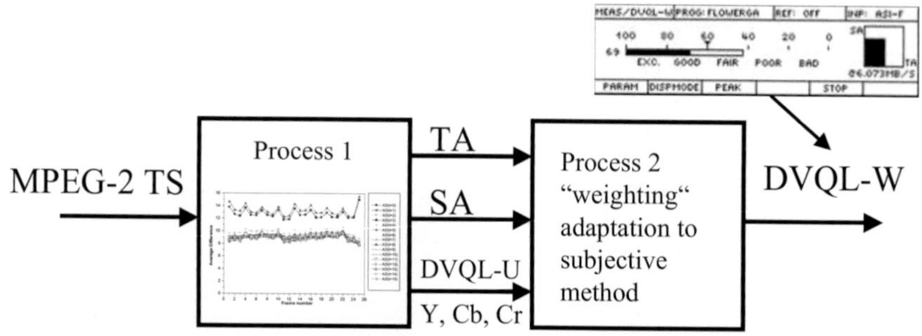

Fig. 12.9. Using the digital video quality analyzer [DVQ]

Weighting is then performed in a second process which thus takes into account subjective factors. The display of the digital video quality analyzer shows both the digital video quality level - weighted or unweighted - and the spatial and temporal activity. The analyzer is also able to detect decoding problems, in addition to the video quality: These problems include

- picture freeze (TA = 0),
- picture loss (TA = 0, SA = 0), and
- sound loss.

Digital video quality analyzers are mainly used close to the MPEG-2 encoding stages since the transmission has no further effect on the video quality itself. Naturally, such an analyzer will also detect the decoding problems caused by bit errors produced by the transmission link. Since in many cases the network operator is not the program provider, as well, digital video quality analyzers are often also found at the network termination end so that objective measurement parameters are available as a basis for any discussion between network operator and program provider. Digital video quality analyzers are also of great importance in the testing of MPEG-2 encoders.

12.3 Summary and Outlook

Artefacts caused by the MPEG encoding process lead to a picture quality which is more or less good. This is apparent as

- blocking (visible block transitions)
- blurring (lack of high frequencies)
- "mosquito" noise (visible DCT structures).

In the case of the more recent image coding methods such as MPEG-4 AVC, however, other approaches must be used in picture quality assessment. E.g., deblocking filters are used here in an attempt to keep blocking as invisible as possible. The only objective information is provided by an analysis of the quantization performed by the encoder. Such measurements are supported by the MPEG analyzer DVM [DVM].

However, the influence of the distribution and transmission paths in the digital TV network remains the same. It leads to bit errors which, in turn, can lead to a "fall-off-the-cliff" at some time (Fig. 12.11.), either it works or it no longer works at some time. In a transition case, slice structures become apparent as can be seen in Fig. 12.11.

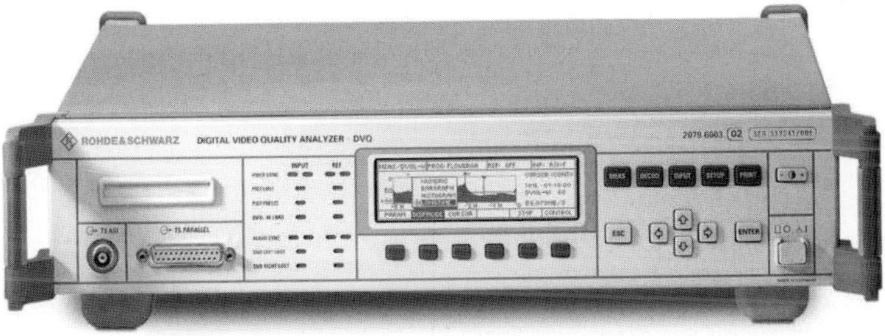

Fig. 12.10. Digital video quality analyzer, Rohde&Schwarz [DVQ]

Fig. 12.11. Picture disturbances caused by interference during the transmission; slice-like structures are recognizable which are caused here by severe rain during satellite reception.

Bibliography: [ITU500], [DVQ], [DVM]

13 Basic Principles of Digital Modulation

To begin with, this chapter quite generally creates a basis for an approach to the digital modulation methods. Following this chapter, it would also be possible to continue e.g. in the field of mobile radio technology (GSM, IS95 or UMTS) as the basic knowledge discussed here applies to the field of communication technology and its applications overall. However, its prime intent is to create the foundation for the subsequent chapters on DVB-S, DVB-C, OFDM/COFDM, DVB-T, ATSC and ISDB-T. Experts, of course can simply skip this chapter.

13.1 Introduction

Analog transmission of information has long been effected by means of amplitude modulation (AM) and frequency modulation (FM). The information to be transmitted is impressed on the carrier by varying either its amplitude or frequency or phase, this process being referred to as modulation.

To transmit data signals, i.e. digital signals, amplitude or frequency shift keying was used in the early times of data transmission. To transmit a data stream of e.g. 10 Mbit/s by means of simple amplitude shift keying (ASK), a bandwidth of at least 10 MHz is required if a non-return-to-zero code (NRZ) is used. According to the Nyquist theorem, a bandwidth corresponding to at least half the data rate is required for the NRZ baseband signal. Using ASK produces two sidebands and that gives a RF signal with a bandwidth which is equal to the data rate of the baseband signal. The bandwidth actually required is even larger because of the signal filtering necessary to suppress adjacent-channel interference.

An analog telephone channel is about 3 kHz wide. Initially, a data rate of 1200 bit/s could be achieved for this channel. Today, 56 kbit/s is no problem any more. We are used to our fax and modem links operating at such data rates. This quantum leap ahead was possible only through the use of modern digital modulation methods known as IQ modulation. IQ modulation is basically a form of amplitude modulation.

W. Fischer, *Digital Video and Audio Broadcasting Technology*, Signals and Communication Technology, 3rd ed., DOI: 10.1007/978-3-642-11612-4_13,

We know the following modulation methods:

- Amplitude modulation
- Frequency modulation
- Phase modulation
- Amplitude shift keying (ASK)
- Frequency shift keying (FSK)
- Phase shift keying (PSK)
- Amplitude and phase shift keying (QAM)

What we want is to reduce the bandwidth for data signal transmission. This is possible only by using modern digital modulation methods. Our aim is to cut the required bandwidth by several factors relative to the data rate of the signal transmitted.

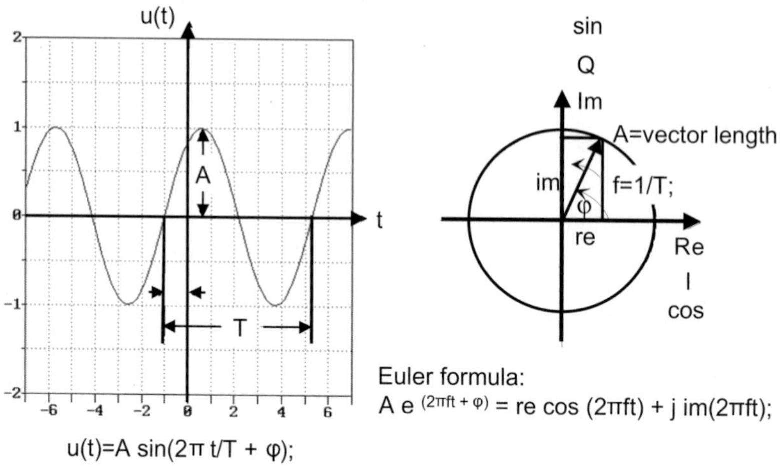

Fig. 13.1. Vector representation of a sinusoidal signal

It is obvious that this will not go without disadvantages, i.e. susceptibility to noise and interference will increase. In the following, the digital modulation methods will be discussed.

Before entering into this subject, we should like to point out that in electrical engineering it is customary to represent sinusoidal quantities by means of vectors (s. Fig. 13.1.). Each sinusoidal quantity can be unambiguously described by its amplitude and zero-phase angle. Moreover, the frequency must be known. In the vector representation, the rotating vector at the time t = 0 is shown. The vector is then at the zero-phase angle and its length corresponds to the amplitude of the sinusoidal quantity.

Fig. 13.1. represents a sine signal in the time domain and in the form of a vector. The rotating vector, whose length corresponds to the amplitude, is shown at zero-phase angle φ. The sine signal is obtained by projecting the rotating vector on the vertical axis (Im) and recording the position of the vector tip versus time. The corresponding cosine signal is obtained by projecting the rotating vector on the horizontal axis (Re).

The vector can be split into its real part and its imaginary part, terms which are derived from the theory of complex numbers in mathematics. The real part corresponds to the projection onto the horizontal axis and is calculated from Re = A • cos φ. The imaginary part corresponds to the projection onto the vertical axis and can be calculated from Im = A • sin φ. The length of the vector is related to the real part and the imaginary part via Pythagoras' theorem

$$A = \sqrt{\mathrm{Re}^2 + \mathrm{Im}^2};.$$

The real part can also be imagined to be the amplitude of a cosine signal and the imaginary part as the amplitude of a sine signal.

Any desired sine or cosine signal can be obtained by the superposition of a sine and a cosine signal of the same frequency and the desired amplitudes.

The real part is also called the I, or in-phase, component and the imaginary part is called the Q, or quadrature, component, where in-phase stands for 0^0 phase angle to a reference carrier and quadrature stands for 90^0 phase angle. The terms real part, imaginary part, cosine and sine component and I and Q component will appear time and again in the sections to follow.

13.2 Mixer

We will see that the mixer is one of the most important electronic components that make up an IQ modulator. A mixer is basically a multiplier. The modulation signal is usually converted to the IF by means of a carrier signal. As a result, two sidebands about the carrier are obtained. This type of modulation is known as double-sideband amplitude modulation with suppressed carrier. The mixer shown in Fig. 13.2. is basically a double switch driven by the carrier. It reverses the polarity of the modulation signal at the carrier frequency.

In the case of a purely sinusoidal modulation signal, two spectral lines are obtained – one above and one below the carrier frequency – each spaced from the carrier at an offset of the modulation frequency. Moreover, sub-harmonics at an offset of the carrier frequency are produced. The latter have to be suppressed by means of a lowpass filter.

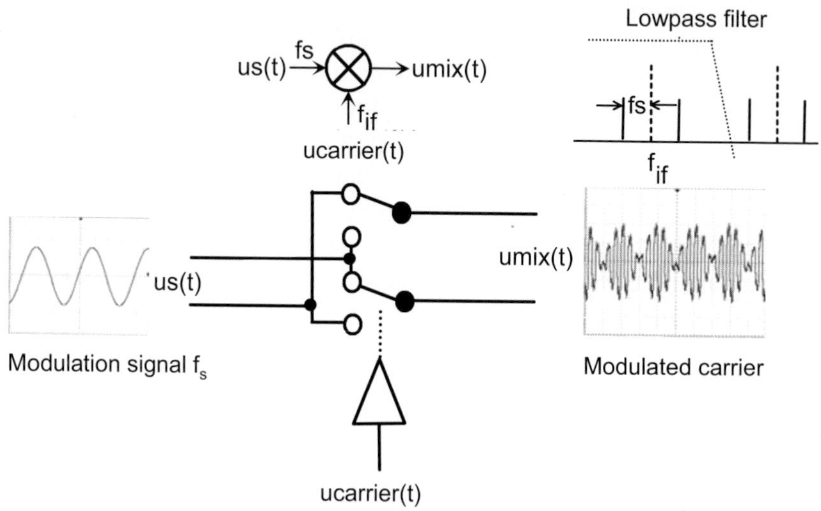

Fig.13.2. Mixer and mixing process. Amplitude modulation with suppressed carrier

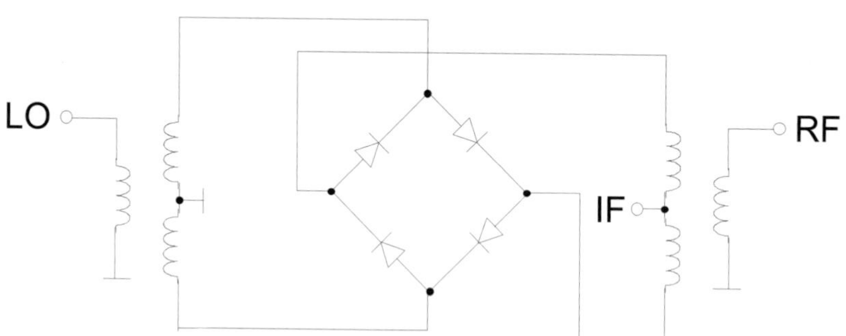

Fig. 13.3. Block diagram of a double-balanced mixer

Fig. 13.3. is a block diagram of a modern analog double-balanced mixer. The polarity of the modulation signal is switched by 4 PIN diodes. The carrier signal (LO = local oscillator) is coupled in via an RF transformer, and the modulation product is coupled out via an RF transformer. The modulation signal is fed DC-coupled.

Mixers are today often implemented in the form of purely digital multipliers which, except for quantization noise and rounding errors, have an ideal behaviour.

If a direct voltage is applied as modulation signal, the carrier itself appears at the output of the mixer. Superimposing a sinusoidal signal on the DC leads to normal amplitude modulation with unsuppressed carrier. (Fig. 13.4.).

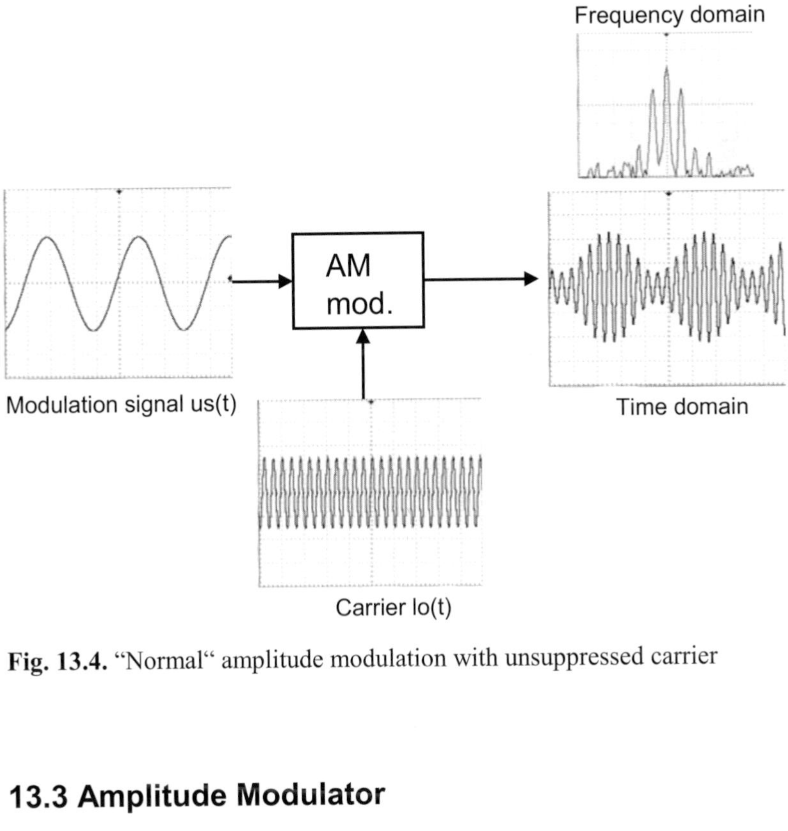

Fig. 13.4. “Normal“ amplitude modulation with unsuppressed carrier

13.3 Amplitude Modulator

In amplitude modulation, the information is contained in the amplitude of the carrier. The modulation signal changes (modulates) the carrier amplitude. This is effected by means of an AM modulator.

Fig. 13.4. illustrates "normal" AM modulation, in which the carrier is not suppressed. A sinusoidal modulation signal varies the carrier amplitude and so is impressed on the carrier as an envelope. In the example in Fig. 13.4., both the carrier and the modulation signal are sinusoidal signals. Looking at the spectrum, we not only find a spectral line at the carrier frequency but also two sidebands spaced from the carrier at an offset of the modulation frequency. For example, if a 1 MHz carrier is amplitude-modulated with a sinusoidal 1 kHz signal, a modulation spectrum with the carrier signal at 1 MHz and two sideband signals at 1 kHz above and below the carrier will be obtained. The bandwidth is 2 kHz in this case.

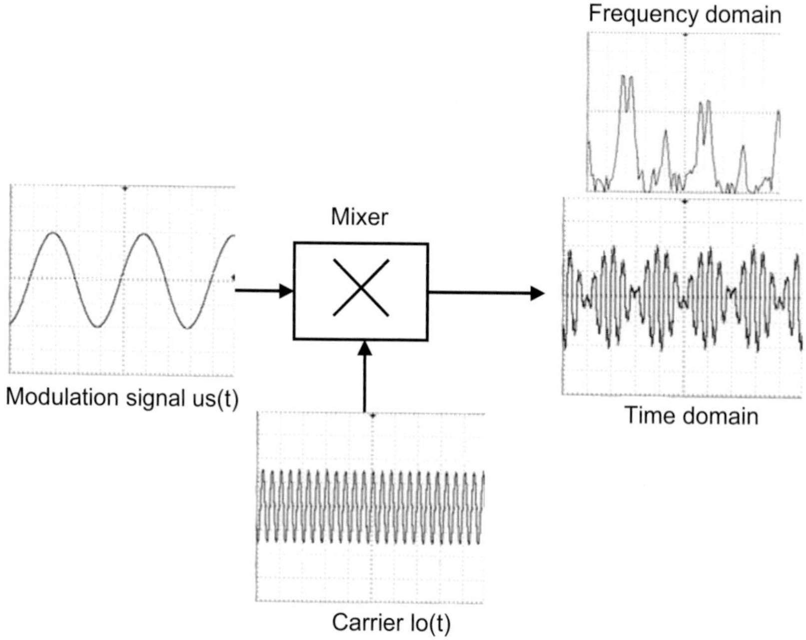

Fig. 13.5. Amplitude modulation with suppressed carrier

As mentioned above, the carrier is suppressed by the mixer. If a mixer is used for amplitude modulation and the modulation signal itself has no DC component, no spectral line at the carrier frequency will be found in the modulation spectrum. There are only the two sidebands. Fig. 13.5. shows amplitude modulation effected by means of a double-balanced mixer. In the modulation spectrum, we find not only the two sidebands but also harmonic sidebands about multiples of the carrier frequency. The latter have to be suppressed by lowpass filters. Fig. 13.5. also shows a typical amplitude-modulated signal in the time domain with suppressed carrier. The bandwidth is the same as with "normal" amplitude modulation, i.e. with unsuppressed carrier.

13.4 IQ Modulator

In colour television, quadrature modulation or IQ modulation has been used for a long time for the transmission of colour information. With PAL or NTSC colour subcarriers, the chrominance information is contained in the phase of the subcarrier and the colour saturation, or colour intensity, in the amplitude of the subcarrier. The colour subcarrier is superimposed on the luminance signal.

The modulated colour subcarrier is generated by means of an IQ modulator or quadrature modulator, where "I" stands for in-phase and "Q" for quadrature phase.

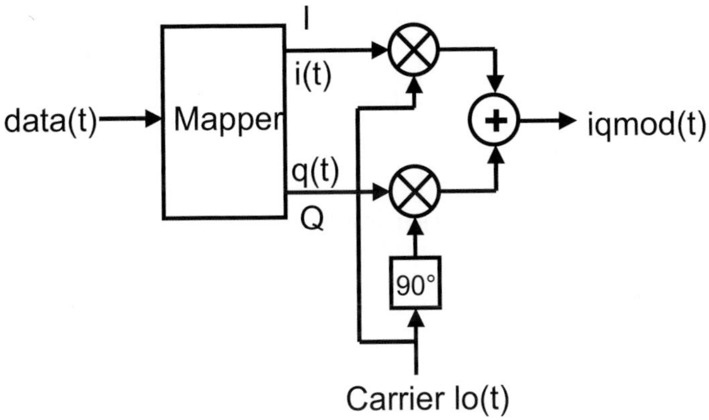

Fig. 13.6. IQ modulator

An IQ modulator (see Fig. 13.6.) has an I path and a Q path. The I path incorporates a mixer which is driven with 0° carrier phase. The mixer in the Q path is driven with 90° carrier phase. This means that I stands for 0° and Q for 90° carrier phase. I and Q are orthogonal to each other. In the vector diagram, the I axis coincides with the real axis and the Q axis with the imaginary axis.

PAL or NTSC modulators, too, incorporate an IQ modulator. For digital modulation, a mapper is connected ahead of the IQ modulator. The mapper is fed with the data stream data(t) to be transmitted; the output signals i(t) and q(t) of the mapper are the modulation signals for the I and the Q mixer. i(t) and q(t) are no longer data signals but signed voltages.

If i(t)=0, the I mixer produces no output signal, if q(t)=0, the Q mixer produces no signal. If i(t) is at 1 V, for example, the I mixer will output a carrier signal with constant amplitude and 0° carrier phase. If q(t), on the other hand, is at 1 V, the Q mixer will output a carrier signal with constant amplitude and 90° carrier phase (s. Fig. 13.10.).

The I and Q modulation products are combined in an adder.

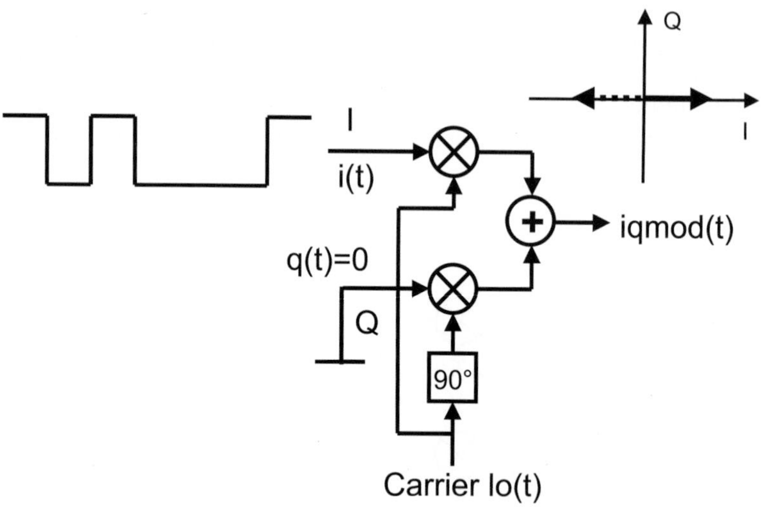

Fig. 13.7. IQ modulator, I path only

The product iqmod(t) is, therefore, the sum of the output signals of the I mixer and the Q mixer. If the Q mixer supplies no output signal, iqmod(t) corresponds to the output signal of the I path and vice versa.

Since the output signals of the I and the Q path are sine and cosine signals of the same frequency (carrier frequency) and differ only in amplitude, a sinusoidal output signal iqmod(t) of variable amplitude and phase is obtained through the superposition of the sinusoidal I output signal and the cosinusoidal Q output signal. Therefore, with the aid of control signals i(t) and q(t), we can vary the amplitude and phase of iqmod(t).

With the IQ modulator, we can generate pure amplitude modulation, pure phase modulation, or combined amplitude and phase modulation. A sinusoidal modulator output signal can thus be controlled in amplitude and phase.

The following applies to the amplitude and phase of iqmod(t):

$$A = \sqrt{(Ai)^2 + (Aq)^2};$$

$$\varphi = \arctan(\frac{Aq}{Ai});$$

where Ai is the amplitude of the I path and Aq the amplitude of the Q path.

From the incoming data stream data(t), the mapper generates the two modulation signals i(t) and q(t). We will see later that bit groups are combined to create certain patterns for i(t) and q(t), i.e. for the modulation signals of the I and the Q path.

Let us first look at the I path only (Fig. 13.7.). The Q path is driven with q(t)=0, i.e. it delivers no output signal and so does not contribute to iqmod(t). We now apply alternately +1 V and -1 V to the I path modulation input, so that i(t)=+1 V or i(t)= -1 V. Looking at the output signal iqmod(t), we see that the carrier lo(t) is present and switched only in phase between 0° and 180°. By varying the amplitude of i(t), we can vary the amplitude of iqmod(t).

For the vector diagram this means that the vector changes between 0° and 180° and varies in length but always remains on the I axis as long as only i(t) is present and being varied (Fig. 13.7).

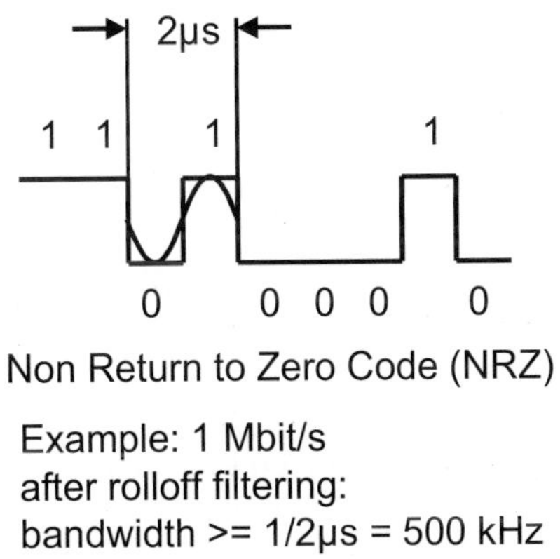

Fig. 13.8. NRZ code

This is a suitable point for discussing fundamentals relating to bandwidth conditions in the baseband and at RF. In the extreme case, the bandwidth of a data signal with an NRZ (non-return to zero) code (Fig. 13.8.) at a data rate of 1 Mbit/s can be cut (filtered) to such an extent that 500 kHz bandwidth are just sufficient to ensure reliable decoding. With using much mathematics, this can be explained quite simply by the fact that 01 alternations represent the highest frequency; i.e. the period has a length of 2 bits and is thus 2 µs in the case of a data rate of 1 Mbit/s. The reciprocal of 2 µs is then 500 kHz and the minimum baseband bandwidth for transmitting an NRZ code is then:

$$f_{baseband\_NRZ}\ [Hz] \geq 0.5 \bullet \text{data rate}_{NRZ}\ [bits/s];$$

If such a filtered NRZ code (Fig. 13.8.) is then supplied, e.g. without DC to a mixer as in the I path of this IQ modulator, two sidebands having each the bandwidth of the input baseband signal (Fig. 13.9.) are produced at RF. The minimum bandwidth required at RF is thus:

$$f_{RF\_NRZ}\ [Hz] \geq \text{data rate}_{NRZ}\ [bits/s];$$

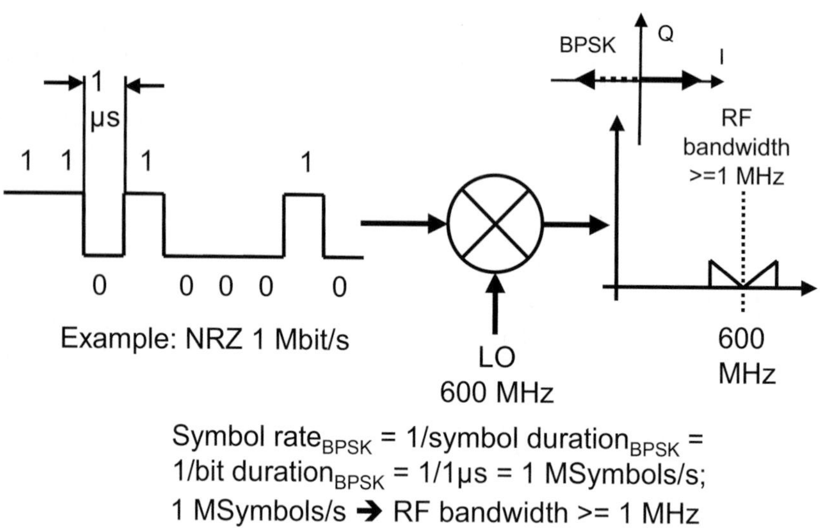

Fig. 13.9. BPSK modulation

In this type of modulation, therefore, the ratio between data rate and minimum bandwidth required at RF is 1:1. This type of modulation is called binary phase shift keying, or biphase shift keying, BPSK. With BPSK, a data rate of 1 Mbit/s requires a minimum bandwidth of 1 MHz at RF level. The duration of one stable state of the carrier is called a symbol and in BPSK, a symbol has exactly the same duration as one bit. The reciprocal of the symbol duration is called the symbol rate.

Symbol rate = 1/symbol duration;

With a data rate of 1 Mbit/s with BPSK (Fig. 13.9.), the symbol rate is 1 MSymbol/s. The minimum bandwidth required always corresponds to the symbol rate, i.e. 1 MSymbol/s requires a minimum bandwidth of 1 MHz.

Now let us assume that i(t) is zero and there is only a q(t) output signal. We now switch q(t) between +1 V and -1 V. iqmod(t) corresponds to the output signal of the Q mixer; there is no contribution from the I path. Again, a sine signal is obtained for iqmod(t), but with phase 90° or 270°. By varying the amplitude of q(t), the amplitude of iqmod(t) can be varied. For the vector diagram this means that the vector changes between 90° and 180° and varies in length along the Q axis (imaginary axis).

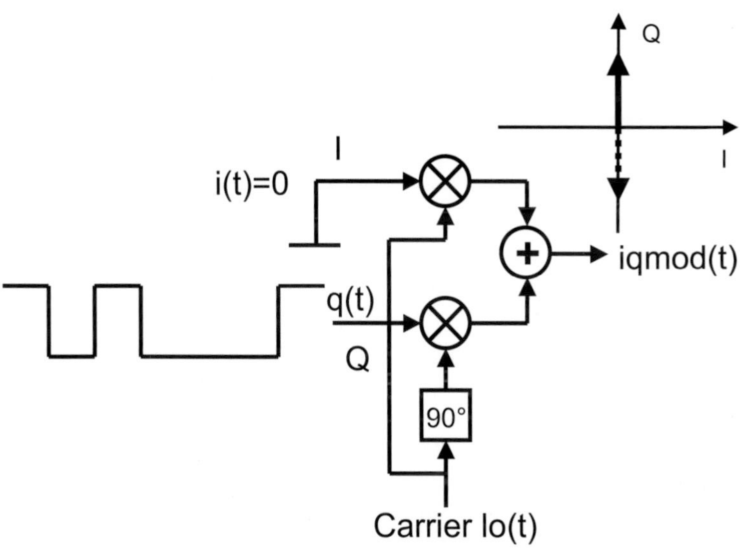

Fig. 13.10. IQ modulator, Q path active only

Next, we want to vary both i(t) and q(t) between +1 V and -1 V. In this case, the modulation products of the I path and the Q path are added up, so we can switch the carrier between 45°, 135°, 225° and 315°. This is referred to as quadrature phase shift keying or QPSK. Allowing any voltages for i(t) and q(t), any desired amplitude and phase can be generated for iqmod(t).

The data stream data(t) is converted to the two modulation signals i(t) for the I path and q(t) for the Q path by means of a mapper. This is shown in Fig. 13.13. for QPSK modulation. The mapping table is the rule according to which the data stream data(t) is converted to modulation signals i(t) and q(t). In the case of QPSK, two bits (corresponding to bit 0 and bit 1 in the mapping table) are combined to form a dibit. For dibit combination 10, for example, the mapper outputs the signals i(t)= -1 V and q(t)= -1 V according to the mapping table shown here.

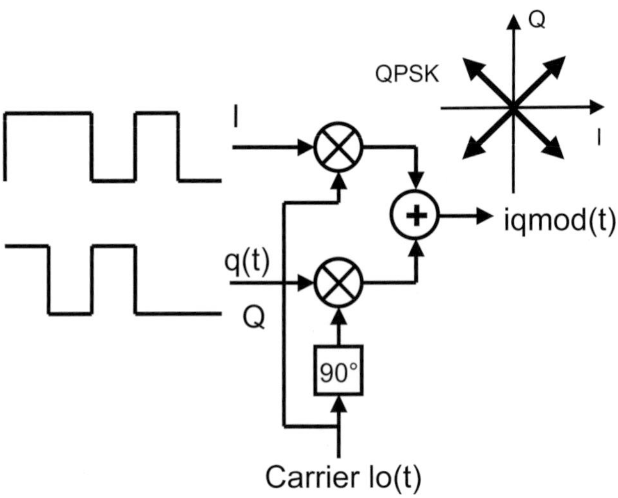

Fig. 13.11. IQ Modulator, I and Q Paths active and identical Amplitudes (QPSK)

The bit combination 11 yields i(t)=+1 V and q(t)= -1 V in this example. The allocation of bits to modulation signals, defining how the bit stream is to be read and converted by the mapper, is merely a matter of definition. It is important that the modulator and the demodulator, i.e. the mapper and the demapper, use the same mapping rules. Fig. 13.12. also shows that in this case the data rate is halved after the mapper. QPSK can transmit two bits per state. Two bits each are combined to form a dibit that determines the state of the mapper output signals i(t) and q(t). Therefore, in this case, i(t) and q(t) have half the data rate of data(t). i(t) and q(t) in turn modulate the carrier signal and, in the case of QPSK, switch it only in phase. There are four possible constellations for iqmod(t): 45°, 135°, 225° and 315°. The information is contained in the phase of the carrier. Now that we can switch the carrier phase at half the data rate relative to the input rate, the required channel bandwidth is reduced by a factor of 2. The time the carrier or vector dwells on a specific phase (dwell time = symbol duration) is referred to as symbol (Fig. 13.12. and 13.14.). The reciprocal of the symbol duration is the symbol rate. The required bandwidth corresponds to the symbol rate. Compared with simple bit transmission, the available bandwidth capacity is now boosted by a factor of 2.

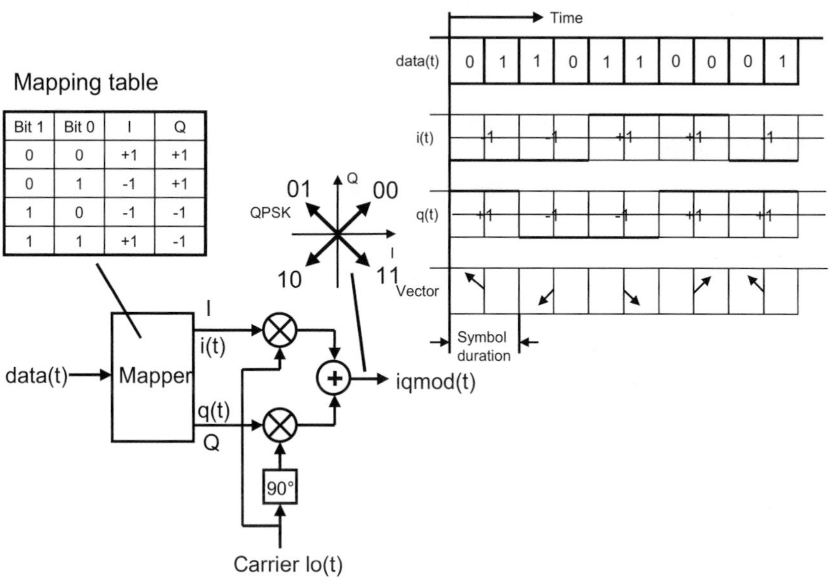

Fig. 13.12. Mapping with QPSK modulation

In practice, higher-order modulation methods are used besides QPSK. Fig. 13.11. shows 16QAM produced by varying the amplitude and phase. The information is in the amplitude, or magnitude, and in the phase. In the case of 16QAM (= 16 quadrature amplitude modulation), four bits are combined in the mapper; one carrier constellation can, therefore, carry four bits, and there are 16 possible carrier constellations. The data rate after the mapper, or the symbol rate, is a fourth of the input data rate. This means that the required channel bandwidth has been reduced by a factor of four.

In vector diagrams for IQ modulation, it is common practice to represent only the end point of the vector. A vector diagram in which all possible vector constellations are entered is referred to as a constellation diagram.

Fig 13.13. shows constellation diagrams of real QPSK, 16QAM and 64QAM signals, i.e. impaired by noise. The decision thresholds of the demapper are shown too.

The number of bits transmitted per symbol is the logarithm to the base of 2 of the constellation.

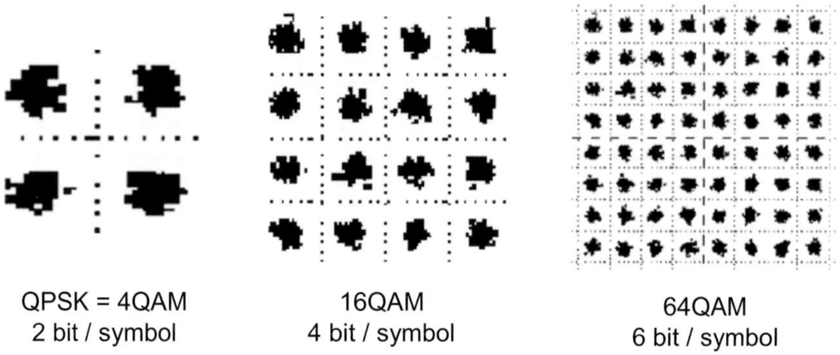

Fig. 13.13. Constellation diagrams of QPSK, 16QAM and 64QAM

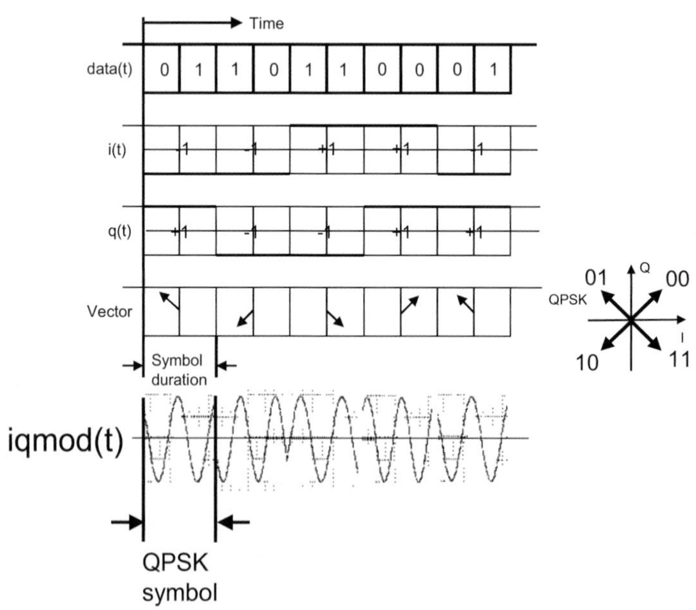

Fig. 13.14. QPSK

Fig. 13.14. shows the original data stream data(t), the resulting constellations of the carrier vector, and the switched, or keyed, carrier signal iqmod(t) in the time domain. Each switching status is referred to as a symbol. The duration of a switching status is called symbol duration. The reciprocal of the symbol duration is the symbol rate.

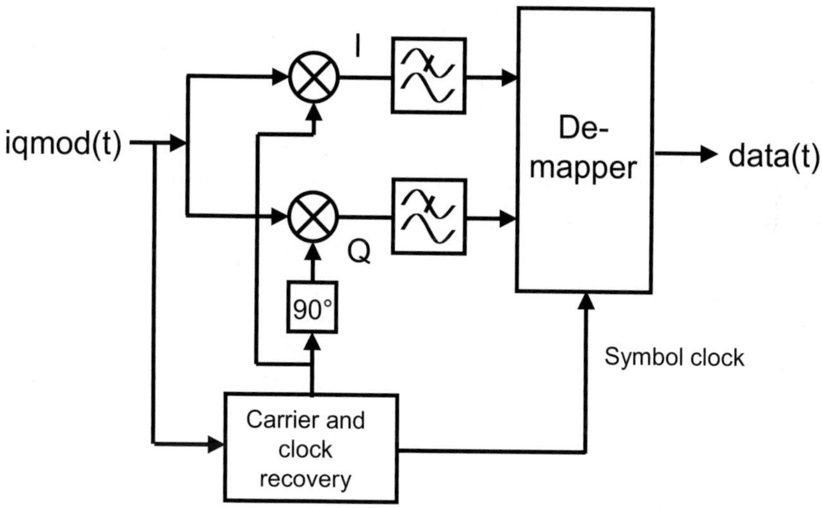

Fig. 13.15. IQ demodulator

13.5 The IQ Demodulator

In this section, IQ demodulation will be discussed briefly (s. Fig. 13.15.). The digitally modulated signal iqmod(t) is fed to the I mixer, which is driven with 0° carrier phase, and to the Q demodulator, which is driven with 90° carrier phase. At the same time, the carrier and the symbol clock are recovered in a signal processing block. To recover the carrier, the input signal iqmod(t) is squared twice. So, a spectral line at the fourfold carrier frequency can be isolated by means of a bandpass filter. A clock generator is locked to this frequency by means of a PLL. Moreover, the symbol clock has to be recovered, i.e. the point in the middle of the symbol has to be determined. Some modulation methods allow carrier recovery only with an uncertainty of multiples of 90°.

By IQ mixing, the baseband signals i(t) and q(t) are retrieved. The carrier harmonics superimposed on these signals have to be eliminated by means of a lowpass filter before the signals are applied to the demapper.

The demapper simply reverses the mapping procedure, i.e. it samples the baseband signals i(t) and q(t) at the middle of the symbol and so recovers the data stream data(t).

Fig. 13.16. illustrates the processes of IQ modulation and demodulation in the time domain and in the form of constellation diagrams for the QPSK method. The signal in the first line represents the input data stream data(t). The second and the third line show the signals i(t) and q(t) at the modulation end. The fourth and the fifth line are the voltage characteristics after the I and the Q mixer of the modulator, the sixth line the characteristic of iqmod(t). The phase steps between the symbols are clearly visible. The amplitude does not change (QPSK). In the last line, the corresponding constellation diagrams are shown. Lines 7 and 8 show the digitally recovered signals i(t) and q(t) at the demodulation end. It can be seen that, in addition to the baseband signals, the traces contain the carrier at double the frequency. The latter has to be eliminated both in the I and the Q path by means of a lowpass filter prior to demapping. In the case of analog mixing, harmonics would be superimposed in addition which would also be suppressed by the lowpass filters.

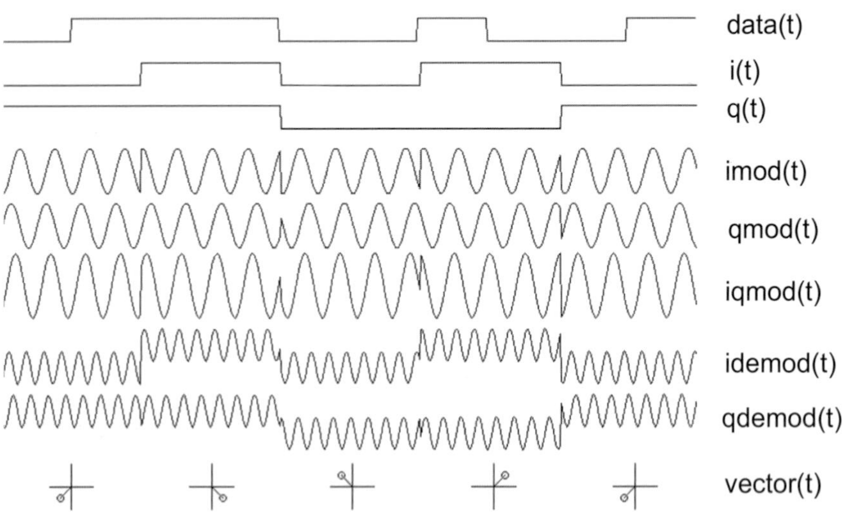

Fig. 13.16. IQ modulation and demodulation (mapping table different to examples before)

Very frequently, however, demodulation is performed using the fs/4 method, which requires a less complex IQ demodulator. The modulated signal iqmod(t) is passed through an anti-aliasing lowpass filter and then sampled by means of an A/D converter which operates at the fourfold IF of the modulated signal iqmod(t). Therefore, if the carrier of iqmod(t) is at f_{IF}, the sampling frequency is 4 • f_{IF}. This means that a complete carrier cycle is sampled four times (see Fig. 13.8.). Provided the A/D converter clock is fully synchronous with the carrier clock, the rotating carrier vector is sampled exactly at the instants shown in Fig. 13.17. The symbol clock is recovered in a carrier and clock recovery block as described above.

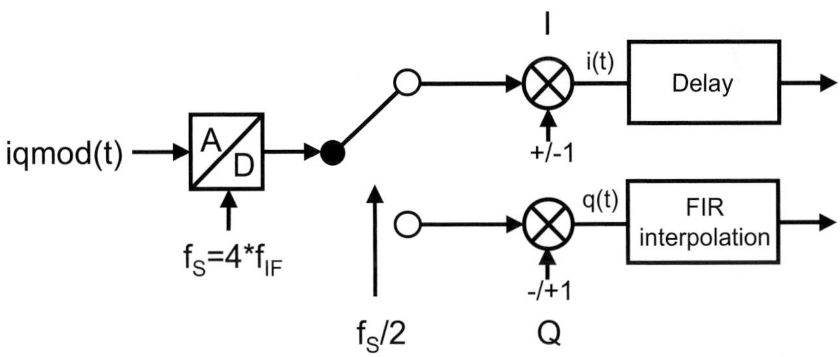

Fig. 13.17. IQ demodulation using f_S/4 method

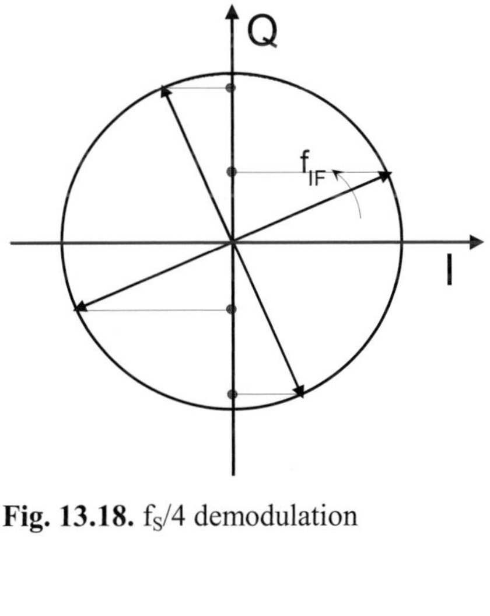

Fig. 13.18. f_S/4 demodulation

After the A/D converter, a switch splits the data stream into two streams of half the data rate. For example, the odd samples are taken to the I path and the even samples to the Q path. This means that only every second sample is taken to the I path or the Q path, respectively, thus halving the data rate in both paths. The multipliers in the two paths only reverse the sign, i.e. they multiply the samples alternately by +1 and -1.

Principle of the fs/4 method:

If the A/D converter operates exactly at the fourfold carrier frequency (IF) and the A/D converter clock and the carrier clock are fully synchronized, the samples correspond alternately to an I and a Q value. This can be seen from Fig. 13.18. Each second sample in the I and the Q path has a negative sign and so has to be multiplied by -1.

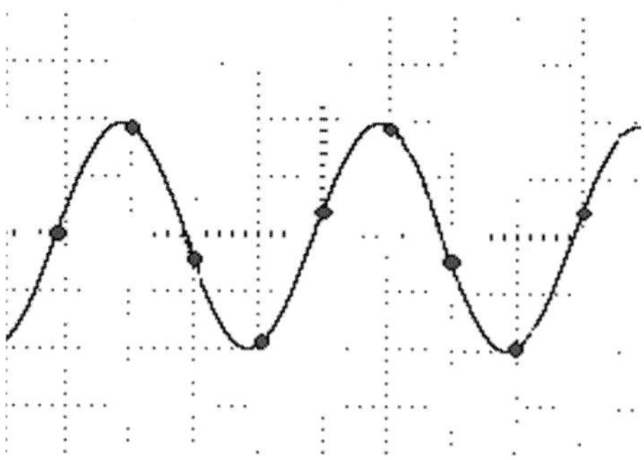

Fig. 13.19. IQ demodulation according to the $f_S/4$ method

The baseband signals i(t) and q(t) are thus recovered in a very simple way. Since the signals i(t) and q(t) have to settle after each symbol change (change of switching status), and settling is delayed by half a clock cycle by the switch after the A/D converter, the signals have to be pulled back into synchronism with the aid of digital filters.

To this effect a signal, for example q(t), is interpolated, so retrieving the sample between two values. This is done with the aid of an FIR filter (finite impulse response filter, digital filter). Each digital filter has a basic delay, however, which has to be compensated by introducing a corresponding delay in the other path, i.e. the I path in this case, by means of a delay line. After the FIR filter and the delay line, the sampled and clock-synchronous signals i(t) and q(t) are available and can be applied to the demapper.

As already mentioned, the less complex fs/4 method is frequently used in practice. In the case of OFDM-(Orthogonal Frequency Division Multiplex)-modulated signals, this circuit is provided directly ahead of the FFT signal processing block. The fs/4 demodulation method is supported by many modern digital circuits.

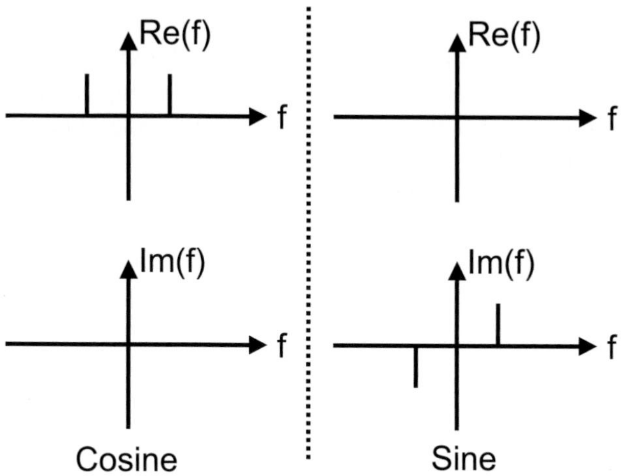

Fig. 13.20. Fourier Transform of a cosine and a sine

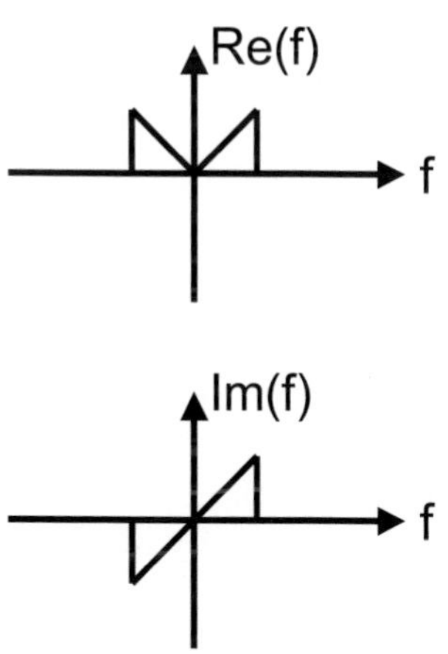

Fig. 13.21. Fourier Transform of a general real time domain signal

13.6 Use of the Hilbert transform in IQ modulation

In this section we will discuss the Hilbert Transform, which plays a major role in some digital modulation methods such as OFDM or 8VSB (c.f. ATSC, the U.S. version of digital terrestrial TV).

Let us start with sine and cosine signals. At the time t=0, the sine signal has the value 0, the cosine signal the value 1. The sine signal is shifted 90° relative to the cosine signal, i.e. it leads the cosine signal by 90°. We will see later that the sine signal is the Hilbert Transform of the cosine signal.

Based on the sine and cosine functions, we can arrive at some important definitions: the cosine function is an even function, i.e. it is symmetrical about t=0, so that cos(x) = cos(-x) applies.

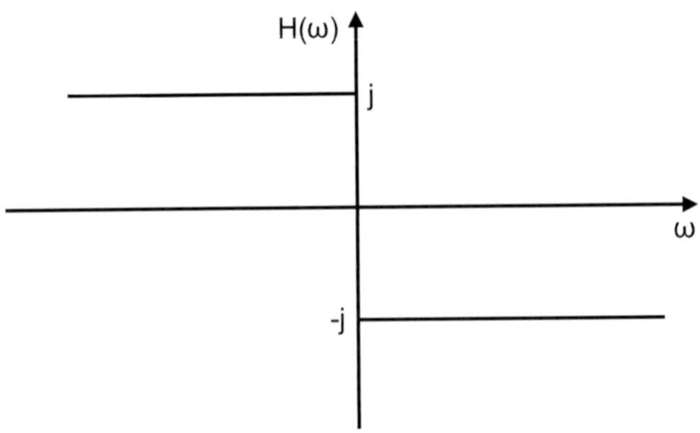

Fig. 13.22. Fourier Transform of the Hilbert Transform

The sine function, on the other hand, is an odd function, i.e. it is half-turn symmetrical about t=0, so that sin(x) = -sin(-x) applies. The spectrum of the cosine, i.e. its Fourier Transform, is purely real and symmetrical about f=0. The imaginary component is zero (s. Fig. 13.20.).

The spectrum of the sine, i.e. its Fourier transform, is purely imaginary and half-turn symmetrical (Fig. 13.20.). The real component is zero. The above facts are important for understanding the Hilbert Transform. For all real time-domain signals, the spectrum of all real components versus f (Re(f)) is symmetrical about f=0, and the spectrum of all imaginary components versus f (Im(f)) is half-turn symmetrical about f=0 (s. Fig. 13.21.).

Any real time-domain signal can be represented as a Fourier series – the superposition of the cosinusoidal and sinusoidal harmonics of the signal. The cosine functions are even and the sine functions odd. Therefore, the characteristics previously stated for a single cosine function or a single sine function also generally apply to a sum of cosine functions or a sum of sine functions. Let us now discuss the Hilbert Transform itself. Fig. 13.22. shows the transfer function of a Hilbert transformer. A Hilbert transformer is a signal processing block with special characteristics. Its main purpose is to phase shift a sine signal by 90°. This means that a cosine is converted to a sine and a sine to a minus cosine. The amplitude remains invariant under the Hilbert Transform. These characteristics apply to any type of sinusoidal signal, i.e. of any frequency, amplitude, or phase. Hence, they also apply to all the harmonics of any type of time-domain signal. This is due to the transfer function of the Hilbert transformer which is shown in Fig. 13.22. – essentially it only makes use of the symmetry characteristics of even and odd time-domain signals referred to above.

Examining the transfer function of the Hilbert transformer, we find:

- All negative frequencies are multiplied by j, all positive frequencies by -j. j is the positive, imaginary square root of -1
- The rule j • j = -1 applies
- Real spectral components, therefore, become imaginary and imaginary components become real
- Multiplication by j or -j may invert the negative or positive part of the spectrum

Applying the Hilbert Transform to a cosine signal, the following is obtained: A cosine has a purely real spectrum symmetrical about zero. If the negative half of the spectrum is multiplied by j, a purely positive imaginary spectrum is obtained for all negative frequencies. If the positive half of the spectrum is multiplied by -j, a purely negative imaginary spectrum is obtained for all frequencies above zero. The spectrum of a sine is obtained.

This applies analogously to the Hilbert Transform of a sine signal:

By multiplying the positive imaginary negative sine spectrum by j, the latter becomes negative real (j • j = -1). By multiplying the negative imaginary positive sine spectrum by -j, the latter becomes purely positive real (-j • -j = -(√-1 • √-1) = 1). The spectrum of a minus cosine is obtained.

The cosine-to-sine and sine-to-minus-cosine mapping by the Hilbert Transform also applies to all the harmonics of any type of time-domain signal.

Summarizing, the Hilbert Transform shifts the phases of all harmonics of any type of time-domain signal by 90°, i.e. it acts as a 90° phase shifter for all harmonics.

13.7 Practical Applications of the Hilbert Transform

Often, a sideband or parts of a sideband have to be suppressed during modulation. With single-sideband modulation (SSB modulation), for example, the upper or lower sideband has to be suppressed, which can be done in a variety of ways. For example, simple lowpass filtering can be used or, as is common practice in analog TV, vestigial sideband filtering. Hard lowpass filtering has the disadvantage that significant group delay distortion is produced. The latter method in any case is technically complex. For a long time, however, an alternative to single-sideband modulation has been available, this alternative being known as the phase method. A single-sideband modulator using the phase method operates as follows: the IQ modulator is fed with a modulation signal which is applied unmodified to the I path and to the Q path with 90° phase shift. A phase shift of plus or minus 90° in the Q path results in suppression of the upper or the lower sideband, respectively.

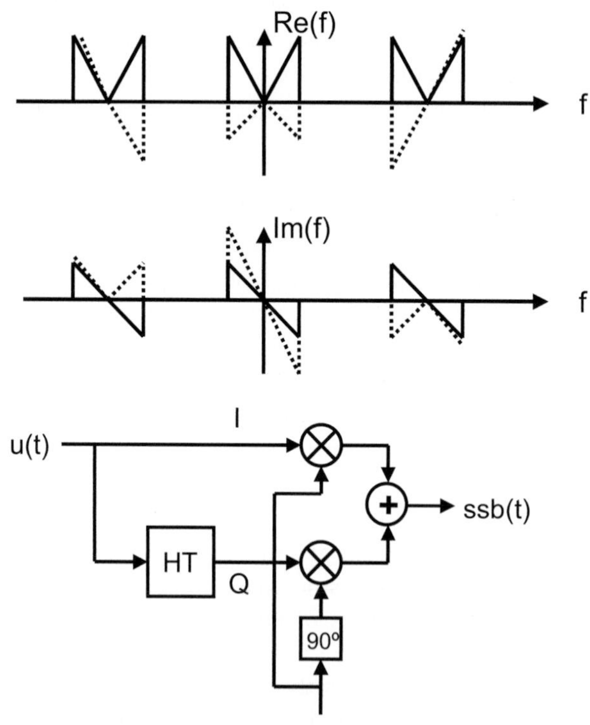

Fig. 13.23. Practical application of the Hilbert Transform to suppressing a sideband in SSB modulation

It is difficult to implement an ideal 90° phase shifter for all harmonics of a baseband signal as an analog circuit. Digital implementation is no problem – thanks to the Hilbert Transform. A Hilbert transformer is a 90° phase shifter for all components of a real time-domain signal.

Fig. 13.23. shows the suppression of a sideband by means of an IQ modulator and a Hilbert transformer. A real baseband signal is directly fed to the I path of an IQ modulator and to the Q path via a Hilbert transformer. The continuous lines at f=0 represent the spectrum of the baseband signal, the dashed lines at f=0 the spectrum of the Hilbert Transform of the baseband signal.

It can be seen clearly that, under the Hilbert Transform, the half-turn symmetrical imaginary component becomes a mirror symmetrical real component and the mirror symmetrical real component becomes a half-turn symmetrical imaginary component at the baseband.

If the unmodified baseband signal is then fed into the I path and the Hilbert Transform of the baseband signal into the imaginary path, spectra about the IQ modulator carrier like those shown in Fig. 13.22. are obtained. It can be seen that in this case the lower sideband is suppressed.

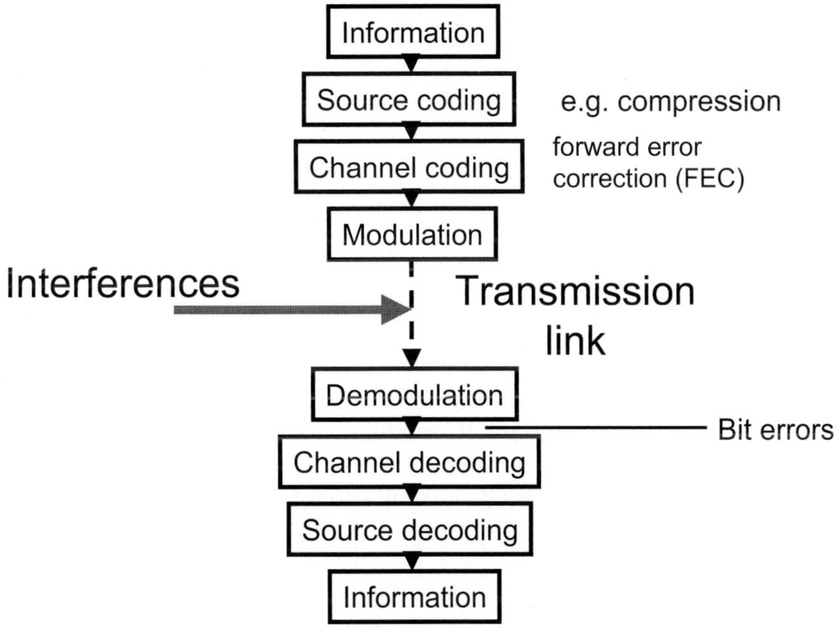

Fig. 13.24. Information transmission

13.8 Channel Coding/Forward Error Correction

In addition to the most suitable modulation method, the most appropriate error protection, i.e. channel coding, is selected from among the characteristics of the respective transmission channel. The current aim is to approach the Shannon-Limit as closely as possible. This section discusses commonly used error protection mechanisms and creates the foundations for the transmission methods in digital television.

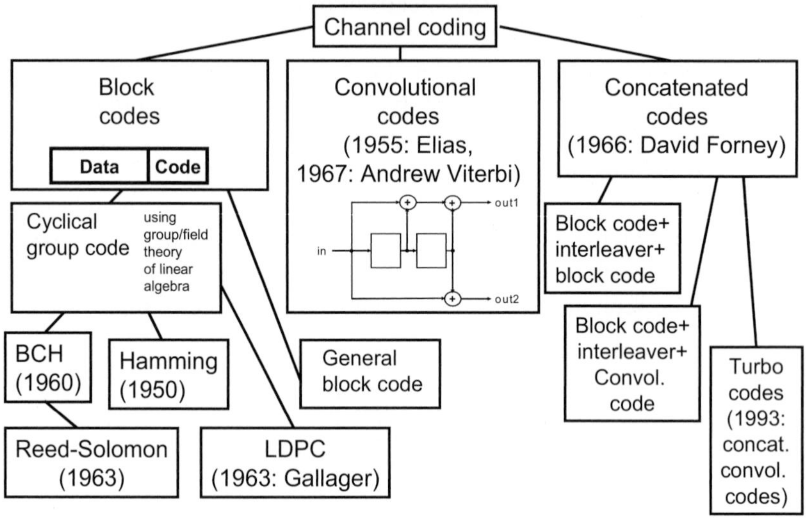

Fig. 13.25. Channel coding

Before information is transmitted, source encoding (Fig. 13.24.) is used for changing it into a form in which it can be transmitted in as little space as possible. This simply means that it is compressed as well as is possible and tolerable. After that, error protection (Fig. 13.24.) is added before the data are sent on their journey. This corresponds to channel coding. The error-protected data are then digitally modulated onto a sinusoidal carrier after which the information is sent on its way, subjected to interference such as noise, linear and nonlinear distortion, discrete and wide-band interferers, intermodulation, multipath propagation etc. Due to the varying degree of signal quality at the receiving end (Fig. 13.24.), this causes bit errors after its demodulation back to a data stream. Using the error protection added in the transmitter (FEC - Forward Error Correction), errors can then be corrected to a certain extent in the channel decoder. The bit error ratio is reduced back to a tolerable amount, or to zero. The information is then processed in such a way that it can be presented. I.e. the data are decompressed, if necessary, which corresponds to source decoding..

The "toolbox" (Fig. 13.25.) which can be used for providing error protection is not as large as one might assume. The essential basics were largely created back in 1950 – 1970. Essentially, there are block codes and convolutional codes. Block codes (Fig. 13.27.) are based on principles of linear algebra and simply protect a block of data with an error protection block. From the data to be transmitted, a type of checksum is basically calculated which can be used to find out if errors have crept in during the transmission or not, and where the errors, if any, are located. Some block codes also allow a certain number of errors to be repaired. Convolutional codes (Fig. 13.28.) delay and randomize the data stream with itself and thus introduce a certain "intelligence" into the data stream to be transmitted. The counterpart to the convolutional coder is the Viterbi-decoder developed by Andrew Viterbi in 1967.

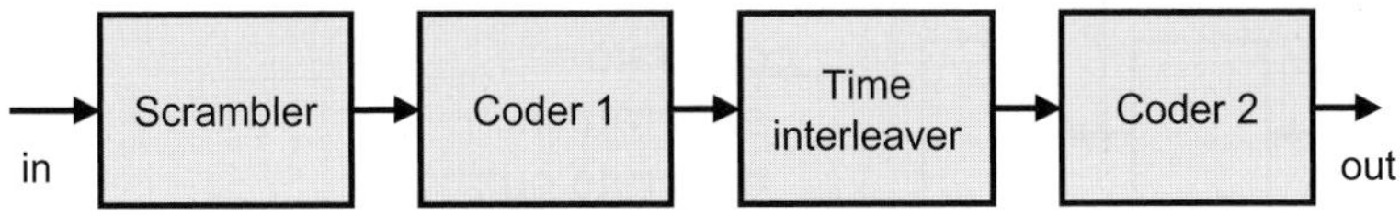

Example: DVB-S, DVB-T:
scrambler = energy dispersal,
coder 1 = Reed-Solomon,
time interleaver = Forney interleaver
coder 2 = convolutional coder

Fig. 13.26. Concatenated Forward Error Correction

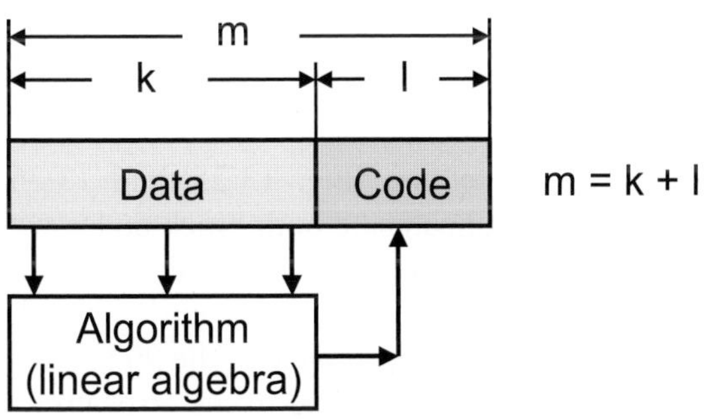

Example: DVB Reed-Solomon code:
k = 188 byte, l = 16 byte, m = 204 byte

Fig. 13.27. Block code

Before the data are supplied to the error protection section (Fig. 13.26.), however, they are first scrambled in order to bring movement into the data stream, to break up any adjoining long strings of zeroes or ones into more or less random data streams. This is done by mixing and EXOR-operations on a pseudo random binary sequence (PRBS). At the receiving end, the data stream now encrypted must be recovered by synchronous descrambling. The scrambling is followed by the first FEC. The data stream is then distributed in time by means of time interleaving. This is necessary so that during the deinterleaving at the receiving end, burst errors can be broken up into individual errors. This can be followed by a second FEC.

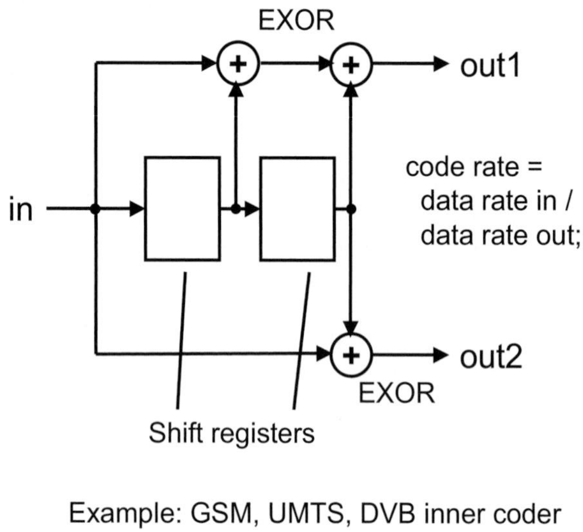

Fig. 13.28. Convolutional coding

There is also concatenated error protection (Fig. 13.25. and 13.26.) (David Forney, 1966). It is possible to concatenate both block codes with block codes and block codes with convolutional codes or also convolutional codes with convolutional codes. Concatenated convolutional codes are called turbo codes. They only made their appearance in the 90's.

It depends on the choice of modulation method and of the error protection how closely the Shannon-Limit is approached. Shannon determined the theoretical limit of the data rate in a distorted channel of a certain bandwidth. The precise formula for this is:

$$C = B \cdot \log_2(1 + \frac{S}{N});$$

If the signal/noise ratio is more than 10 dB, the following formula can also be used:

$$C[Bit/s] \approx \frac{1}{3} \cdot B[Hz] \cdot SNR[dB];$$

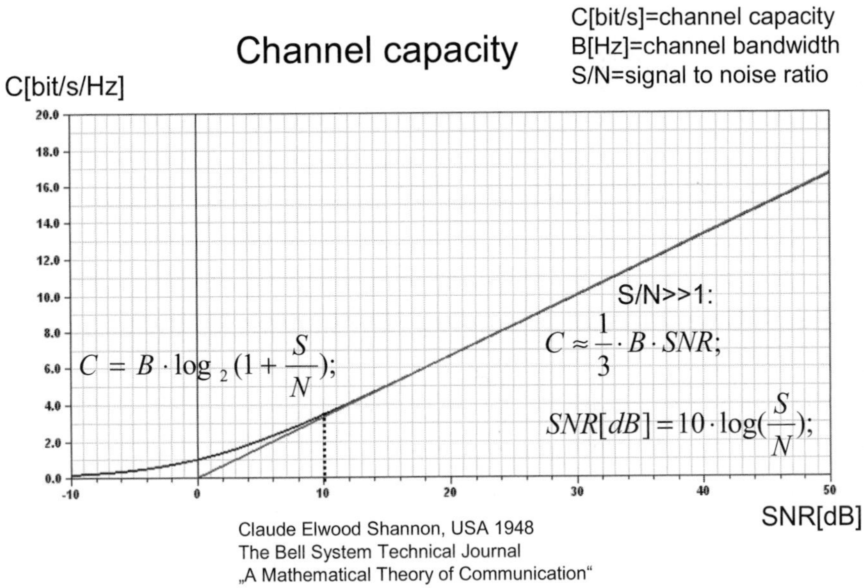

Fig. 13.29. Channel capacity

Depending on the properties of the transmission channel, a certain amount of data can be transmitted within a shorter or longer time period. The available channel bandwidth determines the maximum possible symbol rate. The signal/noise ratio present in the channel then determines the modulation method to be selected, in combination with the appropriate error protection. These relationships are illustrated by Prof. Küpfmüller's so-called information cube (Fig. 13.30.).

The FEC actually used in the transmission process will be discussed in the relevant chapter.

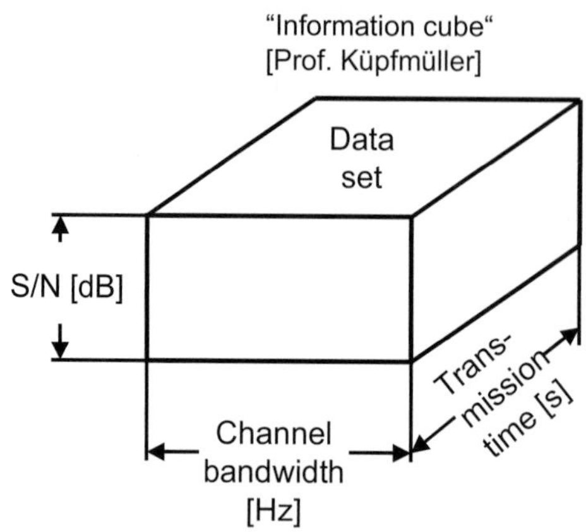

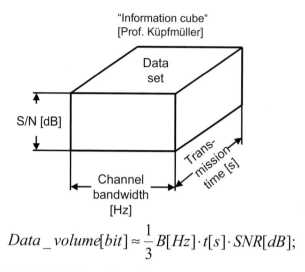

$$Data\_volume[bit] \approx \frac{1}{3} B[Hz] \cdot t[s] \cdot SNR[dB];$$

Fig. 13.30. Information cube

13.9 A Comparison to Analog Modulation Methods

In the age of digital transmission methods, it still pays to look at the traditional analog modulation methods which have been part of our lives for more than 100 years, from the beginning of carrier keying in Morse code, to AM radio and then FM radio which, due to its quality, is still giving digital audio broadcasting a run for its money. Adjacently to a frequency-modulated carrier, however, digitally modulated signals are now also being transmitted (e.g. IBOC, HD radio) which is why it makes sense to acquaint oneself again with the analog modulation methods. In this section, therefore, the special features of the traditional modulation methods of

- amplitude modulation (AM),
- frequency modulation (FM), and
- phase modulation (PM)

will be presented. The relevant experience is also quite applicable to the digital modulation methods. In addition, this chapter is intended to see to it that this traditional knowledge is not completely lost. In the second half of the 19th century, budding communication engineers were confronted with the question of how to convey messages by wire and wirelessly from one place to another. The first variant of message transmission by wire was telephony and the telegram. In telephony, the voice was converted by a carbon microphone into amplitude variations of an electrical voltage and transmitted as a pure baseband signal via two-wire lines. In the case of the telegram, a direct voltage was keyed on and off, making it possible, with the aid of the Morse alphabet, to transmit text messages from point A to point B. The Morse alphabet was thus already virtually a type of source encoding, working with redundancy reduction. Short codes were used for letters occurring frequently in the language and long codes were used for letters occurring less frequently. After the discovery of and research into electromagnetic waves, it was then a matter of applying them in the wireless transmission of messages. Baseband signals (voice, various texts) were then impressed on a sinusoidal carrier of a particular frequency and this carrier then transmitted the information from a transmitting antenna to one or more receiving antennas. The necessity of selecting a suitable transmitting frequency, i.e. carrier frequency, within the correct range of frequencies or wavelengths and modulating it with the information to be sent out arises firstly from the mere fact that electromagnetic waves will emanate from the transmitting antenna only when the wavelength $\lambda = c/f$ reaches the order of magnitude of the antenna dimensions. Depending on the frequency range selected, these messages could then be transmitted over a greater or lesser distance. But above all else, it was possible to select different carrier frequencies and thus to send out many different messages simultaneously, a principle which applies to the present day. In contrast to the past, however, the main problem today is that very many people wish to send out great amounts of information at the same time, with the resultant problem of a lack of frequencies and thus of having to control the availability of frequencies. At the beginning of communication technology it didn't matter whether an entire band was occupied or only a part of it, differently from today where the frequency resource is scarce and must be well managed. There are separate international organisations especially established for this purpose which deal with this problem.

13.9.1 Amplitude Modulation

In amplitude modulation, the information to be transmitted is impressed on the amplitude of a sinusoidal carrier (Fig. 13.31.). This type of modulation can be considered simply as multiplying a modulation signal by a carrier signal. If the carrier signal is multiplied by the value zero, the result is also zero. If the carrier signal is multiplied by a particular, information-dependent value, a particular carrier amplitude is obtained. The simplest variant of amplitude modulation is amplitude shift keying (ASK). In the original Morse-type transmission, a carrier was simply switched on and off. The original characters could be decoded again from the duration of the on- and off-periods. However, we will now consider the case of modulating a sinusoidal or cosinusoidal carrier with a a sinusoidal or cosinusoidal modulation signal. Using the cosine instead of the sine for representing the situation results in simpler addition theorems which can be used for explaining the physics. The modulation signal is described as:

$$u_{signal}(t) = U_{signal} \cdot \cos(2\pi f_{signal} t) = U_{signal} \cdot \cos(\omega_{signal} t);$$

The carrier signal is described as:

$$u_{carrier}(t) = U_{carrier} \cdot \cos(2\pi f_{carrier} t) = U_{carrier} \cdot \cos(\omega_{carrier} t);$$

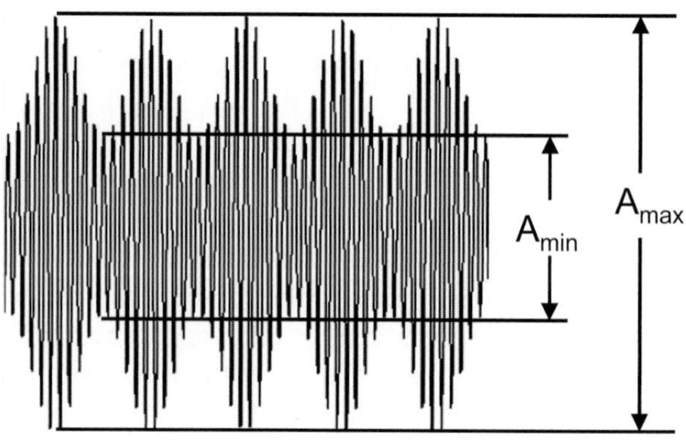

Fig. 13.31. Amplitude Modulation

If a direct voltage component of U_{DC} is added to the modulation signal $u_{signal}(t)$ and then multiplied by the carrier signal by means of a mixer (multiplier) multiplied, the following is obtained

$$u(t) = (u_{signal}(t) + U_{DC}) \cdot u_{carrier}(t) =$$
$$(U_{signal} \cdot \cos(\omega_{signal} t) + U_{DC}) \cdot U_{carrier} \cdot \cos(\omega_{carrier} t);$$

Applying the addition theorems of mathematics /geometry

$$\cos(\alpha) \cdot \cos(\beta) = \frac{1}{2}\cos(\alpha - \beta) + \frac{1}{2}\cos(\alpha + \beta);$$

the result is then:

$$u(t) = U_{DC} \cdot u_{carrier}(t) + u_{signal}(t) \cdot u_{carrier}(t);$$

or

$$u(t) = U_{DC} \cdot U_{carrier} \cos(\omega_{carrier} t)$$
$$+ \frac{1}{2} U_{signal} U_{carrier} \cos((\omega_{carrier} - \omega_{signal})t)$$
$$+ \frac{1}{2} U_{signal} U_{carrier} \cos((\omega_{carrier} + \omega_{signal})t);$$

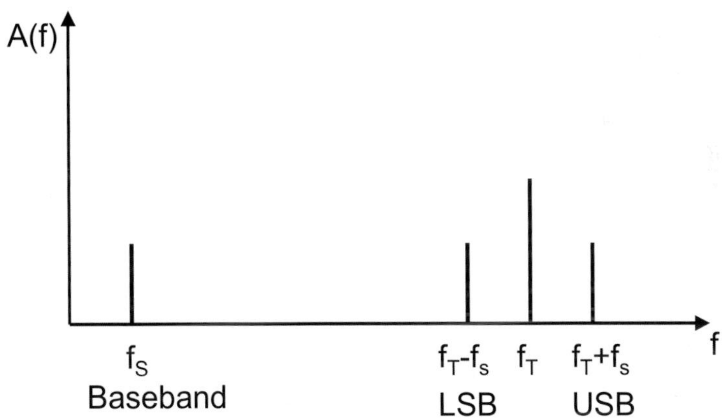

Fig. 13.32. Spectrum of the amplitude modulation (AM) with baseband signal, lower sideband (LSB) and upper sideband (USB)

Reinterpreting this, a carrier component is now produced at the center of the band, plus two sidebands - one lower than the carrier by the modulation frequency and one higher than the carrier by the modulation frequency. Setting the DC component to zero results in an amplitude modulation with suppressed carrier. Depending on the DC component added to the modulation component, a greater or lesser carrier component is produced. It can be seen that two sidebands of the respective highest modulation frequency are created (Fig. 13.32.) and that, as a result, the minimum bandwidth required at the RF domain must be greater than/equal to twice the highest modulation frequency:

$$b_{RFAM} \geq 2 \cdot f_{signal};$$

The amount of amplitude modulation of a sinusoidal or cosinusoidal carrier is determined by the modulation factor m (Fig. 13.31.). If A_{max} corresponds to the maximum amplitude of the modulated carrier and A_{min} corresponds to the minimum amplitude of the modulated carrier signal, the modulation factor m is defined as:

$$m = \frac{A_{\max} - A_{\min}}{A_{\max} + A_{\min}};$$

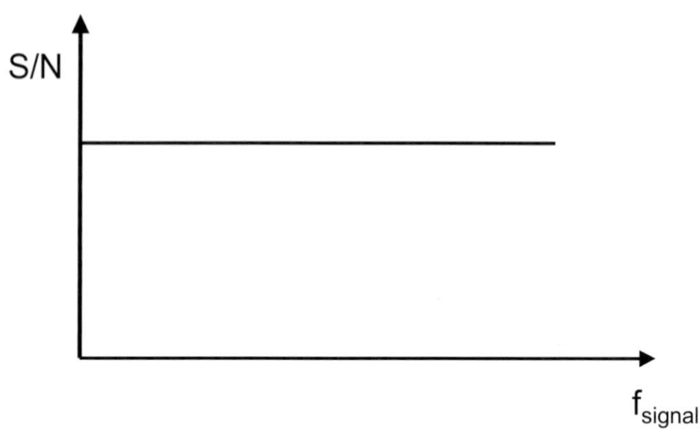

Fig. 13.33. LF signal/noise ratio in AM as a function of the baseband signal frequency

In amplitude modulation, the information to be transmitted is located in the amplitude of the modulated carrier. Nonlinearities in the transmission channel have a direct effect as amplitude distortions in the demodulated baseband signal. which means that AM systems have to be very linear. However, real transmission systems also exhibit interfering effects in the form of noise. In the case of AM, a baseband signal with superimposed additive white Gaussian noise (AWGN) is obtained after the demodulation. The noise power density is here constant and independent of the frequency of the modulation signal (Fig. 13.33.). The resultant baseband signal/noise ratio (Fig. 13.34.) directly corresponds linearly to the RF signal/noise ratio or may possibly be shifted slightly in parallel due to negative demodulation characteristics.

13.9.2 Variants of Amplitude Modulation

In amplitude modulation, various variants have become successful in practice which are:

- traditional AM with unsuppressed carrier and both sidebands,
- AM with suppressed carrier and both sidebands,
- single-sideband modulation with unsuppressed carrier,
- single-sideband modulation with suppressed carrier,
- and vestigial-sideband modulation.

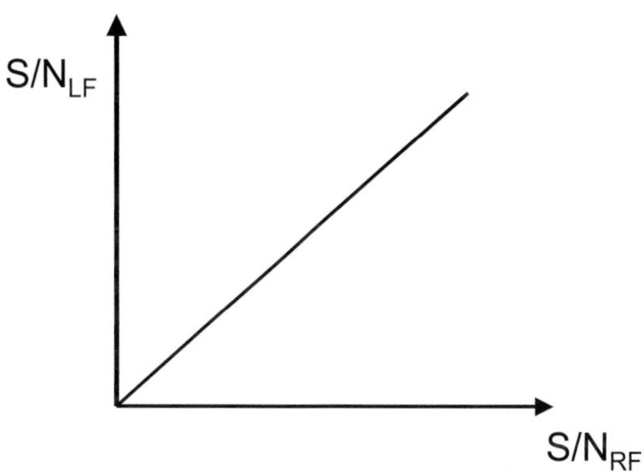

Fig. 13.34. LF signal/noise ratio of AM as a function of the RF signal/noise ratio

In single-sideband modulation, either the upper or the lower sideband is completely suppressed whereas in vestigial-sideband modulation one sideband is only partially suppressed. The practical use is simply the saving in bandwidth since the information is present completely both in the upper sideband and in the lower sideband. The relevant sideband was formerly suppressed completely or partially by analog filtering and later by applying a Hilbert transformer or 90-degree phase shifter, and an IQ modulator. In vestigial sideband modulation (VSB), the analog filters could be made less severe at the transmitting and receiving end.

13.9.3 Frequency Modulation

In frequency modulation (Fig. 13.35.), the information to be transmitted is impressed on the frequency of the carrier, i.e. the frequency of the carrier changes to a certain extent in dependence on the information to be transmitted. The simplest variant of frequency modulation is frequency shift keying (FSK). The principle of frequency modulation can be traced back to Edwin Howard Armstrong (1933) who also invented the superheterodyne receiver. The aim had been to become more insensitive to atmospheric interference. Today, frequency modulation is of great significance mainly in the field of VHF sound broadcasting. Frequency modulation (FM) is quite tolerant of nonlinearities and much more insensitive to noise-like influences.

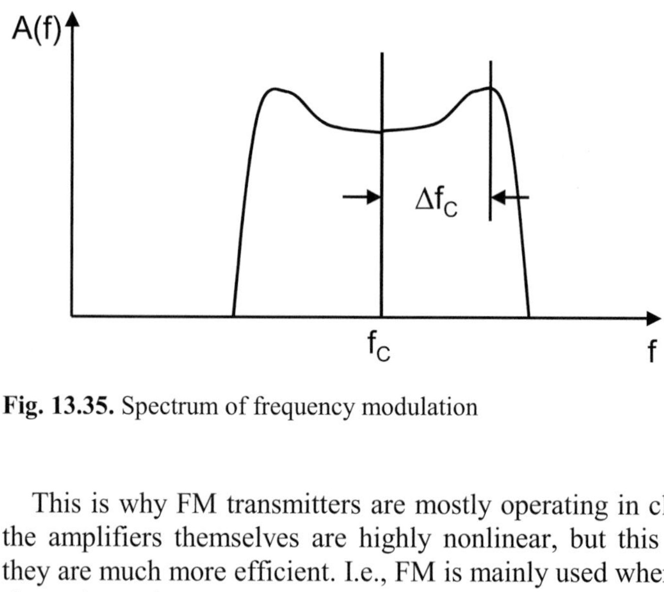

Fig. 13.35. Spectrum of frequency modulation

This is why FM transmitters are mostly operating in class C mode, i.e. the amplifiers themselves are highly nonlinear, but this also means that they are much more efficient. I.e., FM is mainly used where corresponding channel requirements are set (Low SNR, nonlinearities). In analog TV transmission via satellite, travelling tube amplifiers (TWAs), which are quite nonlinear, are used both in the earth terminal and in the satellite. Moreover, the SNR is about 10 dB due to the long distance of about 36000 km between satellite and Earth.

The frequency modulation can be expressed mathematically by:

$$u(t) = U_{carrier} \cdot \cos(2\pi f(t) \cdot t);$$

i.e. the frequency f(t) is a function of time and is influenced by the modulation factor. This involves two parameters, namely

- frequency deviation $\Delta f_{carrier}$ and
- maximum modulation frequency $f_{signal\ max}$

In frequency modulation, the modulation index is

$$M = \frac{\Delta f_{carrier}}{f_{signal\ \max}};$$

From Carson's formula (J.R. Carson, 1922), the approximate minimum FM bandwidth required at the RF level can be specified as:

$$B_{10\%} = 2(\Delta f_{carrier} + f_{signal\ \max});$$

bzw.

$$B_{1\%} = 2(\Delta f_{carrier} + 2 \cdot f_{signal\ \max});$$

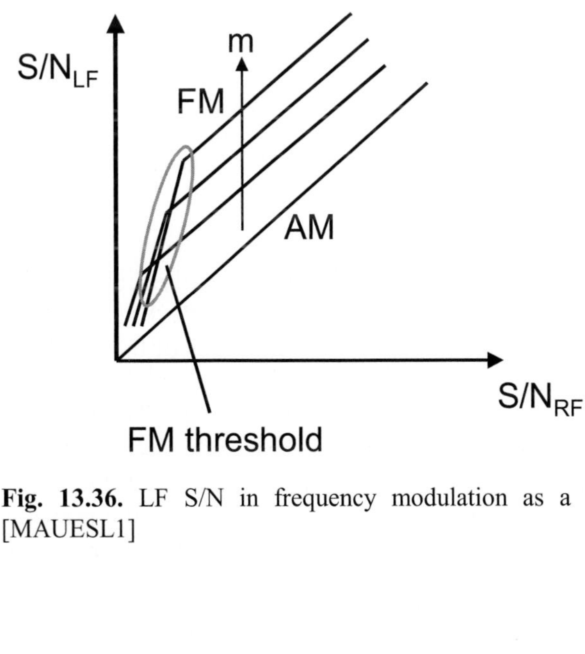

Fig. 13.36. LF S/N in frequency modulation as a function of the RF S/N [MAUESL1]

All signal components in the channel are here below 10% or 1%, respectively. The spectral lines produced can be determined by Bessel functions. Limiting the bandwidth causes nonlinear distortions in the demodulated signal. The spreading of the bandwidth in the channel results in a gain in the LF SNR with respect to the amplitude modulation above the so-called FM threshold (Fig. 13.36.). This gain can be expressed as:

$$FM_{gain\_SN\_LF} = 10 \cdot \log(3 \cdot M^3) dB;$$

This FM gain is only present above the FM threshold which is a SNR in the RF domain of about:

$$FM_{threshold} \approx 7...10dB + 10 \cdot \log(2 \cdot (M+1)) dB;$$

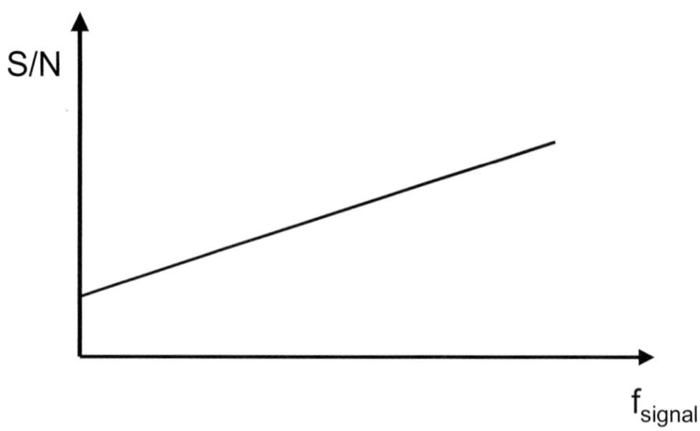

Fig. 13.37. LF signal-to-noise ratio of frequency modulation as a function of the baseband signal frequency

The FM threshold itself is defined as disproportionally high drop-off compared with the FM gain at the 1-dB point. Below the FM threshold, spike-like noise signals occur due to phase discontinuities of the carrier. In the case of the wideband FM normally used in analog television by satellite, small white splashes are then produced in the picture. The LF noise occurring with frequency modulation is called "delta noise" (Fig. 13.37.), i.e. the noise power density is not constant but increases with increasing LF bandwidth. To counteract this, preemphasis is applied at the transmitting end, i.e. higher frequencies are emphasized more. At the receiving end, deemphasis is then applied, decreasing the amplitude of the higher frequencies again in accordance with the preemphasis characteristic so that a linear frequency response is obtained again.

13.9.4 Phase Modulation

Phase modulation is closely related to frequency modulation. In phase modulation, the information to be transmitted is impressed on the phase of the carrier:

$$u(t) = U_{carrier} \cdot \cos(2\pi ft + \varphi(t));$$

Both frequency modulation and phase modulation are known collectively as angle modulation. Like frequency modulation, phase modulation is insensitive to nonlinearities. Technically, phase modulation is used mainly in frequency modulation with preemphasis. To be able to distinguish between frequency modulation and phase modulation, the following relationships must be considered: In frequency modulation, the frequency deviation $\Delta f_{carrier}$ is proportional to the amplitude of the modulating signal U_{signal}:

$$\Delta f_{carrier\_FM} \sim U_{signal};$$

The frequency deviation is not dependent on the modulating signal, i.e. is not a function of the latter:

$$\Delta f_{carrier\_FM} \neq f(f_{signal});$$

The phase deviation $\Delta\varphi_{carrier}$ in frequency modulation corresponds to the modulation index and is inversely proportional to the frequency of the modulating signal f_{signal}:

$$\Delta\varphi_{carrier\_FM} \sim \frac{1}{f_{signal}};$$

In phase modulation, the phase deviation $\Delta\varphi_{carrier}$ is proportional to the amplitude of the modulating signal U_{signal}:

$$\Delta\varphi_{carrier\_PM} \sim U_{signal};$$

The frequency deviation in phase modulation is dependent on the maximum signal frequency and proportional to the signal frequency of the modulating signal:

$$\Delta f_{carrier\_PM} \sim f_{signal};$$

The phase deviation in phase modulation is not dependent on the maximum frequency of the modulating signal, i.e. is not a function of the latter:

$$\Delta\varphi_{carrier\_PM} \neq f(f_{signal});$$

I.e., frequency modulation can be distinguished physically from phase modulation only when the frequency of the modulating signal is changing; in frequency modulation, the frequency deviation does not change then whereas in phase modulation the frequency deviation of the carrier changes in dependence on the signal frequency of the modulating signal. FM and PM can also be distinguished in the LF signal to noise ratio: in FM, the LF SNR increases with increasing signal frequency (delta noise) whereas in PM the LF SNR is not a function of the signal frequency.

13.10 Band Limiting of Modulated Carrier Signals

Modulated carrier signals must only occupy their designated channel; they must not interfere with adjacent or even more remote channels. This applies to any type of modulation, whether amplitude, phase or frequency modulation or whether analog or digital. To this end, measures are taken both on the baseband side and on the RF side. The baseband signal itself must already be band limited. On the RF side, too, precautions for protecting the adjacent channels are taken in most cases by using SAW filters at the intermediate frequency level. In addition, harmonic traps and channel-dependent mask filters are used directly in the RF path.

In the case of digital modulation, in particular, baseband filtering will still be discussed briefly at this point because this is a matter of concern in every digital TV or sound broadcasting transmission standard using single-carrier modulation. It can thus be dealt with here centrally in advance. If sinusoidal carriers are keyed by a square wave as is the case with digital modulation (amplitude and phase shift keying), this results in a multiplication of a square wave by a sinusoidal signal in the time domain and a convolution with the Fourier transform of the square wave signal with the Fourier transform of the sinusoidal signal in the frequency domain. If a single square-wave pulse from minus infinity to plus infinity were present, a continuous sin(x)/x-shaped spectrum with zeroes corresponding to the inverse of the square-wave pulse duration would be obtained (Fig. 13.38.).

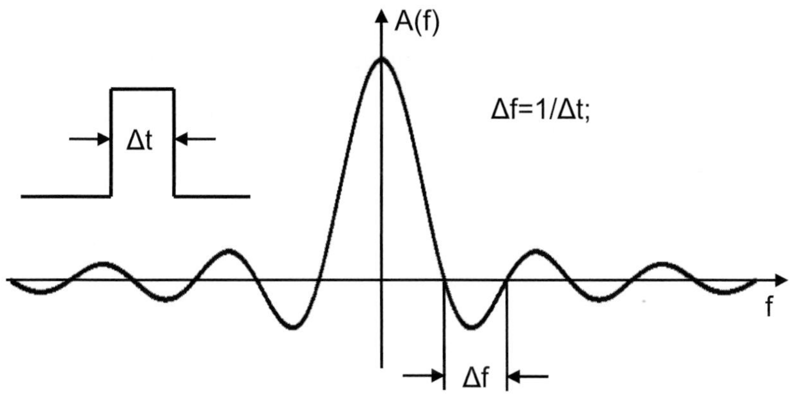

Fig. 13.38. Spectrum of a single rectangular pulse

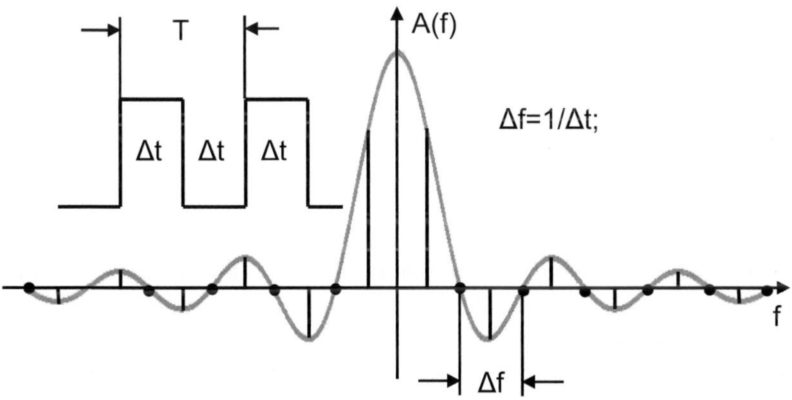

Fig. 13.39. Line spectrum of a symmetric sequence of rectangular pulses

A sequence of rectangular, or square-wave, pulses results in a line spectrum molded to the sin(x)/x function. The spectral lines occur with the spacing of the period T. If the period corresponds to exactly twice the width of the of the square-wave pulse, the line spectrum with the maximum frequency possible is obtained, providing spectral components spaced apart by $1/(2\bullet\Delta t)$ (Fig. 13.39.). If the period has a longer duration, the spectral lines move towards lower frequencies. But there will always be a periodic line spectrum of multiples of the fundamental which corresponds to the inverse of the period of the sequence of square wave pulses. If the periods are fluctuating and the square wave pulse duration is constant, the sin(x)/x form is more or less "modulated out", resulting in practice in a sin(x)/x-shaped overall spectrum in the case of digital modulation with a data signal with good energy dispersal. However, only the fundamental is needed for demodulation, i.e. all harmonics can be suppressed. This is done maximally in rectangular form (Fig. 13.40.).

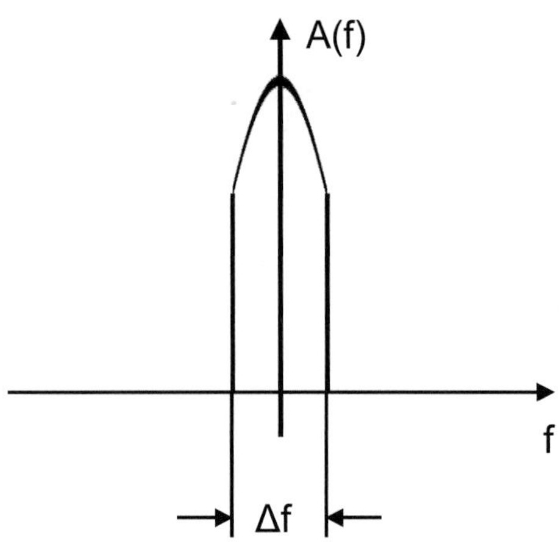

Fig. 13.40. Rectangular suppression of the harmonics (linear representation)

The control signals at the IQ modulator input i(t) and q(t) are initially square wave signals and meet the above-mentioned conditions. They must be band-limited before they are supplied to the IQ modulator. In the case of digital modulation at the baseband level, this band limiting is carried out by special "well-mannered" low-pass filters. These are constructed in most cases, and always digitally today, as root cosine square filters with a special roll-off characteristic. The roll-off factor r describes here where the filtering starts from in relation to the Nyquist bandwidth. The filter curve is symmetric and has its center at the so-called Nyquist point. The same matched filtering is performed again in the receiver, resulting in the total filter curve. The filter characteristic is designed to result in minimum overshoot in the demodulated signals i(t) and q(t). Fig.13.41. shows the resultant RF spectrum. The dashed curve corresponds to the spectrum after the modulator and the continuous curve corresponds to the spectrum after additional filtering (matched filter) in the demodulator. In the case of GSM, Gaussian filtering takes the place of the cosine filtering. In DVB-S, DVB-S2 and DVB-C, however, the cosine square or root cosine square filtering shown in Fig. 13.41, is used. On the baseband side, the spectrum to the left of the vertical axis can be imagined to consist of negative frequencies; the vertical axis corresponds to the band center of the channel on the RF side.

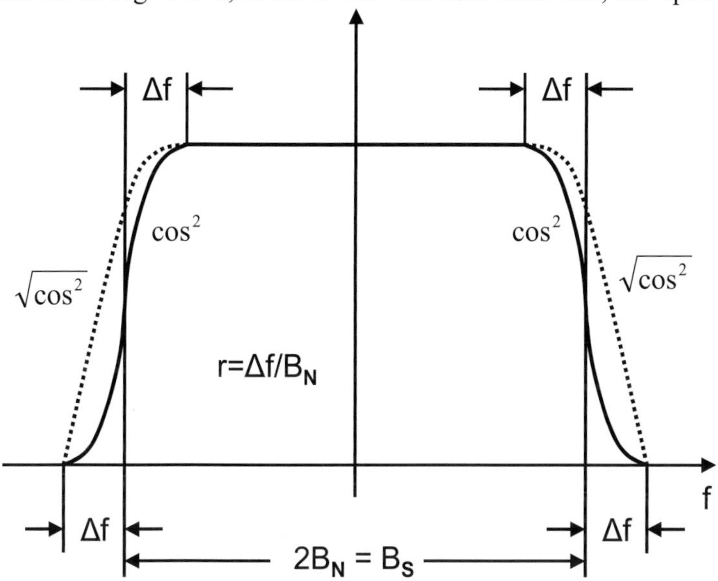

Fig. 13.41. Roll-off filtering of digitally modulated signals

13.11 Summary

In this chapter, many basic principles underlying the video and audio transmission standards have been repeated or recreated. A basic understanding of single carrier modulation (SC modulation) is also the prerequisite for an understanding of multicarrier modulation (MC modulation). Whilst single carrier modulation is the subject of many transmission standards, others employ multicarrier modulation, depending on the characteristics and requirements of the transmission channel.

Bibliography: [MAEUSL1], [BRIGHAM], [KAMMEYER], [LOCHMANN], [GIROD], [KUEPF], [REIMERS], [STEINBUCH]

14 Transmitting Digital Television Signals by Satellite - DVB-S/S2

Today, analog television signals are widely received by satellite since this type of installation has become extremely simple and inexpensive. Thus, in Europe, a simple satellite receiving system complete with dish, LNB and receiver is available for less than 100 Euros and there are no follow-up expenses. In the meantime, analog satellite reception in Europe is being replaced more and more by DVB-S - digital video broadcasting by satellite - and also DVB-S2 since 2005. In this chapter, the method of transmitting MPEG-2 source encoded TV signals via satellite is described.

Centrifugal force:

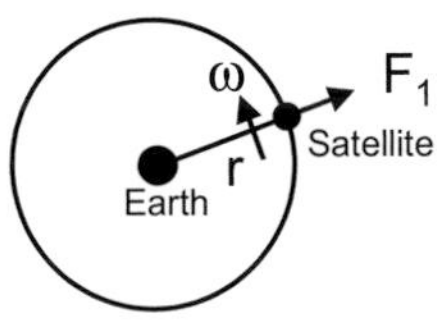

$F_1 = m_{Sat} \cdot \omega^2 \cdot r$;

m_{Sat} = mass of satellite;

$\omega = 2 \cdot \pi / T$ = angular speed;

$\pi = 3.141592654$ = circular constant;

$T = 1 \text{ day} = 24 \cdot 60 \cdot 60 \text{ s} = 86400 \text{ s}$;

Fig. 14.1. Centrifugal force of a geostationary satellite

Every communication satellite is located geostationary (Figs. 14.1., 14.2. and 14.3.) above the equator in an orbit of about 36000 km above the Earth's surface. This means that these satellites are positioned in such a way that they move around the Earth at the same speed as that with which the Earth itself is rotating, i.e. once per day. There is precisely only one orbital position, at a constant distance of about 36,000 km from the Earth's surface, where this can be achieved, the only point at which the centrifugal force of the satellite and the gravitational attraction of the Earth cancel

W. Fischer, *Digital Video and Audio Broadcasting Technology*, Signals and Communication Technology, 3rd ed., DOI: 10.1007/978-3-642-11612-4_14, © Springer-Verlag Berlin Heidelberg 2010

each other. However, the various satellites can be positioned at various degrees of longitude, that is to say angular positions above the Earth's surface. For example, ASTRA is positioned at 19.2° East. It is due to this position of the satellite above the equator that all satellite receiving antennas point to the South in the Northern hemisphere, and to the North in the Southern hemisphere.

Centripetal force:

$F_2 = \gamma \cdot m_{Eearth} \cdot m_{Sat} / r^2$;

m_{Erde} = mass of Earth ;

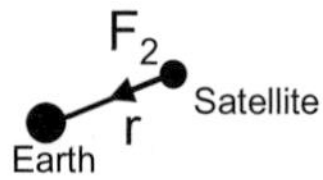

γ = constant of graphitation = $6.67 \cdot 10^{-11}\ m^3/kg\ s^2$;

Fig. 14.2. Centripetal force acting on a geostationary satellite

Balance condition:
centrifugal force = centripetal force:

$F_1 = F_2$;

$m_{Sat} \cdot \omega^2 \cdot r = \gamma \cdot m_{Earth} \cdot m_{Sat} / r^2$;

$r = (\gamma \cdot m_{Earth} / \omega^2)^{1/3}$;

$r = 42220$ km ;

$d = r - r_{Erde} = 42220\ km - 6370\ km = 35850\ km$;

Fig. 14.3. Equilibrium condition

The orbital data of a geostationary satellite can be calculated on the basis of the following relationships: The satellite is moving at a speed of one day per orbit around the Earth. This results in the following centrifugal force: The satellite is attracted by the Earth with a particular gravitational force of attraction due to its orbital height: The two forces, centrifugal force and centripetal force, must be in equilibrium. From this, it is possible to determine the orbit of a geostationary satellite (Fig. 14.1. to 14.3.).

Compared with the orbit of a space shuttle, which is about 400 km above the earth's surface, geostationary satellites are far distant from the Earth, about one tenth of the way to the moon. Geostationary satellites launched, e.g. by the space shuttle or by similar carrier systems, must first be pushed up into this distant orbit by firing auxiliary rockets (apogee motors). From there, they will never pass back into the earth's atmosphere. On the contrary, shortly before their fuel reserves for path corrections are used up they must be pushed out into the so-called "satellite cemetery" orbit which is even farther away. Only satellites close to the earth in a non-stationary orbit can be "collected" again. As a comparison - the orbital time of near-earth satellites which, in principle, also include the International Space Station ISS or the space shuttle, is about 90 minutes per orbit at about 27000 km/h.

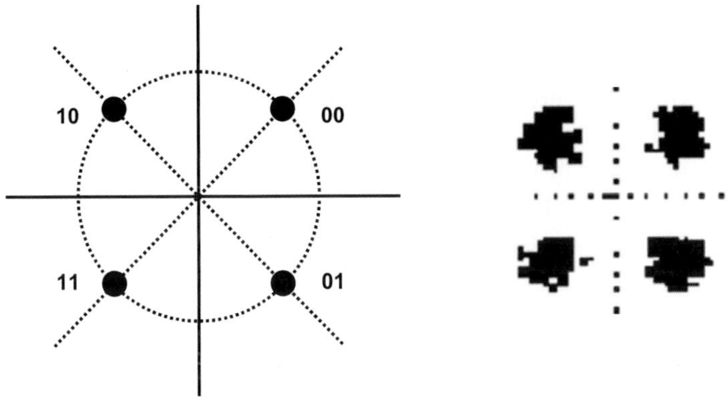

Fig. 14.4. Modulation parameters in DVB-S (QPSK, Gray coded)

But now let us return to DVB-S. In principle, the same satellite systems can be used for transmitting both analog TV signals and digital TV signals. However, in Europe, the digital signals are located in a different frequency band while the previous satellite frequency bands are still occupied with analog television. Hundreds of programs in Europe can be received both as analog signals and as digital signals via satellite and a lot of these are completely free to air.

In the following sections, the techniques for transmitting digital television via satellite will be described. This chapter also forms the basis for understanding digital terrestrial television (DVB-T). Both systems make use of the same error protection mechanisms but in DVB-T, a much more elaborate modulation method is used.

The DVB-S transmission method is defined in the ETSI Standard ETS 300421 "Digital Broadcasting Systems for Television, Sound and Data Services; Framing Structure, Channel Coding and Modulation for 11/12 GHz Satellite Services" and was adopted in 1994.

14.1 The DVB-S System Parameters

The modulation method selected for DVB-S was quadrature phase shift keying (QPSK). For some time, the use of 8PSK modulation instead of QPSK has also been considered in order to increase the data rate. In principle, satellite transmission requires a modulation method which is relatively resistent to noise and, at the same time, is capable of handling severe nonlinearities. Due to the long distance of 36000 km between the satellite and the receiving antenna, satellite transmission is subject to severe noise interference caused by the free-space attenuation of about 205 dB. The active element in a satellite transponder is a traveling wave tube amplifier (TWA) which exhibits severe nonlinearities in its modulation characteristic. These nonlinearities cannot be compensated for since this would be associated with a decrease in energy efficiency. During daylight, the solar cells provide power both to the electronics of the satellite and to the batteries. During the night, the energy for the electronics comes exclusively from the backup batteries. If there are large amounts of nonlinearity, therefore, there must not be any information content in the amplitude of a modulation signal.

Both in QPSK and in 8PSK, the information content is in the phase alone. In the satellite transmission of analog TV, too, frequency modulation was used instead of amplitude modulation for this reason.

A satellite channel of a direct broadcasting satellite usually has a width of 26 to 36 MHz (e.g. 33 MHz in ASTRA 1F, 36 MHz in EUTELSAT Hot Bird 2), the uplink is in the 14 ... 19 GHz band and the downlink is 11 ... 13 GHz. It is then necessary to select a symbol rate which produces a spectrum which is narrower than the transponder bandwidth. The symbol rate selected is, therefore, often 27.5 MS/s. As QPSK allows the transmission of 2 bits per symbol, a gross data rate of 55 Mbit/s is obtained.

gross_data_rate = 2 bits/symbol • 27.5 Megasymbols/s = 55 Mbit/s;

However, the MPEG-2 transport stream now to be sent to the satellite as QPSK-modulated signal must first be provided with error protection before being fed into the actual modulator. In DVB-S, two error protection mechanisms are used, namely a Reed-Solomon block code which is coupled with convolutional (trellis) coding. In the case of the Reed-Solomon error protection, already known from the audio CD, the data are assembled into packets of a certain length and these are provided with a special checksum of a particular length. This checksum allows not only errors to be detected but also a certain number of errors to be corrected. The number of errors which can be corrected is a direct function of the length of the checksum. In Reed-Solomon, the number of repairable errors always corresponds to exactly one half of the error protection bytes (checksum).

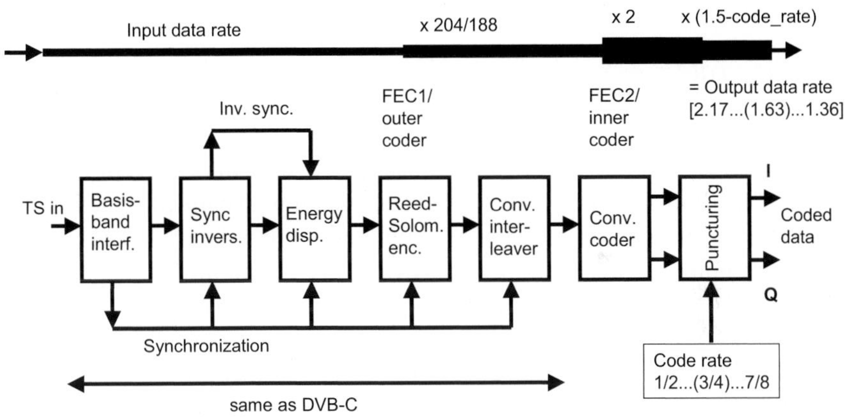

Fig. 14.5. Forward error correction (FEC) in DVB-S and DVB-T. DVB-S modulator Part 1

It is possible then to always consider exactly one transport stream packet as one data block and to protect this block with Reed Solomon error protection. An MPEG-2 transport stream packet has a length of 188 bytes. In DVB-S, it is expanded by 16 bytes Reed Solomon forward error correction to form a data packet of 204 bytes length. This is called RS (204,188) coding. At the receiving end, up to 8 errors can be corrected in this 204-byte-long packet. The position of this/these error/s is not specified. If there are more than 8 errors in a packet, this will still be reliably detected but it is no longer possible to correct these errors. The transport stream packet is then flagged as errored by means of the transport error indicator in the transport stream header. This packet must then be discarded by the

MPEG-2 decoder. The Reed Solomon forward error correction reduces the data rate:

$$\begin{aligned} \text{net\_data\_rate}_{\text{Reed-Solomon}} &= \text{gross\_data\_rate} \bullet 188/204 \\ &= 55 \text{ Mbit/s} \bullet 188/204 = \\ &= 50.69 \text{ Mbit/s;} \end{aligned}$$

However, simple error protection would not be sufficient for satellite transmission which is why further error protection in the form of convolutional coding is inserted after the Reed Solomon forward error correction. This further expands the data stream. This expansion is made controllable by means of a parameter, the code rate. The code rate describes the ratio between the input data rate and the output data rate of this second error correction block:

$$code\_rate = \frac{input\_data\_rate}{output\_data\_rate};$$

In DVB-S, the code rate can be selected within the range of 1/2, 3/4, 2/3,...7/8. If the code rate is 1/2, the data stream is expanded by a factor of 2. The error protection is now maximum and the net data rate has dropped to a minimum. A code rate of 7/8 provides only a minimum overhead but also only a minimum of error protection. The available net data rate is then at a maximum. A good compromise is usually a code rate of 3/4. The code rate can then be used to control the error protection and thus, as a reciprocal of this, also the net data rate.

The net data rate in DVB-S with a code rate of 3/4, after convolutional coding, is then:

$$\begin{aligned} \text{net\_data\_rate}_{\text{DVB-S 3/4}} &= \text{code\_rate} \bullet \text{net\_data\_rate}_{\text{Reed-Solomon}} \\ &= 3/4 \bullet 50.69 \text{ Mbit/s} \\ &= 38.01 \text{ Mbit/s;} \end{aligned}$$

14.2 The DVB-S Modulator

The following description deals with all component parts of a DVB-S modulator in detail. Since this part of the circuit is also found in a DVB-T modulator, it is recommended to read this section also in conjunction with the latter.

The first stage of a DVB-S modulator (Fig. 14.5.) is the baseband interface. This is where the signal is synchronized with the MPEG-2 transport stream. This MPEG-2 transport stream consists of packets with a constant length of 188 bytes, consisting of 4 bytes header and 184 bytes payload., the header beginning with a sync byte. This has a constant value of 0x47 and follows at constant intervals of 188 bytes. In the baseband interface, the signal is synchronized to this sync byte structure. Synchronization occurs within about 5 packets and all clock signals are derived from this.

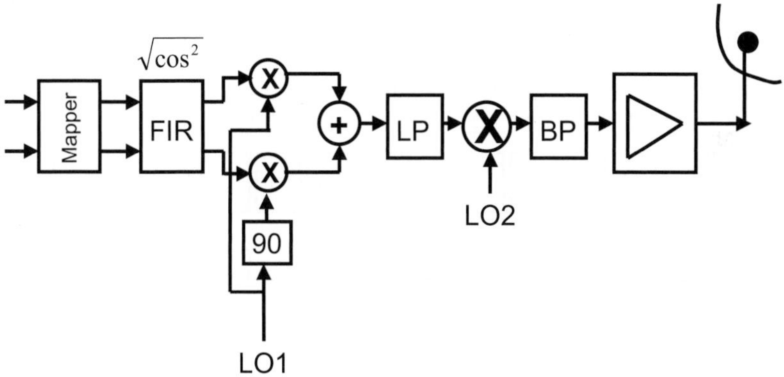

Fig. 14.6. DVB-S modulator, part 2

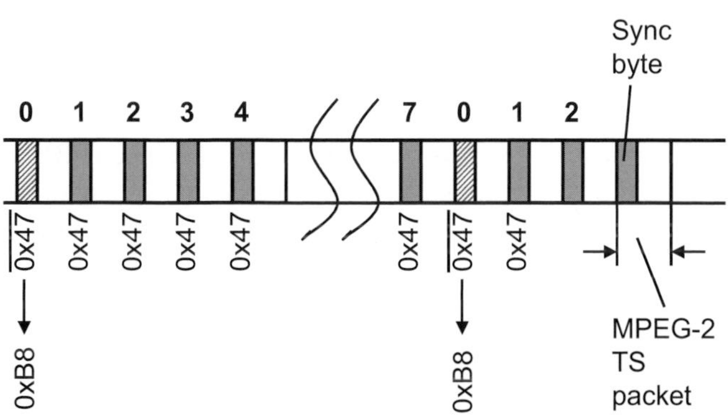

Fig. 14.7. Sync byte inversion

In the next block, the energy dispersal unit, every eighth sync byte is first inverted. I.e., 0x47 then becomes 0xB8 by bit inversion. The other 7 sync bytes between these remain unchanged. Using this sync byte inversion, additional timing stamps are then inserted into the data signal which are certain long-time stamps over 8 packets, compared with the transport stream structure. These time stamps are needed for resetting processes in the energy dispersal block at the transmitting and receiving end. This, in turn means that both the modulator or transmitter and the demodulator or receiver receives this sequence of eight packets of the sync byte inversion transparently in the transport stream and uses them to control certain processing steps. It may happen that relatively long sequences of zeroes or ones occur purely accidentally in a data signal. However, these are unwanted since they do not contain any clock information or cause discrete spectral lines over a particular period. To eliminate them, virtually every digital transmission method applies energy dispersal before the actual modulation.

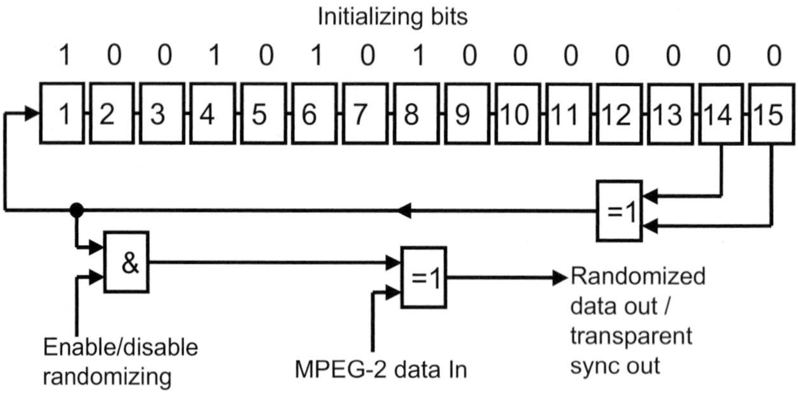

Fig. 14.8. Energy dispersal stage (randomizer)

To achieve energy dispersal, a pseudo random binary sequence (PRBS) (Fig. 14.8.) is first generated which, however, is restarted time and again in a defined way. In DVB-S, the starting and resetting takes place whenever a sync byte is inverted.

The data stream is then mixed with the pseudo random binary sequence (PRBS) by means of an Exclusive OR operation which breaks up long sequences of ones or zeroes. If this energy-dispersed data stream is mixed again with the same pseudo random binary sequence at the receiving end, the dispersal is cancelled again.

The receiving end, therefore, contains the identical circuit, consisting of a 15-stage shift register with feedback which is loaded in a defined way with a start word whenever an inverted sync byte occurs. This means that the two shift registers at the transmitting end and at the receiving end are operating completely synchronously and are synchronized by the sequence of 8 packets of the sync byte inversion block. This synchronization only becomes possible because the sync bytes and the inverted sync bytes are passed through completely transparently and are not mixed with the pseudo random bit sequence.

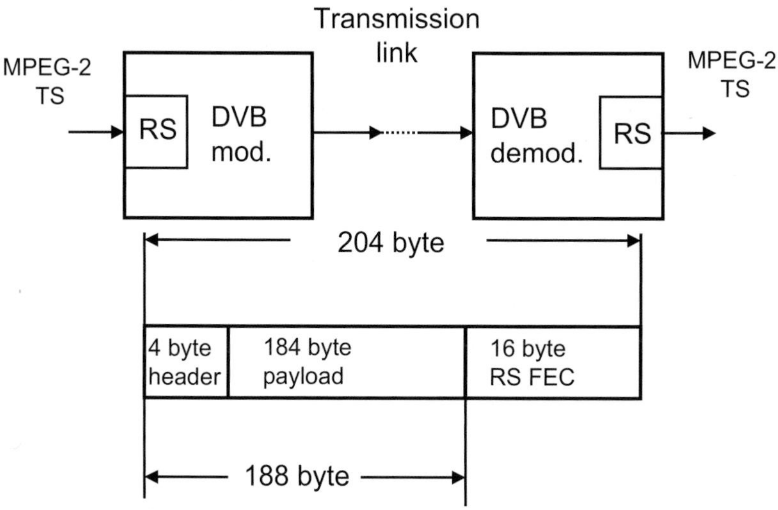

Fig. 14.9. Reed-Solomon coding

The next stage contains the outer coder (Fig. 14.5. and 14.9.), the Reed-Solomon forward error correction. At this point, 16 bytes of error protection are added to the data packets which are still 188 bytes long but are now energy-dispersed. The packets now have a length of 204 bytes which makes it possible to correct up to 8 errors at the receiving end. If there are more errors, the error protection fails and the packet is flagged as errored by the demodulator by the transport error indicator in the transport stream header being set to 'one'.

Frequently, however, burst errors occur during a transmission. If this results in more than 8 errors in a packet protected by Reed-Solomon coding, the block error protection will fail. The data are, therefore, interleaved, i.e. distributed over a certain period of time in a further operating step.

Any burst errors present are then broken up in the de-interleaving (Fig. 14.10.) at the receiving end and are distributed over a number of transport stream packets. It is then easier to correct these burst errors, which have now become single errors, and no additional data overhead is required.

In DVB-S, the interleaving is done in a so-called Forney interleaver (Fig. 14.11.) which is composed of two rotating switches and a number of shift registers. This ensures that the data are scrambled, and thus distributed, as "unsystematically" as possible. Maximum interleaving is over 11 transport stream packets. The sync bytes and inverted sync bytes always precisely follow a particular path. This means that the speed of rotation of the switches corresponds to an exact multiple of the packet length and interleaver and de-interleaver are synchronous with the MPEG-2 transport stream.

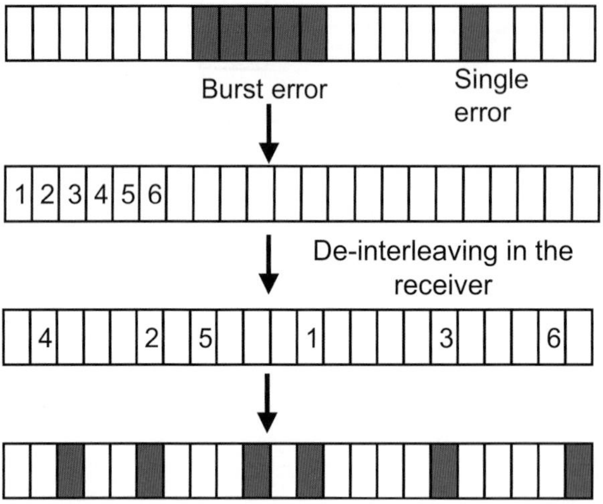

Fig. 14.10. De-interleaving

The next stage of the modulator is the convolutional coder (trellis coder). This stage represents the second, so-called inner error protection. The convolutional coder has a relatively simple structure but understanding it is not quite as simple.

The convolutional coder consists of a 6-stage shift register and two signal paths in which the input signal is mixed with the content of the shift register at certain tapping points. The input data stream is split into 3 data streams. The data first run into the shift register where they influence the upper and lower data stream of the convolutional coder by an Exclusive OR operation lasting 6 clock cycles. This disperses the information of one bit over 6 bits. At specific points both in the upper data branch and in the lower data branch there are EXOR gates which mix the data streams with the contents of the shift register. This provides two data streams at the output of the convolutional coder, each of which exhibits the same data rate as the input signal. In addition, the data stream was only provided with a particular memory extending over 6 clock cycles. The total output data rate is then twice as high as the input data rate which corresponds to a code rate = 1/2. An overhead of 100% has now been added to the data signal.

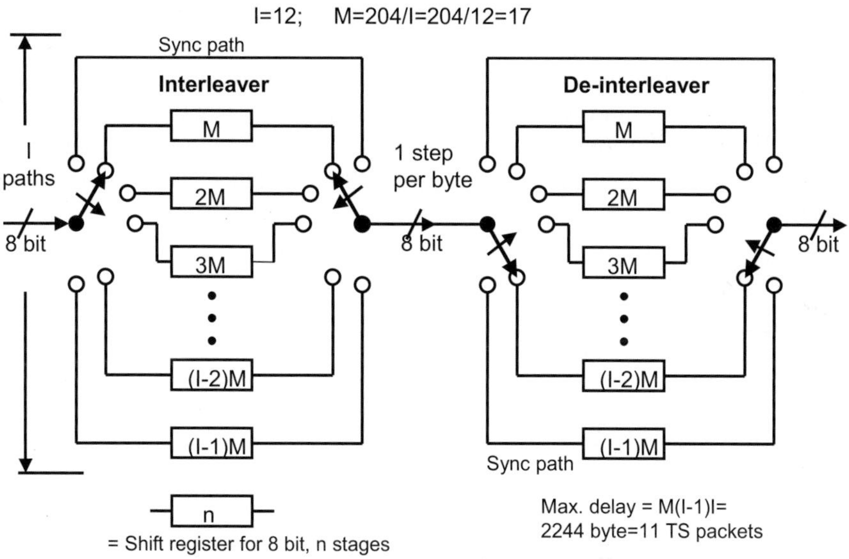

Fig. 14.11. Forney interleaver and de-interleaver

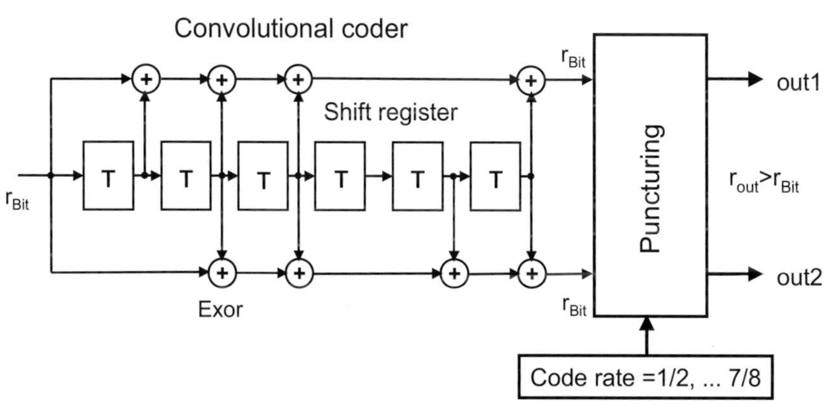

Fig. 14.12. Convolutional coder in DVB-S and DVB-T

14.3 Convolutional Coding

Each convolutional coder (Fig. 14.12.) consists of stages with more or less delay and with memory which, in practice, are implemented by using shift registers. In DVB-S, and also in DVB-T, it was decided to use a six-stage shift register with 5 taps each in the upper and lower signal path. The time-delayed bit streams taken from these taps are Exclusive-ORed with the un-delayed bit stream and thus result in two output data streams, subjected to a so-called convolution, each with the same data rate as the input data rate. A convolution occurs whenever a signal "manipulates" itself, delayed in time.

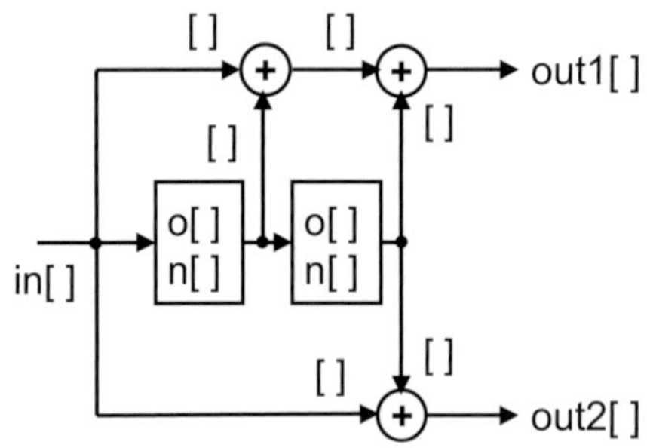

Fig. 14.13. Sample 2-stage convolutional coder

A digital filter (FIR) also performs a convolution. It would take too much time to analyse the convolutional coder used in DVB-S and DVB-T directly since, due to its six stages, it has a memory of $2^6 = 64$. Reducing it, therefore, to a sample encoder having only two stages we only need to look at $2^2 = 4$ states. The shift register can assume the internal states 00, 01, 10 and 11. To test the behaviour of the circuit arrangement it is then necessary to feed a zero and a one into the shift register for each of these 4 states and then to analyse the resulting state and also to calculate the output signals due to the Exclusive OR operations. If, e.g., a zero is fed into the shift register which has a current content of 00, the resultant new content will also be 00 since one zero is shifted out and at the same time a new zero is shifted in. In the upper signal path, the two EXOR operations produce an overall result of 0 at the output. The same applies to the lower signal path.

If a one is fed into the shift register with contents 00, the new state will be 10 and a one is obtained as output signal in the upper signal path as well as in the lower signal path. The other three states can be worked out in the same way by feeding in a one and a zero in each case. The results are shown in Fig. 14.14. The total result of the analysis can be illustrated more clearly in a state diagram (Fig. 14.15.) where the four internal states of the shift register are entered in circles.

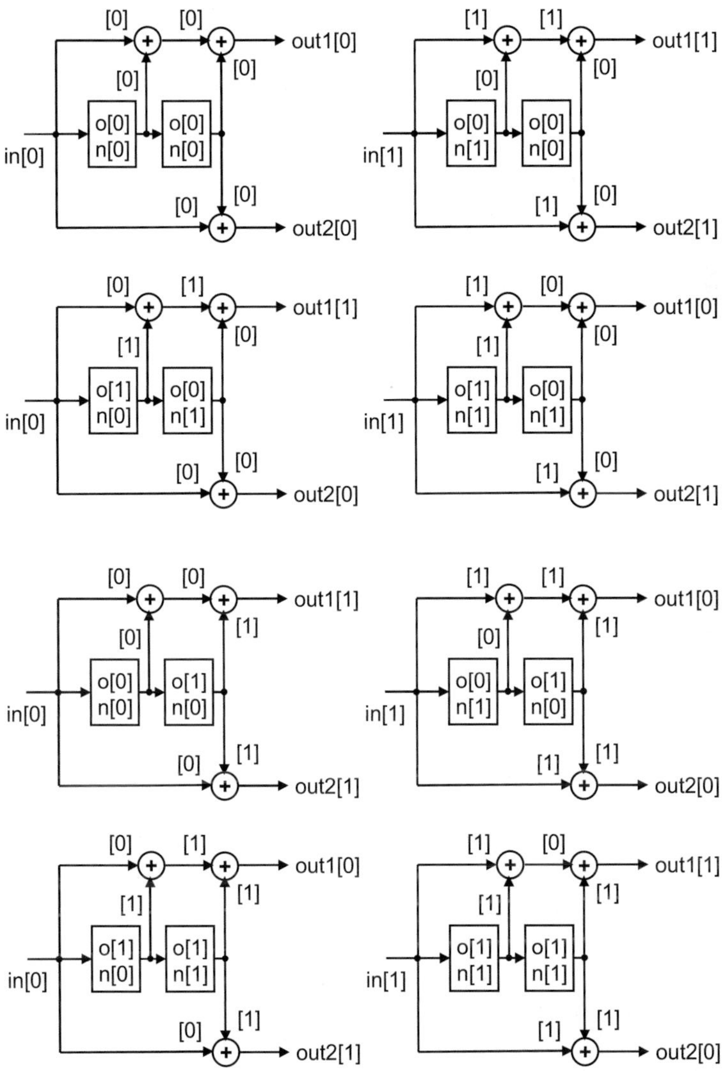

Fig. 14.14. States of the sample convolutional coder (o = old state, n = new state)

The least significant bit is entered on the right and the most significant bit on the left which means that the shift register arrangement has to be imagined upended. The arrows between these circles mark the possible state transitions. The numbers next to the circles describe the respective stimulus bit and the output bits of the arrangement, respectively. It can be seen clearly that not all transitions between the individual states are possible. Thus, it is impossible, for instance, to pass directly from 00 to 11 without first passing, e.g. through the 01 state.

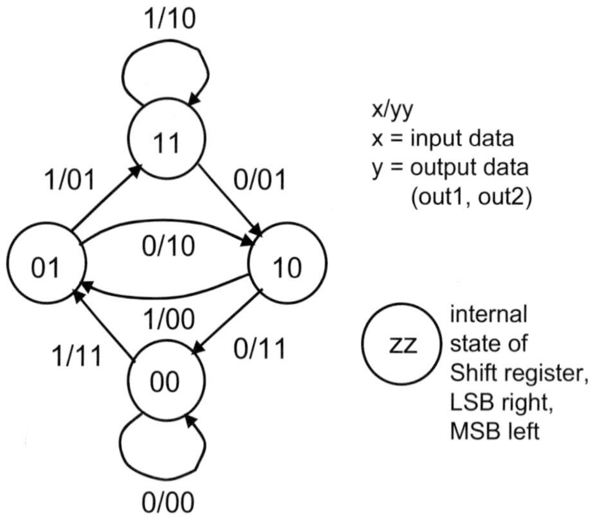

Fig. 14.15. State diagram of the sample convolutional coder

Plotting the permitted state transitions against time results in a so-called trellis diagram. Within the trellis diagram, it is only possible to move along certain paths or branches and not all paths through the trellis are possible. In many country regions, certain plants (fruit trees, wine) are planted to grow along trellises on a wall. They are thus forced to grow in an orderly way in accordance with a particular pattern by being fixed at certain points on the wall. However, it happens sometimes that such a trellis point breaks off due to bad weather, and the trellis is then in disarray. The existing pattern makes it possible, however, to find out where the branch must have been and it can thus be fixed again. The same happens with our data streams after the transmission where the convolutionally encoded data streams can be forced out of the trellis due to bit errors caused, e.g. by noise. But since the history of the data streams, i.e. their course through the trellis diagram is known, bit errors can be corrected on the basis of greatest probability by reconstructing the paths. This is precisely the principle of operation of the so-called Viterbi decoder, named after its inventor. The Viterbi decoder is virtually the counterpart of the convolutional decoder and there is, therefore, no convolutional decoder. The Viterbi decoder is also much more complex than the convolutional coder.

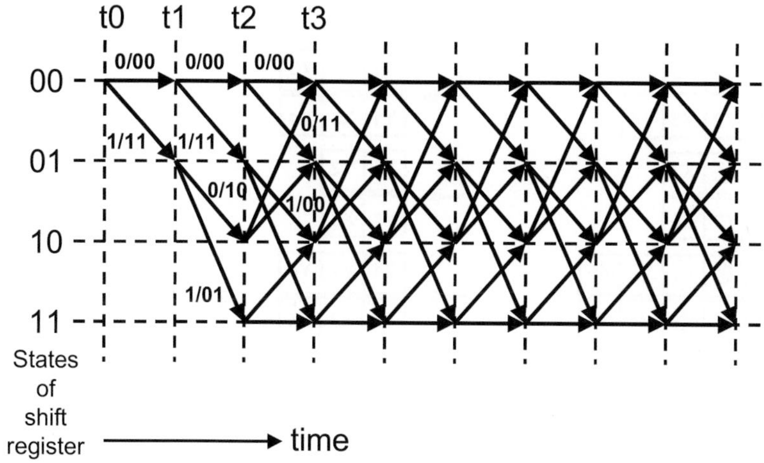

Fig. 14.16. Trellis diagram

After the convolutional coding, the data stream is now inflated by a factor of 2. For example, 10 Mbit/s have now become 20 Mbit/s but the two output data streams together now carry 100% overhead, i.e. error protection. On the other hand, this correspondingly lowers the net data rate available. This overhead, and thus also the error protection, can be controlled in the puncturing unit (Fig. 14.17.), e.g. the data rate can be lowered again by selectively omitting bits. The omitting, i.e. the puncturing, is done in accordance with an arrangement called the puncturing pattern, which is known to the transmitter and the receiver.

This makes it possible to vary the code rate between 1/2 and 7/8. 1/2 means no puncturing, i.e. maximum error protection, and 7/8 means minimum error protection and a maximum net data rate follows correspondingly. At the receiving end, punctured bits are filled up with 'Don't Care' bits and are treated like errors in the Viterbi decoder and thus reconstructed. Up to here the processing stages of DVB-S and DVB-T are 100% alike. In the case of DVB-T, the two data streams are combined to form a common data stream by alternately accessing the upper and lower punctured data stream. In DVB-S, the upper data stream and the lower data stream in each case run directly into the mapper where the two data streams are converted into the corresponding constellation of the QPSK modulation.

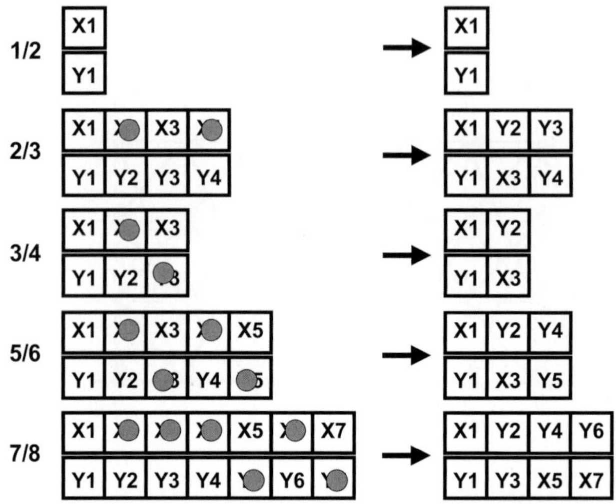

Fig. 14.17. Puncturing in DVB-S

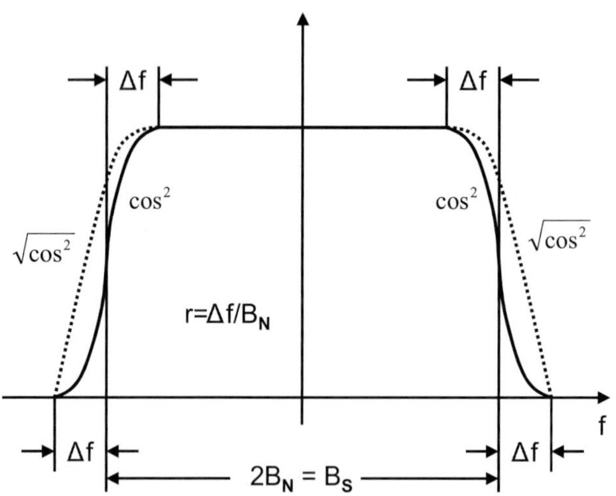

Fig. 14.18. Roll-off filtering

The mapping is followed by digital filtering so that the spectrum “rolls off” gently towards the adjacent channels. This limits the bandwidth of the signal and at the same time optimizes the eye pattern of the data signal. In

DVB-S, the roll-off filtering is carried out with a roll-off factor of $r = 0.35$. The signal rolls off with a root cosine squared shape within the frequency band. The cosine squared shape of the spectrum actually required is only produced by combining the transmitter output filter with the receiver filter because both filters exhibit root cosine squared roll-off filtering. The roll-off factor describes the slope of the roll-off filtering and is defined as $r = \Delta f/f_N$. After the roll-off filtering, the signal is QPSK modulated in the IQ modulator, upconverted to the actual satellite RF and then, after power amplification, fed to the satellite antenna. It is then uplinked to the satellite in the 14...17 GHz band.

14.4 Signal Processing in the Satellite

The geostationary direct broadcasting satellites located permanently above the equator in an orbit of about 36000 km above the Earth's surface receive the DVB-S signal coming from the uplink station and limit it first with a band-pass filter. Since the uplink distance of more than 36000 km results in a free-space loss of over 200 dB and, as a result, the useful signal is correspondingly attenuated, the uplink antenna and the receiving antenna on the satellite must exhibit corresponding gains. In the satellite, the DVB-S signal is converted to the downlink frequency in the 11...14 GHz band and then amplified by means of a TWA (Travelling Wave tube Amplifier). These amplifiers are highly nonlinear and, in practice, can also not be corrected due to the power budget in the satellite. During the day, the satellite is supplied with energy by solar cells and this energy is stored in batteries. During the night, the satellite is then supplied only from its batteries.

Before the signal is sent back to Earth, it is first filtered again in order to suppress out-of-band components. The transmitting antenna of the satellite has a certain pattern so that optimum coverage is obtained in the receiving area to be covered on the ground. This results in a so-called footprint within which the programs can be received. Because of the high free space loss of about 200 dB due to the downlink distance of more than 36000 km, the satellite transmitting antenna must exhibit a correspondingly high gain. The transmitting power is about 60 … 80 W. The signal processing unit for a satellite channel is called a transponder. Uplink and downlink are polarized, i.e. there are horizontally and vertically polarized channels. Polarization is used in order to be able to increase the number of channels.

14.5 The DVB-S Receiver

After the DVB-S signal coming from the satellite has again travelled along its path of 36000 km and, therefore, been attenuated correspondingly by 200 dB and its power has been reduced further by atmospheric conditions such as rain or snow, it arrives at the satellite receiving antenna and is focussed at the focal point of the dish. This is the precise point at which the low noise block (LNB) is mounted. The LNB contains a waveguide with a detector each for the horizontal and vertical polarization. Depending on which plane of polarization has been selected it is either the signal from the horizontal detector or that from the vertical detector which is switched through. The plane of polarization is selected by selection of the amplitude of the supply voltage to the LNB (14/18V). The received signal is then amplified in a low-noise gallium arsenide amplifier and is then down-converted to the first satellite IF in the 900...2100 MHz band.

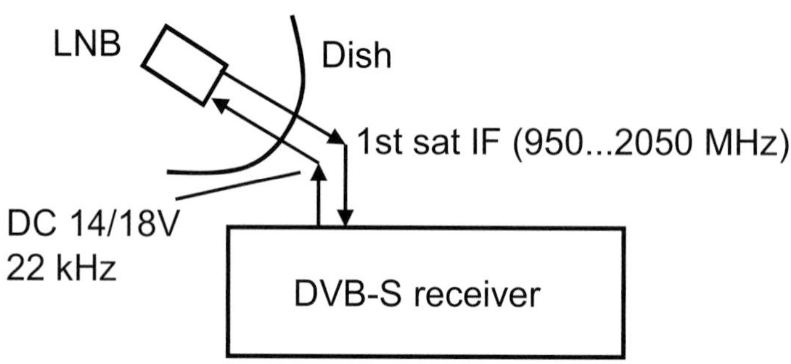

Fig. 14.19. Satellite receiver with LNB and receiver

Modern "universal" LNBs (suitable for receiving digital TV) contain two local oscillators which output a carrier at 9.6 GHz and at 10.6 GHz and the received signal is down-converted by being mixed either with the 9.75 GHz or with the 10.6 GHz depending on whether the received channel is in the upper or lower satellite frequency band. DVB-S channels are usually in the upper band and the 10.6 GHz oscillator is then used.

The phrase "suitable for receiving digital TV" only refers to the presence of a 10.6 GHz oscillator and is thus misleading. The LNB is switched between 9.75 and 10.6 GHz by means of a 22 kHz switching voltage which is superimposed on the LNB supply voltage or not. The LNB is supplied via the coaxial cable which distributes the satellite intermediate frequency in the 900...2100 MHz band now output. During installation work, care should be taken, therefore, to deactivate the satellite receiver since otherwise a possible short circuit could damage the voltage supply for the LNB.

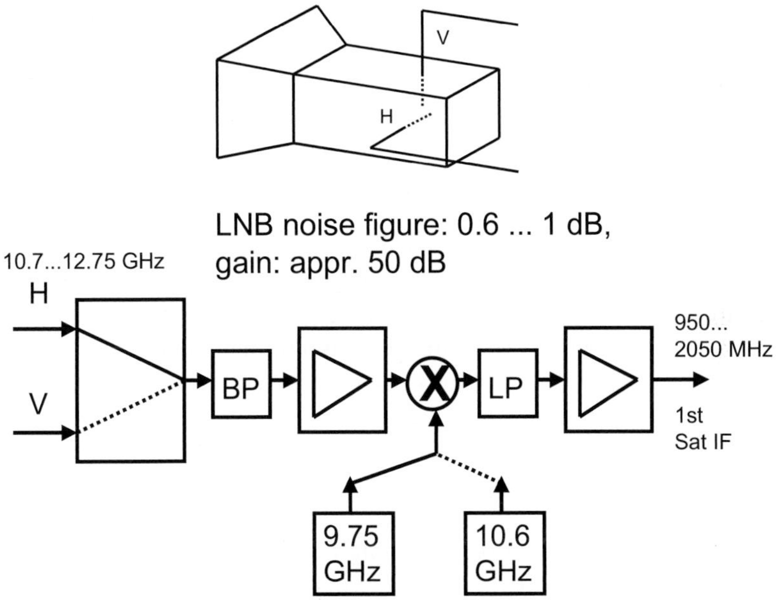

Fig. 14.20. Outdoor unit - LNB

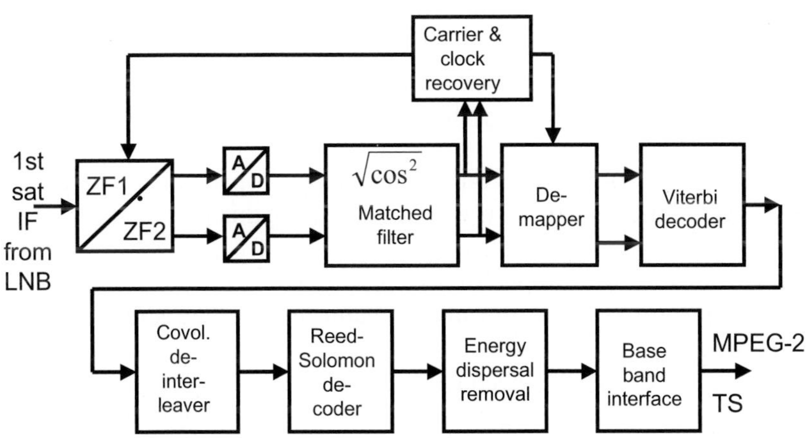

Fig. 14.21. DVB-S receiver (without MPEG-2 decoder)

In the DVB-S receiver, the so-called DVB-S set-top box or decoder, the signal undergoes a second down-conversion to a second satellite IF. This down-conversion is performed with the aid of an IQ mixer which is fed by an oscillator controlled by the carrier recovery circuit. After the IQ conversion, analog I and Q signals are again available. The I and Q signals are then A/D converted and supplied to a matched filter in which the same root cosine squared filtering process as at the transmitting end takes place with a roll-off factor of 0.35. Together with the transmitter filter, this then results in the actual cosine squared roll-off filtering of the DVB-S signal. The filtering process must be matched with respect to the roll-off factor at the transmitting end and at the receiving end.

After the matched filter, the carrier and clock recovery circuit and the demapper tap off their input signals. The demapper again generates a data stream from which the first errors are removed in the Viterbi decoder. The Viterbi decoder is the counterpart of the convolutional coder. The Viterbi decoder must have knowledge of the code rate currently used. The decoder must be informed of this code rate (1/2...2/3...7/8) by operator intervention.

The Viterbi decoder is followed by the convolutional de-interleaving where any burst errors are broken up into individual errors. The bit errors still present then are corrected in the Reed Solomon decoder. The transport stream packets, which had an original length of 188 bytes, had been provided with 16 bytes error protection at the transmitting end. These can be used at the receiving end for correcting up to 8 errors in the packet which now has a length of 204 bytes. Burst errors, i.e. multiple errors in a packet, should have been broken up by the preceding deinterleaving process. However, if an error-protected TS packet with a length of 204 bytes contains more than 8 errors, the error protection will fail. The Transport Error Indicator in the transport stream header is then set to "1" to flag this packet as errored. The packet length is now 188 bytes again. TS packets flagged as errored must not be used by the MPEG-2 decoder and error concealment must be applied.

After the Reed Solomon decoding, the energy dispersal is removed and the inversion of the sync bytes is cancelled. During this process, the energy dispersal unit is synchronized by this sequence of 8 packets of the sync byte inversion. At the output of the following baseband interface, the MPEG-2 transport stream is available again and is then supplied to an MPEG-2 decoder.

Today, the entire DVB-S decoder after the A/D converters is located on one chip which, in turn, is usually integrated in the satellite tuner. I.e., the tuner, which is controlled via the I2C bus, has an F connector input for the signal from the LNB and a parallel transport stream output.

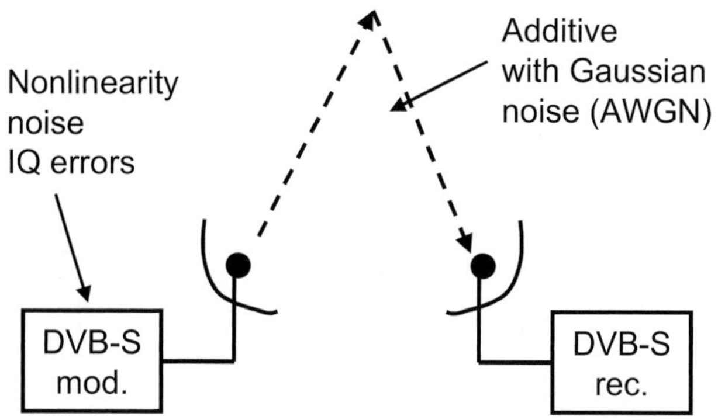

Fig. 14.22. Influences affecting the satellite transmission link

14.6 Influences Affecting the Satellite Transmission Link

This section deals with the influences to be expected on the satellite transmission link (Fig. 14.22.) and it will be seen that these influences are mainly restricted to noise. However, let us first begin with the modulator. This can be assumed to be ideal up to the IQ modulator. The IQ modulator can exhibit different gains in the I and Q branches, a phase error in the 90° phase shifter and a lack of carrier suppression. There can also be noise effects and phase jitter coming from this circuit section. These problems can be ignored, however, because of the rugged nature of the QPSK modulation and will normally never reach an order of magnitude which will noticeably affect the signal quality. In the satellite, the travelling wave tube generates severe nonlinearities but these do not play a part, in practice. In the region of the uplink and the downlink, however, where the DVB-S signal is attenuated severely by more than 200 dB due to the distance of 36000 km each way travelled by the signal, strong noise effects are experienced. It is these noise effects, the additive white gaussian noise (AWGN) becoming superimposed on the signal, which form the only influence to be discussed.

In the part following, the satellite downlink will be analysed by way of an example with respect to the signal attenuation and the resultant noise effects.

The minimum carrier/noise ratios (C/N) necessary and the channel bit error rate needed are known and predetermined from forward error correction (FEC, Reed-Solomon and convolutional coding) (Fig. 14.23.).

To gain an idea about the C/N to be expected, the levels on the satellite downlink will now be considered.

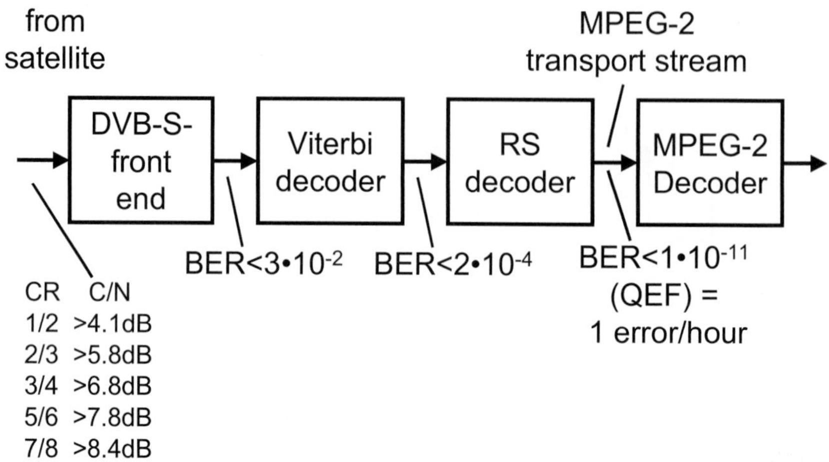

Fig. 14.23. Minimum carrier/noise ratios necessary at the receiving end and bit error ratios

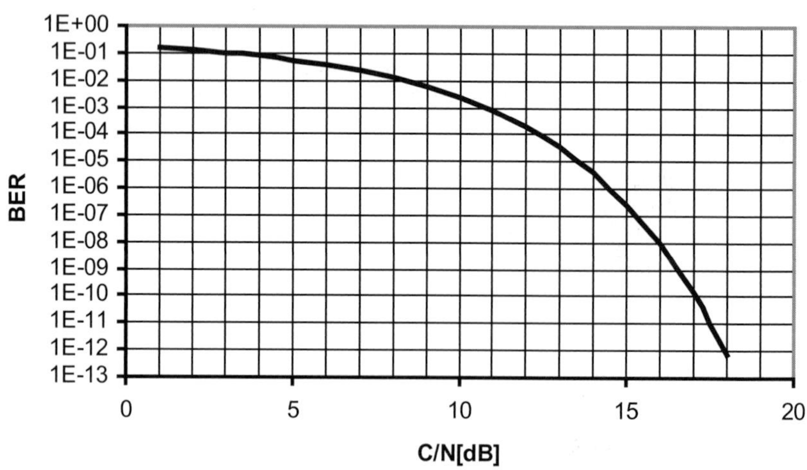

Fig. 14.24. Channel bit error ratio in DVB-S as a function of C/N

A geostationary satellite is "parked" in an orbit of 35800 km above the equator. This is the only orbit in which it can travel around the Earth synchronously. At 45° latitude, the distance from the Earth's surface is then

d = Earth's radius • sin(45°) + 35800 km = 6378 km • sin(45°) + 35800 km = 37938 km;

Transmitted power (e.g. Astra 1F):

Assumed transponder output power: 82 W =	19 dBW
Gain of the transmitting antenna	33 dB
Satellite EIRP (equivalent isotropic radiated power)	52 dBW

Free space attenuation:

Satellite-Earth distance = 37,938 km	91.6 dB
Transmitting frequency = 12.1 GHz	21.7 dB
Loss constant	92.4 dB
Free space attenuation	205.7 dB

Received power:

Satellite EIRP	52.0 dBW
Free space attenuation	205.7 dB
Clear sky attenuation	0.3 dB
Receiver directional error	0.5 dB
Polarisation error	0.2 dB
Received power at the antenna	-154.7 dBW
Antenna gain	37 dB
Received power	-117.7 dBW

Noise power at the receiver:

Boltzmann's constant	-228.6 dBW/K/Hz
Bandwidth = 33 MHz	74.4 dB
Temperature 20 °C = 273K+20K = 293K	24.7 dB
Noise figure of the LNB	1.0 dB
Noise power	-128.5 dBW

Carrier/noise ratio C/N:

Received power C	-117.7 dBW
Noise power N	-128.5 dBW
C/N	**10.8 dB**

Thus, a C/N of about 10 dB can be expected in the example. Actual C/N values can be expected between 9 … 12 dB.

The following equations form the basis for the C/N calculation:

Free space attenuation:

L[dB] = 92.4 + 20•log(f/GHz) + 20•log(d/km);
f = transmission frequency in GHz;
d = Transmitter-receiver distance in km;

Antenna gain of a parabolic antenna:

G[dB] = 20 + 20•log(D/m) + 20•log(f/GHz);
D = antenna diameter in m;
f = transmission frequency in GHz;

Noise power at the receiver input:
N[dBW] = -228.6 + 10•log(b/Hz) + 10•log((T/^{0}C +273)) + F;
B = bandwidth in Hz;
T = temperature in ^{0}C;
F = noise figure of the receiver in dB.

Fig. 14.23. shows the minimum C/N ratios as a function of the code rate used. In addition, the pre-Viterbi, post-Viterbi (= pre-Reed-Solomon) and post-Reed-Solomon bit error rates are plotted. A frequently used code rate is 3/4. With a mimum C/N ratio of 6.8 dB, this results in a pre-Viterbi channel bit error rate of 3^{-2}. The post-Viterbi bit error rate is then 2^{-4} which correponds to the limit at which the subsequent Reed-Solomon decoder still delivers an output bit error rate of 1^{-11} or better. This approximately corresponds to one error per day and is defined as quasi error-free (QEF). At the same time, these conditions also almost correspond to the "fall off the cliff" (or "brickwall effect"). Slightly more noise and the transmission breaks down abruptly.

In the calculated example of the C/N to be expected on the satellite transmission link, there is, therefore, still a margin of about 3 dB available with a code rate of 3/4. The precise relationship between the channel bit error rate, i.e. the pre-Viterbi bit error rate, and the C/N ratio is shown in Fig. 14.24.

14.7 DVB-S2

DVB-S was adopted in 1994, using QPSK as a modulation method and a concatenated error protection system of Reed-Solomon FEC and convolution coding. In 1997, the DVB DSNG standard [ETS301210] was laid down which was created for reporting purposes (DSNG = Digital Satellite News Gathering). Live signals are transmitted by satellite, e.g. from outside broadcast vans at big public events to the studios. DVB DSNG already uses 8PSK and 16QAM. In 2003, new methods were defined, both for direct broadcasting and for professional applications, as "DVB-S2" (s. Fig. 14.25.) in ETSI document [ETS302307] .

Both QPSK, 8PSK (uniform and non-uniform) and 16APSK (16 amplitude phase shift keying) were provided as modulation methods, the latter only being used in the professional field (DSNG). The error protection used is completely new, e.g. LDPC (low density parity check). The standard is quite open for broadcasting, interactive services and DSNG.

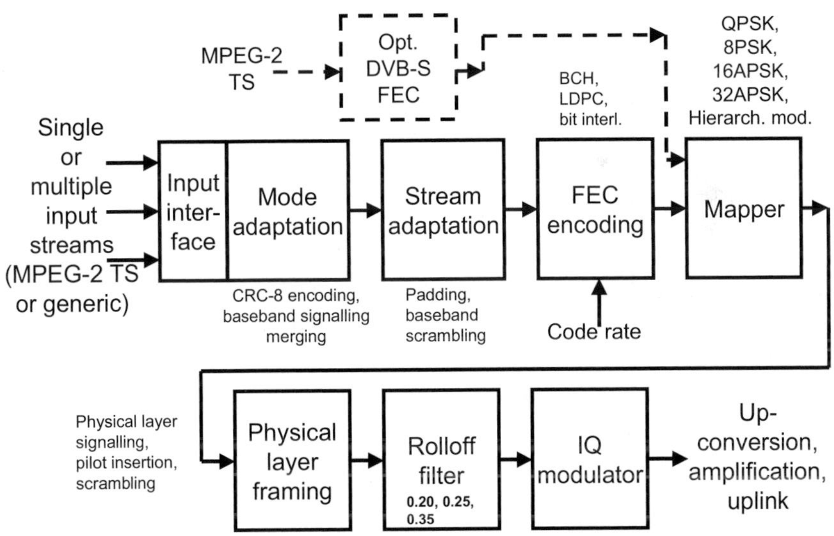

Fig. 14.25. Block diagram of a DVB-S2 modulator

Data streams not conforming to the MPEG-2 transport stream can also be transmitted and it is possible to transmit either one or a number of transport streams. This also applies to generic data streams which can also be divided into packets. Fig. 14.25. shows the block diagram of a DVB-S2 modulator. At the input interface, the data stream or streams appear in the form of an MPEG-2 transport stream or of generic data streams. Following the mode and stream adaptation blocks, the data are fed to the FEC encoding block.

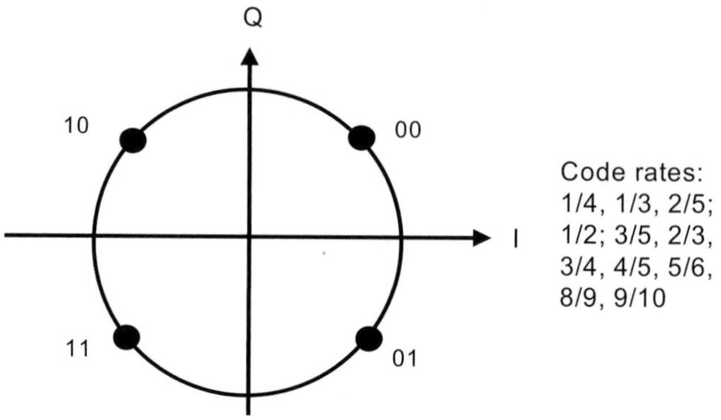

Fig. 14.26. Gray-coded QPSK, absolute mapping (like DVB-S)

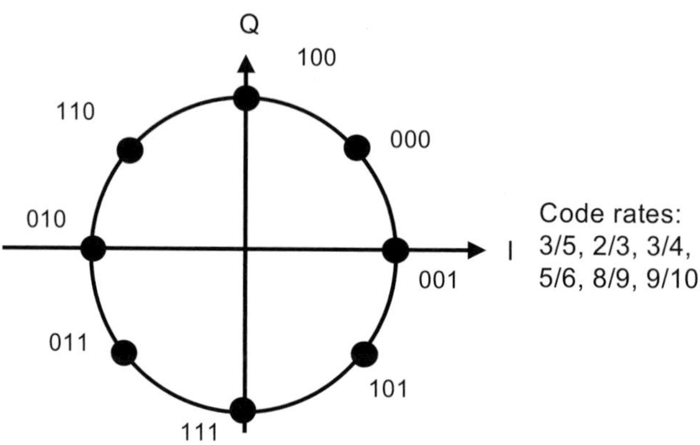

Fig. 14.27. Gray-coded 8PSK

In the downstream mapper, QPSK (Fig. 14.26.), 8PSK (Fig. 14.27.), 16APSK (Fig. 14.28.) or 32APSK (Fig. 14.29.) is then mapped. This is always absolute mapping, i.e. non-differential. Hierarchical modulation is a special case. It is virtually backward compatible with the DVB-S standard, making it possible to transmit a DVB-S stream and an additional DVB-S2 stream. In the hierarchical modulation mode (Fig. 14.30.), the constellation can be interpreted in two different ways.

The quadrant can be interpreted as a constellation point, gaining 2 bits for the high priority path conforming to DVB-S. It is also possible, however, to look for the two discrete points in the quadrant, decoding a further bit for the low priority path in the process. In this case, 3 bits per symbol are transmitted. There is also hierarchical modulation in DVB-T. After the mapping, the signal passes through the physical layer framing and roll-off filtering stages and is then converted into the modulation signal proper in the IQ modulator. The roll-off factor is 0.20, 0.25 or 0.35.

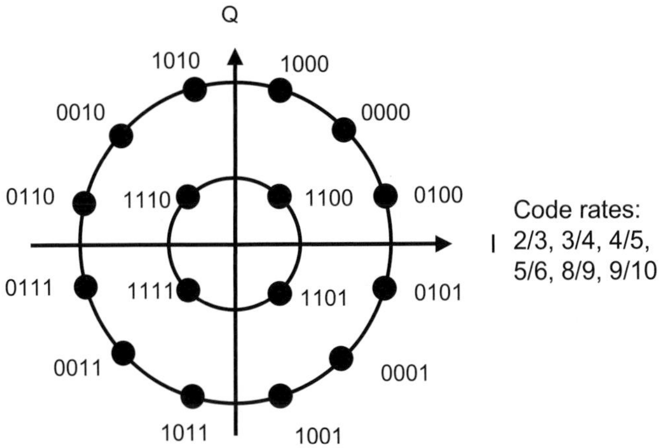

Fig. 14.28. 16APSK

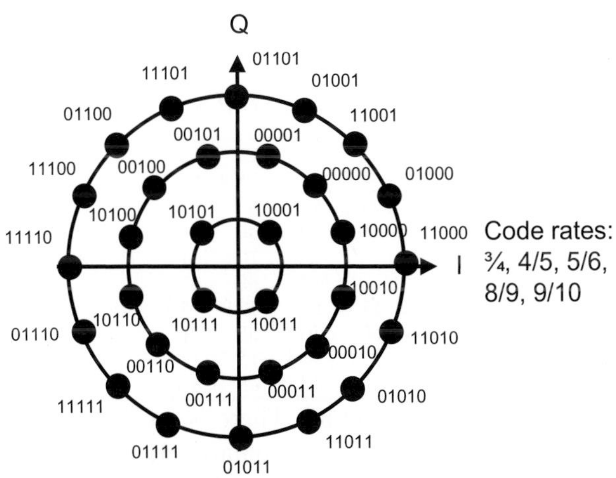

Fig. 14.29. 32 APSK

The error protection (Fig. 14.31.) consists of a BCH (Bose-Chaudhuri-Hocquenghem) coder and an LDPC (low density parity check) encoder followed by the bit interleaver. Possible code rates are 1/4 ... 9/10 and are shown in the figures for the respective constellation diagrams (QPSK ... 32APSK).

Compared with DVB-S, the minimum C/N ratio necessary in DVB-S2 is much more dependent on the modulation method and can also be varied by the code rate.

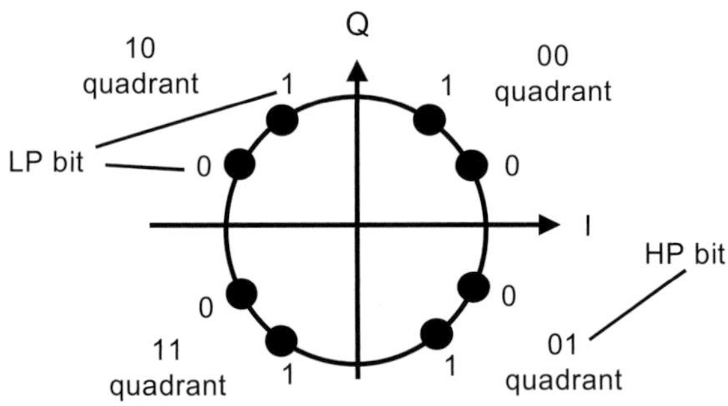

Fig. 14.30. Hierarchical QPSK modulation

Table 14.1 shows the minimum C/N ratios from the DVB-S2 Standard [ETS302307].

Table 14.1. Minimum C/N ratio necessary in DVB-S and DVB-S2

Mod.	CR =1/4	CR =1/3	CR =2/5	CR =1/2	CR =3/5	CR =2/3	CR =3/4	CR =4/5	CR =5/6	CR =8/9	CR =9/10
QPSK	-2.4 dB	-1.2 dB	0 dB	1 dB	2.2 dB	3.1 dB	4 dB	4.6 dB	5.2 dB	6.2 dB	6.5 dB
8PSK	-	-	-	-	5.5 dB	6.6 dB	7.9 dB	-	9.4 dB	10.6 dB	11 dB
16APSK	-	-	-	-	-	9 dB	10.2 dB	11 dB	11.6 dB	12.9 dB	13.1 dB
32APSK	-	-	-	-	-	-	12.6 dB	13.6 dB	14.3 dB	15.7 dB	16.1 dB

Unlike DVB-S, DVB-S2 has a frame structure. There is an FEC frame and a physical layer frame. An FEC frame firstly contains the data to be transmitted which are either data which have an MPEG-2 transport stream structure or data which are quite independent of this, so-called generic data.

In front of this data field there an 80 bit long baseband header. The data block with the baseband header is then padded to a length dependent on the selected code rate of the error protection and then provided with the BCH code plus the LDPC code. Depending on the mode, an FEC frame then has a length of 64800 or 16200 bits.

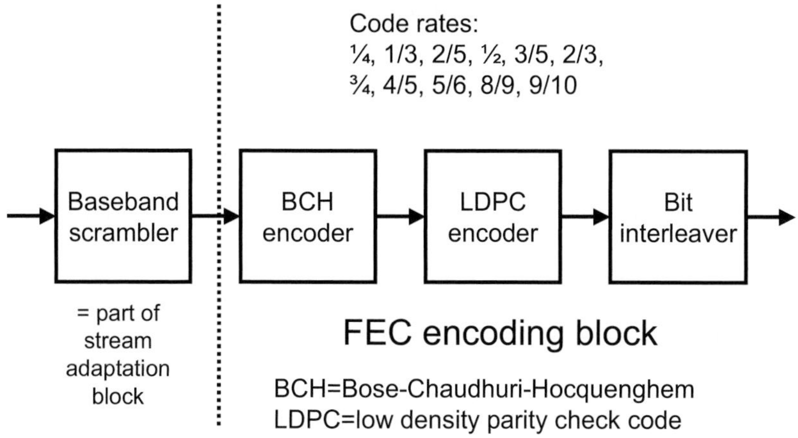

Fig. 14.31. DVB-S2 FEC block

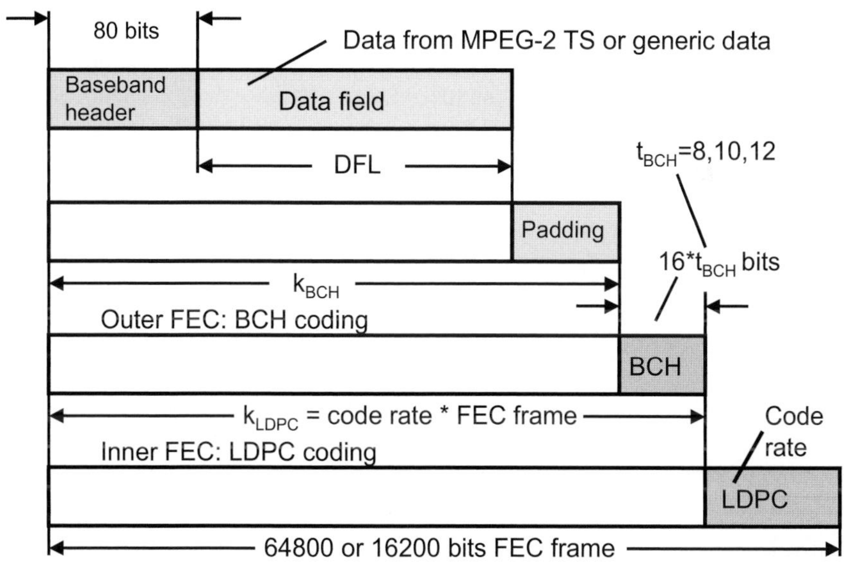

Fig. 14.32. FEC frame in DVB-S2

The FEC frame is then divided into a physical layer frame composed of n slots. The physical layer frame starts with the one-slot-long physical layer header in which the carrier is $\pi/2$-shift BPSK modulated. This is followed by slot 1 ... slot 16.

Slot 17 may be a pilot block if pilots are transmitted (optional). This is followed by another 16 time slots with data and then, after slot 32, possibly another pilot block etc.

Table 14.2. Coding parameters in DVB-S2

LDPC code rate	k_{BCH}	k_{LDPC}	t_{BCH}	FEC frame
¼	16008	16200	12	64800
1/3	21408	21600	12	64800
2/5	25728	25920	12	64800
½	32208	32400	12	64800
3/5	38688	38880	12	64800
2/3	43040	43200	10	64800
¾	48408	48600	12	64800
4/5	51648	51840	12	64800
5/6	53840	54000	10	64800
8/9	57472	57600	8	64800
9/10	58192	58320	8	64800
¼	3072	3240	12	16200
1/3	5232	5400	12	16200
2/5	6312	6480	12	16200
½	7032	7200	12	16200
3/5	9552	9720	12	16200
2/3	10632	10800	12	16200
¾	11712	11880	12	16200
4/5	12432	12600	12	16200
5/6	13152	13320	12	16200
8/9	14232	14400	12	16200
9/10	NA	NA	NA	16200

Table 14.3. Sample data rates in DVB-S and DVB-S2 with a symbol rate of 27.5 MS/s

Standard	Modulation	CR	Pilots	Net data rate [Mbit/s]
DVB-S	QPSK	¾	--	38.01
DVB-S2	QPSK	9/10	On	48.016345
DVB-S2	QPSK	9/10	Off	49.186827
DVB-S2	QPSK	8/9	On	47.421429
DVB-S2	QPSK	8/9	Off	48.577408
DVB-S2	8PSK	9/10	On	72.005046
DVB-S2	8PSK	9/10	Off	73.678193

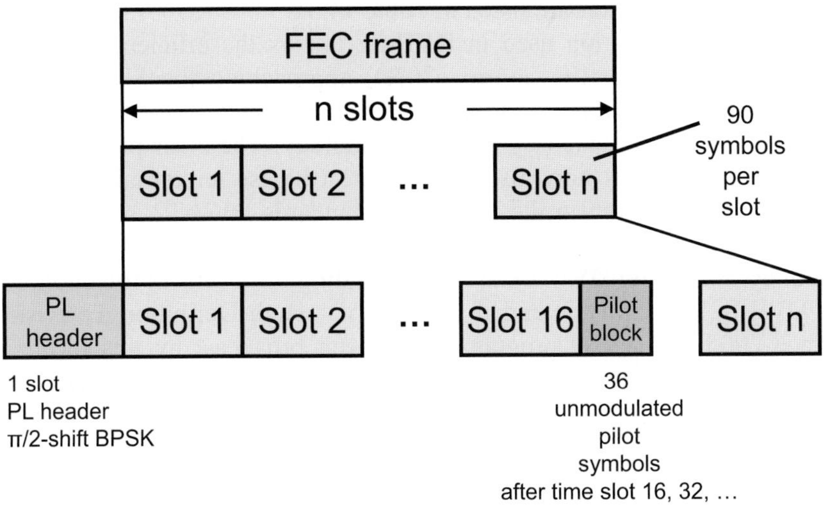

Fig. 14.33. Physical layer frame in DVB-S2

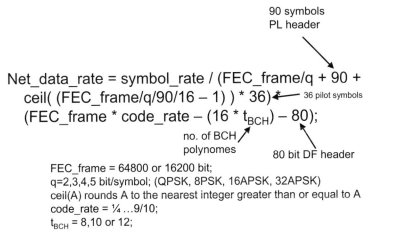

Net_data_rate = symbol_rate / (FEC_frame/q + 90 + ceil((FEC_frame/q/90/16 – 1)) * 36) * (FEC_frame * code_rate – (16 * t_{BCH}) – 80);

(90 symbols PL header; 36 pilot symbols; no. of BCH polynomes; 80 bit DF header)

FEC_frame = 64800 or 16200 bit;
q=2,3,4,5 bit/symbol; (QPSK, 8PSK, 16APSK, 32APSK)
ceil(A) rounds A to the nearest integer greater than or equal to A
code_rate = ¼ ...9/10;
t_{BCH} = 8,10 or 12;

Fig. 14.34. Formula for calculating the net data rate in DVB-S2

A slot has a length of 90 symbols. A pilot block has a length of 36 symbols. Table 14.1. shows the coding parameters of the FEC frame. The data rates in DVB-S2 can be calculated by using the formula shown in Fig.

14.33. In practice (symbol rate of 27.5 MS/s), they are about 49 Mbit/s. Examples of data rates are listed in Table 14.3.

The error protection used in DVB-S2 allows the efficiency to be increased enormously (by approx. 30%), approaching the Shannon limit much more closely. This also requires much greater computing capacity but this can be provided by today's technology. The error protection applied in DVB-S2 is now also used in the Chinese terrestrial digital TV standard DTMB, and in the new DVB standards DVB-T2 and DVB-C2. The other new DVB-SH standard for mobile TV, although largely derived from DVB-S2 (and DVB-T), uses turbo coding for its error protection.

DVB-S2 is mainly intended for HDTV - High Definition Television. Since 2005, some HD programs have been broadcast by this means in Europe, among them Premiere (now Sky), Sat1 and Pro7 in Germany. Sat1 and Pro7 have stopped transmitting HD for the time being, at least until 2010. By then, it will also be possible to receive the public broadcasting stations in HDTV format via satellite. The content will be MPEG-4 AVC coded with a data rate of about 10 Mbit/s per HD program.

Bibliography: [ETS300421], [MÄUSL3], [MÄUSL4], [REIMERS], [GRUNWALD], [FISCHER3], [EN301210], [ETS302307]

15 DVB-S/S2 Measuring Technology

15.1 Introduction

The satellite transmission of digital TV signals has now been discussed in detail. The following sections will deal with DVB-S/S2 measuring technology. Satellite transmission is relatively rugged and, in principle, only subject to noise effects (approx. 205 dB free space attenuation), and possible irradiation due to microwave links. There is also noise interference at the first satellite IF due to cordless telephones (DECT).

The essential tests parameters on a DVB-S signal are:

- Signal level
- Bit error ratio
- C/N (carrier/noise ratio)
- E_b/N_0
- Modulation error ratio (MER)
- Shoulder attenuation

The following are required for measurements on DVB-S signals:

- A modern spectrum analyzer (e.g. Rohde&Schwarz FSP, FSU)
- A professional DVB-S receiver with BER measurement or an antenna test instrument (e.g. Kathrein MSK33) or an MPEG analyzer with corresponding RF interface (Rohde&Schwarz DVM)
- A DVB-S/S2 test transmitter for measurements on set-top boxes and IDTV receivers (e.g. Rohde&Schwarz SFQ, SFU, SFL, SFE)

15.2 Measuring Bit Error Ratios

Due to the inner and outer error protection, there are three different bit error ratios in DVB-S:

W. Fischer, *Digital Video and Audio Broadcasting Technology*, Signals and Communication Technology, 3rd ed., DOI: 10.1007/978-3-642-11612-4_15, © Springer-Verlag Berlin Heidelberg 2010

- Pre-Viterbi bit error ratio
- Pre-Reed-Solomon bit error ratio
- Post-Reed-Solomon bit error ratio

The most interesting bit error ratio providing the most information about the transmission link is the pre- Viterbi bit error ratio. It can be measured by reapplying the data stream after the Viterbi decoder to a convolutional coder with the same configuration as that of the transmitter. If then the data stream before the Viterbi decoder is compared with that after the convolutional coder (Fig. 15.1.) (taking into consideration the delay of the coder), the two are identical if there are no errors. A comparator for the I branch and for the Q branch then determines the differences, and thus the bit errors.

The bit errors counted are then related to the number of bits transmitted in the corresponding period, resulting in the bit error ratio

BER = bit errors / transmitted bits;

The range of the pre-Viterbi bit error ratio is between $1 \bullet 10^{-4}$ to $1 \bullet 10^{-2}$. This means that every ten-thousandth to hundredth bit is errored.

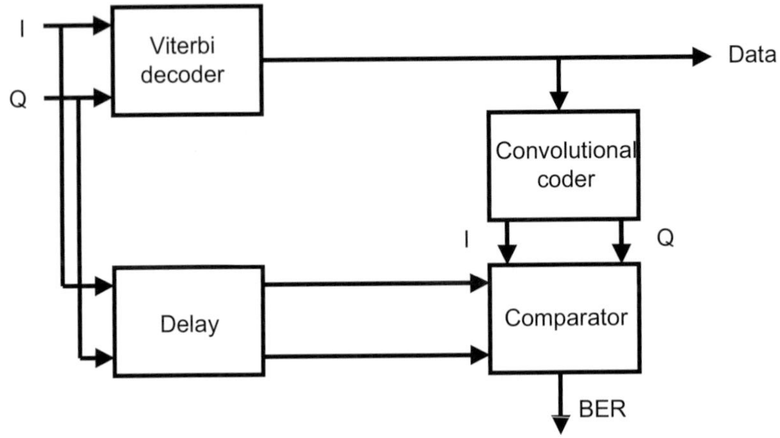

Fig. 15.1. Circuit for measuring the pre-Viterbi bit error ratio

The Viterbi decoder can only correct a proportion of the bit errors. There is thus a residual bit error ratio remaining before the Reed-Solomon decoder. Counting the correction processes of the Reed-Solomon decoder and relating them to the number of bits transmitted within the corresponding period of time provides the pre-Reed-Solomon bit error ratio. The limit pre-Reed-Solomon bit error ratio is about $2 \bullet 10^{-4}$. Up to there, the Reed - Solomon decoder can repair all errors. At the same time, however, the transmission is "on the brink". A little bit more interference, e.g. due to too much attenuation due to rain, and the transmission will break down and the picture will start to show "blocking".

But the Reed-Solomon decoder, too, cannot correct all bit errors, resulting in errored transport stream packets which are then flagged in the TS header (transport error indicator bit = 1). If the errored transport stream packets are counted, the post-Reed-Solomon decoder bit error ratio can be calculated.

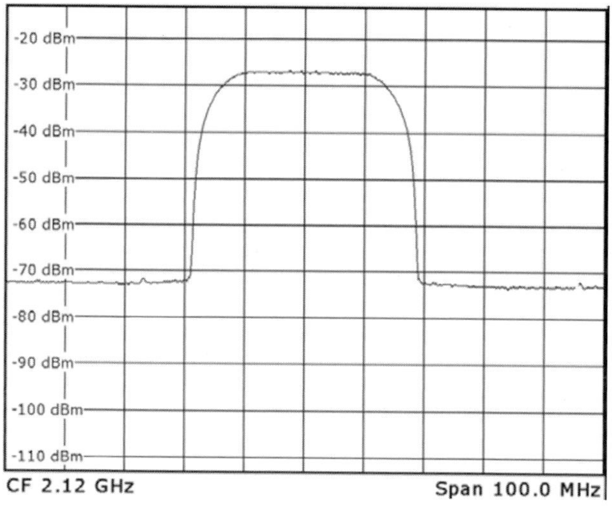

Fig. 15.2. Spectrum of a DVB-S signal (10 dB/Div, 10 MHz/Div, Span 100 MHz)

If very low bit error ratios (e.g. less than 10^{-6}) are measured, long measuring times in the range of minutes or hours must be selected to detect these with any degree of accuracy. Since there is a direct relation between bit error ratio and the carrier/noise ratio, this can be used for determining the latter (see diagram in Section 14.6 “Influences affecting the satellite transmission link”, Fig. 14.24.). Virtually every DVB-S chip or DVB-S receiver contains a circuit for determining the pre-Viterbi bit error ratio because this value can be used for aligning the satellite receiving antenna and for determining the quality of reception. The circuit itself is not very complex. In most cases, DVB-S receivers display two bar graphs in their setup menu, one for the signal strength and one for signal quality. The latter is derived from the bit error ratio.

Due to the altered error protection, the following bit error ratios are defined in DVB-S2:

- Bit error ratios before LDPC,
- Bit error ratios before BCH,
- Bit error ratios after BCH.

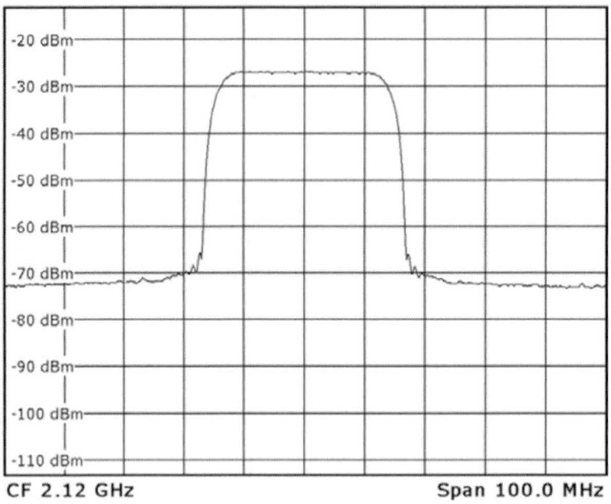

Fig. 15.3. Spectrum of a DVB-S2 signal with rolloff=0.25

15.3 Measurements on DVB-S Signals using a Spectrum Analyzer

A spectrum analyzer is quite suitable for measuring the power in the DVB-S channel, at least in the uplink. Of course, it would also be quite simple to use a thermal power meter but a spectrum analyzer can also be used for determining the carrier/noise ratio in the uplink. Firstly, however, the power of the DVB-S/S2 signal will be determined using the spectrum analyzer. A DVB-S signal has the appearance of noise and has a rather large crest factor. Because of its strong similarity with white Gaussian noise, its power is measured exactly as in the case of noise.

To determine the carrier power, the spectrum analyzer is set as follows: At the analyzer, a resolution bandwidth of 2 MHz and a video band width of 3 to 10 times the resolution bandwidth (10 MHz) are selected. To achieve some averaging, a slow operating time must be set (2000 ms). These parameters are required because of the RMS detector used in the spectrum analyzer. The following settings are used:

- Center frequency at the center of the DVB-S channel,
- Span at 100 MHz,
- Resolution bandwidth at 2 MHz,
- Video bandwidth at 10 MHz (because of RMS detector and log. representation),
- Detector RMS
- Slow operating time (2000 ms)
- Noise marker at channel center (results in C' in dBm/Hz)

This results in a spectrum as shown in Fig. 15.2. The RMS detector calculates the power density of the signal in a window with a bandwidth of 1 Hz, the test window being continuously pushed over the frequency window to be measured (sweep range). In principle, first the RMS (root mean square) value of the voltage is determined from all samples in the signal window of 1 Hz bandwidth:

$$U_{RMS} = \frac{1}{N}\sqrt{u_1^2 + u_2^2 + u_3^2 + \ldots};$$

From this, the power in this signal window is calculated with reference to an impedance of 50 Ω and converted into dBm. This is then the signal power density in a window of 1 Hz bandwidth. The slower the selected sweep time set, the more samples can be accommodated in this window and the smoother and better averaged will be the test result.

Because of the noise-like signal, we use the noise marker measuring power. The noise marker is set to band center for this purpose. The prerequisite is a flat channel but this can always be assumed to be the case in the uplink. If the channel is not flat, other suitable measuring functions must be used for measuring the channel power but these are dependent on the spectrum analyzer.

The analyzer provides us with the value C' as the noise power density at the position of the noise marker in dBm/Hz, and the filter bandwidth and the characteristics of the logarithmic amplifier of the analyzer are automatically taken into consideration. To relate the signal power density C' to the Nyquist bandwidth B_N of the DVB-S signal, it is necessary to calculate the signal power C as follows:

$$C = C' + 10\log B_N = C' + 10\log(\textit{symbol rate/Hz})\ dB; \qquad [dBm]$$

The Nyquist bandwidth of the signal corresponds to the symbol rate of the DVB-S signal.

Example:

Measured value of the noise marker:	-100 dBm/Hz
Correction value at 27.5 MS/s symbol rate:	+ 74.4 dB
Power in the DVB-S channel:	- 25.6 dBm

15.3.1 Approximate Determination of the Noise Power N

If it were possible to switch off the DVB-S signal without changing the noise ratios in the channel, the noise marker at the center of the band would now provide information on the noise ratios in the channel. However, this cannot be done in such a simple way. If not an exact measurement value, then at least a "good idea", is obtained if the noise marker is used on the shoulder of the DVB-S signal for measuring in close proximity of the signal. This is because it can be assumed that the noise fringe in the wanted band continues similarly to its appearance on the shoulder.

The value N' of the noise power density is output by the spectrum analyzer. The noise power N in the channel with the bandwidth B_K of the DVB-S transmission channel is then calculated from the noise power density N' as follows:

$$N = N' + 10\log B_K = N' + 10\log(\textit{channel bandwidth/ Hz})dB; \ [dBm]$$

The channel bandwidth of the signal corresponds to the symbol rate of the DVB-S signal (DVB measurement guidelines).

Example:

Measured value of the noise marker:	-120 dBm/Hz
Correction value at 27.5 MS/s symbol rate:	+ 74.4 dB
Noise power in the DVB-S channel:	- 45.6 dBm

The resultant C/N is:

$$C/N_{[dB]} = C_{[dBm]} - N_{[dBm]};$$

In the example: C/N[dB] = -25.6 [dBm] - (-45.6 dBm) = 20 dB;

In fact, to measure the C/N in the downlink, the noise is measured in the gaps between the individual channels. Other possibilities of measuring C/N would be to use a suitable constellation analyzer for DVB-S/S2 (e.g. the Rohde&Schwarz DVM with RF option) or via the detour of measuring the bit error rate. Naturally, such an analyzer can also be used to measure levels.

15.3.2 C/N, S/N and Eb/N_0

The carrier-to-noise ratio C/N is an important value in assessing the quality of the satellite transmission link. From the C/N, direct conclusion can be drawn with respect to the bit error rate to be expected. The C/N is the result of the power radiated by the satellite (< ~100W), the antenna gain at the transmitting and receiving end (size of the receiving antenna) and the loss in the space between. The alignment of the satellite receiving antenna and the noise figure of the LNB also play a role. DVB-S receivers output the C/N value as an aid for aligning the receiving antenna.

$$C/N[dB] = 10\log(P_{Carrier}/P_{Noise})\ ;$$

In addition to the carrier-to-noise ratio, there is also the signal-to-noise ratio:

$$S/N[dB] = 10\log(P_{Signal}/\ P_{Noise});$$

The signal power is here the power of the signal after roll-off filtering. P_{noise} is the noise power within the Nyquist bandwidth (symbol rate).

The signal-to-noise ratio S/N is thus obtained from the carrier-to-noise ratio as:

$$S/N[dB] = C/N[dB] + 10\log\ (1-r/4);$$

where r is the roll-off factor (= 0.35 in DVB-S); i.e., in DVB-S:

$$S/N[dB] = C/N[dB]\ -0.3977\ dB;$$

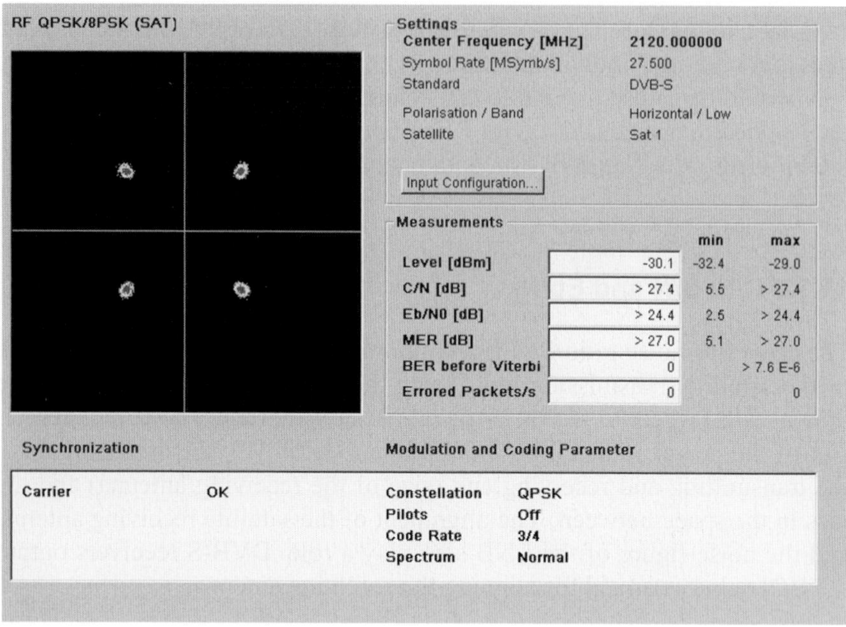

Fig. 15.4. Constellation diagram of an undisturbed DVB-S signal with MER and BER measurement [DVM]

15.3.3 Finding the E_B/N_0 Ratio

In DVB-S, the term E_B/N_0 is often mentioned. This is the energy per bit with respect to the noise power density.

E_B = energy per bit;
N_0 = noise power density in dBm/Hz;

The E_B/N_0 can be calculated from the C/N ratio:

$$E_B/N_0\,[dB] = C/N[dB] + 10\log(188/204) - 10\log(m) - 10\log(\text{code rate});$$

where
m = 2 for QPSK/DVB-S;
m = 4 for 16QAM;
6 for 64QAM and
8 for 256QAM, and

the code rate is 1/2, 2/3, 3/4, 5/6, 7/8.

With a code rate of 3/4 as in the case of the usual QPSK modulation,

$$
\begin{aligned}
E_B/N_0\ [\mathrm{dB}]_{3/4} &= C/N[\mathrm{dB}] + 10\log(188/204) - 10\log(2) - 10\log(3/4) \\
&= C/N[\mathrm{dB}] + 0.3547\ \mathrm{dB} - 3.0103\ \mathrm{dB} + 1.2494\ \mathrm{dB} \\
&= C/N[\mathrm{dB}] - 1.4062\ \mathrm{dB};
\end{aligned}
$$

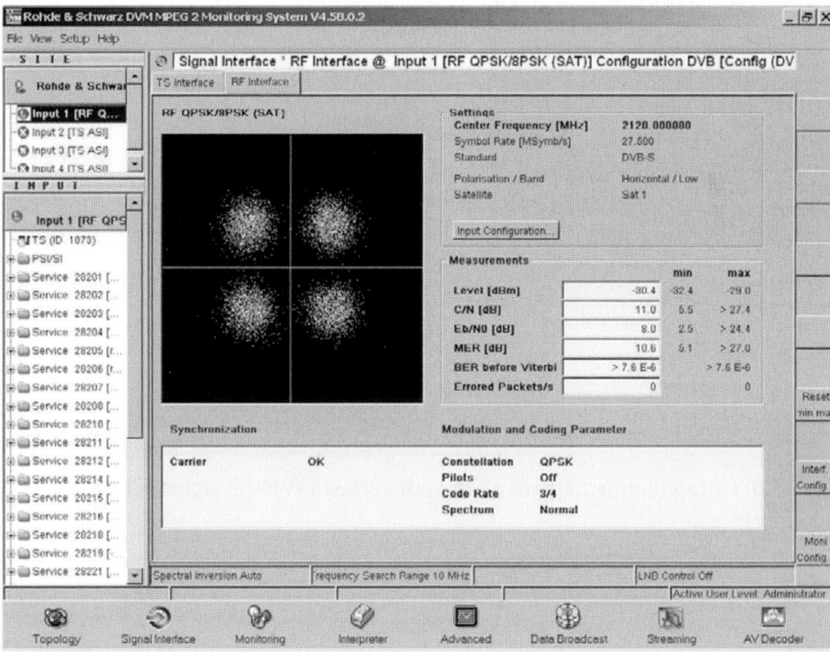

Fig. 15.5. Constellation diagram of an interfered DVB-S signal [DVM]

15.4 Modulation Error Ratio (MER)

The modulation error ratio (MER) is an aggregate parameter in which all the interfering signals affecting a digitally modulated signal are mapped. Each interfering event, or hit, can be described by an error vector which pushes the constellation point out of the ideal center of a decision field. The ratio of the measured RMS value of the signal amplitude to the quadratic mean of the error vectors is then the MER. This is defined in detail in the chapters on DVB-C and DVB-T measuring technology. In the case of

DVB-S/S2, the MER is almost identical to the S/N value since there are virtually only noise effects.

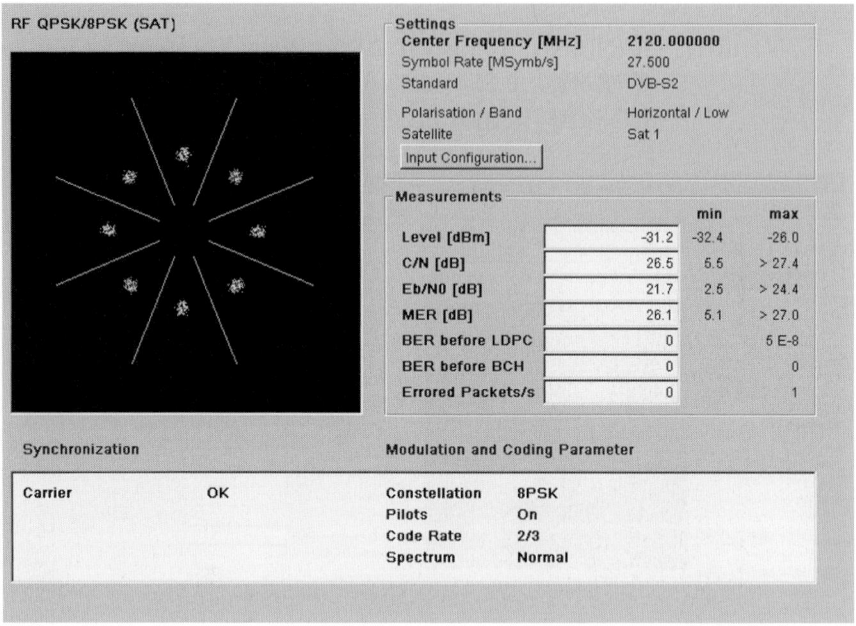

Fig. 15.6. Constellation diagram of an undisturbed DVB-S2 signal (8PSK) [DVM]

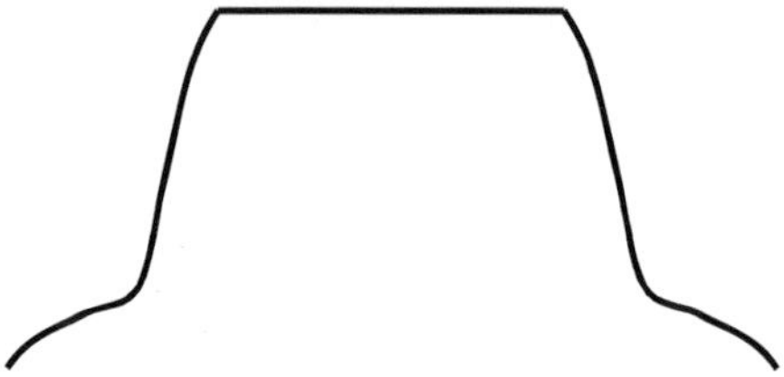

Fig. 15.7. DVB-S spectrum with "shoulders"

15.5 Measuring the Shoulder Attenuation

The DVB-S/2 signal within the wanted DVB-S/2 channel should be as flat as possible, i.e. it should not exhibit any ripple or tilt. Toward the edges of the channel, the DVB-S/2 spectrum drops off filtered with a smooth roll-off. There are, however, still signal components outside the actual wanted band and these are called the 'shoulders' of the DVB-S/S2 signal. The aim is to achieve the best possible shoulder attenuation of at least 40 dB. [ETS 300421] specifies a tolerance mask for the DVB-S signal spectrum but, in principle, the satellite operator can define a particular tolerance mask for the shoulder attenuation.

The signal spectrum is analyzed using a spectrum analyzer and simple marker functions.

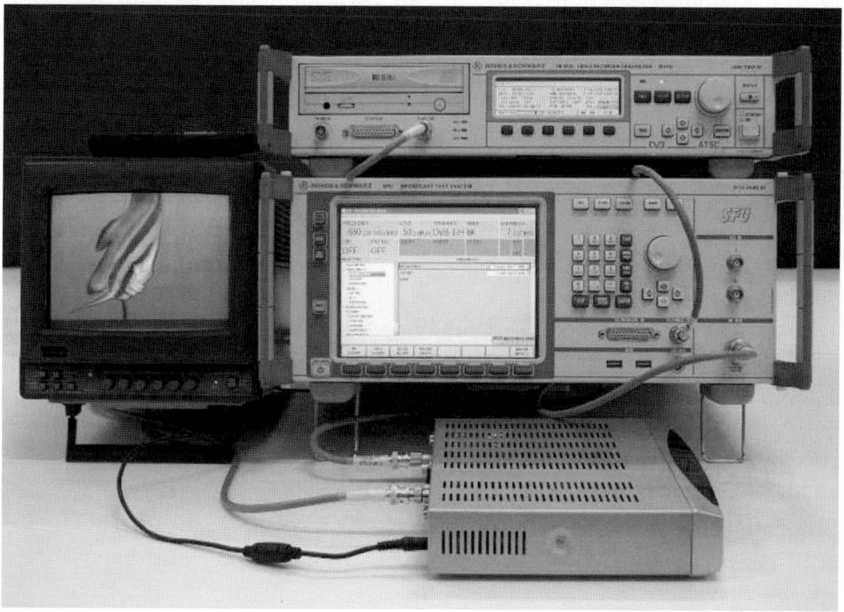

Fig. 15.8. Testing DVB receivers using an MPEG-2 generator (Rohde&Schwarz DVRG) and a test transmitter (Rohde&Schwarz SFU): The MPEG-2 generator (top) supplies the MPEG-2 transport stream with test contents and feeds the DVB test transmitter (center) which, in turn, generates a DVB-conformal IQ modulated RF signal for the DVB receiver (bottom). The video output signal from the DVB receiver is displayed on the TV monitor (left).

15.6 DVB-S/S2 Receiver Test

The testing of DVB-S/S2 receivers (set-top boxes, s. Fig. 15.4., and IDTVs) is accorded great significance. For these tests, DVB-S/S2 test transmitters are used which can simulate the satellite transmission link and the modulation process. Such a test transmitter (e.g. Rohde&Schwarz TV Test Transmitter SFQ or SFU) includes, in addition to the DVB-S/S2 modulator and upconverter, an add-on noise source and possibly even a channel simulator. The test transmitter is fed with an MPEG-2 transport stream from an MPEG-2 generator. The test transmitter then supplies a DVB-S/S2 signal within the range of the first satellite IF (900 - 2100 MHz). This signal can be fed directly to the input of the DVB-S/S2 receiver. It is then possible to create various adverse signal conditions for the DVB-S/S2 receiver by changing numerous parameters in the test transmitter. It is also possible to measure the bit error ratio as a function of the C/N ratio. Such test transmitters are used both in the development and in the production and quality assurance of DVB-S/S2 receivers.

Bibliography: [ETS300421], [ETR290], [REIMERS], [GRUNWALD], [FISCHER3], [SFQ], [SFU], [DVM]

16 Broadband Cable Transmission According to DVB-C

In many countries, good radio and TV coverage is provided via broadband cable, especially in densely populated areas. These cable links have either a bandwidth of about 400 MHz (approx. 50 - 450 MHz) or about 800 MHz (approx. 50 - 860 MHz). In addition to the VHF and UHF band known from terrestrial television, special channels are occupied. Analog television programs can be received easily with a conventional TV set without additional complexities which is why this type of TV coverage is of such great interest to many. The only obstacle in comparison with analog satellite TV reception is the additional monthly line charge with which a satellite receiving system would pay for itself within one year in many cases. If the satellite dish is large enough, the picture quality is often better than via broadband cable since intermodulation products sometimes result in visible interference due to the multiple channel allocation in broadband cable.

The decision between cable and satellite reception simply depends on the following considerations:

- Convenience
- Cabel reception charges
- Single- and multiple-channel reception
- Picture quality
- Personal requirements/preferences

In many areas of Europe, purely terrestrial reception has dropped to below 10%. Naturally, this does not apply to the rest of the world.

Since about 1995, many cable networks are also carrying digital TV signals according to the DVB-C standard and many others in the higher frequency bands above about 300 MHz. This section is intended to explain the methods for transmitting digital TV signals via broadband cable in greater detail. The chosen transmission methods and parameters were selected with reference to the typical characteristics of a broadband cable. Cable exhibits a much better signal/noise ratio than in satellite transmission and there are not many problems with reflections, either, all of which

W. Fischer, *Digital Video and Audio Broadcasting Technology*, Signals and Communication Technology, 3rd ed., DOI: 10.1007/978-3-642-11612-4_16,

permits digital modulation methods of higher quality to be used, from 64QAM (coax) to 256QAM (optical fiber). A broadband cable network consists of the cable head end, of the cable distribution links consisting of coaxial cables and cable amplifiers, of the 'last mile' from the distributor to the house connection of the subscriber and of the subscriber's in-house network itself. Special technical terms such as 'network level' are deliberately avoided here since these terms can be specific to cable operators or countries. The cable distribution links from the head end to the last distribution box can also be run as optical fibers. This broadband cable system then distributes radio programs and analog and digital TV programs. More and more frequently there are also return channel links in the frequency band below about 65 MHz.

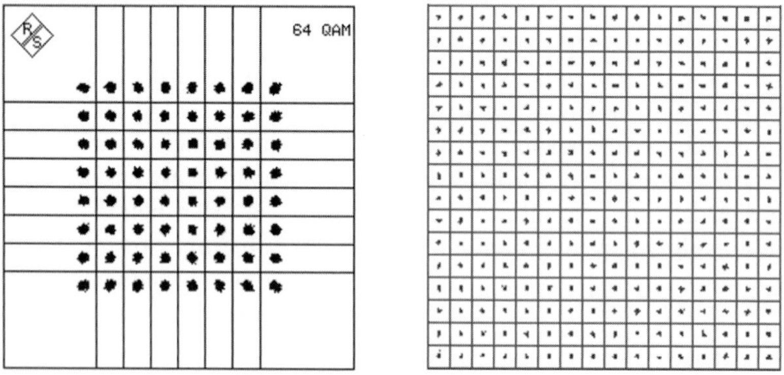

Fig. 16.1. 64QAM (left) and 256QAM (right)

16.1 The DVB-C Standard

Digital video broadcasting for cable applications had been specified in about 1994 in the standard [ETS 300429]. This service has been available in the cable networks since then, or shortly after. We will see that in the DVB-C modulator, the MPEG-2 transport stream passes through almost the same stages of conditioning as in the DVB-S satellite standard. It is only the last stage of convolutional coding which is missing here: it is simply not needed because the medium of propagation is so much more robust. This is followed by the 16, 32, 64, 128 or 256QAM quadrature amplitude modulation. In coax cable systems, 64QAM is used virtually ways whereas optical fibre networks frequently use 256QAM.

Considering then a conventional coax system with a channel spacing of 8 MHz. It normally uses a 64QAM-modulated carrier signal with a symbol rate of, for example, 6.9 MS/s. The symbol rate must be lower than the system bandwidth of 8 MHz in the present case. The modulated signal is rolled off smoothly towards the channel edges with a roll-off factor of r = 0.15. Given 6.9 MS/s and 64 QAM (6 bits/symbol), a gross data rate of

$$\text{Gross\_data\_rate}_{DVB\text{-}C} = 6 \text{ bits/Symbol} \bullet 6.9 \text{ MSymbols/s} = 41.4 \text{ Mbit/s};$$

is obtained. In DVB-C, only Reed-Solomon error protection is used which is the same as in DVB-S, i.e. RS(188,204). Thus, an MPEG-2 transport stream packet of 188 bytes length is provided with 16 bytes of error protection, resulting in a total packet length of 204 bytes during the transmission.

The resultant net data rate is:

$$\text{Net\_data\_rate}_{DVB\text{-}C} = \text{Gross\_data\_rate} \bullet 188/204 = 38.15 \text{ Mbit/s};$$

Thus, a 36-MHz-wide satellite channel with a symbol rate of 27.5 MS/s and a code rate of 3/4 has the same net data rate, i.e. the same transport capacity as this DVB-C channel with a width of only 8 MHz.

The following generally applies for DVB-C:

$$\text{Net\_data\_rate}_{DVB\text{-}C} = \text{ld}(m) \bullet \text{symbol\_rate} \bullet 188/204;$$

As well, however, the DVB-C channel has a much better signal/noise ratio (S/N) with about >30 dB compared with about 10 dB in the case of DVB-S.

The constellations provided in the DVB-C standard are 16QAM, 32QAM, 64QAM, 128QAM and 256QAM. According to DVB-C, the spectrum is roll-off filtered with a roll-off factor of r = 0.15. The transmission method specified in DVB-C is also known as the international standard ITU-T J83A. There is also the parallel standard ITU-T J83B used in North America, which will be described later, and ITU-T J83C which is used in 6 MHz-wide channels in Japan. In principle, J83C has the same structure as DVB-C but it uses a different roll-off factor for 128QAM (r = 0.18) and for 256QAM (r = 0.13). Everything else is identical. ITU-T J83B, the method found in the US and in Canada, has a completely different FEC and is described in a separate Section.

16.2 The DVB-C Modulator

The DVB-C modulator does not need to be described in so much detail since most of the stages are completely identical with the DVB-S modulator. The modulator locks to the MPEG-2 transport stream fed to it at the baseband interface and consisting of 188 byte-long transport stream packets. The TS packets consist of a 4 byte header, beginning with the sync byte (0x47), followed by 184 bytes of payload. Following this, every sync byte is inverted to 0xB8 to carry long-term time markers in the data stream to the receiver for energy dispersal and its cancellation. This is followed by the energy dispersal stage (or randomizer) proper, and then the Reed-Solomon coder which adds 16 bytes of error protection to each 188 byte-long TS packets. The packets, which are then 204 bytes long, are then supplied to the Forney interleaver to make the data stream more resistant to error bursts. The error bursts are broken up by the cancellation of the interleaving in the DVB-C demodulator which makes it easier for the Reed Solomon block decoder to correct errors.

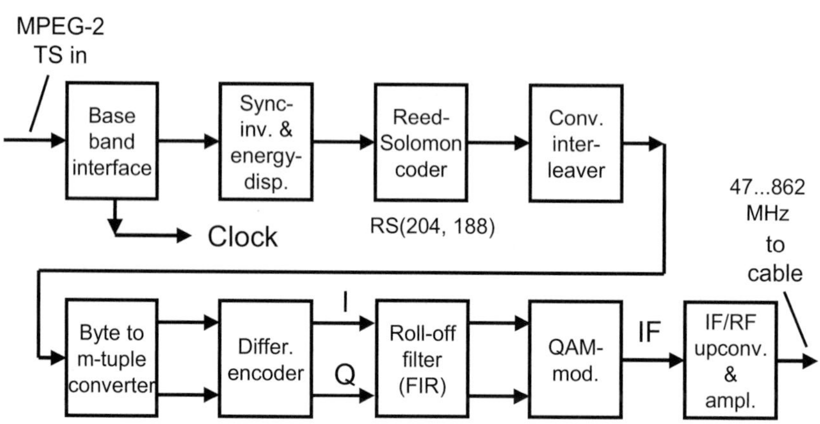

Fig. 16.2. DVB-C modulator

The error-protected data stream is then fed into the mapper where the QAM quadrant must be differentially coded, in contrast to DVB-S and DVB-T. This is because the carrier can only be recovered in multiples of 90° in the 64QAM demodulator and the DVB-C receiver can lock to any multiples of 90° carrier phase. The mapper is followed by the quadrature amplitude modulation which is now done digitally. Usually, 64QAM is selected for coaxial links and 256QAM for fiber-optical links. The signal is roll-off filtered with a roll-off factor of $r = 0.15$. This gradual roll-off towards the band edges optimizes the eye opening of the modulated signal. After power amplification, the signal is then injected into the broadband cable system.

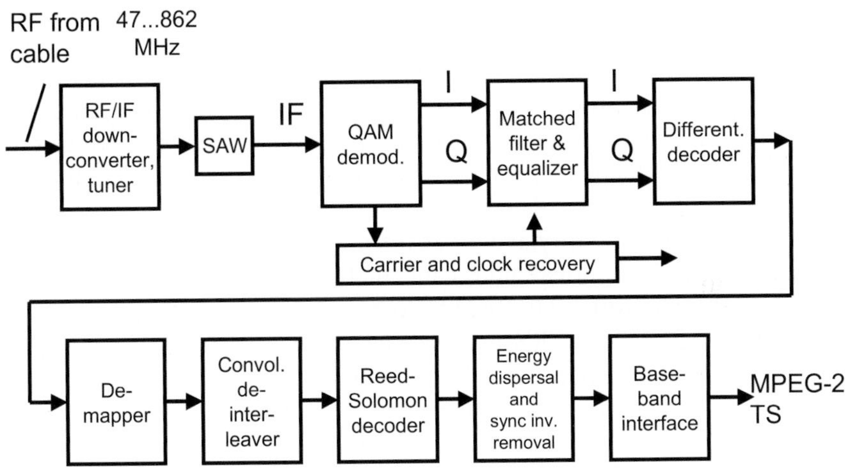

Fig. 16.3. DVB-C receiver

16.3 The DVB-C Receiver

The DVB-C receiver - set-top box or integrated - receives the DVB-C channel in the 50 - 860 MHz band. The transmission has added effects due to the transmission link such as noise, reflections and amplitude and group delay distortion. These effects will be discussed later in a separate Section.

The first module of the DVB-C receiver is the cable tuner which is essentially identical with a tuner for analog television. The tuner converts the 8 MHz-wide DVB-C channel down to an IF with a band center at about 36 MHz. These 36 MHz also correspond to to the band center of an analog TV IF channel according to ITU standard BG/Europe. Adjacent channel components are suppressed by a downstream SAW filter which has a bandwidth of exactly 8 MHz. Where 7 or 6 MHz channels are possible, the filter must be replaced accordingly. This band-pass filtering to 8, 7 or 6 MHz is followed by further downconversion to a lower intermediate frequency in order to simplify the subsequent analog/digital conversion. Before the A/D conversion, however, all frequency components above half the sampling rate must be removed by means of a low-pass filter. The signal is then sampled at about 20 MHz with a resolution of 10 bits. The IF, which is now digitized, is supplied to an IQ demodulator and then to a root cosine squared matched filter operating digitally. In parallel with this, the carrier and the clock are recovered. The recovered carrier with an uncertainty of multiples of 90 degrees is fed into the carrier input of the IQ demodulator. This is followed by a channel equalizer, partly combined with the matched filter, a complex FIR filter in which it is attempted to correct the channel distortion due to amplitude response and group delay errors. This equalizer operates in accordance with the maximum likelihood principle, i.e. it is attempted to optimize the signal quality by "tweeking" digital "setscrews" which are the taps of the digital filter. The signal, thus optimized, passes into the demapper where the data stream is recovered. This data stream will still have bit errors and, therefore, error protection is added. Firstly, the interleaving is cancelled and error bursts are turned into single errors. The following Reed Solomon decoder can eliminate up to 8 errors per 204-byte-long RS packet. The result is again transport stream packets with a length of 188 bytes which, however, are still energy-dispersed. If there are more than 8 errors in a packet, they can no longer be repaired and the transport error indicator in the TS header is then set to 'one'. After the RS decoder, the energy dispersal and the inversion of every 8th sync byte are cancelled and the MPEG-2 transport stream is again present at the physical baseband interface. In practice, all modules from the A/D converter to the transport stream output are implemented in one chip. The essential components in a DVB-C set-top box are the tuner, some discrete components, the DVB-C demodulator chip and the MPEG-2 decoder chip, all of which are controlled by a microprocessor.

16.4 Interference Effects on the DVB-C Transmission Link

Since, in practice, DVB-C modulators only use digital IQ modulators, IQ errors such as amplitude imbalance, phase errors and carrier leakage can be neglected today. These effects are simply no longer present in contrast to first-generation transmission. The effects occurring during transmission are essentially noise, intermodulation and cross-modulation interference and echoes and amplitude and group delay effects. If a cable amplifier is saturated and, at the same time, is occupied with a large number of channels, frequency conversion products are produced which will appear in the useful signal range. Every amplifier, therefore, needs to be operated at the correct operating point. It is, therefore, of importance that the levels on the transmission link are correct. A too high level produces intermodulation in the amplifiers, a too low level reduces the signal/noise ratio, both of which result in noise. The levels in a house installation, for example, should be adjusted in such a way that a maximum signal/noise ratio is obtained for DVB-C. An amplifier which may be present is calibrated in such a way that the signal/noise ratio is at the point of inversion at the most distant antenna socket. DVB-C signals are also very sensitive to amplitude and group delay response.

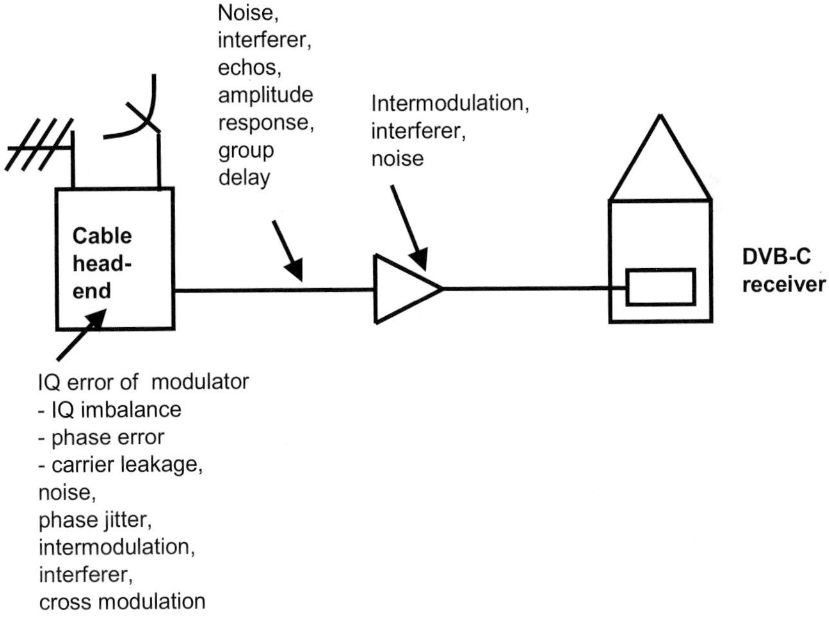

Fig. 16.4. Interference effects on the DVB-C transmission link

A slightly defective connecting line between the TV cable socket and the DVB-C receiver is often sufficient to render correct reception impossible. Operation of a DVB-C transmission link which is still quasi error free (QEF) requires a signal/noise ratio S/N of more than 26 dB for 64 QAM. The channel bit error ratio, i.e. the bit error ratio before Reed Solomon, is then $2 \bullet 10^{-4}$. The Reed Solomon decoder then corrects errors up to a residual bit error ratio after Reed Solomon of $1 \bullet 10^{-11}$. This corresponds to quasi error free operation (1 error per hour) but is also close to the "brick wall" (or "fall off the cliff"). A little more noise and the transmission will break down abruptly. The S/N ratio required for the QEF case depends on the degree of modulation. The higher the degree of quadrature amplitude modulation, the more sensitive the transmission system. Fig. 16.5. shows the variation of the bit error ratio with respect to the S/N ratio for QPSK, 16QAM, 64QAM and 256QAM.

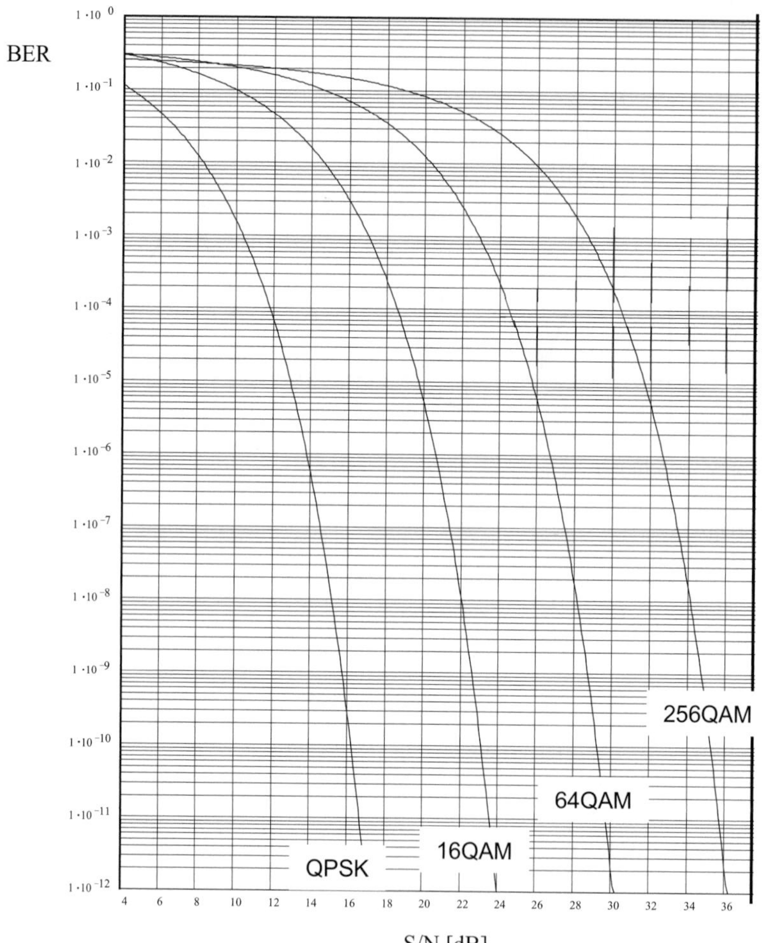

Fig. 16.5. Bit error ratio as a function of S/N in DVB-C [EFA]

Currently, the most widely used signals in coax networks are 64QAM-modulated signals. These require a S/N ratio of more than 26 dB which then corresponds to operation close to the "brick wall" (or "fall-off-the-cliff").

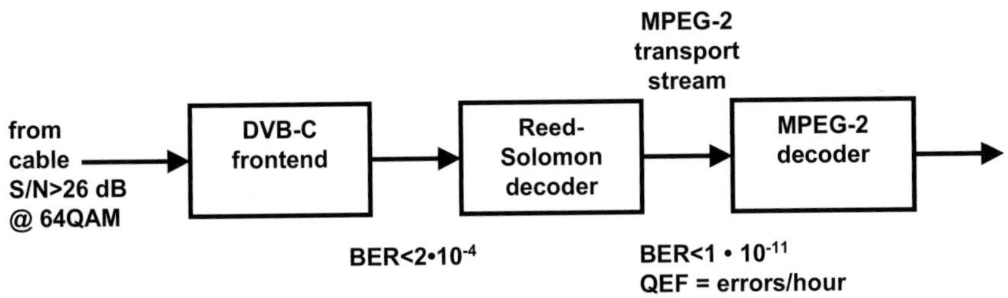

Fig. 16.6. DVB-C Bit error ratios

Bibliography: [ETS300429], [ETR290], [EFA], [GRUNWALD], [ITU-T J83]

17 Broadband Cable Transmission According to ITU-T J83B (US)

In North America, a different standard is used for transmitting digital TV signals over broadband cable which is ITU-T J83B.

"J83B" is a part of a document which describes a total of 4 system proposals for broadband cable standards:

- System A corresponds fully to DVB-C (8 MHz bandwidth)
- System B is the currently used US cable standard (6 MHz bandwidth), = J83B
- System C is the Japanese cable standard (largely identical with DVB-C, different roll-off, 6 MHz bandwidth)
- System D is "ATSC for cable" (16VSB, 6 MHz bandwidth, not in use)

In principle, J83B is comparable to J83A, C (Europe, Japan) but there are great differences in detail, especially in the FEC. The channel bandwidth in J83B is 6 MHz as in J83C (Japan). The modulation methods used are only 64QAM and 256QAM with a roll-off factor of $r = 0.18$ (64QAM) and $r = 0.12$ (256QAM). The error protection (FEC) is much more elaborate than in J83A or C. This begins with the MPEG framing.

The sync byte in the MPEG-2 transport stream is replaced by a special checksum which is also calculated continuously in parallel at the receiving end as in ATM (asynchronous transfer mode) and is used as criterion for synchronization if they agree.

J83B makes it possible to transmit both an MPEG-2 transport stream and ATM. After replacing the sync byte with a CRC, this is followed by a Reed Solomon block encoder RS(128,122) which, in contrast to J83A, is not set up for the MPEG-2 block structure. In the RS encoder, 6 RS symbols are added to every 122 7-bit long symbols. This makes it possible to repair 3 symbols within 128 symbols at the receiving end, forming a frame of several RS(128,122) packets to which a 42-bit- or 40-bit-long sync trailer is added in which, among other things, the adjustable interleaver length is signalled. The RS encoder is followed by an interleaver which condi-

W. Fischer, *Digital Video and Audio Broadcasting Technology*, Signals and Communication Technology, 3rd ed., DOI: 10.1007/978-3-642-11612-4_17, © Springer-Verlag Berlin Heidelberg 2010

conditions a data stream in order to prevent error bursts. A randomizer provides for an advantageous spectral distribution and breaks up long sequences of zeroes and ones in the data stream. The last stage in the FEC is a trellis encoder (convolutional coder) which inserts additional error protection and, naturally, overhead into the data stream. The data stream conditioned in this way is then 64QAM or 256QAM modulated and then transmitted in the coaxial or fiber-optical broadband cable.

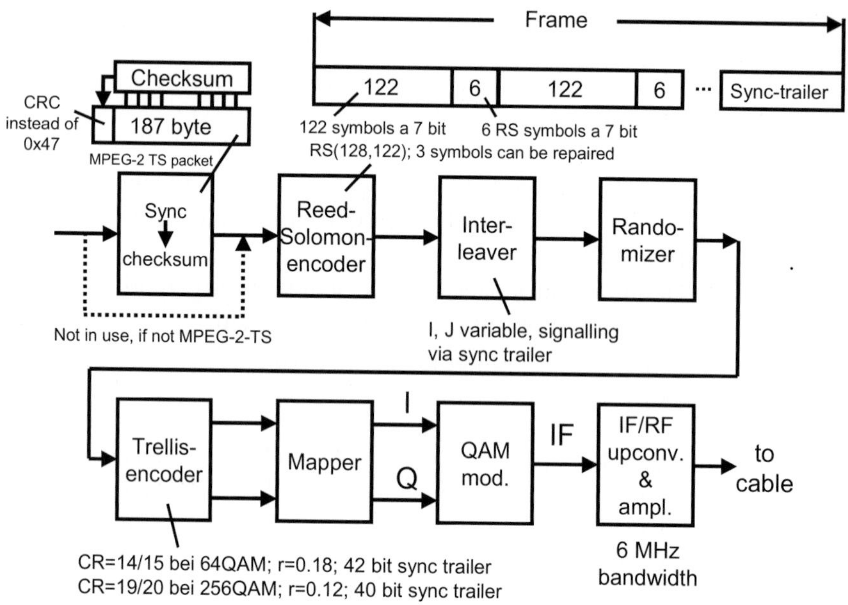

Fig. 17.1. Block diagram of an ITU-T J83B modulator

In addition to J83A, B and C, there is also the J83D Standard described in the same ITU document but this is not being used in practice. J83D corresponds to ATSC (discussed in Chapter 23), the only difference being that 16VSB modulation is proposed here instead of 8VSB.

17.1 J83B Transmission Parameters

The transmission parameters provided in J83B are:

- 64QAM, symbol rate = 5.05641 MS/s, r=0.18, net data rate = 26.970352 Mbit/s
- 256QAM, symbol rate = 5.360537 MS/s, r=0.12, net data rate = 38.810701 Mbit/s.

Fig. 17.2. shows the constellation diagrams used in J83B.

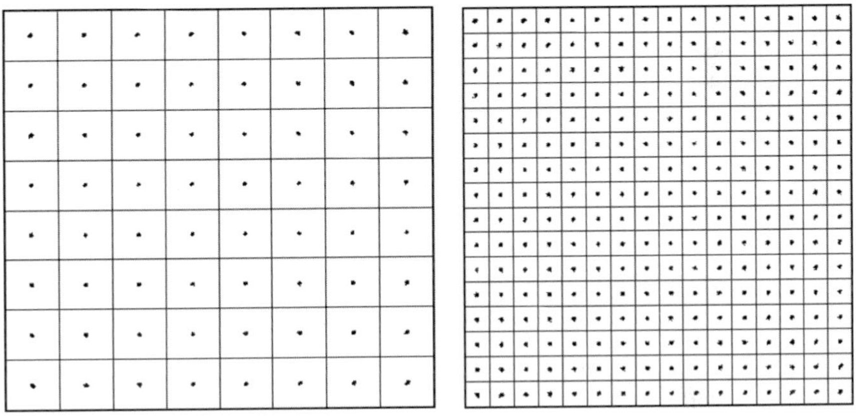

Fig. 17.2. Constellation diagrams in J83B (64QAM left and 256QAM right)

The gross data rates in J83B are calculated as follows:

Gross data rate = Symbolrate • bits/symbol;

64QAM: Gross data rate = 5.05641 MS/s • 6 bits/symbol
= 30.34 Mbit/s;

256QAM: Gross data rate = 5.360537 MS/s • 8 bits/symbol
= 42.88 Mbit/s;

Because the symbol rate is higher with 256QAM, a smaller roll-off-factor of r=0.12 is used there.

17.2 J83B Baseband Input Signals

In contrast to DVB-C, J83B is not tied to the MPEG-2 transport stream as input signal. J83B provides both MPEG-2 transport streams and, e.g., ATM data. There is no coupling between the J83B FEC layer and the input data signal. Should there be an MPEG-2 transport stream, however, the sync byte 0x47 is replaced by a checksum calculated over the entire transport stream packet (similar to the ATM case).

17.3 Forward Error Correction

The error protection in J83B consists of a Reed Solomon RS(128,122) error protection which is composed of 7-bits-long Reed Solomon symbols. It is possible to repair 3 symbols per RS block. An FEC frame starts with a sync trailer which has the following structure:

- 64QAM: 42-bit sync trailer, 0x752C0D6C (28 bits), 4-bit control-word, 10 reserved bits (set to zero)
- 256QAM: 40-bit sync trailer, 0x71E84DD4 (32 bits), 4-bit control word, 4 reserved bits (set to zero).

The FEC frame in J83B consists of

- 60 RS-blocks of 128 RS symbols with a width of 7 bits with 64QAM
- 88 RS-blocks of 128 RS symbols with a width of 7 bits with 256QAM.

The Reed Solomon encoder is followed by the time interleaver (Fig. 17.3.). This has the task of breaking up burst errors into individual errors in the deinterleaver in the receiver. Following that, the data stream is interleaved by mixing with a pseudo random sequence (Fig. 17.4.). The randomizer is reset during the sync trailer. The data stream is then fed to the trellis coder. Trellis coding is a special type of convolutional coding. The trellis coder has one input and N outputs, the number of outputs being matched to the subsequent mapping or to the modulation, respectively. If the selected modulation scheme allows the transmission of N bits per symbol, the trellis coder will have N outputs. Trellis coding can be traced back to Gottfried Ungerböck (IBM, 1982) and was used for the first time in telephone modems.

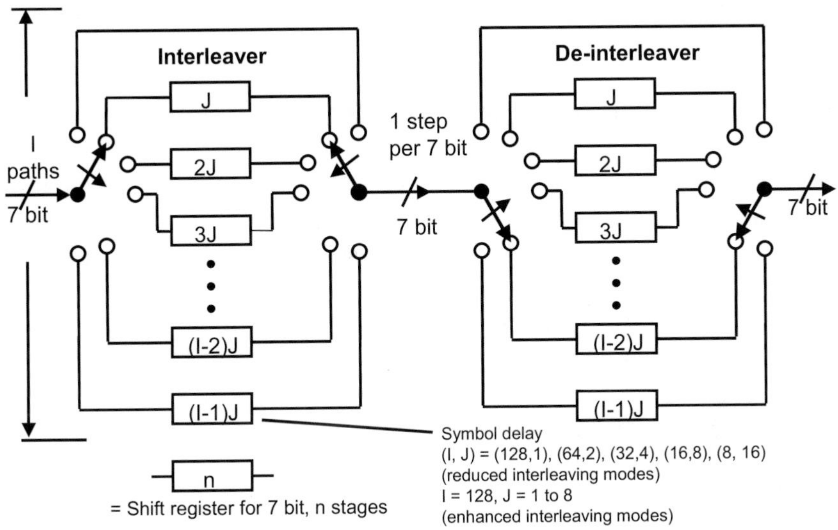

Fig. 17.3. Time interleaver

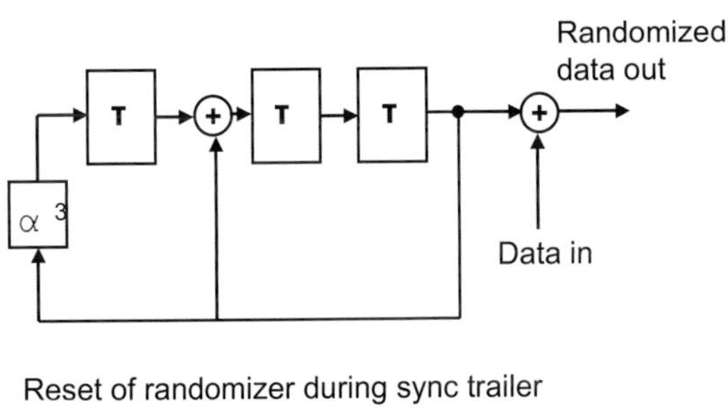

Fig.17.4. Randomizer

The subsequent modulation is a combination of coherent and differential modulation. Due to the N x 90^0 uncertainty in the QAM modulation, the quadrant is mapped differentially. The trellis coder in J83B consists of a parser which supplies 4 or 6 bits directly to the mapper, and a differential precoder followed by convolutional coders which process the two bits contained in the quadrants. A total of 6 or 8 bits, respectively, are then mapped. The input signal of the parser is formed by groups of 7 bits each (a total of 28 or 38 bits, resp.). The parser is responsible for forming the groups and also alters their sequential order in accordance with a defined arrangement.

17.4 Calculation of the Net Data Rate

In J83B, the net data rate is calculated with knowledge of the gross data rate, the error protection used and the J83B frame structure. There is no frame structure in DVB-S and DVB-C which is why the net data rate can be calculated there in a relatively simple way. It is even more complex in the case of standards like DVB-T and especially with the new DVB standards. The formula for calculating the net data rate with J83B is:

Net data rate = Gross data rate • f1 • f2 • f3;

where
f1 = 122/128 = 0.953125 = Reed Solomon factor;
f2 = frame_size_factor = f4/(f4+f5);
f3 = code_rate;
f4 = bits_per_frame_without_trailer;
f5 = bits_per_trailer;

With 64QAM, this results in:

f4 = 60 • 128 • 7 bits = 53760 bits;
f5 = 42 bits;
f2 = 53760/(53760 + 42) = 0.9992;
f3 = 14/15 = 0.93333;

Net data rate = 30.34 Mbit/s • 0.9992 • 0.93333 = 26.97 Mbit/s;

and with 256QAM:

f4 = 88 • 128 • 7 bits = 78848 bits;
f5 = 40 bits;
f2 = 78848/(78848 + 40) = 0.9995;
f3 = 19/20 = 0.95;

Net data rate = 42.88 Mbit/s • 0.953125 • 0.9995 = 38.81 Mbit/s;

17.5 Roll-off Filtering

In J83B, two different roll-off factors are possible depending on the type of modulation selected, which are:

- with 64QAM, r=0.18
- and with 256QAM, r=0.12.

Fig. 17.8. shows the spectra with 64QAM and 256QAM. The different roll-off factors can be clearly seen.

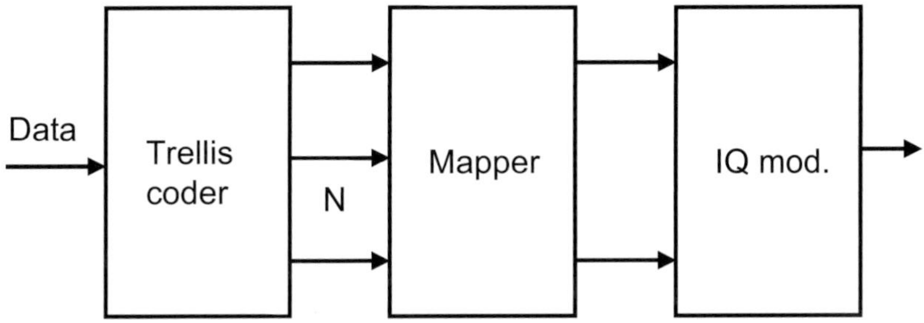

Fig. 17.5. Principle of trellis coded modulation (TCM)

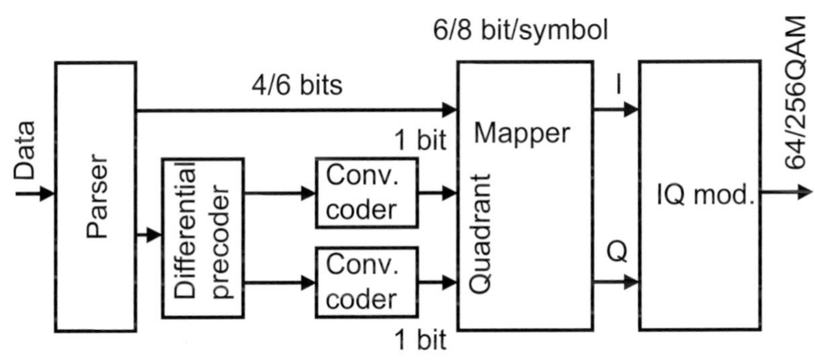

Fig. 17.6. Trellis-coded modulation in J83B

17.6 Fall-off-the-Cliff

Due to the fact that J83B uses two error protection mechanisms, there are, in principle, three possible bit error ratios in the receiver which are

- the bit error ratio before Viterbi,
- the bit error ratio before Reed-Solomon
- and the bit error ratio after Reed-Solomon.

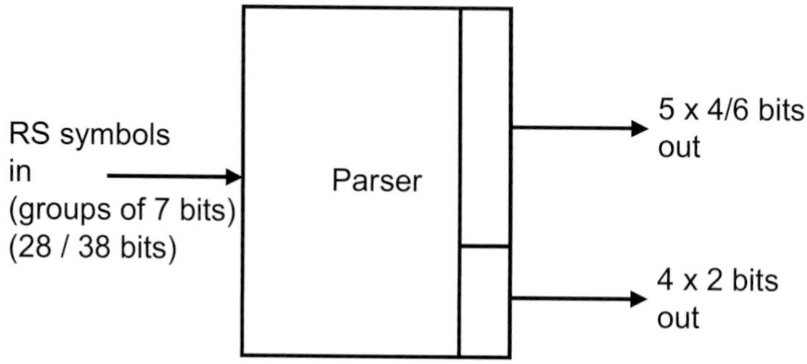

Fig. 17.7. Parser

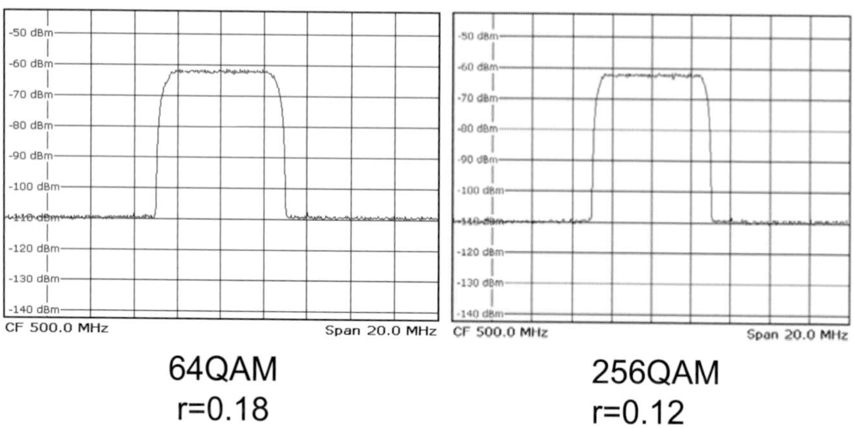

Fig. 17.8. Spectra of 64QAM and 256QAM

With trellis coding, the bit error ratio before Viterbi cannot be measured technically because of ambiguities in the receiver in the Viterbi decoding. With J83B, the transmission system is "at-the-cliff" (approx.)

- 64QAM with a C/N of 22 dB
- 256QAM with a C/N of 28 dB.

This corresponds to a bit error ratio before Reed-Solomon of

$1.4 \cdot 10^{-4}$ (experimental values; not from the standard)

Bibliography: [ITUJ83], [EFA], [SFQ], [SFU], [ETL]

18 Measuring Digital TV Signals in the Broadband Cable

In contrast to the measuring techniques used on digital TV signals transmitted via satellite, a wider range of measuring techniques is provided for broadband cable testing and is also necessary. The influences acting on the broadband cable signal, which can be modulated with up to 256QAM, are more varied by far, and more critical than in the satellite domain. In this section, the test instruments and measuring methods for measurements on DVB-C and J83A, B, C signals will be discussed. A large amount of space is reserved for the so-called constellation analysis of I/Q modulated signals which is also encountered in DVB-T. The influences or parameters to be considered in cable transmission are:

- Signal level
- C/N and S/N ratio
- I/Q modulator errors
- Interferers
- Phase jitter
- Echoes in the cable
- Frequency response
- Bit error ratio
- Modulation error ratio and error vector magnitude

To be able to detect and evaluate these influences, the following test instruments are used:

- An up-to-date spectrum analyzer
- A test receiver with constellation analysis
- A test transmitter with integrated noise generator and/or channel simulator for stress testing DVB-C and J83A, B, C receivers

W. Fischer, *Digital Video and Audio Broadcasting Technology*, Signals and Communication Technology, 3rd ed., DOI: 10.1007/978-3-642-11612-4_18,

18.1 DVB-C/J83A, B, C Test Receivers with Constellation Analysis

The most important test instrument for measuring digital TV signals in broadband cable networks is a DVB-C/J83A, B, C test receiver with an integrated constellation analyzer. Such a test receiver operates as follows:

The digital TV signal is received by a high-quality cable tuner which converts it to IF. The TV channel to be received is then band limited to 8, 7 or 6 MHz by an SAW (surface acoustic wave) filter, thus suppressing adjacent channels. The TV channel is then usually down-converted to an even lower 2nd IF in order to be able to use more inexpensive and better A/D converters. The IF signal, which has been filtered with an anti-aliasing-type low-pass filter, is then sampled with an A/D converter and demodulated in the DVB-C/J83A, B, C demodulator. During this process, a signal processor accesses the demodulator at the I/Q level and detects the constellation points as hit frequencies in the I and Q direction in the decision fields of the QAM constellation diagram. This provides frequency distributions ('clouds') around the individual constellation points - there are 64QAM clouds in the case of 64QAM. The individual QAM parameters are then determined by mathematical analyses of the frequency distributions. In addition, the constellation diagram itself is displayed graphically and can then be assessed visually. The signal is then also demodulated to become the MPEG-2 transport stream which can be supplied to an MPEG-2 test decoder for further analyses.

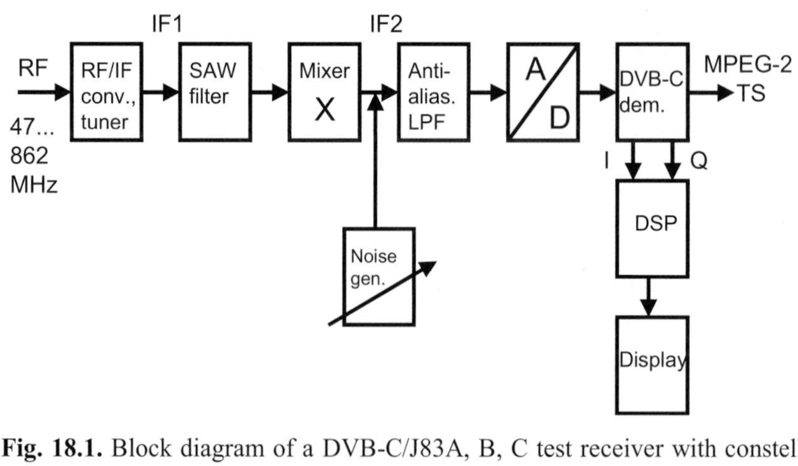

Fig. 18.1. Block diagram of a DVB-C/J83A, B, C test receiver with constellation analyzer

If the correct DVB-C or J83A, B, C signal is present at the test receiver and all settings at the receiver have been selected so that it can correctly lock to the QAM signal, a constellation diagram with constellation points of varying size (Fig. 18.2.) and the appearance of noise clouds is obtained. The size of the constellation points depends on the magnitude of the interference effects. The smaller the constellation points, the better the signal quality.

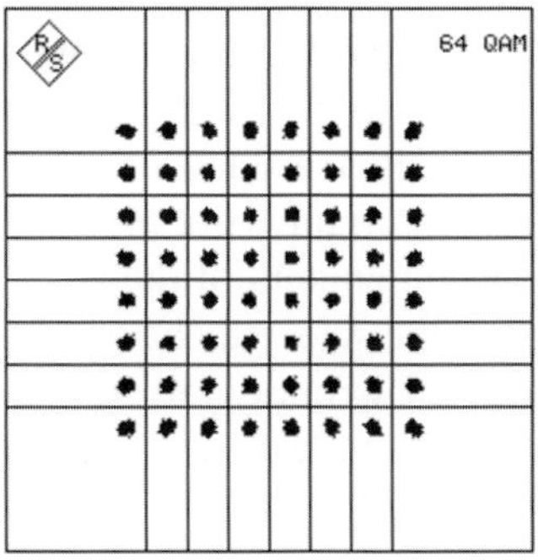

Fig. 18.2. Correctly locked 64QAM constellation diagram with noise

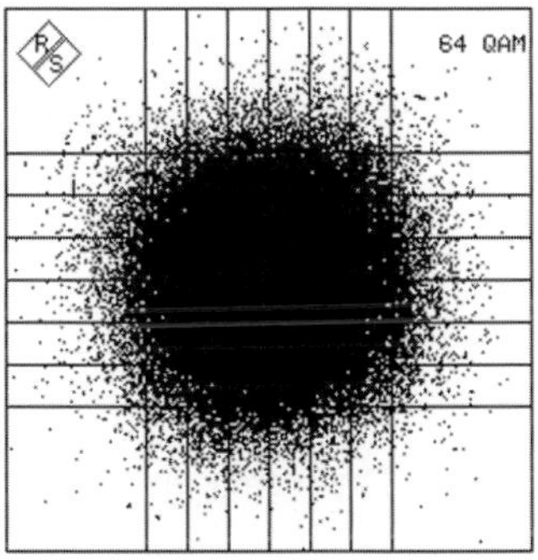

Fig. 18.3. No QAM signal in the selected channel, only noise

For the receiver to lock up in DVB-C and J83A, B, C, the following adjustment parameters of the test receiver must be selected correctly:

- Channel frequency: channel band center, approx. 47 to 862 MHz
- Standard: DVB-C/J83A, J83B or J83C

- Channel bandwidth: 6, 7, 8 MHz
- QAM level: 16QAM, 32QAM, 64QAM, 128QAM, 256QAM
- Symbol rate: approx. 2 to 7 MS/s
- SAW filter: ON with adjacent channels occupied
- Input attenuation control: to AUTO if provided

If there is simply no signal in the selected RF channel, the constellation analyzer of the test receiver will display a completely noisy constellation diagram (Fig. 18.3.) which exhibits no regular features whatever. It appears like a giant constellation point in the center of the display, but without sharp contours.

If accidentally an analog channel has been selected instead of, e.g. a DVB-C channel, constellation diagrams like Lissajou figures are produced which change continuously depending on the current content of the analog TV channel. If, however, there is a QAM signal in the selected channel but some of the receiver parameters have been selected wrongly (RF not exactly right, maybe the wrong symbol rate, wrong QAM level etc), a giant constellation point with much sharper contours appears.

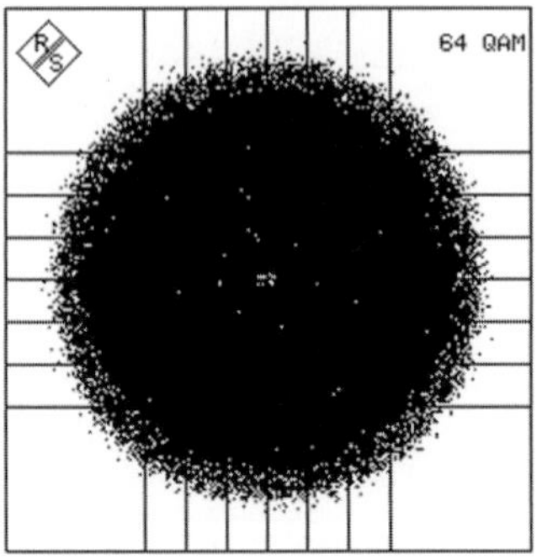

Fig. 18.4. Constellation diagram with the wrong carrier frequency and wrong symbol Rate selected (completely unsynchronized)

If all parameters have been selected correctly and only the carrier frequency is still divergent, the constellation diagram will rotate. It is then possible to see concentric circles.

An ideal, completely undistorted constellation diagram would show only a single constellation point per decision field in the exact center of the fields (Fig. 18.6.). However, such a constellation diagram can only be generated in a simulation.

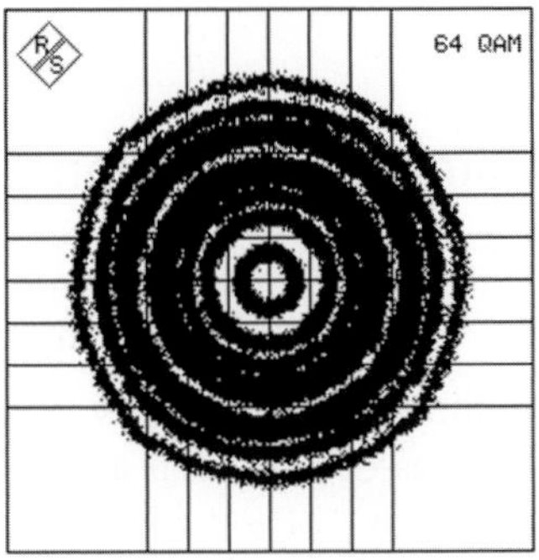

Fig. 18.5. QAM signal with the carrier out of sync

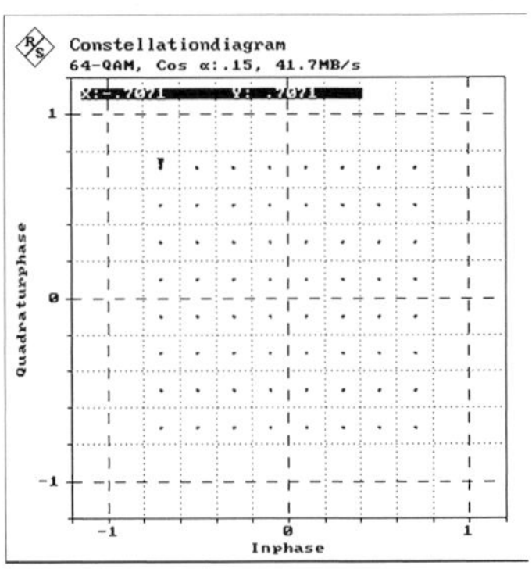

Fig. 18.6. Completely undistorted constellation diagram of an ideal undisturbed 64QAM signal

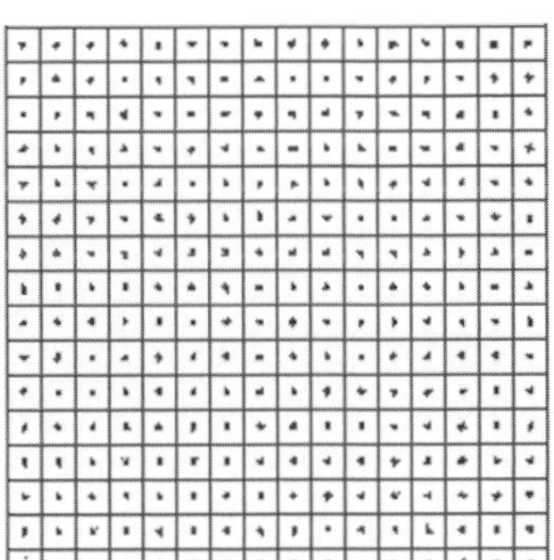

Fig. 18.7. 256QAM modulated DVB-C signal

Today, transmissions of up to 256 QAM are also encountered, mainly in HFC (hybrid fiber coax) networks. Such a constellation diagram is shown in Fig. 18.7.

18.2 Detecting Interference Effects Using Constellation Analysis

In this section, the most important interference effects on the broadband cable transmission link are discussed, and how they are analysed by using the constellation diagram. The following influences can be seen and distinguished directly by means of constellation analysis:

- Additive white Gaussian noise
- Phase jitter
- Interference
- Modulator I/Q errors

Apart from assessing the constellation diagram purely visually, the following parameters can also be calculated directly from it:

- Signal level
- C/N and S/N ratio
- Phase jitter
- I/Q amplitude imbalance
- I/Q phase error
- Carrier suppression
- Modulation error ratio (MER)
- Error vector magnitude (EVM)

18.2.1 Additive White Gaussian Noise (AWGN)

One interference effect which affects all types of transmission links in the same way is the so-called white Gaussian noise (AWGN). This effect can emanate more or less from virtually any point along the transmission link. In the constellation diagram, noise-like effects are recognized from the constellation points which are now of varying size (Fig. 18.8.). To measure the RMS value of the noise-like interferer, the hits in the individual areas are counted within the individual constellation fields, i.e., the frequency with which the center and the areas around it are hit at ever increasing distance is detected. If these hits or counts within a constellation field were to be displayed multi-dimensionally, a two-dimensional bell-shaped Gaussian curve would be obtained (Fig. 18.9.).

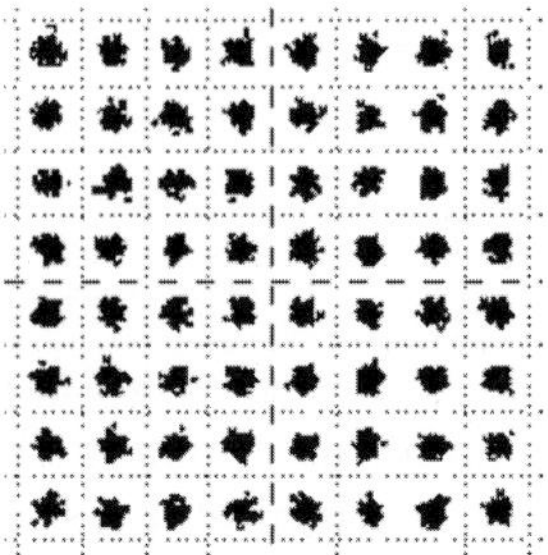

Fig. 18.8. Constellation diagram of a 64QAM signal with additive noise

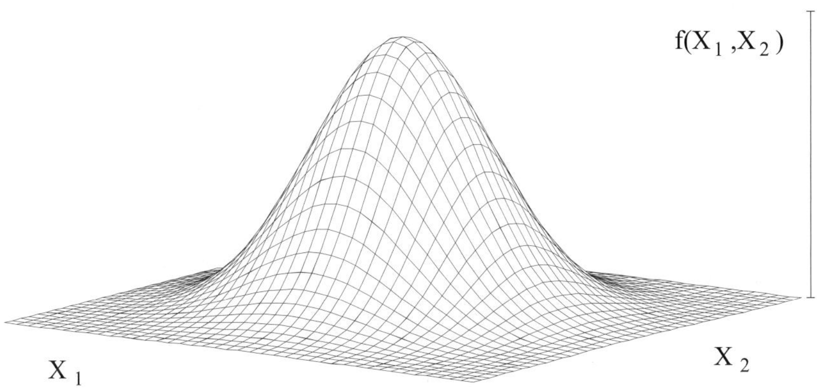

Fig. 18.9. Two-dimensional bell-shaped Gaussian curve [EFA]

This two-dimensional distribution will then be found similarly in every constellation field. To find the RMS value of the noise effect, the standard deviation is then simply calculated from these hit results. The standard deviation corresponds directly to the RMS value of the noise signal. Relating this RMS value N to the amplitude of the QAM signal S allows the logarithmic signal/noise ratio S/N in dB to be calculated by taking the logarithm.

A normal frequency distribution can be described by the Gaussian normal distribution function as:

$$y(x) = \frac{1}{\sigma \cdot \sqrt{2\pi}} e^{-0.5 \cdot (\frac{x-\mu}{\sigma})^2};$$

where σ = standard deviation, μ = mean value. The standard deviation can be calculated from the counter results as:

$$\sigma = \sqrt{\int_{-\infty}^{\infty} (x-\mu)^2 f(x)dx};$$

It can be seen clearly that, in principle, the formula for determining the standard deviation corresponds to the mathematical relationship for calculating the RMS value.

It must be noted, however, that it is not only noise but also impulse interferers or intermodulation and cross-modulation products which, due to non-linearities on the transmission link, can cause comparable noise-cloud-like distortions in the constellation diagram and thus can not be distinguished from actual noise.

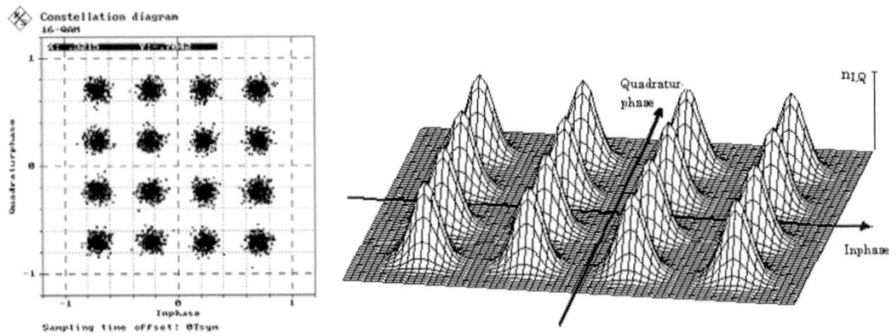

Fig. 18.10. Two-dimensional hit-rates in the 16QAM constellation diagram [HOFMEISTER]

In principle, there are two definitions for the signal-to-noise level: the signal-to-noise ratio S/N and the carrier-to-noise ratio C/N. Each can be converted to the other one. The C/N ratio is always referred to the actual bandwidth of the channel which is 6, 7 or 8 MHz on broadband cable networks. S/N refers to the conditions after roll-off filtering and to the actual Nyquist bandwidth of the signal. The symbol rate of the signal should be used for signal bandwidth and for noise bandwidth. But basically it is recommended to use the symbol rate also as reference bandwidth for the noise bandwidth in measuring C/N, providing an unambiguous definition for the C/N ratio as proposed, e.g., in the DVB Measurement Guidelines [ETR290].

The signal power S is obtained from the carrier power C, as

$$S = C[\text{dBm}] + 10 \log(1 - r/4);$$

where r is the roll-off factor.

The logarithmic signal-to-noise ratio S/N is, therefore:

$$S/N[\text{dB}] = C/N[\text{dB}] + 10 \log(1-r/4);$$

Example:
Channel bandwidth: 8 MHz
Symbol rate: 6.9 MS/s
Roll-off factor: 0.15

$$S/N[\text{dB}] = C/N[\text{dB}] + 10 \log(1-0.15/4)$$
$$= C/N[\text{dB}] - 0.1660 \text{ dB};$$

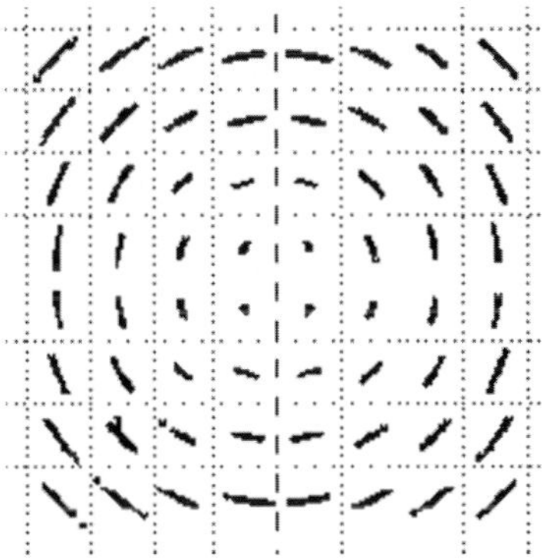

Fig. 18.11. State diagram of a 64QAM signal with phase jitter

18.2.2 Phase Jitter

Phase jitter or phase noise in the QAM signal is caused by converters in the transmission path or by the I/Q modulator itself. In the constellation diagram, phase jitter produces smear distortion of greater or lesser magnitude (Fig. 18.11.). The constellation diagram 'totters' in rotation around the center point.

To find the phase jitter, the smear distortions of the outermost constellation points are measured which is where the phase jitter has the greatest effect. Then the frequency distribution within the decision field is considered along the circular path whose center point is at the origin of the state diagram. Again, the standard deviation which is still affected by additional noise can be calculated here. This noise effect must then still be calculated out.

18.2.3 Sinusoidal Interferer

A sinusoidal interferer (Fig. 18.12.) produces circular distortions of the constellation points. These circles are the result of the interference vector rotating around the center of the constellation point. The diameter of the circles corresponds to the amplitude of the sinusoidal interferer.

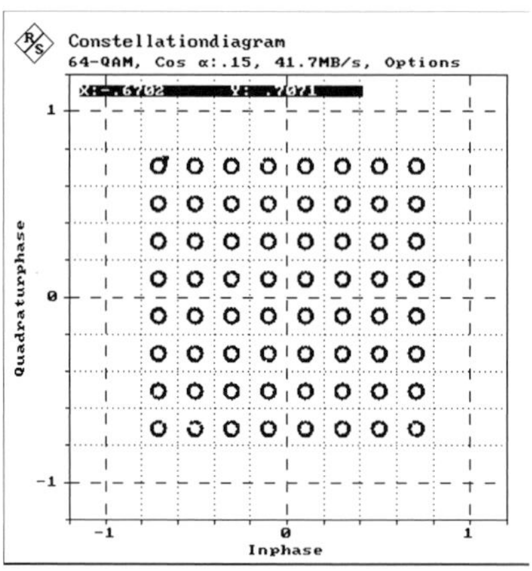

Fig. 18.12. Effect of a sinusoidal interferer

18.2.4 Effects of the I/Q Modulator

In the first generation of DVB-C modulators, analog I/Q modulators were used. Errors in the I/Q modulator (Fig. 18.13.) then resulted in I/Q errors in the QAM-modulated signal. If, e.g., the I branch has a different gain than the Q branch of the I/Q modulator, I/Q amplitude imbalance is produced. If the 90° phase shifter in the carrier feed to the Q modulator is not exactly 90 degrees, an I/Q phase error is produced. Lack of carrier suppression was an even more frequent problem. This is caused by carrier cross-talk or by some DC component in the I or Q modulation signal. Today, the I/Q modulators in broadband cable are exclusively digital and the problems of the I/Q modulator described are no longer relevant. They will only be mentioned briefly here for the sake of completeness.

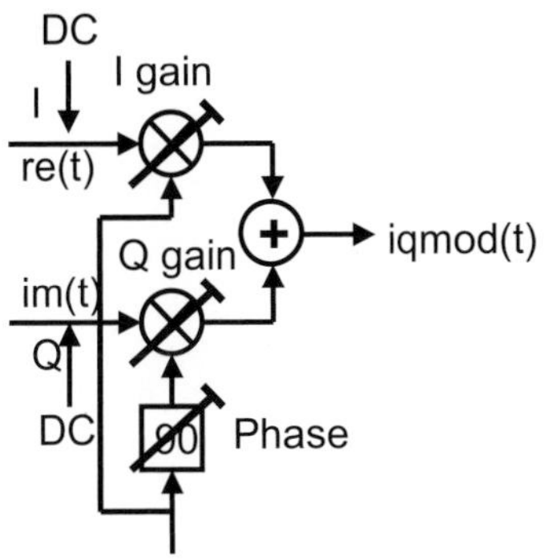

Fig. 18.13. IQ modulator with IQ errors

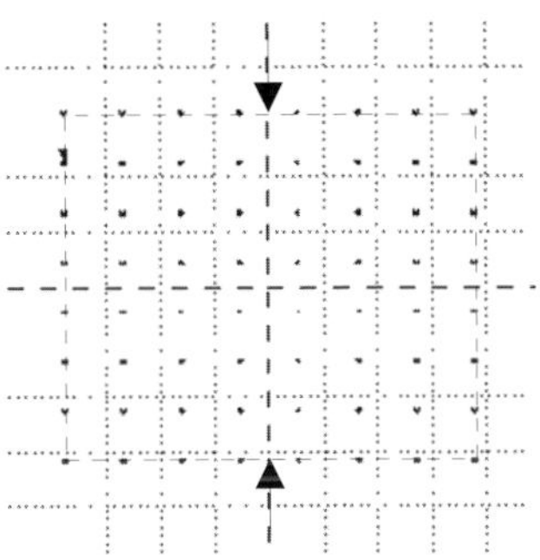

Fig. 18.14. Constellation diagram with IQ imbalance

18.2.4.1 I/Q Imbalance

In the case of an I/Q imbalance, the constellation diagram is squashed in the I or Q direction resulting in a rectangular diagram instead of a square one (Fig. 18.14.). The amplitude imbalance can be determined by measuring the lengths of the sides of the rectangle. It is defined as

$$AI = (v_2/v_1 - 1) \bullet 100\%;$$

where v_I is the gain in the I direction or I side of the rectangle, and v_2 is the gain in the Q direction or Q side of the rectangle.

18.2.4.2 I/Q Phase Error

An I/Q phase error (PE) (Fig. 18.15.) leads to a diamond-shaped constellation diagram. The phase error in the 90° phase shifter of the I/Q modulator can be determined from the angles of the diamond in the constellation diagram. The acute angle then has a value of 90° - PE and the obtuse angle has a value of 90° + PE.

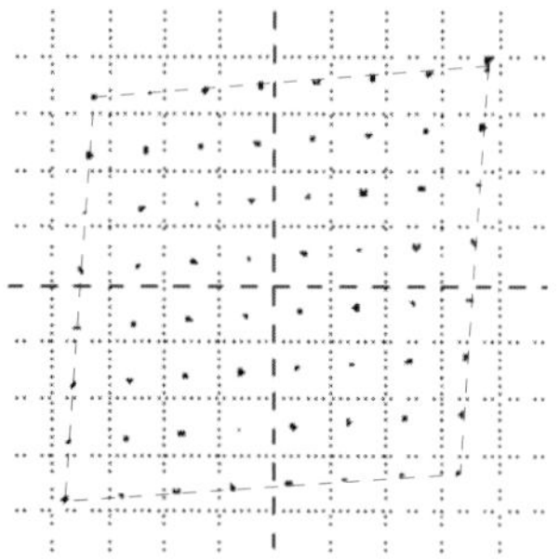

Fig. 18.15. Constellation diagram with an IQ phase error

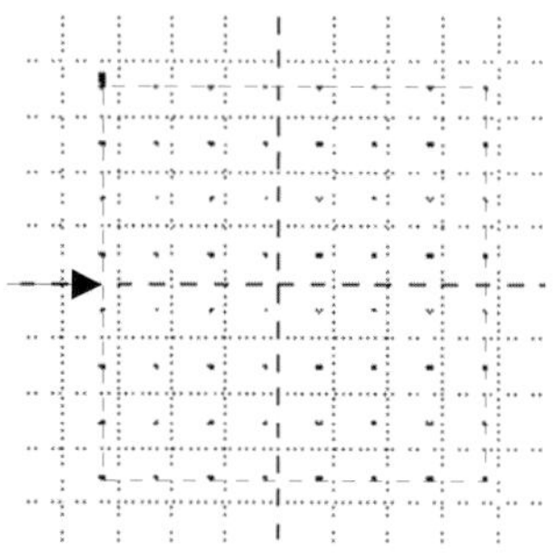

Fig. 18.16. Constellation diagram with insufficient carrier suppression

18.2.4.3 Carrier Suppression

In the case of insufficient carrier suppression (Fig. 18.16.), the constellation diagram is pushed away from the center in some direction. The degree of carrier suppression can be calculated from the magnitude of the displacement.

It is defined as:

$$CS = -10 \log(P_{RC}/P_{SIG}); \text{ [dB]}$$

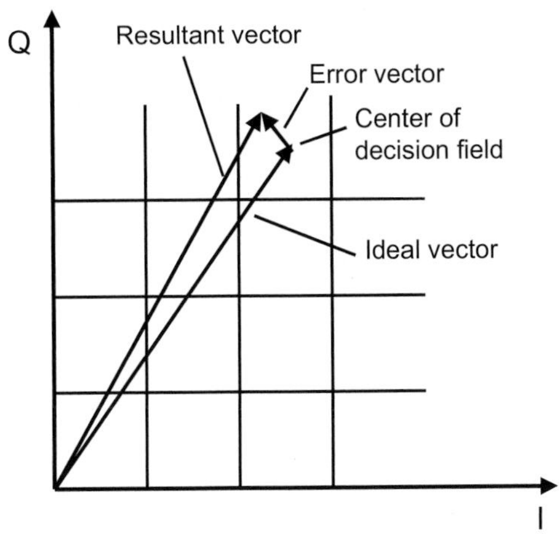

Fig. 18.17. Error vector for determining the modulation error ratio (MER)

18.2.5 Modulation Error Ratio (MER)

All the interference effects on a digital TV signal in broadband cable networks previously explained cause the constellation points to exhibit deviations from their nominal position in the center of the decision fields. If the deviations are too great, the decision thresholds will be exceeded and bit errors are produced. However, the deviations from the decision field center can also be considered to be measurement parameters for the size of any interference quantity. Which is precisely the object of an artificial measurement parameter like the modulation error ratio (MER). The MER measurement assumes that the actual hits in the constellation fields have been pushed out of the center of the respective field by interference quantities (Fig. 18.17.). The interference quantities are given error vectors and the error vector points from the center of the constellation field to the point of the actual hit in the constellation field. Then the lengths of all these error vectors are measured against time in each constellation field and the quadratic mean is formed or the maximum peak value is acquired in a time window. The exact definition of MER can be found in DVB Measurement Guidelines [ETR290].

$$MER_{PEAK} = \frac{\max(|error\_vector|)}{U_{RMS}} \cdot 100\%;$$

$$MER_{RMS} = \frac{\sqrt{\frac{1}{N}\sum_{n=0}^{N-1}(|error\_vector|)^2}}{U_{RMS}} \cdot 100\%;$$

The reference U_{RMS} is here the RMS value of the QAM signal. Usually, however, a logarithmic scale is used:

$$MER_{dB} = 20\lg\left(\frac{MER[\%]}{100}\right); \quad [dB]$$

The MER value is thus an aggregate quantity which includes all possible individual errors and thus completely describes the performance of the transmission link.

In principle:

MER [dB] ≤ S/N [dB];

Table 18.1. MER and EVM

QAM	MER>EVM [%]	EVM>MER [%]	MER>EVM [dB]	MER>EVM [dB]
4	EVM=MER	MER=EVM	|EVM|=MER	MER=|EVM|
16	EVM= MER/1.342	EVM= MER•1.342	|EVM|= MER +2.56dB	MER= |EVM| -2.56dB
32	EVM= MER/1.304	EVM= MER•1.304	|EVM|= MER +2.31dB	MER= |EVM| -2.31dB
64	EVM= MER/1.527	EVM= MER•1.527	|EVM|= MER +3.68dB	MER= |EVM| -3.68dB
128	EVM= MER/1.440	EVM= MER•1.440	|EVM|= MER +3.17dB	MER= |EVM| -3.17dB
256	EVM= MER/1.627	EVM= MER•1.627	|EVM|= MER +4.23dB	MER= |EVM| -4.23dB

18.2.6 Error Vector Magnitude (EVM)

The error vector magnitude (EVM) is closely related to the modulation error ratio (MER), the only difference being the different reference used. Whereas in the MER, the reference is the RMS value of the QAM signal, it is the peak value of the QAM signal which is used as reference for the EVM.

EVM and MER can be converted from one to the other with the aid of table 18.1.

18.3 Measuring the Bit Error Ratio (BER)

In DVB-C and in J83A, C, the transmission is protected by Reed Solomon forward error correction RS(204,188). Using 16 error protection bytes per transport stream packet, this protection allows 8 single errors per TS packet to be corrected at the receiving end. Counting the correction events performed by the Reed Solomon decoder at the receiving end and assuming that these are attributable to single errors, and relating them to the incoming bitstream in the comparable period (a transport stream packet has 188•8 useful bits and a total of 204•8 bits), provides the bit error rate, a value between $1 \bullet 10^{-4}$ and $1 \bullet 10^{-11}$.

However, not all the errors can be corrected by the Reed Solomon decoder. Errors in TS packets which are no longer correctable lead to errored packets which are then marked by the transport error indicator in the MPEG-2 transport stream header. Counting the non-correctable errors and relating them to the corresponding data volume allows the post-Reed Solomon bit error ratio to be calculated.

Thus, there are two bit error ratios in DVB-C and J83A, C:

- Bit error ratio before Reed Solomon - the channel bit error ratio
- Bit error ratio after Reed Solomon

The bit error ratio (BER) is defined as:

BER = bit errors / transmitted bits;

The bit error ratio has a fixed relationship to the signal/noise ratio if only noise is involved. This relationship is shown in the figure below. In addition, the figure includes the equivalent noise degradation (END) and the noise margin for an example.

Equivalent Noise Degradation (END):

The equivalent noise degradation is a measure of the 'insertion loss' of the entire system from the modulator via the cable link up to the demodulator. It specifies the deviation of the real S/N ratio from the ideal for a BER of $1 \bullet 10^{-4}$ in dB. In practice, values of around 1 dB are achieved.

Noise Margin:

The noise margin is the margin between the C/N ratio leading to a BER of $1 \bullet 10^{-4}$, and the C/N value of the cable system. When the C/N value is measured in the cable, the channel bandwidth of the QAM signal is used as the noise bandwidth.

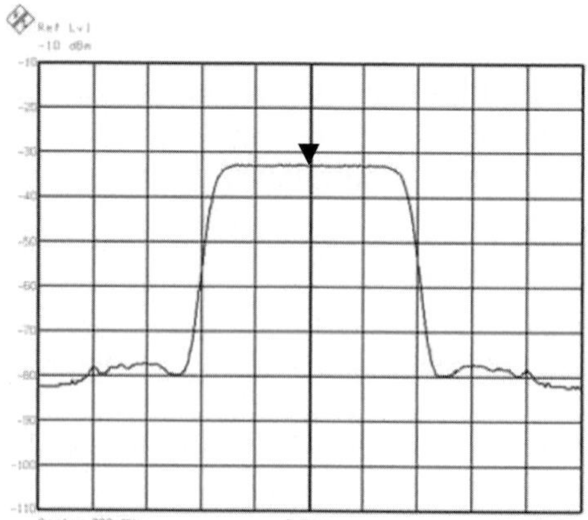

Fig. 18.18. Spectrum of a DVB-C signal

18.4 Using a Spectrum Analyzer for Measuring DVB-C Signals

A spectrum analyzer is a good instrument for measuring the power of the DVB-C channel, at least at the modulation end. A DVB-C signal looks like noise and has quite a high crest factor. Due to its similarity with white Gaussian noise, the power is measured the same way as in a noise power measurement.

To find the DVB-C/J83A,B,C carrier power, the spectrum analyzer is set up as follows:

At the analyzer, a resolution bandwidth of 300 kHz and a video bandwidth of 3 to 10 times the width of the resolution bandwidth (3 MHz) is selected. To achieve some averaging, a slow sweep time (2000 ms) must be set. These parameters are required because we are using the RMS detector of the spectrum analyzer. The following settings are then used:

- Center frequency: center of the cable channel
- Span: 10 MHz
- Resolution BW: 300 MHz
- Video BW: 3 MHz (due to RMS detector and log. display)
- Detector: RMS
- Sweep: slow (2000 ms)
- Noise marker: channel center (C' in dBm/Hz)

To measure power, the noise marker is used because of the noise-like signal. The noise marker is set to band center for this. The prerequisite is a flat channel which, however, can always be assumed at the modulator. If the channel is not flat, other suitable but analyzer-dependent measuring functions must be used for measuring the channel power.

The analyzer provides the C' value as noise power density at the position of the noise marker in dBm/Hz, automatically taking into consideration the filter bandwidth and the characteristics of the logarithmic amplifier of the analyzer. To relate the signal power density C' to the Nyquist bandwidth B_N of the cable signal. the signal power C must be calculated as follows:

$$C = C' + 10\log B_N$$
$$= C' + 10\log(symbol\_rate / Hz)\ \text{dB}; \qquad [\text{dBm}]$$

The Nyquist bandwidth of the signal corresponds to the symbol rate of the cable signal.

Example:

Measurement value of the noise marker:	-100.0 dBm/Hz
Correction value at 6.9 MS/s symbol rate:	+ 68.4 dB
Power in the channel:	- 31.6 dBm

Finding the Noise Power N by Approximation:

If it were possible to switch off the DVB-C/J83A,B,C signal without changing the noise conditions in the channel, the noise marker in the center of the channel would provide information about the noise conditions in the channel. However, this can not be done so easily. A 'good idea'at least, if not an exact measurement value, about the noise power in the channel is obtained if the noise marker is used for measuring quite near to the signal on the 'shoulder' of the DVB-C/J83A, B, C signal. This is because it can be assumed that the noise fringe within the useful band continues in a similar way to how it appears on the shoulder.

The value N' of the noise power density is output by the spectrum analyzer. To calculate the noise power in the channel having the bandwidth B_K of the cable transmission channel from the noise power density N', the noise power N must be found as follows:

$$N = N' + 10\log(B_N)$$
$$= N' + 10\log(signal\_bandwidth / Hz)\ dB; \qquad [dBm]$$

The noise bandwidth is recommended by [ETR290] which is, the symbol rate.

Example:

Measurement value of the noise marker:	-140.0 dBm/Hz
Correction value at 8 MHz bandwidth:	+ 68.4 dB
Noise power in the cable channel:	- 71.6 dBm

The resultant C/N value is then:

$$C/N[\text{dB}] = C[\text{dBm}] - N[\text{dBm}];$$

i.e. $C/N_{[dB]} = -31.6\ \text{dBm} - (-71.6\ \text{dBm})$
$= 40\ \text{dB};$

18.5 Measuring the Shoulder Attenuation

Out-of-band components close to the wanted DVB-C/J83A,B,C band are recognized from the 'shoulders' of the QAM signal (s. Fig. 18.18. and 18.19.). These shoulders should be suppressed as well as possible so as to cause the least possible interference to the adjacent channels. This is defined as required minimum shoulder attenuation (e.g. 43 dB). The shoulder attenuation is measured by using simple marker functions of the spectrum analyzer.

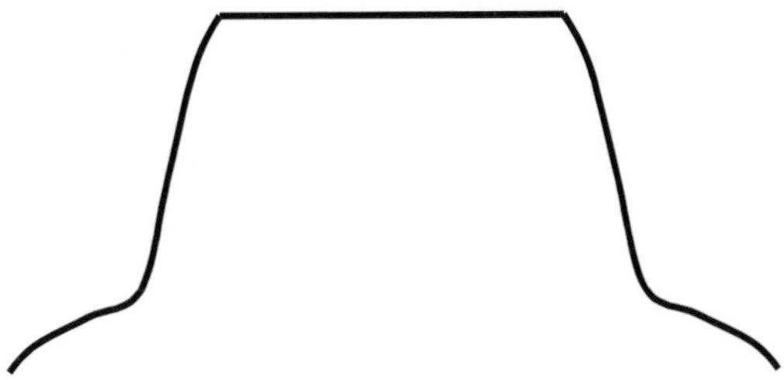

Fig. 18.19. Shoulders on the DVB-C signal

18.6 Measuring the Ripple or Tilt in the Channel

The ripple in the amplitude response of a digital TV channel should be as low as possible (less than 0.4 dB_{PP}). Moreover, the tilt of this channel should not be greater than this value, either. The ripple and tilt of the channel can be measured by using a spectrum analyzer. The correction data of the channel equalizer in the test receiver can also be used for this measurement. Some cable test receivers allow the channel frequency response to be calculated from this.

18.7 DVB-C/J83A,B,C Receiver Test

As in DVB-S and also in DVB-T, the testing of receivers (set-top boxes and integrated) is very important. The test transmitters can simulate a cable transmission link and the modulation process. Apart from the cable modulator and upconverter, such a test transmitter (e.g. Rohde&Schwarz TV Test Transmitter SFQ, SFU) also contains an add-on noise source and possibly even a channel simulator. The test transmitter is fed with an MPEG-2 transport stream from an MPEG-2 generator. The output signal of the test transmitter can be supplied directly to the input of the cable receiver. It is then possible to generate various stress conditions for the receiver by altering numerous parameters. It is also possible to measure the bit error rate as a function of the C/N ratio.

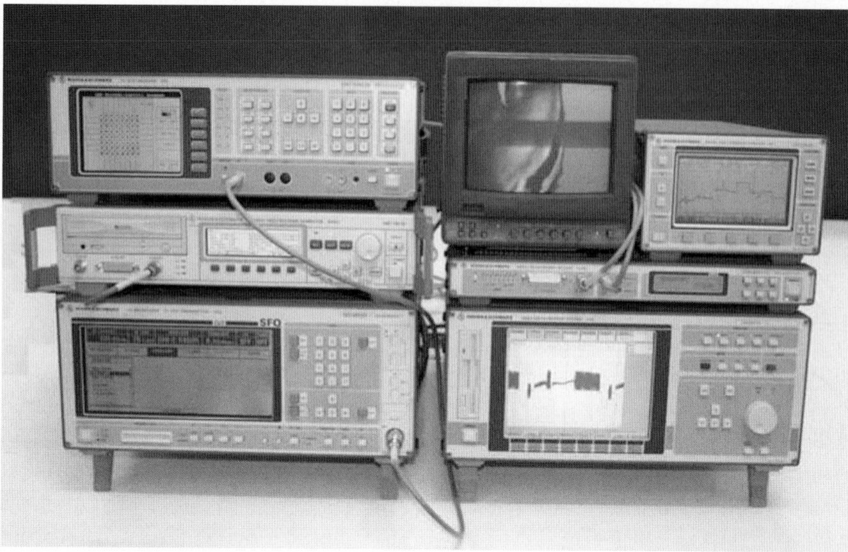

Fig. 18.20. Constellation analysis on a DVB-C signal from a test transmitter (Rohde&Schwarz SFQ, bottom left) using a test transmitter (Rohde&Schwarz EFA, top left): An MPEG-2 generator (Rohde&Schwarz DVRG, center left) supplies an MPEG-2 transport stream with test contents which is fed into the test transmitter. The DVB-C receiver EFA displays the DVB-C Signal back into the MPEG-2 transport stream which can then be decoded with an MPEG-2 test decoder (Rohde&Schwarz DVMD, center right). The picture also shows the video analyzer VSA (bottom right), the TV monitor (top center) and a "601" analyzer VCA (top right).

Bibliography: [ETR290], [EFA], [SFQ], [HOFMEISTER], [ETS300429], [REIMERS], [GRUNWALD], [JAEGER], [FISCHER3], [SFU]

19 Coded Orthogonal Frequency Division Multiplex (COFDM)

Almost from the beginning of the electrical transmission of messages about 100 years ago, single-carrier methods have been used for transmitting information. The message to be transmitted is impressed on a sinusoidal carrier by applying analog amplitude, frequency or phase modulation techniques. Since the eighties, single-carrier transmission is more and more by digital methods in the form of frequency shift keying (FSK) and in many cases also by vector modulation (QPSK, QAM). The main applications for this are fax, modem, mobile radio, microwave links and satellite transmission and the transmission of data over broadband cables. However, the characteristics of many transmission paths are such that single-carrier methods prove to be sensitive to interference, complex or inadequate. Since the days of Marconi and Hertz, however, it is precisely these transmission links which are used most frequently. Today, every child knows of transistor radios, television receivers and mobiles or the simple walkie-talkies, all of which operate with a modulated carrier in a terrestrial environment. And every car driver knows the effect of reception of the radio program he is listening to suddenly ceasing when he stops at a red light - he is in a 'dead spot'. Due to multi-path reception, fading occurs which is frequency- and location-selective. In terrestrial radio transmission, narrowband or wideband sinusoidal or impulse-type interferers must also be expected which can adversely affect reception. Location, type and orientation and mobility, i.e. movement, all play a role. This applies both to radio and TV reception and to reception via mobile radios. Terrestrial conditions of reception are the most difficult types of reception of all. This similarly applies to the old two-wire line in the telecommunications field. There can be echoes, crosstalk from other pairs, impulse interferers and amplitude and group delay response. However, the demand for data links with higher bit rates from PC to Internet is increasing more and more. The usual single-carrier methods and also data transmission systems such as ISDN are already reaching their limits. For many people, 64 kbit/s or 128 kbit/s with grouped channels in ISDN is not enough. In the terrestrial radio link it is now the broadcasting services, which have always had a wide

W. Fischer, *Digital Video and Audio Broadcasting Technology*, Signals and Communication Technology, 3rd ed., DOI: 10.1007/978-3-642-11612-4_19, © Springer-Verlag Berlin Heidelberg 2010

bandwidth such as television with normally up to 8 MHz bandwidth, which are 'crying out' for reliable digital transmission methods. Using a multi-carrier method is one reliable approach to this. The information is transmitted digitally not via one carrier but via many - in some cases thousands of subcarriers with multiple error protection and data interleaving. These methods, which have been known since the seventies, are:

- Coded Orthogonal frequency division multiplexing (COFDM),
- Discrete multitone (DMT).

They are used in

- Digital Audio Broadcasting (DAB),
- Digital Video Broadcasting (DVB-T),
- Asymmetrical Digital Subscriber Line (ADSL),
- Transmission of data signals via power lines.
- ISDB-T
- DTMB
- DVB-T2
- DVB-C2

In this Section, the background, characteristics and generation of multi-carrier modulation methods such as coded orthogonal frequency division multiplexing (COFDM) or discrete multitone (DMT) are described.

The concept of multi-carrier modulation goes back to investigations in the Bell Laboratories in the U.S. [CHANG] and to ideas in France in the seventies. In those days, however, chips which were fast enough to implement these ideas were nowhere in sight. It wasn't until many years later, at the beginning of the nineties, that the concept was turned into reality and applied for the first time in Digital Audio Broadcasting (DAB). Although DAB cannot really be called an absolute marketing success, this is certainly not due to the technology but rather the inappropriate marketing (or complete lack of it) and this, in principle, is attributable to industry and politics. The technology itself is first class. Even today, it is difficult to convince the consumer in many fields that this product or the other is better. It is certainly correct to leave many decisions to the consumer but he or she must then have knowledge of the new possibilities and underlying principles or even be able to purchase the new type of product. In the case of DAB, this has only been possible since 2001 (availability of DAB receivers) which is unfortunate for this excellent method of transmitting audio virtually in CD quality via terrestrial channels. The situation is different in the case of ADSL in telecommunications where ADSL is increasingly accepted and demanded for Internet access because of its speed, and in the case of DVB-T which is spreading in countries where it is politically promoted and prescribed as an appropriate technology. If applied correctly, DAB and DVB-T make a good contribution to the conservation of energy and frequencies whilst at the same time delivering better performance.

Fig. 19.1. The terrestrial radio channel

19.1 Why Multi-Carrier?

Multi-carrier methods belong to the most complicated transmission methods of all and are in no way inferior to the code division multiple access (CDMA) methods. But why this complexity? The reason is simple: the transmission medium is an extremely difficult medium (Fig. 19.1.) to deal with.

The terrestrial transmission medium involves

- terrestrial transmission paths,
- difficult line-associated transmission conditions.

The terrestrial transmission paths, in particular, exhibit the following characteristic features:

- Multipath reception via various echo paths caused by reflections from buildings, mountains, trees, vehicles;
- Additive white Gaussian noise (AWGN);

- Narrow-band or wide-band interference sources caused by internal combustion engines, streetcars or other radio sources;
- Doppler effect, i.e. frequency shift in mobile reception.

Multipath reception leads to location- and frequency-selective fading phenomena (Fig. 19.2.), an effect known as "red-light effect" in car radios. The car stops at a red stop light and radio reception ceases. If one were to select another station or move the car slightly forward, reception would be restored. If information is transmitted by only one discrete carrier precisely at one particular frequency, echoes will cause cancellations of the received signal at particular locations at exactly this frequency. This effect is a function of the frequency, the intensity of the echo and the echo delay.

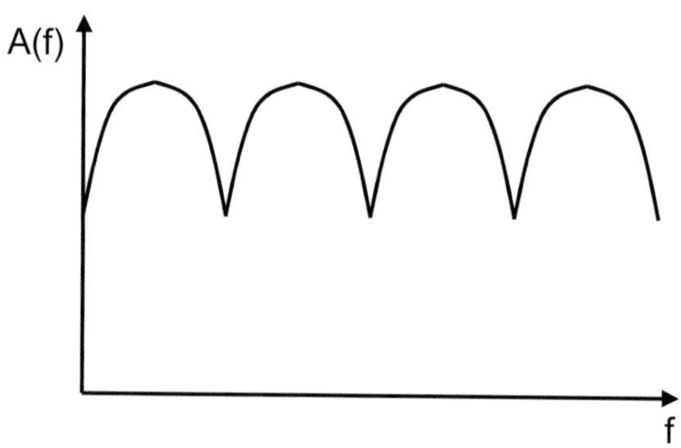

Fig. 19.2. Transfer function of a radio channel with multipath reception, frequency-selective Fading

If high data rates of digital signals are transmitted by vector modulated (I/Q modulated) carriers, they will exhibit a bandwidth which corresponds to the symbol rate.

The available bandwidth is usually specified. The symbol rate is obtained from the type of modulation and the data rate. However, single-carrier methods have a relatively high symbol rate, often within a range of more than 1 MS/s up to 30 MS/s. This leads to very short symbol periods of 1 µs and shorter (inverse of the symbol rate). However, echo delays can easily be within a range of up to 50 µs or more in terrestrial transmission channels. Such echoes would lead to inter-symbol interference between adjacent symbols or even far distant symbols and render transmission more or less impossible. An obvious trick would now be to make the symbol period as long as possible in order to minimize inter-symbol interference and, in addition, pauses could be inserted between the symbols, so-called guard intervals.

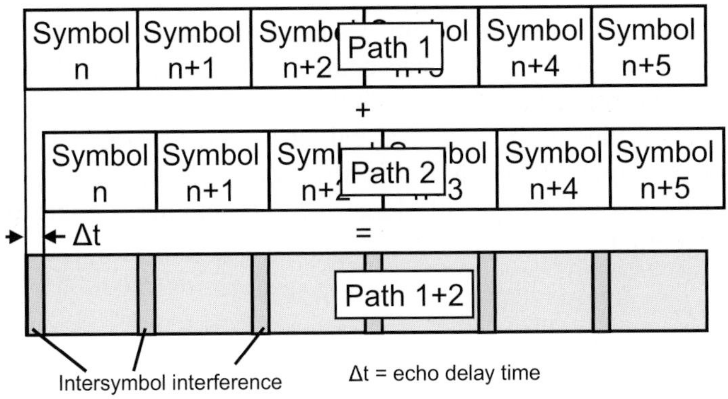

Fig. 19.3. Inter-symbol interference / intersymbol crosstalk with multipath reception

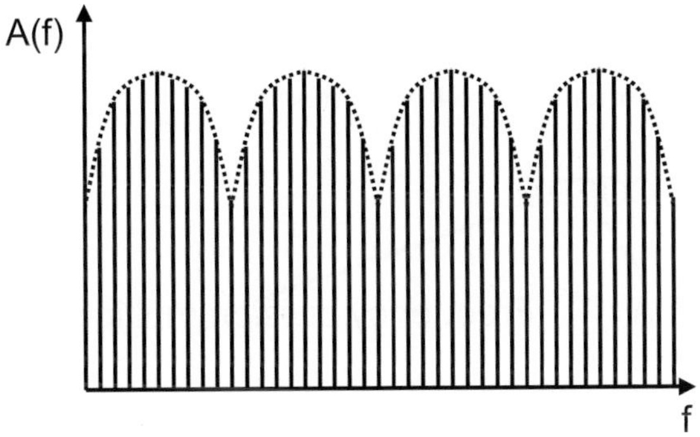

Fig. 19.4. COFDM: multicarrier in radio channel with fading

However, there is still the problem of the location- and frequency-selective fading phenomena. If then the information is not transmitted via a single carrier but is distributed over many, up to thousands of subcarriers and a corresponding overall error protection is built in, the available channel bandwidth remaining constant, individual carriers or carrier bands will be affected by the fading, but not all of them.

At the receiving end, sufficient error-free information could then be recovered from the relatively undisturbed carriers to be able to reconstruct an error-free output data stream by means of the error protection measures taken. If, however, many thousands of subcarriers are used instead of one carrier, the symbol rate is reduced by the factor of the number of subcarriers and the symbols are correspondingly lengthened several thousand times up to a millisecond. The fading problem is solved and, at the same time, the problem of inter-symbol interference is also solved due to the longer symbols and the appropriate pauses between them.

A multi-carrier method is born and is called Coded Orthogonal Frequency Division Multiplex (COFDM). It is now only necessary to see that the many adjacent carriers do not interfere with one another, i.e. are orthogonal to one another.

19.2 What is COFDM?

Orthogonal frequency division multiplex is a multi-carrier method with up to thousands of subcarriers, none of which interfere with each other because they are orthogonal to one another. The information to be transmitted is distributed interleaved to the many subcarriers, first having added the appropriate error protection, resulting in coded orthogonal frequency division multiplex (COFDM). Each of these subcarriers is vector modulated, i.e. QPSK, 16QAM and often up to 64QAM modulated.

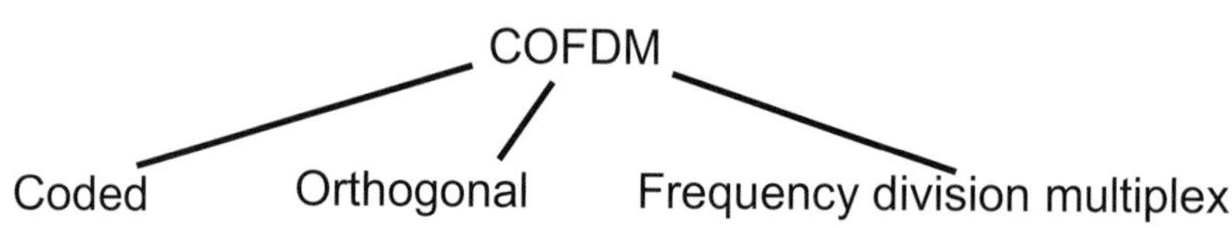

COFDM is a composite of orthogonal (at right angles to one another or, in other words, not interfering with one another) and frequency division multiplex (division of the information into many subcarriers in the frequency domain).

In a transmission channel, information can be transmitted continuously or in time slots. It is then possible to transport different messages in the various time slots, e.g. data streams from different sources. This timeslot method has long been applied, mainly in telephony for the transmission of different calls on one line, one satellite channel or also one mobile radio channel. The typical impulse-type interference caused by a mobile telephone conforming to the GSM standard with irradiation into stereo systems and TV sets has its origin in this timeslot method, also called time division multiple access (TDMA) in this case. However, it is also possible to subdivide a transmission channel of a certain bandwidth in the frequency domain, resulting in subchannels into each one of which a subcarrier can be placed. Each subcarrier is modulated independently of the others and carries its own information independently of the other subcarriers. Each of these subcarriers can be vector modulated, i.e. QPSK, 16QAM and often up to 64QAM modulated.

All subcarriers are spaced apart by a constant interval Δf. A communication channel can contain up to thousands of subcarriers, each of which could carry the information from a source which would have nothing at all to do with any of the others. However, it is also possible first to provide a common data stream with error protection and then to divide it into the many subcarriers. This is then frequency division multiplex (FDM). Thus, in FDM, a common data stream is split up and transmitted in one channel, not via a single carrier but via many, up to thousands of subcarriers, digitally vector modulated. Since these subcarriers are very close to one another, e.g. with a spacing of a few kHz, great care must be taken to see that these subcarriers do not interfere with one another. The carriers must be orthogonal to each other. The term orthogonal normally stands for 'at 90 degrees to each other' but in communications engineering quite generally means signals which do not interfere with one another due to certain characteristics. When will adjacent carriers of an FDM system then influence each other to a greater or lesser extent? Surprisingly, one has to start with a rectangular pulse and its Fourier transform (Fig. 19.5.). A single rectangular pulse of duration Δt provides a sin(x)/x-shaped spectrum in the frequency domain, with nulls spaced apart by a constant Δf = 1/Δt in the spectrum. A single rectangular pulse exhibits a continuous spectrum, i.e. instead of discrete spectral lines there is a continuous sin(x)/x-shaped curve.

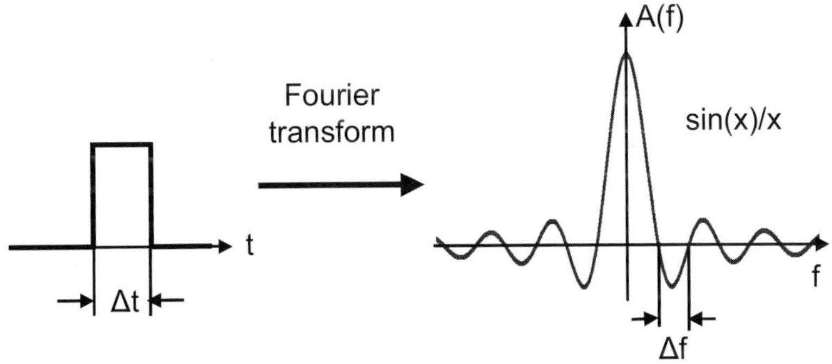

Fig. 19.5. Fourier Transform of a rectangular pulse

Varying the period Δt of the rectangular pulse varies the spacing Δf of the nulls in the spectrum. If Δt is allowed to tend towards zero, the nulls in the spectrum will tend towards infinity. This results in a Dirac pulse which has an infinitely flat spectrum which contains all frequencies. If Δt tends towards infinity, the nulls in the spectrum will tend towards zero. This results in a spectral line at zero frequency which is DC. All cases in between simply correspond to

$$\Delta f = 1/\Delta t;$$

A train of rectangular pulses of period T_p and pulse width Δt also corresponds to this sin(x)/x-shaped variation but there are now only discrete spectral lines spaced apart by $f_P = 1/T_P$ which, however, conform to this sin(x)/x-shaped variation.

What then is the relationship between the rectangular pulse and orthogonality? The carrier signals are sinusoidal. A sinewave signal of frequency $f_S = 1/T_S$ results in a single spectral line at frequency f_S and $-f_S$ in the frequency domain. However, these sinusoidal carriers carry information by amplitude- and frequency-shift keying.

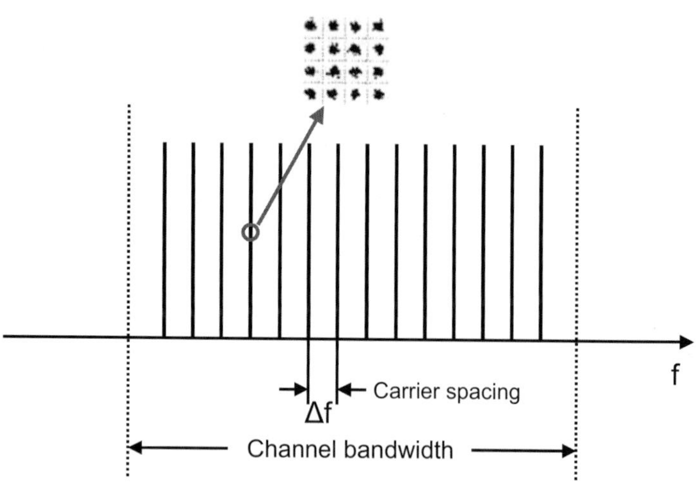

Fig. 19.6. Coded Orthogonal Frequency Division Multiplex (COFDM)

I.e., these sinusoidal carrier signals do not extend continuously from minus infinity to plus infinity but change their amplitude and phase after a particular time Δt. Thus one can imagine a modulated carrier signal to be composed of sinusoidal sections cut out rectangularly, so-called burst packets. Mathematically, a convolution occurs in the frequency domain, i.e. the spectra of the rectangular window pulse and of the sinewave become superimposed. In the frequency domain there is then a sin(x)/x-shaped spectrum at the f_s and $-f_s$ position instead of a discrete spectral line. The nulls of the sin(x)/x spectrum are described by the length of the rectangular window Δt. The space between the nulls is $\Delta f = 1/\Delta t$.

If then many adjacent carriers are transmitted simultaneously, the sin(x)/x-shaped tails produced by the bursty transmission will interfere with the adjacent carriers.

However, this interference is minimized if the carrier spacing is selected in such a way that a carrier peak always coincides with a null of the adjacent carriers. This is achieved by selecting the subcarrier spacing Δf to correspond to the inverse of the length of the rectangular window, i.e. the burst period or symbol period. Such a burst packet with many and often thousands of modulated subcarriers is called a COFDM symbol.

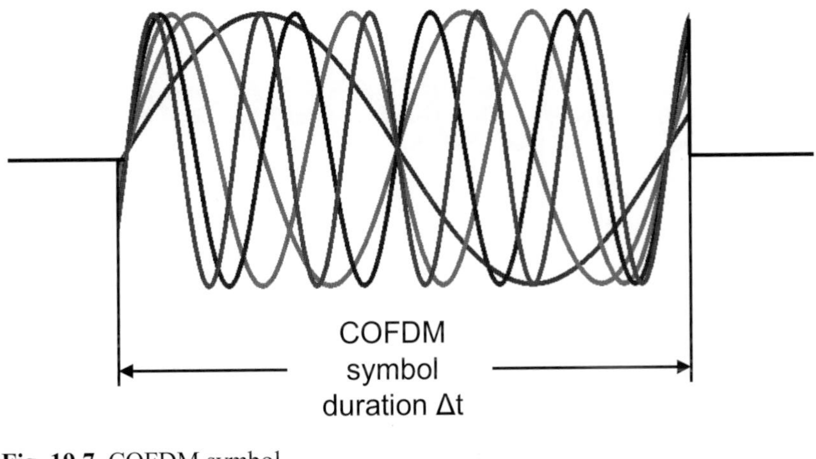

Fig. 19.7. COFDM symbol

The following holds true as COFDM orthogonality condition (Fig. 19.8.):

$$\Delta f = 1/\Delta t;$$

where Δf is the subcarrier spacing and Δt is the symbol period.

If, for example, the symbol period of a COFDM system is known, the subcarrier spacing can be inferred directly, and vice versa.

In DVB-T, the following conditions apply for the so-called 2k and 8k mode (Table 19.1.):

Table 19.1. COFDM modes in DVB-T

Mode	2k	8k
No. of subcarriers	2048	8192
Subcarrier spacing Δf	~ 4 kHz	~ 1 kHz
Symbol duration Δt	1/Δf = ~ 250 µs	~ 1 ms

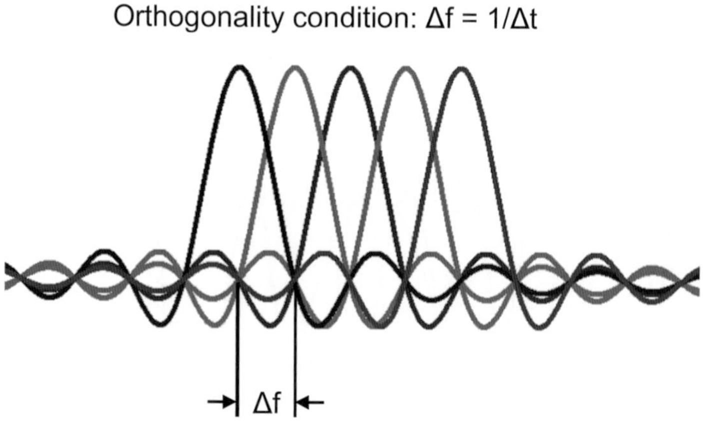

Fig. 19.8. Orthogonality condition in COFDM

19.3 Generating the COFDM Symbols

In COFDM, the information to be transmitted is first error protected, i.e. a considerable overhead is added before this data stream consisting of payload and error protection is impressed on the large number of subcarriers. Each one of these often thousands of subcarriers must then transmit a portion of this data stream. As in the single-carrier method, each subcarrier requires mapping by which the QPSK, 16QAM or 64QAM is generated. Each subcarrier is modulated independently of the others. In principle, a COFDM modulator could be imagined to be composed of up to thousands of QAM modulators, each with a mapper. Each modulator receives its own, precisely derived carrier. All the modulation processes are synchronized with one another in such a manner that in each case a common symbol is produced which has the exact length of Δt = 1/Δf. However, this procedure is pure theory: in practice, its costs would be astronomical and it would be unstable but, nevertheless, it serves to illustrate the principle of COFDM.

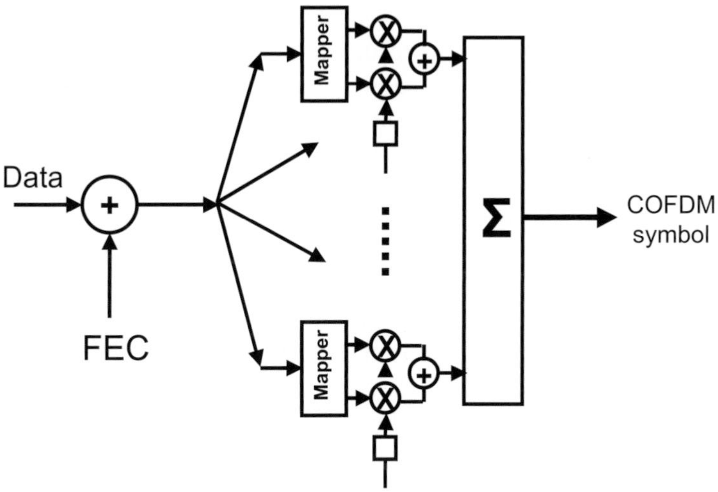

Fig. 19.9. Theoretical block diagram of a COFDM modulator

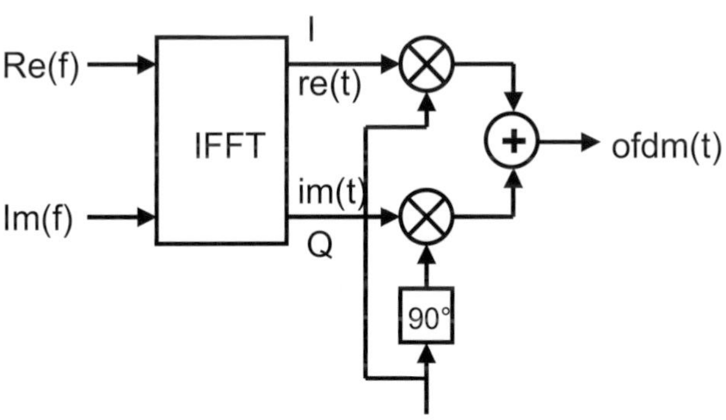

Fig. 19.10. Practical implementation of a COFDM modulator by IFFT

In reality, a COFDM symbol is generated by a multiple mapping process in which two tables are produced, followed by an Inverse Fast Fourier Transform (IFFT). I.e., COFDM is simply the result of applying numerical mathematics in a high-speed computer (Fig. 19.10.).

The COFDM modulation process is as follows: The error-protected data stream, thus provided with an overhead, is split up and divided as randomly as possible into a large number of up to thousands of substreams, a process called multiplexing and interleaving. Each substream passes packet by packet into a mapper which generates the description of the respective subvector, divided into real and imaginary parts. Two tables are generated with up to many thousands of entries, resulting in a real-part table and an imaginary-part table. This results in the description of the time domain section in the frequency domain. Each subcarrier, which is now modulated, is described as x-axis section and y-axis section or, expressed mathematically, as cosinusoidal and sinusoidal component, or real and imaginary part. These two tables - real table and imaginary table - are now the input signals for the next signal processing block, the Inverse Fast Fourier Transform (IFFT). After the IFFT, the symbol is now available in the time domain. The signal shape has a purely random, stochastic, appearance due to the many thousands of independently modulated subcarriers it contains. From experience, many people find it difficult to visualize how the many carrier are produced which is why the process of modulation with the aid of IFFT will now be described step by step.

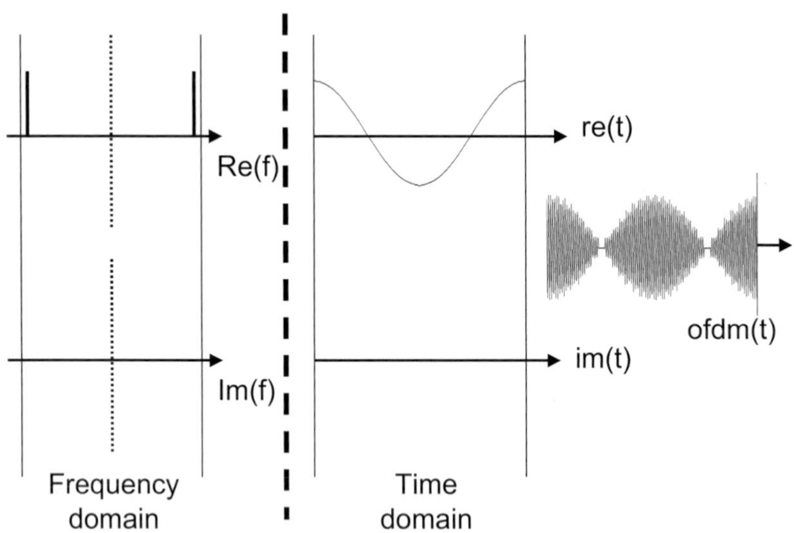

Fig. 19.11. IFFT of a symmetric spectrum

The COFDM modulator shown in Fig. 19.10., which consists of the IFFT block followed by a complex mixer (I/Q modulator), is fed one by one with various real-part and imaginary-part tables in the frequency domain after which the Inverse Fast Fourier Transform is performed and the result is considered at the outputs re(t) and im(t) after the IFFT, i.e. in the time domain, and after the complex mixer.

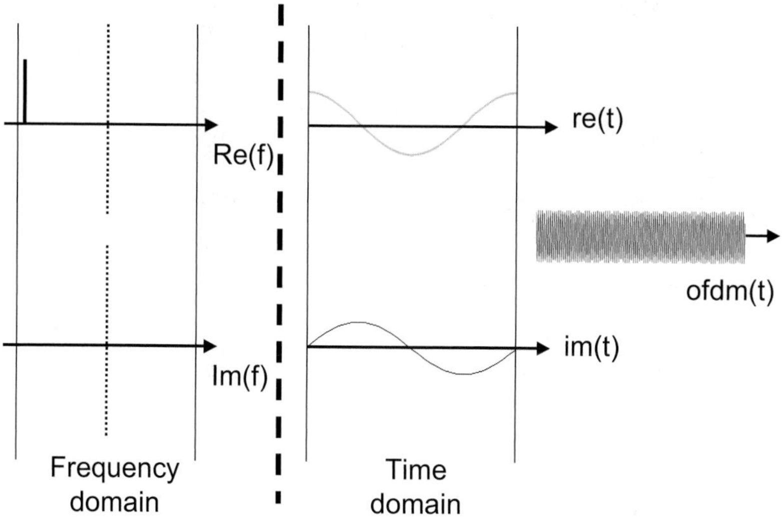

Fig. 19.12. IFFT of a asymmetric spectrum

This begins with a spectrum which is symmetric with respect to the band center of the COFDM channel (Fig. 19.11.). simply consisting of carrier No.1 and N. After the IFFT, an output signal is produced at output re(t) which is purely cosinusoidal. At output im(t), u(t) = 0V is present. A purely real time-domain signal is expected since the spectrum meets the conditions of symmetry required for this. After the I/Q modulator, an amplitude modulated signal with suppressed carrier is produced which is only generated by the real time-domain component (see Fig. 19.11.).

If, however, e.g. the spectral line in the upper range of the band, that is to say the carrier at N, is suppressed and only the component at carrier No.1 is left, a complex time-domain signal is obtained due to the asymmetric spectrum (Fig, 19.12. and 19.13.). At output re(t) after the IFFT, a cosinusoidal signal with half the amplitude as before is now present. In addition, the IFFT now supplies a sinusoidal output signal of the same frequency and the same amplitude at output im(t). This produces a complex signal in the time domain. If this, i.e. re(t) and im(t), is fed into the following I/Q modulator, the modulation disappears, resulting in a single sinusoidal oscillation converted into the carrier frequency band. A single-sideband modulated signal is produced and the arrangement now represents an SSB modulator. Changing the frequency of the stimulating quantity at the frequency level only changes the frequency of the cosinusoidal and sinusoidal output signals at re(t) and im(t). re(t) and im(t) have exactly the same amplitude and frequency and a phase difference of 90 degrees as before. The decisive factor in understanding this type of COFDM implementation is that, in principle, this mutual relationship applies to all subcarriers. For every subcarrier, im(t) is always at 90 degrees to re(t) and has the same amplitude.

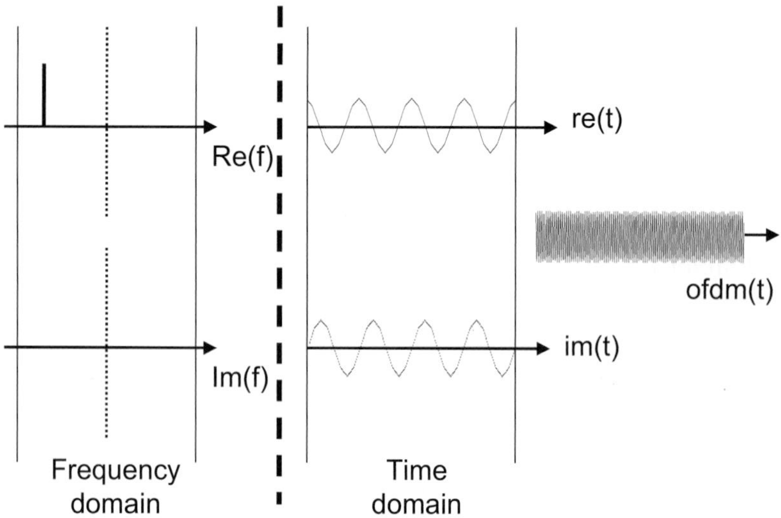

Fig. 19.13. IFFT with altered frequency

Including more and more carriers produces a signal with ever more random appearance for re(t) and im(t), the real and imaginary part-signals having a 90° phase relation to one another in the time domain.

im(t) is said to be the Hilbert Transform of re(t). This transform can be imagined to be a 90° phase shifter for all spectral components. If both time domain signals are fed into the I/Q modulator following, the actual COFDM symbol is produced. In each case, the corresponding upper or lower COFDM subband is suppressed by this type of modulation, providing a thousandfold phase-shift-type single-sideband modulator. Many references, some of which date back more than 20 years, contain notes regarding single-sideband modulators of this phase-shifting type. It is only due to the fact that each subcarrier at re(t) and im(t) has the same amplitude and they are at precisely 90 degrees to one another that the upper COFDM sideband does not produce crosstalk in the lower one, and vice versa, with respect to the center frequency.

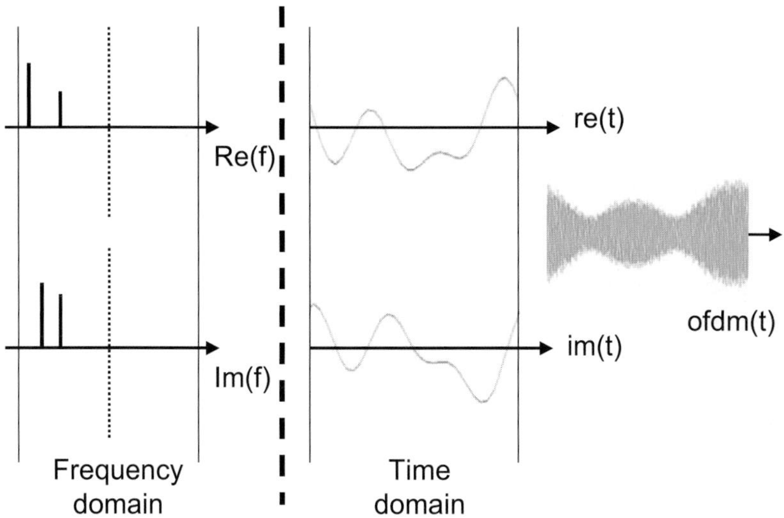

Fig. 19.14. COFDM with 3 carriers

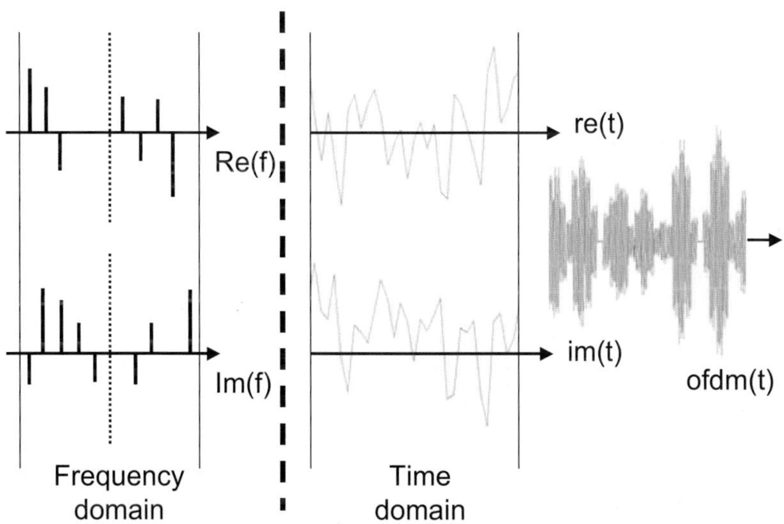

Fig. 19.15. COFDM with 12 carriers

Since nowadays analog, i.e. non-ideal, I/Q modulators are very often used because of the direct modulation method, the effects arising can only be explained in this way.

The more carriers (Fig. 19.14.), the more random the appearance of the corresponding COFDM symbol. Even just 12 single carriers placed in relatively random order with respect to one another result in a COFDM symbol with stochastic appearance. The symbols are calculated and generated section by section in pipeline fashion. The same number of data bits are always combined and modulated onto a large number of up to thousands of COFDM subcarriers. Firstly, real- and imaginary-part tables are produced in the frequency domain and then, after the IFFT, tables for re(t) and im(t) which are stored in downstream memories. Period by period, a COFDM symbol of the exact constant length of $\Delta t = 1/\Delta f$ is then generated. Between these symbols, a guard interval of defined but often adjustable length is maintained.

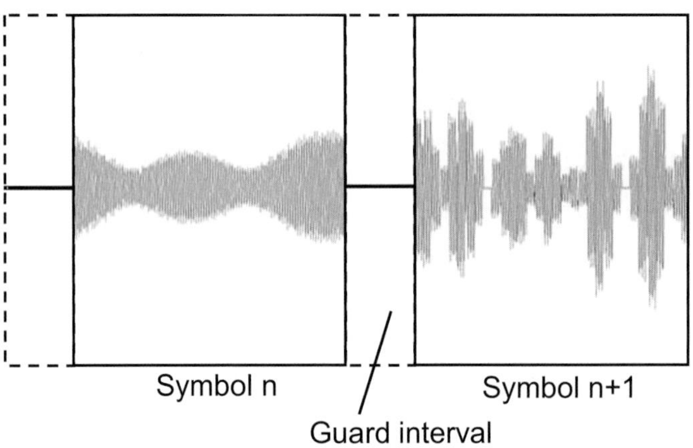

Fig. 19.16. COFDM symbols with guard interval

Inside this guard interval, transient events due to echoes can decay which prevents inter-symbol interference. The guard interval must be longer than the longest echo delay time of the transmission system. At the end of the guard interval, all transient events should have decayed. If this is not the case, additional noise is produced due to the inter-symbol interference which, in turn, is a simple function of the intensity of the echo.

However, the guard intervals are not simply set to zero. Usually, the end of the next symbol is keyed precisely into this time interval (Fig. 19.17.)

and the guard intervals can thus not be seen in any oscillogram. Purely from the point of view of signal processing, these guard intervals can be generated quite easily. The signals produced after the IFFT are first written into a memory in any case and are then read out alternately in accordance with the pipeline principle. The guard interval is then simply created by first reading out the end of the respective complex memory content in corresponding guard interval length (Fig. 19.18.).

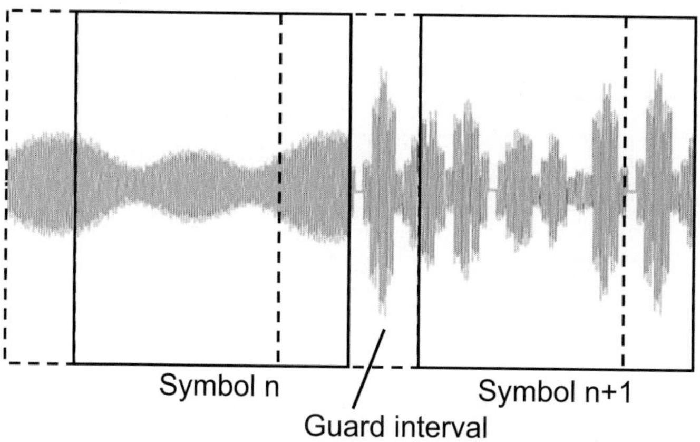

Fig. 19.17. Guard interval filled up with the end of the next symbol (CP = cyclic prefix)

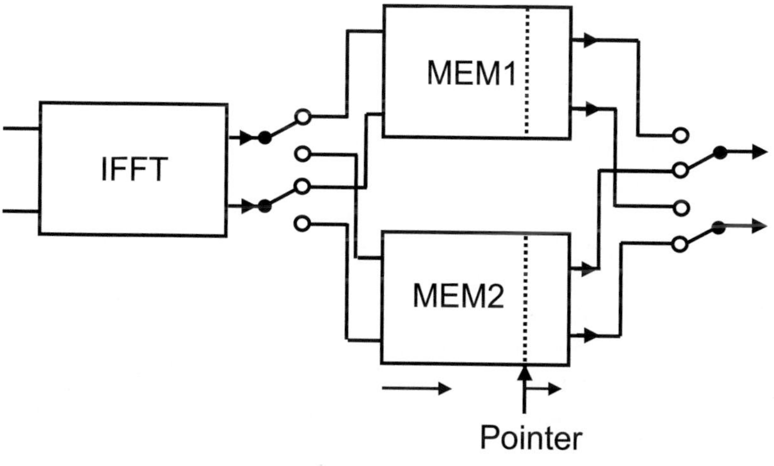

Fig. 19.18. Generating the guard interval

But why not simply leave the guard interval empty instead of filling it up with the end of the next symbol as is usually done? The reason is based on the way in which a COFDM receiver locks onto the COFDM symbols. If the guard interval (also often called CP = cyclic prefix) were not occupied with payload information, the receiver would have to hit the COFDM symbols exactly at the right spot which, however, is no longer possible in practice due to their rounding off due to multiple echoes during the transmission.

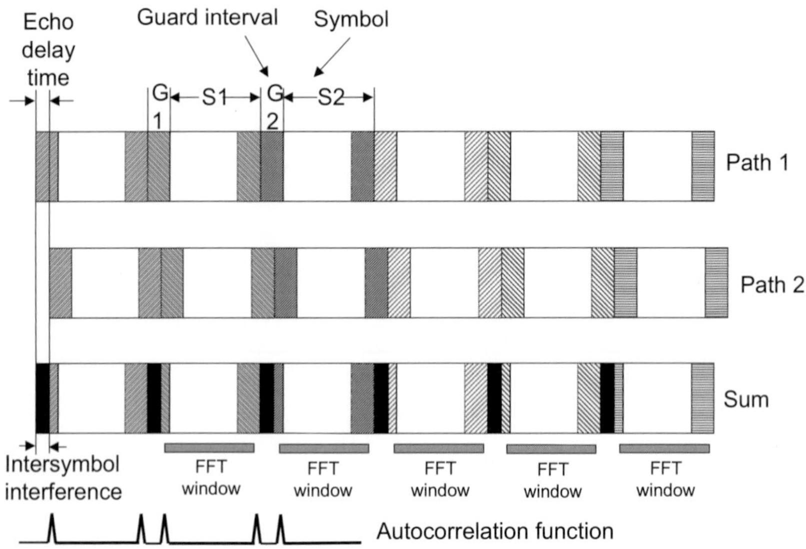

Fig. 19.19. Multipath reception in COFDM

The beginning and the end of the symbols could only be detected with difficulty in this case. If, however, e.g. the end of the next symbol is repeated in the preceding guard interval, the signal components existing several times in the signal can be easily found by means of the autocorrelation function in the receiver. This makes it possible to find the beginning and the end of the area within the symbols not affected by inter-symbol interference due to echoes. Fig. 19.19. shows this for the case of two receiving paths. Using the autocorrelation function, the receiver positions its FFT sampling window, which has the exact length of one symbol, within the symbols in such a way that it always lines up with the undisturbed area (Fig. 19.20. and 19.21.). Thus, the sampling window is not positioned precisely over the actual symbol but this only results in a phase error which produces a turning of all constellation diagrams and must be eliminated in subsequent processing steps. However, this phase error produces a rotation of all constellation diagrams.

It should not be thought, however, that the guard interval can be used for eliminating fading. This is not so. There is nothing that can be done against fading apart from adding error protection to the data stream by means of upstream FEC (forward error correction) and distributing the data stream as uniformly as possible over all COFDM subcarriers in the transmission channel.

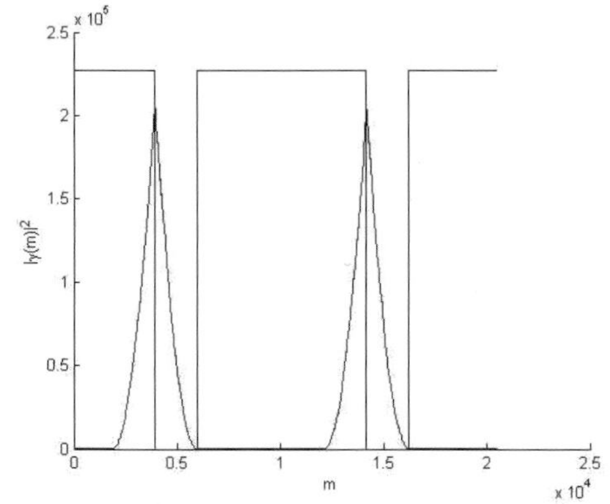

Fig. 19.20. Practical example: autocorrelation function and FFT window position, receiving only one signal path [VIERACKER]

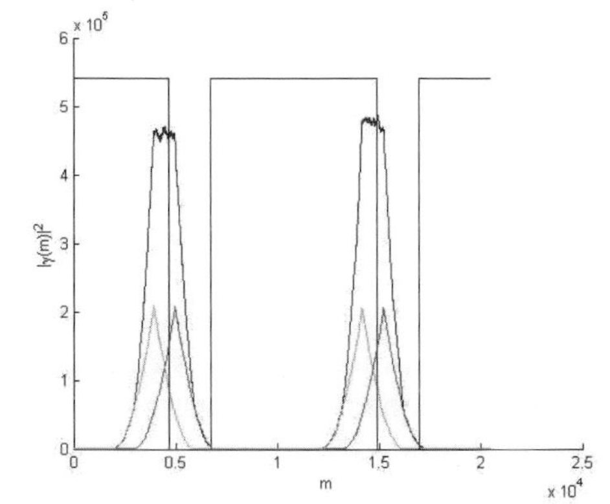

Fig. 19.21. Practical example: autocorrelation function and FFT window position, receiving two paths with 0dB attenuation (0dB echo); sum autocorrelation function and autocorrelation function for both signal paths [VIERACKER]

19.4 Supplementary Signals in the COFDM Spectrum

Up until now it has only been said that in orthogonal frequency division multiplex, the information plus error protection is distributed over the many subcarriers and these are then vector modulated and transmitted. This gives the impression that every carrier is carrying payload. However, this is not so, in fact. In all familiar COFDM transmission methods (DAB, DVB-T, ISDB-T, WLAN, ADSL), the following categories of COFDM carriers can be found to a greater or lesser extent, or not at all:

- Payload carriers,
- Unused carriers set to zero,
- Fixed pilots,
- Scattered pilots which are not fixed,
- Special data carriers for supplementary information.

The term 'supplementary signals' has been deliberately kept general since, although they have the same function everywhere, they have different designations.

In this section, the function of these supplementary signals in the COFDM spectrum will be discussed in greater detail.

The payload carriers have already been described. They transmit the actual payload data plus error protection and are vector modulated in various ways. Among others, coherent QPSK, 16QAM or 64QAM modulation is often used as modulation and the combined 2, 4 or 6 bits per carrier are then mapped directly onto the respective carrier. In the case of the non-coherent differential coding which is also frequently used, the information is contained in the difference of the carrier constellation from one symbol to the next. The main methods are DQPSK or DBPSK. Differential coding has the advantage that it is 'self-healing', i.e. any phase errors which may be present are corrected automatically, saving channel correction facilities in the receiver which thus becomes simpler. However, this is at the cost of twice the bit error ratio in comparison with coherent coding.

Thus, the data carriers can be coded as follows:

- Coherent,
- Differentially coded.

The edge carriers, that is to say the top and bottom carriers, are not used in most cases and are set to zero and do not carry any information at all.

They are called zero-information carriers and there are two basic reasons for the existence of these unused zero-information carriers.

- Preventing adjacent channel crosstalk by facilitating the filtering of the shoulders of the COFDM spectrum, and
- Adapting the bit capacity per symbol to the input data structure.

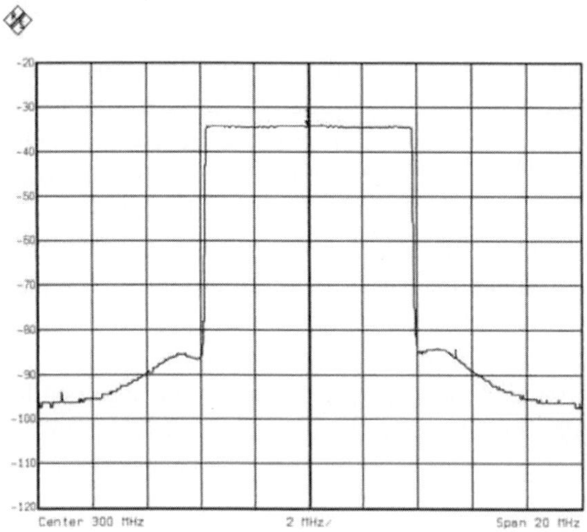

Fig. 19.22. Real DVB-T COFDM spectrum with shoulders

A COFDM spectrum (Fig. 19.22.) has so-called 'shoulders' which are simply the result of the sin(x)/x-shaped tails of each individual carrier. These shoulders cause interference in the adjacent channels and it is, therefore, necessary to improve the so-called shoulder attenuation by applying suitable filtering measures. These filtering measures, in turn, are made easier by simply not using the edge carriers because the filters do not need to be so steep in this case.

Following an integral multiple of symbols, it is also often necessary to join up with the input data structure which is also often structured in blocks. A symbol can carry a certain number of bits due to the data carriers present in the symbol. The data structure of the input data stream can also supply a certain number of bits per block. The number of payload carriers in the symbol is then selected to be only such that the calculation comes out exactly after a certain number of complete data blocks and symbols.

Because IFFT is used, however, it is necessary to select a power of two as the number of carriers which, after subtracting all data and pilot carriers, still leaves carriers, namely the zero-information carriers.

There are then also the following pilot carriers:

- Pilot carriers with a fixed position in the spectrum, and
- Pilot carriers with a variable position in the spectrum.

Pilot carriers with a fixed position in the spectrum are used for automatic frequency control (AFC) in the receiver, i.e. to lock it to the transmitted frequency. These pilot carriers are usually cosinusoidal signals and are thus located on the real axis at fixed amplitude positions. There is usually a number of such fixed pilots in the spectrum. If the receive frequency is not tied to the transmit frequency, all constellation diagrams will rotate also within one symbol. At the receiving end, these fixed pilots within a symbol are simply missed out and the receive frequency is corrected in such a way that the phase difference from one fixed pilot to the next within a symbol becomes zero.

The pilots with variable position in the spectrum are used as measuring signal for channel estimation and channel correction at the receiving end in the case of coherent modulation. One could say they represent a sweep signal for the channel estimation in order to be able to measure the channel.

Special data carriers with supplementary information are very often used as a fast information channel from transmitter to receiver in order to inform the receiver of changes made in the type of modulation, e.g. switching from QPSK to 64QAM. In this way, frequently all current transmission parameters are transmitted from transmitter to receiver, e.g. in DVB-T. It is then only necessary to set the approximate receiving frequency at the receiver.

19.5 Hierarchical Modulation

Digital transmission methods often exhibit a hard ‘fall off the cliff’ or ‘brickwall' effect when the reception abruptly ceases because the signal/noise ratio limit has been exceeded. Naturally, this also applies to COFDM. In some COFDM transmission methods (DVB-T, ISDB-T), so-called ‘hierarchical modulation’ is used to counteract this effect. When hierarchical modulation is switched on, the information is transmitted by means of two different transmission methods within one COFDM spec-

trum. One of the transmission methods is more robust but cannot support such a high data rate. The other one is less robust but is capable of handling a higher data rate, making it possible to transmit e.g. the same video signal with poorer signal quality and with better signal quality in the same COFDM stream. At the receiving end, one or the other method can then be selected with an eye on the conditions of reception. Hierarchical modulation will not be discussed in greater detail at this point because there are several approaches and these depend on the relevant standard.

19.6 Summary

Coded orthogonal frequency division multiplex (COFDM) is a transmission method which, instead of one carrier, uses a large number of subcarriers in one transmission channel. It is especially designed for the characteristics of a terrestrial transmission channel containing multiple echoes. The information to be transmitted is provided with error protection (coded orthogonal frequency division multiplex - COFDM) and distributed over all these subcarriers. The subcarriers are vector modulated and in each case transmit a part of the information. COFDM produces longer symbols than single-carrier transmission and, as a result, and with the aid of a guard interval, intersymbol interference due to echoes can be eliminated. Due to the error protection and the fact that the information is distributed over the many subcarriers it is possible to recover the original data stream free of errors in spite of any fading due to echoes. A final note: Many references mention both COFDM and OFDM. In practice, there is no difference between the two methods. OFDM is a part of COFDM. OFDM would never work without the error protection contained in COFDM.

Bibliography: [REIMERS], [HOFMEISTER], [FISCHER2], [DAMBACHER], [CHANG], [VIERACKER]

20 Terrestrial Transmission of Digital Television Signals (DVB-T)

The particular characteristics of a terrestrial radio channel have already been explained in the previous chapter on COFDM (Coded Orthogonal Frequency Division Multiplex). They are mainly determined by multipath reception which leads to location- and frequency-selective fading. In DVB-T, i.e. in the terrestrial transmission of digital TV signals according to the Digital Video Broadcasting standard, it was decided that the most appropriate modulation method to cope with this problem would be COFDM, the principles of which are explained in the previous chapter. Fig. 20.1. shows a block diagram of the DVB-T modulator, consisting at its heart of the COFDM modulator with the IFFT block followed by the I/Q modulator which can be a digital or an analog type. The position of the I/Q modulator in the circuit can vary depending on how the DVB-T modulator is implemented in practice. The COFDM modulation is preceded by the channel coding, i.e. the error correction, which is exactly the same in DVB-T as in DVB-S satellite transmission.

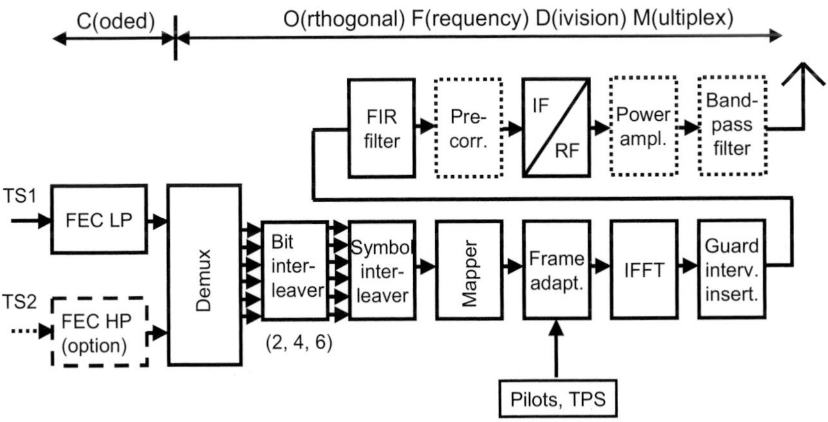

Fig. 20.1. Block diagram of the DVB-T modulator – part 1

W. Fischer, *Digital Video and Audio Broadcasting Technology*, Signals and Communication Technology, 3rd ed., DOI: 10.1007/978-3-642-11612-4_20, © Springer-Verlag Berlin Heidelberg 2010

As can also be seen from the block diagram, two MPEG-2 transport stream inputs are possible which then provides for the so-called hierarchical modulation. However, hierarchical modulation is provided as an option in DVB-T and has not as yet been put into practical use. Hierarchical modulation was originally provided for transmitting the same TV programs with different data rate, different error correction and different quality in one DVB-T channel. The HP (high priority) path transmits a data stream with a low data rate, i.e. a poorer picture quality due to higher compression, but allows better error protection or a more robust type of modulation (QPSK) to be used. The LP (low priority) path is used for transmitting the MPEG-2 transport stream with a higher data rate, a lower error protection and a type of modulation of higher order (16QAM, 64QAM). At the receiving end, HP or LP can be selected in dependence on the conditions of reception. Hierarchical modulation is intended to lessen the impact, as it were, of the "fall off the cliff". But it is also quite conceivable to transmit two totally independent transport streams. Both HP and LP branches contain the same channel coder as in DVB-S but, as already mentioned, this is an option in the DVB-T modulator, not the receiver where this involves very little additional expenditure.

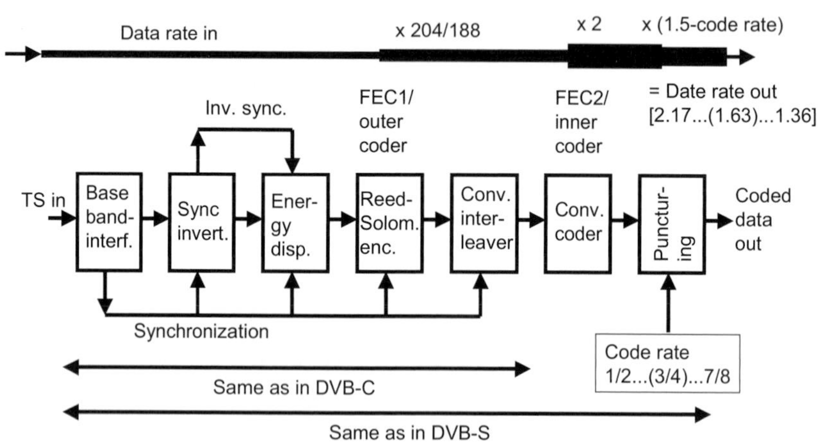

Fig. 20.2. Block diagram of the DVB-T modulator – part 2, FEC

Not every COFDM carrier in DVB-T is a payload carrier. There is also a large number of pilot carriers and special carriers. These special carriers are used for frequency synchronization, channel estimation and channel correction, and for implementing a fast information channel. They are inserted into their locations in the DVB-T spectrum before the IFFT.

Before discussing the DVB-T standard in greater detail, let us first ask: "Why DVB-T?"

There are fully operational scenarios for supplying digital television via satellite and cable, both of which paths are accessible to many households throughout the world. Why then the need for yet another, terrestrial path, e.g. via DVB-T which, in addition, is complex and expensive and may require a large amount of maintenance? The additional coverage with digital terrestrial television is necessary for reasons of

- Regional requirements (historical infrastructures, no satellite reception)
- Regional geographic situations
- Portable TV reception
- Mobile TV reception
- Local supplementary municipal services (regional/urban television)

Many countries in the world do not have satellite TV coverage, or only inadequately so, for the most varied reasons of a political, geographic or other nature. In many cases, substitute coverage by cable is not possible, either, because of e.g. permafrost and also often cannot be financed because of sparse population density. This leaves only the terrestrial coverage. Countries which are far away from the equator such as those in Scandinavia naturally have more problems with satellite reception since, e.g. the satellite receiving antennas are almost pointing at the ground. There are also many countries which have not previously had analog satellite reception as a standard such as Australia where satellite reception plays only a minor role. Population centers there are covered terrestrially and via cable or satellite. In many countries, it is not permitted to gather in an uncontrollable variety of TV programs from the sky for political reasons. Even regions in Central Europe with good satellite and cable coverage require additional terrestrial TV coverage, mainly for local TV programs which are not being broadcast via satellite. And portable and mobile reception is virtually only possible via the terrestrial path.

20.1 The DVB-T Standard

In 1995, the terrestrial standard for the transmission of digital TV programs was defined in ETS 300744 in connection with the DVB-T project. A DVB-T channel can have a bandwidth of 8, 7 or 6 MHz. There are two different operating modes: the 2K mode and the 8K mode where 2K stands for a 2046-points IFFT and 8K stands for an 8192-points IFFT. As is already known from the chapter on COFDM, the number of COFDM subcarriers must be a power of two. In DVB-T, it was decided to use symbols with a length of about 250 μs (2K mode) or 1 ms (8K mode). Depending on requirements, one or the other mode can be selected. The 2K mode has greater subcarrier spacing of about 4 kHz but the symbol period is much shorter. Compared with the 8K mode with a subcarrier spacing of about 1 kHz, it is much less susceptible to spreading in the frequency domain caused by doppler effects due to mobile reception and multiple echoes but much more susceptible to greater echo delays. In single frequency networks, for example, the 8K mode will always be selected because of the greater transmitter spacing possible. In mobile reception, the 2K mode is better because of the greater subcarrier spacing. The DVB-T standard allows for flexible control of the transmission parameters.

Apart from the symbol length, which is a result of the use of 2K or 8K mode, the guard interval can also be adjusted within a range of 1/4 to 1/32 of the symbol length. It is possible to select the type of modulation (QPSK, 16QAM or 64QAM)). The error protection (FEC) is designed to be the same as in the DVB-S satellite standard. The DVB-T transmission can be adapted to the respective requirements with regard to robustness or net data rates by adjusting the code rate (1/2 ... 7/8).

In addition, the DVB-T standard provides for hierarchical coding as an option. In hierarchical coding, the modulator has two transport stream inputs and two independently configurable but identical FECs. The idea is to apply a large amount of error correction to a transport stream with a low data rate and then to transmit it with a very robust type of modulation. This transport stream path is then called the high priority (HP) path. The second transport stream has a higher data rate and is transmitted with less error correction and, e.g. 64QAM modulation and the path is called the low priority (LP) path. It would then be possible, e.g. to subject the identical program packet to MPEG-2 coding, once at the higher data rate and once at the lower data rate, and to combine the two packets in two multiplex packets transported in independent transport streams. Higher data rate automatically means better (picture) quality. The data stream with the lower data rate and correspondingly lower picture quality is fed to the high priority path and that with the higher data rate is supplied to the low priority path. At the receiving end, the high priority signal is demodulated more easily than the low priority one. Depending on the conditions of reception, the HP path or the LP path will be selected at the receiving end. If the reception is poor, there will at least still be reception due to the lower data rate and higher compression, even if the quality of the picture and sound is inferior.

In DVB-T, coherent COFDM modulation is used, i.e. the payload carriers are mapped absolutely and are not differentially coded. However, this requires channel estimation and correction for which numerous pilot signals are provided in the DVB-T spectrum and are used as test signal for the channel estimation.

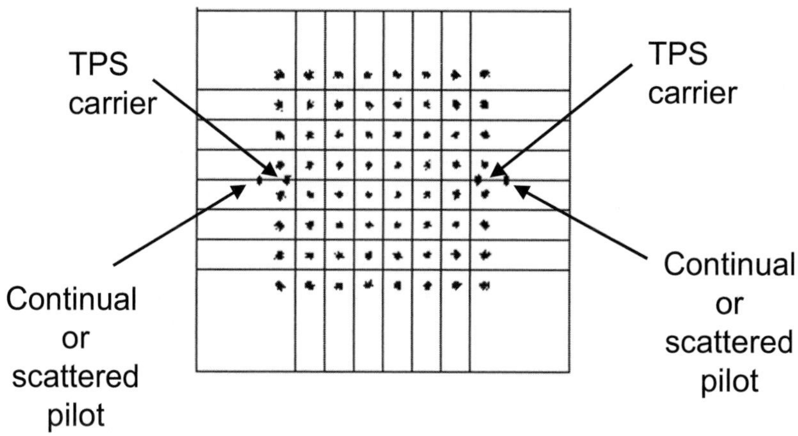

Fig. 20.3. DVB-T carriers: payload carriers, Continual and Scattered Pilots, TPS carriers

20.2 The DVB-T Carriers

In DVB-T, an IFFT with 2048 or 8192 points is used. In theory, 2048 or 8192 carriers would then be available for the data transmission. However, not all of these carriers are used as payload carriers. In the 8K mode, there are 6048 payload carriers and in the 2K mode there are 1512. The 8K mode thus has exactly four times as many payload carriers as the 2K mode but since the symbol rate is higher by a factor of exactly 4 in the 2K mode, both modes will always have the same data rate, given the same conditions of transmission. DVB-T contains the following types of carrier:

- Inactive carriers with fixed position (set to zero amplitude)
- Payload carriers with fixed position
- Continual pilots with fixed position
- Scattered pilots with changing position in the spectrum
- TPS carriers with fixed position

The meaning of the words 'payload carrier' is clear: these are simply the carriers used for the actual data transmission. The edge carriers at the upper and lower channel edge are set to zero, i.e. they are inactive and carry no modulation at all, i.e. their amplitudes are zero. The continual pilots are located on the real axis, i.e. the I (in-phase) axis, either at 0 degrees or at 180 degrees and have a defined amplitude. The continual pilots are boosted by 3 dB compared with the average signal power and are used in the receiver as phase reference and for automatic frequency control (AFC), i.e. for locking the receive frequency to the transmit frequency. The scattered pilots are scattered over the entire spectrum of the DVB-T channel from symbol to symbol and virtually constitute a sweep signal for the channel estimation. Within each symbol, there is a scattered pilot every 12th carrier. Each scattered pilot jumps forward by three carrier positions in the next symbol, i.e. in each case two intermediate payload carriers will never become a scattered pilot whereas those at every 3rd position in the spectrum are sometimes payload carriers and sometimes scattered pilots. The scattered pilots are also on the I axis at 0 degrees and 180 degrees and have the same amplitude as the continual pilots.

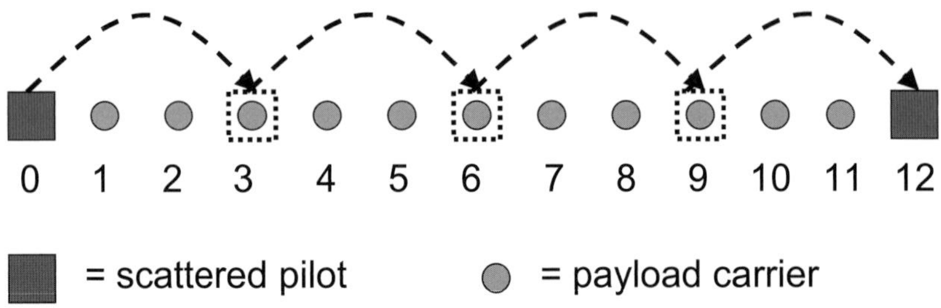

Fig. 20.4. Change of position of the Scattered Pilots

The TPS carriers are located at fixed frequency positions. For example, carrier No. 50 is a TPS carrier. TPS stands for Transmission Parameter Signalling. These carriers represent virtually a fast information channel via which the transmitter informs the receiver about the current transmission parameters. They are DBPSK (differential bi-phase shift keying) modulated and are located on the I axis either at 0 degrees or at 180 degrees. They are differentially coded, i.e. the information is contained in the difference between one and the next symbol. All the TPS carriers in one symbol carry the same information, i.e. they are all either at 0 degrees or all at 180 degrees on the I axis. At the receiving end, the correct TPS carrier position of 0 degrees or 180 degrees is then determined by majority voting for each symbol and is then used for the demodulation. DBPSK means that a zero is transmitted when the state of the TPS carriers changes from one symbol to the next, and a one if the TPS carrier phase does not change from one symbol to the next. The complete TPS information is broadcast over 68 symbols and comprises 67 bits. This segment over 68 symbols is called a frame and the scattered pilots within this frame also jump over the DVB-T channel from the start of the channel right to the end of the channel.

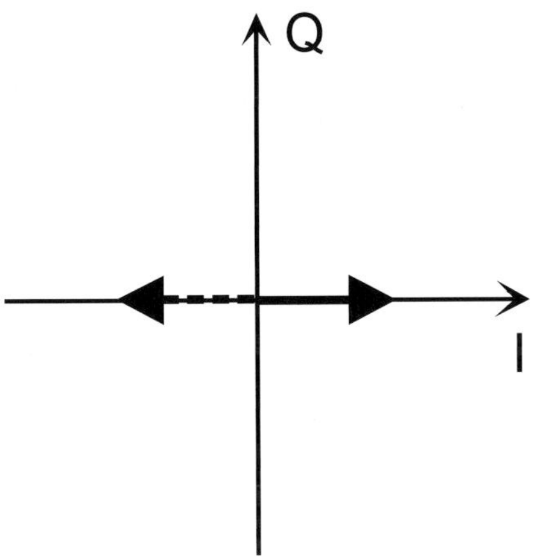

Fig. 20.5. DBPSK modulated TPS carriers

17 of the 68 TPS bits are used for initialization and synchronization, 13 bits are error protection, 22 bits are used at present and 13 bits are reserved for future applications. Table 20.1. explains how the TPS carriers are utilized.

Thus, the TPS carriers keep the receiver informed about:

- Mode (2K, 8K)
- Length of the guard interval (1/4, 1/8, 1/16, 1/32)
- Type of modulation (QPSK, 16QAM, 64QAM)
- Code rate (1/2, 2/3, 3/4, 5/6, 7/8)
- Use of hierarchical coding

However, the receiver should have already determined the mode (2K, 8K) and the length of the guard interval which are thus actually meaningless as TPS information.

Table 20.1. Bit allocation of the TPS carriers

Symbol number	Format	Purpose/content
s_0		Initialization
s_1 - s_{16}	0011010111101110 or 1100101000010001	Synchronization word
s_{17} - s_{22}	010 111	Length indicator
s_{23}, s_{24}		Frame number
s_{25}, s_{26}		Constellation 00=QPSK/01=16QAM/10=64QAM
s_{27}, s_{28}, s_{29}		Hierarchy information 000=Non hierarchical, 001=α=1, 010=α=2, 011=α=4
s_{30}, s_{31}, s_{32}		Code rate, HP stream 000=1/2, 001=2/3, 010=3/4, 011=5/6, 100=7/8
s_{33}, s_{34}, s_{35}		Code rate, LP stream 000=1/2, 001=2/3, 010=3/4, 011=5/6, 100=7/8
s_{36}, s_{37}		Guard interval 00=1/32, 01=1/16, 10=1/8, 11=1/4
s_{38}, s_{39}		Transmission mode 00=2K, 01=8K
s_{40} - s_{53}	all set to "0"	Reserved for future use
s_{54} - s_{67}	BCH code	Error protection

In Fig. 20.3., the position of the pilots and TPS carriers can be seen clearly in a 64QAM constellation diagram. The two outer points on the I axis correspond to the positions of the continual pilots and the scattered pilots. The two inner points on the I axis are the TPS carriers.

The position of the continual pilots and of the TPS carriers in the spectrum can be seen from the tables 20.2. and 20.3. In these tables, the carrier numbers are listed at which the continual pilots and the TPS carriers can be found. Counting begins at carrier number zero which is the first non-zero carrier at the beginning of the channel.

The various types of carriers used in DVB-T are briefly summarized as follows. Of the 2048 carriers in the 2K mode, only 1705 carriers are used and all others are set to zero. Within these 1705 carriers there are 1512 payload carriers which can be QPSK, 16QAM or 64QAM modulated, 142 scattered pilots, 45 continual pilots and 17 TPS carriers.

Table 20.2. Carrier positions of the Continual Pilots

2K mode	8K mode
0 48 54 87 141 156	0 48 54 87 141 156 192 201 255 279 282 333 432 450
192 201 255 279 282	483 525 531 618 636 714 759 765 780 804 873 888 918
333 432 450 483 525	939 942 969 984 1050 1101 1107 1110 1137 1140 1146
531 618 636 714 759	1206 1269 1323 1377 1491 1683 1704 1752 1758 1791
765 780 804 873 888	1845 1860 1896 1905 1959 1983 1986 2037 2136 2154
918 939 942 969 984	2187 2229 2235 2322 2340 2418 2463 2469 2484 2508
1050 1101 1107 1110	2577 2592 2622 2643 2646 2673 2688 2754 2805 2811
1137 1140 1146 1206	2814 2841 2844 2850 2910 2973 3027 3081 3195 3387
1269 1323 1377 1491	3408 3456 3462 3495 3549 3564 3600 3609 3663 3687
1683 1704	3690 3741 3840 3858 3891 3933 3939 4026 4044 4122
	4167 4173 4188 4212 4281 4296 4326 4347 4350 4377
	4392 4458 4509 4515 4518 4545 4548 4554 4614 4677
	4731 4785 4899 5091 5112 5160 5166 5199 5253 5268
	5304 5313 5367 5391 5394 5445 5544 5562 5595 5637
	5643 5730 5748 5826 5871 5877 5892 5916 5985 6000
	6030 6051 6054 6081 6096 6162 6213 6219 6222 6249
	6252 6258 6318 6381 6435 6489 6603 6795 6816

Table 20.3. Carrier positions of the TPS Carriers

2K mode	8K mode
34 50 209 346 413 569 595	34 50 209 346 413 569 595 688 790 901
688 790 901 1073 1219 1262	1073 1219 1262 1286 1469 1594 1687 1738
1286 1469 1594 1687	1754 1913 2050 2117 2273 2299 2392 2494
	2605 2777 2923 2966 2990 3173 3298 3391
	3442 3458 3617 3754 3821 3977 4003 4096
	4198 4309 4481 4627 4670 4694 4877 5002
	5095 5146 5162 5321 5458 5525 5681 5707
	5800 5902 6013 6185 6331 6374 6398 6581
	6706 6799

Some of the scattered pilots occasionally coincide with positions of continual pilots which is why the number 131 should be used for calculating the actual payload carriers in the case of the scattered pilots in 2K mode. The conditions in the 8K mode are comparable. Here, too, not all the 8192 carriers are being used but only 6817 of which, in turn, only 6048 are actual payload carriers. The rest are scattered pilots (568), continual pilots (177) and TPS carriers (68). As before, the number 524 must be used for the scattered pilots in calculating the payload carriers since sometimes a scattered pilot will coincide with a continual pilot. Every 12th carrier in a symbol is a scattered pilot. It is thus easy to calculate the number of scattered pilots by dividing the number of carriers actually used by 12 ($1705/12 = 142$, $6817/12 = 568$).

Table 20.4. Number of different carriers in DVB-T

2K mode	8K mode	
2048	8192	Carrier
1705	6817	used carrier
142/131	568/524	Scattered pilots
45	177	Continual pilots
17	68	TPS carrier
1512	6048	payload carrier

The payload carriers are either QPSK, 16QAM or 64QAM modulated and transmit the error-protected MPEG-2 transport stream. Fig. 20.6. shows the constellation diagrams for QPSK, 16QAM and 64QAM with the positions of the special carriers in the case of non-hierarchical modulation.

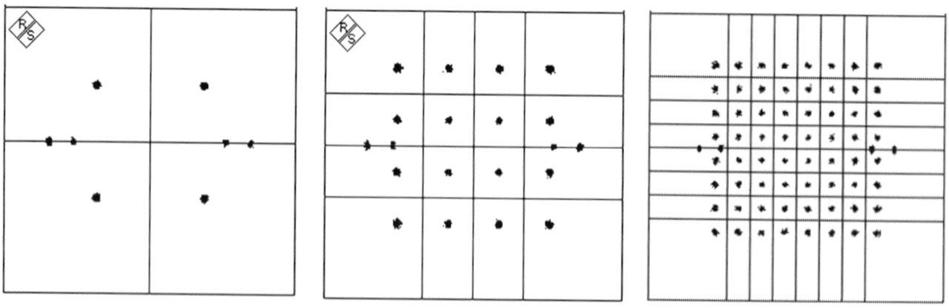

Fig. 20.6. DVB-T constellation diagrams for QPSK, 16QAM and 64QAM

20.3 Hierarchical Modulation

To ensure that reliable reception is still guaranteed even in poor conditions, hierarchical modulation is provided as an option in DVB-T. Without it, e.g. a signal/noise ratio which is too bad will lead to a hard "fall off the cliff", otherwise known as the 'brick wall effect'. In the case of the frequently used DVB-T transmission with 64QAM modulation and a code rate of 3/4 or 2/3, the limit of stable reception is at a signal/noise ratio of just under 20 dB. In this section, the details of hierarchical modulation/coding will be explained more fully. If hierarchical modulation is used, the DVB-T modulator has two transport stream inputs and two FEC blocks. One transport stream with a low data rate is fed into the so-called high priority path (HP) and provided with a large amount of error protection, e.g. by selecting the code rate 1/2. A second transport stream with a higher data rate is supplied in parallel to the low priority path (LP) and is provided with less error protection, e.g. with the code rate 3/4.

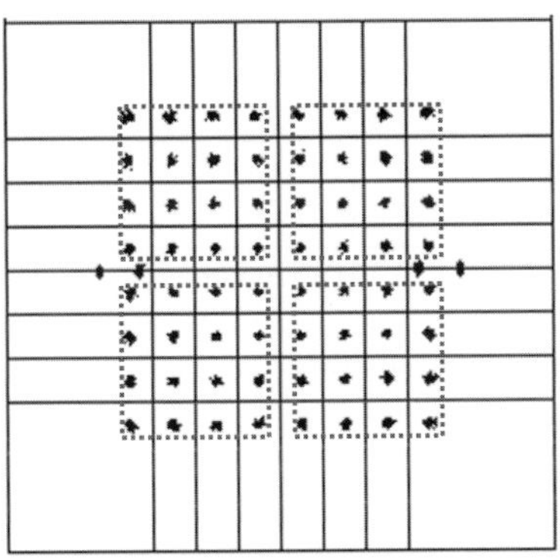

Fig. 20.7. Embedded QPSK in 64QAM with hierarchical modulation

In principle, both HP and LP transport streams can contain the same programs but at different data rates, i.e. with different amounts of compression. However, the two can also carry completely different payloads. On the high priority path, QPSK is used which is a particularly robust type of modulation. On the low priority path, a higher level of modulation is needed due to the higher data rate. In DVB-T, the individual payload carriers are not modulated with different types of modulation. Instead, each payload carrier transmits portions both of LP and of HP. The high priority path is transmitted as so-called embedded QPSK in 16QAM or 64QAM.

Fig. 20.7. shows the case of QPSK embedded in 64QAM. The LP information is carried by the discrete constellation point and the HP is described by the quadrant. A cloud of 8 times 8 points in a quadrant as a whole thus corresponds virtually to the total constellation point of the QPSK in this quadrant.

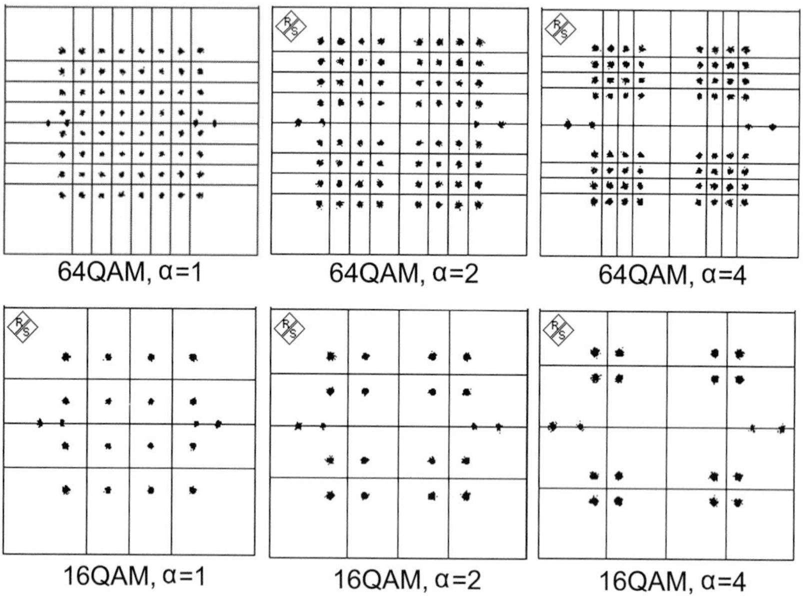

Fig. 20.8. Possible Constellations with hierarchical modulation

A 64QAM modulation enables 6 bits per symbol to be transmitted. However, since the quadrant information, as QPSK, diverts 2 bits per symbol for the HP stream, only 4 bits per symbol remain for the transmission of the LP stream. The gross data rates for LP and HP thus have a fixed ratio of 4:2 to one another. In addition, the net data rates are dependent on the code rate used. QPSK embedded in 16QAM is also possible. The ratio between the gross data rates of LP and HP is then 2:2. To make the QPSK of the high priority path more robust, i.e. less susceptible to interference, the constellation diagram can be spread at the I axis and the Q axis. A factor α of 2 or 4 increases the distance between the individual quadrants of the 16QAM or 64QAM diagrams. The higher α is, the more insensitive the high priority path becomes and the more sensitive the low priority path becomes since the discrete constellation points move closer together. Fig. 20.8. shows the 6 possible constellations with hierarchical modulation, i.e.

64QAM with $\alpha = 1, 2, 4$ and 16QAM with $\alpha = 1, 2, 4$. The information about the presence or absence of hierarchical modulation and the α factor and the code rates for LP and HP are transmitted in the TPS carriers. This information is evaluated in the receiver which automatically adjusts its demapper accordingly. The decision to demodulate HP or LP in the receiver can be made automatically in dependence on the current conditions of reception (channel bit error rate) or left to the user to select manually. Hierarchical modulation is provided as an option in modern DVB-T chipsets and set-top boxes since, in practice, no additional hardware is required. In many DVB-T receivers, however, no software is provided for this option since it is currently not used in any country. At the beginning of 2002, hierarchical modulation was tested in field tests in Australia but it is currently not used there, either.

20.4 DVB-T System Parameters of the 8/7/6 MHz Channel

In the following paragraphs, the system parameters of DVB-T will be derived and explained in detail. These parameters are:

- IFFT sampling frequencies
- DVB-T signal bandwidths
- Spectrum occupied by the 8/7 and 6 MHz DVB-T channel
- Data rates
- Signal levels of the individual carriers

The basic system parameter in DVB-T is the IFFT sampling frequency of the 8-MHz channel which is defined as:

$$f_{\text{sample IFFT 8MHz}} = 64/7 \text{ MHz} = 9.142857143 \text{ MHz}.$$

From this basic parameter, all other system parameters can be derived, i.e. those of the 8/7 and 6 MHz channel. The IFFT sampling frequency is the sampling rate of the COFDM symbol or, respectively, the bandwidth within which all 2K (= 2048) and 8K (= 8192) subcarriers can be accommodated. However, many of these 2048 or 8192 subcarriers are set to zero and the bandwidth of the DVB-T signal must be narrower than that of the actual 8, 7 or 6 MHz wide channel. As will be seen, the signal bandwidth of the 8 MHz channel is only about 7.6 MHz and there is thus a space of approx. 200 kHz between the top and the bottom of this channel and its adjacent channels.

These 7.6 MHz contain the 6817 or 1705 carriers actually used. In the case of the 7 or 6 MHz channel, the IFFT sampling frequency of these channels can be calculated from the IFFT sampling frequency of the 6 MHz channel by simply multiplying it by 7/8 or 6/8, respectively.

$f_{sample\ IFFT\ 7MHz} = 64/7\ MHz \bullet 7/8 = 8\ MHz;$

$f_{sample\ IFFT\ 6MHz} = 64/7\ MHz \bullet 6/8 = 48/7\ MHz = 6.857142857\ MHz;$

All 2048 or 8192 IFFT carriers in the 8/7 and 6 MHz channel can be found within these IFFT bandwidths. From these bandwidths or sampling frequencies, the respective subcarrier spacing can be easily derived by dividing the bandwidth $f_{sample\ IFFT}$ by the number of IFFT subcarriers:

$\Delta f = f_{sample\ IFFT} / N_{total\_carriers};$
$\Delta f_{2k} = f_{sample\ IFFT} / 2048;$
$\Delta f_{8k} = f_{sample\ IFFT} / 8192;$

Therefore, the COFDM subcarrier spacing in an 8, 7 or 6 MHz-wide DVB-T channel in 2K and 8K mode is:

Table 20.5. Subcarrier spacing in 2K and 8K mode

Channel bandwidth	Δf of 2K mode	Δf of 8K mode
8 MHz	4.464285714 kHz	1.116071429 kHz
7 MHz	3.90625 kHz	0.9765625 kHz
6 MHz	3.348214275 kHz	0.8370535714 kHz

From the subcarrier spacing, the symbol length Δt_{symbol} can be determined directly. Due to the orthogonality condition, it is:

$\Delta t_{symbol} = 1/\Delta f;$

Therefore, the symbol lengths in the various modes and channel bandwidths in DVB-T are:

Table 20.6. Symbol durations in 2K and 8K mode

Channel bandwidth	Δt_{symbol} of 2K mode	Δt_{symbol} of 8K mode
8 MHz	224 us	896 ms
7 MHz	256 us	1.024 ms
6 MHz	298.7 us	1.1947 ms

The DVB-T signal bandwidths are obtained from the subcarrier spacing Δf of the respective channel (8, 7, 6 MHz) and the number of carriers actually used in 2K and 8K mode (1705 and 6817).

$$f_{\text{signal DVB-T}} = N_{\text{used\_carriers}} \bullet \Delta f;$$

Table20.7. Signal Bandwidths in DVB-T

Channel bandwidth	$f_{\text{signal DVB-T}}$ of 2K mode	$f_{\text{signal DVB-T}}$ of 8K mode
8 MHz	7.612 MHz	7.608 MHz
7 MHz	6.661 MHz	6.657 MHz
6 MHz	5.709 MHz	5.706 MHz

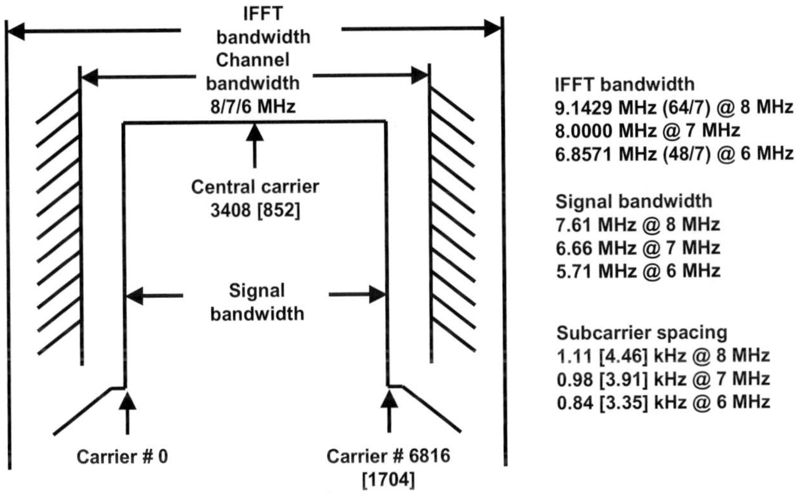

Fig. 20.9. Spectrum of a DVB-T signal in 8K and [2K] mode for the 8/7/6 MHz channel

In principle, there are two ways of counting the COFDM subcarriers of the DVB-T channel. The carriers can either be counted through from 0 to 2047 or from 0 to 8192 in accordance with the number of IFFT carriers or counting can begin with carrier number zero at the first carrier actually used in the respective mode. The latter counting method is the more usual one, counting from 0 to 1704 in 2K mode and from 0 to 6816 in 8K mode. In Fig. 20.9. then the position of the spectrum of the DVB-T channel is shown and the most important DVB-T system parameters are summarized again.. Fig. 20.9. also shows the center carrier numbers which are of particular importance in testing. This carrier number of 3408 in the 8K mode and 852 in the 2K mode corresponds to the exact center of the DVB-T channel. Some effects which can be caused by the DVB-T modulator can only be observed at this point. The values provided in square brackets in the Figure apply to the 2K mode (e.g.: 3408 [852]) and the others apply to the 8K mode.

The gross data rate of the DVB-T signal is derived from, among other things, the symbol rate of the DVB-T COFDM signal. The symbol rate is a function of the length of the symbol and of the length of the guard interval. as follows:

$$symbol\_rate_{COFDM} = \frac{1}{symbol\_duration + guard\_duration};$$

The gross data rate is then the result of the symbol rate, the number of actual payload carriers and the type of modulation (QPSK, 16QAM, 64QAM). In 2K mode, there are 1512 payload carriers and in 8K mode there are 6048. In QPSK, 2 bits per symbol are transmitted, in 16QAM it is 4 bits per symbol and in 64QAM it is 6 bits per symbol. Since the symbols are longer by a factor of 4 in the 8K mode but, on the other hand, there are four times as many payload carriers in the channel, this factor cancels out again which means that the data rates are independent of the mode (2K or 8K). The gross data rate of the DVB-T channel is thus:

$$\text{gross\_data\_rate} = \text{symbol\_rate}_{\text{COFDM}} \bullet \text{no\_of\_payload\_carriers} \bullet \text{bits\_per\_symbol};$$

The total length of the COFDM symbols is composed of the symbol length and the length of the guard intervals:

Table 20.8. Total symbol durations in DVB-T

	total_symbol_duration = symbol_duration + guard_duration [us]							
Chan. bandw. [MHz]	2K 1/4	2K 1/8	2K 1/16	2K 1/32	8K 1/4	8K 1/8	8K 1/16	8K 1/32
8	280	252	238	231	1120	1008	952	924
7	320	288	272	264	1280	1152	1088	1056
6	373.3	336	317.3	308	1493.3	1344	1269.3	1232

total_symbol_duration = symbol_duration + guard_duration;

The symbol rate of the DVB-T channel is calculated as:

symbol_rate = 1 / total_ symbol_duration;

The DVB-T symbol rates are listed in Table 20.9. as a function of mode and channel bandwidth.

Table 20.9. Symbol rates in DVB-T

	Symbol rate [kS/s]							
Channel band-width	2K guard 1/4	2K guard 1/8	2K guard 1/16	2K guard 1/32	8K guard 1/4	8K guard 1/8	8K guard 1/16	8K guard 1/32
8 MHz	3.5714	3.9683	4.2017	4.3290	0.8929	0.9921	1.04504	1.0823
7 MHz	3.1250	3.4722	3.6760	3.7888	0.7813	0.8681	0.9191	0.9470
6 MHz	2.6786	2.9762	3.1513	3.2468	0.6696	0.7440	0.7878	0.8117

The gross data rate is then determined from:

gross_data_rate = symbol_rate • no_of_payload_carriers
• bits_per_symbol;

The gross DVB-T data rates are listed in Table 20.10. as a function of channel bandwidth and the guard interval length.

Table 20.10. Gross data rates in DVB-T

	Gross data rate [MBit/s]							
Channel band-width, modula-tion	2K guard 1/4	2K guard 1/8	2K guard 1/16	2K guard 1/32	8K guard 1/4	8K guard 1/8	8K guard 1/16	8K guard 1/32
8 MHz QPSK	10.800	12.000	12.706	13.091	10.800	12.000	12.706	13.091
8 MHz 16QAM	21.6	24.0	25.412	26.182	21.6	24.0	25.412	26.182
8 MHz 64QAM	32.4	36.0	38.118	39.273	32.4	36.0	38.118	39.273
7 MHz QPSK	9.45	10.5	11.118	11.455	9.45	10.5	11.118	11.455
7 MHz 16QAM	18.9	21.0	22.236	22.91	18.9	21.0	22.236	22.91
7 MHz 64QAM	28.35	31.5	33.354	34.365	28.35	31.5	33.354	34.365
6 MHz QPSK	8.1	9.0	9.530	9.818	8.1	9.0	9.530	9.818
6 MHz 16QAM	16.2	18.0	19.06	19.636	16.2	18.0	19.06	19.636
6 MHz 64QAM	24.3	27.0	28.59	29.454	24.3	27.0	28.59	29.454

The net data rate additionally depends on the code rate of the convolutional coding used and on the Reed Solomon error protection RS(188, 204) as follows:

net_data_rate = gross_data_rate • 188/204 • code_rate;

Since the factor of 4 cancels out, the overall formula for determining the net data rate of DVB-T signals is independent of the mode (2K or 8K) and is:

net_data_rate = 188/204 • code_rate • $\log_2(m)$ • 1/(1 + guard) • channel • const1;

where
m = 4 (QPSK), 16 (16-QAM), 64 (64-QAM);
$\log_2(m)$ = 2 (QPSK), 4 (16-QAM), 6 (64-QAM);
code rate = 1/2, 2/3, 3/4, 5/6, 7/8;
guard = 1/4, 1/8, 1/16, 1/32;
channel = 1 (8MHz), 7/8 (7MHz), 6/8 (6 MHz);
const1 = $6.75 \cdot 10^6$ bits/s;

Table 20.11. Net data rates with non-hierarchical modulation in the 8 MHz DVB-T channel

Modulation	Code rate	Guard 1/4	Guard 1/8	Guard 1/16	Guard 1/32
		Mbit/s	Mbit/s	Mbit/s	Mbit/s
QPSK	1/2	4.976471	5.529412	5.854671	6.032086
	2/3	6.635294	7.372549	7.806228	8.042781
	3/4	7.464706	8.294118	8.782007	9.048128
	5/6	8.294118	9.215686	9.757785	10.05348
	7/8	8.708824	9.676471	10.24567	10.55617
16QAM	1/2	9.952941	11.05882	11.70934	12.06417
	2/3	13.27059	14.74510	15.61246	16.08556
	3/4	14.92941	16.58824	17.56401	18.09626
	5/6	16.58824	18.43137	19.51557	20.10695
	7/8	17.41765	19.35294	20.49135	21.11230
64QAM	1/2	14.92941	16.58824	17.56401	18.0926
	2/3	19.90588	22.11765	23.41869	24.12834
	3/4	22.39412	24.88235	26.34602	27.14439
	5/6	24.88235	27.64706	29.27336	30.16043
	7/8	26.12647	29.02941	30.73702	31.66845

From this, the net data rates of the 8, 7 and 6 MHz channel in the various operating modes can be determined:

The net data rates in DVB-T vary between about 4 and 31 Mbit/s in dependence on the transmission parameters and channel bandwidths used. In the 7 MHz and 6 MHz channels, the available net data rates are lower by a factor of 7/8 or 6/8, respectively. in comparison with the 8 MHz channel.

Table 20.12. Net data rates with non-hierachical modulation in the 7 MHz DVB-T Channel

Modulation	Code rate	Guard 1/4	Guard 1/8	Guard 1/16	Guard 1/32
		Mbit/s	Mbit/s	Mbit/s	Mbit/s
QPSK	1/2	4.354412	4.838235	5.122837	5.278075
	2/3	5.805882	6.450980	6.830450	7.037433
	3/4	6.531618	7.257353	7.684256	7.917112
	5/6	7.257353	8.063725	8.538062	8.796791
	7/8	7.620221	8.466912	8.964965	9.236631
16QAM	1/2	8.708824	9.676471	10.245675	10.556150
	2/3	11.611475	12.901961	13.660900	14.074866
	3/4	13.063235	14.514706	15.368512	15.834225
	5/6	14.514706	16.127451	17.076125	17.593583
	7/8	15.240441	16.933824	17.929931	18.473262
64QAM	1/2	13.063235	14.514706	15.368512	15.834225
	2/3	17.417647	19.352941	20.491350	21.112300
	3/4	19.594853	21.772059	23.052768	23.751337
	5/6	21.772059	24.191177	25.614187	26.390374
	7/8	22.860662	25.400735	26.894896	27.709893

In hierarchical modulation, the gross data rates in 64QAM modulation are distributed at a ratio of 2:4 between HP and LP and in16QAM the ratio between HP and LP gross data rates is exactly 2:2. In addition, the net data rates in the high priority and low priority paths depend on the code rates used there.

The formulas for determining the net data rates of HP and LP are:

$$net\_data\_rate_{HP} = 188/204 \bullet code\_rate_{HP} \bullet bits\_per\_symbol_{HP} \bullet 1/(1 + guard\_duration) \bullet channel \bullet const1;$$

$$net\_data\_rate_{LP} = 188/204 \bullet code\_rate_{LP} \bullet bits\_per\_symbol_{LP} \bullet 1/(1 + guard\_duration) \bullet channel \bullet const1;$$

where
bits_per_symbol$_{HP}$ = 2:
bits_per_symbol$_{LP}$ = 2 (16QAM) or 4 (64QAM);
code_rate $_{HP/LP}$ = 1/2, 2/3, 3/4, 5/6, 7/8;
guard_duration = 1/4, 1/8, 1/16, 1/32;
channel = 1 (8MHz), 7/8 (7MHz), 6/8 (6 MHz);
const1 = $6.75 \bullet 10^6$ bits/s;

Table 20.13. Net data rates with non-hierarchical modulation in the 6 MHz DVB-T Channel

Modulation	Code rate	Guard 1/4	Guard 1/8	Guard 1/16	Guard 1/32
		Mbit/s	Mbit/s	Mbit/s	Mbit/s
QPSK	1/2	3.732353	4.147059	4.391003	4.524064
	2/3	4.976471	5.529412	5.854671	6.032086
	3/4	5.598529	6.220588	6.586505	6.786096
	5/6	6.220588	6.911765	7.318339	7.540107
	7/8	6.531618	7.257353	7.684256	7.917112
16QAM	1/2	7.464706	8.294118	8.782007	9.048128
	2/3	9.952941	11.058824	11.709343	12.064171
	3/4	11.197059	12.441177	13.173010	13.572193
	5/6	12.441176	13.823529	14.636678	15.080214
	7/8	13.063235	14.514706	15.368512	15.834225
64QAM	1/2	11.197059	12.441177	13.173010	13.572193
	2/3	14.929412	16.588235	17.564014	18.096257
	3/4	16.795588	18.661765	19.759516	20.358289
	5/6	18.661765	20.735294	21.955017	22.620321
	7/8	19.594853	21.772059	23.052768	23.751337

This brings us to the last details of the DVB-T standard related to actual field experience: the constellation and the levels of the individual carriers. Depending on the type of constellation (QPSK, 16QAM or 64QAM, hierarchical with $\alpha = 1$, 2 or 4), a mean signal value of the payload carriers is obtained which can be calculated simply by means of the quadratic mean (RMS value) of all possible vector lengths in their correct distribution. This mean is then defined as 100% or simply as One. In the case of the 2K mode, there are 1512 or 6048 payload carriers, the mean power of which is 100% or One. The TPS carrier levels are set in the same way exactly in relation to the individual payload carriers. The continual and scattered pilots are differently arranged. Due to the need for easy detectability, these pilots are boosted by 2.5 dB with respect to the mean signal level of the payload carriers. I.e., the voltage level of the continual and scattered pilots is higher by 4/3 compared with the mean level of the payload carriers and the power level is higher by 16/9.

$20 \log(4/3) = 2.5$ dB; voltage ratio of continual and scattered pilots with respect to the payload carrier signal average;

and

$10 \log(16/9) = 2.5$ dB; power ratio of continual and scattered pilots with respect to the payload carrier signal average;

In summary, it can be said that the position of the TPS carriers in the constellation diagram always corresponds to the 0 dB point of the mean value of the payload carriers and that the position of the continual pilots and of the scattered pilots always corresponds to the 2.5 dB point, regardless of the DVB-T constellation involved at the time.

Test instruments are often calibrated for carrier-to-noise ratio (C/N) and not for signal-to-noise ratio (S/N). The signal-to-noise ratio, however, is relevant to the calculation of the bit error ratio (BER) caused by pure noise interference in the channel. The C/N must then be converted into S/N. When converting C/N to S/N, the energy in the pilots must be taken into consideration. The energy in the pure payload carrier without pilots can be determined as follows, both for the 2K mode and for the 8K mode:

$$\text{payload\_to\_signal} = (\text{payload} / (\text{payload} + (\text{scattered} + \text{continual}) \bullet (4/3)^2 + \text{TPS} \bullet 1));$$

$$\text{payload\_to\_signal}_{2k} = 10 \log(1512/(1512 + (131 + 45) \bullet 16/9 + 17 \bullet 1)) = -0.857 \text{ dB};$$

$$\text{payload\_to\_signal}_{8k} = 10 \log(6048/(6048 + (524 + 177) \bullet 16/9 + 68 \bullet 1)) = -0.854 \text{ dB};$$

The level of the DVB-T payload carriers alone is thus about 0.86 dB below the total carrier level.

Mapping of the constellation diagrams for QPSK, 16QAM and 64QAM is another DVB-T system parameter. The mapping tables describe the bit allocation to the respective constellation diagrams. The mapping tables below are layed out with the LSB (bit 0) on the left and the respective MSB on the right. Therefore, the order from left to right is bit 0, bit 1 for QPSK, bit 0, bit 1, bit 2, bit 3 for 16QAM and bit 0, bit 1, bit 2, bit 3, bit 4, bit 5 for 64QAM.

QPSK

10 •	00 •
11 •	01 •

16QAM

1000 •	1010 •	0010 •	0000 •
1001 •	1011 •	0011 •	0001 •
1101 •	1111 •	0111 •	0101 •
1100 •	1110 •	0110 •	0100 •

64QAM

100000 •	100010 •	101010 •	101000 •	001000 •	001010 •	000010 •	000000 •
100001 •	100011 •	101011 •	101001 •	001001 •	001011 •	000011 •	000001 •
100101 •	100111 •	101111 •	101101 •	001101 •	001111 •	000111 •	000101 •
100100 •	100110 •	101110 •	101100 •	001100 •	001110 •	000110 •	000100 •
110100 •	110110 •	111110 •	111100 •	011100 •	011110 •	010110 •	010100 •
110101 •	110111 •	111111 •	111101 •	011101 •	011111 •	010111 •	010101 •
110001 •	110011 •	111011 •	111001 •	011001 •	011011 •	010011 •	010001 •
110000 •	110010 •	111010 •	111000 •	011000 •	011010 •	010010 •	010000 •

Fig. 20.10. DVB-T mapping tables

20.5 The DVB-T Modulator and Transmitter

Having dealt with the DVB-T standard and all its system parameters in detail, the DVB-T modulator and transmitter can now be discussed. A DVB-T modulator can have one or two transport stream inputs followed by forward error correction (FEC) and this only depends on whether this modulator supports hierarchical modulation or not. If hierarchical modulation is used, both FEC stages are completely independent of one another but are completely identical as far as their configuration is concerned. One transport stream path with FEC is called the high priority path (HP) and the other one is the low priority path (LP). Since the two FEC stages are completely identical with the FEC of the DVB-S satellite standard, discussed in the relevant chapter (Chapter 14), they do not need to be discussed in detail here.

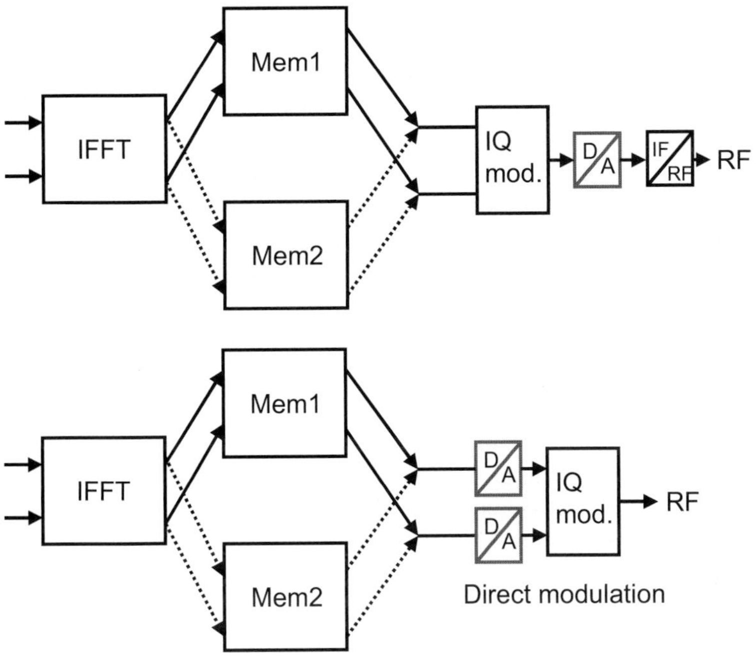

Fig. 20.11. Possible implementations of a DVB-T modulator

The modulator locks to the transport stream, present at the transport stream input, in the baseband interface. It uses for this the sync byte which has a constant value of 0x47 at intervals of 188 bytes. To carry also long-term time stamps in the transport stream, every eighth sync byte is then inverted and becomes 0xB8. This is followed by the energy dispersal stage which is synchronized by these inverted sync bytes both at the transmitting end and at the receiving end. Following this, initial error control is performed in the Reed Solomon encoder. The TS packets are now expanded by 16 bytes error protection. After this block coding, the data stream is interleaved in order to be able to break up error bursts during the deinterleaving at the receiver end. In the convolutional encoder, additional error protection is added which can be reduced again in the puncturing stage.

Up to this point, both HP and LP paths are absolutely identical but may have different code rates. The error-controlled data of the HP and LP paths, or the data of the one TS path in the case of non-hierarchical modulation, then pass into the de-multiplexer where they are then divided into 2, 4 or 6 outgoing data streams depending on the type of modulation (2 paths for QPSK, 4 for 16QAM and 6 for 64QAM). The divided data streams then pass into a bit interleaver where 126-bit-long blocks are formed which are then interleaved on each path. In the symbol interleaver following, the blocks are then again mixed block by block and the error-controlled data stream is distributed uniformly over the entire channel. Adequate error control and good distribution over the DVB-T channel are the prerequisites for COFDM to function correctly. Together, this is then COFDM - Coded Orthogonal Frequency Division Multiplex. After that, all the payload carriers are then mapped depending on whether hierarchical or non-hierarchical modulation is used, and on the factor α being = 1, 2 or 4. This results in two tables, namely that for the real part Re(f) and that for the imaginary part Im(f). However, they also contain gaps into which the pilots and the TPS carriers are then inserted by the frame adaptation block. The complete tables, comprising 2048 and 8192 values, respectively, are then fed into the heart of the DVB-T modulator, the IFFT block.

After that, the COFDM signal is available separated into real and imaginary part in the time domain. The 2048 and 8192 values, respectively, for real and imaginary part in the time domain are then temporarily stored in buffers organized along the lines of the pipeline principle. I.e., they are alternately written into one buffer whilst the other one is being read out. During read-out, the end of the buffer is read out first as a result of which the guard interval is formed. To obtain a better understanding of this section, special reference is made to the chapter on COFDM. The signal is then usually digitally filtered at the temporal I/Q level (FIR filter) to provide for better attenuation of the shoulders.

The signal is now pre-equalized in a power transmitter in order to compensate for nonlinearities in the output stage. At the same time it is clipped in order to limit the DVB-T signal with respect to its crest factor since otherwise the output stages could be destroyed because of the very high crest factor of the COFDM signal due to its very high and very low amplitudes.

The position of the I/Q modulator depends on how the DVB-T modulator or transmitter is implemented in practice. The signal is either digital/analog converted separately for I and Q at the I/Q level and then supplied to an analog I/Q modulator which allows direct mixing to RF in accordance with the principle of direct modulation, a principle commonly used at present. The other approach is to remain at the digital level up to and including the I/Q modulator and then to perform the D/A conversion.

This, however, requires a further converter stage from a lower intermediate frequency to the final RF which is more complex and costs more and, therefore, is usually avoided today. On the other hand, this advantage is gained at the expense of the possibly unpleasant characteristics of an analog I/Q modulator, the presence of which can virtually always be detected in the output signal. Given the correct implementation, however, it is possible to manage direct modulation from baseband to RF (Fig. 20.11.).

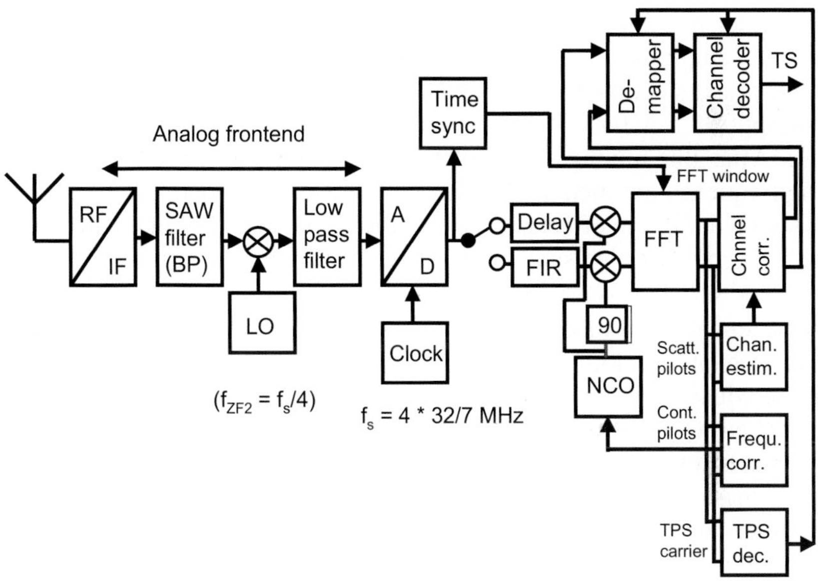

Fig. 20.12. Block diagram of a DVB-T receiver (part 1)

20.6 The DVB-T Receiver

One may think that the DVB-T modulator is a rather complex device but the receiving end is even more complicated. Due to the high packing density of modern ICs, however, most of the modules of the DVB-T receiver (Fig. 20.12.) can be accommodated in a single chip today.

The first module of the DVB-T receiver is the tuner. It is used for converting the RF of the DVB-T channel down to IF. In its construction, a DVB-T tuner only differs in being required to have a much better phase noise characteristic. The tuner is followed by the DVB-T channel at 36 MHz band center. This also corresponds to the band center of an analog

TV channel with a bandwidth of 8 MHz. However, In analog television, everything is referred to the vision carrier frequency which is 38.9 MHz at intermediate frequency. In digital television, i.e. in DVB-S, DVB-C and also in DVB-T, it is the channel center frequency which is considered to be the channel frequency. At intermediate frequency, the signal is bandpass filtered to a bandwidth of 8, 7 or 6 MHz, using surface acoustic wave (SAW) filters. In this frequency range, the filters can be implemented easily with the characteristics required for DVB-T. Following this bandpass filtering, the adjacent channels are suppressed to an acceptable degree. An SAW filter has minimum phase shift, i.e. there is no group delay distortion, only amplitude and group delay ripple.

In the next step, the DVB-T signal is converted down to a lower, second IF at approx. 5 MHz. This is frequently an IF of 32/7 MHz = 4.571429 MHz. After this mixing stage, all signal components above half the sampling frequency are then suppressed with the aid of a low-pass filter in order to avoid aliasing effects. This is followed by analog/digital conversion. The A/D converter is usually clocked at exactly four times the second IF, i.e. at 4 • 32/7 = 18.285714 MHz. This is necessary in order to be able to use the so-called fs/4 method for I/Q demodulation in the DVB-T modulator (see chapter on I/Q modulation). Following the A/D converter, the data stream, which is now available with a data rate of about 20 Megawords/s, is supplied to the time synchronization stage, among others. In this stage, autocorrelation is used to derive synchronization information. Using autocorrelation, signal components are detected which exist in the signal several times and in the same way. Since in the guard interval, the end of the next symbol is repeated before each present symbol, the autocorrelation function will supply an identification signal in the area of the guard intervals and in the area of the symbols. The autocorrelation function is then used to position the FFT sampling window into the area of guard interval plus symbol free of inter-symbol interference and this positioning control signal is fed into the FFT processor in the DVB-T receiver.

In parallel with the time synchronization, the data stream coming from the A/D converter is split into two data streams by a changeover switch. E.g., the odd-numbered samples pass into the upper branch and the even-numbered ones pass into the lower branch, producing two data streams with half the data rate in each case. However, these streams are offset from one another by half a sampling clock cycle. To eliminate this offset, the intermediate values are interpolated by means of an FIR filter, e.g. in the lower branch. This filter, in turn, causes a basic delay of, e.g. 30 clock periods or more which must be replicated in the upper branch by using simple shift registers. The two data streams are then fed to a complex mixer which is supplied with carriers by a numerically controlled oscillator

(NCO). This mixer and the NCO are then used for correcting the frequency of the DVB-T signal but because the oscillators lack accuracy, the receiver must also be locked to the transmitted frequency by means of automatic frequency control (AFC). This is done by the AFC evaluating the continual pilots after the Fast Fourier Transform (FFT). If the receiver frequency differs from the transmitted frequency, all the constellation diagrams will rotate more or less quickly clockwise or anticlockwise. The direction of rotation simply depends on whether the deviation is positive or negative and the speed depends on the magnitude of the error. It is then only necessary to measure the position of the continual pilots in the constellation diagram. The only factor of interest with respect to the frequency correction is the phase difference of the continual pilots from symbol to symbol, the aim being to reduce this phase difference to zero. The phase difference is a direct controlled variable for the AFC, i.e. the NCO frequency is changed until the phase difference becomes zero. The rotation of the constellation diagrams is then stopped and the receiver is locked to the transmitted frequency.

The FFT signal processing block, the sampling window of which is controlled by the time synchronization, transforms the COFDM symbols back into the frequency domain, providing again 2048 or 8192 real and imaginary parts. However, these do not as yet correspond directly to the carrier constellations. Since the FFT sampling window is not placed precisely over the actual symbol, there exists a phase shift in all COFDM subcarriers, i.e. all constellation diagrams are twisted. This means that the continual and scattered pilots are no longer located on the real axis, either, but somewhere on a circle, the radius of which corresponds to the amplitude of these pilots. Furthermore, channel distortions must be expected due to echoes or amplitude response or group delay. This, in turn, means that the constellation diagrams can also be distorted in their amplitude and can be additionally twisted to a greater or lesser extent. However, the DVB-T signal carries a large quantity of pilot signals which can be used as measuring signal for channel estimation and channel correction in the receiver. Over the period of twelve symbols, scattered pilots will have come to rest at every third carrier position, i.e. information about the distortion in the channel is available at every third carrier position. Measuring the amplitudes and phase distortion of the continual and scattered pilots enables the correction function for the channel to be calculated, rotating the constellation diagrams back to their nominal position. In addition, the amplitude distortion is removed and the constellation diagrams are compressed or expanded in such a way that the pilots come to rest at the correct position at their nominal position on the real axis.

Knowing about the operation of channel estimation and correction is of importance to understanding test problems in DVB-T. From the channel estimation data, it is possible to deduce both a large amount of test information in the DVB-T test receiver (channel transfer function, impulse response etc) and problems in the DVB-T modulator (I/Q modulator, center carrier).

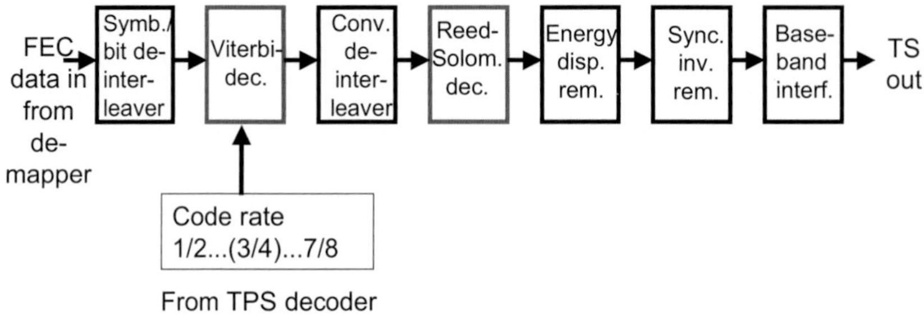

Fig. 20.13. Block diagram of the DVB-T receiver (part 2), channel decoding

In parallel with the channel correction, the TPS carriers are decoded in the uncorrected channel. The transmission parameter signalling carriers do not require channel correction since they are differentially encoded, the modulation of the TPS carriers being DBPSK (differential bi-phase shift keying). Each symbol contains a large number of TPS carriers and each carrier carries the same information. The respective bit to be decoded is determined by differential decoding with respect to the previous symbol and by majority voting within a symbol. In addition, the TPS information is error-protected. Therefore, the TPS information can be evaluated correctly for the DVB-T transmission before the threshold to the "fall off the cliff" is reached. The TPS information is needed by the de-mapper following the channel correction, and also by the channel decoder. The TPS carriers make it possible to derive the currently selected type of modulation (QPSK, 16QAM or 64QAM) and the information about the presence of hierarchical modulation. The de-mapper is then correspondingly set to the correct type of modulation, i.e. the correct de-mapping table is loaded. If hierarchical modulation is provided, a decision about which path (high priority (HP) or low priority (LP)) is to be decoded must be made in dependence on the channel bit error ratio, either manually or automatically. Following the demapper, the data stream is available again and is provided for channel decoding.

Apart from the symbol and bit de-interleaver, the channel decoder (Fig. 20.13.) is configured exactly the same as that for the DVB-S satellite TV standard. The de-mapped data pass from the de-mapper into the symbol and bit de-interleaver where they are resorted and fed into the Viterbi decoder. At the locations where bits have been punctured, dummy bits are inserted again. These are dealt with similarly to errored bits by the Viterbi decoder which then attempts to correct the first errors in accordance with methods known from the trellis decoder.

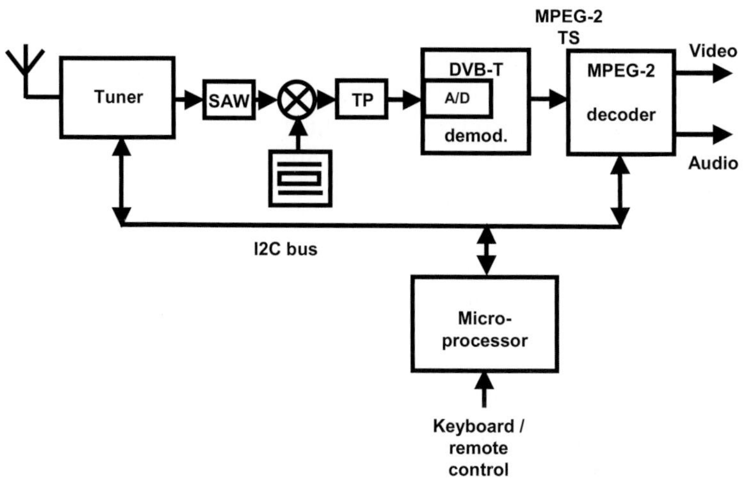

Fig. 20.14. Block diagram of a DVB-T set-top box

The Viterbi decoder is followed by the convolutional deinterleaver which breaks up error bursts by undoing the interleaving. This makes it easier for the Reed Solomon decoder to correct bit errors. The Reed Solomon decoder corrects up to 8 bit errors per packet with the aid of the 16 error control bytes. If there are more than 8 errors per packet, the 'transport error indicator' is set to one and then this transport stream packet cannot be processed further in the MPEG-2 decoder and error masking must be carried out. As well, the energy dispersal must then be undone. This stage is synchronized by the inverted sync bytes and this sync byte inversion must also be undone, after which the MPEG-2 transport stream is available again.

A practical DVB-T receiver (Fig. 20.14.) has only a few discrete components such as the tuner, SAW filter, the mixing oscillator for the 2nd IF and the low-pass filter. These are followed by a DVB-T demodulator chip which contains all modules of the DVB demodulator after the A/D converter. The transport stream coming out of the DVB-T demodulator is fed into the downstream MPEG-2 decoder where it is decoded back into video and audio. All these modules are controlled by a microprocessor via an I^2C bus.

20.7 Interference on the DVB-T Transmission Link and its Effects

Terrestrial transmission paths are subject to numerous influences (Fig. 20.15.). Apart from additive white Gaussian noise, these are mainly the many echoes, i.e. the multi-path reception which makes this type of transmission so very problematic. Terrestrial reception is easy or difficult depending on the echo situation.

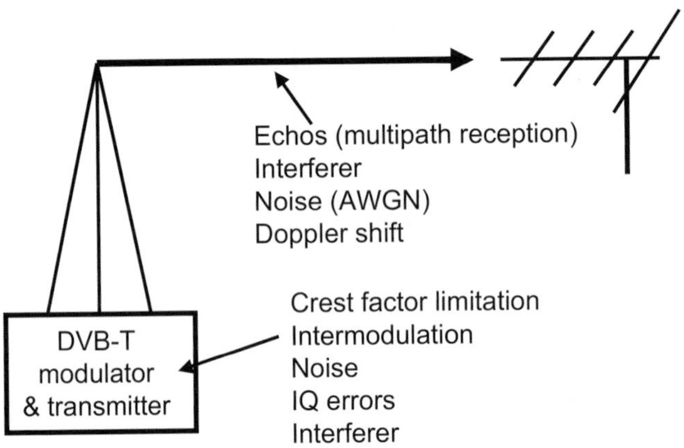

Fig. 20.15. Interferences on the DVB-T transmission link

The quality of the transmission link is also determined by the DVB-T modulator and transmitter. The high crest factor of COFDM transmissions results in special requirements even at the transmitting end. In theory, the crest factor, i.e. the ratio between the maximum peak amplitude and the RMS value of DVB-T signals is of the order of magnitude of 35 to 41 dB but it would not be possible to operate any practical power amplifier with these crest factors. Sooner or later, they would lead to its destruction. In practice, therefore, the crest factor is limited to about 12 to 13 dB before the DVB-T signal is fed into the power amplifier. However, this leads to a poor shoulder attenuation in the DVB-T signal and, in addition, in-band noise of the same order of magnitude as the shoulder attenuation is produced due to intermodulation and cross-modulation. The shoulder attenuation is then about 38...40 dB. To bring this shoulder attenuation back to a reasonable order of magnitude, passive band-pass filters tuned to the DVB-T channel are connected downstream (Fig. 20.16.). This again provides a shoulder attenuation of better than 50 dB (critical mask). But there is nothing that can be done against the in-band carrier/noise ratio of about 38...40 dB now present. These interference products are the result of the clipping required for reducing the crest factor and will now determine the performance of the DVB-T transmitter. I.e., every DVB-T transmitter will exhibit a C/N ratio of the order of about 38...40 dB.

Today, direct modulation is used in virtually every DVB-T modulator, i.e. the signal is converted directly from the digital baseband into RF as a result of which analog I/Q modulators are used. In consequence, this circuit section, too, which is now no longer operating with theoretical perfection, has adverse effects on the signal quality, resulting in I/Q errors such as amplitude imbalance, I/Q phase errors and lack of carrier suppression. It is the art of the makers of modulators to keep these influences to a minimum. However, the presence of an analog I/Q modulator in the DVB-T transmitter is always detectable by measuring instruments as will be seen later in the chapter on test engineering. As well, the finite quality of the signal processing in the DVB-T modulator also results in the creation of noise-like interferers. Further noise occurs on the transmission link in dependence on the conditions of reception. Similarly, multiple echoes and sinusoidal or impulse-like interferers can be expected and echoes, in turn, can lead to frequency- and location-selective fading.

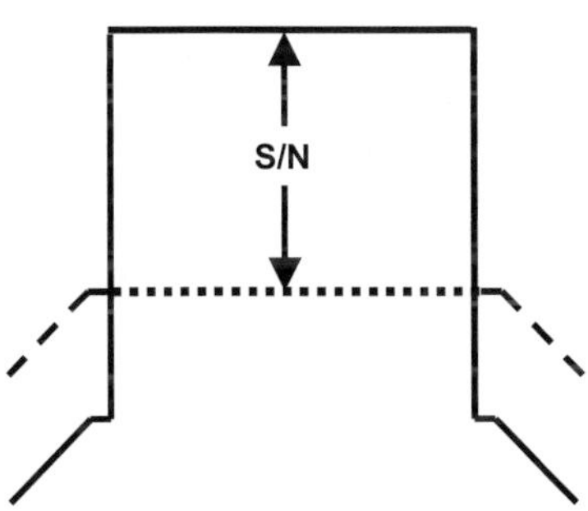

Fig. 20.16. Shoulder attenuation after clipping and after bandpass filtering

To calculate the crest factor in COFDM signals:

The crest factor is usually defined as:

$c_{fu} = 20 \log(U_{peak}/U_{RMS})$;

Power meters and spectrum analyzers are sometimes also calibrated to the following definition:

$c_{fp} = 10 \log(PEP)/P_{AVG}$;

where PEP is the peak envelope power $(U_{peak}/\sqrt{2})^2/ z_o$;
and $P_{AVG} = U_{RMS}^2/ z_o$;

The two crest factor definitions thus differ by 3 dB:

$c_{fu} = c_{fp}.+ 3$ dB;

The crest factor of COFDM signals is calculated as follows:

The maximum peak voltage is obtained by adding together the peak amplitudes of all single carriers:

$U_{peak} = N \bullet U_{peak0}$;

where U_{peak0} is the peak amplitude of a single COFDM carrier
and N is the number of COFDM carriers used.

The RMS value of a COFDM signal is calculated from the quadratic mean as:

$$U_{RMS} = \sqrt{N \cdot U_{RMS0}^2};$$

where U_{RMS} is the RMS voltage of a single COFDM carrier

$$U_{RMS0} = \frac{U_{peak0}^2}{\sqrt{2}};$$

The RMS value of the COFDM signal is then:

$$U_{RMS} = \sqrt{N \cdot \frac{U_{peak0}^2}{2}};$$

Inserting into the equation the maximum peak value occurring when all individual carriers are superimposed and the RMS value of the total signal provides:

$$cf_{COFDM} = 20\log(\frac{U_{peak}}{U_{RMS}}) = 20\log(\frac{N \cdot U_{peak0}}{\sqrt{N \cdot \frac{U_{peak0}^{2}}{2}}});$$

This, in turn, can be transformed and simplified to become:

$$cf_{COFDM} = 20\log(\frac{N}{\sqrt{\frac{N}{2}}} = 20\log\sqrt{2N} = 10\log(2N);$$

The theoretical crest factors in DVB-T are then

$$cf_{DVB\text{-}T2K} = 35 \text{ dB};$$

in 2K mode with 1705 carriers used, and

$$cf_{DVB\text{-}T8K} = 41 \text{ dB};$$

in 8K mode with 6817 carriers used.

It must be noted that these are theoretical values which, due to the limited resolution of the signal processing and the clipping, cannot occur in practice. Practical values are of the order of magnitude of 13 dB (DVB-T power transmitter) to about 15 dB (with modulators without clipping).

In the following paragraphs, the DVB-T transmission path itself will be considered in greater detail. In the ideal case, exactly one signal path arrives at the receiving antenna. The signal is then only attenuated to a greater or lesser extent and is merely subjected to additive white Gaussian noise (AWGN). This channel with a direct view of the transmitter is called a Gaussian channel and provides the best conditions of reception for the receiver (Fig. 20.17.).

If multiple echoes are added to this direct signal path, the conditions of reception become much more difficult. This channel with a direct line of sight and a defined number of multiple echoes, which can be simulated as a mathematical channel model, is called a Ricean channel (Fig. 20.18.).

If then the direct line of sight to the transmitter, i.e. the direct signal path, is also blocked, the channel is called a Rayleigh channel (Fig. 20.19.). This represents the worst conditions of stationary reception.

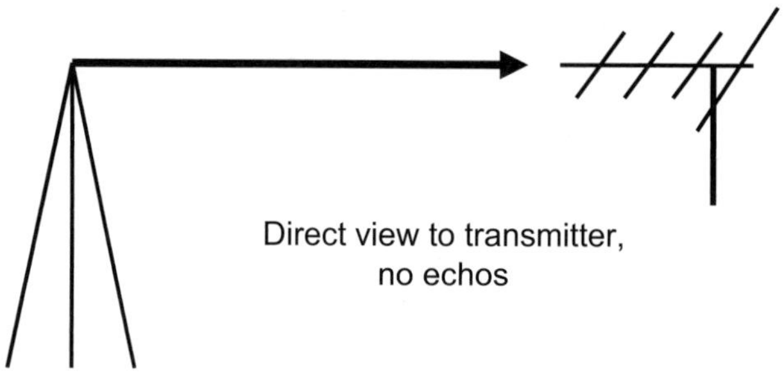

Fig. 20.17. Gaussian channel

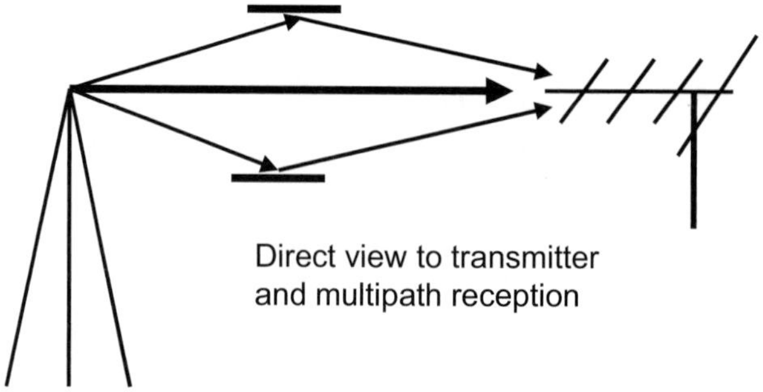

Fig. 20.18. Ricean channel

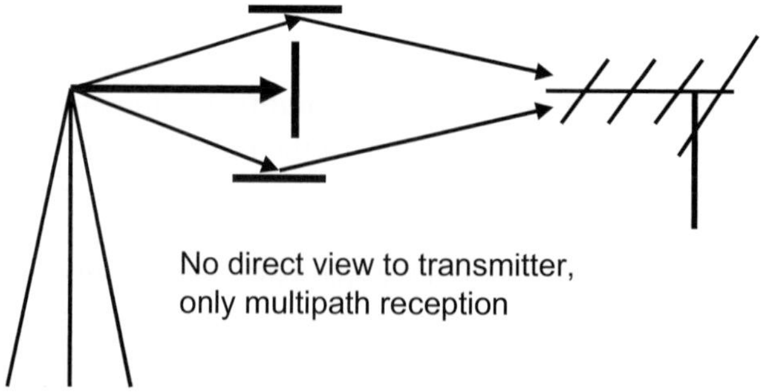

Fig. 20.19. Rayleigh channel

If, for instance, the receiver is moving at a certain speed away from the transmitter or towards the transmitter (Fig. 20.20.), a negative or positive frequency shift Δf will occur due the Doppler effect. This frequency shift by itself does not present any problems to the DVB-T receiver which will compensate for it by means of its AFC. It can be calculated from the speed of movement, the transmitting frequency and the velocity of light.

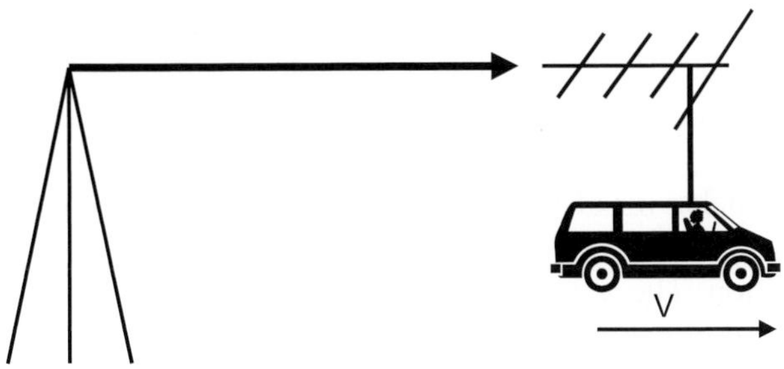

Fig. 20.20. Doppler effect

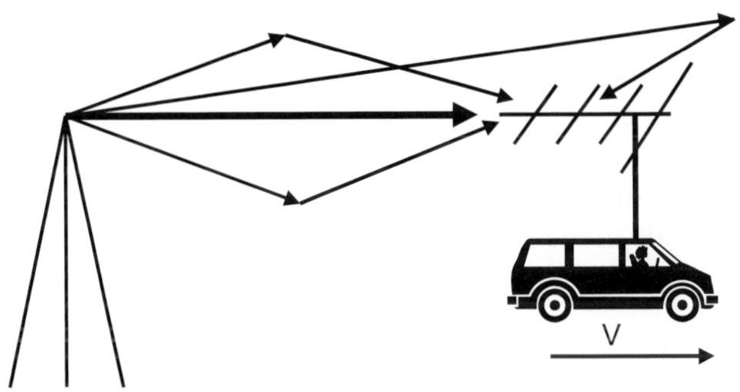

Fig. 20.21. Doppler effect in combination with multipath reception

The following applies:

$$\Delta f = v \bullet (f/c) \bullet \cos(\varphi);$$

where
v is the speed,
f the transmitting frequency,
c the velocity of light (299792458 m/s) and
φ the angle of incidence of the echo in relation to the direction of movement.

Example: At a transmitting frequency of 500 MHz and a speed of 200 km/h, the Doppler shift is 94 Hz.

If, however, multiple echoes are added (Fig. 20.21.), the COFDM spectrum becomes smeared. This smearing is due to the fact that the mobile receiver is both moving towards signal paths and moving away from other sources. I.e. there are now spectral COFDM combs which are shifting upward and downward. Due to its subcarrier spacing, which is narrower by a factor of 4, the 8K mode is much more sensitive to such smearing in the frequency domain than the 2K mode. The 2K mode is thus the better choice for mobile reception, although DVB-T was originally not intended for mobile reception.

Considering then the behavior of the DVB-T receiver in the presence of noise. More or less noise in the DVB-T channel leads to more or fewer bit errors during the reception. The Viterbi decoder can correct more or fewer of these bit errors depending on the code rate selected in the convolutional encoder. In principle, the same rules apply to DVB-T as do for a single carrier method (DVB-C or DVB-S), i.e. the same "waterfall" curves of bit error ratio vs. signal/noise ratio apply. The only caution is advised with respect to the signal/noise ratio which is also often called carrier/noise ratio. The two differ slightly in DVB-T, the reason being the power in the pilot carriers and auxiliary carriers (continual and scattered pilots and TPS carriers). To determine the bit error rate in DVB-T, only the power in the actual payload carriers can be used as the signal power. In DVB-T, the difference between the overall carrier power and the power in the pure payload carriers is 0.857 dB in the 2K mode and 0.854 dB in the 8K mode but the noise bandwidth of the pure payload carriers is reduced with respect to the overall signal.

The reduced noise bandwidth of the payload carriers is:

$$10 \log(1512/1705) = -0.522 \text{ dB; in 2K mode}$$

and

$$10 \log(6048/6917) = -0.520 \text{ dB; in 8K mode.}$$

Thus, the difference between C/N and S/N in DVB-T is:

$$C/N - S/N = -0.522 \text{ dB} - (-0.857 \text{ dB}) = 0.34 \text{ dB; in 2K mode, and}$$

$$C/N - S/N = -0.52 \text{ dB} - (-0.854 \text{ dB}) = 0.33 \text{ dB; in 8K mode.}$$

From the S/N in Fig. 20.22., the bit error ratio before Viterbi, i.e. the channel bit error ratio, can be determined. Fig. 20.22. only applies to non-hierarchical modulation since the constellation pattern can be expanded with hierarchical modulation.

The theoretical minimum carrier-to-noise ratios for quasi error-free operation depend on the code rate both in DVB-T and in DVB-S. In addition, the type of modulation (QPSK, 16QAM, 64QAM) and the type of channel (Gaussian, Ricean, Rayleigh) have an influence. The theoretical minimum C/Ns are listed below for the case of non-hierarchical coding.

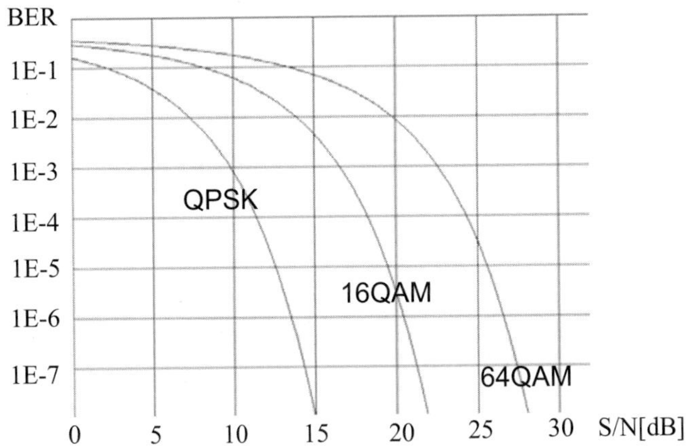

Fig. 20.22. Bit error ratio in DVB-T as a function of S/N in QPSK, 16QAM and 64QAM with non-hierarchical modulation

Table 20.14. Minimum C/N required with non-hierachical modulation

Modulation	Code rate	Gaussian channel [dB]	Rice channel [dB]	Rayleigh channel [dB]
QPSK	1/2	3.1	3.6	5.4
	2/3	4.9	5.7	8.4
	3/4	5.9	6.8	10.7
	5/6	6.9	8.0	13.1
	7/8	7.7	8.7	16.3
16QAM	1/2	8.8	9.6	11.2
	2/3	11.1	11.6	14.2
	3/4	12.5	13.0	16.7
	5/6	13.5	14.4	19.3
	7/8	13.9	15.0	22.8
64QAM	1/2	14.4	14.7	16.0
	2/3	16.5	17.1	19.3
	3/4	18.0	18.6	21.7
	5/6	19.3	20.0	25.3
	7/8	20.1	21.0	27.9

Thus, the demands for a minimum C/N fluctuate within a wide range from about 3 dB for QPSK with a code rate of 1/2 in a Gaussian channel up to about 28 dB for 64QAM with a code rate of 7/8 in a Rayleigh channel. Practical values are about 18 to 20 dB (64QAM, code rate 2/3 or 3/4) for stationary reception and about 11 to 17 dB (16QAM, code rate 2/3 or 3/4) for mobile reception.

Table 20.15. Theoretical minimum C/N with hierarchical modulation (QPSK, 64QAM, $\alpha = 2$); low priority path (LP)

Modulation	Code rate	Gaussian channel [dB]	Rice channel [dB]	Rayleigh channel [dB]
QPSK	1/2	6.5	7.1	8.7
	2/3	9.0	9.9	11.7
	3/4	10.8	11.5	14.5
64-QAM	1/2	16.3	16.7	18.2
	2/3	18.9	19.5	21.7
	3/4	21.0	21.6	24.5
	5/6	21.9	22.7	27.3
	7/8	22.9	23.8	29.6

20.8 DVB-T Single-Frequency Networks (SFN)

COFDM is well suited to single frequency operation. As the name indicates, in single frequency operation, all transmitter operate at the same frequency which makes for great economy with regard to frequency resources. All transmitters radiate the identical signal and have to operate in complete synchronism with each other. Signals from adjacent signals are seen by a transmitter as if they were simply echoes. Frequency synchronization is the easiest condition because frequency accuracy and stability had to meet high demands even in analog television. In DVB-T, the transmitter RF is locked to the best reference available: the signal from the GPS (Global Positioning System) which is available throughout the world and is now also used for synchronizing the transmitting frequencies of a DVB-T single-frequency network. The GPS satellites radiate a 1 pps (pulse per second) signal to which a 10 MHz oscillator in professional GPS receivers is locked which, in turn, acts as reference signal for the DVB-T transmitters.

However, there is also a strict requirement with respect to the maximum distance between transmitters (Fig. 20.23. and Tables 20.16, 20.17. and 20.18.). This distance is related to the length of the guard interval and the velocity of light, i.e. the associated signal delay. Intersymbol interference can only be avoided if in the case of multipath reception, the delay on any path is no longer than the length of the guard interval. The question about what would happen if the signal received from a more distant transmitter violates the guard interval is easily answered: it results in intersymbol interference which becomes noticeable as noise in the receiver.

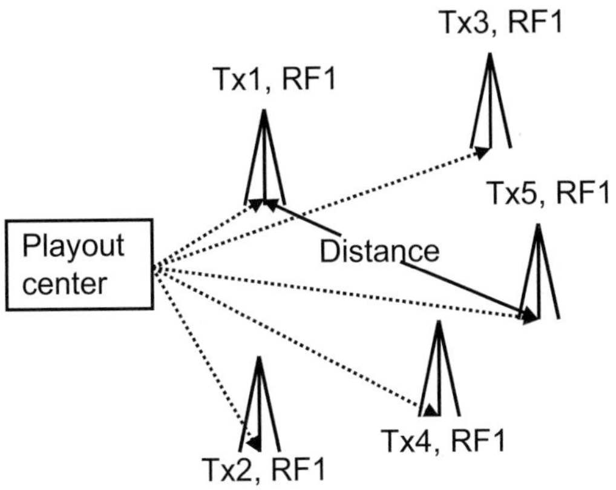

Fig. 20.23. DVB-T single-frequency network (SFN)

Table 20.16. Guard interval lengths for 8K, 2K modes and transmitter distances (8 MHz channel)

Mode	Symbol duration µs	Guard interval ratio	Guard interval µs	Transmitter distance km
2K	224	1/4	56	16.8
2K	224	1/8	28	8.4
2K	224	1/16	14	4.2
2K	224	1/32	7	2.1
8K	896	1/4	224	67.1
8K	896	1/8	112	33.6
8K	896	1/16	56	16.8
8K	896	1/32	28	8.4

Signals from transmitters at greater distances must simply be attenuated sufficiently. The threshold for quasi error free operation is formed by the same conditions as for pure noise. It is, therefore, of particular importance that the levels in a single-frequency network are calibrated correctly. It is not the maximum transmitting power at every transmitting site which is required but the correct one. Planning of the network requires topographical information.

In many cases, however, network planning is relatively simple since mostly only small regional single-frequency networks with only very few transmitters are set up.

Table 20.17. Guard interval lengths for 8K, 2K modes and transmitter distances (7 MHz channel)

Mode	Symbol duration µs	Guard interval ratio	Guard interval µs	Transmitter distance km
2K	256	1/4	64	19.2
2K	256	1/8	32	9.6
2K	256	1/16	16	4.8
2K	256	1/32	8	2.4
8K	1024	1/4	256	76.7
8K	1024	1/8	128	38.4
8K	1024	1/16	64	19.2
8K	1024	1/32	32	9.6

Table 20.18. Guard interval lengths for 8K, 2K modes and transmitter distances (6 MHz channel)

Mode	Symbol duration µs	Guard interval	Guard interval µs	Transmitter distance km
2K	299	1/4	75	22.4
2K	299	1/8	37	11.2
2K	299	1/16	19	5.6
2K	299	1/32	9	2.8
8K	1195	1/4	299	89.5
8K	1195	1/8	149	44.8
8K	1195	1/16	75	22.4
8K	1195	1/32	37	11.2

The velocity of light is c = 299792458 m/s which results in a signal delay per kilometer transmitter distance of t_{1km} = 1000 m/c = 3.336 µs. Since in the 8K mode, the guard interval is longer in absolute terms, it is mainly this mode which is provided for single frequency operation.

Long guard intervals are provided for single frequency networks. Medium-length guard intervals are used in regional networks. The short guard intervals, finally, are provided for local networks or used outside of single frequency networks.

In a single frequency network, all the individual transmitters must be synchronized with one another. The program contribution is injected from the playout center in which the MPEG-2 multiplexer is located, e.g. via satellite, optical fiber or microwave link. It is clear that the MPEG-2 transport streams are subject to different feed line delays due to different path lengths. However, it is necessary that in each DVB-T modulator in an SFN network the same transport stream packets are processed into COFDM symbols. Every modulator must perform every operating step completely synchronously with all the other modulators in the network. The same packets, the same bits and the same bytes must all be processed at the same time. Every DVB-T transmitter site must broadcast absolutely identical COFDM symbols at exactly the same time.

The DVB-T modulation is structured in frames, one frame being composed of 68 DVB-T COFDM symbols. Within a frame, the complete TPS information is transmitted and the scattered pilots are scattered over the entire DVB-T channel. Four such frames, in turn, make up one superframe.

Frame structure of DVB-T:

- 68 COFDM symbols = 1 frame
- 4 frames = 1 superframe

One superframe in DVB-T accommodates an integer number of MPEG-2 transport stream packets, as follows:

Table 20.19. Number of transport stream packets per superframe

Code-rate	QPSK 2K	QPSK 8K	16QAM 2K	16QAM 8K	64QAM 2K	64QAM 8K
1/2	252	1008	504	2016	756	3024
2/3	336	1344	672	2688	1008	4032
3/4	378	1512	756	3024	1134	4536
5/6	420	1680	840	3360	1260	5040
7/8	441	1764	882	3528	1323	5292

In consequence, a superframe in a single frequency network must be composed of absolutely identical transport stream packets and each modulator in the SFN must generate and broadcast the superframe at the same time.

These modulators must, therefore, be synchronized with one another and, in addition, the differences in the feed line delays must be equalized statically and dynamically. To achieve this, packets with time stamps are inserted into the MPEG-2 transport stream in the playout center. These packets are special transport stream packets which are configured similarly to an MPEG-2 table (PSI/SI). For this purpose, the transport stream is divided into sections, the lengths of which are selected to be approximately a half second because they must correspond to a certain integral number of transport stream packets fitting into a certain integral number of superframes. These sections are called megaframes.

A megaframe is composed of an integral number of superframes, as follows:

- 1 megaframe = 2 superframes in 8K mode
- 1 megaframe = 8 superframes in 2K mode

The 1 pps signals of the GPS satellites are also used for synchronizing the timing of the DVB-T modulators. In the case of a single frequency network, there is a professional GPS receiver outputting both a 10 MHz reference signal and this 1 pps time signal at every transmitter site and at the playout center (Fig. 20.24.) where the multiplexed stream is assembled.

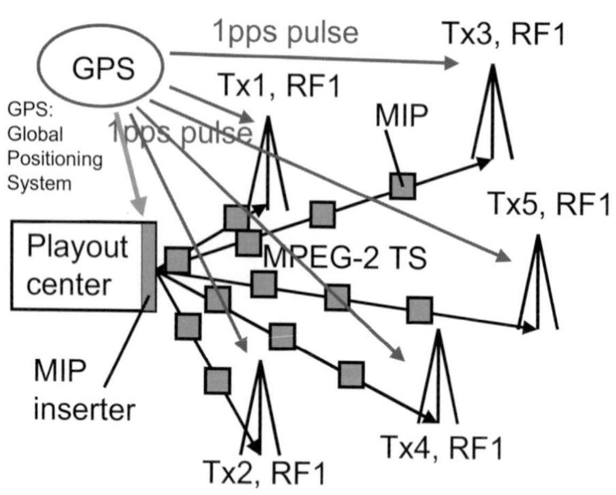

Fig. 20.24. DVB-T distribution network with MIP insertion

At the multiplexer site there is a so-called MIP inserter which inserts this special transport stream packet into one megaframe in each case, which is why this packet is called the megaframe initializing packet (MIP). The MIP has a special PID of 0x15 so that it can be identified and it contains time reference and control information for the DVB-T modulators. Among other things, it contains the time counting back to the time the last 1 pps pulse was received at the MIP inserter. This time stamp with a resolution of 100 ns steps is used for automatically measuring the feed distance. This time information is evaluated by the SFN adapter which automatically corrects the delay from the playout center to the transmitter site by means of a buffer store. It also requires information about the maximum delay in the network. Given this information, which can either be input manually at every transmitter site or is carried in the MIP packet, each SFN adapter adjusts itself to this time. The MIP packet also contains a pointer to the start of the next megaframe in numbers of TS packets. Using this pointer information, each modulator is then able to start a megaframe at the same time.

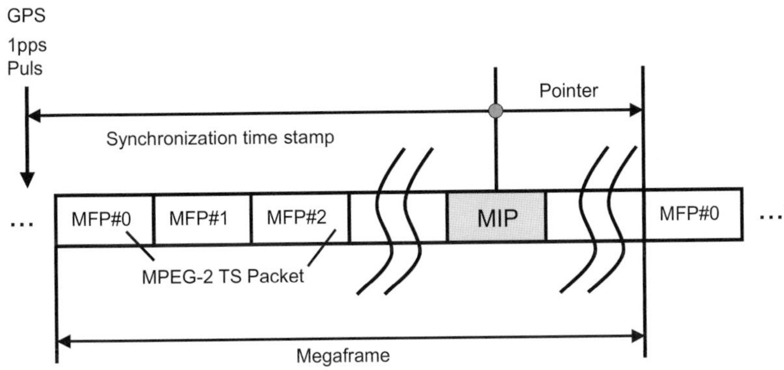

Fig. 20.25. Megaframe structure at transport stream level

The length of a megaframe depends on the length of the guard interval and on the bandwidth of the channel. The narrower the channel (8, 7 or 6 MHz), the longer the COFDM symbols since the subcarrier spacing becomes less. Every DVB-T modulator can now be synchronized by means of the information contained in the MIP packet. The MIP packet can always be transmitted at a fixed position in the megaframe but this position is also allowed to vary. Table 20.20. contains a list of the exact lengths of one megaframe.

Table 20.20. Duration of a megaframe

Guard interval	8 MHz channel	7 MHz channel	6 MHz channel
1/32	0.502656 s	0.574464 s	0.670208 s
1/16	0.517888 s	0.598172 s	0.690517 s
1/8	0.548352 s	0.626688 s	0.731136 s
1/4	0.609280 s	0.696320 s	0.812373 s

An MIP can also be used for transmitting additional information such as the DVB-T transmission parameters which makes it possible to control and configure the entire DVB-T SFN from one center. For example, it can be used for changing the type of modulation, the code rate, the guard interval length etc. However, although this is possible, it may not be supported by every DVB-T modulator.

If the transmission of the MIP packets stops for some reason or if the information in the MIP packets is corrupted, the single frequency network will lose synchronization. If a DVB-T transmitter detects that it has dropped lock or that it has not received a GPS signal for some time and the 1 pps reference and the 10 MHz reference have, therefore, drifted, it has to go off air or it will only be a source of noise in the single frequency network. Reliable reception is then only possible with directional reception close to the transmitter. For this reason, the MIPs in the transport stream arriving at the transmitter are often monitored using an MPEG-2 test decoder (see Fig. 20.26.).

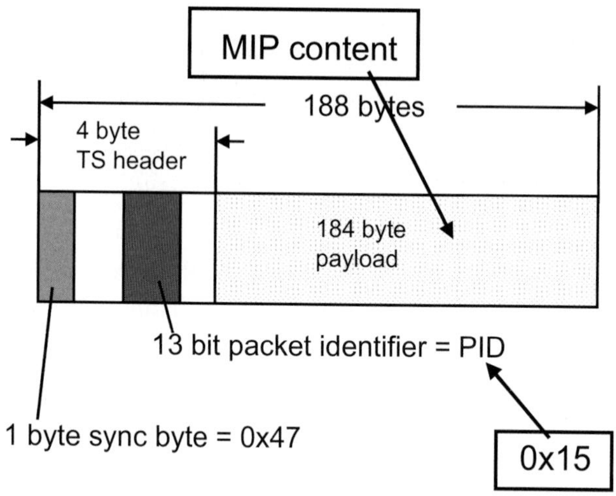

Fig. 20.26. Megaframe initializing packet

Fig. 20.27. (MIP = Megaframe Initializing Packet) clearly shows that the multiplexed MPEG-2 stream is now carrying a further table-like packet, namely the MIP packet, containing the synchronization time stamp, the pointer and the maximum delay. It also contains the transmission parameters. It can also be seen that every transmitter in the link-up can be addressed. Like a table, the content of the MIP packet is protected by a CRC checksum.

In addition, each transmitter can also be "pushed", i.e. it is possible to change the time when the COFDM symbol is broadcast. This will not push the single frequency network out of synchronization but only vary the delay of the signals of the transmitters with respect to each other and can thus be used for optimizing the SFN network. These time offsets are found in the 'TX time offset' functions in Fig. 20.27. Shifting the broadcasting time makes it appear to the receiver as if the geographic position of the respective transmitter has changed. This may be of interest if two transmitter in an SFN are very far apart and are approaching the limit of the guard interval (e.g. DVB-T network Southern Bavaria with the Olympic tower in Munich and the Mount Wendelstein transmitter at a distance d of 63 km) or if the guard interval has been chosen to be very short for reasons of data rates (e.g. Sydney, Australia, with g=1/16).

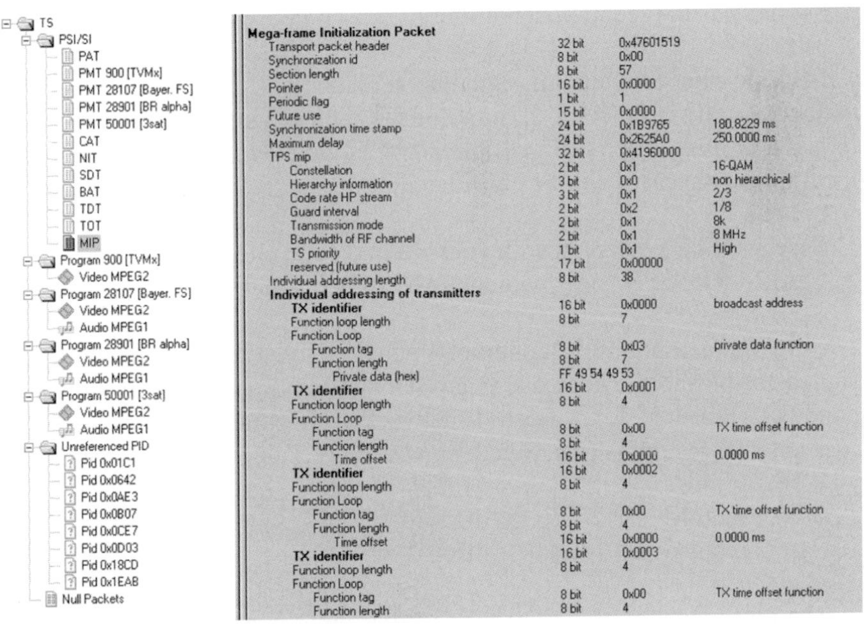

Fig. 20.27. MIP packet analysis [DVMD]

20.9 Minimum Receiver Input Level Required with DVB-T

To obtain error-free reception of a DVB-T signal, the minimum receive level required must be present at the DVB-T receiver input. Below a certain signal level, reception breaks off and at the threshold blocking and freezing effects will occur, above which reproduction is faultless. This section discusses the principles for determining this minimum level.

The minimum level in DVB-T is dependent on:

- Type of modulation (QPSK, 16QAM, 64QAM)
- Error protection used (code rate 1/2, 2/3, 3/4,... 7/8)
- Channel model (Gaussian, Ricean, Rayleigh)
- Bandwidth (8, 7, 6 MHz)
- Ambient temperature
- Actual receiver characteristics (noise figure of the tuner etc)
- Multipath reception conditions

In principle, a minimum signal/noise ratio S/N is required which is mathematically a function of some of the factors listed above. The theoretical S/N limits are listed in Table 20.14. in Section 20.7. As an example, the 2 following cases will be considered:

Case 1: Ricean channel with 16QAM and code rate = 2/3, and
Case 2: Ricean channel with 64QAM and code rate = 2/3.

Case 1 corresponds to conditions adapted for a DVB-T network designed for portable indoor use (e.g. Germany) and Case 2 corresponds to conditions adapted for a DVB-T network with parameters designed for roof antenna reception (e.g. Sweden, Australia). Table 20.14. shows that

Case 1 requires a S/N of 11.6 dB, and
Case 2 requires a S/N of 17.1 dB.

The noise level N present at the receiver input is obtained from the following physical relation:

$$N[dBW] = -228.6 + 10 \log(B/Hz) + 10 \log((T/^0C + 273)) + F;$$

where
B = bandwidth in Hz;

T = temperature in ^{0}C;
F = noise figure of the receiver in dB;

The constant -228.6 dBW/K/Hz is the so-called Boltzmann's constant. Assuming that:

ambient temperature T = 20^{0}C;
noise figure of the tuner = 7 dB;
receiver bandwidth B = 8 MHz;

then

$$N[dBW] = -228.6 + 10 \log(8000000/Hz) + 10 \log((20/^0C+273)) + 7;$$

0 dBm @ 50 ohms = 107 dBµV;
0 dBm @ 75 ohms = 108.8 dBµV;

N = -98.1 dBm = -98.1 dBm + 108.8 dB = 10.7 dBµV; (at 75 ohms)

Thus, the noise level present at the receiver input under these conditions is 10.7 dBµV.

For Case 1 (16QAM), the minimum receiver input level required is, therefore:

S = S/N [dB] + N [dBµV] = (11.6 + 10.7)[dBµV] = 22.3 dBµV;

For Case 2 (64QAM), the minimum receiver input level required is:

S = S/N [dB] + N [dBµV] = (17.1 + 10.7)[dBµV] = 27.8 dBµV;

In practice it is found that these values can be met quite confortably with only one signal path but as soon as several signal paths are present at the receiver input (multipath reception), the required level is often higher by up to 10 to 15 dB and varies greatly with different types of receiver.

The received level actually present is the result of:

- Received signal strength present at the receiving location
- Antenna gain
- Polarization losses
- Losses in the feed line from the antenna to the receiver

The following applies to the conversion of the antenna output level from the field strength present at the receiving site:

E[dBμV/m] = U[dBμV] + k[dB];

k[dB] = (-29.8 + 20 log(f[MHz]) - g[dB];

where:

E = electrical field strength,
U = antenna output level,
k = k factor of the antenna,
f = received frequency,
g = antenna gain.

The level present at the receiver input as a result is then:

S[dBμV] = U[dBμV] - loss[dB];

Where 'loss' designates the implementation losses (antenna feeder etc.)

Considering now Case 1 (16QAM) and Case 2 (64QAM) at 3 frequencies:

a) f = 200 MHz,
b) f = 500 MHz,
c) f = 800 MHz.

The antenna gain is assumed to be g = 0 dB in each case (non-directional rod antenna).
k factors of the antenna:

a) k = (-29.8 + 48) dB = 16.2 dB;
b) k = (-29.8 + 54) dB = 24.2 dB;
c) k = (-29.8 + 58.1) dB = 28.3 dB;

Field strengths for Case 1 (16QAM; minimum required level
U = S - loss = 22.3 dBμV - 0dB = 22.3 μV:

a) E = (22.3 + 16.2) dBμV/m = 38.5 dBμV/m;
b) E = (22.3 + 24.2) dBμV /m = 46.5 dBμV/m;
c) E = (22.3 + 28.3) dBμV/m = 50.6 dBμV/m;

If a directional antenna with gain is used, e.g. a roof antenna, the following conditions are obtained.

a) with f = 200 MHz (assuming g = 6 dB), E = 32,5 dBµV/m;
b) with f = 500 MHz (assuming g = 10 dB), E = 36,5 dBµV/m;
c) with f = 800 MHz (assuming g = 10 dB), E = 40.6 dBµV/m;

Field strengths for Case 2 (64QAM; minimum required level
U = S - loss = 27.8 dBµV - 0dB = 27.8 µV):

a) E = (27.8 + 16.2) dBµV/m = 44.0 dBµV/m;
b) E = (27.8 + 24.2) dBµV /m = 52.0 dBµV/m;
c) E = (27.8 + 28.3) dBµV/m = 56.1 dBµV/m;

If a directional antenna with gain is used, e.g. a roof antenna, the following conditions are obtained.

a) with f = 200 MHz (assuming g = 6 dB), E = 38.0 dBµV/m;
b) with f = 500 MHz (assuming g = 10 dB), E = 42.0 dBµV/m;
c) with f = 800 MHz (assuming g = 10 dB), E = 46.1 dBµV/m;

Under free space conditions, the field strength at the receiving site can be calculated as:

E[dBµV/m] = 106.9 + 10 log(ERP[kW]) - 201g(d[km]);

where:
E = electrical field strength
ERP = effective radiated power, i.e. the transmitter power
plus antenna gain;
d = transmitter - receiver distance;

Under real conditions, however, much lower field strengths must be assumed because this formula does not take into consideration shading, multipath reception etc. The reduction depends on the topological conditions (hills, mountains, buildings etc) and can be up to about 20 - 30 dB, but also much more with complete shading.

Example (without reduction; a reduction of at least 20 dB is recommended):

ERP = 50 kW;

d = 1 km; E = (106.9 + 10 log(50) – 20 log(1)) dBµV/m
= 123.9 dBµV/m;

d = 10 km; E = (106.9 + 10 log(50) – 20 log(1)) dBµV/m
= 103.9 dBµV/m;
d = 30 km; E = 94.4 dBµV/m;

d = 50 km; E = 89.9 dBµV/m;

d = 100 km; E = 83.9 dBµV/m;

Since the plane of polarization was frequently changed from horizontal to vertical at the transmitter site as part of the DVB-T conversion. There may also be polarization losses of about 10…20 dB at the receiving antenna if this has not also been changed from horizontal to vertical.

If the DVB-T signal is received with an indoor antenna inside the house attenuation due to the building must also be considered which amounts to another 10 to 20 dB.

In Germany, the following field strength values were assumed with 16QAM, CR=2/3 as limit values for the field strength outside the building during the simulation of the conditions of reception:

- Reception with a roof antenna: approx. 55 dBµV/m
- Reception with an outdoor antenna: approx. 65 dBµV/m
- Reception with an indoor antenna: approx. 75…85 dBµV/m

Bibliography: [ETS300744], [REIMERS], [HOFMEISTER], [EFA], [SFQ], [TR101190], [ETR290].

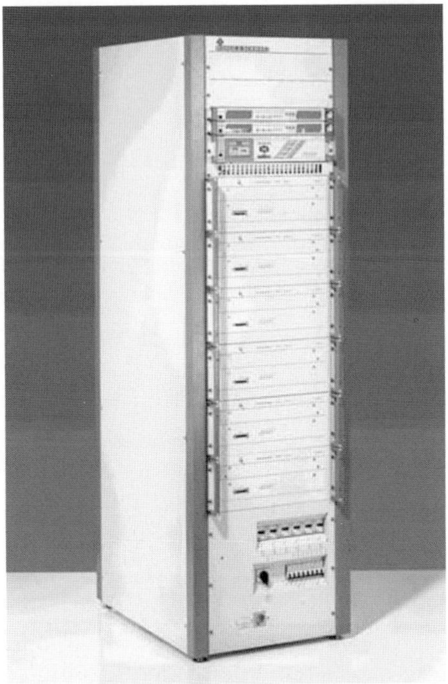

Fig. 20.28. Medium-power DVB-T transmitter (Rohde&Schwarz)

Fig. 20.29. DVB-T mask filter, critical mask (dual mode filter, manufacturer's photo Spinner)

21 Measuring DVB-T Signals

The DVB-T standard and its complicated COFDM modulation method have now been discussed thoroughly. The present chapter deals with methods for testing DVB-T signals in accordance with the DVB Measurement Guidelines ETR 290 and also beyond these. The requirement for measurements is much greater in DVB-T than in the other two transmission path systems DVB-C and DVB-S due to the highly complex terrestrial transmission path, the much more complicated DVB-T modulator and the analog IQ modulator used there in most cases. DVB-T measuring techniques must cover the following interference effects:

- Noise (AWGN)
- Phase jitter
- Interferers
- Multipath reception
- Doppler effect
- Effects in the single-frequency network
- Interference with the adjacent channels (shoulder attenuation)
- I/Q errors of the modulator:
    - I/Q amplitude imbalance
    - I/Q phase errors
    - lack of carrier suppression

The test instruments used in DVB-T measuring techniques are essentially comparable to those used in broadband cable measuring techniques. The following are required for measuring DVB-T signals:

- A modern spectrum analyzer
- a DVB-T test receiver with constellation analyzer
- a DVB-T test transmitter for measurements on DVB-T receivers

The DVB-T test receiver is by far the most important measuring means in DVB-T. Due to the pilot signals integrated in DVB-T, it allows the most extensive analyses to be performed on the signal without using other aids,

W. Fischer, *Digital Video and Audio Broadcasting Technology*, Signals and Communication Technology, 3rd ed., DOI: 10.1007/978-3-642-11612-4_21, © Springer-Verlag Berlin Heidelberg 2010

the most important one of these being the analysis of the DVB-T constellation pattern. Although extensive knowledge in the field of DVB-C constellation analysis has been gathered since the 90s, simply copying it across into the DVB-T world is not sufficient. This chapter deals mainly with the special features of DVB-T constellation analysis, points out problems and provides assistance in interpreting the results of the measurements.

In comparison with DVB-C constellation analysis, DVB-T constellation analysis is not simply a constellation analysis on many thousands of subcarriers and many things do not lend themselves to being simply copied across.

Fig. 21.1. shows the constellation diagram of a 64QAM modulation in DVB-T. The positions of the scattered pilots and of the continual pilots (on the left and on the right outside the 64QAM constellation diagram on the I axis) and of the TPS carriers (constellation points inside the constellation diagram, also on the I axis) can be easily seen. The scattered pilots are used for channel estimation and correction and thus represent a checkpoint in the constellation diagram which is always corrected to the same position. The transmission parameter signalling carriers (TPS) serve as a fast information channel from the transmitter to the receiver. Apart from noise, there are no further influences acting on the constellation diagram shown (Fig. 21.1.).

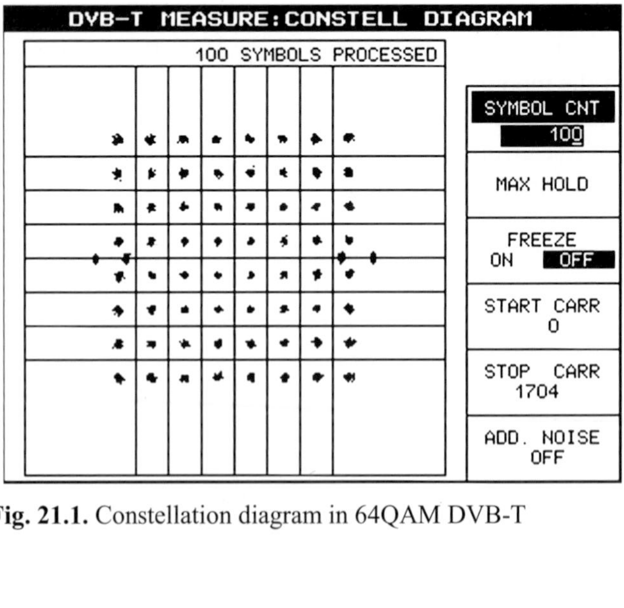

Fig. 21.1. Constellation diagram in 64QAM DVB-T

A DVB-T test receiver (Fig. 21.2.) can be used for detecting all influences acting on the transmission link. A DVB-T test receiver basically differs from a set-top box in the analog signal processing being of a much higher standard and the I/Q data and the channel estimation data being accessed by a signal processor (DSP). The DSP then calculates the constellation diagram and the measurement values. In addition, the DVB-T signal can be demodulated down to the MPEG-2 transport stream level.

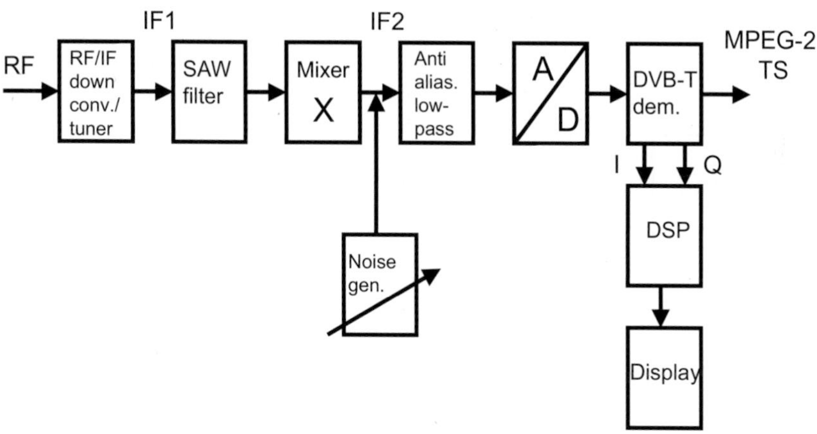

Fig. 21.2. Block diagram of a DVB-T test receiver

21.1 Measuring the Bit Error Ratio

In DVB-T, as in DVB-S, there are 3 bit error ratios due to the inner and outer error protection:

- Bit error ratio before Viterbi
- Bit error ratio before Reed Solomon
- Bit error ratio after Reed Solomon

The error ratio of greatest interest and providing the most information is the pre-Viterbi bit error ratio. It can be determined by applying the post-Viterbi data stream to another convolutional encoder of the same configuration as that at the transmitter end. If the data stream before Viterbi is compared with that after the convolutional encoder - taking into consideration the delay of the convolutional encoder - the two are identical provided there are no errors. The differences, and thus the bit errors, are then determined by a comparator for the I and Q branch.

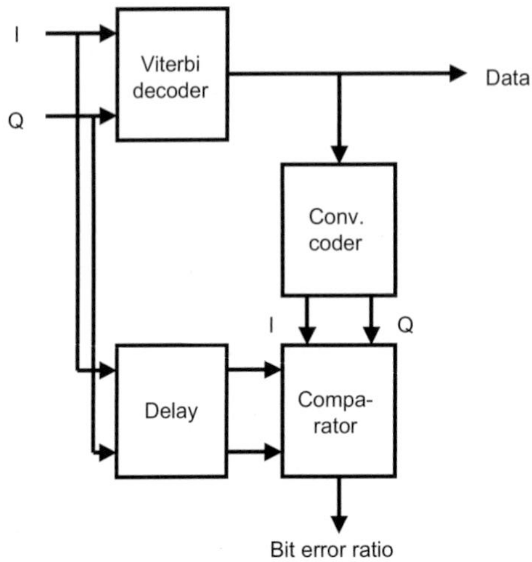

Fig. 21.3. Circuit for determining the pre-Viterbi bit error ratio

The bit errors counted are then related to the number of bits transmitted within the corresponding period, providing the bit error ratio (BER)

BER = bit errors/transmitted bits;

The pre-Viterbi bit error ratio range is between 10^{-9} (transmitter output) and 10^{-2} (receiver input with poor receiving conditions).

The Viterbi decoder can only correct some of the bit errors, leaving a residual bit error ratio before Reed Solomon. Counting the corrections of the Reed Solomon decoder and relating them to the number of bits transmitted within the corresponding period provides the pre-Reed Solomon bit error ratio.

However, the Reed Solomon decoder is not able to correct all bit errors, either, and this then results in errored transport stream packets. These are flagged in the TS header (transport error indicator bit = 1). Counting the errored transport stream packets enables the post-Reed Solomon bit error ratio to be calculated.

A DVB-T test receiver will detect all 3 bit error ratios and indicate them in one of the main measurement menus. It must be noted that with the relatively low bit error ratios usually available after the Viterbi and Reed-

Solomon decoders, measuring times of corresponding length in the range of minutes up to hours must be selected.

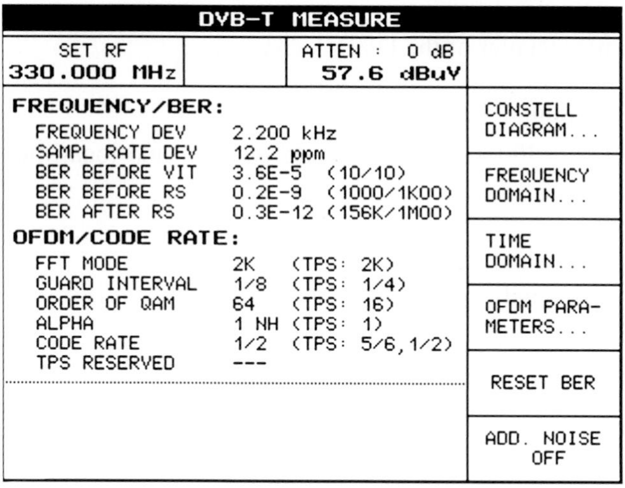

Fig. 21.4. Bit error ratio measurement [EFA]

The measurement menu example [EFA] shows that all the important information about the DVB-T transmission is combined here. Apart from the RF selected, it also shows the received level, the frequency deviation, all 3 bit error ratios and the decoded TPS parameters.

21.2 Measuring DVB-T Signals Using a Spectrum Analyzer

A spectrum analyzer is very useful for measuring the power of the DVB-T channel, at least at the DVB-T transmitter output. Naturally, one could simply use a thermal power meter for this purpose but, in principle, it is also possible to use a spectrum analyzer which will provide a good estimate of the carrier/noise ratio. Firstly, however, the power of the DVB-T signal will now be determined. A COFDM signal looks like noise and has a crest factor which is rather high. Due to its similarity with white Gaussian noise, its power is measured in a comparable way.

To determine the carrier power, the spectrum analyzer is set as follows: On the analyzer, a resolution bandwidth of 30 kHz and a video bandwidth of 3 to 10 times the resolution bandwidth, i.e. 300 kHz, is selected. To achieve a certain amount of averaging, a slow sweep time of 2000 ms is set. These parameters are needed because we are using the RMS detector of the spectrum analyzer. The following settings are then used:

- Center frequency: center of the DVB-T channel
- Span: 20 MHz
- Resolution bandwidth: 30 kHz
- Video bandwidth: 300 kHz (due to RMS detector and logarithmic scale)
- Detector: RMS
- Sweep: slow (2000 ms)
- Noise marker: channel center (resultant C' value in dBm/Hz)

$$C[dBm]-10\lg\left(\frac{DVB-T-signal\_bandwidth}{resolution\_bandwidth}\right)[dB];$$

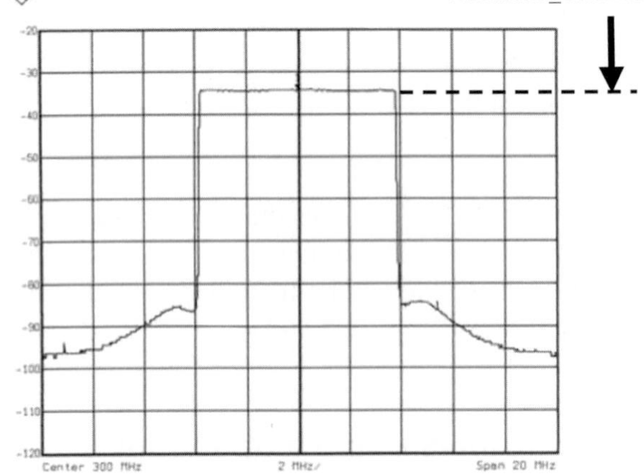

Fig. 21.5. Spectrum of a DVB-T signal

The level indicated in the useful band of the DVB-T spectrum (Fig. 21.5.) depends on the choice of resolution bandwidth (RBW) of the spectrum analyzer (e.g. 1, 4, 10, 20, 30 kHz) with respect to the bandwidth of the DVB-T signal (7.61 MHz, 6.66 MHz, 5.71 MHz). In the literature (DVB-T standard, systems specifications), 4 kHz is often quoted as reference bandwidth but is not always supported by spectrum analyzers. At 4 kHz reference bandwidth, the level shown in the useful band is 38.8 dB (7.61 MHz) or 32.2 dB, respectively, below the level of the DVB-T signal.

Table 21.1. Level of the useful band shown by the spectrum analyzer vs. signal level

Resolution bandwidth [kHz]	Attenuation [dB] in useful band vs. DVB-T signal level in the 7 MHz channel	Attenuation [dB] in useful band vs. DVB-T signal level in the 8 MHz channel
1	38.8	38.2
4	32.8	32.2
5	31.8	31.2
10	28.8	28.2
20	25.8	25.8
30	24.0	24.0
50	21.8	21.8
100	18.8	18.8
500	11.8	11.8

To measure the power, the noise marker is used because of the noise-like signal. The noise marker is set to band center for this but the prerequisite is a flat channel which can always be assumed to exist at the transmitter. If the channel is not flat, other suitable measuring functions must be used for measuring channel power but these depend on the spectrum analyzer.

The analyzer provides the C' value as noise power density at the position of the noise marker in dBm/Hz, automatically taking into consideration the filter bandwidth and the characteristics of the logarithmatic amplifier of the analyzer. To bring the signal power density C' into relation with the Nyquist bandwidth B_N of the DVB-T signal it is necessary to calculate the signal power C as follows:

$$C = C' + 10\log(\text{signal bandwidth/Hz})\ [\text{dBm}]$$

The signal bandwidth of the DVB-T signal is

- 7.61 MHz in the 8 MHz channel,
- 6.66 MHz in the 7 MHz channel,
- 5.71 MHz in the 6 MHz channel.

Example (8 MHz channel):

Measurement value of the noise marker:	-100 dBm/Hz
Correction value at 7.6 MHz bandwidth:	<u>+ 68.8 dB</u>
Power in the DVB-T channel:	- 31.2 dBm

Approximate Determination of the Noise Power N:

If it were possible to switch off the DVB-T signal without changing the noise ratios in the channel, the noise marker in the center of the band would provide information about the noise ratios in the channel. However, this cannot be done so easily. Using the noise marker for measuring very closely to the signal on the shoulder of the DVB-T signal will provide at least a "good idea" about the noise power in the channel, if not a precise measurement value. This is because it can be assumed that in the useful band, the noise fringe continues similarly to that found on the shoulder.

The spectrum analyzer outputs the value N' of the noise power density. The noise power N in the channel with the bandwidth B_C of the DVB-T transmission channel is calculated from the noise power density N' as follows:

$$N = N' + 10\log(B_C/Hz); \text{ [dBm]}$$

The channel bandwidth to be used is the actual bandwidth of the DVB-T channel, e.g. 8 MHz.

Example:

Measurement value of the noise marker:	-140 dBm/Hz
Correction value at 8 MHz bandwidth:	+ 69.0 dB
Noise power in the DVB-T channel:	- 71.0 dBm

From this, the C/N value is obtained as:

$$C/N_{[dB]} = C_{[dBm]} - N_{[dBm]}$$

In the example: $C/N_{[dB]}$ = -31.2 dBm - (-71.2 dBm) = 40 dB.

In estimating C/N in this way by means of the shoulders of the DVB-T signal it is important that this measurement is made directly at the output interfaces after the power amplifier and before any passive bandpass filters. Otherwise, only the shoulders lowered by the bandpass filter will be seen. The author has repeatedly verified the validity of this measuring method in comparisons with measurement results from a DVB-T test receiver.

21.3 Constellation Analysis of DVB-T Signals

The great difference between the constellation analysis of DVB-T signals and DVB-C is that in DVB-T, many thousands of COFDM subcarriers are analyzed. The carrier range must be selectable. Often, displaying all the constellation diagrams (carriers No. 0 to 6817 or 0 to 1705, resp.) as constellation diagrams written on top of each other is of interest. The carrier ranges can be selected in 2 ways:

- start/stop carrier no.,
- center/span carrier no.

Apart from the pure payload carriers, the pilot carriers and the TPS carriers can also be considered but no mathematical constellation analysis will be performed on these carriers. In the following paragraphs, the individual influences and measurement parameters will be discussed.

The following measurement values can be detected by using constellation analysis:

- Signal/noise ratio S/N
- Phase jitter
- I/Q amplitude imbalance
- I/Q phase error
- Modulation error ratio MER

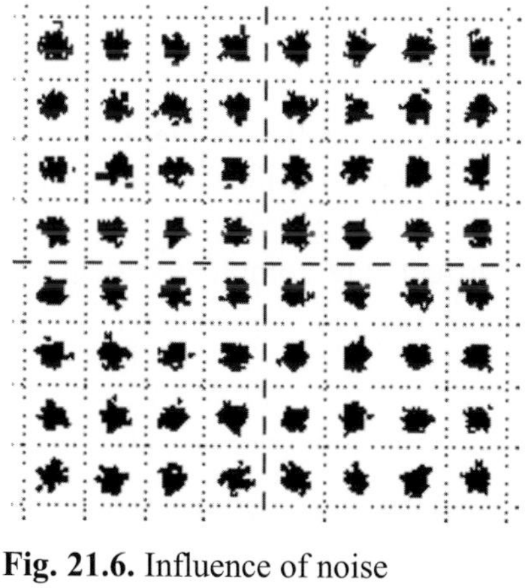

Fig. 21.6. Influence of noise

21.3.1 Additive White Gaussian Noise (AWGN)

White noise (AWGN, Additive White Gaussian Noise) leads to cloud-shaped constellation points. The larger the constellation point, the greater the noise effect. The signal/noise parameter S/N can be determined by analyzing the distribution function (normal Gaussian distribution) in the decision field. The RMS value of the noise component corresponds to the standard deviation. Noise effects affect every DVB-T subcarrier and can also be found on every subcarrier. The effects and measuring methods are completely identical with the DVB-C methods.

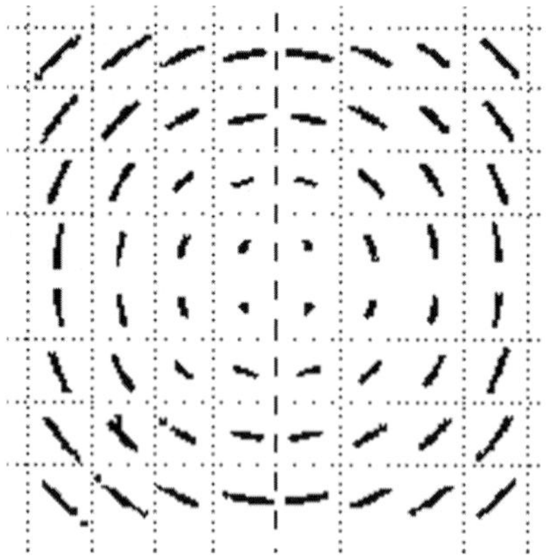

Fig. 21.7. Effect of phase jitter

21.3.2 Phase Jitter

Phase jitter leads to a striated distortion in the constellation diagram. It is caused by the oscillators in the modulator and also affects every carrier and can also be found on every carrier.

Here, too, the measuring methods and effects are completely identical. with those in DVB-C.

21.3.3 Interference Sources

Interference sources affect individual carriers or carrier ranges. They can be noise-like and the constellation points become noise clouds, but they can also be sinusoidal when the constellation points appear as circles.

21.3.4 Echoes, Multipath Reception

Echoes, i.e. multipath reception, lead to frequency-selective fading. There is interference in individual carrier ranges but the information lost as a result can be restored again due to the interleaving across the frequency and the large amount of error protection (Reed Solomon and convolutional coding) provided in DVB-T. Of course, COFDM (Coded Orthogonal Frequency Division Multiplex) was developed precisely for this purpose, namely to cope with the effects of multipath reception in terrestrial transmission.

21.3.5 Doppler Effect

In mobile reception, a frequency shift occurs over the entire DVB-T spectrum due to the Doppler effect. By itself, the Doppler effect does not present a problem in DVB-T transmission because a shift of a few hundred Hertz at motor vehicle speeds can be handled easily. It is when Doppler effect and multipath reception are combined that the spectrum becomes smeared. Echoes moving towards the receiver will shift the spectrum into a different direction from those moving away from the receiver and, as a result, the signal/noise ratio in the channel deteriorates.

21.3.6 I/Q Errors of the Modulator

The focus of this discussion will now shift to the I/Q errors of the DVB-T modulator, the effects of which differ from those in DVB-C.

The COFDM symbol is produced by means of the mapper, the real parts and imaginary parts of all subcarriers being set in the frequency domain before the IFFT (Inverse Fast Fourier Transform). Each carrier is independently QAM modulated (QPSK, 16QAM, 64QAM) in accordance with the information to be transmitted. The spectrum has no symmetries or centro-symmetries and thus is not conjugate with respect to the IFFT band center.

According to system theory, therefore, a complex time-domain signal must be produced after the IFFT. Considering then the real time-domain signal re(t) and the imaginary time-domain signal im(t) carrier by carrier, it is found that for each carrier, re(t) has exactly the same amplitude as im(t) and that im(t) is always shifted by exactly 90 degrees in phase with respect to re(t). All re(t) superimposed in time are fed into the I branch of the complex I/Q mixer and all im(t) superimposed in time are fed into the Q branch. The I mixer is fed with 90 degrees carrier phase and the Q mixer is fed with 90 degrees carrier phase and the two modulation products added together result in the COFDM signal cofdm(t).

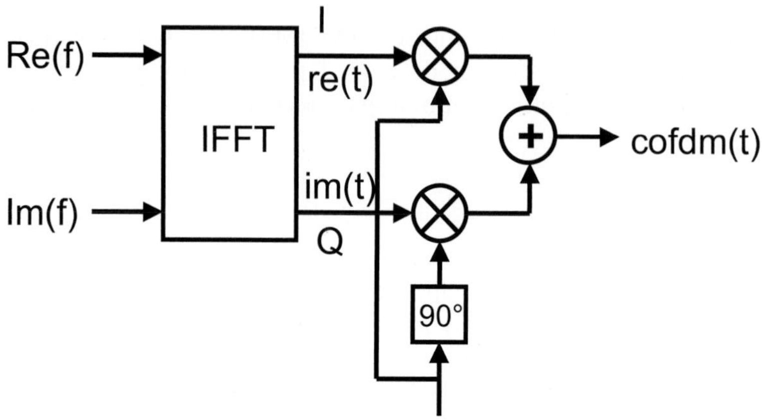

Fig. 21.8. COFDM modulator

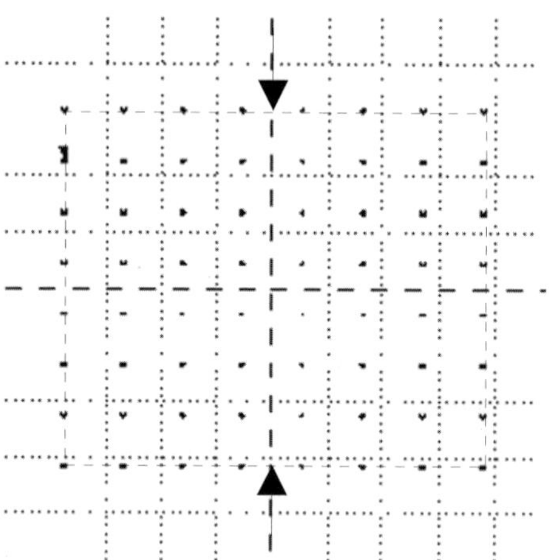

Fig. 21.9. I/Q imbalance

The signal branches re(t) and im(t) must exhibit exactly the right ratios of levels with respect to one another. The 90^0 phase shifter must also be set correctly. And there must not be any DC component superimposed on the re(t) and im(t) signals. Otherwise, so-called I/Q errors will occur. The resultant phenomena appearing in the DVB-T signal will be shown in Fig. 21.9.

Fig. 21.9. shows the constellation diagram with an I/Q amplitude imbalance in the I/Q mixer of the modulator. The pattern is rectangularly distorted, i.e. compressed in one direction (horizontal or vertical). This effect can be observed easily in DVB-C but can only be verified on the center carrier (band center) in DVB-T where all the other carriers display noise-like interference.

An I/Q phase error leads to a rhomboid-like distortion of the constellation diagram (Fig. 21.10.). This effect can be observed without problems in the DVB-C cable standard but can only be verified on the center carrier (band center) in DVB-T where all the other carriers also display noise-like interference due to this effect.

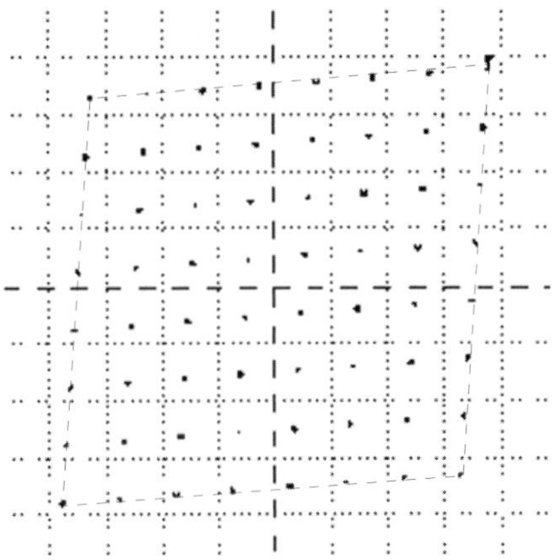

Fig. 21.10. I/Q phase error

A residual carrier present at the I/Q mixer (Fig. 21.11.) shifts the constellation diagram out of the center in some direction. The pattern itself remains undistorted. This effect can only be observed on the center carrier and only affects this carrier.

Today, virtually most of the DVB-T modulators operate in accordance with the direct modulation method. An analog I/Q modulator used in this mode usually exhibits problems in the suppression of the carrier, among others. Although the manufacturers have managed to overcome problems of I/Q amplitude imbalance and I/Q phase errors in most cases, there is a remaining problem of carrier suppression to be found more or less with every DVB-T modulator of this type and has been observed to a greater or lesser extent at many DVB-T transmitter sites around the globe by the author. The residual-carrier problem can only be verified at the center carrier (3408 or 852, resp.) in the center of the band and only causes interference there or in the areas around the center carrier. A lack of carrier suppression can be detected right away as a dip in the display of the modulation error ratio over the range of DVB-T subcarriers in the center of the band and an expert in DVB-T measuring techniques can immediately tell that there is a DVB-T modulator operating in direct modulation mode.

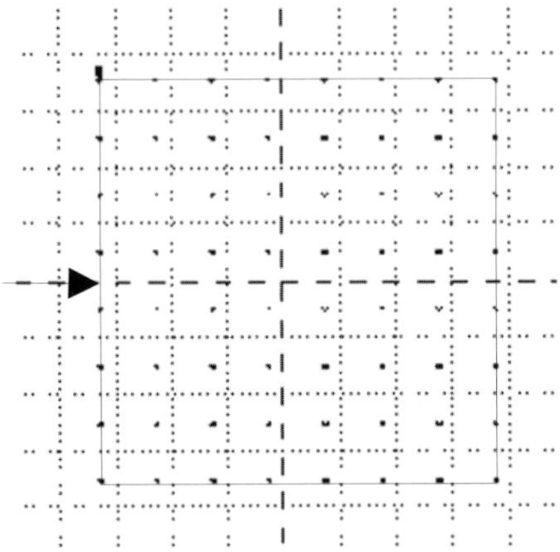

Fig. 21.11. Effect of residual carrier

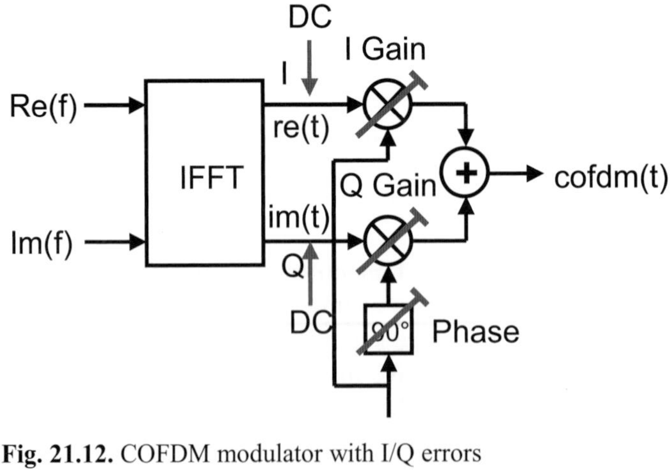

Fig. 21.12. COFDM modulator with I/Q errors

21.3.7 Cause and Effect of I/Q Errors in DVB-T

What then is the cause of I/Q errors, why can these effects be observed only at the center carrier and why do all other carriers exhibit noise-like interference in the presence of any I/Q amplitude imbalance and I/Q phase error?

Fig. 21.12. shows the places in the I/Q modulator at which these errors are produced. A DC component in re(t) or im(t) after the IFFT will lead to a residual carrier in the I or Q branch or in both branches. Apart from a corresponding amplitude, the residual carrier will, therefore, also exhibit a phase angle.

Different gains in the I and Q branches will result in an I/Q amplitude imbalance. If the phase angle at the I/Q mixer differs from 90 degrees, an I/Q error (quadrature error) is produced.

The disturbances in DVB-T caused by the I/Q errors can be explained quite clearly without much mathematics by using vector diagrams. Let us begin with the vector diagram of a normal amplitude modulation. An AM can be represented as a rotating carrier vector and by superimposed vectors of the two sidebands, one sideband vector rotating counterclockwise and one sideband vector rotating clockwise. The resultant vector is always located in the plane of the carrier vector, i.e. the carrier vector is varied (modulated) in amplitude.

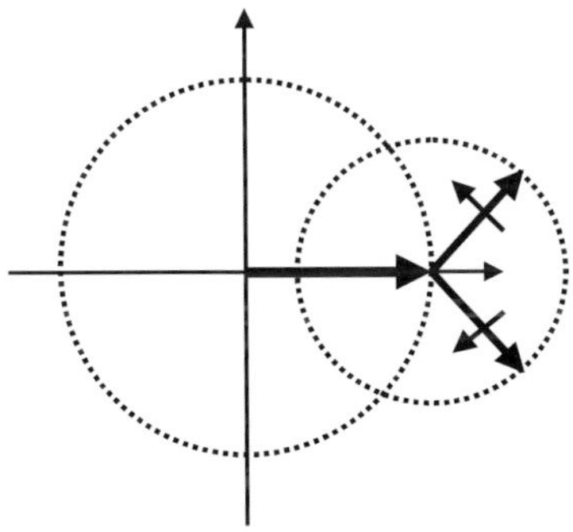

Fig. 21.13. Vector diagram of an amplitude modulation

Suppressing the carrier vector results in amplitude modulation with carrier suppression.

Correspondingly, the behaviour of an I/Q modulator can also be represented by superimposing 2 vector diagrams (Fig. 21.15.). Both mixers operate with suppressed carrier.

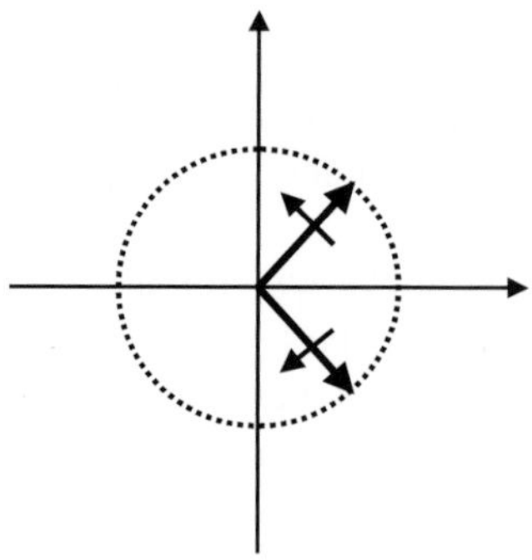

Fig. 21.14. Vector diagram of an amplitude modulation with suppressed carrier

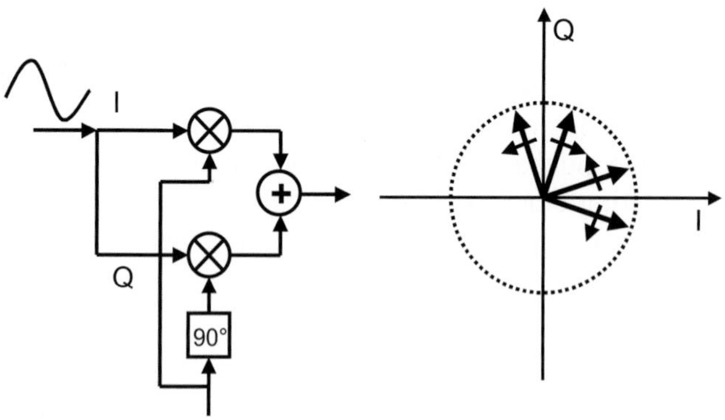

Fig. 21.15. I/Q modulation

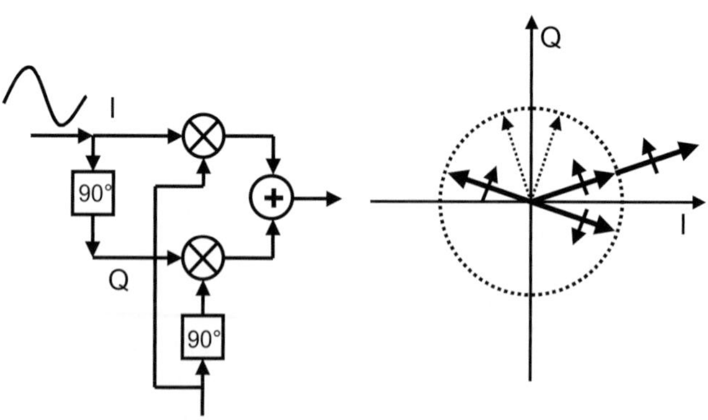

Fig. 21.16. Single sideband modulation

If the same signal is fed into the I branch and into the Q branch, but with a phase difference of 90 degrees from one another, a vector diagram as shown in Fig. 21.16. ('single sideband modulation') is obtained. It can be seen clearly that two sideband vectors are added and two sideband vectors cancel (are subtracted). One sideband is thus suppressed, resulting in single sideband amplitude modulation. A COFDM modulator can thus be interpreted as being a single sideband modulator for many thousands of subcarriers. In an ideal COFDM modulator, there is no crosstalk from the upper COFDM band to the lower one and vice versa.

Since the IFFT is a purely mathematical process, it can be assumed to be ideal. The I/Q mixer, however, can be implemented as a digital (ideal) mixer or as an analog mixer and there are and will be in future analog I/Q mixers in DVB-T modulators (direct modulation).

If then an I/Q amplitude imbalance exists, this means that the upper or lower sideband no longer cancel completely, leaving an interference component. The same applies to an I/Q phase error. It is clear, therefore, that all the subcarriers are subject to noise-like interference, with the exception of the center carrier. It is also clear why a residual carrier will push the constellation pattern away from the center at the center carrier and only interferes with the latter.

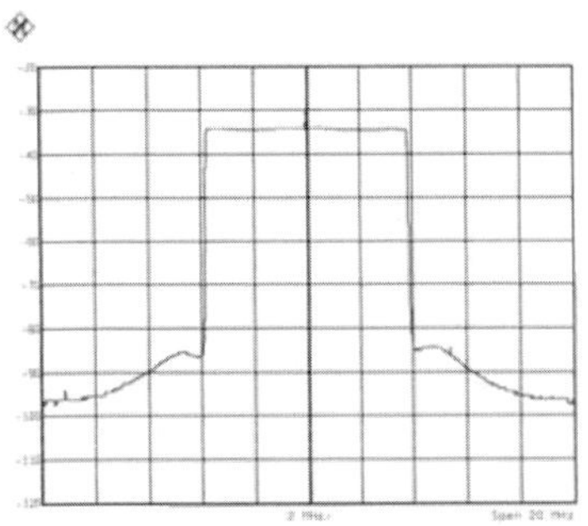

Fig. 21.17. Spectrum of a DVB-T signal

This can also be shown impressively in the spectrum of the DVB-T signal if the DVB-T modulator has the test function of switching off e.g. the lower carrier band in the spectrum. This can be done, for example, with a DVB test transmitter. In the center of the band (center carrier), an existing residual carrier can be seen clearly. If the I/Q modulator is then adjusted to produce an amplitude imbalance, crosstalk from the upper to the lower sideband is clearly apparent. The same applies to an I/Q phase error.

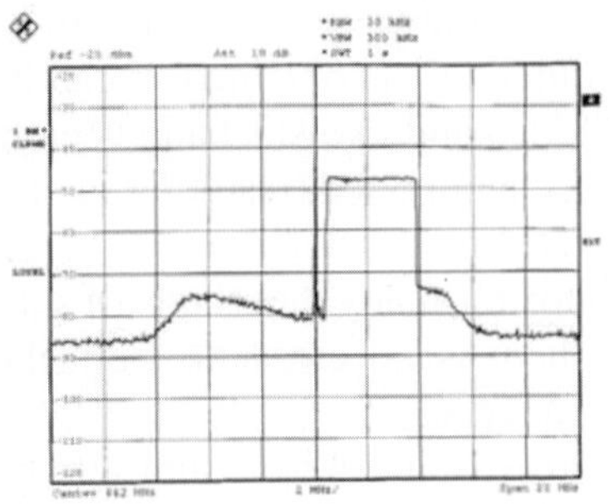

Fig. 21.18. Lower band switched off

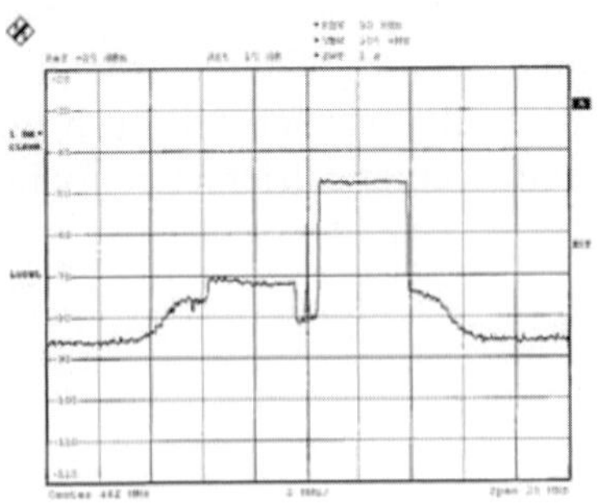

Fig. 21.19. 10% amplitude imbalance

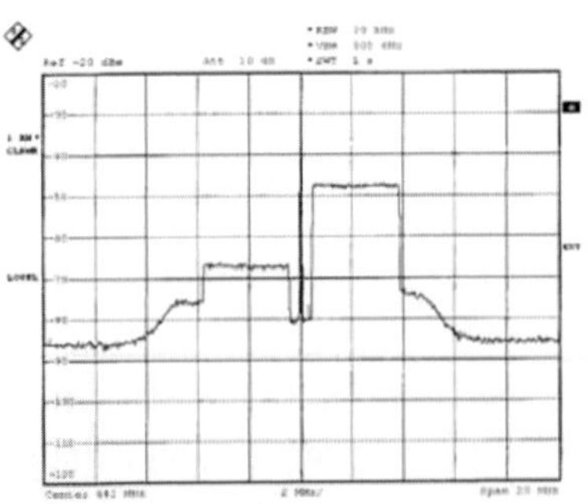

Fig. 21.20. 10° phase error

The process of noise-like crosstalk can be described easily by means of simple trigonometric operations which can be derived from the vector diagram.

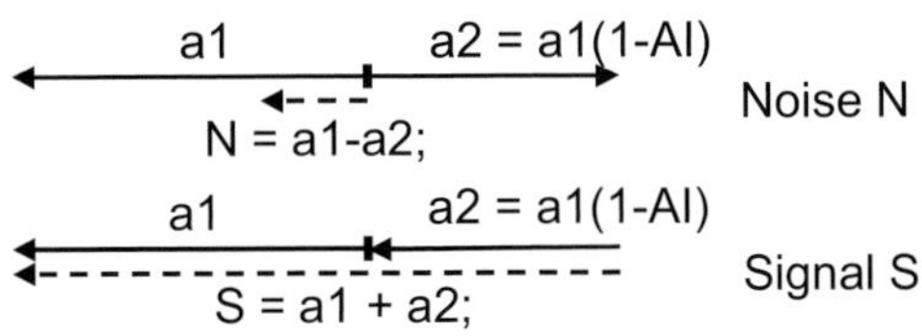

$$S/N = (a1+a2)/(a1-a2) = (a1+a1(1-AI)/(a1-a1(1-AI) = (2-AI)/AI;$$

$$S/N[dB] = 20lg((2-AI[\%]/100)/(AI[\%]/100));$$

Fig. 21.21. Determining the S/N with amplitude imbalance

In the case of amplitude imbalance, the opposing vectors no longer cancel completely (Fig. 21.21.), resulting in a noise vector causing crosstalk from the upper DVB-T band to the lower band and vice versa. The actual useful signal amplitude decreases by the same amount by which the crosstalk increases.

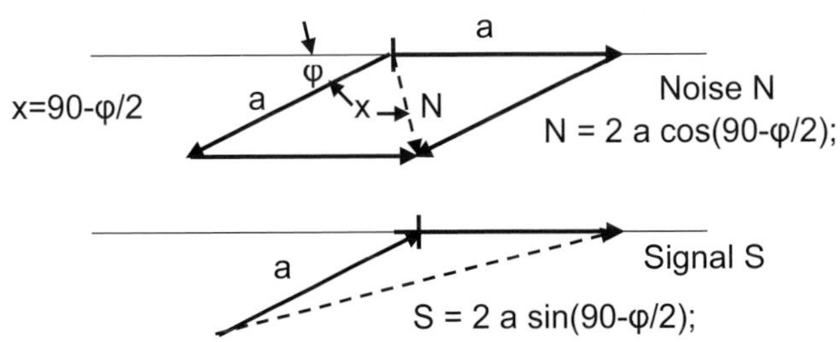

$$S/N = (2a)/(2a)\ (\sin(90-\varphi/2)/\cos(90-\varphi/2)) = \tan(90-\varphi/2);$$

$$S/N[dB] = 20lg(\tan(90-\varphi/2));$$

Fig. 21.22. Determining the S/N in the presence of an I/Q phase error

A phase error will result in a noise vector the length of which can be determined from the vector parallelogram. The useful signal amplitude also decreases by the same amount. Fig. 21.22. shows the conditions for the signal/noise ratio S/N in the presence of amplitude imbalance and of a phase error, respectively, which have now been derived as formulae. In the practical implementation of a DVB-T modulator, the aim is an amplitude imbalance of less than 0.5% and a phase error of less than 0.5 degrees.

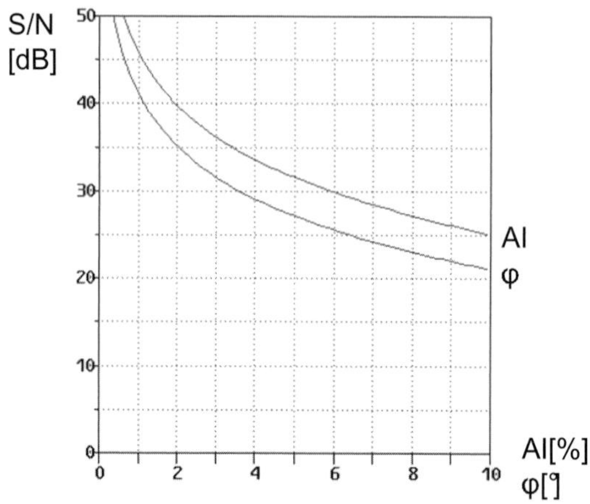

Fig. 21.23. DVB-T signal/noise ratios with amplitude imbalance (AI) and phase error (PE) of the IQ modulator

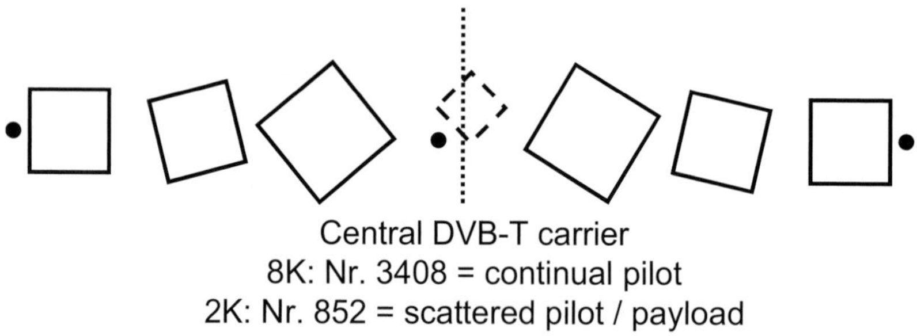

Fig. 21.24. Distortions in the vicinity of the center carrier due to the channel correction in the DVB-T receiver

Thus, I/Q errors of the DVB-T modulator can only be identified by observing the center carrier but may interfere with the entire DVB-T signal. In addition, it will be found that in each case at least the two upper and lower carriers adjacent to the center carrier are also distorted. This is caused by the channel correction in the DVB-T receiver where channel estimation and correction is performed on the basis of the evaluation of the scattered pilots. But these are only available in intervals of 3 carriers and between them it is necessary to interpolate.

In 2K mode, the center carrier is No. 852 which is a payload carrier or sometimes a scattered pilot. Verifying I/Q errors will not present any problems, therefore. The situation is different in 8K mode where the center carrier is No. 3408 and is always a continual pilot. I/Q errors can only be extrapolated in this case by observing the adjacent upper and lower carriers.

Each of the effects described has its own measurement parameters. In the DVB-C cable standard, these parameters have been combined to form an additional aggregate parameter called the modulation error ratio.

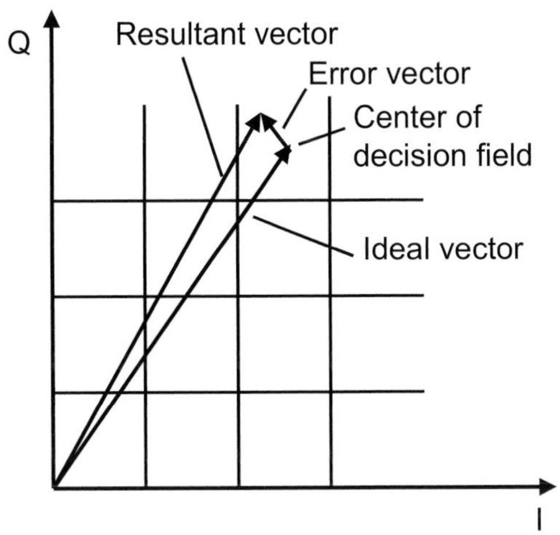

Fig. 21.25. Error vector for determining the modulation error ratio (MER)

The modulation error ratio (MER) is a measure of the sum of all interference effects occurring on the transmission link. Like the signal/noise ratio, it is usually specified in dB. If only one noise effect is present, MER and S/N are equal.

The result of all the interference effects on a digital TV signal in broadband cable networks, explained above, is that the constellation points exhibit deviations with respect to their nominal position in the center of the decision errors. If the deviations are too great, the decision boundaries are crossed and bit errors occur. However, the deviations from the center of the decision field can also be considered to be measurement parameters for the magnitude of an arbitrary interferer. Which is precisely the aim of an artificial measurement parameter like the MER. When measuring the MER, it is assumed that the actual hits in the constellation fields have been pushed away from the center of the respective error by interferers. The interferers are allocated error vectors, the error vector pointing from the center of the constellation field to the point of the actual hit in the constellation field. Then the lengths of all these error vectors are measured with respect to time and the quadratic mean is formed or the maximum peak value is measured in a time window. The definition of MER can be found in the DVB Measurement Guidelines [ETR290].

The MER is calculated from the error vector length, using the following relation:

$$MER_{PEAK} = \frac{\max(|error\_vector|)}{U_{RMS}} \cdot 100\%;$$

$$MER_{RMS} = \frac{\sqrt{\frac{1}{N}\sum_{n=0}^{N-1}(|error\_vector|)^2}}{U_{RMS}} \cdot 100\%;$$

The reference U_{RMS} is here the RMS value of the QAM signal.

Usually, however, the logarithmic scale is used:

$$MER_{dB} = 20 \cdot \lg\left(\frac{MER[\%]}{100}\right); \quad [dB]$$

The MER value is, therefore, an aggregate quantity which includes all the possible individual errors. The MER value thus completely describes the performance of this transmission link.

In principle,

MER [dB] ≤ S/N [dB];

The representation of MER as a function of the subcarrier number MER(f) is of particular significance in DVB-T because it allows the overall situation in the channel to be observed. It is easy to see areas with disturbed carriers. Often only an averaged single MER measurement value is mentioned in connection with DVB-T measurements but this value does not provide much practical information. A graphical representation of the MER versus frequency is always of importance.

In summary, it can be said that noise and phase jitter affect all carriers to the same extent, interferers affect carriers or ranges of carriers like noise or sinusoidally. Echoes also affect only carrier ranges.

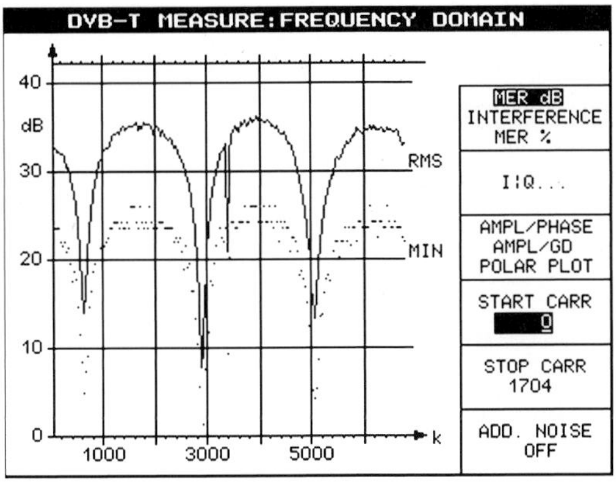

Fig. 21.26. Modulation error ratio (MER) vs. COFDM subcarriers MER(f) [EFA]

Table 21.2. DVB-T interference effects

Interference effect	Effect	Verification
Noise	all carriers	all carriers
Phase jitter	all carriers	all carriers
Interferer	single carriers	carriers affected
Echoes	carrier ranges	carriers affected, impulse response
Doppler	all carriers	frequency deviation, smearing
IQ amplitude imbalance	all carriers	center carrier
IQ phase error	all carriers	center carrier
Residual carrier carrier leakage	center carrier and adjacent carriers	center carrier

I/Q errors of the modulator partially affect the carriers as noise-like disturbance and as such can only be identified by observing the center carrier.

All the influences on the DVB-T transmission link described can be observed easily by constellation analysis in a DVB-T test receiver. In addition, a DVB-T test receiver also allows measurement of the received level, measurement of the bit error rate, calculation of the amplitude and group delay response and of the impulse response from the channel estimation data. The impulse response is of great importance in detecting multipath reception in the field, particularly in single-frequency networks (SFNs). Apart from the I/Q analysis discussed, therefore, a DVB-T test receiver also enables for a large number of significant measurements to be made on the DVB-T transmission link.

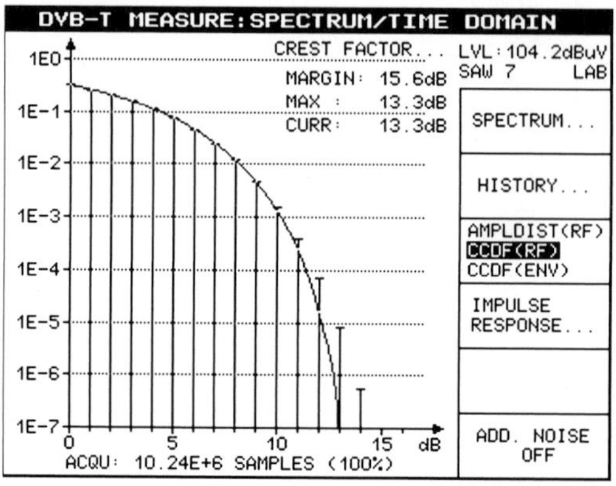

Fig. 21.27. Crest factor measurement [EFA]

21.4 Measuring the Crest Factor

DVB-T signals have a large crest factor which can be up to 40 dB in theory. In practice, however, the crest factor is limited to about 13 dB in power transmitters. The crest factor can be measured by using a DVB-T test receiver. For this purpose, the test receiver picks up the data stream immediately after the A/D converter and calculates from it both the RMS value and the maximum peak signal value occurring in a time window. According to the definition, the crest factor is then

$$c_f = 20 \log(U_{max\ peak}/U_{RMS});$$

21.5 Measuring the Amplitude, Phase and Group Delay Response

Although DVB-T is quite tolerant with respect to linear distortion such as amplitude, phase and group delay distortion, it is, on the other hand, no great problem to measure these parameters. A DVB-T test receiver is easily able to analyse the pilot carrriers (scattered pilots and continual pilots) contained in the signal and to calculate from these the linear distortions. The linear distortion is thus determined from the channel estimation data.

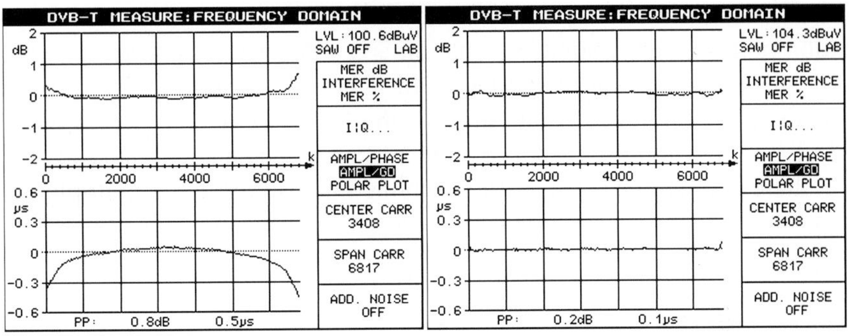

Fig. 21.28. Amplitude and group delay response measurement using the scattered pilots

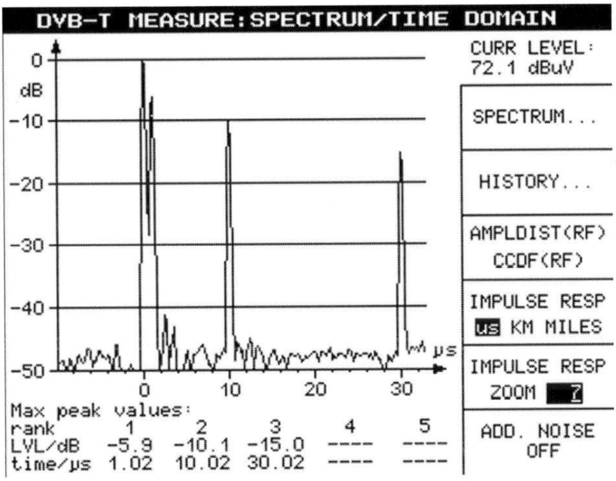

Fig. 21.29. Impulse response measurement via an IFFT of the channel impulse response CIR

21.6 Measuring the Impulse Response

Transforming the channel estimation data, which are available in the frequency domain and from which the representation of the amplitude and phase response was derived, into the time domain by means of an inverse fast Fourier transform provides the impulse response. The maximum length of the calculable impulse response depends on the samples provided by the channel estimation. Every third subcarrier supplies a contribution to the channel estimation at some time, i.e. the distance between two interpolation points of the channel estimation is 3 • Δf, where Δf corresponds to the subcarrier spacing of the COFDM. The calculable impulse response length is thus 1/3 Δf, i.e. one third of the COFDM symbol period. In the ideal case, the impulse response only consists of one main impulse at t = 0, i.e. there is only one signal path. From the impulse response, multiple echoes can be easily classified in accordance with delay and path attenuation.

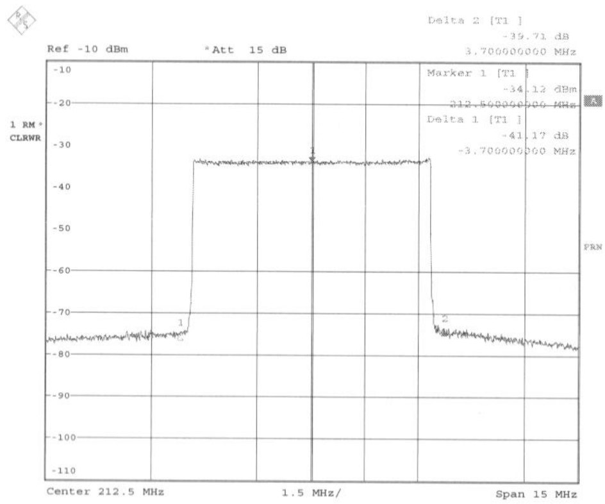

Fig. 21.30. Spectrum of a DVB-T signal at the transmitter output before the mask filter

21.7 Measuring the Shoulder Attenuation

The system does not utilize the full channel bandwidth, i.e. some of the 2K or 8K subcarriers are set to zero so that no interference to adjacent channels will be caused. Due to nonlinearities, however, there are still outband components and the effect on the spectrum and its shape has given rise to the term 'shoulder attenuation'.

In the Standard, the permissible shoulder attenuation is defined as a tolerance mask. Fig. 21.30. the spectrum of a DVB-T signal at the power amplifier output, i.e. before the mask filter. To determine the shoulder attenuation, different methods are defined and especially a relatively elaborate method in the Measurement Guidelines [ETR290]. In practice, the DVB-T spectrum is in most cases simply measured by using three markers, setting one marker to band center and the others to +/- (DVB-T channel bandwidth/2 + 0.2 MHz). With an 8 MHz channel, this results in test points at +/- 4.2 MHz relative to band center and +/- 3.7 MHz for the 7 MHz channel. Fig. 21.31, shows the spectrum of a DVB-T signal after the mask filter (critical mask). The DVB-T standard [ETS 300 744] defines various tolerance masks for various adjacent channel allocations.

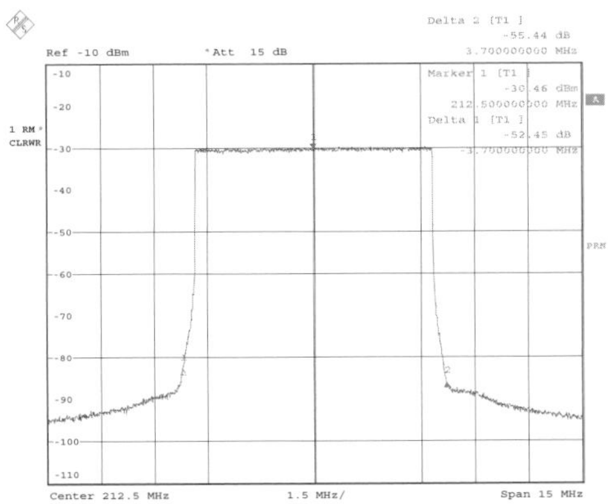

Fig. 21.31. Spectrum of a DVB-T signal measured after the mask filter (critical mask)

In practice, the following shoulder attenuations are achieved:

- Power amplifier, undistorted: approx. 28 dB
- Power amplifier, equalized: approx. 38 dB
- After the output BPF: approx. 52 dB (critical mask)

Usually the tolerance masks listed in Table 21.3. (uncritical mask) and 21.4. (critical mask) are used for evaluating a DVB-T signal (7 and 8 MHz bandwidth). In the corresponding documents (DVB-T Standard [ETS300744], System Specifications), the ratio with respect to channel power at 4 kHz reference bandwidth is usually specified. If the spectrum analyzer does not support this resolution bandwidth, it is possible to select a different one (e.g. 10, 20 or 30 kHz) and the values can be converted.

10lg(4/7610) = -32.8 dB and 10lg(4/6770) = -32.2 dB correspond to the attenuation with respect to the total signal power of the DVB-T signal with 4 kHz reference bandwidth in the useful DVB-T band. If another resolution bandwidth of the analyzer is used, corresponding values must be inserted into the formula. The tables also show the relative attenuation compared with the useful channel independently of the reference bandwidth.

The important factor in choosing the resolution bandwidth of the spectrum analyzer is that it be not too small and not too large. Usually, 10, 20 or 30 kHz is selected.

Table 21.3. DVB-T tolerance mask (uncritical) in the 7 and 8 MHz channel

f_{rel}[MHz] at 7MHz channel bandwidth	f_{rel}[MHz] at 8MHz channel bandwidth	Attenuation [dB] vs. channel power at 4 kHz reference bandwidth	Attenuation [dB] at 7 MHz channel bandwidth	Attenuation [dB] at 8 MHz channel bandwidth
+/-3.4	+/-3.9	-32.2 (7 MHz) -32.8 (8 MHz)	0	0
+/-3.7	+/-4.2	-73	-40.8	-40.2
+/-5.25	+/-6.0	-85	-52.8	-52.2
+/-10.5	+/-12.0	-110	-77.8	-77.2
+/-13.85		-126	-93.8	

Table 21.4. DVB-T tolerance mask (critical) in the 7 and 8 MHz channel

f_{rel}[MHz] at 7MHz channel bandwidth	f_{rel}[MHz] at 8MHz channel bandwidth	Attenuation [dB] vs. channel power at 4 kHz reference bandwidth	Attenuation [dB] at 7 MHz channel bandwidth	Attenuation [dB] at 8 MHz channel bandwidth
+/-3.4	+/-3.9	-32.2 (7 MHz) -32.8 (8 MHz)	0	0
+/-3.7	+/-4.2	-83	-50.8	-50.2
+/-5.25	+/-6.0	-95	-62.8	-62.2
+/-10.5	+/-12.0	-120	-87.8	-87.2
+/-13.85		-126	-93.8	

Fig. 21.32. DVB-T mask filter (uncritical Mask, low power, manufactured by Spinner) with directional test coupler at input and output

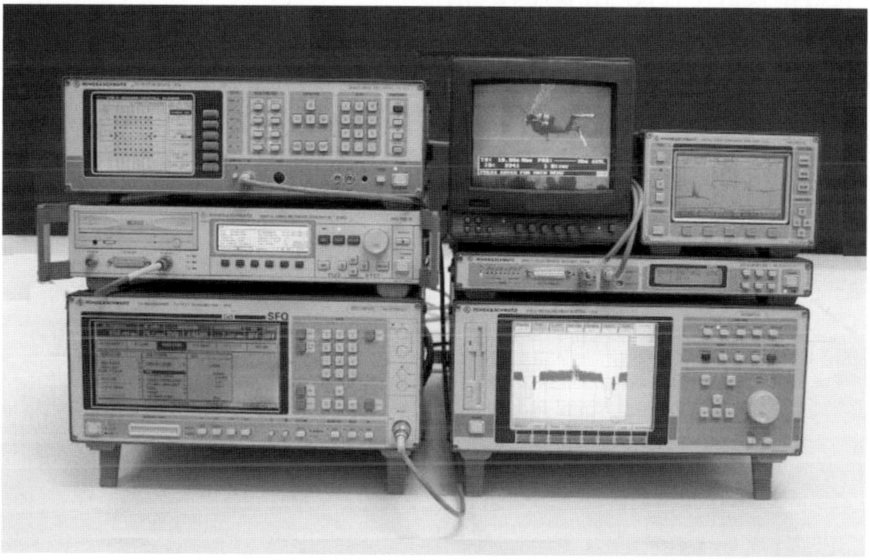

Fig. 21.33. DVB-T transmission link with MPEG-2 test generator DVRG (center left), DVB-T test transmitter SFQ (bottom left), DVB-T test receiver EFA (top left), MPEG-2 test decoder DVMD (center right) and TV monitor, video analyzer VSA and "601" analyzer VCA (Rohde&Schwarz).

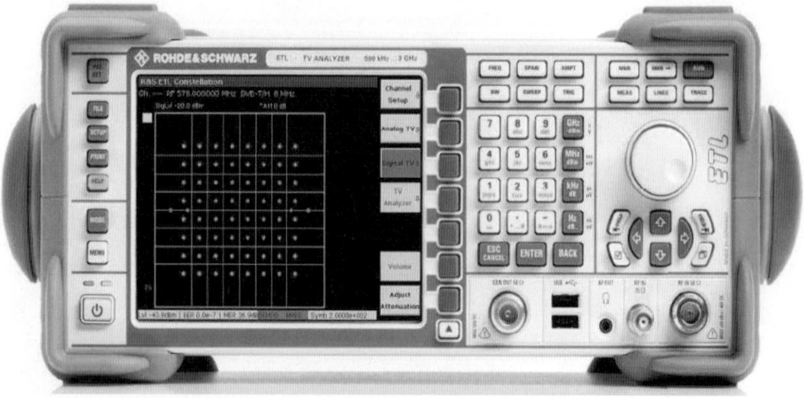

Fig. 21.34. TV Analyzer ETL showing a DVB-T 64QAM constellation diagram display [ETL]

Bibliography: [ETS300744], [ETR290], [HOFMEISTER], [EFA], [SFQ], [SFU], [FISCHER2], [ETL]

22 DVB-H/DVB-SH - Digital Video Broadcasting for Handhelds

22.1 Introduction

The introduction of 2G GSM (2nd generation Global System for Mobile Communication) has triggered quite a boom for this wireless type of communication. If the possession of car telephones or similar mobile-type telephones was the prerogative of mostly special circles of people at the beginning of the nineties, at least every second person had his own personal mobile telephone by the end of the nineties and in most cases it was used only either for telephoning or for sending and receiving short messages - SMS - until then. By then, however, people also wanted to be able to send and receive data via a mobile telephone, e.g. from a PC. To be able to check one's e-mail database was initially a pleasant way of keeping oneself up-to-date, especially in the professional field; today, this is standard usage. In the GSM standard, developed mainly for mobile telephony, however, the data rates are about 9600 bit/s. This is quite adequate for simple text e-mails without attachments but becomes rather troublesome when long files are attached to the original message. It can also be used for surfing the Internet but is a cumbersome and expensive way of doing this. With the introduction of 2.5G mobile telephony, the GPRS (General Packet Radio System), the data rate was increased to 171.2 kbit/s by forming packets, i.e. combining time slots of the GSM system. It was only with the 3rd generation, the UMTS (Universal Mobile Telecommunication System), that the data rate could be increased to 144 - 384 kbit/s and 2 Mbit/s, respectively which, however, greatly depends on the respective conditions of reception and coverage. Using higher-level modulation (8PSK), the EDGE (Enhanced Data Rates for GSM Evolution) standard, too, allows higher data rates of up to 345.6 kbit/s (ECSD) and 473.6 kbit/s (EGPRS), repectively.

Due to their nature, all mobile radio standards are designed for bidirectional communication between the terminal and the base station. The

W. Fischer, *Digital Video and Audio Broadcasting Technology*, Signals and Communication Technology, 3rd ed., DOI: 10.1007/978-3-642-11612-4_22, © Springer-Verlag Berlin Heidelberg 2010

modulation methods such as, e.g. GMSK (Gaussian Minimum Shift Keying) in GSM or WCDMA (Wideband Code Division Multiple Access) in UMTS have been designed for these "rough" receiving conditions in mobile reception.

Today, mobile telephones are no longer mere telephones but can be used as cameras or as games consoles or organizers, evolving more and more to become multimedia terminals. Equipment manufacturers and network operators are continuously searching for more new applications.

In parallel with the evolution of mobile radio, the transition of analog television to digital television took place. If at the end of the eighties, it still appeared to be impossible to be able to send moving pictures digitally via existing transmission paths such as satellite, cable TV or the old-fashioned, terrestrial way, this is an accepted fact today. It is made possible by modern compression methods such as MPEG (Moving Picture Experts Group) and modern modulation methods and matching error protection (FEC). A key event in this area can be considered to be the first use of the so-called DCT (Discrete Cosine Transform) in the JPEG (Joint Photographic Experts Group) standard. JPEG is a method for compressing still pictures used in digital cameras. At the beginning of the nineties, experience gained with the DCT was also applied to the compression of moving pictures in the MPEG standard. First, the MPEG-1 standard developed for CD data rates and applications was produced. With MPEG-2 it became possible to compress SDTV (Standard Definition TV) moving pictures from originally 270 Mbit/s to less than 5 Mbit/s), the data rate of the associated lip-synchronized audio channel being in most cases 200 - 400 kbit/s. Even HDTV (High Definition TV) signals could now be reduced to tolerable data rates of about 15 Mbit/s. Knowledge of the anatomy of the human eye made it possible to perform a so-called irrelevance reduction in combination with redundancy reduction. Signal components, information, not perceived by the eye and ear are removed from the signal before transmission.

The compression methods have been refined further in MPEG-4 (H.264, MPEG-4 Part 10 AVC) and today even lower data rates are possible, with better picture and sound quality.

During the development of DVB, three different transmission methods were developed: DVB-S (Satellite), DVB-C (Cable) and DVB-T (Terrestrial). In DVB-T, digital television is broadcast terrestrially in 6, 7 or 8 MHz-wide radio channels in a gapped frequency band of about 47 MHz to 862 MHz at net data rates of either approx. 15 Mbit/s or 22 Mbit/s. In some countries such as the UK, Sweden or Australia, DVB-T is designed for pure roof aerial reception and the possible data rate is correspondingly high at about 22 Mbit/s. Countries like Germany have selected the "Portable Indoor" option providing the possibility of receiving more than 20 programmes "free to air" via a (passive or active) indoor antenna. Due to the higher degree of error protection (FEC) required and the more rugged modulation method (16QAM instead of 64QAM), only lower data rates such as, e.g. about 15 Mbit/s are possible.

If DVB-T is operated as a network which can be received by portable receivers, the data rates are about 15 Mbit/s and correspondingly only about 4 programmes or services can be accomodated in a DVB-T channel. This is still four times as many as could previously be received in a comparable analog TV channel. The data rates available per programme are, therefore, about 2.5 to 3.5 Mbit/s, present as variable data rates in a so-called statistical multiplex in most cases.

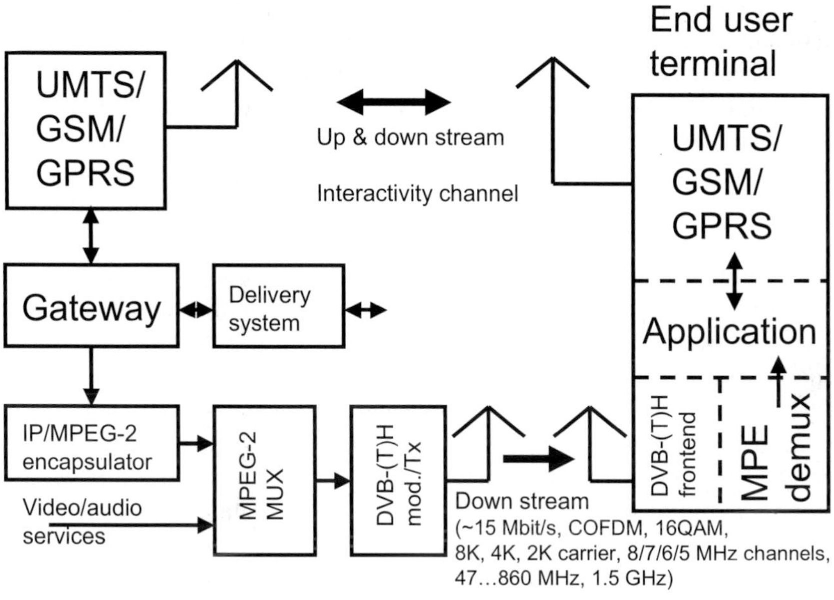

Fig. 22.1. Convergence between mobile radio and DVB

22.2 Convergence between Mobile Radio and Broadcasting

Mobile radio networks are networks in which bi-directional (point-to-point) connections are possible at relatively low data rates. Modulation methods, error protection and hand-over procedures are correspondingly adapted to the mobile environment. Billing etc. are also found to be system related in the standard. The type of services to be selected, be it a telephone call, an SMS or a data link, is determined by the end user and he is billed accordingly.

Broadcasting networks are unidirectional networks in which contents are distributed point-to-multipoint jointly to a large number of parties at relatively high data rates. On-demand contents are relatively rare, a predetermined content being distributed to many parties from one transmitter site or nowadays also by single-frequency networks ever a number of transmitters. This content is usually a radio or television program. The data rates are much higher than in mobile radio networks. Modulation methods and error protection are often designed only for portable or roof antenna reception. Mobile reception is provided for only as part of DAB (Digital Audio Broadcasting) in the Standard. DVB-T has been developed only for stationary or portable reception.

As part of DVB-H (Digital Video Broadcasting for Hand-held mobile terminals), attempts are now being made to merge the mobile radio world with that of broadcasting (Fig. 22.1.) and to combine the advantages of both network systems, combining the bi-directionality of mobile radio networks at relatively low data rates with the uni-directionality of broadcast networks at relatively high data rates. If the same services, such as certain video/audio services on demand are demanded by many subscribers, the data service is mapped from the mobile radio network onto the point-to-multipoint broadcast rail, depending on the demand and on the amount of information involved.

The type of information diverted from the mobile radio network to the broadcasting network depends only on the current requirements. It is currently still undecided what services will be offered to the mobile telephones in future over the DVB-H service. They can be purely IP-based services or also video/audio over IP. In every case, however, DVB-H will be a UDP/IP-based service in connection with MPEG/DVB-T/-H. Conceivable applications are current sports programs, news and other services which could be of interest to the public using mobile telephones. It is certain that, given the right conditions of reception, a DVB-H-enabled mobile will also be able to receive pure, non-chargeable DVB-T transmissions.

22.3 Essential Parameters of DVB-H

The essential parameters of DVB-H correspond to those of the DVB-T Standard. The physical layer of DVB-T has been expanded only slightly. In addition to the 8K and 2K mode, already present in DVB-T, the 4K mode was introduced as a good compromise between the two, allowing single-frequency networks of reasonable size to be formed whilst at the same time being more suitable for mobile use. The 8K mode is not very suitable for mobile use because of its small subcarrier spacing and the 2K mode allows for only short distances of about 20 km between transmitters. The 8K mode requires more memory space for data interleaving and de-interleaving than do the 4K and 2K modes. Space becoming available in 4K and 2K mode can now be used for deeper interleaving in DVB-H, i.e. the interleaver can be selected between 'native' and 'in-depth' in 4K and 2K mode. For signalling additional parameters, TPS (Transmission Parameter Signalling) bits, reserved or already used otherwise, are used in DVB-H.

The parameters additionally introduced in DVB-H are listed as an appendix in the DVB-T Standard [ETS300744]. All other changes or extensions relate to the MPEG-2 transport stream. These, in turn, can be found in the DVB Data Broadcast Standard [ETS301192]. The MPEG-2 transport stream, as DVB baseband signal, is the input signal for a DVB-H modulator. In DVB-H, Multiprotocol Encapsulation (MPE), already defined in the context of DVB Data Broadcasting before DVB-H, is used as a time-slicing method so that energy can be saved in the mobile part. Both length and spacing of time slots must be signalled. The IP packets packaged in MPE time slots can be optionally provided with additional FEC (forward error correction) in DVB-H. This is a Reed Solomon error protection at IP packet level. Everything else corresponds directly to DVB-T or MPEG-2, respectively. DVB-H is a method for time slot IP packet transmission over an MPEG-2 transport stream. The physical layer used is DVB-T with some extensions. Its aim is the convergence between a mobile radio network and a DVB-H broadcast network. Data services are transmitted to the mobile either via the mobile radio network or via the DVB-H network, depending on traffic volume.

22.4 DSM-CC Sections

In the MPEG-2 Standard ISO/IEC 13818 Part 6, mechanisms for transmitting data, data services and directory structures were created early on. There are the so-called DSM-CC sections, where DSM-CC stands for Digital Storage Media Command and Control. In principle, DSM-CC Sections have a comparable structure to the PSI/SI Tables. They begin with a Table ID which is always within a range from 0x3A to 0x3E. DSM-CC Sections have a length of up to 4 kbytes and are also divided into transport stream packets and broadcast multiplexed into the transport stream. Using object carousels (cyclically repeated broadcasting of data), entire directory trees with different file-scan be transmitted to the DVB receiver via DSM-CC sections. This is done, e.g. in MHP (Multimedia Home Platform), where HTML and Java files are transmitted which can then be executed in the MHP-enabled DVB receiver.

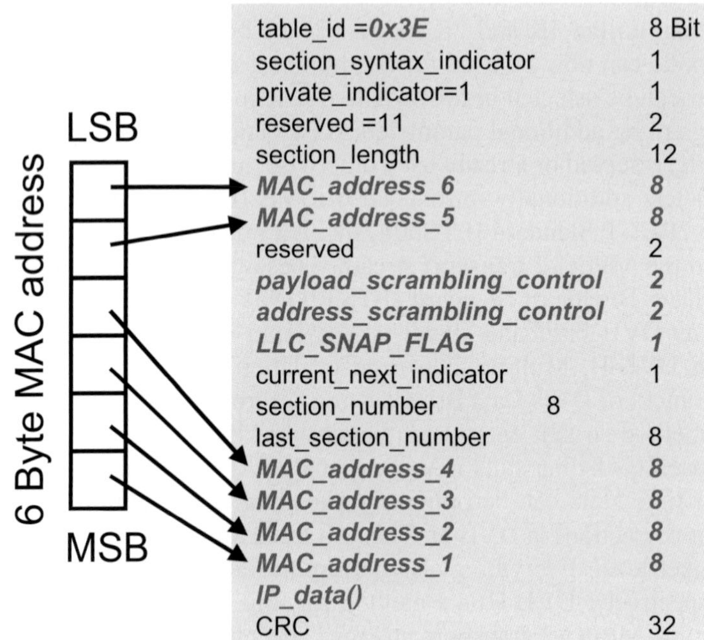

Fig. 22.2. DSM-CC section for IP transmission (Table_ID=0x3E)

DSM-CC sections with Table ID=0x3E (Fig. 22.2.) can be used for transmitting Internet (IP) packets in the MPEG-2 transport stream. In an IP packet, a TCP (Transport Control Protocol) packet or a UDP (User Datagram Protocol) packet is transmitted. TCP packets carry out a controlled transmission between transmitter and receiver via a handshake procedure. In contrast, UDP packets are sent out without any return message. Since in most cases, there is no return channel in broadcast operation (hence the term 'broadcasting'), TCP packets do not make any sense. For this reason, only UDP protocols are used in DVB during the IP transmission in the so-called Multiprotocol Encapsulation (MPE). Although there is a return channel in DVB-H via the mobile radio network, an IP packet cannot be newly requested since the messages must go simultaneously to many addressees in DVB-H.

22.5 Multiprotocol Encapsulation

In DVB Multiprotocol Encapsulation, contents such as, e.g. HTML files or even MPEG-4 video and audio streams are transported in UDP (User Datagram Protocol) packets. Windows Media 9 applications can also be transmitted by this means and can also be reproduced in devices equipped accordingly. The UDP packets contain the port address of the destination (DST Port) (Fig. 22.3.), a 16-bit-wide numerical value via which the destination application is addressed. For example, the World Wide Web (WWW) always communicates via port No. 0x80. Ports are blocked and controlled by a firewall.

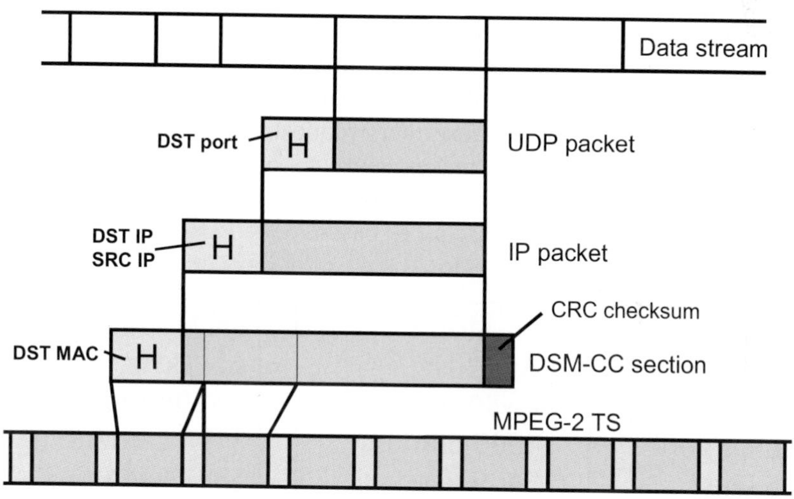

Fig. 22.3. Multiprotocol Encapsulation (MPE)

The UDP packets, in turn, are then embedded into the payload part of IP packets. The header of the IP packets then contains the source and destination (SRC and DST) IP address via which an IP packet is looped through the network from transmitter to receiver in controlled fashion.

If the IP packets are transmitted via a normal computer network, they are mostly transported in Ethernet packets. The header of the Ethernet packets again contains the hardware addresses of the network components communicating with one another, the so-called MAC (Media Access Command) addresses.

When IP packets are transmitted via DVB networks, the Ethernet layer is replaced by the MPEG-2 transport stream and the physical DVB layer

(DVB-C, -S, -T). The IP packets are first packaged in DSM-CC sections which are divided into many transport stream packets, in turn. This is called Multiprotocol Encapsulation: UDP divided into IP, IP into DSM-CC, DSM-CC into TS packets. The header of the DSM-CC sections contains the destination (DST) MAC address. It has a length of 6 bytes as in the Ethernet layer. There is no source MAC address.

22.6 DVB-H Standard

DVB-H stands for "Digital Video Broadcasting for Handheld mobile terminals" and is an attempt at convergence between mobile radio networks and broadcasting networks. The downstream from the mobile radio network (GSM/GPRS, UMTS) is remapped onto the broadcasting network in dependence on the traffic volume. If, e.g., only a single subscriber requests a service via UMTS, for example, this downstream continues to pass via UMTS. If a large number of subscribers request the same service at approximately the same time, it makes sense to offer this service, e.g. a video, point to multipoint via the broadcast network. The services intended to be implemented via DVB-H are all IP based.

2K Mode	**4K Mode**	**8K Mode**
Δf~4kHz,	**Δf~2kHz,**	**Δf~1kHz,**
t_s~250us	**t_s~500us**	**t_s~1000us**
2048 carriers	4096 carriers	8192 carriers
1705 used carriers	3409 used carriers	6817 used carriers
continual pilots	continual pilots	continual pilots
scattered pilots	scattered pilots	scattered pilots
TPS carrier	TPS carrier	TPS carrier
1512 data carrier	3024 data carrier	6048 data carrier
in-depth interleaving on/off	in-depth interleaving on/off	

Fig. 22.4. Overview of the 2K, 4K and 8K modes in DVB-H

DVB-H is intended to provide the framework for a modified DVB-T network to broadcast IP services in time slots in an MPEG-2 transport stream. The physical modulation parameters are very similar or almost identical to those of a DVB-T network. The MPEG-2 transport stream requires greater modifications.

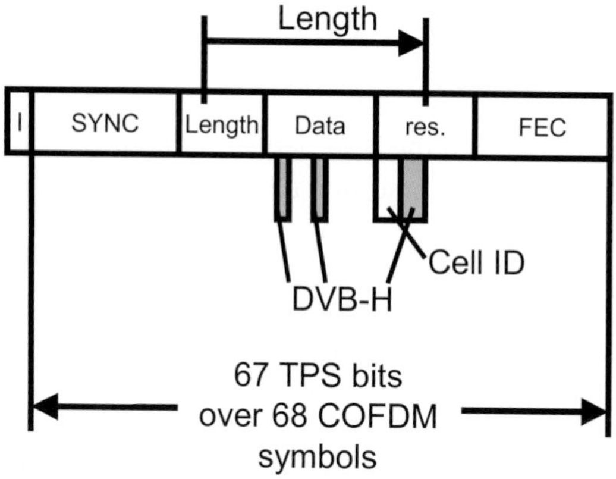

Bit 27...29: hierarchical mode 000, 001, 010
Bit 27: 0 = native Interleaver, 1 = in-depth
interleaver (only in 2K and 4K mode)
Bit 38, 39: 00 = 2K, 01 = 8K, 10 = 4K mode
Bit 40...47: Cell ID
2 new TPS bits:
Bit 48: DVB-H (time slicing) on/off
Bit 49: IP FEC on/off

Fig. 22.5. TPS bits in a DVB-T frame (Transmission Parameter Signalling)

A system overview of DVB-H is provided in ETSI document [TM2939]. The relevant details are described in the DVB Data Broadcasting Standard [ETS301192] and in the DVB-T Standard [ETS300744].

The physical layer DVB-T has been modified or influenced least. In addition to the 8K mode especially well suited to single-frequency (SFN) networks and the 2K mode which more suitable for mobile reception, the 4K mode was introduced additionally as optional compromise. Using the 4K mode, twice the transmitter spacing can be achieved compared with the 2K mode and the mobile capability is distinctly improved compare with the 8K mode. Memory capacities becoming available in the interleaver and de-interleaver should provide for in-depth interleaving which, in turn, would make DVB-H more resistant to burst errors, i.e. multibit errors and the data stream is distributed better over time.

Some additional parameters must also be signalled via TPS carriers in DVB-H.

These are:

- Time slicing on/off in the MPEG-2 transport stream (=DVB-H)
- IP FEC on/off
- In-depth interleaving on/off
- 4K mode

For this purpose, 2 additional bits from the reserved TPS (Transmission Parameter Signalling) bits, bits 42 and 43, and bits already used are used. The details can be found in Fig. 22.5.

Using the 4K mode and in-depth interleaving in the 4K and 2K mode allows a better RF performance to be achieved in the mobile channel. At the same time, the achievable transmitter spacing in 4K mode (approx. 35 km) is greater by a factor of 2 compared with the 2K mode (approx. 17 km) in an SFN network.

Apart from the 8, 7 or 6 MHz channel known from DVB-T, a bandwidth of 5 MHz (L band, USA) can now be selected in DVB-H.

The other modifications are found in the structure of the MPEG-2 transport stream.

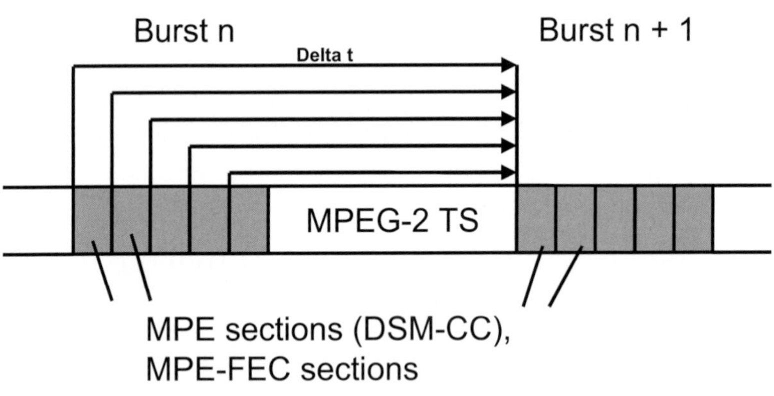

Fig. 22.6. Time slicing in DVB-H

In DVB-H, IP transmission is achieved via the MPEG-2 transport stream by means of the Multiprotocol Encapsulation (MPE) already described. Compared with conventional MPE, however, there are some special features in DVB-H: the IP packets can be protected with an additional Reed Solomon FEC (Fig. 22.7.). The Reed Sololomon FEC of an IP datagram is transmitted in its own MPE FEC sections. These sections have the value 0x78 as Table_ID. The header of these FEC sections has the same structure as that of the MPE sections. Due to the separate transmission of the FEC, a receiver is capable to retrieve the IP packet even without FEC evaluation if there are no errors. Furthermore, the IP information to be transmitted is combined into time slots in the MPEG-2 transport stream. In the time slots, the time Δt until the beginning of the next time slot is signalled in the DSM-CC header. After receiving a time slot, the mobile telephone can then "go to sleep" again until shortly before the next time slot in order to save battery power. On average, the data rates in the time slots will be up to about 400 kbit/s, depending on application. This is IP information requested simultaneously by many users. To signal the time Δt until the next time slot, 4 of the total of 6 bytes provided for the destination MAC address in the DSM-CC header are used. The end of a time slot is signalled via the frame boundary and table boundary bit in the MPE and FEC sections (Fig. 22.8.). The mobile receiver is notified about where an IP service can be found by means of a new SI table, the IP MAC Notification Table (INT) in the MPEG-2 transport stream. The time slot parameters are also transmitted there (Fig. 22.8.).

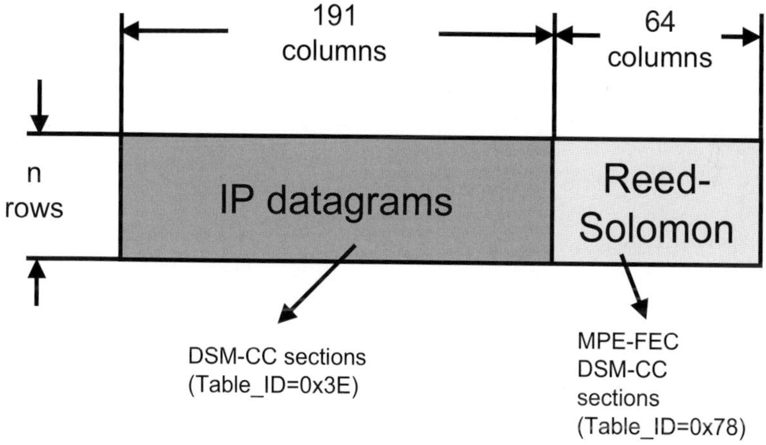

Fig. 22.7. MPE and FEC sections in DVB-H

Instead of the least significant 4 MAC address bytes, the MPE section in DVB-H contains the time slot parameters, the time Δt until the beginning of a new time slot in 10-ms steps, and the two "table_boundary" and "frame_boundary" bits. "table_boundary" marks the last section within a time slice and "frame_boundary" marks the real end of a time slice, especially when MPE FEC sections are used.

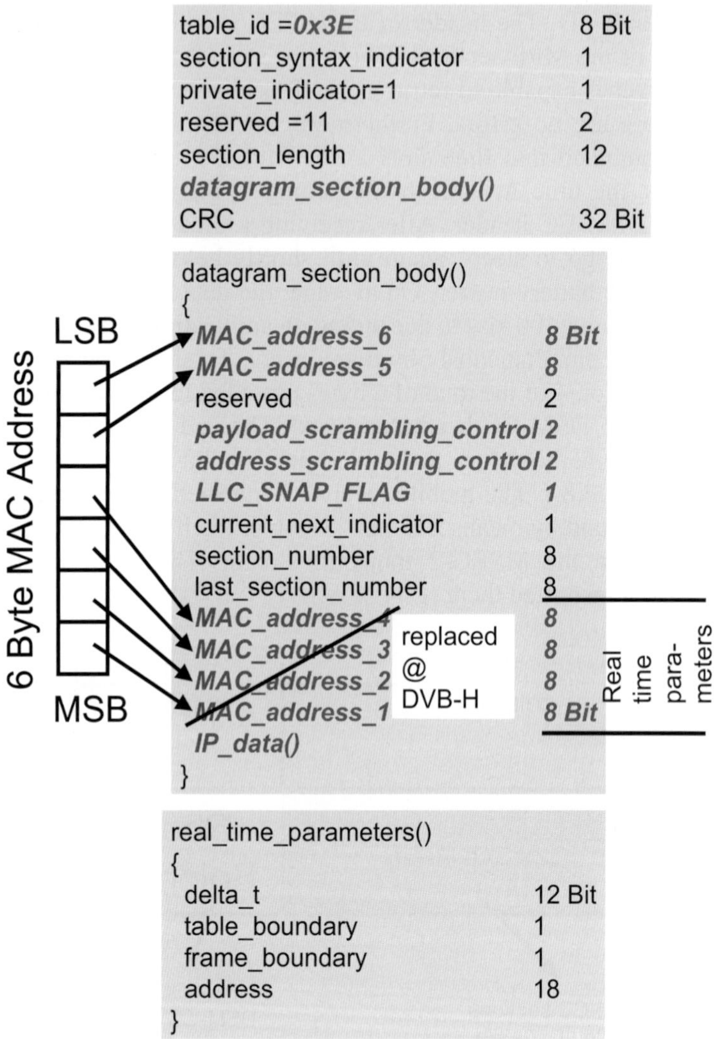

Fig. 22.8. Structure of an MPE section with time slot parameters according to DVB-H

22.7 Summary

DVB-H represents a convergence between GSM/UMTS and DVB. The GSM/UMTS mobile radio network is used as the interactive channel via which high-rate services such as, e.g. video streaming (H.264/MPEG-4

Part 10 AVC Advanced Video Coding or Windows Media 9) are requested which are then transmitted either via the mobile radio network (UMTS) or are remapped onto the DVB-H network. In DVB-H, a DVB-T network is virtually used physically, with some modifications of the DVB-T Standard. As part of DVB-H, additional operating modes were introduced:

- The 4K mode as a good compromise between the 2K and 8K mode with 3409 carriers now used.
- In-depth interleaving is possible in the 4K and 2K modes
- 2 new TPS bits for additional signalling and additional signalling via TPS bits already used,
- Time slicing to save power
- IP packets with FEC protection
- Introduction of a 5 MHz channel (US L band)

table_id = 0x78 = MPE-FEC

table_id =***0x78***	8 Bit
section_syntax_indicator	1
private_indicator=1	1
reserved =11	2
section_length	12
MPE_FEC_section_body()	
CRC	32 Bit

MPE_FEC_section_body()	
{	
padding_columns	***8 Bit***
reserved_for_future_use	***8***
reserved	2
reserved_for_future_use	***5***
current_next_indicator	1
section_number	8
last_section_number	8
real_time parameters()	***42 Bit***
RS_data()	
}	

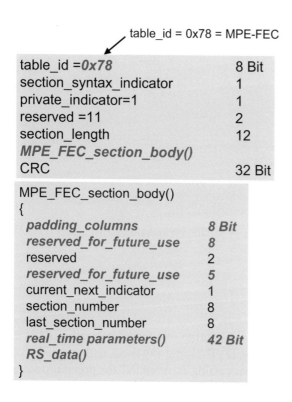

Fig. 22.9. Structure of a DVB-H MPE FEC section with time slot parameters

In the MPEG/2 transport stream, Multiprotocol Encapsulation is applied in a time slicing method. The IP packets to be transmitted can be protected by an additional Reed Solomon FEC code. The end user terminal is notified via a new DVB-SI table about where it can find the IP service.

At the end of 2003, a first prototype of a DVB-H-enabled terminal was presented which has a DVB-H receiver integrated in a modified battery pack.

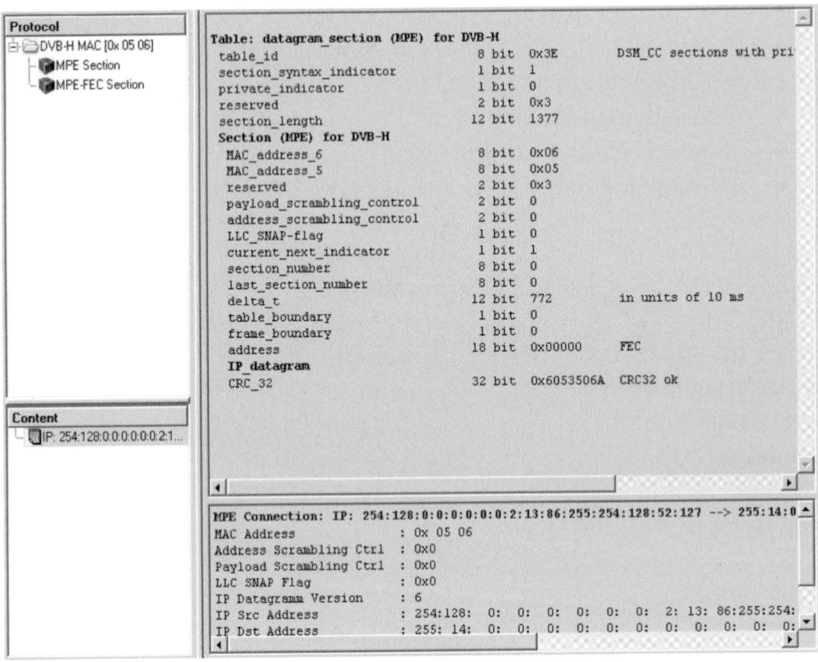

Fig. 22.10. Representation of a DVB-H Section on an MPEG-2 Analyzer [DVM]

22.8 DVB-SH

DVB-SH stands for Mobile TV via satellite and terrestrial paths and is ultimately a virtual combination of DVB-H and DVB-S2 with altered technical parameters. Terrestrial broadcasting is intended for population centers and satellite coverage in the S band (21270 to 2200 GHz) is for rural areas. This MSS (mobile satellite services) band is next to the UMTS band. Terrestrial broadcasting uses the COFDM multicarrier technology, known from DVB-T/H, in the UHF band and the satellite link uses a single-carrier modulation method in the near GHz range where mobile GPS is also working very well. Satellite reception is more difficult in the population centers because of the buildings but terrestrial coverage from the TV towers is cost-effective; the opposite holds true in rural areas. The DVB-SH proposal originated with Alcatel is now an ETSI standard [EN302583] which was published in August 2007. In the technical parameters, changes were made in the error protection; the time interleaver was extended to about 300 ms and the convolutional coding was replaced by turbo coding.

On the direct and indirect path from the satellite or via repeaters or TV towers (Fig. 22.11.), the technical parameters of DVB-SH are:

- Derived from DVB-T/H
- COFDM mode
- QPSK
- 16QAM
- FEC = 3GPP2 turbo encoder (modified and extended)
- 1k, 2k, 4k, 8k mode
- Hierarchical modulation via 16QAM, α=1, 2, 4
- Bandwidths of 8, 7, 6, 5, 1.7 MHz.

On the direct path from the satellite to the terminal alone, the technical parameters can also be (Fig. 22.11.):

- Derived from DVB-S2
- Single carrier TDM mode
- QPSK
- 8PSK
- 16APSK
- FEC = 3GPP2 turbo encoder (modified and extended)
- Bandwidths of 8, 7, 6, 5, 1.7 MHz.

As in DVB-H, the baseband signal used is the MPEG-2 transport stream with time slicing but generic streams are not excluded (are considered as an option). In DVB-SH, two operating modes are provided:

- Type 1: SH-A-SFN: The same frequency is used both for satellite and terrestrial paths.
- Type 2: SH-B or SH-A-MFN: Different frequencies are used for satellite and terrestrial paths.

With regard to the time interleaver depth, two receiver types have been defined in DVB-SH:

- Class 1 receiver: 4 Mbit interleaver memory only
- Class 2 receiver: 512 Mbit interleaver memory.

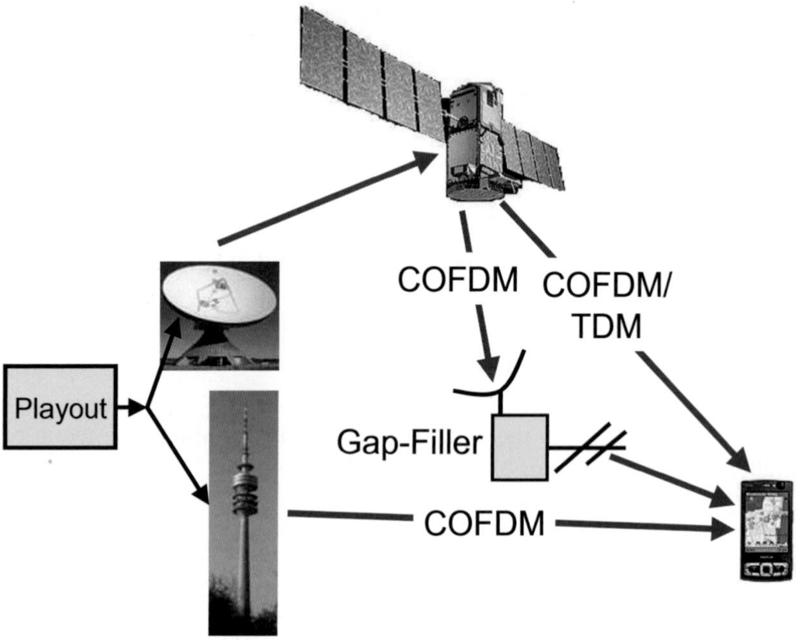

Fig. 22.11. DVB-SH transmission scenario

Bibliography: [ETS300744], [TM2939], [ETSA301192], [ISO/IEC13818-6], [R&S_APPL_1MA91], [EN302583]

23 Digital Terrestrial TV to North American ATSC Standard

Although terrestrial radio transmission poses a variety of problems due to multipath reception and is best handled using multicarrier methods (Coded Orthogonal Frequency Division Multiplex - COFDM), North America opted in favour of a single carrier method under the Advanced Television Systems Committee (ATSC). In the years 1993 to 1995, the Advanced Television Systems Committee – with the participation of AT&T, Zenith, General Instruments, MIT, Philips, Thomson and Sarnoff – developed a method for the terrestrial, and also cable, transmission of digital TV signals. The cable transmission method proposed by ATSC was not put into practice, and the J.83/B Standard was introduced instead. As in all other digital TV transmission methods, the baseband signal is in the form of an MPEG-2 transport stream. The video signal is MPEG-2 coded (MPEG: Moving Picture Experts Group); the audio signal is Dolby digital AC-3 coded. In contrast to DVB, high definition television (HDTV) was favoured in ATSC. The input signal to an ATSC modulator, therefore, is a transport stream with MPEG-2 coded video and Dolby AC-3 coded audio information (AC-3: digital audio compression). Video signals are either SDTV (Standard Definition Television) or HDTV signals. The modulation mode used is eight-level trellis-coded vestigial sideband (8VSB). This is a single-carrier method based on IQ modulation using only the I axis. Eight equidistant constellation points are distributed along the I axis. The 8VSB baseband signal has eight discrete amplitude modulation levels. First, however, an 8ASK signal is generated (ASK: amplitude shift keying).

The ASK signal is a staircase signal (Fig. 23.2.). The bit information to be transmitted is contained in the step height. One step width corresponds to one symbol or symbol duration; three bits can be transmitted per symbol. The reciprocal of the step width is the symbol rate. The ASK staircase signal is amplitude-modulated on a sinusoidal carrier. As a result, a double-sideband spectrum is obtained.

To reduce bandwidth, one sideband is partially suppressed in 8VSB modulation, same as with analog TV. In other words, the amplitude-modulated signal is subjected to vestigial sideband filtering, hence the des-

W. Fischer, *Digital Video and Audio Broadcasting Technology*, Signals and Communication Technology, 3rd ed., DOI: 10.1007/978-3-642-11612-4_23, © Springer-Verlag Berlin Heidelberg 2010

ignation 8VSB. The upper sideband and a vestigial lower sideband remain. Vestigial sideband filtering at the transmitter end calls for Nyquist filtering at the receiver end. The 8VSB signal is subjected to soft Nyquist filtering at the original band center at the receiver end.

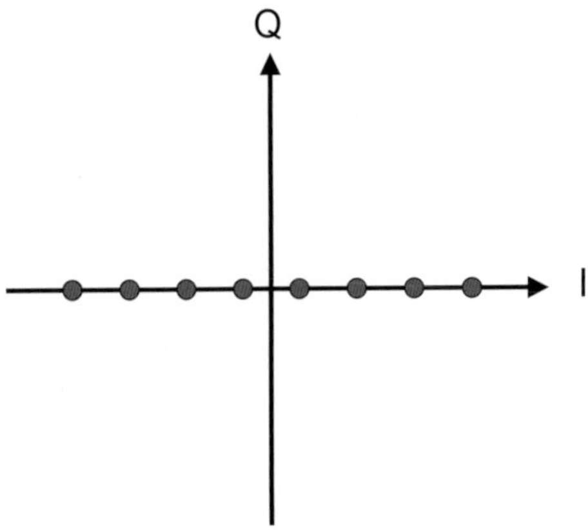

Fig. 23.1. Constellation diagram of an 8ASK signal

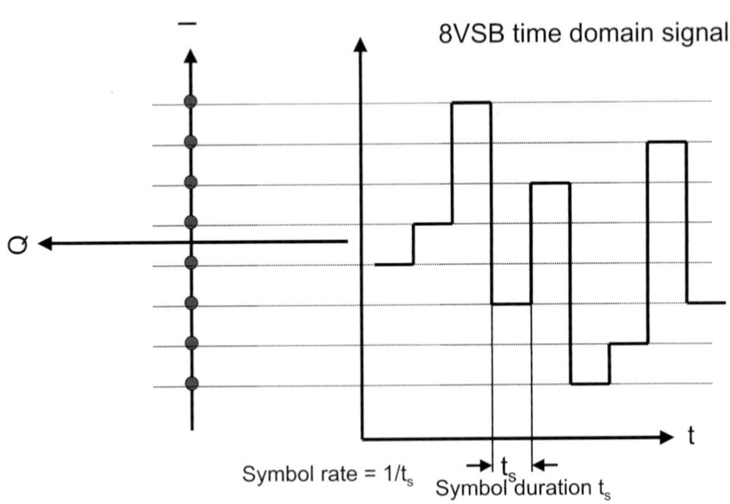

Fig. 23.2. 8VSB/8ASK baseband signal

The area below the Nyquist slope left of the previous band center (Fig. 23.5.) corresponds exactly to the area above the Nyquist edge right of the previous band center, and so compensates for the missing part to complete the upper sideband. As a result, a flat amplitude frequency response is obtained. If the Nyquist slope is not properly adjusted, amplitude frequency response at low frequencies will be the consequence.

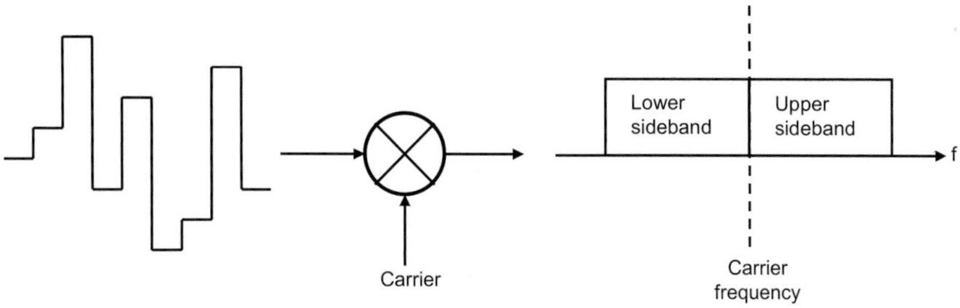

Fig. 23.3. 8ASK modulation at RF domain

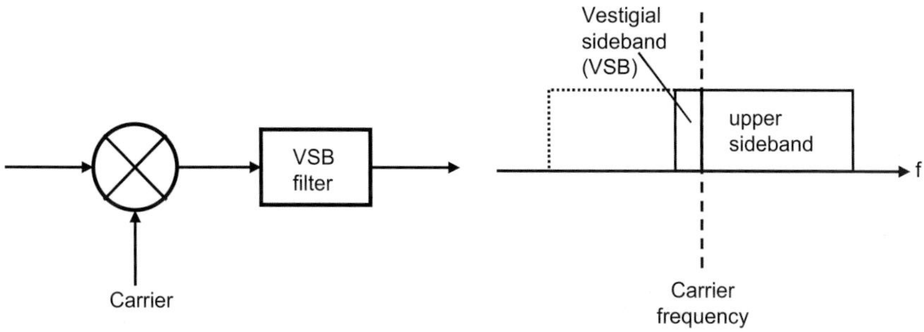

Fig. 23.4. Vestigial sideband filtering

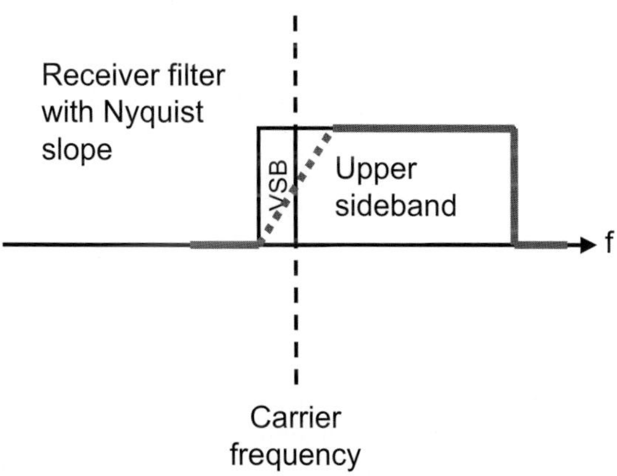

Fig. 23.5. IF filtering with Nyquist slope

With a double-sided spectrum, the vectors representing the upper and the lower sideband (each starting from the tip of the carrier vector) rotate in opposite directions, thus varying the length of the carrier vector, i.e. modulating the carrier. The carrier vector itself remains on the I axis. Even if the carrier is suppressed, the sum vector yielded by the upper and the lower sideband still remains on the I axis (Fig. 23.6.).

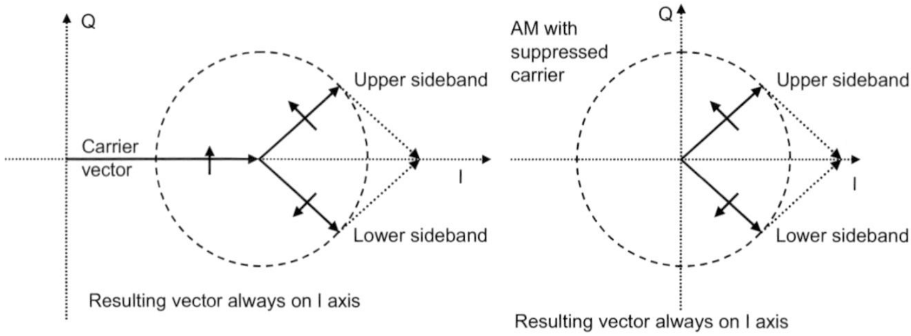

Fig. 23.6. Vector diagram showing amplitude modulation with and without carrier

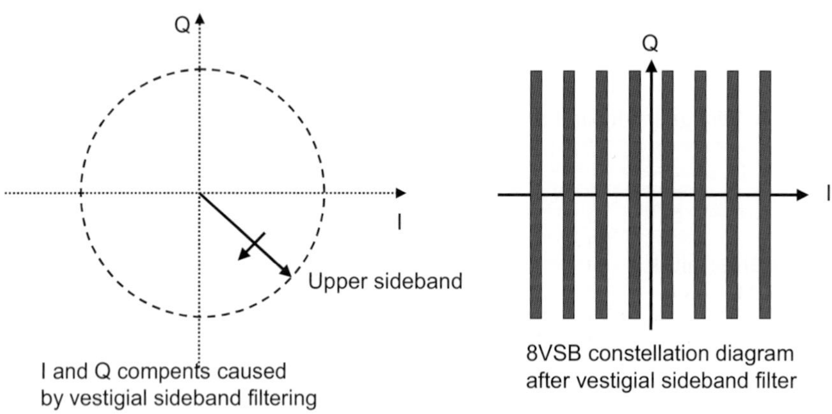

Fig. 23.7. Vector diagram and constellation diagram of an 8VSB signal

However, if one sideband is suppressed in part or completely, the resulting vector will swing about the I axis. Vestigial sideband filtering produces a Q component. Such a Q component is also contained in analog vestigial sideband filtered TV signals (Fig. 23.7.). Analog TV test receivers usually have a Q output in addition to the video output (I output). The Q output is used for measuring incidental carrier phase modulation (ICPM). Due to vestigial sideband filtering, the constellation diagram of an 8VSB signal also includes a Q component, and modulation is no longer shown by points, but by vertical lines. The 8VSB constellation diagram output by an ATSC test receiver, therefore, exhibits vertical lines (Fig. 23.7., 23.8.).

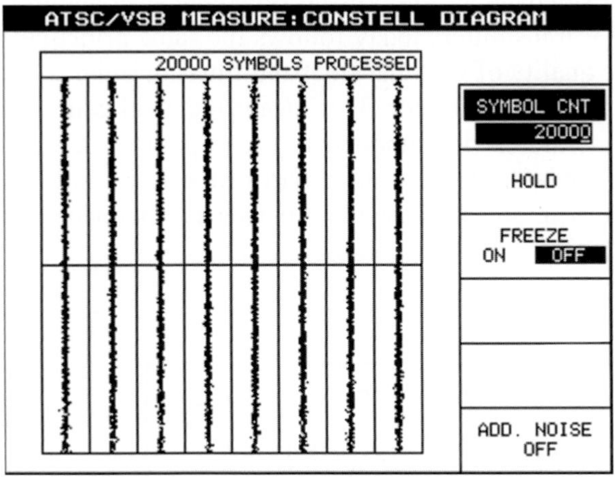

Fig. 23.8. Constellation diagram produced by an ATSC test receiver [EFA]

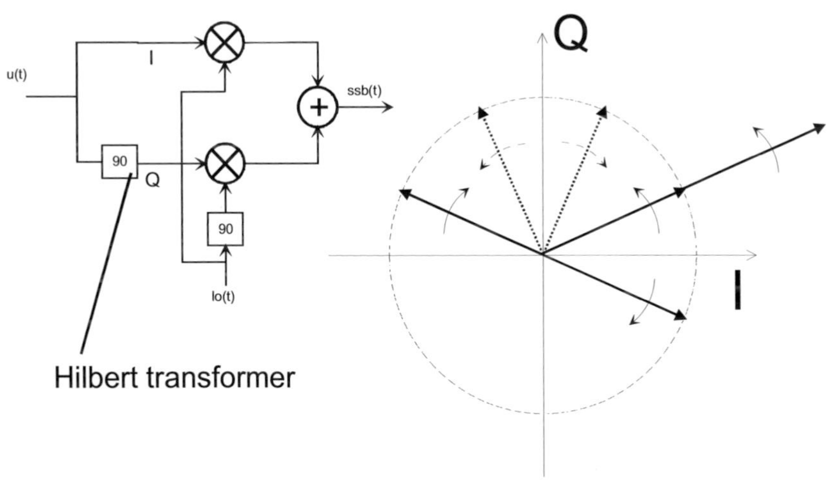

Fig. 23.9. Vestigial sideband or single sideband modulation by means of a Hilbert transformer

8VSB is no longer effected by means of a simple analog vestigial sideband filter as it used to be in analog TV. Today a Hilbert transformer and an IQ modulator are used (Fig. 23.9.). The 8VSB baseband signal is split into two paths. One path is directly applied to the I mixer, the other one is taken via a Hilbert transformer to the Q mixer. A Hilbert transformer is a 90° phase shifter for all frequencies of the band to be filtered. Together with the IQ modulator, it acts as a single sideband modulator; part of the frequencies of the lower sideband are suppressed. Vestigial sideband filtering of modern analog TV transmitters today follows the same principle. A vital prerequisite for the quality of vestigial sideband filtering is the correct setting and operation of the IQ modulator. This means identical gain in the I and Q paths; moreover, the carrier supplied to the Q path must have a phase of exactly 90°. Otherwise the unwanted part of the lower sideband will not be fully suppressed, so that a residual carrier is obtained at the band center.

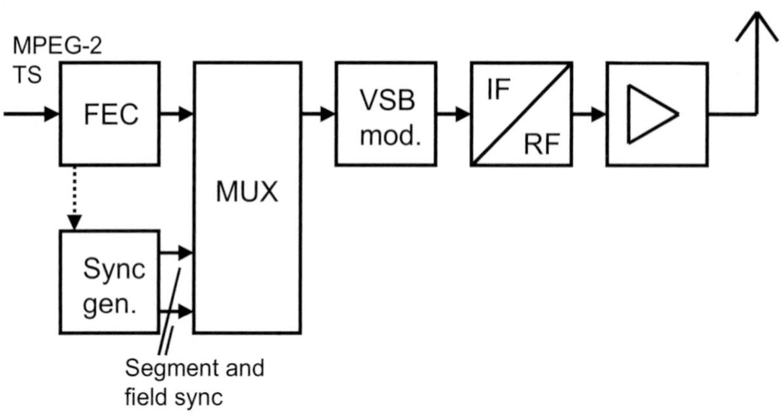

Fig. 23.10. 8VSB modulator and transmitter

23.1 The 8VSB Modulator

After discussing the principle of ATSC modulation, let us take a closer look at the 8VSB modulator (Fig. 23.10.). The ATSC-conformant MPEG-2 transport stream, including PSIP tables, MPEG-2 video elementary streams and Dolby digital AC-3 audio elementary streams, is fed to the forward error correction (FEC) block of the 8VSB modulator at a data rate of exactly 19.3926585 Mbit/s. In the baseband interface, the input transport stream synchronizes to the MPEG-2 188-byte packet structure by means of a sync byte.

The 188 bytes include the transport stream packet header with the sync byte, which has a constant value of 0x47. The transport stream packet clock and the byte clock, which are derived in the baseband interface, are used in the FEC block and also taken to the sync generator for the 8VSB modulator. From the transport stream packet clock and the byte clock, the sync generator generates the data segment sync and the data field sync.

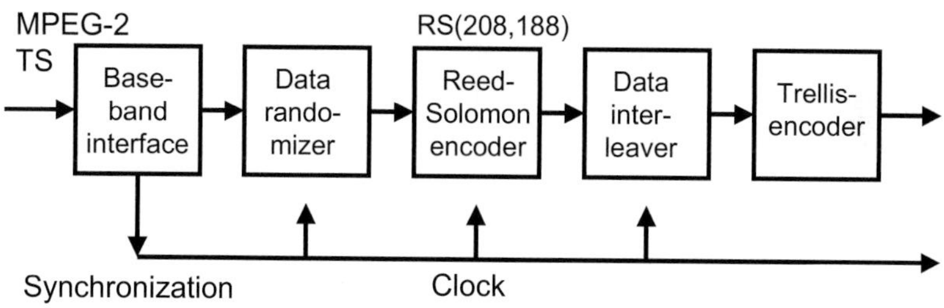

Fig. 23.11. ATSC 8VSB FEC

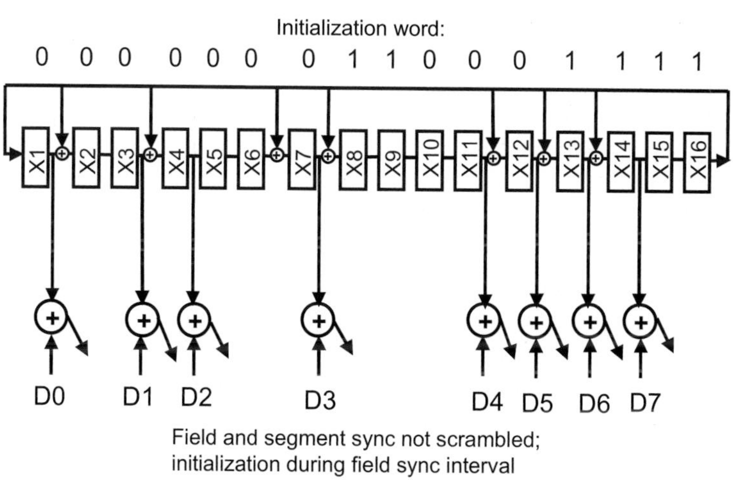

Fig. 23.12. Shift register for randomization

In the FEC block (Fig. 23.11.), the data is fed to a randomizer (Fig. 23.12.) to break up any long sequences of 1s or 0s that may be contained in the transport stream. The data randomizer XORs the incoming data bytes with a pseudo random binary sequence (PRBS). The PRBS generator consists of a 16-bit feedback shift register; it is reset to a defined initialization word at a defined time during the field sync interval. The sync information (e.g. data field sync, data segment sync), which will be discussed in greater detail below, is not randomized and is used, among other things, for receiver-to-modulator coupling. At the receiver end, there is a complementary PRBS generator and randomizer, i.e. of exactly the same design and running exactly in synchronism with the generator/randomizer at the transmitter end.

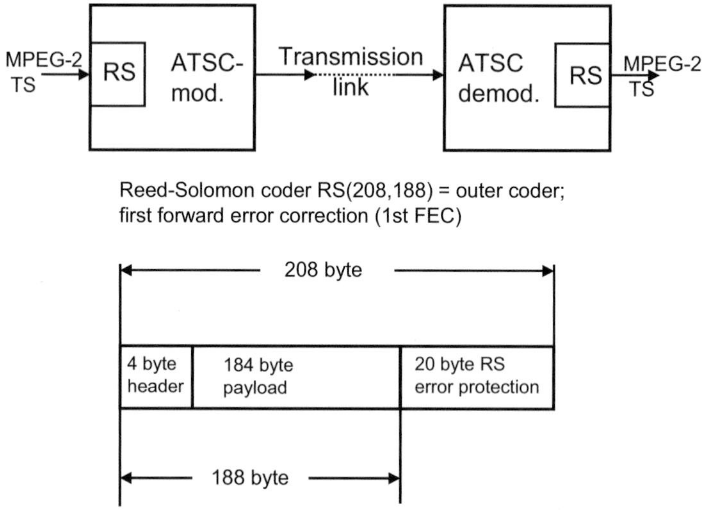

Fig. 23.13. Reed-Solomon FEC

The randomizer at the receiver end reverses the procedure that takes place at the transmitter end, i.e. it restores the original data stream. Randomizing is necessary since long sequences of 1s or 0s may occur. During such sequences, there is no change in the 8VSB symbols and therefore no clock information. This would cause synchronization problems in the receiver and, during the transmission of long sequences of 1s or 0s, produce discrete spectral lines in the transmission channel. This effect is cancelled by randomizing, which causes energy dispersal, i.e. it creates an evenly distributed power density spectrum. The randomizer is followed by the Reed-Solomon block encoder. An ATSC RS encoder (Fig. 23.12.) adds 20 error protection bytes to the 188-byte transport stream (TS) packet (compared with 16 bytes in DVB), yielding a total packet size of 208 bytes. The 20 error protection bytes allow up to 10 errored bytes per TS packet (including FEC part) to be corrected at the receiver end. If more than 10 errored bytes are contained in a TS packet, Reed-Solomon error correction fails, and the transport stream packet concerned is marked as errored.

To mark a TS packet as errored, the transport error indicator bit (Fig. 23.14.) in the TS packet header is set to 1. The packet in question will then be discarded by the MPEG-2 decoder following the 8VSB demodulator in the receiver, and the error will be concealed.

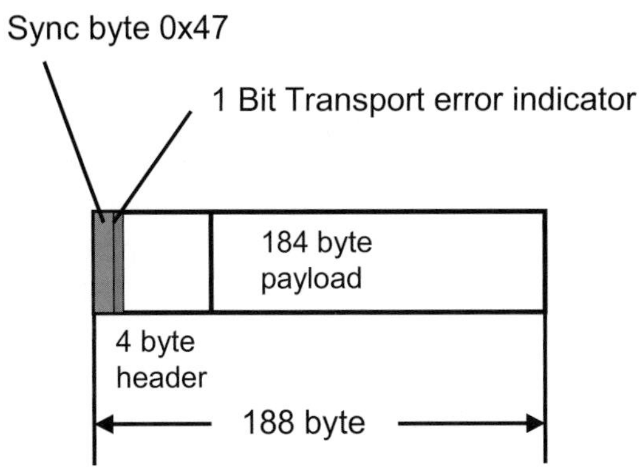

Fig. 23.14. Transport error indicator in TS header

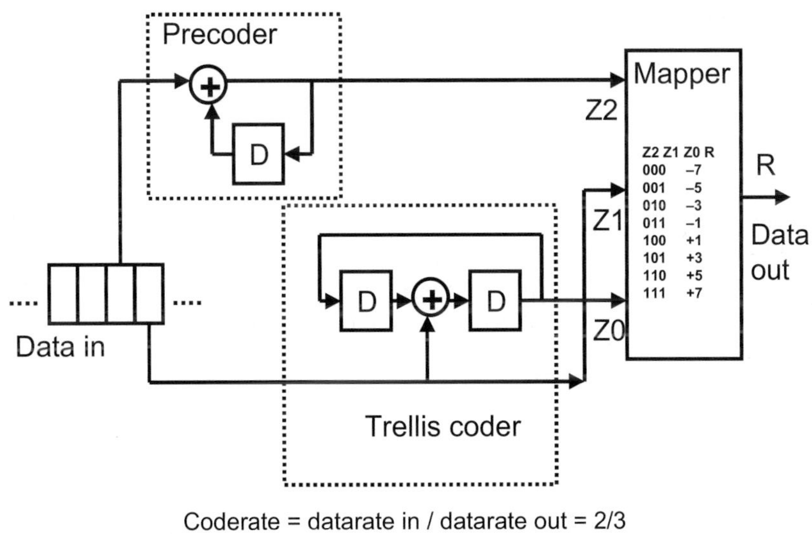

Fig. 23.15. Trellis encoder

The Reed-Solomon encoder RS(188,208) is followed by a data interleaver, which changes the time sequence of the data, i.e. it scrambles the data. At the receiver end, the de-interleaver restores the original time sequence of the data. With interleaving, even long burst errors can be corrected as they are distributed over several frames and can thus be handled more easily by the Reed-Solomon decoder. Interleaving is followed by a second error correction in the form of a trellis encoder. The trellis encoder can be compared to the convolutional encoder used in DVB-S and DVB-T.

The ATSC system employs a trellis encoder (Fig. 23.15.) with two signal paths. From the incoming bit stream, one bit is taken to a precoder with a code rate of 1, and the second bit to a trellis encoder with a code rate of 1/2. This yields an overall code rate of 2/3. The three data streams generated by the precoder and the trellis encoder are fed to a symbol mapper, which outputs the 8-level VSB baseband signal. The counterpart of the trellis encoder at the receiver end is the Viterbi decoder.

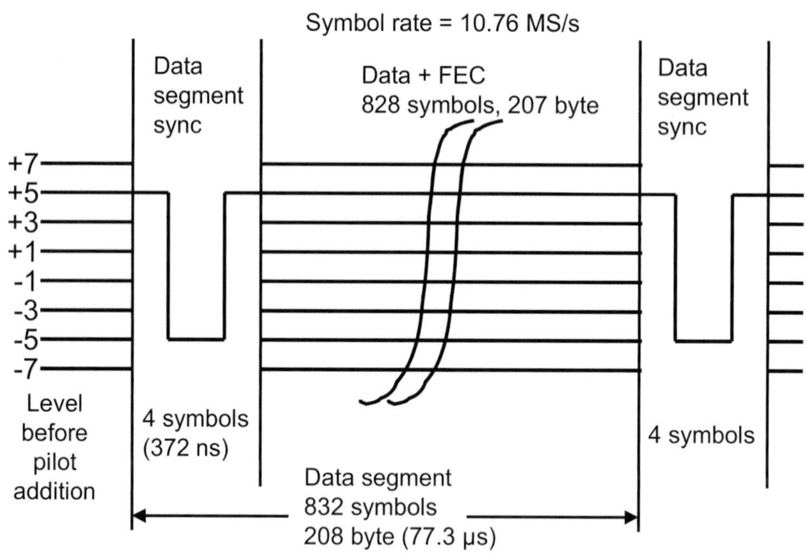

Fig. 23.16. 8VSB data segment

The Viterbi decoder corrects bit errors by retracing the path through the trellis diagram that has with the highest probability been followed through the encoder (see also the chapter on DVB-S). Parallel to the FEC block, a sync generator is provided in the 8VSB modulator. This generator produces, at defined intervals, special sync patterns that are transmitted instead of data in the 8VSB signal as sync information for the receiver. The FEC encoded data and the segment sync and field sync produced by the sync generator are combined in the multiplexer. The 8VSB signal is divided into data segments. Each data segment starts with a data segment sync.

The data segment sync consists of 4 symbols which are assigned defined 8VSB signal levels: The first symbol is at signal level +5, the two middle symbols at signal level –5, and the last symbol at +5. The data segment sync can be compared to the analog TV sync pulse. It marks the beginning of a data segment consisting of 828 symbols and carrying a total of 207 data bytes. A complete data segment – including the sync – comprises 832 symbols and has a length of 77.3µs (s. It is followed by the next data segment, which likewise starts with the 4-symbol data segment sync. A total of 313 data segments each combine to form a field. In 8VSB transmission, a distinction is made between field 1 and field 2. With either field comprising 313 data segments, field 1 and field 2 contain a total of 626 data segments. Each field starts with a field sync. This is a special data segment that likewise starts with a 4 symbol data segment sync but contains special data. Each field of 313 data segments is 24.2 ms long, yielding an overall length of 48.4 ms for field 1 and field 2.

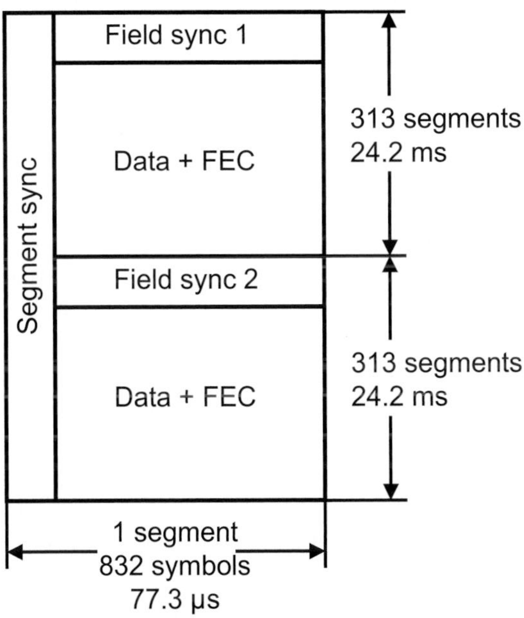

Fig. 23.17. 8VSB data frame with two fields

The field sync, like a data segment, starts with a data segment sync. Instead of normal data, however, this data segment sync contains a number of pseudo random sequences, the VSB mode information, and some special, reserved symbols. The VSB mode bits carry the 8VSB/16VSB mode information. 16VSB was intended for cable transmission, but has not been implemented in practice.

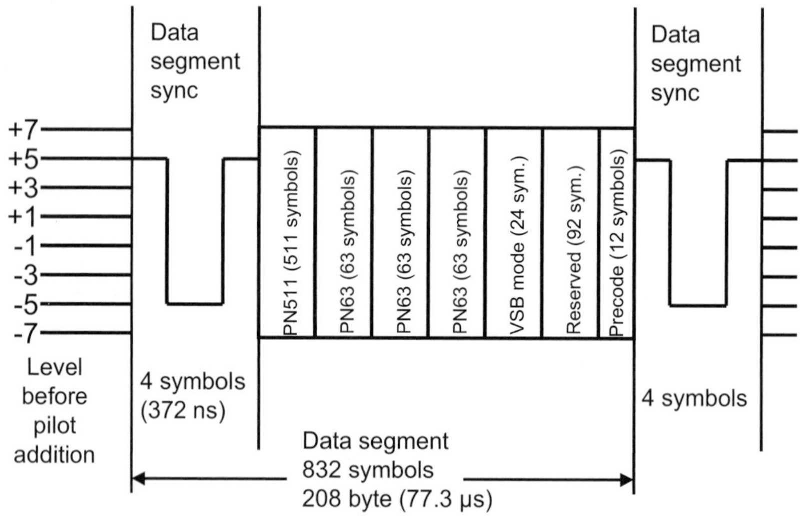

Fig. 23.18. 8VSB field sync

Terrestrial transmission employs the 8VSB mode. The pseudo random sequences contained in the field sync are used as training sequences by the channel equalizer in the receiver. Moreover, it is the pseudo random sequences by which the receiver detects the field sync and is thus able to synchronize to the frame structure. During the field sync, the randomizer block is reset in the modulator and in the receiver. The resulting 8VSB baseband signal, consisting of field syncs and data segments, is taken to the 8VSB modulator. Prior to amplitude modulation, a relative DC component of +1.25 is added to the 8 level signal. Prior to this addition, the 8VSB signal has discrete amplitude stages of -7, -5, -3, -1, +1, +3, +5 and +7. Adding the DC component shifts all 8VSB levels by a relative value of +1.25.

Amplitude modulation of a baseband signal, no longer free of DC but with a mixer signal actually free of carrier, however, produces a signal with a carrier component. This carrier component is referred to as an

8VSB pilot signal, and is found exactly at the center of the 8VSB modulation product before it is subjected to vestigial sideband filtering. As a double-sided spectrum, the modulation product would occupy bandwidth at least as wide as the symbol rate. The symbol rate is 10.76 MS/s, so the minimum required bandwidth is 10.76 MHz. The channel bandwidth in the North American ATSC TV system is, however, only 6 MHz. As in analog TV, therefore, the 8VSB signal is vestigial sideband filtered after amplitude modulation, i.e. the major part of the lower sideband is suppressed. This could be done by means of a conventional analog vestigial sideband filter; this method is today no longer employed however, not even by modern analog TV transmitters. Instead, the 8VSB baseband signal with its pilot DC component is split into two signals: One is taken directly to an I mixer and the other first to a Hilbert transformer and then to a Q mixer (Fig. 23.20.).

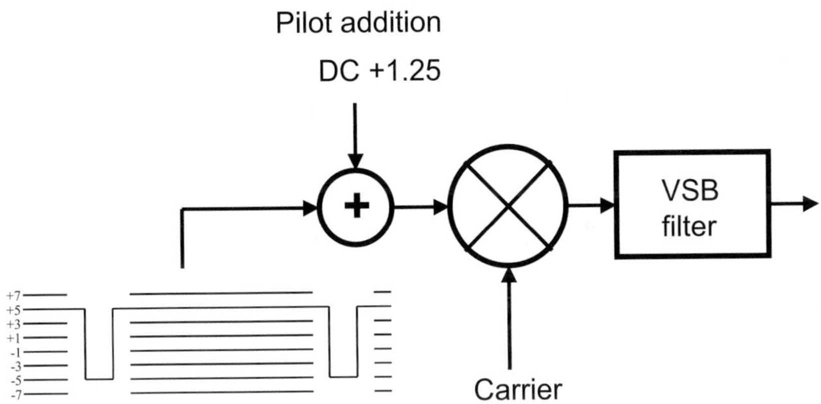

Fig. 23.19. 8VSB modulation with pilot

A Hilbert transformer is a 90° phase shifter for all frequencies of a band. The Hilbert transformer in conjunction with the IQ modulator causes partial suppression of the lower sideband, which is obtained due to the configuration of the amplitudes and phases involved. The resulting 8VSB spectrum only contains the upper sideband and a vestigial lower sideband. Moreover, a spectral line is found at the previous band center, i.e. the band center before vestigial sideband filtering. This spectral line results from the added DC component and is referred to as pilot carrier. The 8VSB spectrum is Nyquist filtered with a roll-off factor of r = 0.115. After VSB modulation, the signal is converted to RF. This conversion is today usually effected by direct modulation simultaneously with VSB modulation. An analog IQ modulator is therefore normally used in VSB modulation which directly converts the baseband signal to RF. As an analog component, the IQ modulator no longer operates as perfectly as does a digital device. It must therefore be ensured that the gain in the I and Q paths is identical, and that the phase of the carrier supplied to the Q path is exactly 90 °. Otherwise the unwanted part of the lower sideband will be inadequately suppressed. After RF conversion, the signal passes through the pre-equalization and power amplifier stages and is then taken to the antenna. A passive bandpass filter in the antenna feeder line suppresses out-of-band components.

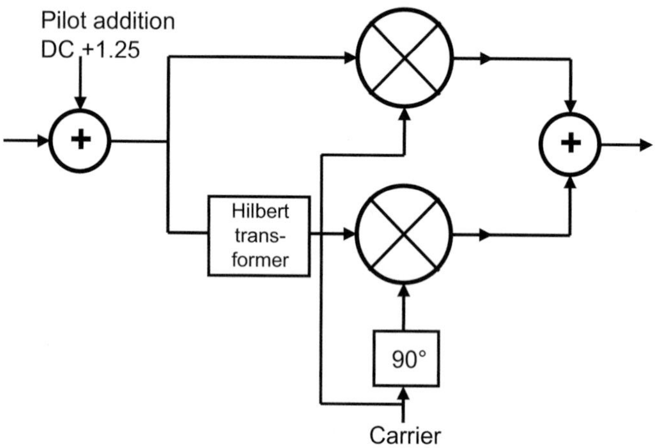

Fig. 23.20. Typical 8VSB modulator with Hilbert transformer

23.2 8VSB Gross Data Rate and Net Data Rate

The symbol rate employed in 8VSB is calculated as follows:

symbol_rate = 4.5/286 • 684 MS/s = 10.76223776 MS/s;

This yields the following gross data rate:

gross_data_rate = 3 bit/symbol • 10.76 MS/s = 32.2867 Mbit/s;

The net data rate is then:

net_data_rate = 188/208 • 2/3 • 312/313 • gross_data_rate
= 19.39265846 Mbit/s;

The above equations are based on the following parameter values:

- 8VSB = 3 bit/symbol
- Reed-Solomon = 188/208
- Code rate = 2/3 (trellis)
- Field sync = 312/313

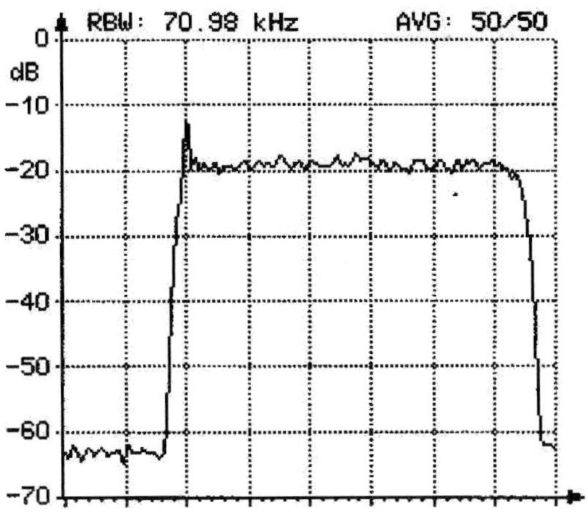

Fig. 23.21. 8VSB spectrum (roll-off filtered with r=0.115)

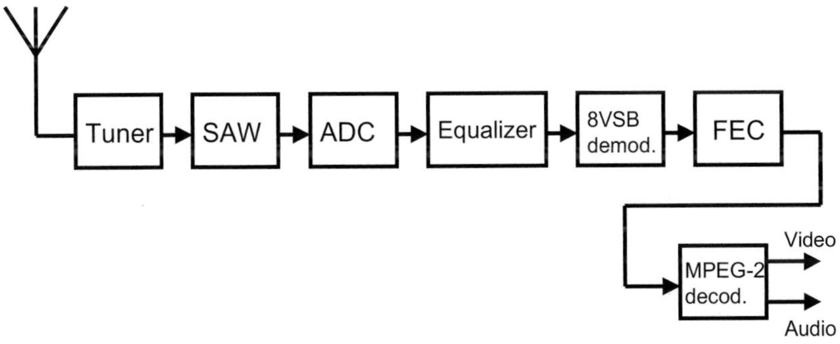

Fig. 23.22. ATSC receiver

23.3 The ATSC Receiver

In the ATSC receiver, a tuner converts the signal from RF to IF. Then the adjacent channels are suppressed by a SAW filter with a Nyquist slope. The band-limited ATSC signal is converted to a second, lower IF for simplified A/D conversion after the anti-aliasing lowpass filter. A/D conversion is followed by a digital channel equalizer that corrects transmission errors. The channel equalizer block also includes a matched filter which performs roll-off filtering with a roll-off factor of $r = 0.115$. The 8VSB signal is then demodulated, and errored bytes are corrected in the FEC block. This again yields the original transport stream, which is applied to the MPEG-2 decoder to restore the original video and audio signals.

23.4 Causes of Interference on the ATSC Transmission Path

ATSC transmission paths are subject to the same types of interference as DVB-T transmission paths. Terrestrial transmission channels are characterized by interference as follows:

- Noise
- Interferers
- Multipath reception (echoes)
- Amplitude response, group delay
- Doppler effect in mobile reception (not considered in ATSC/8VSB)

Of the above types of interference, noise is the only one that can be well predicted and relatively easily handled in ATSC transmission. All other effects, especially multipath reception, are difficult to manage. This is due to the principle of single carrier transmission employed by ATSC. While the equalizer in 8VSB/ATSC is capable of correcting echo, 8VSB is more susceptible to interference compared with COFDM. Mobile reception is virtually impossible.

The "brickwall effect" occurs at an S/N of about 14.9 dB in ATSC. This corresponds to about 2.5 segment errors per second or to a segment error rate of $1.93 \bullet 10^{-4}$, respectively. The pre-Reed Solomon bit error rate is then $2 \bullet 10^{-3}$ and the post-Reed Solomon bit error rate is $2 \bullet 10^{-6}$.

Assuming that the noise power at the tuner input is about 10 dBµV (see chapter on DVB-T), the minimum required receiver input voltage is about 25 dBµV in ATSC.

23.5 ATSC-M/H Mobile DTV

Two proposals by the companies Samsung/Rohde&Schwarz ("A-VSB" - Advanced VSB) and Harris/LG ("ATSC-MHP") resulted in the creation of a so-called Candidate Standard ATSC M/H as an extension to the ATSC standard for mobile TV in 2008. The ATSC M/H standard is called [A/153] and consists of a number of documents. Part 2 describes the transmission part (Transmission System Characteristics). The intention is to make ATSC more portable and receivable by mobile by employing new technologies. The extensions are backward compatible and, therefore, do not interfere with existing ATSC receivers. The services inserted for the use of mobiles are virtually invisible to normal ATSC receivers. In the MPEG-2 data stream supplied to the ATSC M/H modulator, the mobile services run on a PID which is not signalled via the PSI tables, a so-called unreferenced PID. However, this PID is known to an ATSC M/H modulator which inserts the contents into the ATSC frame in a special way. The "normal" ATSC services and the mobile services then share the constant total data rate of 19.39 Mbit/s. The MPEG-2 transport stream is correspondingly edited by an ATSC M/H multiplexer which inserts the mobile DTV contents into the data stream. The mobile DTV services are MPEG-4 AVC and MPEG-4 AAC coded video and audio at total data rates of approx. 0.5 Mbit/s net per service at display resolutions corresponding to mobile telephones (416 pixels x 40 lines (16:9)). For reasons of compatibility, the contents are embedded in UDP and IP protocols similar to DVB-H. In addition, the ATSC M/H compatible modulator is supplied with signalling and control signals. The possibility of forming SFNs (single-frequency networks) does not form a part of the proposed ATSC but is also provided for in parallel. There is a relevant standard [A/110B] which is surrently not used for reasons of expenditure and licensing, and also proprietary solutions for SFN synchronization (e.g. Rohde&Schwarz).

23.5.1 Compatibility with the Existing Frame Structure

The key in ATSC M/H is the backward compatibility with normal ATSC. Normal ATSC receivers must not sense any disturbance from the additional mobile contents. To achieve this, the format must be completely matched to the existing ATSC segment, field and frame structure. An ATSC segment contains a time interleaved and doubly error protected (Reed Solomon and trellis) MPEG-2 transport stream packet. The original MPEG-2 sync bytes which have not been displaced in time have been replaced by the segment sync (4 symbols). The 187 bytes - without sync byte - of an MPEG-2 transport stream packet now become 2484 bytes of an ATSC segment by means of RS(188, 208) and 2/3 trellis coding. Due to the time interleaving, however, the error protected data can no longer be found transparently in a segment but are distributed over 52 segments or transport stream packets, respectively.

ATSC has allocated 313 segments to one field and 2 fields to one frame. If it is intended to incorporate new mobile contents compatibly into ATSC, this frame structure must be adhered to. The following considerations are of importance for the adaptation of ATSC M/H:

- 52 transport stream packets or 52 adjacent segments contain interleaved coherent data
- 3 x 52 transport stream packets result in a number of 156 packets or 156 segments, respectively
- 2 x 156 = 312 segments result in one ATSC field
- Together with the field sync, the ATSC field has a total length of 313 segments

If single frequency networks are to be formed, the basic prerequisite is that all modulators operate completely synchronously with respect to frequency, time and data. I.e. it is necessary to have synchronization of the frame structure, starting with the ATSC M/H multiplexer, the multiplexer informing the ATSC modulator of the ARSC frame start and the emission time of the ATSC symbols.

In ATSC M/H - "Mobile DTV" - the content for the mobile service,

- MPEG-4 AVC - H.264 coded (video), image resolution 426 pixels x 240 lines (16:9)
- and MPEG-4 AAC coded (audio)
- embedded in an IP environment
- provided with additional Reed Solomon and convolutional code
- with additional TPC (transmission parameter channel) information (about the physical layer, error protection)
- extended with additional FIC (fast information channel) information (number of services)

- rendered transparently operable with additional SSC (service signalling channel)
- better received with additional training sequences for the mobile receiver
- optionally equipped with ESG (Electronic Service Guide) via OMA BCAST (Open Mobile Alliance Broadcast)

thus prepared, is keyed into the MPEG-2 transport stream supplied to the ATSC modulator, in such a way that the interleaving taking place there and the error protection are virtually "outwitted", i.e. are precalculated. The additional contents are running on a special PID (0x1FF9 or user-defined) known to the ATSC M/H multiplexer and modulator. This also includes any type of signalling. The data for ATSC M/H are pre-interleaved (52 transport stream packets) in advance in such a manner that the correct order is restored by the interleaving in the modulator (52 segments). The main tasks such as additional error protection, inserting of additional training sequences etc. have already been accomplished in the ATSC M/H multiplexer.

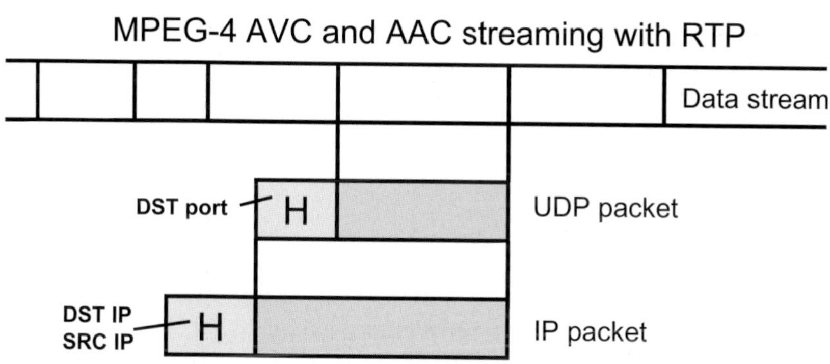

Fig. 23.23. MPEG-4 AVC and AAC streaming via IP

23.5.2 MPEG-4 Video and Audio Streaming

In ATSC M/H, as in other mobile TV standards, the content transmitted for the mobile applications is embedded via IP protocols. MPEG-4 AVC and MPEG-4 AAC streaming material is initially inserted into the MPEG-2 transport stream in UDP (User Datagram Protocol) packets by way of DSM-CC sections. This variant of data transmission provides for the compatibility with other transmission variants (s.a. the DVB-H chapter). MPEG-4 AVC (= H.264) is currently the most modern and most effective type of video compression. The same applies to MPEG-4 AAC+ for the audio coding.

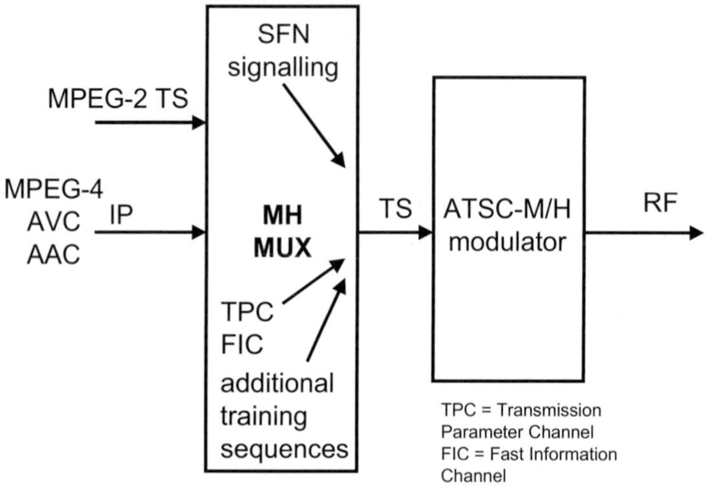

Fig. 23.24. ATSC-M/H multiplexer and ATSC-M/H modulator

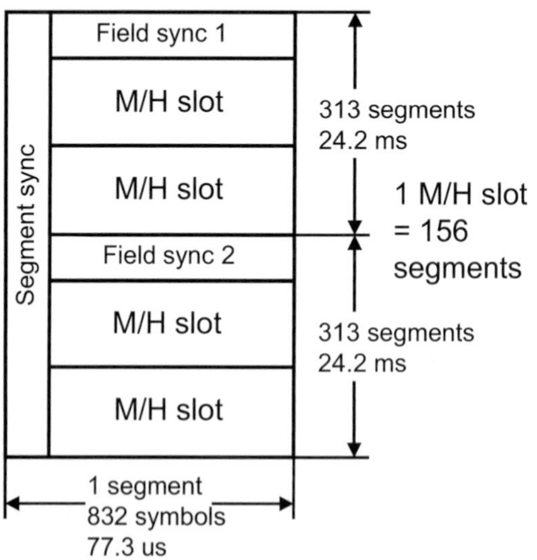

Fig. 23.25. Forming ATSC-M/H slots

23.5.3 ATSC M/H Multiplexer

The ATSC M/H multiplexer is actually the key element for the ATSC M/H signal processing. It conditions the supplementary ATSC M/H data in a "digestible" way for the ATSC M/H modulator. It provides for the IP encapsulation and for the error protection and for the pre-interleaving and inserting of the additional training sequences for the M/H receivers.

Considered in a simplified way, the ATSC M/H slots begin precisely at a field boundary. In reality, however, they have an offset of 37 transport stream packets or segments before the field sync. This is not shown in the Figure and initially might only cause confusion.

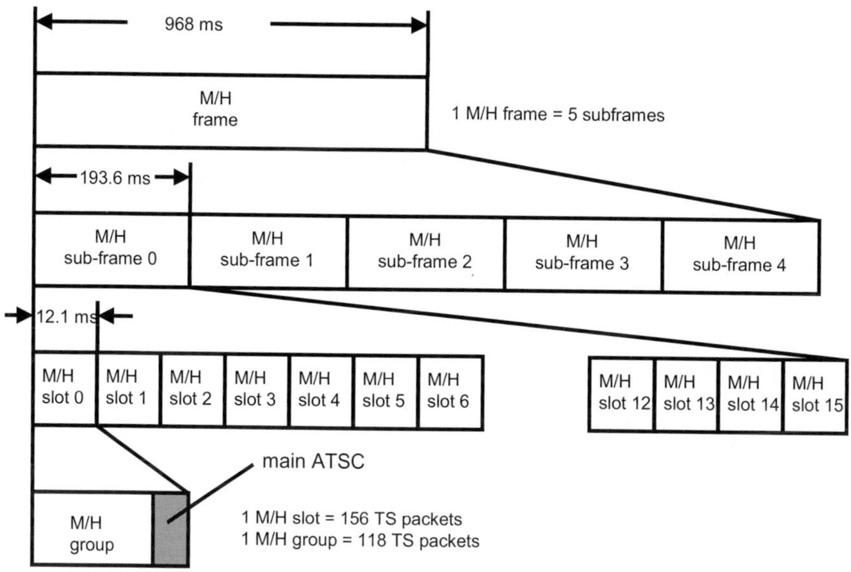

Fig. 23.26. ATSC-M/H framing

23.5.3.1 Compatible ATSC M/H Framing

An ATSC M/H field is here divided into two M/H slots. Each M/H slot has a length of 156 segments (to recall: 3 x 52 interleaved TS packets = 156 packets). The actual ATSC M/H component is located in the first 118 segments whilst the rest of the 156 segments is always used for the ATSC main stream. One to eight M/H slots make up a parade. A parade can carry one or two ATSC M/H ensembles. A parade is thus nothing else but a series of M/H slots transmitting the same M/H contents. The data rate share of a parade in the total of 19.38 Mbit/s of the net ATSC channel is 0.9 to

7.3 Mbit/s. However, the actual useful data rate of a parade is lower due to the additional error protection.

In ATSC M/H, an M/H subframe consisting of 16 adjacent M/H slots is first formed. 5 M/H subframes make up one M/H frame. It depends on the number of slots (1 to 8) allocated to a parade which M/H slots are really occupied with M/H content and this is defined in detail in the standard. If an M/H slot is occupied with M/H content, only the first 118 segments or transport stream packets are used for this purpose, the rest is reserved for the main ATSC. The intention is that the MPEG buffers for main ATSC in the receiver should not be empty.

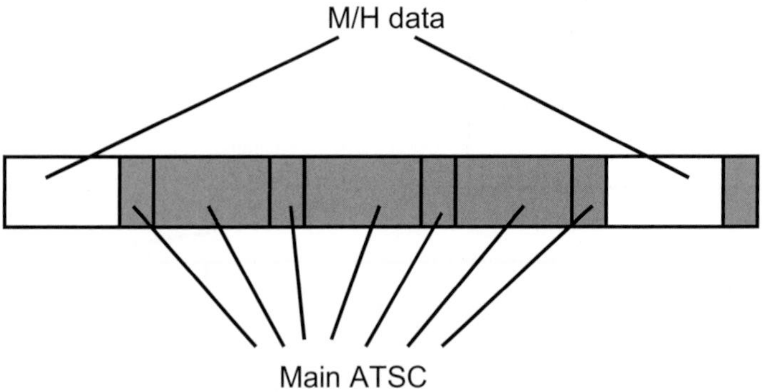

Fig. 23.27. M/H data and main ATSC data

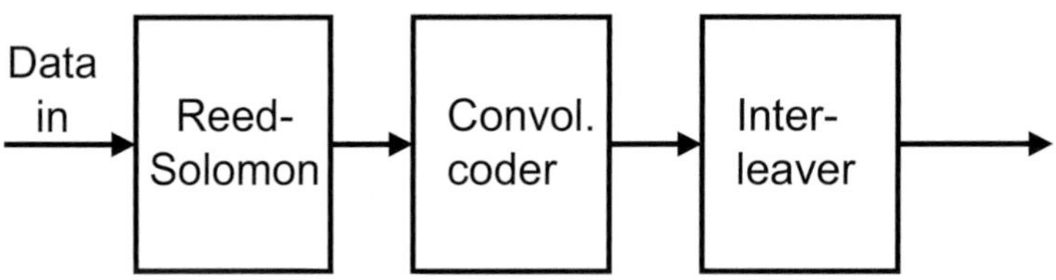

Fig. 23.28. Additional Error Protection

23.5.3.2 Additional Error Protection

The ATSC M/H content is additionally error-protected, the error protection already being added in the ATSC M/H multiplexer. The additional error protection consists of a Reed Solomon error protection and of convolutional coding. The additional convolutional coding together with the trellis coding with the ATSC modulator then results in a turbo code. After the convolutional coding, the ATSC M/H content is additionally interleaved.

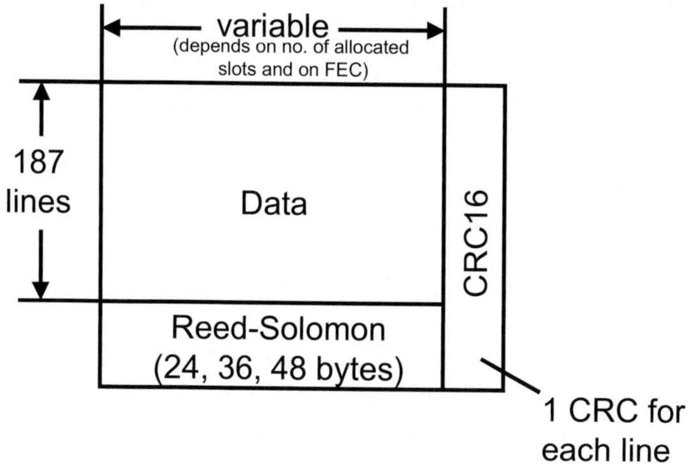

Fig. 23.29. Additional Reed-Solomon FEC

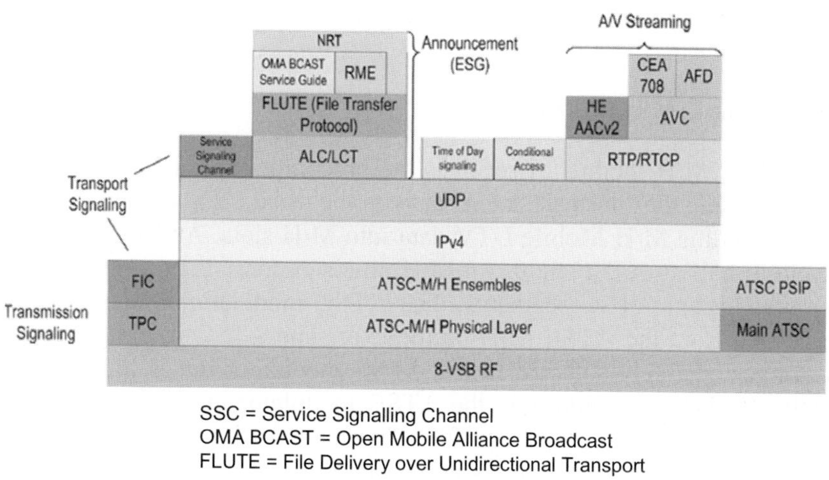

Fig. 23.30. ATSC-M/H layer (source: Rohde&Schwarz)

23.5.3.3 Additional Training Sequences

The ATSC multiplexer also inserts 6 additional trainng sequences for the equalizer in the receiver. This enables the receiver to adapt itself better to unfavourable portable and mobile receiving conditions. It also enables a receiver to handle single-frequency network conditions more easily.

23.5.3.4 Supplementary Data

Apart from the MPEG-4 streaming data, additional information of importance for the receiver is also transmitted in different layers in ATSC M/H, which is

- TPC - Transmission Parameter Channel,
- FIC - Fast Information Channel,
- SSC - Service Signalling Channel,
- OMA BCAST - Open Mobile Alliance Broadcast.

The Transmission Parameter Channel and the Fast Information Channel are included relatively closely at the lowest physical layer; both of them being entered permanently in certain bytes in the M/H slots. The Transmission Parameter Channel signals the mapping of the M/H slots and the additional error protection. Via the Fast Information Channel, the number of services and their IDs are transmitted. The actual service names are found in the Service Signalling Channel which is transmitted at the IP level. In addition, EPG data can be emitted in the OMA BCAST.

23.5.4 ATSC M/H Modulator

The ATSC M/H modulator receives the preconditioned data on a special PID including the "normal" ATSC data on the usual PIDs signalled via PSI and inserts the M/H Mobile DTV data into M/H slots. An M/H slot has a length of half an ATSC field. An M/H time slot consists of a certain number of adjacent ATSC segments. The ATSC modulator is controlled by signalling from the ATSC M/H multiplexer. The ATSC framing can now no longer be freely selected in the ATSC modulator but is linked to the ATSC M/H data. In addition, the ATSC modulator must now perform a trellis reset at certain times. But for a "normal" ATSC receiver, the result after the modulation only looks as if a part of the transmitted data were contained in unknown, i.e. unreferenced PIDs. However, ATSC M/H receivers are well able to utilize these data, too.

23.5.5 Forming Single Frequency Networks

Due to the scarcity of frequencies, single frequency networks provide a very economic way of reusing the same frequency over a relatively large area. This possibility is used very intensively mainly in COFDM-based transmission standards. ATSC is now attempting to use the same approach, applying same rule that all transmitters in a single frequency ATSC network must be

- synchronous in frequency,
- synchronous in time,
- synchronous in their data and
- meeting emission times.

There are no guard intervals in single-carrier modulation. It is necessary that appropriate equalizers in the receiver assist in recovering as much as is possible from the conditions of reception. The essential factor in ATSC with respect to single frequency networks is the frame synchronization of all modulators involved and the emission time of the symbols. The modulators are synchronized via the multiplexer.

23.5.6 Summary

ATSC M/H "Mobile DTV" is an extension in the existing ATSC standard to make ATSC more portable and useful in mobile applications. Numerous backward compatible extension, simply ignored by a "normal" ATSC receiver, have been inserted into the ATSC signal. To be able to broadcast ATSC M/H, an ATSC M/H multiplexer is required in addition to an ATSC M/H-compatible modulator. New ATSC receivers can compensate for delay differences of approximately up to 40 μs within certain limits in the case of multipath reception and thus also provide for single-frequency networks in this order of magnitude; however this is mainly dependent on the level differences of the receiving paths and on the number of paths, and mainly on the receiver.

Fig. 23.31. Closed Captioning insert on a TV screen

23.6 Closed Captioning

It is, or was, normal practice in NTSC to transmit closed captioning (CC) data in the vertical blanking interval in line 21 of the first and second field of a TV frame. In principle, CC is used for sending subtitles, possibly in various languages, in two bytes per field each to the TV receiver; i.e. texts which are related to the current program. The relevant data rate can be calculated as:

- 2 bytes per field at 60 fields per second are
- 120 bytes per second per 8 bits = 820 bits/sec.

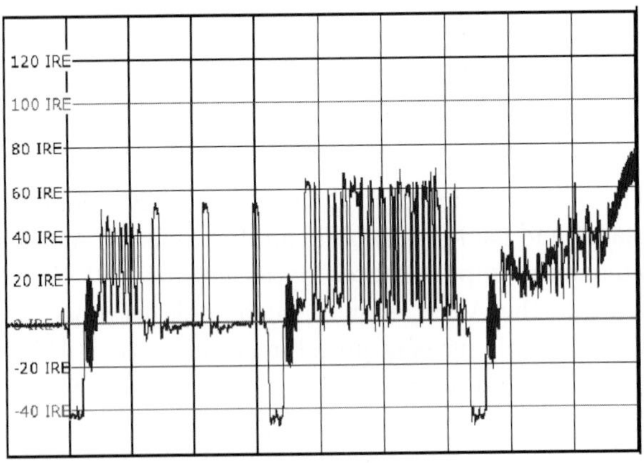

Fig. 23.32. Closed Captioning data in line 21 of an analog TV signal

Fig. 23.31. shows an example of the insertion of closed captioning on a TV screen. The insert can be switched on or off by the viewer via the CC key on their remote control. Fig. 23.32. shows line 21 with closed captioning data.

The relevant standard is EIA 608 A. Broadcasting of closed captioning data is regulated by law in the US. The compatible transmission of such data has also been standardized in ATSC in the EIA 708 B standard. The CC data are not sent out in private PES streams as in DVB but via the optional user data after the picture header in the video PES steam. The data are thus automatically synchronous with the video stream. The data rate is ten times that of analog television. The data rate of the CC data in ATSC is thus

- 9600 bits,
- or 20 bytes per picture user data field at 60 frames per second.

Fig. 23.3. shows the position of the picture user data in the video PES stream. A CC decoder, which can be a component of an MPEG decoder chip, fetches these data from the data field after the picture header and inserts the corresponding text information into the decoded image.

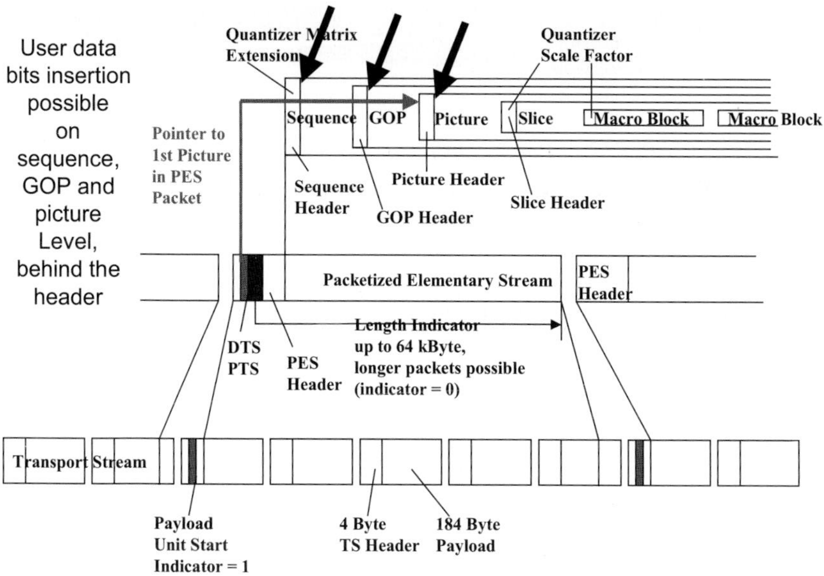

Fig. 23.33. Position of the picture user data in the video PES stream

23.7 Analog Switch-off

The original plan was to switch off the analog high-power transmitters in the US on 17th January 2009. However, the legislators moved this date to 12th June 2009 because of a scarcity of available digital TV receivers. From this date on, analog terrestrial television by high-power transmitter was past history and only low-power transmitters and analog services by cable continued in operation.

Canada is planning its complete switch-over to ATSC for the 31st of August 2011. In Mexico the date for the planned switch-off of analog television is in 2022 and South Korea wants to switch its analog TV services from NTSC to ATSC by 31st December 2012, which is also the year for the proposed analog switch-off in the UK and Australia.

Fig. 23.34. ATSC-M/H multiplexer Rohde&Schwarz AEM100

Bibliography: [A53], [EFA], [SFQ], [SFU], [A153], [A110B], [7EB01_APP], [EIA608A], [EIA708B]

24 ATSC/8VSB Measurements

In the following section, the measurements required at the air interface to the North American terrestrial digital TV transmission system will be discussed in detail. The ATSC – Advanced Television Systems Committee – standard employs a modulation method with a single carrier, that is 8VSB, which stands for 8-level vestigial sideband modulation. The 8VSB constellation diagram does not exhibit points but lines. Due to the Q component resulting from vestigial sideband filtering, eight lines are formed from the originally eight points. As a basic rule in 8VSB, it can be said that the narrower the eight lines, the better the signal quality. While 8VSB modulation appears relatively simple compared to the COFDM multicarrier method, it exhibits correspondingly higher susceptibility to the various types of interference from the terrestrial environment.

The following causes of interference will, therefore, be discussed below:

- Additive white Gaussian noise
- Echoes
- Amplitude and group-delay distortion
- Phase jitter
- IQ errors of modulator
- Insufficient shoulder attenuation
- Interferers

All of the above types of interference manifest themselves as bit errors in the demodulated 8VSB signal. Bit errors can be corrected to a certain extent by means of forward error correction (FEC). Vital in this context are measurement of the bit error ratio and a detailed analysis of the causes of bit errors.

24.1 Bit Error Ratio (BER) Measurement

In ATSC/8VSB, three different bit error ratios are known. These result from two error protection methods being combined, i.e. Reed-Solomon

W. Fischer, *Digital Video and Audio Broadcasting Technology*, Signals and Communication Technology, 3rd ed., DOI: 10.1007/978-3-642-11612-4_25, © Springer-Verlag Berlin Heidelberg 2010

block coding, and convolutional coding. The bit error ratios (BER) are as follows:

- Bit error ratio before Viterbi
- Bit error ratio before Reed-Solomon
- Bit error ratio after Reed-Solomon

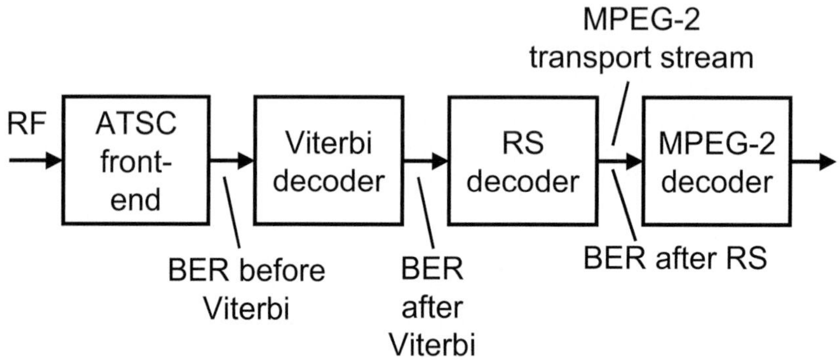

Fig. 24.1. Bit error ratios in ATSC

The most significant BER is the BER before Viterbi as it represents the channel bit error ratio. But there is a problem: with trellis coding, the bit error ratio before Viterbi cannot be measured technically because of ambiguities in the receiver in the Viterbi decoding.

The BER after Viterbi, i.e. before Reed-Solomon, is derived directly from the Reed-Solomon decoder. The BER after Reed-Solomon, then, indicates non-correctable bit errors, i.e. more than 10 bit errors occurring in a 208-byte RS block coded transport stream packet. The BER after Reed Solomon is likewise derived from the Reed-Solomon decoder. Non-correctable bit errors are marked by transport error indicator bits (set to 1) in the MPEG-2 transport stream. Bit error ratio measurement is performed by means of an ATSC/8VSB test receiver.

24.2 8VSB Measurements Using a Spectrum Analyzer

By means of a spectrum analyzer, both in-band and – most importantly – out-of-band measurements can be performed on the 8VSB signal. The parameters to be measured with a modern spectrum analyzer are as follows:

- Shoulder attenuation
- Amplitude frequency response
- Pilot carrier amplitude
- Harmonics

Make the following settings on a modern spectrum analyzer:

- Center frequency at center of band
- Span 20 MHz
- RMS detector
- Resolution bandwidth 20 kHz
- Video bandwidth 200 kHz
- Slow sweep time (>1 s) to allow averaging by RMS detector
- No averaging function activated

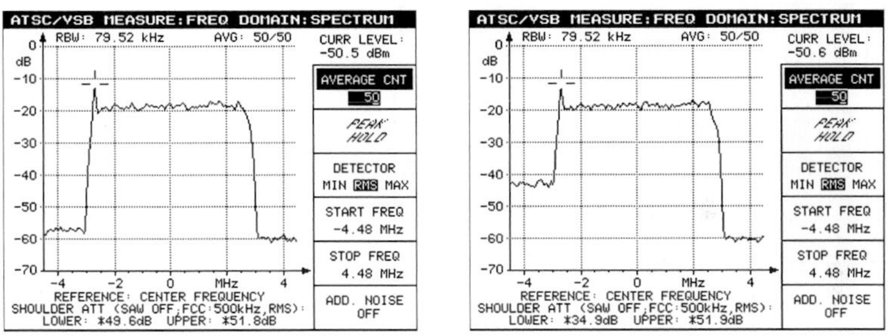

Fig. 24.2. 8VSB signal spectrum with appropriate (left) and poor (right) vestigial sideband suppression

Then the shoulder attenuation and, most importantly, the suppression of the unwanted part of the lower sideband can be measured, as well as the pilot amplitude and the amplitude distortion in the passband.

24.3 Constellation Analysis on 8VSB Signals

In contrast to a quadrature amplitude modulation (QAM) diagram, which shows points, the constellation diagram of an 8VSB signal exhibits lines. An ATSC test receiver usually comprises a constellation analyzer, which displays the 8VSB diagram by 8 parallel vertical lines that should in the ideal case be extremely narrow.

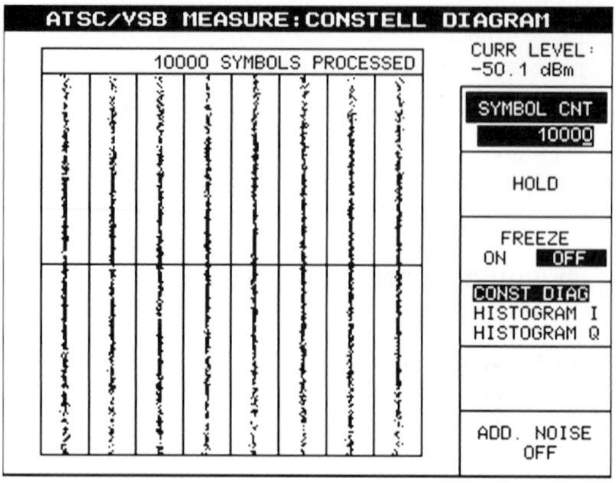

Fig. 24.3. Undistorted constellation diagram of an ATSC/8VSB signal

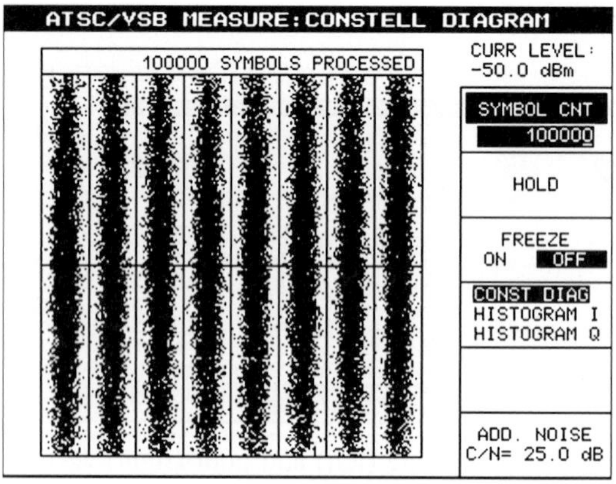

Fig. 24.4. 8VSB constellation diagram revealing noise impairment

The constellation diagram in Fig. 24.3. with very narrow lines reveals only a slight impairment by noise, such as caused already in the ATSC modulator or transmitter. As a basic rule, it can be said that the narrower the lines, the less significant the signal distortion. In the event of pure noise distortion, the lines are uniformly widened over their entire length. The wider the lines, the greater the impairment due to noise. In the constellation analysis, the RMS value of the noise is determined. Based on a statistical function, i.e. the Gaussian distribution (normal distribution), the standard deviation of the I/Q points obtained in the decision fields of the constellation diagram is determined. From the RMS noise value, the test receiver calculates the signal-to-noise ratio (S/N ratio) in dB referenced to the signal power, which is likewise calculated by the test receiver.

In the event of phase jitter, the lines in the decision fields of the constellation diagram are trumpet-shaped, i.e. they become increasingly wider as the distance from the horizontal center line increases (Fig. 24.5.).

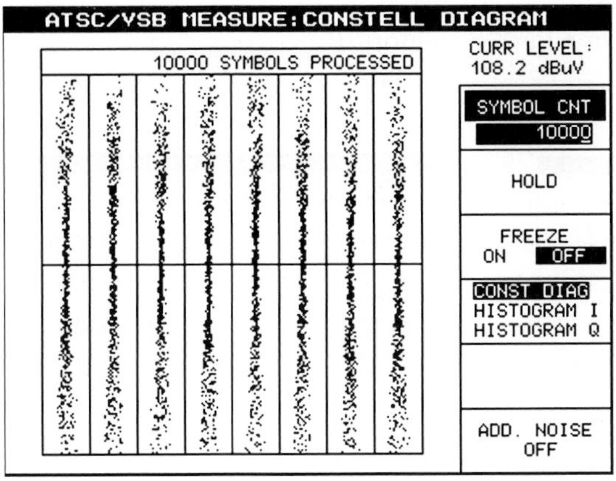

Fig. 24.5. 8VSB constellation diagram revealing phase jitter

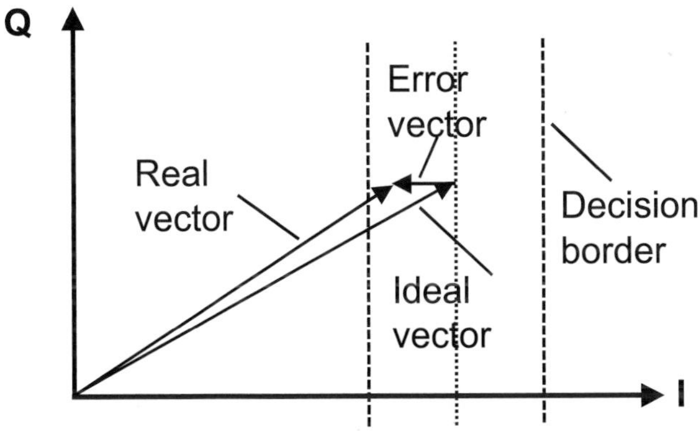

Fig. 24.6. Determining the MER of an 8VSB signal

The Modulation Error Ratio (MER) parameter summarizes all errors that can be measured within a constellation diagram. For each type of error (interference), an error vector is continually calculated. The sum of the squares (RMS value) of all error vectors is calculated. The ratio of the error-vector RMS value and the signal amplitude yields the MER, which is usually specified in dB. In the event of pure noise impairment, the MER is equal to the S/N ratio.

The following applies:

$$MER[dB] <= S/N[dB];$$

$$MER_{RMS}[dB] = -10 \log(1/n \bullet (|error\_vector|^2/P_{signal\_without\_pilot});$$

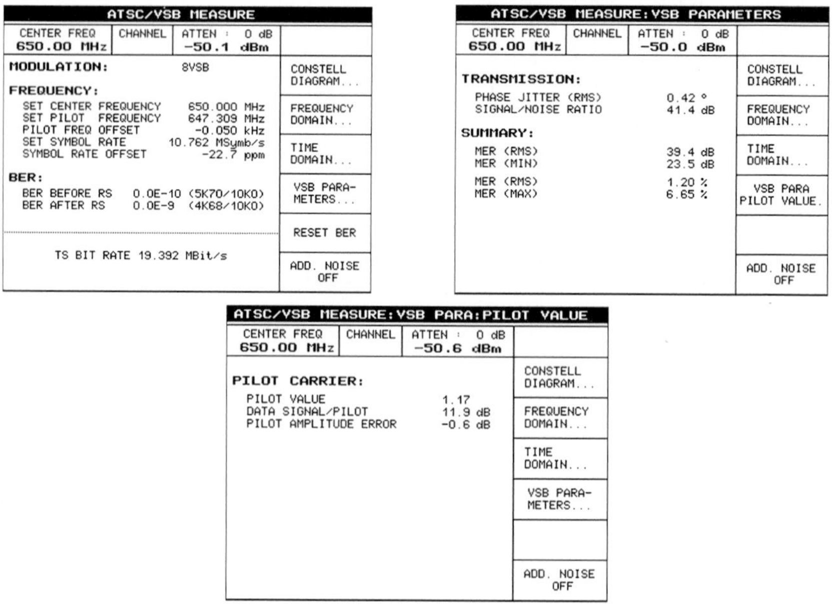

Fig. 24.7. Numerical results output by an 8VSB test receiver

Many test parameters are also output as numerical results by the 8VSB test receiver. These include the signal amplitude, bit error rate, pilot amplitude, symbol rate, phase jitter, S/N ratio, and MER.

24.4 Measuring Amplitude Response and Group Delay Response

Although the ATSC/8VSB signal carries no pilot signals that would provide information on channel quality, the amplitude, group-delay and phase response can be roughly determined – with the aid of the test receiver equalizer – from the PRBS sequences contained in the 8VSB signal. The signal characteristics output by the 8VSB test receiver can be used to align an ATSC modulator or transmitter, for example. The equalizer data also provides information on echoes in the transmission channel and allows calculation of the impulse response.

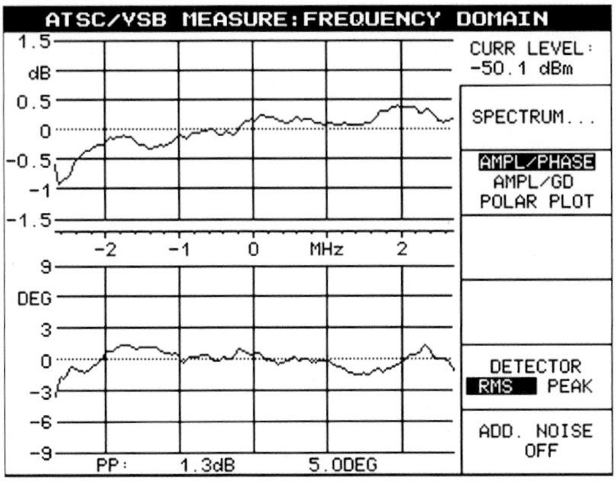

Fig. 24.8. Amplitude and phase response measurement using an 8VSB test receiver [EFA]

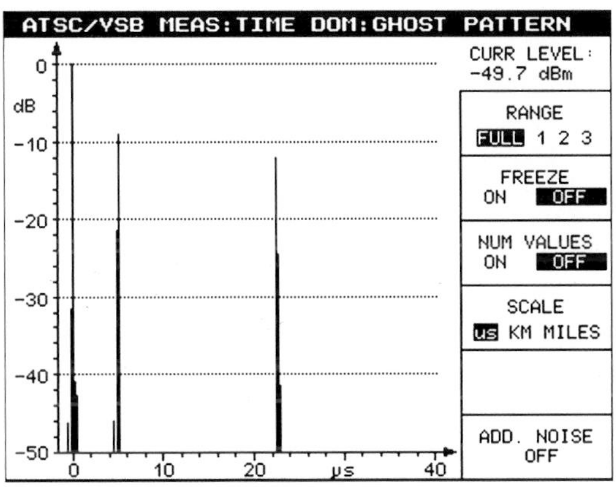

Fig. 24.9. Ghost pattern/impulse response

Bibliography: [A53], [EFA], [SFQ], [SFU], [ETL]

responses can be equally determined with the aid of the test receiver …

25 Digital Terrestrial Television according to ISDB-T

25.1 Introduction

The Japanese standard for digital terrestrial television is called ISDB-T, i.e. Integrated Services Digital Broadcasting – Terrestrial, which was adopted in 1999, quite a long time after DVB-T and ATSC. This delay made it possible to take into account also the experience gained with the older standards. Unlike ATSC, where a single-carrier method is used, it was decided (correctly) for ISDB-T to use a COFDM multicarrier system as in DVB-T. ISDB-T is even more complex than DVB-T; it is also more robust because of the greater interleaving with time. The first pilot station was installed on the Tokyo Tower and overall, ISDB-T started with eleven pilot stations throughout Japan.

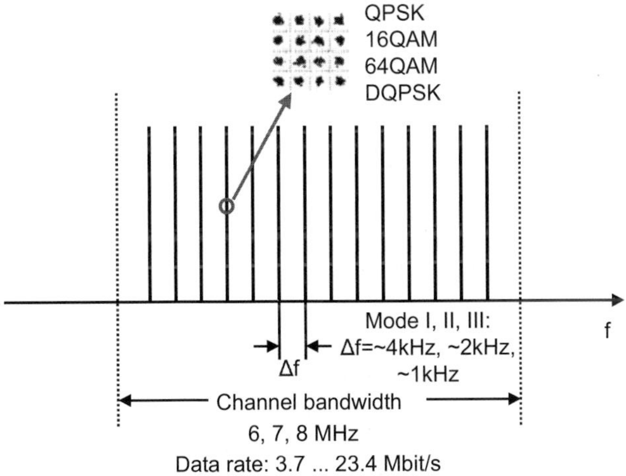

Fig. 25.1. COFDM in ISDB-T

W. Fischer, *Digital Video and Audio Broadcasting Technology*, Signals and Communication Technology, 3rd ed., DOI: 10.1007/978-3-642-11612-4_26,

25.2 ISDB-T Concept

In ISDB-T, COFDM (coded orthogonal frequency division multiplex) is used in 2K, 4K and 8K mode (Fig. 25.1.). The 6 MHz-wide channel can be subdivided into 13 subbands (Fig. 25.2.) in which different modulation parameters can be selected and contents transmitted. Time interleaving can be optionally switched on in various stages. With an actual channel bandwidth of 6 MHz, the useful band only has a width of 5.57 MHz, i.e. there is a guard band of about 200 kHz each for the upper and lower adjacent channels. One subband of the ISDB-T channel has a width of 430 kHz.

It is possible to select different types of modulation in ISDB-T:

- QPSK with channel correction
- 16QAM with channel correction
- 64QAM with channel correction
- DQPSK without channel correction (not required with DQPSK).

There are 3 possible modes (6 MHz channel as example):

- Mode I, with
  108 carriers per subband
  3.968 kHz subcarrier spacing
  1404 carriers within the channel
  2048-points IFFT
- Mode II, with
  216 carriers per subband
  1.9841 kHz subcarrier spacing
  2808 carriers within the channel
  4196-points IFFT
- Mode III, with
  432 carriers per subband
  0.99206 kHz subcarrier spacing
  5616 carriers within the channel
  8192-points IFFT

As already mentioned, the full 6 MHz channel can be subdivided into 13 subbands of precisely 3000/7 kHz = 428.7 kHz (Fig. 25.2.) each.

Not all of the 2048, 4192 or 8192 COFDM carriers in mode I, II or III are actually used as payload carriers. In ISDB-T, there are

- Zero carriers, i.e. those which are not used,

- Data carriers, i.e. real payload,
- Scattered pilots (but not with DQPSK),
- Continual pilots
- TMCC (Transmission and Multiplexing Configuration Control) carriers,
- AC (Auxiliary Channels).

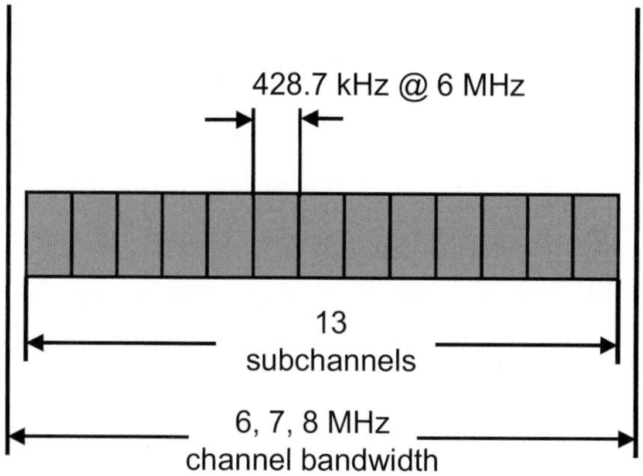

Fig. 25.2. Subchannels in ISDB-T

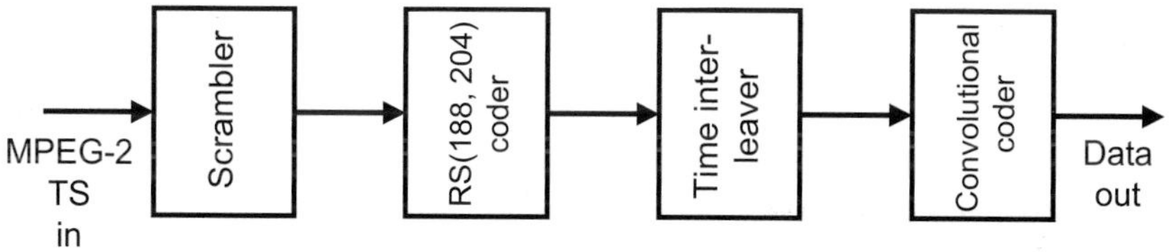

Fig. 25.3. ISDB-T FEC

The net data rates are between 280.85 kbit/s per segment or 3.7 Mbit/s per channel and 1787.28 kbit/s per segment or 23.2 Mbit/s per channel.

Due to the subband or segment concept, it is possible to build both narrow-band receivers which receive only one or a number of subbands and broadband receivers which receive the entire 6-MHz-wide channel.

In principle, the ISDB-T modulator configuration is similar to that of a DVB-T modulator. It has outer error protection, implemented as Reed

Solomon RS(204,188) coder, an energy dispersal unit, an interleaver, an inner coder implemented as convolutional coder, a configurable time interleaver which can be switched on or off, a frequency interleaver, the COFDM frame adapter, the IFFT etc.. The basic configuration of the error protection corresponds directly to that of DVB-T. The selectable code rates are like those of DVB-T:

- 1/2
- 2/3
- 3/4
- 5/6
- 7/8

but the time interleaving is much deeper and can also be configured in stages:

- 0
- 1
- 2
- 4

The following guard interval lengths can be set:

- 1/4
- 1/8
- 1/16
- 1/32

25.1 Forming Layers

The individual segments in ISDB-T can be combined to form a total of 3 layers in which different transmission parameters (type of modulation and error protection) can be selected. In the 3 hierarchical layers, different contents can then be error-protected to different degrees and transmitted with modulation of different robustness. The number of segments to be combined in one layer is selectable but the same transmission parameters are used in each segment of a layer. In the case of 3 layers, in principle, 3 associated, mutually independent data streams must then be supplied. However, by using a "trick", this can be done by supplying one common transport stream (see also Chapter 25.3).

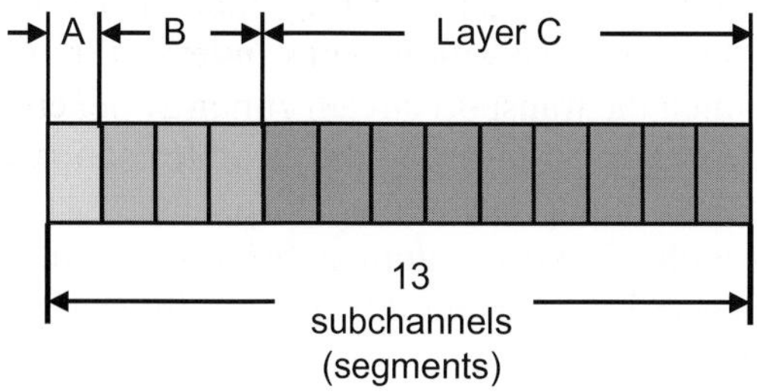

Fig. 25.4. Forming layers

25.2 Baseband Encoding

In the baseband encoding/source encoding area, of course, ISDB-T is just as open as any other standard, too. MPEG-2 video and audio are just as possible here as are the new, more optimal MPEG-4 video and audio codecs. Currently, MPEG-2 video is used for baseband coding (SDTV and HDTV), and MPEG-2 AAC for audio in Japan.

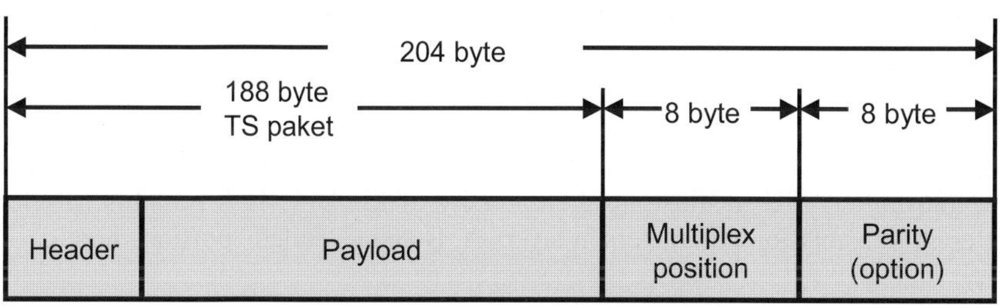

Fig. 25.5. Layer signalling in the 16 overhead bytes in the transport stream packet in the transport stream supplied

25.3 Changes in the Transport Stream Structure

In addition to activating the 3 layers via the transport stream structure, there are also extensions and new tables which, although they largely correspond to those of DVB-SI, differ in their detail or are completely different in some cases. These tables are called ARIB tables (Association of Radio Industries and Business) in ISDB-T. Control information for assigning data to the individual layers of ISDB=T is found in the transport stream in a 16-bit extension in the transport steam packets which are usually 188 bytes long (Fig. 25.5.). The "multiplex position" informs the ISDB-T modulator about the layer for which the transport stream currently contains information. This extension bears no relationship to the error protection in DVB or ISDB-T. However, it has already been common practice to provide the 188-byte-long packets with 16 bytes of dummy information at the baseband level in DVB. The possibility of using this dummy information for layer signalling has been taken up in ISDB-T.

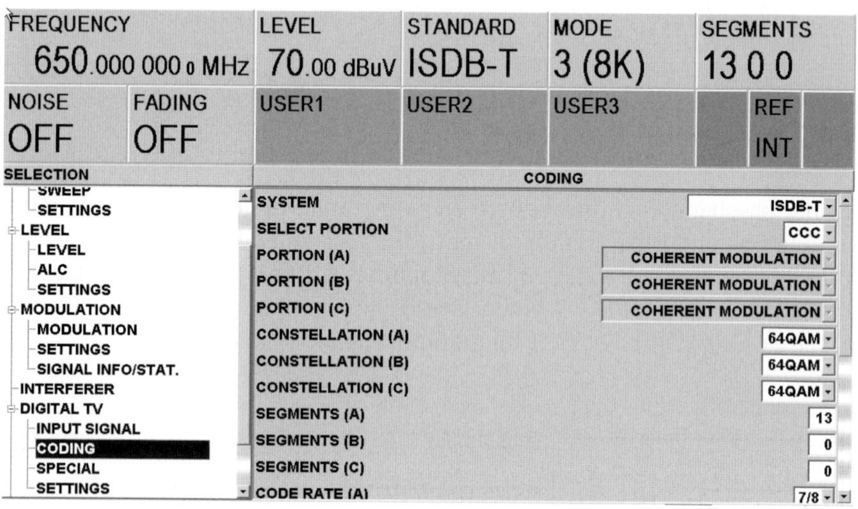

Fig. 25.6. Possible adjustments on the ISDB-T encoder of the Rohde&Schwarz test transmitter SFU to illustrate the diversity of ISDB-T; It is possible to select different transmission parameters in layers A, B and C, called "portion" here.

The order of segments is counted not from left to right, i.e. from channel start to channel end, but starts in the center with segment S0. Segment S1 is then on the left of S0 and S2 is on the right of S0. S0 is the segment used in the 1-segment mode. ISDB-Tsb (Integrated Services Digital Broadcast Terrestrial – sound broadcast) is a narrowband version where only 1 … 3 segments are used. And this is the order before the frequency interleaver. After the frequency interleaver the carriers from the different segments are distributed over the complete channel to avoid frequency selective problems. Only in case of the narrow band ISDB-T versions the subbands are placed in the center.

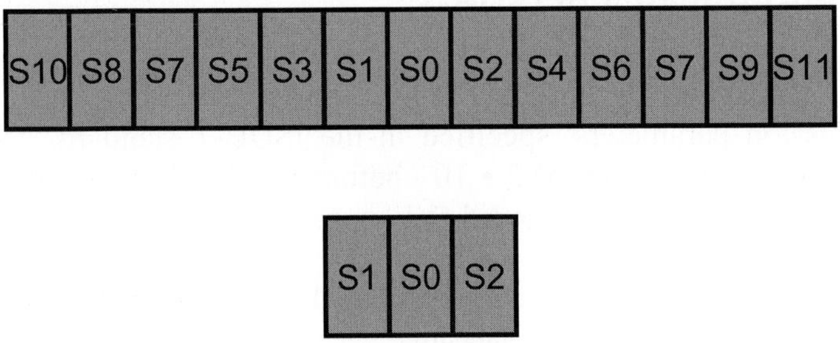

Fig. 25.7. Order of segments in ISDB-T (top) and ISDB-Tsb (bottom)

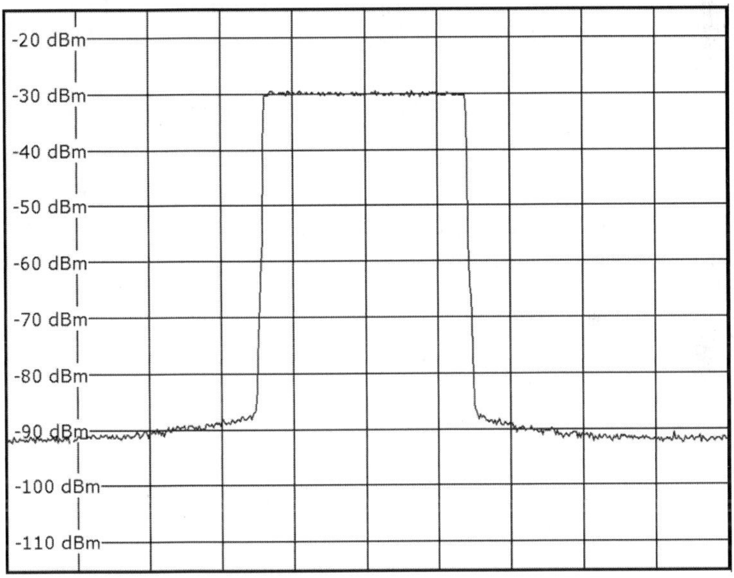

Fig. 25.8. ISDB-T spectrum

25.4 Channel Tables

In ISDB-T, the TV channels have been shifted upward by 1/7 MHz. As an example, Channel 7 is originally located at 177 MHz. The new center frequency is here now 177 + 1/7 MHz = 177.143 MHz.

25.5 Performance of ISDB-T

Table 25.1 shows the minimum signal/noise ratios as a function of the transmission parameters, specified in the ISDB-T standard. They correspond to a bit error rate of 2 • 10^{-4} before Reed-Solomon and to a quasi-error-free data stream after Reed-Solomon.

Table 25.1. Transmission parameters of ISDB-T and theoretical minimum signal/noise ratio according to the standard

Modulation	CR=1/2 [dB]	CR=2/3 [dB]	CR=3/4 [dB]	CR=5/6 [dB]	CR=7/8 [dB]
DQPSK	6.2	7.7	8.7	9.6	10.4
QPSK	4.9	6.6	7.5	8.5	9.1
16QAM	11.5	13.5	14.6	15.6	16.2
64QAM	16.5	18.7	20.1	21.3	22.0

Table 25.2 shows the data rates reproduced in the ISDB-T standard as a function of the transmission parameters.

Table 25.2. Data rates of ISDB-T as a function of the ISDB-T transmission parameters according to the standard (at 6 MHz bandwidth)

Modulation	Code rate	g=1/4 [Mbit/s]	g=1/8 [Mbit/s]	g=1/16 [Mbit/s]	g=1/32 [Mbit/s]
DQPSK QPSK	½	3.651	4.056	4.295	4.425
DQPSK QPSK	2/3	4.868	5.409	5.727	5.900
DQPSK QPSK	¾	5.476	6.085	6.443	6.638
DQPSK QPSK	5/6	6.085	6.761	7.159	7.376
DQPSK QPSK	7/8	6.389	7.099	7.517	7.744
16QAM	½	7.302	8.133	8.590	8.851
16QAM	2/3	9.736	10.818	11.454	11.801
16QAM	¾	10.953	12.170	12.886	13.276
16QAM	5/6	12.170	13.522	14.318	14.752
16QAM	7/8	12.779	14.198	15.034	15.489
64QAM	½	10.953	12.170	12.886	13.276
64QAM	2/3	14.604	16.227	17.181	17.702
64QAM	¾	16.430	18.255	19.329	19.915
64QAM	5/6	18.255	20.284	21.477	22.128
64QAM	7/8	19.168	21.298	22.551	23.234

Table 25.3. Data rates of ISDB-T per segment in dependence of the ISDB-T transmission parameters according to the standard (at 6 MHz bandwidth)

Modulation	Code rate	g=1/4 [kbit/s]	g=1/8 [kbit/s]	g=1/16 [kbit/s]	g=1/32 [kbit/s]
DQPSK QPSK	½	280.25	312.06	330.42	340.43
DQPSK QPSK	2/3	374.47	416.08	440.56	453.91
DQPSK QPSK	¾	421.28	468.09	495.63	510.65
DQPSK QPSK	5/6	468.09	520.10	550.70	567.39
DQPSK QPSK	7/8	491.50	546.11	578.23	595.76
16QAM	½	561.71	624.13	660.84	680.87
16QAM	2/3	748.95	832.17	881.12	907.82
16QAM	¾	842.57	936.19	991.26	1021.30
16QAM	5/6	936.19	1040.21	1101.40	1134.78
16QAM	7/8	983.00	1092.22	1156.47	1191.52
64QAM	½	842.57	936.19	991.26	1021.30
64QAM	2/3	1123.43	1248.26	1321.68	1361.74
64QAM	¾	1263.86	1404.29	1486.90	1531.95
64QAM	5/6	1404.29	1560.32	1652.11	1702.17
64QAM	7/8	1474.50	1638.34	1734.71	1787.28

25.6 Other ISDB Standards

Apart from ISDB-T, there are some other ARIB standards which are:

- ISDB-S (satellite)
- ISDB-Tsb (terrestrial sound broadcast)
- ISDB-C (cable)
- ISDB-Tmm (terrestrial mobile multi-media).

ISDB-S is more effective than DVB-S by a factor of 1.5 and also uses, among other things, 8PSK in single-carrier modulation.

ISDB-C corresponds to a 6 MHz variant of DVB-C and is ITU-T J83C, to put it precisely. Apart from its bandwidth and the roll-off factor, ITU-T J83C is virtually identical with DVB-C and reference is made here to the corresponding chapter.

ISDB-Tsb and ISDB-Tmm are the narrowband versions of ISDB-T, occupying only 1 to 3 segments. There, too, the more effective MPEG-4 baseband coding is provided.

25.7 ISDB-T measurements

ISDB-T measurements are very similar to DVB-T measurements. But there are different constellation diagrams possible in the different layers or portions. Fig. 25.9. shows the different constellation diagrams in the different layers. In case of differential modulation the MER is about 3 dB lower in comparison to coherent modulation (Fig 25.10. center subchannels).

ISDB-T measurements is

- Constellation analysis (Fig. 25.9.)
- MER measurement (Fig. 25.10. and Fig. 25.11.)
- BER measurement (in all 3 layers)
- IQ impairment measurement
- Impulse response measurement (especially in SFN's)
- MPEG transport stream analysis

For more details please see chapter 21 – DVB-T measurements.

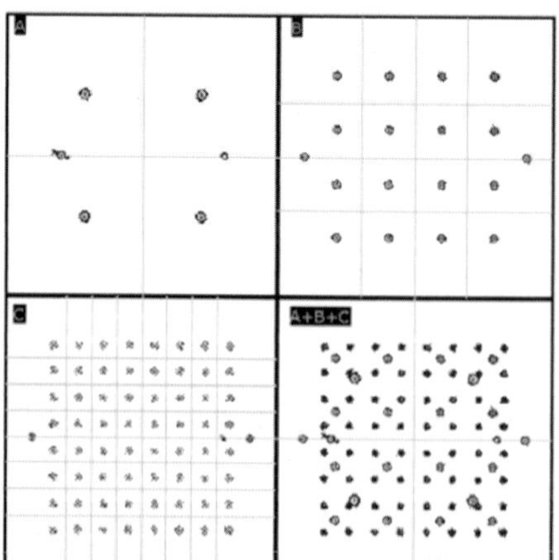

Fig. 25.9. ISDB-T constellation diagrams in the 3 different layers [ETL]

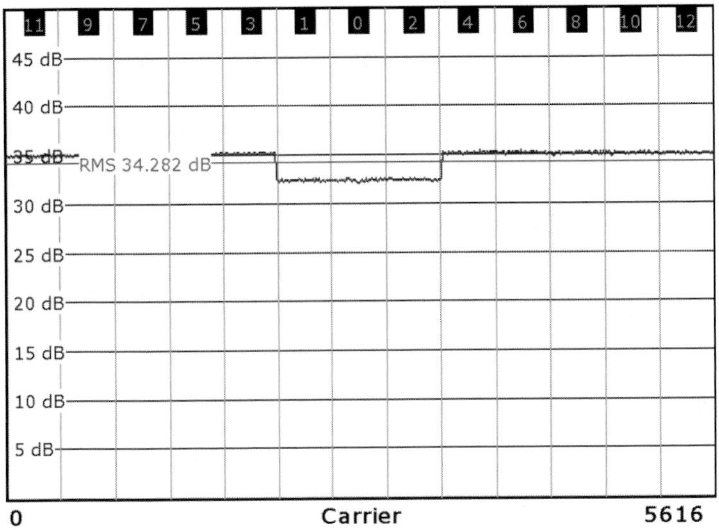

Fig. 25.10. MER(f) measurement in ISDB-T

Pass	Limit <	Results			Unit
		Layer A	Layer B	Layer C	
MER (rms)	24.0	32.3	35.0	35.0	dB
MER (Peak)	10.0	21.5	23.2	23.2	dB
MER (total,rms)	24.0	34.6			dB
MER (total,peak)	10.0	18.1			dB
MER (TMCC,rms)	24.0	36.0			dB
MER (AC,rms)	24.0	35.7			dB
	Limit <	Results <		Limit	Unit
Carrier Suppression	10.0	-----			dB
Carrier Phase		-----			deg
Amplitude Imbalance	-2.00	-----		2.00	%
Quadrature Error	-2.00	-----		2.00	deg

Fig. 25.11. ISDB-T measurement list [ETL]

25.8 Summary

In addition to the 6 MHz channel normally used in Japan, ISDB-T is also defined for 7 and 8 MHz channels. However, it is doubtful that it will be widely used in 7 MHz and 8 MHz countries because DVB-T has found wide-spread acceptance, in the meantime.

ISDB-T is certainly the more flexible standard and, because of the possibility of long time interleaving, also the more robust standard in some applications.

By its very nature, ISDB-T has SFN capability, of course, and single frequency networks are also being formed.

Brazil is the first country outside Japan which has decided to adopt ISDB-T. In Brazil, MPEG-4 AVC is used as baseband coding, plus MPEG-4 LC AAC or MPEG-4 HE AAC. The Brazilian terrestrial digital TV standard is called SBTVD, i.e. Sistema Brasileiro de Televisao Digital.

Bibliography: [ISDB-T], [SFU], [ETL]

26 Digital Audio Broadcasting - DAB

Although DAB (Digital Audio Broadcasting) was introduced back in the early days of the nineties, well before DVB, it is still relatively unknown to the public in many countries and it is only in a few countries such as, e.g. the UK that some measure of success of DAB in the market can be registered. This chapter deals with the principles of the digital sound radio standard DAB.

At first, however, let us consider the history of sound radio. The age of the transmission of audio signals for broadcasting purposes began in the year 1923 with medium-wave broadcasting (AM). In 1948, the first FM transmitter was taken into operation, developed and manufactured by Rohde&Schwarz. The first FM home receivers were also developed and produced by Rohde&Schwarz. 1983 was the year when everyone took the step from analog audio to digital audio with the introduction of the compact disk, the audio CD. In 1991, digital audio signals intended for the public at large were broadcast for the first time via satellite in Europe, DSR (Digital Satellite Radio). This method, operating without compression, did not last long, however, and was little known in public. 1993 then, ADR (Astra Digital Radio) started operation which is broadcast on subcarriers of the ASTRA satellite system on which analog TV programs are also transmitted. The MUSICAM method, used up to the present for audio compression in MPEG-1 and MPEG-2 layer II and is also used in DAB or, to put it more precisely, was developed for DAB as part of the DAB project, was laid down in 1989. Digital Audio Broadcasting, DAB, was developed at the beginning of the nineties and used the then revolutionary new techniques of MPEG-1 and MPEG-2 audio and the COFDM (Coded Orthogonal Frequency Division Multiplex) modulation method. In the mid-nineties then the DVB-S, DVB-C and DVB-T standards for digital television were finalized and thus the age of digital television had also begun. Since 2001, there is a further digital sound radio standard DRM (Digital Radio Mondiale), intended for digital short- and medium wave use, which is also based on COFDM but uses MPEG-4 AAC audio coding.

The first DAB pilot test was carried out in 1991 in Munich. Germany currently has a DAB coverage of about 80%, mainly in Band III. There are also L-Band transmitters for local programs. As ever, DAB is almost un-

W. Fischer, *Digital Video and Audio Broadcasting Technology*, Signals and Communication Technology, 3rd ed., DOI: 10.1007/978-3-642-11612-4_26, © Springer-Verlag Berlin Heidelberg 2010

known to the public in Germany, one of the reasons being that there have been no receivers available for a long time, and also that the contents broadcast do not really cover the variety obtainable in FM radio. None of this has anything to do with technical reasons, however. In the UK, DAB was greatly expanded in 2003/2004, Singapore has 100% coverage, Belgium 90%. DAB is broadcast in France, Spain, Portugal and Canada. DAB activities are taking place in 27 countries and DAB frequencies are available in 44 countries.

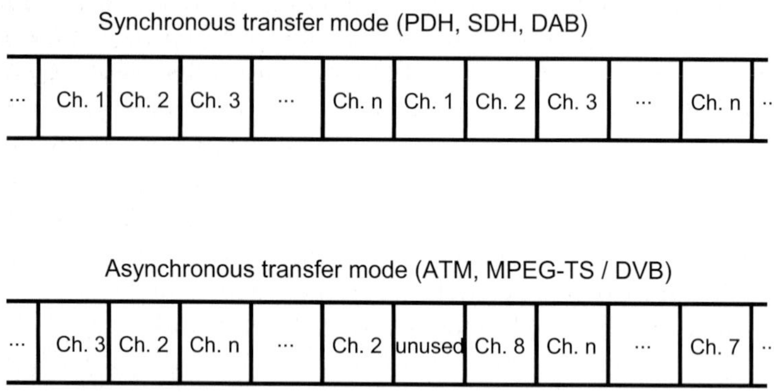

Fig. 26.1. Synchronous and asynchronous transfer mode

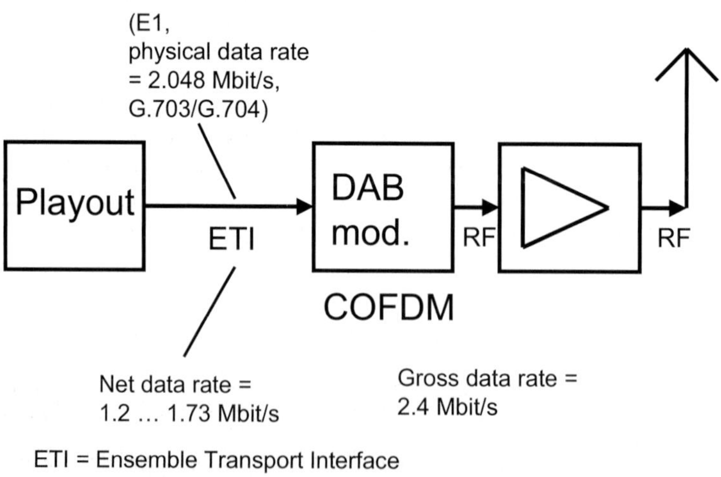

Fig. 26.2. DAB transmission link

26.1 Comparing DAB and DVB

In a comparison of DAB and DVB, the basic characteristics of both methods will first be compared, pointing out properties and differences. In principle, it is possible to transmit data synchronously or asynchronously (Fig. 16.1.). In synchronous transmission, the data rate is constant for each data channel and the time slots of the individual data channels are fixed. In asynchronous transmission, the data rate of the individual data channels can be constant or it can vary. The time slots have no fixed allocation. They are allocated as required and their order in the individual channels can thus be completely random. Examples of synchronous data transmission are PDH (Plesiochronous Digital Hierarchy), SDH (Synchronous Digital Hierarchy) and DAB (Digital Audio Broadcasting). Examples of asynchronous data transmission are ATM (Asynchronous Transfer Mode) and the MPEG-2 transport stream/Digital Video Broadcasting (DVB).

DAB is a completely synchronous system, a completely synchronous data stream being produced right back in the playout center, i.e. at the point where the DAB multiplex signal is generated. The data rates of the individual contents are constant and are always a multiple of 8 kbit/s. The time slots in which the contents from the individual sources are transmitted are permanently allocated and vary only when there is a complete change in the multiplex, i.e. in the composition of the data stream. The data signal coming from the multiplexer which is supplied to the DAB modulator and transmitter is called ETI (Ensemble Transport Interface) (Fig. 26.2.). The multiplexed data stream, or multiplex, itself is called ensemble. The ETI signal uses E1 transmission paths known from telecommunication which have a physical data rate of 2.048 Mbit/s. E1 would correspond to 30 ISDN channels and 2 signalling channels of 64 kbit/s each, also called G.703 and G.704 interface. Physically, these are PDH interfaces but DAB uses a different protocol. Although the physical data rate is 2048 kbit/s, the actual net data rate of the DAB signal transported across it is between (0.8) 1.2 ... 1.73 Mbit/s. The ETI signal is transmitted either without error protection, or with a Reed Solomon error protection code which, however, is removed again at the input of the DAB modulator. The error protection of the DAB system itself is added only in the DAB modulator although this is often wrongly shown to be different in various references. The modulation method used in DAB is COFDM (Coded Orthogonal Frequency Division Multiplex) and the subcarriers are $\pi/4$-shift DQPSK modulated. After the error protection has been added the gross data rate of the DAB signal is 2.4 Mbit/s. A special feature of DAB consists in that the different contents can be error protected to a different degree (unequal FEC).

MPEG-2, and thus DVB, is a completely asynchronous system.

The MPEG-2 transport stream is a baseband signal forming the input signal to a DVB modulator. The MPEG-2 transport stream is generated in the playout center by encoding and multiplexing the individual programs (services) and is then supplied to the modulator via various transmission paths (Fig. 26.3.). In the DVB modulator, it must be decided how, i.e. by which transmission path, the MPEG-2 transport stream is to be emitted: terrestrial (DVB-T), by cable (DVB-C) or by satellite (DVB-S). Naturally, the transmission rates and modulation methods differ for the individual transmission methods. In DVB-T, COFDM is used in conjunction with QPSK, 16QAM or 64QAM. In DVB-C, it is either 64QAM or 256QAM depending on the type of cable link (coaxial cable or optical fibre). In DVB-S, the modulation method of choice has been QPSK because of the poor signal/noise ratio in the channel.

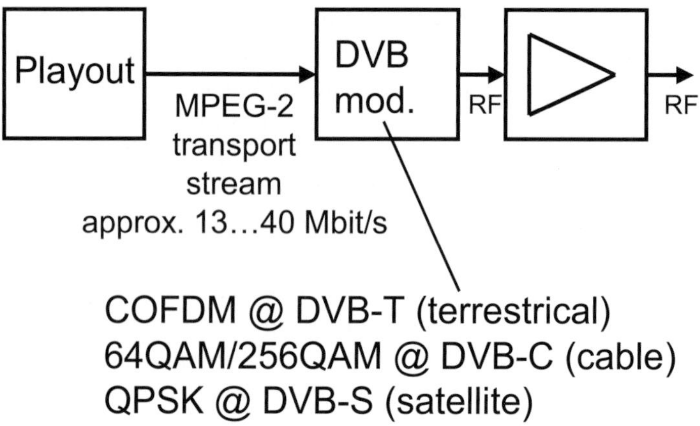

Fig. 26.3. DVB transmission link

In DVB, all contents transmitted carry the same degree of error protection (equal FEC).

As a rule, the data rate in DVB-S is about 38 Mbit/s. It only depends on the symbol rate selected and on the code rate, i.e. the error protection. Using QPSK 2 bits/symbol can be transmitted. The symbol rate is mostly 27.5 Msymbols/s. If 3/4 is selected as code rate, the resultant data rate is 38.01 Mbit/s.

If, e.g., 64QAM (coax networks) is selected in DVB-C, and a symbol rate of 6.9 Msymbols/s, the resultant net data rate is 38.15 Mbit/s.

In DVB-T, the possible data rate is between 4 and about 32 Mbit/s depending on operating mode (type of modulation - QPSK, 16QAM,

64QAM, error protection, guard interval, bandwidth). The usual data rate is, however, approx. 15 Mbit/s in applications allowing portable reception and approx. 22 Mbit/s in stationary applications with a roof antenna. A DVB-T broadcast network is designed either for portable reception or for roof antenna reception, i.e. if a roof antenna is used in a DVB-T network designed for portable reception, this will not produce an increase in data rate.

The MPEG-2 transport stream is the data signal supplied to the DVB modulators. It consists of packets with a constant length of 188 bytes. The MPEG-2 transport stream represents asynchronous transmission, i.e. the individual contents to be transmitted are keyed into the payload area of the transport stream packets purely randomly as required. The contents contained in the transport stream can have completely different data rates which do not need to be absolutely constant, either. The only rule relating to data rates is that the aggregate data rate provided by the channel must not be exceeded. And, naturally, the data rate of the MPEG-2 transport stream must correspond absolutely to the input data rate of the DVB modulators resulting from the modulation parameters.

Table 26.1. DAB and DVB comparison

	Digital Audio Broadcasting - DAB	Digital Video Broadcasting - DVB
Transfer mode	synchronous	asynchronous
Forward error correction (FEC)	unequal	equal
Modulation	COFDM mit $\pi/4$-shift DQPSK	single carrier QPSK, 64QAM, 256QAM or COFDM with QPSK, 16QAM, 64QAM
Transmission link	terrestrical	satellite, cable, terrestrical

In summary: DAB is a completely synchronous transmission system and DVB is a completely asynchronous one. Remembering this will make it easier to gain a better understanding of the characteristics of both systems

The error protection in DAB is unequal, meaning it can be selected to be different for different contents, whereas it is equal for all contents to be transmitted in DVB and, because of the asynchronous mode, could not even be selected to be different since it is not known what content is being transmitted when.

The DAB modulator demultiplexes the current content in the ETI signal and takes it into consideration. The DVB modulator is not interested in the current content transmitted. In DAB, the modulation method is COFDM with π/4-shift DQPSK. DVB uses single-carrier transmission or COFDM depending on transmission path. DAB is intended for terrestrial applications whereas DVB provides terrestrial, cable and satellite transmission standards. Satellite transmission is provided for in DAB but currently not used.

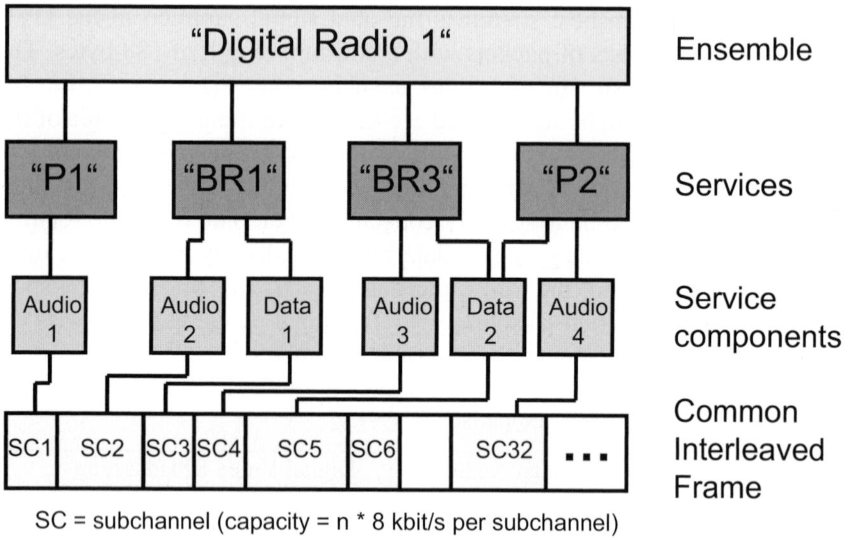

Fig. 26.4. DAB ensemble

26.2 An Overview of DAB

The following sections will provide a brief overview of DAB - Digital Audio Broadcasting. The DAB Standard is the ETSI Standard ETS300401. In the Standard, the data structure, the FEC and the COFDM modulation of the DAB Standard is described. In addition, the ETI (Ensemble Transport Interface) supply signal is described in ETS300799, and in ETS300797 the supply signals for the ensemble multiplexer STI (Service Transport Interface) are described. A further important document is TR101496 which contains guidelines and rules for the implementation and operation of DAB. Furthermore, ETS301234 describes how multimedia objects (data broadcasting) can be transmitted in DAB.

Fig. 26.4. shows an example of the composition of a multiplexed DAB data stream. The term "Ensemble" covers several programs which are combined to form one data stream. In the present case, the ensemble given the exemplary name "Digital Radio 1" is composed of 4 programs, the so-called services, here having the designations "P1", "BR1", "BR3" and "P2". These services, in turn, can be composed of a number of service components. A service component can be, e.g. an audio stream or a data stream. In the example, the service "P1" contains an audio stream, "Audio1". This audio stream is physically transmitted in subchannel SC1. "BR" is composed of an audio stream "Audio2" and a data stream "Data1" which are broadcast in subchannels SC2 and SC3. Each subchannel has a capacity of n • 8 kbit/s. The transmission in the subchannels is completely synchronous, i.e. the order of the subchannels is always the same and the data rates in the subchannels are always constant. All subchannels together - up to a possible maximum of 64 - result in the so-called Common Interleaved Frame. Service components can be associated with a number of services, e.g. as in the example "Data2".

During their transmission in the DAB system, the different subchannels can be provided with different degrees of error protection (unequal FEC).

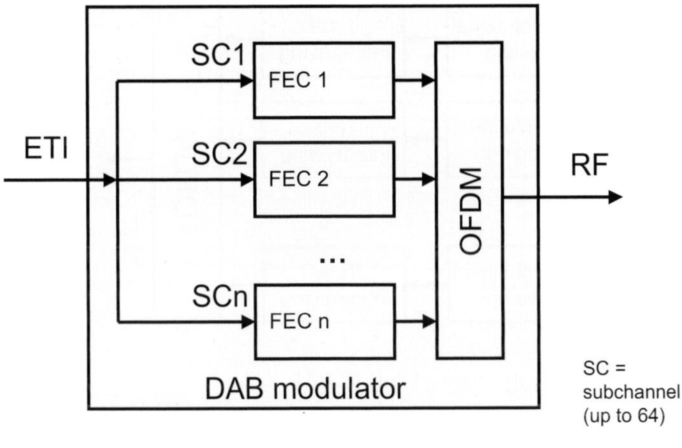

Fig. 26.5. DAB modulator

The data stream generated in the DAB multiplexer is called ETI (Ensemble Transport Interface). It contains all programs and contents to be broadcast later via the DAB transmitter. The ETI signal can be supplied to the modulator from the playout center, e.g. via optical fiber links via existing telecommunication networks or by satellite. A suitable link for this purpose is an E1 link having a data rate of 2.048 Mbit/s.

In the DAB modulator, the COFDM is carried out (Fig. 26.5.). The data stream is first provided with error protection and then COFDM-modulated. After the modulator, the RF signal power is amplified and then radiated via the antenna.

In DAB, all subchannels are error-protected individually and to different degrees. Up to 64 subchannels are possible. The FEC is provided in the DAB modulator. In many block diagrams, FEC is often described in conjunction with the DAB multiplexer which, although it is not wrong in principle, does not correspond to reality. The DAB multiplexer forms the ETI data signal in which the subchannels are transmitted synchronously and unprotected.

The ETI, however, carries the information about how much protection is to be provided for the individual channels. The ETI data stream is then split up in the DAB modulator and each subchannel is then error-protected to a different degree in accordance with the signalling in the ETI. The subchannels provided with FEC are then supplied to the COFDM modulator.

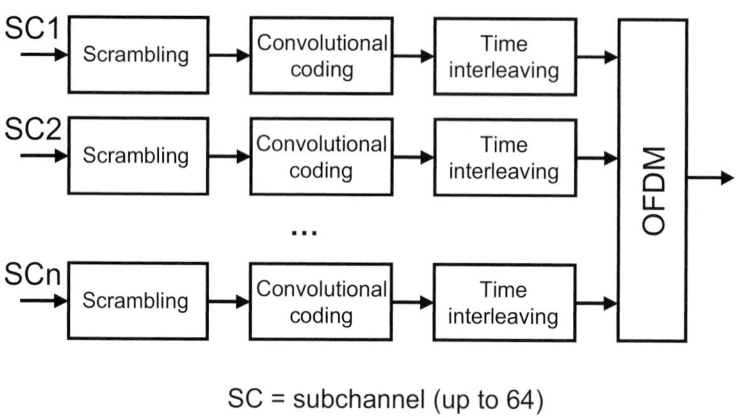

Fig. 26.6. Forward error correction (FEC) in DAB

The error protection in DAB (Fig. 16.6.) is composed of scrambling followed by convolutional coding. In addition, the DAB signal is then subjected to long time interleaving, i.e. the data are interleaved over time so that they are more resistant to block errors during the transmission. Each subchannel can be error-protected to a different degree (unequal forward error correction). The data from all subchannels are then supplied to the COFDM modulator which first carries out frequency interleaving and then modulates them onto a large number of COFDM subcarriers.

There are 4 different selectable modes in DAB. These modes are provided for different applications and frequency bands. Mode I is used in the VHF band and Mode II to IV are used in the L band, depending on frequency and application. The number of carriers is between 192 and 1536 and the bandwidth of the DAB signal is always 1.536 MHz. The difference between the modes is simply the symbol length and the number of subcarriers used.

Mode I has the longest symbol and the most subcarriers and thus the smallest subcarrier spacing. This is followed by Mode IV, Mode II and finally Mode III with the shortest symbol period and the least carriers and thus the largest subcarrier spacing. In principle, however, it holds true that the longer the COFDM symbol, the better the echo tolerance and the smaller the subcarrier spacing, the poorer the suitability for mobile applications.

The modes actually used in practice are Mode I for the VHF band and Mode II for the L band

Table 26.2. DAB modes

Mode	Frequency range	Subcarrier spacing [kHz]	No. of COFDM carriers	Used for	Symbol duration [µs]	Guard interval duration [µs]	Frame length
I	Band III VHF	1	1536	single-frequency network (SFN)	1000	246	96 ms 76 symbols
II	L band (<1.5 GHz)	4	384	multi-frequency network (MFN)	250	62	24 ms 76 symbols
III	L band (<3 GHz)	8	192	satellite	125	31	24 ms 152 symbols
IV	L band (<1.5 GHz)	2	768	small single-frequency network (SFN)	500	123	48 ms 76 symbols

The audio signals in DAB are coded to MPEG-1 or MPEG-2 (Layer II), i.e. compressed from about 1.5 Mbit/s to 64 ... 384 kbit/s. During this process, the audio signal is divided into 24 or 48 ms long sections which are then individually compressed, using a type of perceptual coding in which audio signal components inaudible to the human ear are omitted. These methods are based on the MUSICAM (Masking pattern adapted Universal Subband Integrated Coding And Multiplexing) principle described in ISO/IEC Standards 11172-3 (MPEG-1) and 13818-3 (MPEG-2) and actually developed for DAB as part of the DAB project. In MPEG-1 and -2 it is possible to transmit audio in mono, stereo, dual sound and joint stereo modes. The frame length is 24 ms in MPEG-1 and 48 ms in MPEG-2. These frame lengths are also found in the DAB Standard and also affect the length of the COFDM frames. The same applies as before: DAB is a completely synchronous transmission system where all processes are synchronized with one another.

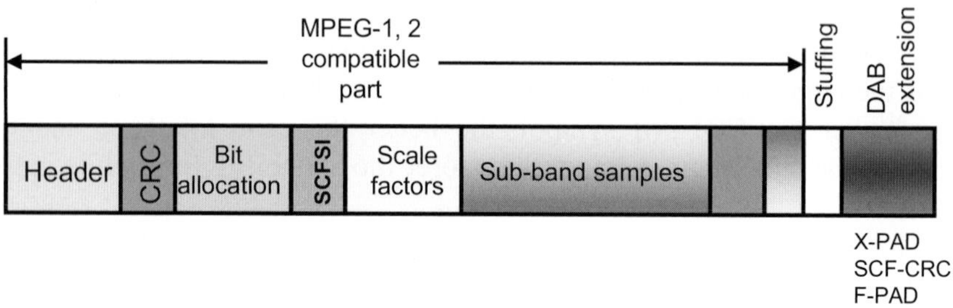

Fig. 26.7. DAB audio frame

Fig. 26.7. shows the structure of a DAB audio frame. An MPEG-1-compatible frame has a length of 24 ms. The frame begins with a header containing 32 bits of system information. The header is protected by a 16 bit long CRC checksum. This is followed by the block with the bit allocation in the individual sub-bands, followed by the scale factors and sub-band samples. In addition, ancillary data can be optionally transmitted. The sampling rate of the audio signal is 48 kHz in MPEG-1 and thus does not correspond to the 44.1 kHz of the audio CD. The data rates are between 32 and 192 kbit/s for a single channel or between 64 and 384 kbit/s for stereo, joint stereo or dual sound. The data rates are multiples of 8 kbit/s. In MPEG-2, the MPEG-1 frame is supplemented by an MPEG-2 extension. In MPEG-2 Layer II, the frame length is 48 ms and the sampling rate of the audio signal is 24 kHz.

This audio frame structure of the MPEG-1 and -2 Standards is repeated in DAB. The MPEG-1- and MPEG-2-compatible part is supplemented by a DAB extension in which program-associated data (PAD) are transmitted. Between these, stuffing bytes (padding) are used, if necessary. In the PAD, a distinction is made between the extended PAD "X-PAD" and the fixed PAD "F-PAD". Among other things, the PAD include an identifier for music/voice, program-related text and additional error protection.

DAB audio data rates normally used in practice are:

- Germany:
  mostly 192 kbit/s, PL3
  60 kbit/s or 192 kbit/s in some cases, PL4 (one additional per gram)
- UK:
  256 kbit/s, classical music
  128 kbit/s, popular music,
  64 kbit/s, voice

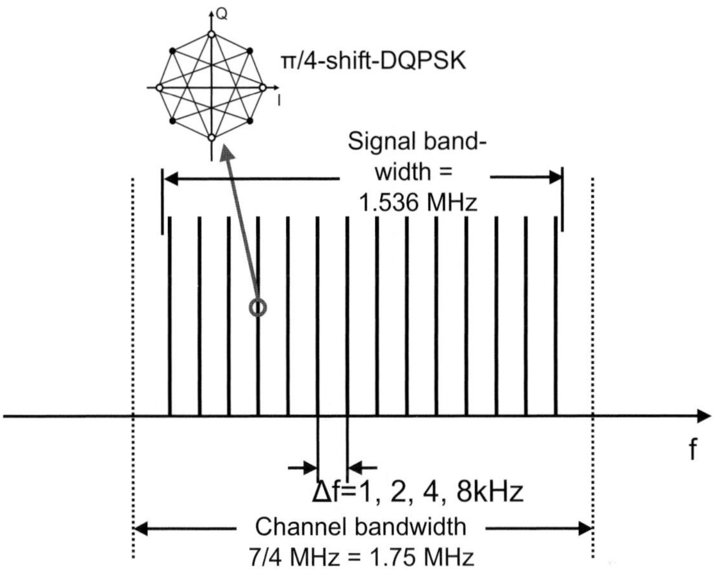

Fig. 26.8. DAB COFDM channel

26.3 The Physical Layer of DAB

In the section following, the implementation of COFDM in Digital Audio Broadcasting DAB will be discussed in detail. The main item of concern are the DAB details at the modulation end. COFDM is a multicarrier transmission method in which, in the case of DAB, between 192 and 1536 carriers are combined to form one symbol. Due to DQPSK, each carrier can carry 2 bits in DAB. A symbol is the superposition of all these individual carriers. A guard interval with a length of about 1/4 of the symbol length is added to the symbol which has a length of between 125 μs and 1 ms. In the guard interval, the end of the following symbol is repeated where echoes due to multipath reception can "wear themselves down". This prevents intersymbol interference as long as a maximum echo interval is not exceeded.

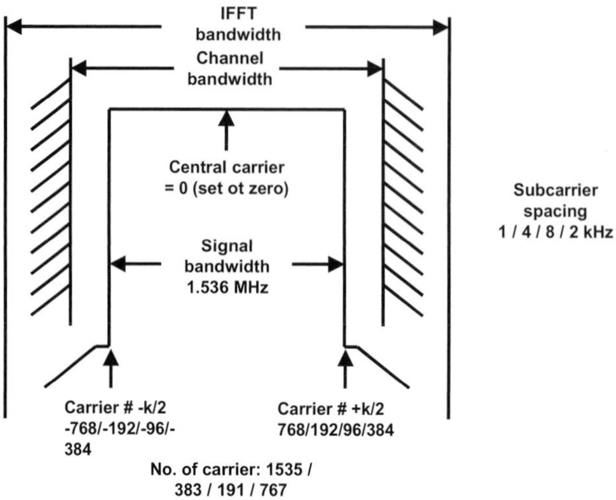

Fig. 26.9. DAB spectrum

Instead of one carrier, COFDM involves hundreds to thousands of subcarriers in one channel (Fig. 26.8.). The carriers are equidistant from one another. All carriers in DAB are $\pi/4$-shift DQPSK (Differential Quadrature Phase Shift Keying) modulated. The bandwidth of a DAB signal is 1.536 MHz, the channel bandwidth available, e.g. in VHF Band 12 (223 ... 230 MHz) is 1.75 MHz which corresponds to exactly 1/4 of a 7 MHz channel.

Firstly, however, let us turn to the principle of differential QPSK: The vector can take up four positions, which are 45, 135, 225 and 315 degrees. However, the vector is not mapped in absolute values but differentially. I.e., the information is contained in the difference between one symbol and the next. The advantage of this type of modulation lies in the fact that no channel correction is necessary. It is also irrelevant how the receiver is locked in phase, the decoding will always operate correctly. There is also a disadvantage, however: the arrangement requires a signal to noise ratio which is better by about 3 dB than in the case of absolute mapping (coherent modulation) since in the case of an errored symbol, the difference with respect to the preceding symbol and the following symbol is false and will lead to bit errors. Any interference event will then cause 2 bit errors.

In reality, however, DAB does not use DQPSK but π/4-shift DQPSK, which will be discussed in detail later. Many references wrongly mention only DQPSK in DAB. If the DAB standard is analyzed in detail, however, and especially the COFDM frame structure, this special type of DQPSK is encountered automatically via the phase reference symbol (TFPR).

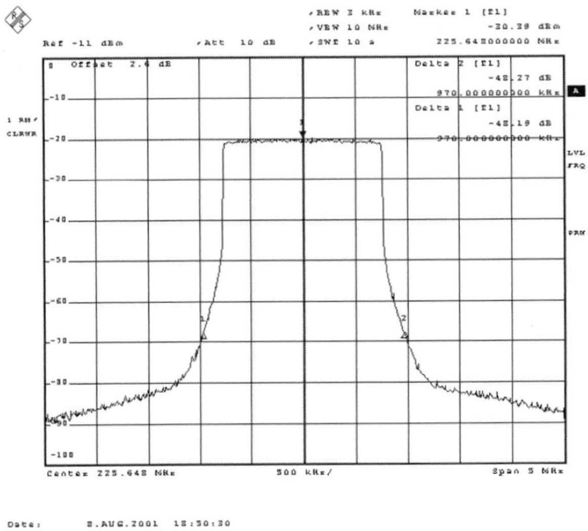

Fig. 26.10. Real DAB spectrum after the mask filter

COFDM signals are generated with the aid of an Inverse Fast Fourier Transform (IFFT) (s. COFDM chapter) which requires a number of carriers corresponding to a power of two. In the case of DAB, either a 2048-point IFFT, a 512-point IFFT, a 256-point IFFT or a 1024-point IFFT is performed. The cumulative IFFT bandwidth of all these carriers is greater than the channel bandwidth but the edge carriers are not used and are set to zero (guard band), making the actual bandwidth of DAB 1.536 MHz. The channel bandwidth is 1.75 MHz. The subcarrier spacing is 1, 4, 8 or 2 kHz depending on DAB mode (Mode I, II, II or IV) (see Fig. 26.8. and 26.9.).

Fig. 26.10. shows a real DAB spectrum as it would be measured with a spectrum analyzer at the transmitter output after the mask filter. The width of the spectrum is 1.536 MHz. There are also signal components which extend into the adjacent channels, the relevant terms being shoulders and shoulder attenuation. The shoulders are lowered by using mask filters.

In DAB, a COFDM frame (Fig. 26.11.) consists of 77 COFDM symbols. The length of a COFDM symbol depends on the DAB mode and is between 125 μs and 1 ms, to which the guard interval is added which is about 1/4 of the symbol length. The total length of a symbol is thus between about 156 μs and 1.246 ms. Symbol No. 0 is the so-called null symbol. During this time, the RF carrier is completely gated off. The null symbol starts the DAB frame and is followed by the time frequency phase reference (TFPR) used for frequency and phase synchronisation in the receiver. It does not contain any data.

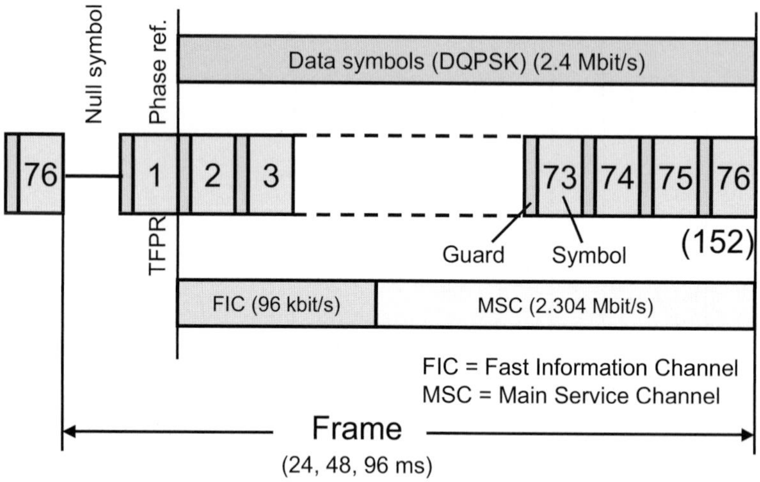

Fig. 26.11. DAB frame

All COFDM carriers are set to defined amplitude and phase values in the phase reference symbol. The actual data transmission starts with the second symbol. In contrast to DVB, the data stream in DAB is completely synchronous with the COFDM frame. In the first symbols of the DAB frame, the Fast Information Channel (FIC) is transmitted, the length of which is dependent on the DAB mode. The data rate of the FIC is 96 kbit/s. In the FIC, important information for the DAB receiver is transmitted. Following the FIC, the transmission of the Main Service Channel (MSC) starts in which the actual payload data are found. The data rate of the MSC is a constant 2.304 Mbit/s and is mode-independent. Both FIC and MSC additionally contain FEC gated in by the DAB COFDM modulator. The FEC in DAB is very flexible and can be configured differently for the various subchannels, resulting in net data rates of (0.8) 1.2 to 1.73 Mbit/s for the actual payload (audio and data). The type of modulation used in DAB is differential QPSK. The aggregate gross data rate of FIC and MSC is 2.4 Mbit/s. The length of a DAB frame is between 24 and 96 ms (mode-dependently).

In the further description, the details of the COFDM implementation in DAB will be discussed in greater detail. In DAB, a COFDM frame starts with a null symbol. All carriers are simply set to zero in this symbol. However, Fig. 26.12. only shows a single carrier over a number of symbols. The first symbol shown at the left-hand edge of the picture is the null symbol where the vector has the amplitude zero. This is followed by the phase reference symbol to which the phase of the first data symbol (symbol no. 2) i referred. The difference between the phase reference symbol and symbol no. 2 and, continuing from there, the difference between two adjacent symbols provides the coded bits. I.e., the information is contained in the phase change.

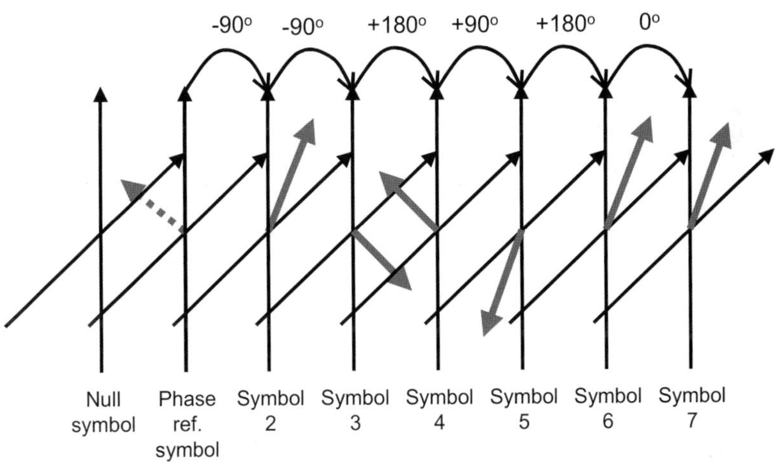

Fig. 26.12. DQPSK sequence with null symbol and phase reference symbol

The principle shown in Fig. 26.12. does still not correspond to the precise reality in DAB which, however, we are approaching step by step.

Fig. 26.12. shows the mapping and the state transitions in the case of simple QPSK or simple DQPSK. It can be seen clearly that phase shifts of +/-90 degrees and +/-180 degrees are possible. In the case of +/-180 degree phase shifts, however, the voltage curve passes through zero which leads to the envelope curve being pinched in. In single carrier methods it is usual, therefore, to carry out so-called π/4-shift DQPSK instead of DQPSK, thus avoiding this problem. In this type of modulation, the carrier phase is shifted by 45 degrees from phase to phase, i.e. by π/4. The receiver is informed about this and cancels out this process. An example of π/4-shift DQPSK is the TETRA mobile radio standard. In DAB, too, this modulation method was adopted, but in this case in conjunction with the COFDM multicarrier method.

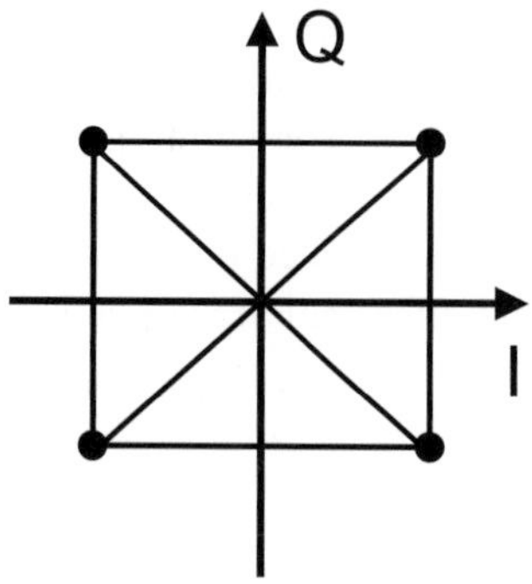

Fig. 26.13. Mapping of an "normal" QPSK or a "normal" DQPSK, with state transitions which also pass trough the zero point

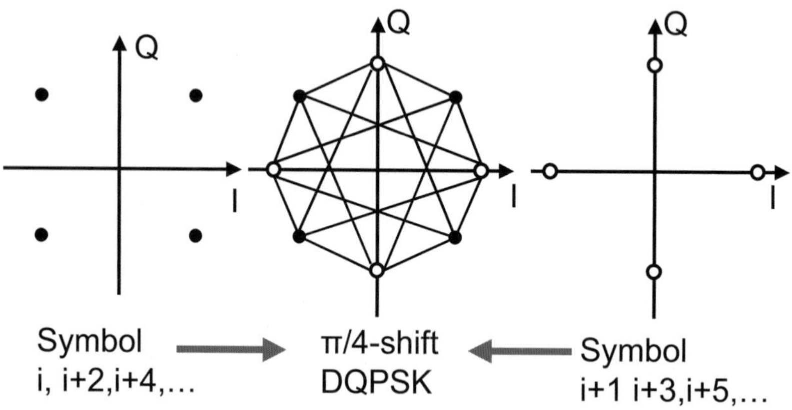

Fig. 26.14. Transition from DQPSK to π/4-shift DQPSK

Considering now the transition from DQPSK to π/4-shift DQPSK (Fig. 26.14.). On the left, the constellation pattern of simple QPSK is shown. On the right, QPSK rotated by 45 degrees, i.e. by π/4, is shown. π/4-shift DQPSK is composed of both. The carrier phase is shifted on by 45 degrees from symbol to symbol. If only 2 bits per vector transition are to be represented, the 180-degree phase shifts can be avoided. It can be shown that phase shifts of +/-45 degrees (+/-π/4) and +/-135 degrees (+/-3/4π) are sufficient for transmitting 2 bits per symbol difference by differential mapping. The constellation pattern of π/4-shift DQPSK (Fig. 26.14., center) shows the state transitions used. It can be seen that there is no 180-degree shift.

In DAB, π/4-shift DQPSK is used in conjunction with COFDM. The COFDM frame starts with the null symbol in DAB. During this time, all carriers are set to zero, i.e. u(t) = 0 for the period of a COFDM symbol. This is followed by the phase reference symbol, or more precisely by the time frequency phase reference (TFPR) symbol where all carriers are mapped onto n•90 degrees corresponding to the so-called CAZAC (Constant Amplitude Zero Autocorrelation) sequence. This means that the carriers are mapped onto the I or Q axis differently for each carrier according to a particular pattern, i.e. assume the phase space of 0, 90, 180, 270 degrees. The phase reference symbol is the reference for the π/4-shift DQPSK of the first data symbol, i.e. symbol no. 2. The carriers in symbol No. 2 thus accupy the phase space of n•45 degrees. Symbol no. 3 gets its phase reference from symbol no. 2 and occupies the phase space of n•90 degrees etc. The same applies to all other carriers.

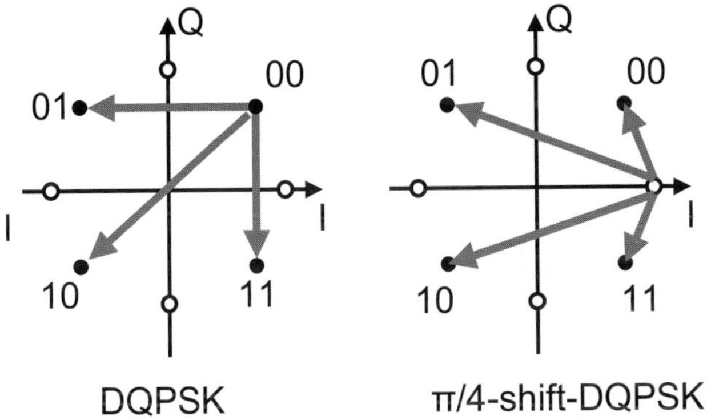

Fig. 26.15. Constellation pattern of DQPSK compared with π/4-shift DQPSK

Fig. 26.15. shows the comparison of a DQPSK with a π/4-shift DQPSK. The selected mapping rule has been selected arbitrarily here and could easily be selected differently.

If it is intended to transmit the bit combination 00 by using the DQPSK in the example, the phase angle will not change. Bit combination 01 is signalled by a +45 degrees phase shift, bit combination 11 corresponds to a –45 degrees phase shift. A 10, in turn, corresponds to 180 degrees phase shift.

In the right-hand drawing of Fig. 26.15., the state transitions of a $\pi/4$-shift DQPSK are shown with phase shifts of +/-45 degrees and +/-135 degrees. The carrier never dwells on a constant phase, neither are there any 180-degree phase shifts.

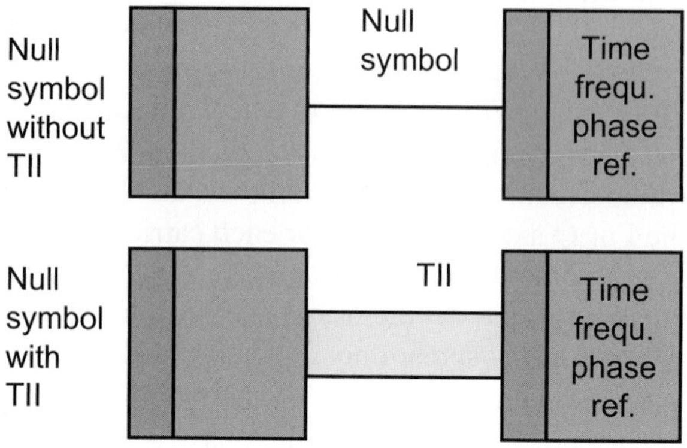

Fig. 26.16. Null symbol with and without TII

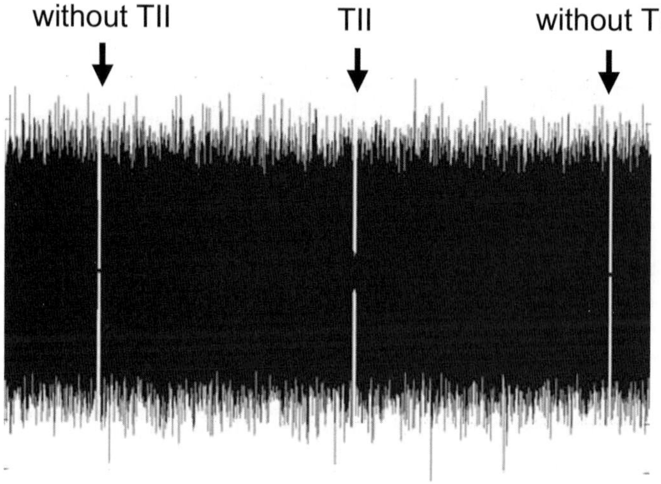

Fig. 26.17. Oscillogram of a DAB frame with null symbols (only each second null symbol includes TII = Transmitter Indentification Information)

The null symbol is the very first symbol of a DAB frame, called symbol no. 0 in numerical order. During this time, the amplitude of the COFDM signal is zero. The length of a null symbol corresponds approximately to the length of a normal symbol plus guard interval. In reality, however, it is slightly longer because it is used for adjusting the DAB frame length to exactly 14, 48 or 96 ms to match the frame length of the MPEG-1 or –2 audio layer II. The null symbol marks the beginning of a DAB COFDM frame. It is the first symbol of this frame and can be easily recognized since all carriers are zeroed during this time (Fig. 26.16., Fig. 26.17., Fig. 26.19. and Fig. 26.20.). It is thus used for roughly synchronizing the receiver timing. During the null symbol, a transmitter ID, a so-called TII (transmitter identification information) (Fig. 26.16. and Fig. 26.17.), can also be transmitted. In the case of a TII, certain carrier pairs in the null symbol are set and can be used for signalling the transmitter ID (Fig. 26.18.).

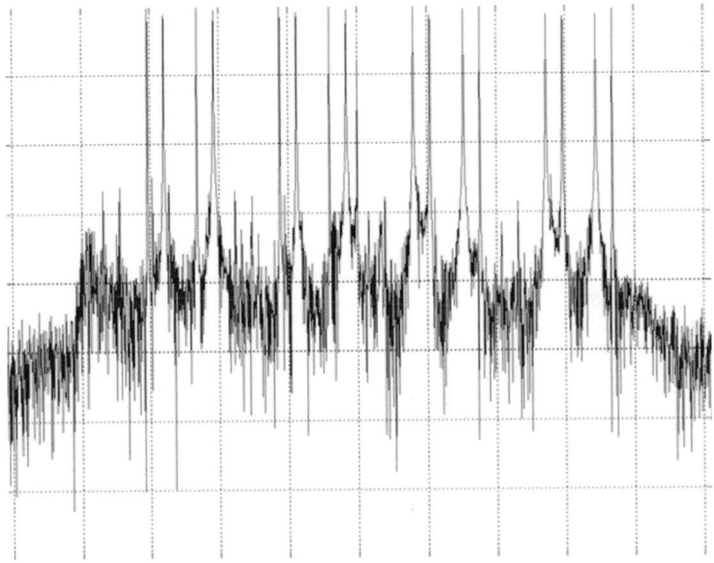

Fig. 26.18. FFT of a null symbol with TII; carrier pairs are set for signalling the TII Main ID and TII Sub-ID

The frame lengths, the symbol lengths and thus also the zero symbol lengths depend on the DAB mode and are listed in Table 26.2.

The phase reference symbol or TFPR (Time Frequency Phase Reference) symbol is the symbol following directly after the null symbol. Within this symbol, all carriers are set to certain fixed phase positions according to the CAZAC (Constant Amplitude Zero Autocorrelation) sequence. This symbol is used for receiver AFC (automatic frequency control), on the one hand, and, on the other hand as starting phase reference for the π/4-shift DQPSK.

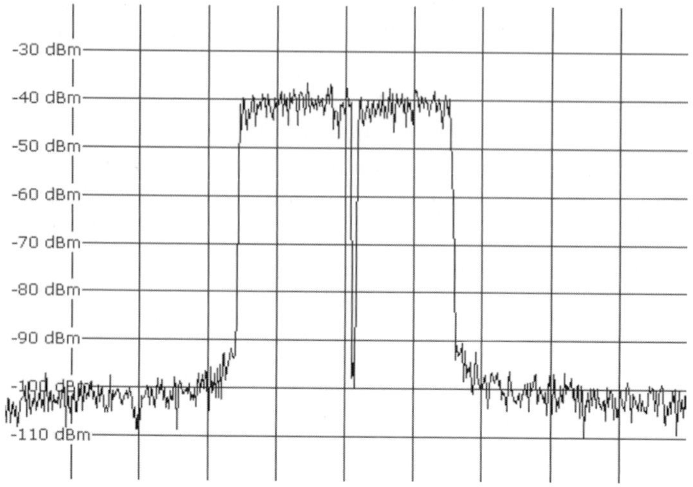

Fig. 26.19. Spectrum of DAB signal with the null symbol running through

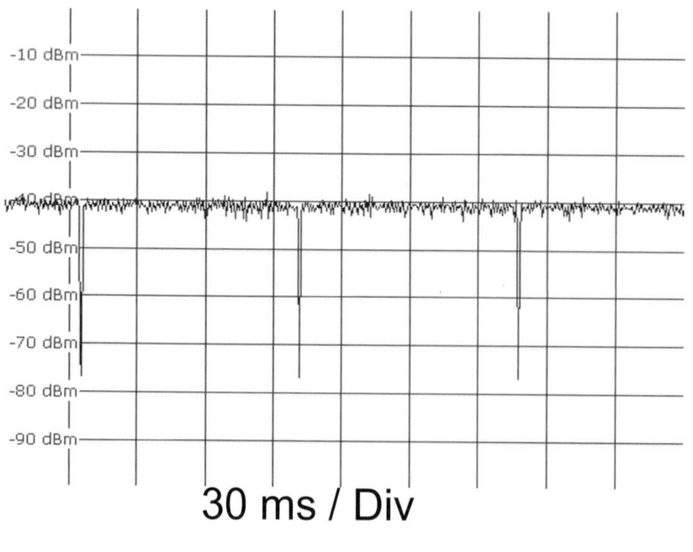

Fig. 26.20. Spectrum of a DAB signal with zero span; the DAB frame with the null symbol can easily be seen (Mode I)

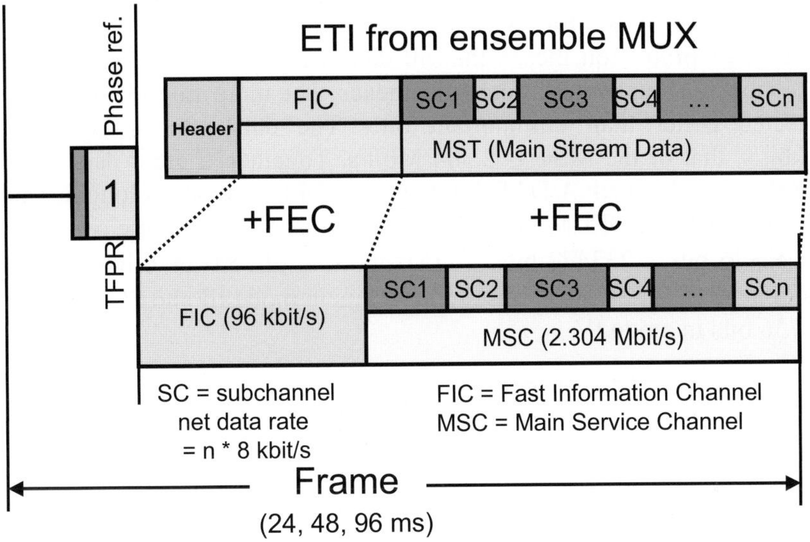

Fig. 26.21. DAB frame

The receiver can also use this symbol for calculating the impulse response of the channel in order to carry out accurate time synchronisation, among other things for positioning the FFT sampling window in the receiver. The impulse response allows the individual echo paths to be identified. During the TFPR symbol, the carriers are set to 0, 90, 180 or 270 degrees, differently for each carrier. The relevant rule is defined in tables in the standard (CAZAC Sequence).

Returning now to the DAB data signal, the gross data rate of a DAB channel is 2.4 Mbit/s. Subtracting the FIC (Fast Information Channel) which is used for receiver configuration, and the error protection (convolutional coding), a net data rate of (0.8) 1.2 … 1.73 Mbit/s is obtained. In contrast to DVB, DAB operates completely synchronously. Whereas in DVB-T, no COFDM frame structure can be recognized in the data signal, the MPEG-2 transport stream, the DAB data signal also consists of frames. A DAB COFDM frame (Fig. 26.21.) begins with a null symbol.

During this time, the RF signal is zeroed. This is followed by the reference symbol. There is no data transmission during the time of the null symbol and the reference symbol. Data transmission starts with COFDM symbol no. 2 with the transmission of the FIC (Fast Information Channel), followed by the MSC, the Main Service Channel. The FIC and MSC already contain error protection (FEC) inserted by the modulator. The error protection used in the FIC is equal and that used in the MSC is unequal. Equal error protection means that all data are provided with equal error protection, unequal error protection means that more important data are protected better than unimportant ones. The data rate of the FIC is 96 kbit/s, that of the MSC is 2.304 Mbit/s. Together, a gross data rate of 2.4 Mbit/s is obtained. A DAB frame is 77 COFDM symbols long in mode I, II, IV and 153 COFDM symbols long in mode III. The frame consists of 1536•2•76 bits = 233472 bits in DAB mode I, of 384•2•76 bits = 58638 bits in mode II, 152•2•151 bits = 57984 bits in mode III and 768•76 bits = 116736 bits in mode IV.

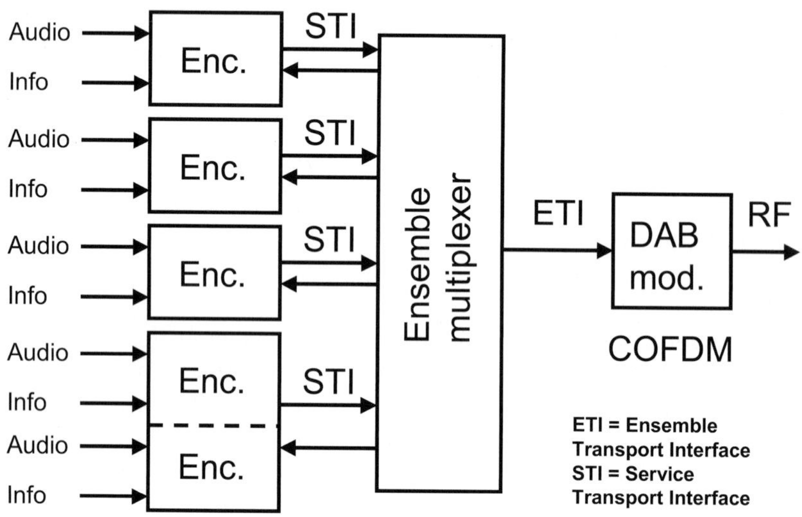

Fig. 26.22. DAB feed via ETI

The DAB data are fed from the ensemble multiplexer to the DAB modulator and transmitter via a data signal called ETI (Ensemble Transport Interface) (Fig. 27.22.). The data rate of the ETI signal is lower than that of the DAB frame since it does not yet contain error protection. Error protection is only added in the modulator (convolutional coding and interleaving). However, the ETI signal already contains the frame structure of DAB (Fig. 26.21.). An ETI frame starts with a header. This is followed by the data of the Fast Information Channel (FIC). After that comes the mainstream (MST). The mainstream is sub divided into subchannels. Up to 64 subchannels are possible. The information about the structure of the mainstream and the error protection to be added in the modulator is found in the Fast Information Channel (FIC). The FIC is intended for automatic configuration of the receiver.

The modulator obtains its information for the composition and configuration of the multiplexed data stream from the ETI header however.

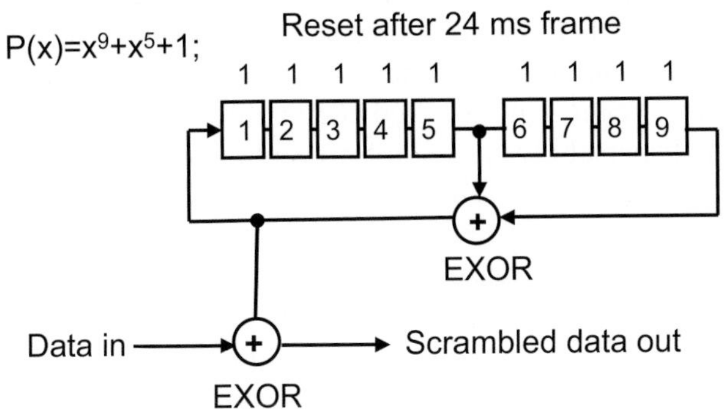

Fig. 26.23. Data scrambling

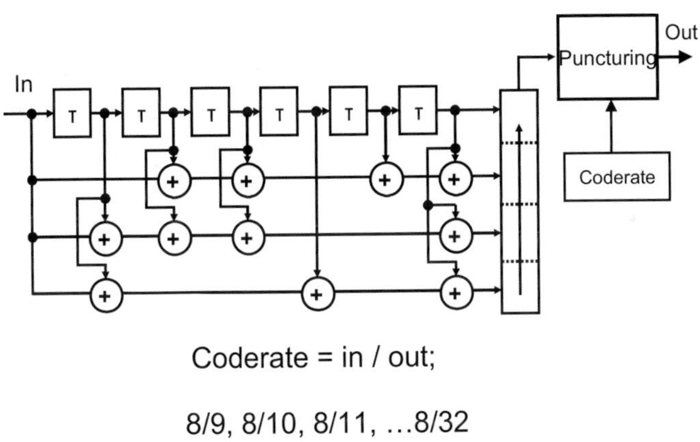

Fig. 26.24. DAB convolutional coding with puncturing

26.4 DAB – Forward Error Correction

In this section, error protection, the Forward Error Correction (FEC) used in DAB will be discussed in greater detail.

In DAB, all subchannels are error protected individually and to different degrees (Fig. 26.5. and 26.6.). Up to 64 subchannels are possible. Error protection (FEC) is carried out in the DAB modulator.

Before the data stream is provided with error protection, it is scrambled (Fig. 26.23.). This is done by mixing with a pseudo random binary sequence (PRBS). The PRBS is generated with the aid of a shift register with feedback. The data stream is then mixed with this PRBS by using an exclusive-OR gate. This breaks up long sequences of ones and zeroes which maybe present in the data stream. This is called energy dispersal. In single-carrier methods, energy dispersal is required for preventing the carrier vector from staying at constant positions. This would lead to discrete spectral lines. But error protection, too, only operates correctly if there is movement in the data signal. This is the reason why this scrambling is carried out at the beginning of the FEC also in the COFDM method. Every 24 ms, the shift register arrangement is loaded with all ones and thus reset.

Such an arrangement is also found in the receiver and must be synchronised with the transmitter. Mixing again with the same PRBS in the receiver restores the original data stream.

This is followed by the convolutional coding. The convolutional coder used in DAB (followed by puncturing) is shown in Fig. 26.24. The data signal passes through a 6-stage shift register. In parallel to this, it is exclusive-OR-ed with the information stored in the shift registers at different delay times in three branches. The shift register content delayed by six clock cycles and the three data signals manipulated by EXOR operations are serially combined to form a new data stream now having four times the data rate of the input data rate.

This is called ¼ code rate. The code rate is the ratio of input data rate to output data rate.

After the convolutional coding, the data stream had been expanded by a factor of four. However, the output data stream now carries 300% overhead, i.e. error protection. This lowers the available net data rate. This overhead, and thus the error protection, can be controlled in the puncturing unit. The data rate can be lowered again by selectively omitting bits. Omitting, i.e. puncturing, is done in accordance with a scheme known to the transmitter and the receiver: a puncturing scheme. The code rate describes the puncturing and thus provides a measure for the error protection. The code rate is simply calculated from the ratio of input data rate to output data rate. In DAB, it can be varied between 8/9, 8/10, 8/11...8/32. 8/32 provides the best error protection at the lowest net data rate, 8/9 provides the lowest error protection at the highest net data rate. Various data contents are protected to a different degree in DAB. Frequently, however, burst errors occur during a transmission. If the burst errors last longer, the error protection will fail. For this reason, the data are interleaved in a further operating step, i.e. distributed over a certain period of time. Long interleaving over 384 ms makes the system very robust and suitable for mobile use. During the de-interleaving at the receiving end, any burst errors which maybe present are then broken up and distributed more widely in the data stream. It is now easier to repair these burst errors, which have become single errors, and this without any additional data overhead. In DAB there are two types of error protection being used, namely equal error protection and unequal error protection.

Equal error protection means that all components are provided with the same FEC overhead. This applies to the Fast Information Channel (FIC) and to the case of pure data transmission.

Audio contents, i.e. the components of an MPEG-1 or -2 audio frame carry unequal protection. Some components in the audio frame are more important because bit errors would cause greater disruption there and these parts are protected more therefore. These different components in the audio frame are provided with different code rates.

In many transmission methods, constant equal error protection is used. An example of this is DVB. In DAB, only parts of the information to be transmitted are provided with equal error protection. This includes the following data: the FIC is protected equally with a mean code rate of 1/3. The data of the packet mode can be provided with a code rate of 2/8, 3/8, 4/8 or 6/8.

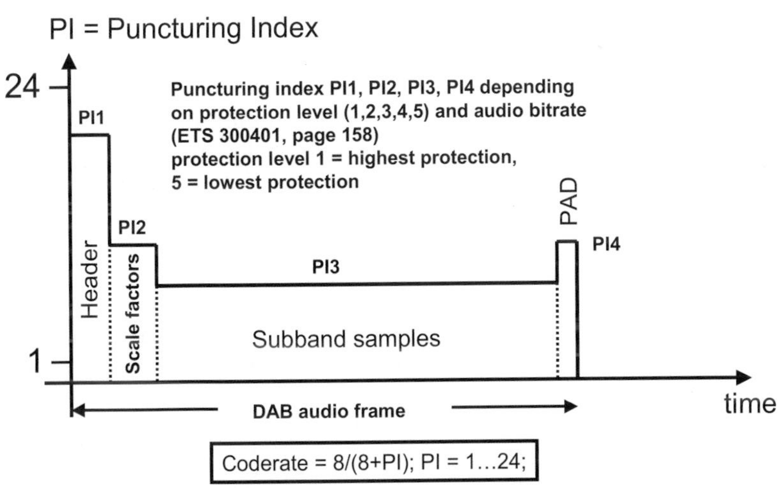

Fig. 26.25. Unequal forward error correction of a DAB audio frame

The MPEG audio packets are protected with unequal error protection which is also controllable in DAB. Some components of the MPEG audio packet are more sensitive to bit errors than other ones.

The components in the DAB audio frame which are provided with different error protection are:

- Header
- Scale factors
- Subband samples
- Program-associated data (PAD)

The header must be protected particularly well. If errors occur in the header, this will lead to serious synchronization problems. The scale factors must also be well protected since bit errors in this area would make for very unpleasant listening. The subband samples are less sensitive and their error protection is correspondingly lower.

Fig. 26.25. shows an example of the unequal error protection within a DAB audio frame. The puncturing index describes the quality of the error protection. From the puncturing index, the code rate in the relevant section can be easily calculated using the following formula:

code_rate = 8/(8+PI);

where PI = 1 … 24, the puncturing index.

The puncturing index, in turn, is obtained from the protection level, which is in the range 1, 2, 3, 4 or 5, and the audio bit rate. Table 26.3. lists the mean code rates as a function of protection level and the audio bit rates. PL1 offers the highest error protection and PL5 offers the lowest error protection.

Table 26.3. DAB protection levels and mean code rates

Audio bitrate [kbit/s]	Mean code rate protection level 1	Mean code rate protection level 2	Mean code rate protection level 3	Mean code rate protection level 4	Mean code rate protection level 5
32	0.34	0.41	0.50	0.57	0.75
48	0.35	0.43	0.51	0.62	0.75
56	X	0.40	0.50	0.60	0.72
64	0.34	0.41	0.50	0.57	0.75
80	0.36	0.43	0.52	0.58	0.75
96	0.35	0.43	0.51	0.62	0.75

112	X	0.40	0.50	0.60	0.72
128	0.34	0.41	0.50	0.57	0.75
160	0.36	0.43	0.52	0.58	0.75
192	0.35	0.43	0.51	0.62	0.75
224	0.36	0.40	0.50	0.60	0.72
256	0.34	0.41	0.50	0.57	0.75
320	X	0.43	X	0.58	0.75
384	0.35	X	0.51	X	0.75

Table 26.4. shows the minimum signal/noise ratio S/N needed and the number of programs which can be accommodated in a multiplexed DAB data stream on the basis of a data rate of 192 kbit/s per program, in dependence on the protection level. If, e.g. PL3 is used, 6 programs of 196 kbit/s each can be accommodated in a multiplexed DAB data stream and the minimum signal/noise ratio then needed is 11 dB. The gross data rate of the DAB signal (including error protection) is 2.4 Mbit/s and the net data rate is between (0.8) 1.2 and 1.7 Mbit/s depending on the error protection selected.

Table 26.4. DAB channel capacity and minimum S/N

Protection level (FEC)	No. of programs at 196 kbit/s	S/N [dB]
PL1 (highest)	4	7.4
PL2	5	9.0
PL3	6	11.0
PL4	7	12.7
PL5 (lowest)	8	16.5

Table 26.5. DAB parameters and quality

Program type	Format	Quality	Sampling rate [kHz]	Protection level	Bitrate [kbit/s]
music/voice	mono	broadcast	48	PL2 oder 3	112...160
music/voice	2-channel stereo	broadcast	48	PL2 oder 3	128...224
music/voice	multichannel	broadcast	48	PL2 oder 3	384...640
voice	mono	acceptable	24 oder 48	PL3	64...112
news	mono	intelligible	24 oder 48	PL4	32 or 64
data	--	--		PL4	32 or 64

The unequal error protection in DAB has the effect that the DAB receivability does not abruptly break off when the signal drops below a certain minimum S/N ratio. At first, audible disturbances arise and receivability ceases only about 2 dB later. Table 26.5. shows frequently selected protection levels and audio rates in DAB [HOEG_LAUTERBACH].

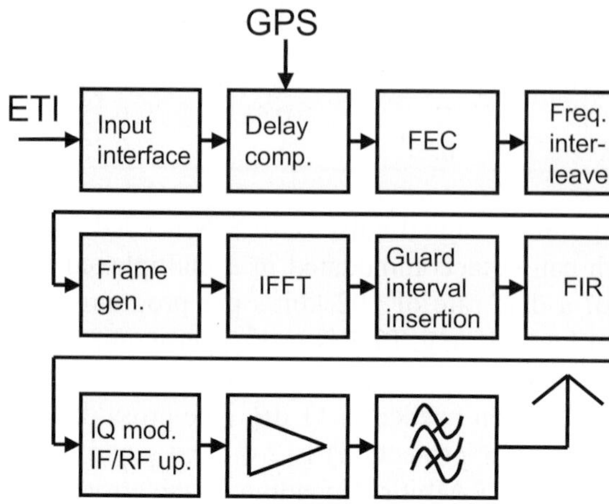

Fig. 26.26. Block diagram of a DAB modulator and transmitter

26.5 DAB Modulator and Transmitter

Let us now consider the overall block diagram of a DAB modulator (Fig. 26.26.) and transmitter. The ETI (Ensemble Transport Interface) is present at the input interface where the modulator synchronizes itself to the ETI signal. In the case of a single-frequency network, delay compensation is carried out in the modulator controlled via the TIST (Time Stamp) in the ETI signal. This is followed by the error protection (FEC) which is different for each signal content.

The error-protected data stream is then frequency interleaved, i.e. distributed. Each COFDM carrier is assigned a part of the data stream which is always 2 bits per carrier in DAB. In the differential mapper the real-and imaginary-part table is then formed, i.e. the current vector position is determined for each carrier. Following this the DAB frame with null symbol, TFPR symbol and data symbols is formed and the completed real-and imaginary-part tables are then supplied to the IFFT, the Inverse Fast Fourier Transform. After that, we are back in the time domain where the guard interval is added to the symbol by repeating the end of the symbol following.

After FIR filtering, pre-correction is carried out in the power transmitter for compensating for the non-linearities of amount and phase of the amplifier characteristic. The IQ modulator following is then usually the IF/RF up converter at the same time. Today, direct modulation is normally used, i.e. direct conversion from baseband up into RF. This is followed by power amplification in transistor output stages. The remaining non-linearities and the necessary clipping of voltage peaks to about 13 dB result in the so-called shoulders of the DAB signal. These are out-off-band components which would interfere with the adjacent channels.

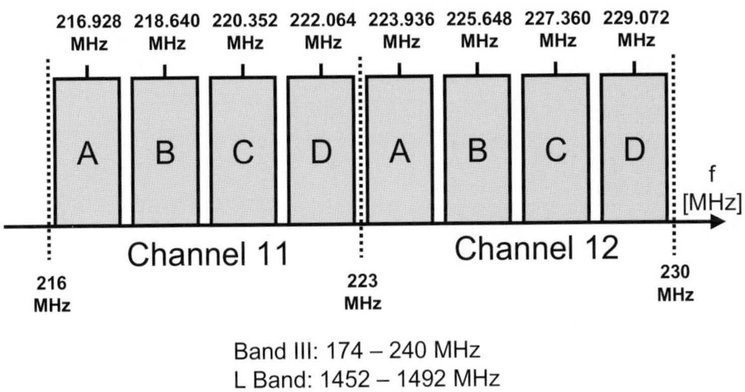

Fig. 26.27. DAB channel allocation with channel 11 and 12 as example

Table 26.6. DAB channel table band III VHF

Channel	Center frequency [MHz]
5A	174.928
5B	176.640
5C	178.352
5D	180.064
6A	181.936
6B	183.648
6C	185.360
6D	187.072
7A	188.928
7B	190.640
7C	192.352
7D	194.064
8A	195.936
8B	197.648
8C	199.360
8D	201.072
9A	202.928
9B	204.640
9C	206.352
9D	208.064

10A	209.936
10N	210.096
10B	211.648
10C	213.360
10D	215.072
11A	216.928
11N	217.088
11B	218.640
11C	220.352
11D	222.064
12A	223.936
12N	224.096
12B	225.648
12C	227.360
12D	229.072
13A	230.784
13B	232.496
13C	234.208
13D	235.776
13E	237.488
13F	239.200

Table 26.7. DAB channel table L band

Channel	Center frequency [MHz]
LA	1452.960
LB	1454.672
LC	1456.384
LD	1458.096
LF	1461.520
LG	1463.232
LH	1464.944
LI	1466.656
LJ	1468.368
LK	1470.080
LL	1471.792
LM	1473.504
LN	1475.216
LO	1476.928
LP	1478.640
LQ	1480.352
LR	1482.064
LS	1483.776
LT	1485.488
LU	1487.200
LV	1488.912
LW	1490.624

Table 26.8. DAB channel table L band, Canada

Channel	Center frequency [MHz]
1	1452.816
2	1454.560
3	1456.304
4	1458.048
5	1459.792

6	1461.536
7	1463.280
8	1465.024
9	1466.768
10	1468.512
11	1470.256
12	1472.000
13	1473.744
14	1475.488
15	1477.232
16	1478.976
17	1480.720
18	1482.464
19	1484.464
20	1485.952
21	1487.696
22	1489.440
23	1491.184

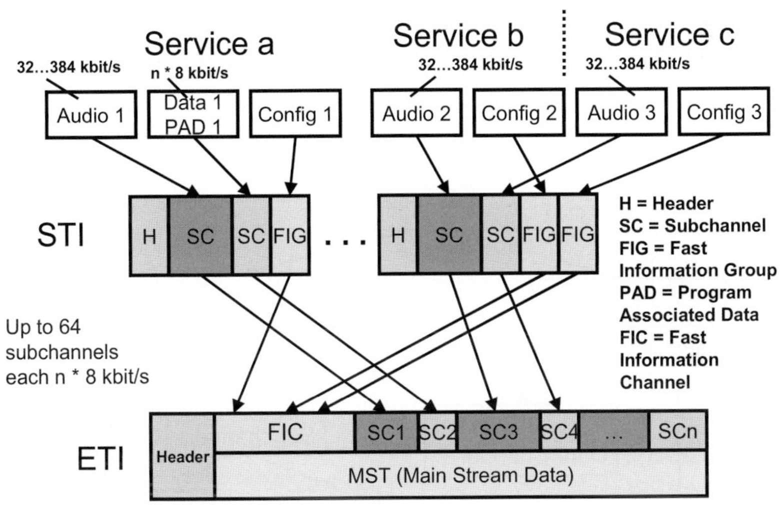

Fig. 26.28. Composition of the ETI data stream

For this reason there is another passive bandpass filter (mask filter). Without pre-correction a DAB signal would have a shoulder attenuation of about 30 dB. If the pre-correction has been properly set the shoulder attenuation will be about 40 dB. This would still interfere with the adjacent channels and would not be authorised by the Authorities. Following the mask filter the shoulders are then lowered by another 10 dB.

Fig. 26.27. shows frequently used DAB blocks. A VHF channel (7 MHz bandwidth) is divided into 4 DAB blocks. The blocks are then called e.g. block 12A, 12B, 12C or 12D.

Tables 26.6., 26.7. and 26.8. list the channel tables used in DAB. Each DAB channel has a width of 7/4 MHz = 1.75 MHz. However, the COFDM signal bandwidth is only 1.536 MHz and there is thus a guard band for the adjacent channels.

26.6 DAB Data Structure

In the section following the essential features of the data structure of DAB will be explained. In DAB, a number of MPEG-1 or –2 Audio Layer II coded audio signals (MUSICAM) combined to form an ensemble (Fig. 26.28.) are transmitted in a 1.75 MHz-wide DAB channel. The maximum net data rate of the DAB channel is about 1.7 Mbit/s and the gross data rate is 2.4 Mbit/s. The data rate of an audio channel is between 32 and 384 kbit/s.

The details described in the following section can be found in the [ETS300401] (DAB), [ETS300799] (ETI) and [ETS300797] (STI) standards.

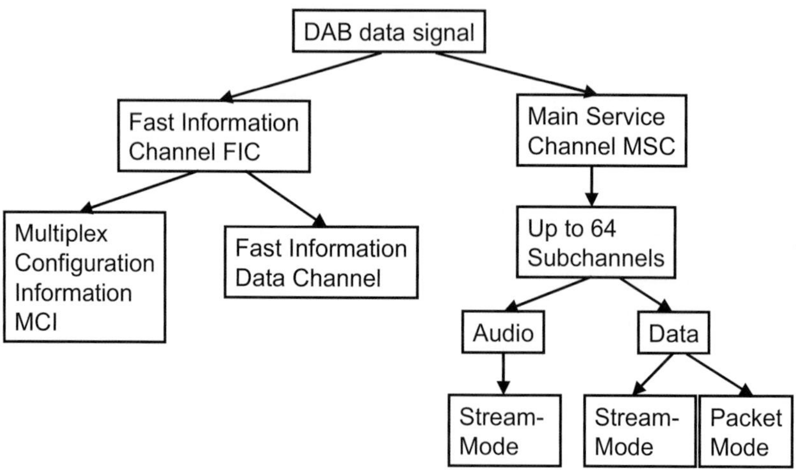

Fig. 26.29. DAB data structure

A DAB data signal (ETI) is composed of the Fast Information Channel (FIC) and the Main Service channel (MSC) (Fig. 26.29.). In the Fast Information Channel, the modulator and the receiver are informed about the composition of the multiplexed data stream by means of the Multiplex

Configuration Information (MCI). The Main Service Channel contains up to 64 subchannels with a data rate of n•8 kbit/s each (Fig. 26.30.). In the subchannels audio signals and data are transmitted. The modulator and receiver obtain the information about the composition of the Main Service Channel from the Multiplex Configuration Information (MCI).

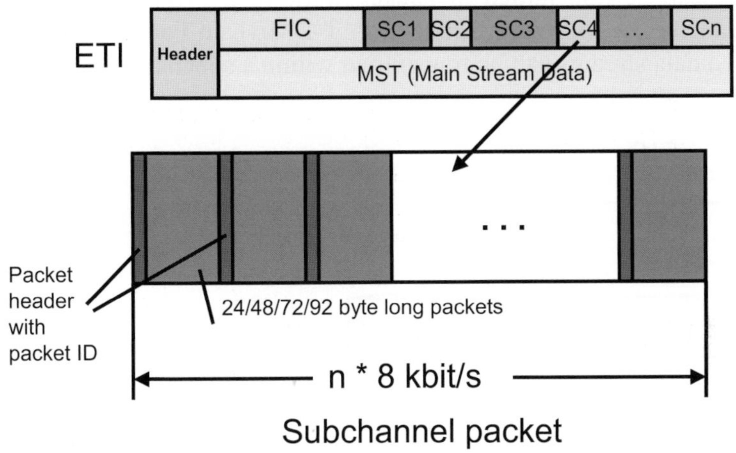

Fig. 26.30. DAB data structure in packet mode

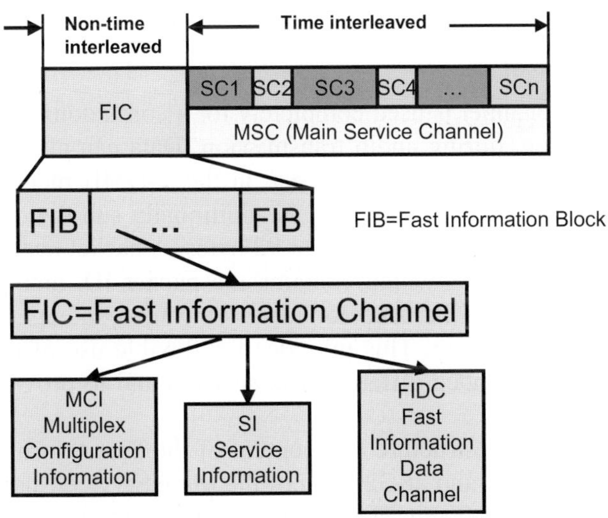

Fig. 26.31. Structure of the DAB Fast Information Channel (FIC)

The transmission in the subchannels can be carried out in Stream Mode and in packet mode. In Stream Mode, data are transmitted continuously. In Packet Mode, the subchannel is additionally sub divided into sub-packets with a constant length. Audio is always transmitted in Stream Mode. The data structure is here pre-determined by the audio coding (24/48 ms pattern). Data can be transmitted in Packet Mode (e.g. MOT - Multimedia Object Transfer) or in Stream Mode (e.g. T-DMB). In Packet Mode, the most varied data streams can be transmitted within a subchannel.

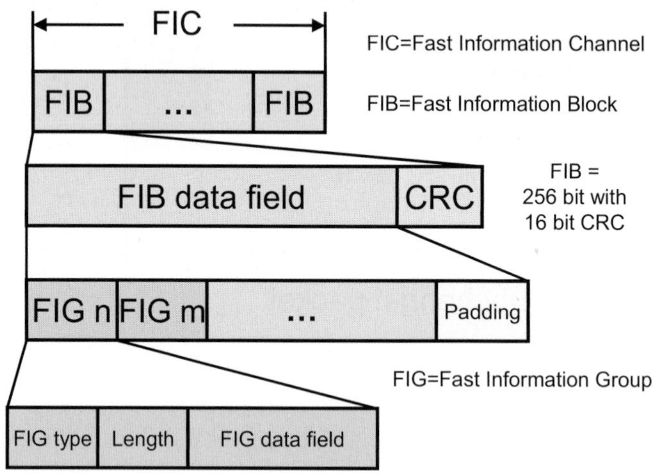

Fig. 26.32. Structure of the DAB FIC

In Stream Mode, a subchannel is used completely for a continuous data stream. This is the case e.g. during audio transmission. Data can also be transmitted in stream mode. This is the case e.g. in the T-DMB method (South Korea). In Packet Mode, a subchannel is additionally sub divided into packets of a constant length of 24, 48, 72 or 92 bytes. A packet begins with a 5-byte-long packet header which contains the packet ID, among other things. The packet ID can be used for identifying the contents. A packet ends with a CRC checksum. This provides for flexible use of the subchannel. It is possible to embed different data services and to provide variable data rates.

In the following section, the structure and content of the Fast Information Channel (FIC) (Fig. 26.31. and 26.32.) and of the Main Service Channel (MSC) will be considered in greater detail. The information transmitted in the fast information channel and main service channel come from the Main Stream Data (MST) from the Ensemble Transport Interface (ETI).

FIC and MSC are provided with error protection (FEC) in the modulator, the FIC being given the strongest protection. The error protection in the MSC is configurable. The strength of the error protection in the MSC is signalled to the receiver in the FIC.

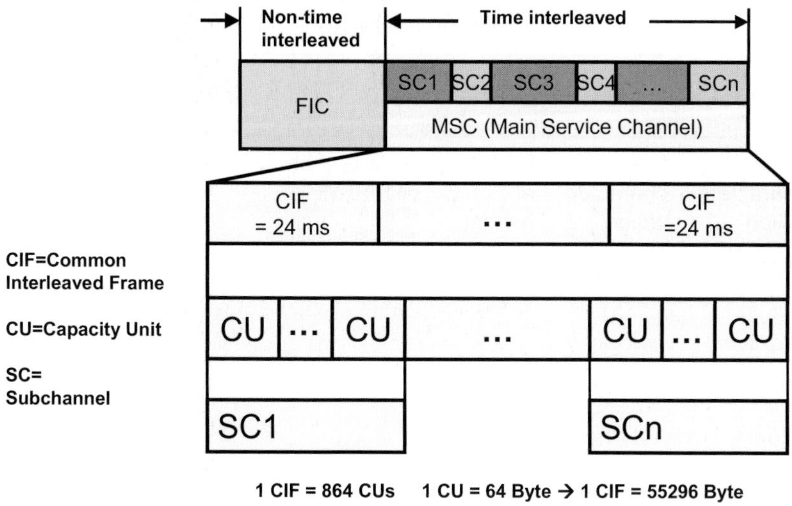

Fig. 26.33. DAB main service channel

In the MSC, the individual subchannels are transmitted, a total of 64 subchannels being possible. Each subchannel can be error-protected to a different degree which is also signalled in the FIC. The subchannels are combined or more precisely allocated to services.

The Fast Information Channel is not time interleaved but transmitted error protected in so-called Fast Information Blocks (FIB).

In the FIC the Multiplex Configuration Information (MCI) is transmitted which is information about the composition of the multiplex data stream, as well as the Service Information (SI) and the Fast Information Data Channel (FIDC).

The SI transmits information about the programs transmitted, the services. In the FIDC, fast supplementary multi-program data are transmitted.

The Fast Information Channel (FIC) is composed of Fast Information Blocks with a length of 256 bits. An FIB consists of an FIB data field and a 16-bit-wide CRC checksum. In the data area of the FIB the messages are transmitted in so-called Fast Information Groups (FIG). Each FIG is identified by its FIG type. An FIG is composed of the FIG type, of the length and the FIG data field in which the actual messages are transmitted.

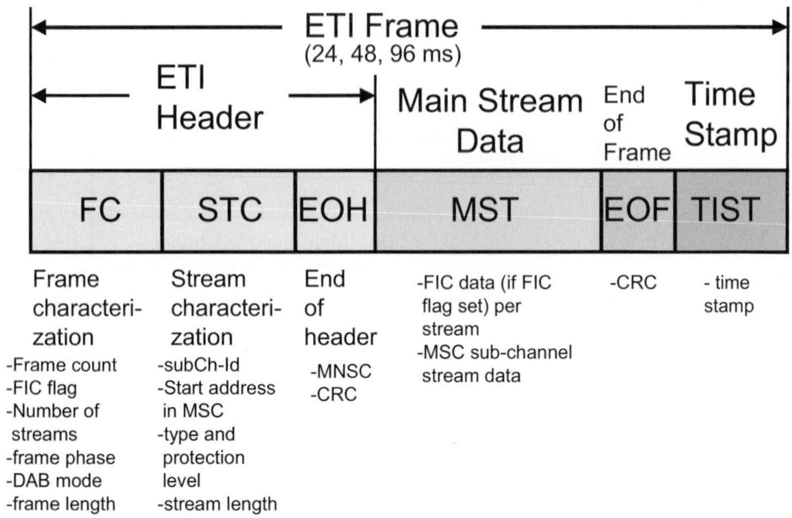

Fig. 26.34. DAB ETI frame structure

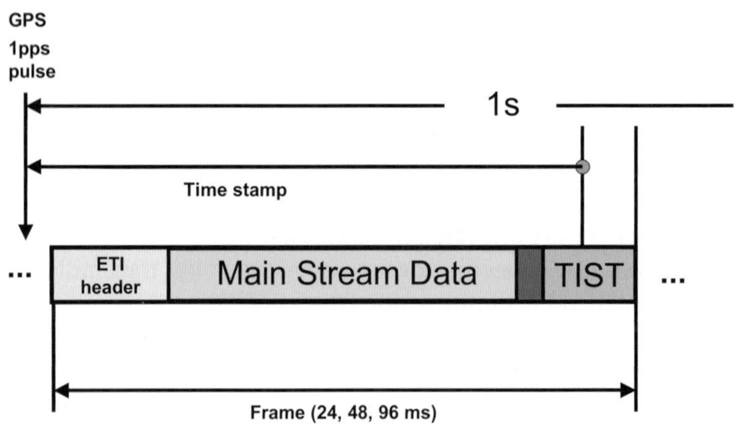

Fig. 26.35. Synchronization of DAB modulators via the TIST in the ETI frame

In the main service channel (Fig. 26.33.), the individual subchannels are broadcast. A total of 64 subchannels are possible. Each subchannel has a data rate of n•8 kbit/s. The subchannels are associated with services (programs). The MSC is composed of so-called Common Interleaved Frames (Fig. 26.33.) which have a length of 24 ms and consist of Capacity Units (CU) with a length of 64 bytes. Overall, 864 CUs result in one CIF which then has a length of 55296 bytes. A number of CUs make up one sub-channel in which the audio frames or data are transmitted.

An ETI frame (Fig. 26.34.) is composed of the header, the Main Stream Data (MST), and End of Frame (EOF) and the Time Stamp (TIST). An ETI frame has a length of 24, 48 or 96 ms.

26.7 DAB Single-Frequency Networks

In the further text, DAB single-frequency networks (SFN) and their synchronisation will be discussed.

COFDM is optimally suited to single-frequency operation. In single-frequency operation, all transmitters are operating at the same frequency which is why single-frequency operation is very economical with regard to frequencies. All transmitters are broadcasting an absolutely identical signal and must operate completely synchronously for this reason. Signals from adjacent transmitters look to a DAB receiver as if they were simply echoes.

The condition which can be met most simply is the frequency synchronisation because frequency accuracy and stability already had to meet high requirements in analog terrestrial radio. In DAB, the RF of the transmitter is tied to the best possible reference. Since the signal of the GPS (Global Positioning System) satellites is available throughout the world, it is used as reference for synchronizing the transmitting frequency of a DAB single-frequency network.

The GPS satellites radiate a 1pps signal to which a 10 MHz oscillator is tied in professional GPS receivers which is used as reference signal for the DAB transmitters.

However, there is also a strict requirement with regard to the maximum transmitter spacing. The maximum possible transmitter spacing is a result of the length of the guard interval and the velocity of light and the associated propagation time. Inter-symbol interference can only be avoided if in multi-path reception no path has a longer propagation time than the guard interval length. The question about what would happen if a signal of a more remote transmitter violating the guard interval is received can be easily answered. Inter-symbol interference is produced which becomes noticeable as disturbing noise in the receiver. Signals from more remote transmitters must simply be attenuated sufficiently well. The threshold for virtually error-free operation is set by the same conditions as in the case of pure noise. It is of particular importance therefore that a single-frequency network has the correct levels. It is not the maximum transmitting power which is required at every site but the correct one. Network planning requires topographical information.

With the velocity of light of C=299792458 m/s, a signal delay of 3.336 µs per kilometer transmitter distance is obtained.

The maximum distances between adjacent transmitters possible with DAB in a single-frequency network are shown in Table 26.9.

Table 26.9. SFN parameters in DAB

	Mode I	Mode IV	Mode II	Mode III
Symbol duration	1 ms	500 µs	250 µs	125 µs
Guard interval	246 ms	123 µs	62 µs	31 µs
Symbol+guard	1246 µs	623 µs	312 µs	15 6µs
Max. transmitter distance	73.7 km	36.8 km	18.4 km	9.2 km

In a single-frequency network, all individual transmitters must operate synchronised with one another. The contributions are supplied by the playout center in which the DAB multiplexer is located, e.g. via satellite, optical fibre or microwave link. It is obvious that due to different path lengths the ETI signals fed in will carry different delays.

However, in each DAB modulator in a single-frequency network the same data packets must be processed to form COFDM symbols. Each modulator must perform all operating steps in complete synchronism with all other modulators in the network. The same packets, the same bits and the same bytes must be processed at the same time. At each DAB transmitter site, absolutely identical COFDM symbols must be radiated at the same time.

The DAB modulation is organized in frames.

To carry out delay compensation in the DAB SFN, Time Stamps (TIST) (Fig. 26.34.) derived from the GPS signal are added to the ETI signal in the multiplexer.

At the end of an ETI frame the TIST is transmitted which is derived by the DAB ensemble multiplexer by GPS reception and is keyed into the ETI signal. It specifies the time back to the reception of the last GPS 1pps signal (Fig. 26.35.). The time information in the TIST is then compared in the modulator with the GPS signal also received at the transmitter site and used for performing a controlled ETI signal delay.

26.8 DAB Data Broadcasting

In the following section, the possibility of data broadcasting in DAB will be briefly discussed. In DAB data broadcasting (Fig. 26.36.), a distinction is made between the MOT (Multimedia Object Transfer) standard as defined in the [ETS301234] Standard, and the IP transmission via DAB. In both cases, a DAB subchannel is operated in packet mode, i.e. the data packets to be transmitted are divided into short constant-length packets. Each of these packets has a packet ID in the header section by means of which the transmitted content can be identified.

In the Multimedia Object Transfer (MOT) according to [ETS301234], a distinction is made between file transmission, a slide show and the "Broadcast Web Page" operation. In file transmission, only files are fed out cyclically. A slide show can be configured with respect to its display speed. It is possible to transmit JPEG or GIF files.

In the "Broadcast Web Page", a directory of HTML pages is cyclically transmitted and a starting page can be defined. The resolution corresponds to ¼ VGA.

Fig. 26.37. shows the MOT data structure. The files to be transmitted, the slide show or the HTML data are transmitted in the payload segment of an MOT packet. The MOT packet plus header is inserted into the payload segment of an MSC data group, the MOT header coming first followed by a CRC checksum. The entire MOT packet is divided into short constant-length packets of the packet mode. These packets are then transmitted in subchannels.

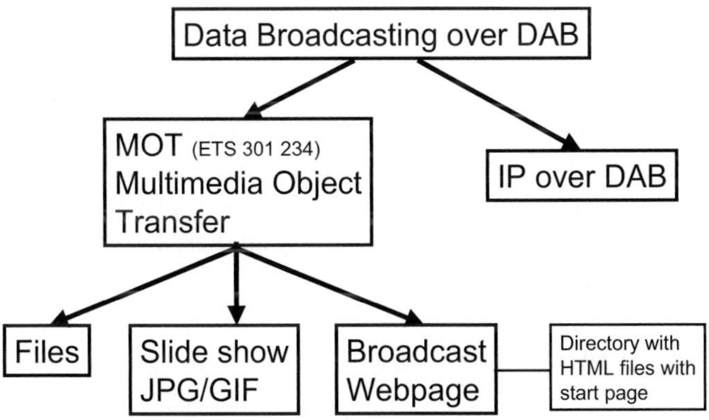

Fig. 26.36. Data broadcasting over DAB

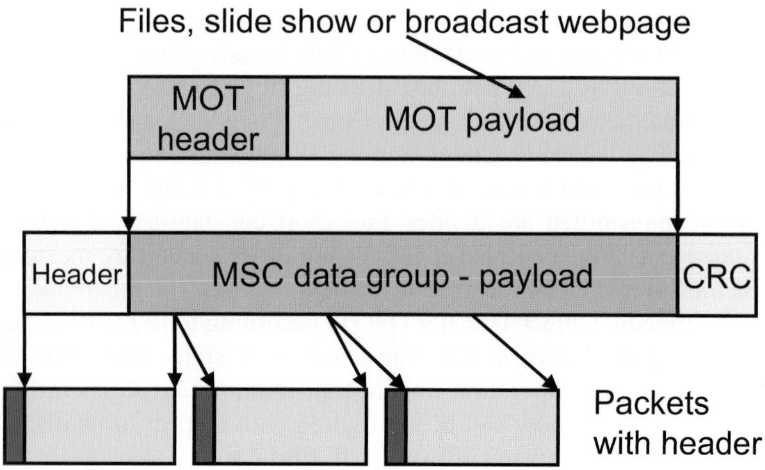

Fig. 26.37. MOT data structure

The category DAB Data Broadcasting should also include T-DMB (Terrestrial Digital Multimedia Broadcasting). In this South Korean method, DAB is operated in the data stream mode.

26.9 DAB+

One of the more recent developments in audio broadcasting is DAB+. In DAB+, MPEG-4 AAC Plus is used instead of MPEG-1 or -2 Layer II Audio. The unequal error protection originally provided in DAB is then no longer possible since it is coupled directly to the MPEG-1 or MPEG-2 Layer II frame. However, it is now possible to accommodate three times as many services, i.e. programs, per DAB multiplex. Similar to a T-DMB transmission, there are no changes in the physical layer of DAB. Australia has decided to adopt DAB+ and is now also setting up these networks.

26.10 DAB Measuring Technology

The DAB measuring technology can be copied directly from the world of DVB-T. It is necessary both to test DAB receivers and to measure DAB transmitters. For these purposes, test transmitters are now available which deliver a DAB signal [SFU] and test receivers which are capable of analyzing DAB signals [ETL].

26.10.1 Testing DAB Receivers

In the DAB receiver test, the reality of DAB reception must be simulated for the DAB receiver. This requires multi-path reception, noise, minimum receiver input level, interferers etc. as necessary test scenarios. Similar to DVB-T, the source of these inputs is provided by a corresponding test transmitter with fading simulator [SFU]. This can also be used for T-DMB and DAB+ since the physical layer is the same.

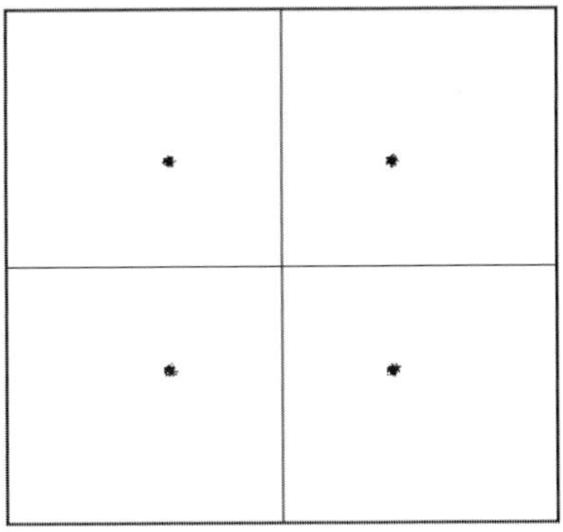

Fig. 26.38. Relatively undisturbed differentially demodulated DAB constellation diagram [ETL]

26.10.2 Measuring the DAB Signal

In DAB, as in DVB-T, the following measurements can be performed on the DAB signal:

- Detecting the bit error ratios,
- Measurements on the DAB spectrum,
- Constellation analysis.

Due to the unequal error protection, it is more difficult to measure the bit error ratios. Measuring the bit errors is relatively simple only at the Fast

Information Channel (FIC) since a constant error protection with a code rate of 1/3 is present there.

In the DAB constellation analysis [ETL], the constellation diagram is first differentially demodulated and produces 4 points again. The smaller the appearance of these points in the constellation diagram, the more undisturbed was their transmission (Fig. 26.38.).

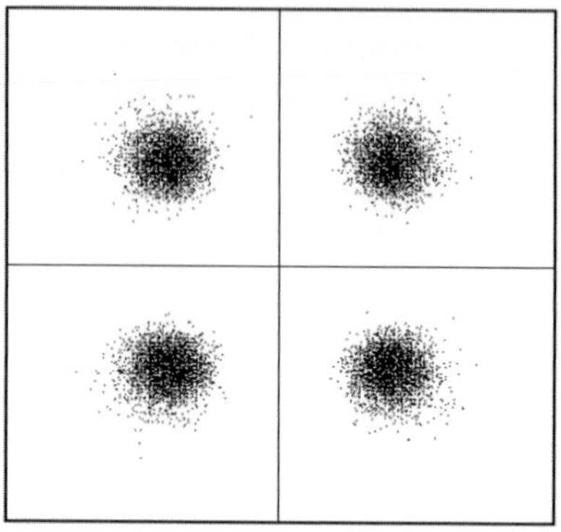

Fig. 26.39. DAB constellation diagram with noise

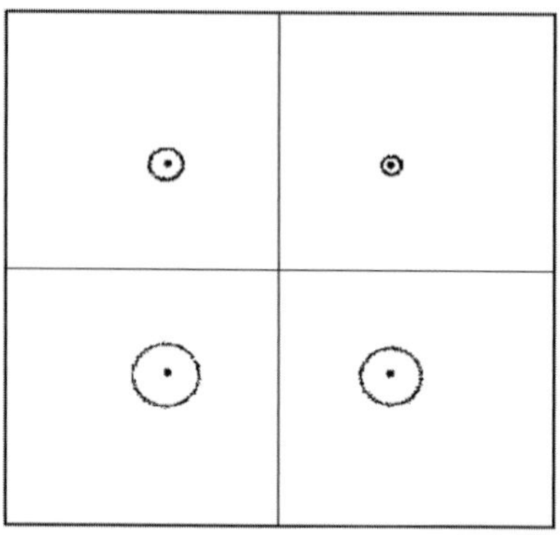

Fig. 26.40. DAB constellation diagram with superimposed sinusoidal interferer

If noise effects are affecting a DAB signal, a DAB constellation diagram will appear as shown in Fig. 26.39. Similar to DVB-T, phase jitter will result in striation-like distortions of the constellation diagram. Sinusoidal interferers will generate circular constellation points (Fig. 26.40.). A wrongly calibrated IQ modulator will produce carrier cross-talk from the lower DAB sub-band into the upper one and conversely and will lead to a poorer S/N ratio just as in DVB-T. In DAB, a modulation error ratio

(MER) can also be defined (see also the chapter on DVB-T Measuring Technology). In DAB, too, an MER can also be defined and measured as a function of the subcarriers (Figs. 26.41. and 26.42.).

A further measurement necessary in DAB is measuring the channel impulse response (Fig. 26.43.). The channel impulse response, which can be calculated by analyzing the TFPR symbol, can be used for verifying if a DAB single-frequency network is running synchronously and that there are no guard interval violations.

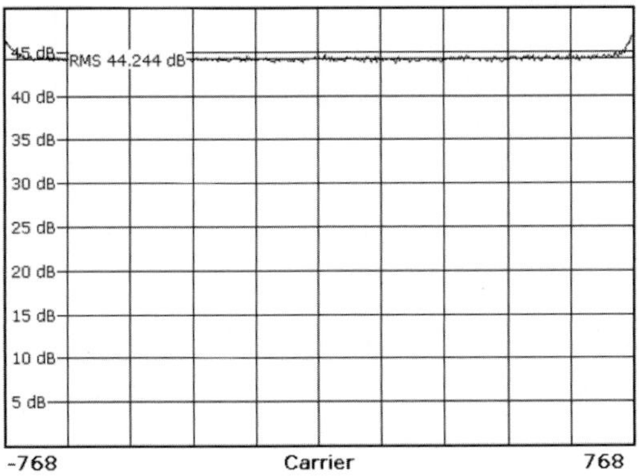

Fig. 26.41. MER(f) in undisturbed DAB (Mode I)

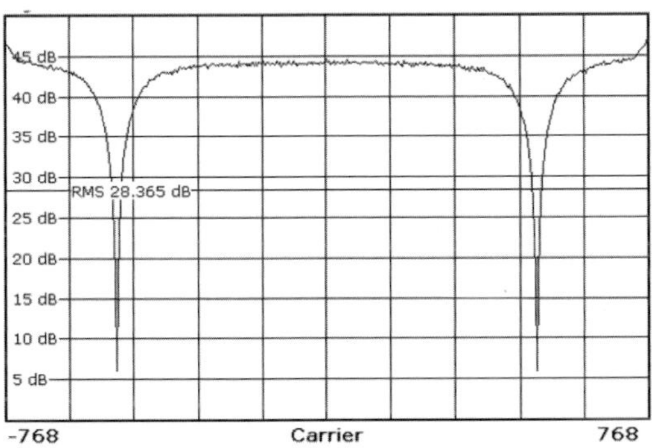

Fig. 26.42. MER(f) in DAB with fading

Naturally, the evaluation of the data contents in the DAB signal is also of interest. The analysis of an ETI signal at the output of the playout center or at the transmitter input, respectively, corresponds to the MPEG-2 analysis in DVB. By now there are analysis tools available also for this purpose. This delay in the availability of DAB test technology is one of the consequences of the DAB market delays and does not have any basis in technical reasons.

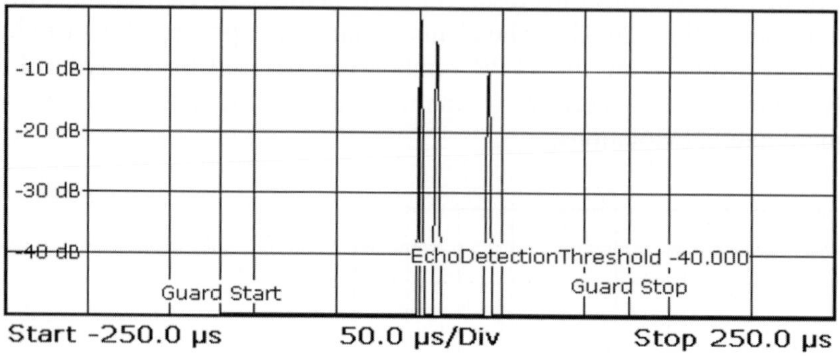

Fig. 26.43. DAB channel impulse response (3 paths) [ETL]

Bibliography: [FISCHER7], [HOEG_LAUTERBACH], [ETS300401], [ETS300799], [ETS300797], [TR101496], [ETS301234], [ETL], [SFU]

27 DVB Data Services: MHP and SSU

Apart from DVB-H, there are also other DVB data services. These are the Multimedia Home Platform, or MHP in short, and the System Software Update (SSU) for DVB receivers. In parallel with these, there is also MHEG (the Multimedia and Hypermedia Information Coding Experts Group) running over DVB-T in the UK. All these data services have in common that they are broadcast via so-called object carousels in DSM-CC sections. Applications are transmitted to the receiver via MHP and MHEG and can be stored and run by a receiver especially equipped for this purpose. In the case of MHP, these are HTML files and Java applications transmitted to the terminal in complete directory structures. MHEG allows HTML and XML files to be transmitted and started.

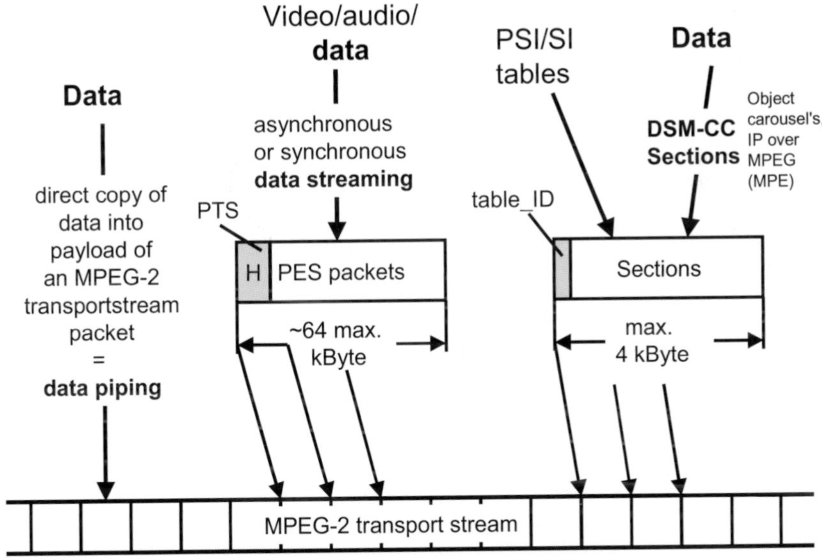

Fig. 27.1. Data transmission via an MPEG-2 transport stream: data piping, data streaming and DSM-CC sections

W. Fischer, *Digital Video and Audio Broadcasting Technology*, Signals and Communication Technology, 3rd ed., DOI: 10.1007/978-3-642-11612-4_27, © Springer-Verlag Berlin Heidelberg 2010

27.1 Data Broadcasting in DVB

In MPEG-2/DVB, data transmission can take place as (Fig. 27.1.):

- Data piping
- Asynchronous or synchronous data streaming
- via object carousels in DSM-CC sections
- as datagram transmission in DSM-CC sections
- as IP transmission in DSM-CC sections

In data piping, the data to be transmitted are copied directly into the payload part of MPEG-2 transport stream packets asynchronously to all other contents and without any other defined intermediate protocol. In data streaming, in contrast, the familiar PES (packetized elementary stream) packet structures are used which allow the contents to be synchronized with one another through the presentation time stamps (PTS). Another mechanism for asynchronous data transmission, defined in MPEG-2, are DSM-CC (Digital Storage Media Command and Control) sections (Fig. 27.2.).

```
table_id (=0x3A ...0x3E)                      8 Bit
section_syntax_indicator                      1
private_indicator=1                           1
reserved =11                                  2
section_length                                12
{
 table_id_extension                           16
 reserved                                     2
 version_number                               5
 current_next_indicator                       1
 section_number                               8
 last_section_number                          8
 switch(table_id)
 {
   case 0x3A: LLCSNAP(); break;
   case 0x3B: userNetworkMessage(); break;
   case 0x3C: downloadDataMessage(); break;
   case 0x3D: DSMCC_descriptor_list(); break;
   case 0x3E: for (i=0; i<dsmcc_section_length-9;i++)
                    private_data_byte;                     8 Bit
 }
}
CRC                                                        32 Bit
```

Fig. 27.2. Structure of a DSM-CC section

27.2 Object Carousels

DSM-CC sections have already been discussed in detail in the DVB-H section. DSM-CC sections are table-like structures and are considered to be private sections according to MPEG-2 Systems. The basic structure of a DSM-CC (Fig. 27.2.) section corresponds to the structure of a so-called long section with a checksum at the end. A DSM-CC section has a length of up to 4 kbytes and begins with a table_ID in the range of 0x3A ... 0x3E. This is followed by the section header with version administration, already discussed in detail in other chapters. Data services such as object carousels or general datagrams or IP packets as in DVB-H (MPE, multiprotocol encapsulation) are then transmitted in the actual trunk of the section. The table_ID shows the type of data services involved.

Table_ID's:

- 0x3A and 0x3C provide for the broadcasting of object/data carousels
- 0x3D provides for the signalling of stream events
- 0x3E provides for the transmission of datagrams or IP packets

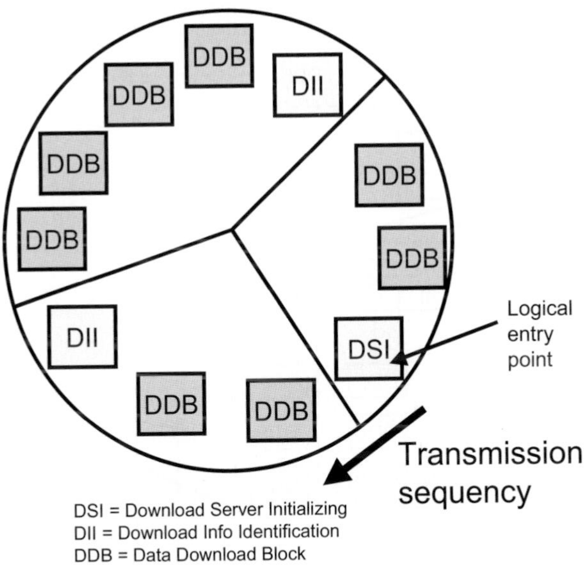

Fig. 27.3. Principle of an object carousel

Object carousels (Fig. 27.3.) allow complete file and directory structures to be transmitted from a server to the terminal via the MPEG-2 transport stream. A restriction imposed by the data carousels is that they only allow a relatively flat directory structure and flat logical structure. Object and data carousels are described both in the standard [ISO/IEC 13818-6] (a part of MPEG-2) and in the DVB data broadcasting document [EN301192].

Firstly, data/object carousels have a logical structure which owes nothing to the content actually to be transmitted (directory tree plus files). The entry point into the carousel is via the DSI (Download Server Initializing) message, or via a DII (Download Information Identification) message in the case of the data carousel. It is retransmitted cyclically with a table_ID=0x3B in a DSM-CC section. Cyclically because this is broadcasting and it must be possible to reach a large number of terminals time and again and the terminals are unable to request messages from the server. The DSI packet then uses IDs to refer to one or more DII messages (Fig. 27.4.) which are also retransmitted cyclically in DSM-CC sections with a table_ID = 0x3B. The DII messages, in turn, refer to modules in which the actual data are then repeatedly broadcast cyclically via many data download blocks (DBB) with a table_ID=0x3C in DSM-CC sections.

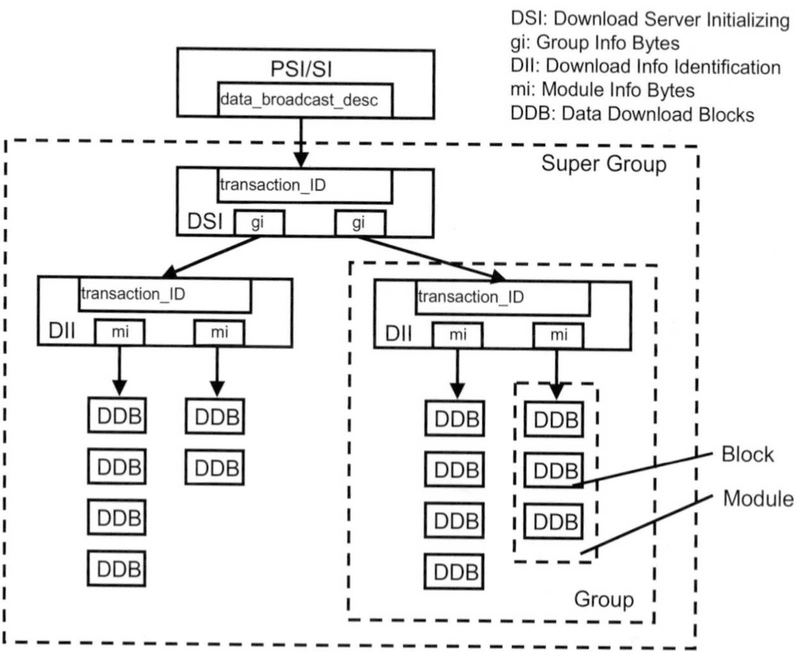

Fig. 27.4. Logical structure of an object carousel

The transmission of a directory tree can take up to several minutes depending on the volume of data and the available data rate.

The presence of an object/data carousel must be announced via PSI/SI tables. Such a data service is allocated to a program service and entered in the respective program map table (PMT) where the PIDs of the object/data carousels are to be found. In the case of a data carousel, entry takes place directly by DII.

Additional items such as a more detailed description of the contents in the carousels are broadcast in separate, new SI tables like the AIT (Application Information Table) and the UNT (Update Notification Table). The AIT belongs to the Multimedia Home Platform and the UNT belongs to the System Software Update and both - AIT and UNT - must also be announced via PSI/SI. The AIT is entered in the PMT of the associated program and the UNT is entered in the NIT.

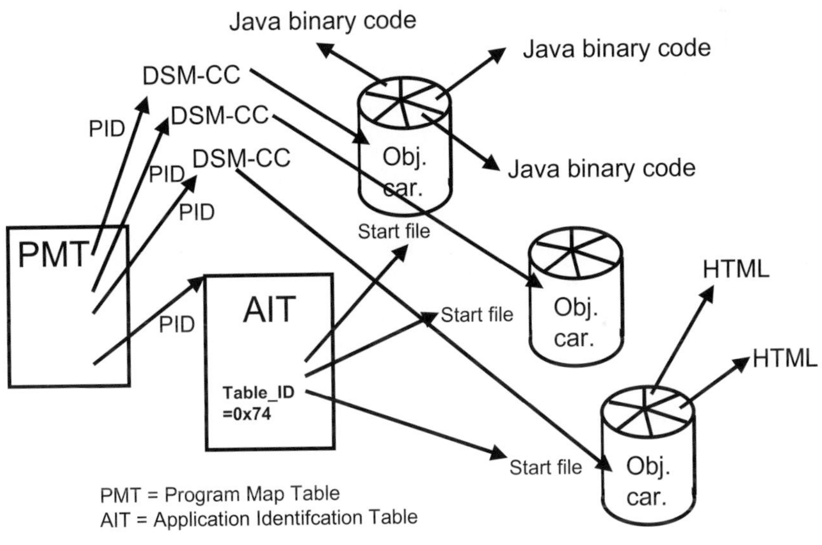

Fig. 27.5. MHP structure

27.3 The Multimedia Home Platform MHP

The Multimedia Home Platform has been provided in DVB as supplementary service for MHP-enabled receivers. The standard, with about 1000 pages, is [ETS101812] and has been released in the year 2000. There are two versions which are MHP 1.1. and MHP 1.2. MHP is used for transmitting HTML (Hypertext Multimedia Language) files familiar from the Internet, and Java applications. Starting the HTML and Java applications requires special software (or middleware) in the receiver. MHP-capable receivers are more expensive and not available in great numbers on the market. MHP applications are broadcast in many countries but are currently really successful only in Italy.

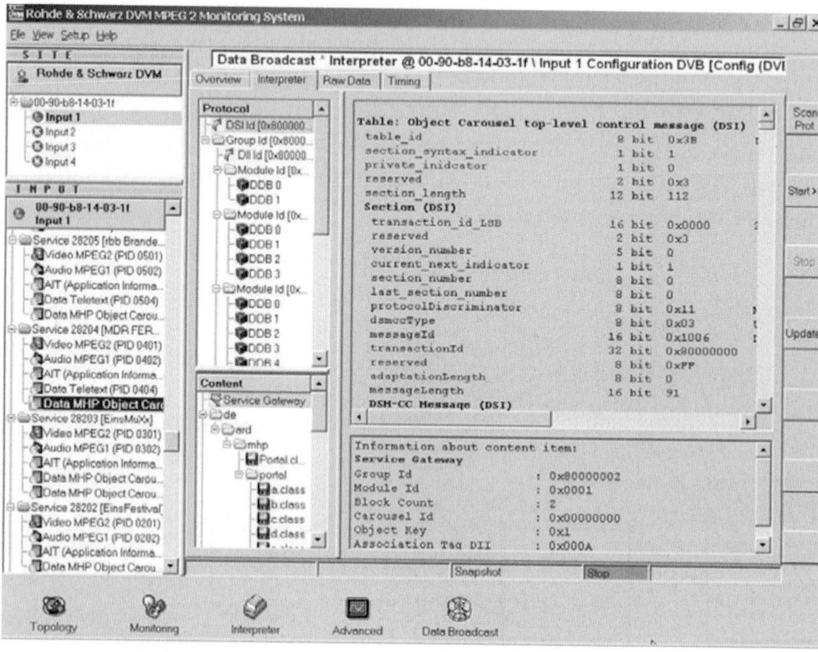

Fig. 27.6. MHP file structure of an object carousel as analyzed on an MPEG analyzer [DVM]

The contents broadcast by MHP are:

- Games
- Electronic programme guides
- News
- Interactive program-associated services
- "Modern" teletext

The entry point into the MHP directory structure (Fig. 27.5., 27.6., 27.7.), the starting file and the name and type of the MHP application are signalled via the AIT (Application Information Table, Fig. 27.5.). The AIT is entered in a PMT as PID with the value of 0x74 as table_ID.

27.4 System Software Update SSU

Since the software of DVB receivers is also subject to continuous updates, it makes sense to provide these to the customer in a relatively simple manner. This can be done "by air" in the case of DVB-S and DVB-T and, of course, via cable in the case of DVB-C. If the software is transmitted in object carousels embedded in the MPEG-2 transport stream according to DVB it is called SSU (System Software Update) and is defined in the [TS102006] standard. Currently, however, mainly proprietary software updates are used.

In SSU, the available software updates are announced via another table, the Update Notification Table (UNT). The PID of the UNT is entered in the NIT and the table_ID of the UNT is 0x4B.

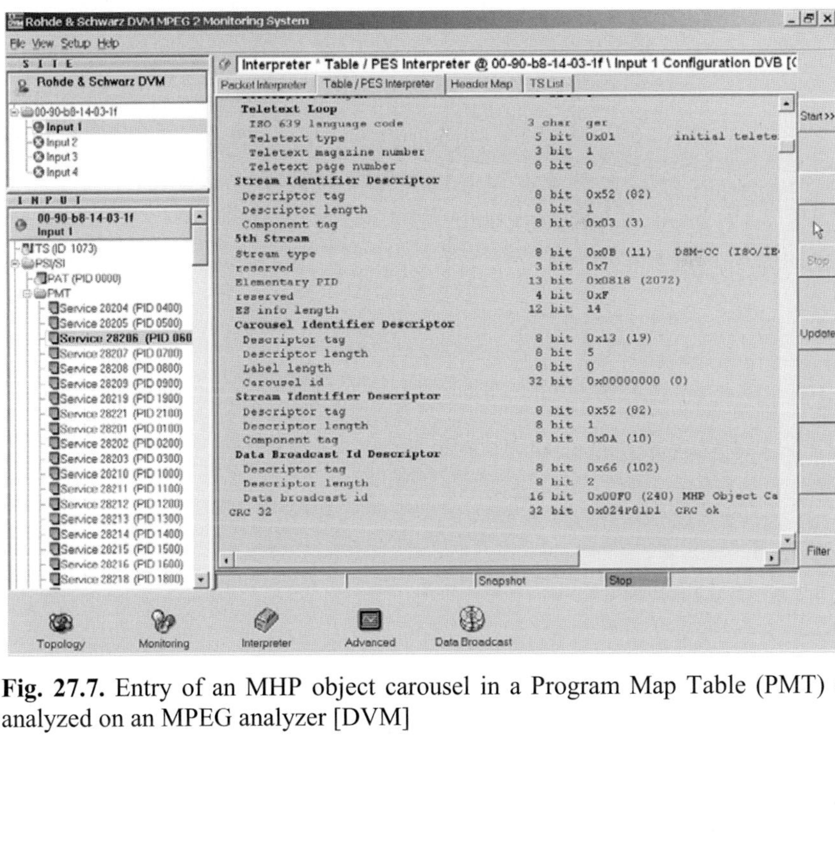

Fig. 27.7. Entry of an MHP object carousel in a Program Map Table (PMT) as analyzed on an MPEG analyzer [DVM]

Bibliography: [ISO/IEC13818/6], [EN301192], [ETS101812], [TS102006]

28 T-DMB

The idea for T-DMB - Terrestrial Digital Multimedia Broadcasting - comes from Germany, it was developed in South Korea, and its physical parameters are identical to the European DAB (Digital Audio Broadcasting) standard. T-DMB is intended for the mobile reception of broadcasting services similar to DVB-H. T-DMB corresponds wholly to DAB which itself supports the data stream mode also used in T-DMB (Fig. 28.1.). However, the "unequal forward error correction" possible in DAB is no longer possible in this case because the entire subchannel used for the T-DMB channel must be equally error protected.

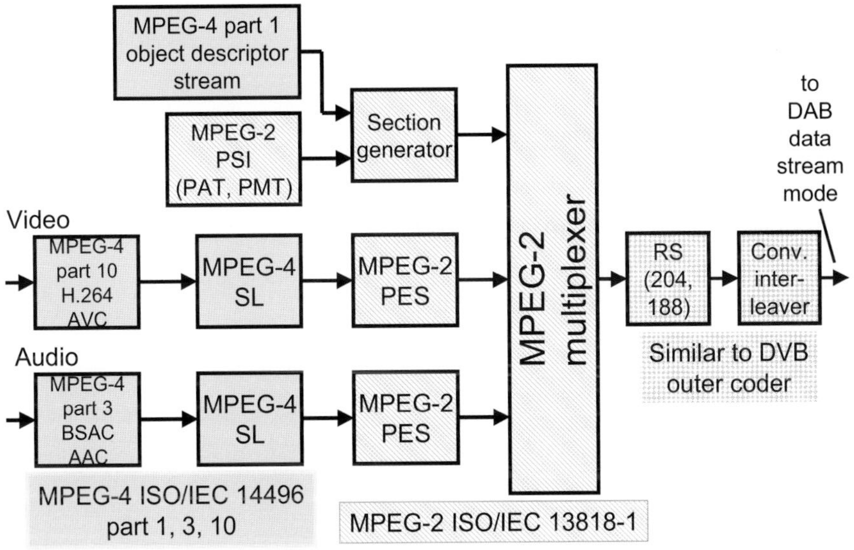

Fig. 28.1. T-DMB modulator block diagram

In T-DMB, the video and audio contents are MPEG-4-AVC- and AAC-coded. The video coding uses the new H.264 method. Video and audio are then packaged in PES packets and are then assembled to form an MPEG-2 transport stream (Fig. 28.1.) which also contains the familiar PSI/SI tables.

W. Fischer, *Digital Video and Audio Broadcasting Technology*, Signals and Communication Technology, 3rd ed., DOI: 10.1007/978-3-642-11612-4_28, © Springer-Verlag Berlin Heidelberg 2010

The transport stream is then error protected similarly to DVB-C, i.e. with Reed Solomon RS(204, 188) error protection plus Forney interleaving, after which the data stream is docked onto DAB in data stream mode (Fig. 28.2.).

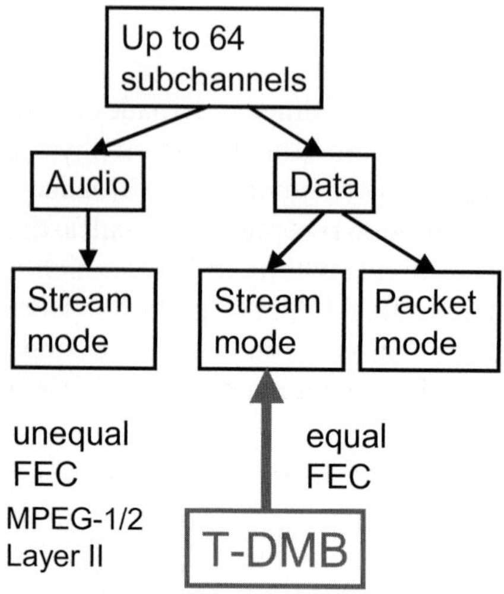

Fig. 28.2. DAB data structure

Bibliography: [ETS300401], [T-DMB]

29 IPTV – Television over the Internet

Thanks to new technologies, the traditional transmission paths for television of terrestrial, broadband cable and satellite transmission have been joined by an additional propagation path, the two-wire line, conventionally known as telephone cable. VDSL (Very-high-bit-rate Digital Subscriber Line, [ITU-T G.993]) now provides for data rates on these lines which also allow television, IPTV – Internet Protocol Television, i.e. television over the Internet. IPTV is now provided, e.g. by the German T-COM/Deutsche Telekom or the Telekom Austria under the new slogan "Triple Play". "Triple Play" is telephone, Internet and television out of one socket. The term has also been applied for some time to broadband cable where all 3 media are also available from one socket.

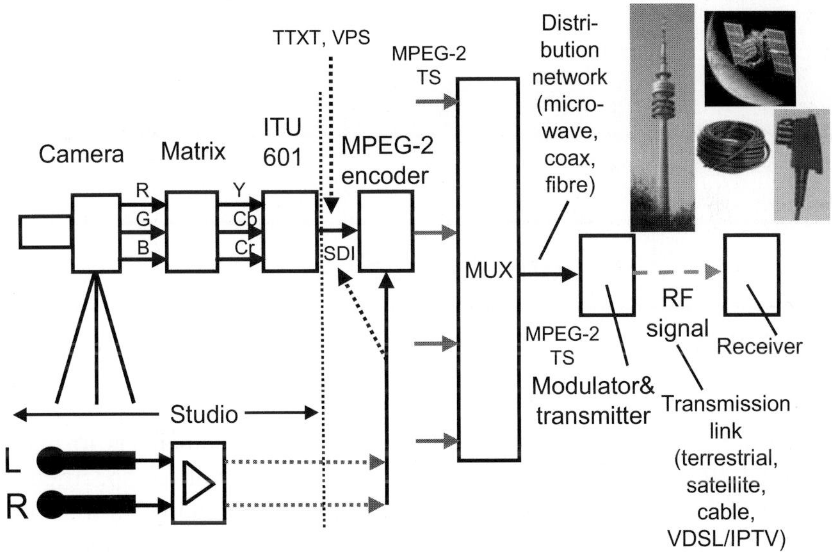

Fig. 29.1. Distribution paths for digital television

The contents are here MPEG-4-coded to compress the input material optimally to the lowest possible data rates, using MPEG-4 AVC (or possibly

W. Fischer, *Digital Video and Audio Broadcasting Technology*, Signals and Communication Technology, 3rd ed., DOI: 10.1007/978-3-642-11612-4_29, © Springer-Verlag Berlin Heidelberg 2010

VC-1 (Windows Media 9)) and AAC. However, MPEG-2-coded video streams and MPEG-1-coded audio streams are also still transmitted via IP. Currently, four possibilities exist for transmitting DTV over IP. The first one of these is proprietary where MPEG-4 Video or possibly. Windows Media 9 (VC-1) is simply embedded, together with MPEG-4 Audio (AAC), in UDP-packets (i.e. without handshake). The UDP-packets, in turn, are then placed in IP-packets and are then transmitted via Ethernet, WLAN, WiMAX or xDSL.

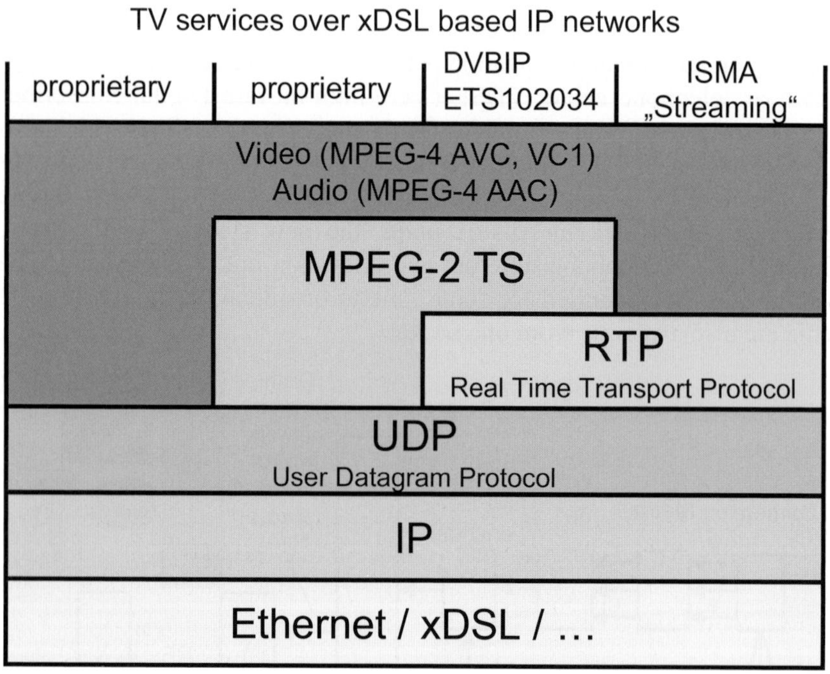

Fig. 29.2. IPTV-Protocols

A further approach, also not standardized at the moment, is to insert video and audio-streams into an MPEG-2-transport stream as specified in the MPEG-2 and MPEG-4-standards, and then to transport this transport stream in UDP- and IP-packets, e.g. also via xDSL. In the method specified as part of DVB-IP in the ETS 102034 standard, the RTP (Real Time Transport Protocol) is additionally inserted between the transport stream and the UDP layer. In ISMA (Internet Streaming Media Alliance) streaming the transport stream layer is missing but the RTP is used here, also. All methods have in common that in each case only one program is transmitted on-demand. In the MPEG-2-transport stream, PAT and PMT-tables are inserted for signalling purposes.

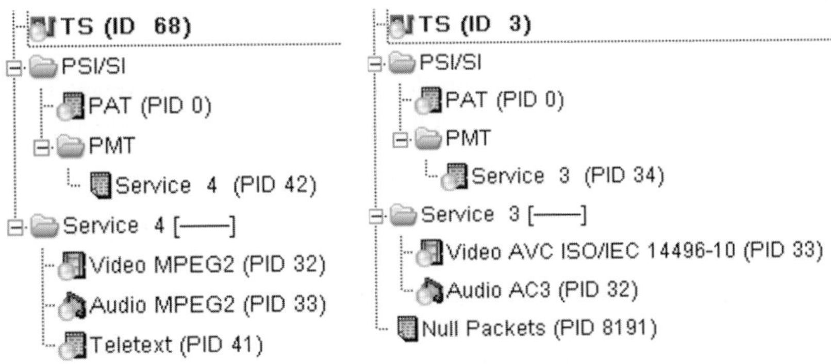

Fig. 29.3. Examples of DVB-IP-compliant transport streams, recorded in the network of Telekom Austria; on the left – MPEG-2 contents, on the right – MPEG-4 AVC and Dolby Digital

29.1 DVB-IP

In DVB-IP [ETS 102034], the MPEG-2 transport stream with either MPEG-4- or MPEG-2-coded video signals and MPEG-4-, MPEG-2- or MPEG-1-coded audio signals (Fig. 29.3.) is embedded via RTP (Real Time Protocol) in UDP packets and then transmitted in an IP network via DXL with instantaneous data rates of 8 or 16 Mbit/s. The main purpose of the RTP is to assist in the restoration of the original order of the packets in an IP network. The RTP also contains mechanisms for managing the timing (s.a. PCR jitter). The DVB IP has provisions for sending the MPEG-2 transport stream either with all PSI/SI tables or only sending the PSI tables along. On registration, the delivery system sends to the DVB-IP receiver a list of available services with the associated socket. A socket consists of an IP address consisting of 4 bytes, and the associated UDP port. This address has the following syntax:

a.b.c.d:port

where a,, b, c and d have a value of between 0 ... 255 and port comprises a range of from 0 ... 65535. "Normal" IP TV runs on multicast addresses within a range of

224.0.0.0 ... 239.255.255.255.

It is only in the case of video on demand that unicast addresses make sense which, with exceptions, can comprise almost the entire address range of from

0.0.0.0 ... 255.255.255.255

According to the Internet Protocol, exceptions are:
127.0.0.1 (= local host),
x.x.x.0 (= current network),
x.x.x.255 (= broadcast),
244.0.0.0 ... 239.255.255.255 (= multicast).

On registering, an IP address, by means of which it can then receive both multicast and unicast services, is assigned to the IPTV receiver. When the service or program is selected by the user, the receiver then signals the corresponding socket to the nearest network node (DSLAM) and is then fed the MPEG-2 transport stream via precisely this address plus UDP port via UDP protocol.

29.2 IP Interface Replaces TS-ASI

It has been noted that the TS-ASI interface is being replaced more and more by a Gigabit Ethernet interface, especially in the head end and in the playout center. This is being mentioned because it fits into the present chapter. If head-end components are connected to one another via gigabit IP, the transport stream, which is otherwise distributed via TS-ASI, must be embedded fully compatibly, i.e. completely, in IP and be provided with all associated PSI/SI tables. In the IP network , the required transport stream is also addressed via a socket, i.e. via the 4-byte-long IP address and the UDP port.

29.3 Summary

Replacing the TS-ASI interface with a Gigabit Ethernet interface appears to be more and more prevalent. It remains to be seen how successful the new IPTV offer via two-wire line will be compared with the other three previous TV propagation paths. However, Triple Play - telephone, Internet and television from one "connector", is a very interesting alternative to what has hitherto been on offer with respect to communications, either by broadband cable or by the two-wire line being discussed here. Like Mobile TV, IPTV had been an important subject in the broadcasting exhibitions in 2007/2008.

Bibliography: [ITU-T G.993], [ETS102034]

30 DRM – Digital Radio Mondiale

In 2000, a further digital broadcasting standard called DRM - Digital Radio Mondiale [ETS 101980] was created. DRM is intended for the frequency band from 30 kHz ... 30 MHz, in which the AM service was normally transmitted. The broadcasting frequency bands were basically divided in accordance with their propagation characteristics, as follows:

- LW (Long Wave) ~30 kHz ... 300 kHz
- MW (Medium Wave) ~300 kHz ... 3 MHz
- SW (Short Wave) ~3 MHz ... 30 MHz
- VHF: ~30 MHz ... 300 MHz
- UHF: ~300 MHz ... 3 GHz

VHF is split into three bands:

- VHF I: 47 ... 85 MHz
- VHF II: 87.5 ... 108 MHz
- VHF III: 174 … 230 MHz

UHF has two frequency bands, which are:

- UHF IV: 470 ... 606 MHz
- UHF V: 606 ... 826 MHz

In the frequency band below 30 MHz, very long-range reception is sometimes possible which, however, is greatly dependent on diurnal (day/night) variations and on solar activity. The channel bandwidths specified here are 9 kHz (ITU-Region 1 (Europe, Africa) and Region 3 (Asia/Pacific)) and 10 kHz (ITU-Region 2 (North and South America)).

DRM is the attempt to replace more and more unused frequency bands in which amplitude modulation has hitherto been used, with modern digital transmission methods. The modulation method applied is COFDM, using MPEG-4 AAC for compressing the audio signals. The net data rates are usually approx. 10 to 20 kbit/s.

W. Fischer, *Digital Video and Audio Broadcasting Technology*, Signals and Communication Technology, 3rd ed., DOI: 10.1007/978-3-642-11612-4_30, © Springer-Verlag Berlin Heidelberg 2010

The channel bandwidths specified for DRM are derived from the bandwidths normally used in the frequency bands provided. The DRM bandwidths are between 4.5 kHz und 20 kHz (Fig. 30.2.) and are defined via the parameter of "spectrum occupancy". Table 30.1 shows the possible bandwidths. As in other standards, too, which define COFDM as the modulation method, modes are defined here. The DRM modes are designated as Robustness Mode A, B, C and D. The mode determines the carrier spacing and the symbol duration. The physical parameters of the DRM-modes can be seen in Table 30.2. The number of carriers in an COFDM-symbol depends on the mode and on the DRM bandwidth. The number of carriers which can be accommodated in a symbol is listed in Table 30.3.

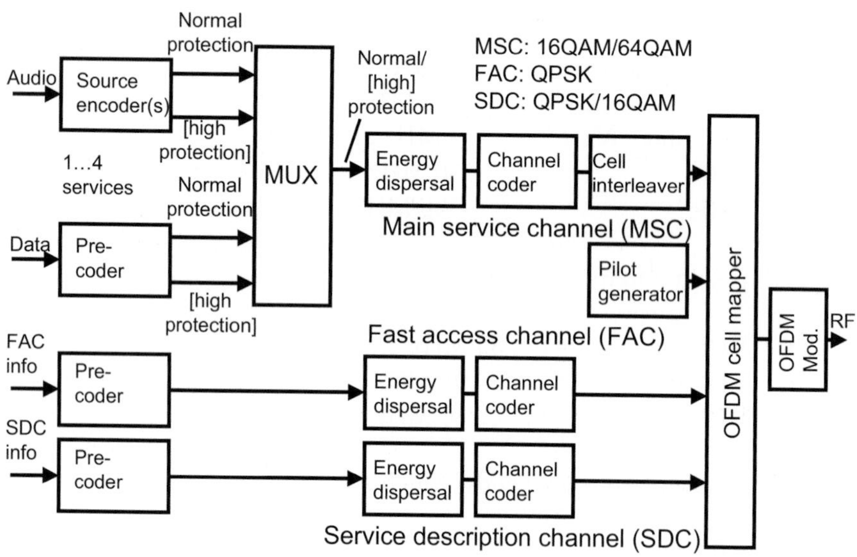

Fig. 30.1. Block diagram of a DRM modulator

Table 30.1. DRM bandwidths

Spectrum occupancy	0	1	2	3	4	5
Channel bandwidth [kHz]	4.5	5	9	10	18	20

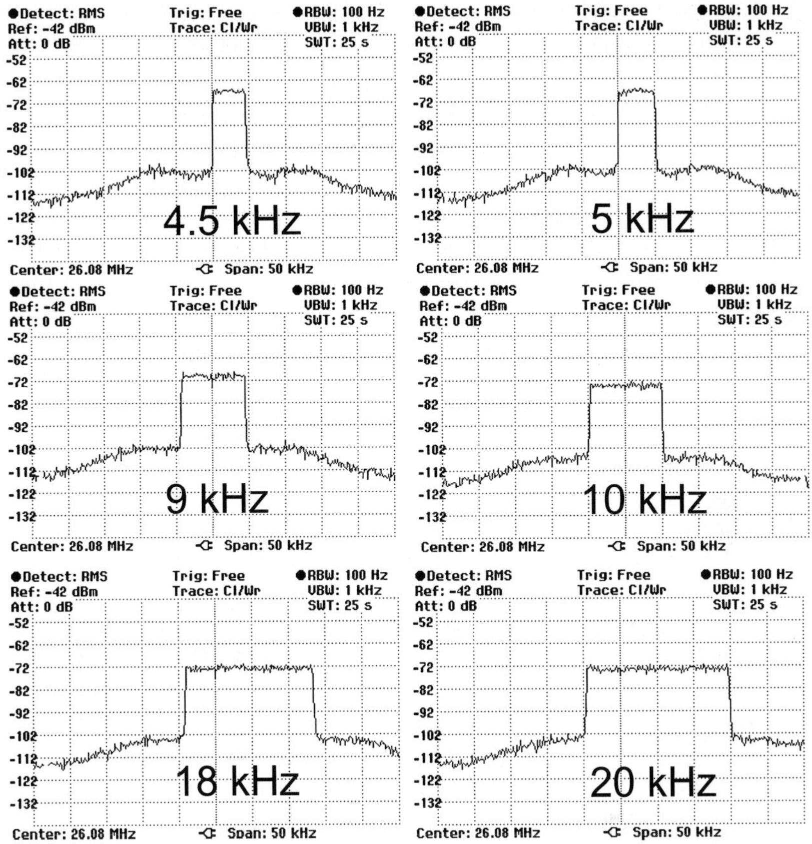

Fig. 30.2. DRM spectra at 4.5, 5, 9, 10, 18 and 20 kHz bandwidth with the same channel frequency in each case; it must be noted that the channel frequency does not always correspond to the band center of the DRM spectrum; compare also Table 30.3. (K_{min}/K_{max}).

Table 30.2. DRM modes and their physical parameters

DRM robustness mode	Symbol duration [ms]	Carrier spacing [Hz]	t_{guard} [ms]	t_{guard}/t_{symbol}	No. of symbols per frame
A	24	41 2/3	2.66	1/9	15
B	21.33	46 7/8	5.33	1/4	15
C	14.66	68 2/11	5.33	4/11	20
D	9.33	107 1/7	7.33	11/14	24

Table 30.3. Number of DRM carrier per COFDM symbol (K_{min} = lowest carrier no., K_{max} = highest carrier no., K_{unused} = unused carrier numbers, SO = spectrum occupancy)

Robustness mode	Carrier	SO 0 4.5 kHz	SO 1 5 kHz	SO 2 9 kHz	SO 3 10 kHz	SO 4 18 kHz	SO 5 20 kHz
A	K_{min}	2	2	-102	-114	-98	-110
A	K_{max}	102	114	102	114	314	350
A	K_{unused}	-1,0,1	-1,0,1	-1,0,1	-1,0,1	-1,0,1	-1,0,1
B	K_{min}	1	1	-91	-103	-87	-99
B	K_{max}	91	103	91	103	279	311
B	K_{unsed}	0	0	0	0	0	0
C	K_{min}	-	-	-	-69	-	-67
C	K_{max}	-	-	-	69	-	213
C	K_{unused}	-	-	-	0	-	0
D	K_{min}	-	-	-	-44	-	-43
D	K_{max}	-	-	-	44	-	135
D	K_{unused}	-	-	-	0	-	0

Fig. 30.1. shows the block diagram of a DRM modulator. Up to 4 services (audio or data) can be combined to form one DRM multiplex and to be transmitted in the so-called MSC (Main Service Channel). A DRM signal contains the following subchannels:

- the MSC = Main Service Channel (16QAM/64QAM modulation)
- the FAC = Fast Information Channel (QPSK)
- the SDC = Service Description Channel (QPSK/16QAM)

The FAC is used for signalling the following information to the receiver:

- Robustness mode
- Spectrum occupancy
- Interleaving depth
- MSC mode (16QAM/64QAM)
- SDC mode (QPSK/16QAM)
- Number of services

The SDC is used for transmitting information such as

- Protection level of the MSC
- Stream description

- Service label
- Conditional access information
- Audio coding information
- Time and date

30.1 Audio source encoding

DRM transmits MPEG-4-coded audio signals which can be compressed with the following algorithms:

- MPEG-4 AAC (= Advanced Audio Coding),
- MPEG-4 CELP speech coding (= Code Excited Linear Prediction),
- MPEG-4 HVXC speech coding (= Harmonic Vector Excitation Coding)

30.2 Forward Error Correction

The forward error correction in DRM is composed of the following:

- an energy dispersal block
- a convolutional coder
- a puncturing block

In DRM, it is possible to chose between

- Equal FEC and
- Unequal FEC

Parts of the audio frame can be error-protected to different degrees by this means. The degree of error protection is determined via the protection level and can be chosen as

- PL = 0 (maximum error protection)
- PL = 1
- PL = 2
- PL = 3 (lowest error protection)

The PL then results in a particular code rate.

30.3 Modulation Method

The Fast Access Channel (FAC) is permanently QPSK-modulated (Fig. 30.3.) since it is virtually the first "entry point" for the DRM receiver and must, therefore, be modulated firmly and very robustly.

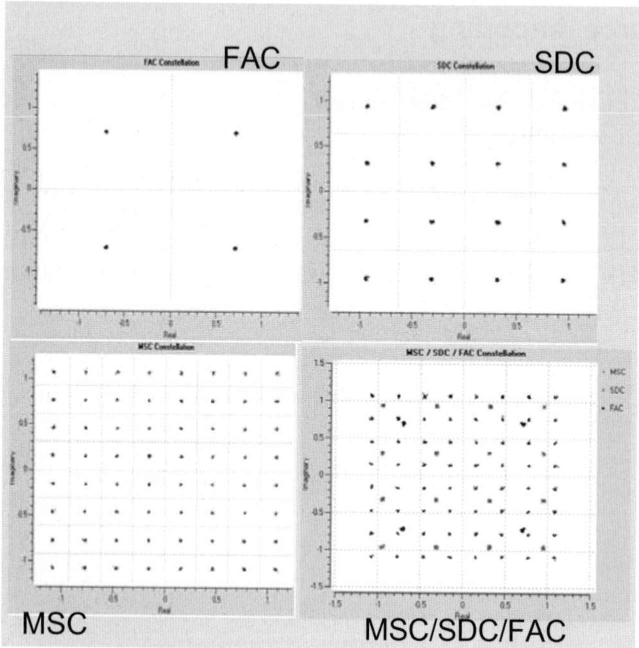

Fig. 30.3. Modulation methods in DRM

In the case of the Service Description Channel (SDC) it is possible to chose between QPSK and 16QAM as modulation method which is again signalled to the receiver via the FAC. The types of modulation possible in the MSC are either 16QAM or 64QAM (Fig. 30.3.) which is also signalled to the receiver via the FAC. Apart from the modulated data carriers which transmit the information of the MSC, FAC and SDC, there are also pilots which are not responsible for any information transport. They have special tasks and are mapped onto fixed constellation schemes known to the modulator and receiver. These pilots are used for:

- Frame, frequency and time synchronization
- Channel estimation and correction
- Robustness-mode-signalling

In DRM it is possible to chose, apart from "simple modulation“ (SM), also "hierarchical modulation“ (HM), similar to DVB-T. Different levels of error protection can then be used on the two paths of the hierarchical modulation.

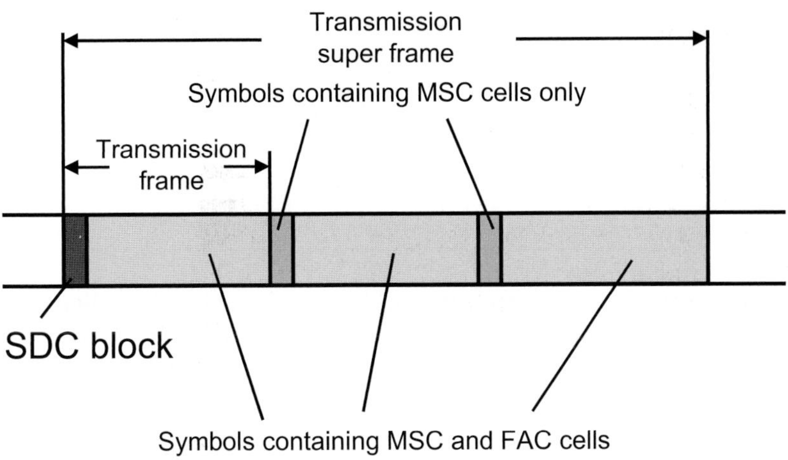

Fig. 30.4. Frame structure in DRM

30.4 Frame structure

Like other transmission standards such as DVB-T or DAB, DRM, too, has a frame structure (Fig. 30.4.) for arranging the COFDM symbols which is organized as follows:

- a certain number of COFDM symbols N_s results in an COFDM-transmission frame
- 3 transmission frames produce one transmission superframe

An COFDM frame, in turn, is composed of:

- Pilot cells
- Control cells (FAC, SDC)
- Data cells (MSC)

In this context, cells are understood to be carriers allocated to various uses. Control cells are used for transmitting the FAC and the SDC. Data cells are used for transporting the MSC.

The pilot cells are simply the pilots already mentioned. Table 30.4. shows how many CODFM symbols make up a transmission frame.

Table 30.4. Number of symbols N_s per frame

Robustness mode	Number of symbols N_s per transmission frame
A	15
B	15
C	20
D	24

At the beginning of a transmission super frame, the so-called SDC-block is transmitted in symbol no. 0 and 1 in Mode A and B and in symbol no. 0, 1 and 2 in Mode C and D. After that, only MSC and FAC cells are transported until the beginning of the next super frame (Fig. 30.4.).

Pilot carriers or pilot cells are distributed over the entire range of COFDM carriers. Depending on the mode, they are spaced apart by 20, 6, 4 or 3 carriers from one another and skip forward by 4, 2 or 1 carrier from symbol to symbol.

Table 30.5. Pilot Carriers

Mode	Pilot carrier spacing in the symbol	Carrier skip distance from symbol to symbol
A	20	4
B	6	2
C	4	2
D	3	1

30.5 Interference on the transmission link

DRM is operated in a frequency band in which atmospheric disturbances and diurnal fluctuations of the transmission characteristics (ground and sky wave) are particularly pronounced. In the frequency band below 30 MHz there is mainly also the presence of man-made noise to be considered.

According to the standard, DRM has a bit error ratio of $1 \bullet 10^6$ in the MSC after the channel decoder with a signal/noise ratio (S/N) of 14.9 dB with 64QAM and a code rate of 0.6. In practice, the "fall off-the cliff“

phenomenon (also known as "brickwall effect") was actually observed with an approximate S/N of 16 dB at a CR=0.5. With 16QAM modulation, this effect occurred with an SNR of about 5 dB (receiver: mixer DRT1 by Sat Schneider and DREAM Software).

Table 30.6. "Fall-off-the-Cliff" (Receiver: Mixer DRT1 by Sat Schneider, Germany and DREAM Software from the Technical University of Darmstadt, Germany

Transmission parameters	S/N at "fall-off-the-cliff"
MSC=64QAM, CR=0.5	16 dB
MSC=16QAM, CR=0.5	5 dB

30.6 DRM data rates

The DRM data rates depend on the DRM bandwidth (spectrum occupancy), on the mode, on the selected type of modulations and on the forward error correction. They are between about 5 and 72 kbit/s.

Table 30.7. MSC net data rates at a code rate of CR=0.6 (equal FEC, simple modulation) with 64QAM

Robustness mode	SO 0 4.5 kHz [kbit/s]	SO 1 5 kHz [kbit/s]	SO 2 9 kHz [kbit/s]	SO 3 10 kHz [kbit/s]	SO 4 18 kHz [kbit/s]	SO 5 20 kHz [kbit/s]
A	11.3	12.8	23.6	26.6	49.1	55.0
B	8.7	10.0	18.4	21.0	38.2	43.0
C	-	-	-	16.6	-	34.8
D	-	-	-	11.0	-	23.4

Table 30.8. MSC net data rates at a code rate of CR=0.62 (equal FEC, simple modulation) with 16QAM

Robustness mode	SO 0 4.5 kHz [kbit/s]	SO 1 5 kHz [kbit/s]	SO 2 9 kHz [kbit/s]	SO 3 10 kHz [kbit/s]	SO 4 18 kHz [kbit/s]	SO 5 20 kHz [kbit/s]
A	7.8	8.9	16.4	18.5	34.1	38.2
B	6.0	6.9	12.8	14.6	26.5	29.8
C	-	-	-	11.5	-	24.1
D	-	-	-	7.6	-	16.3

The lowest possible data rate (CR=0.5, 16QAM, Mode B, 4.5 kHz) 4.8 kbit/s. The highest possible data rate (CR=0.78, 64QAM, Mode A, 20 kHz) is 72 kbit/s.

30.7 DRM transmitting stations and DRM receivers

Numerous transmitting stations throughout the world have already been converted from AM to DRM. Relevant information is available from the Internet. Apart from software-based DRM receivers, compact receivers are also available now. Software-based solutions are in most cases based on a DRM signal down-converted at 12 kHz which is fed into the line-in socket of a PC. A suitable example which can be mentioned is the DREAM software from the Technical University of Darmstadt (see also Fig. 30.5.).

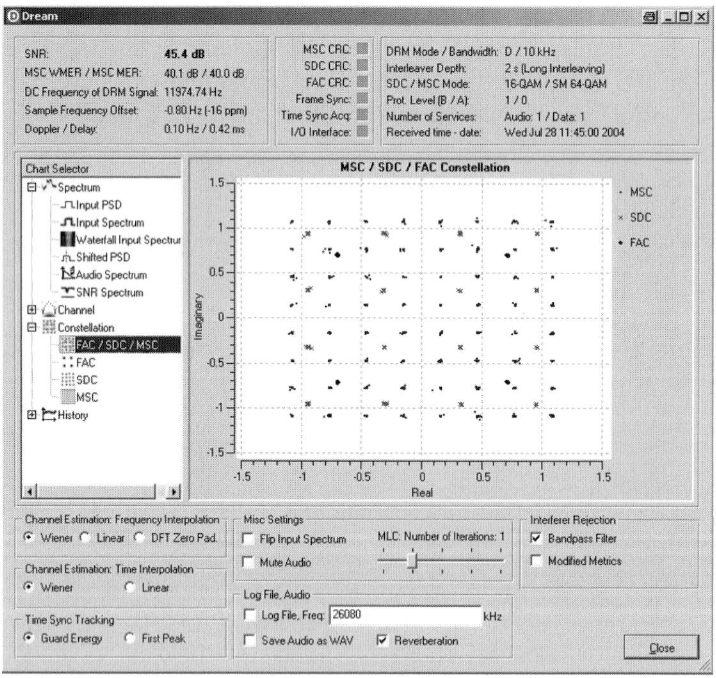

Fig. 30.5. Constellation diagram of a DRM signal (MSC, FAC and SDC superimposed), recorded using the DREAM software

30.8 DRM+

DRM+, an extension of DRM, is being developed and is intended for frequencies above 30 MHz. DRM+ could be a possible alternative to DAB. Like DRM, DRM+ would work with the latest AAC+ codec and could be used both in VHF band II, where VHF FM technology is currently employed, and in VHF band I which, as now, is empty.

Bibliography: [ETS101980], [DREAM]

31 Digital Terrestrial TV Networks in Practice

This chapter is intended to provide the practical engineer with an overview of the configuration of TV transmitting stations and of the structure of single-frequency DVB-T networks, using as examples the DVB-T SFN networks of Southern and Eastern Bavaria with some TV transmitting stations of the Bayerischer Rundfunk (BR) and T-Systems/Deutsche Telekom (now Media Broadcast GmbH). The author has been able to experience the commissioning of both networks at close quarters, both during the initial training of some of the operating personnel, during visits in the installation phase, and when the networks were being switched on. In addition, both networks are located in the region in which the author has grown up and is still living. And all the TV transmitting stations are completely equipped throughout with "Bavarian Technology" by the companies Rohde&Schwarz, Spinner and Kathrein. Beginning with the playout center, the entire feed link of the single-frequency networks (SFN) and especially the transmitting stations themselves, from the mask filter and the combiner to the transmitting antenna will be described in this chapter. The author was even granted the privilege of experiencing the "flight" of the antenna by helicopter at the Olympic Tower, Munich, and at the Mt. Wendelstein transmitter during the installation. This chapter also describes measurements in a single-frequency network and explains how coverage is measured in an SFN. In addition, the practical requirements for an SFN-enabled receiver are discussed and how they can be verified. All this information is derived from practical experience and cannot be found in some standards or documents: information from practical engineer to practical engineer.

31.1 The DVB-T SFNs Southern and Eastern Bavaria

The DVB-T single-frequency networks (SFNs) used as the example are networks in the South of Germany, in Germany's largest federal state which has a geography of high (Alps) and low mountain ranges and in-

W. Fischer, *Digital Video and Audio Broadcasting Technology*, Signals and Communication Technology, 3rd ed., DOI: 10.1007/978-3-642-11612-4_31, © Springer-Verlag Berlin Heidelberg 2010

cludes the gentle foothills. This topography was of the greatest significance in the planning of the networks.

The Southern Bavaria DVB-T network consists of the two transmitters Olympic Tower Munich and Mt. Wendelstein. The Olympic Tower is a typical telecommunications tower to the North-West of Munich with a height of 292 m at about 450 m above sea level. It was originally used as microwave tower for telephoning and was built in 1968. Microwave has been largely replaced by optical fiber today and is no longer of the same significance as before. Thus there are hardly any more microwave dishes in operation on the Olympic Tower. At the upper end of the Olympic Tower there are the transmitting antennas for FM radio, DAB and now also for DVB-T.

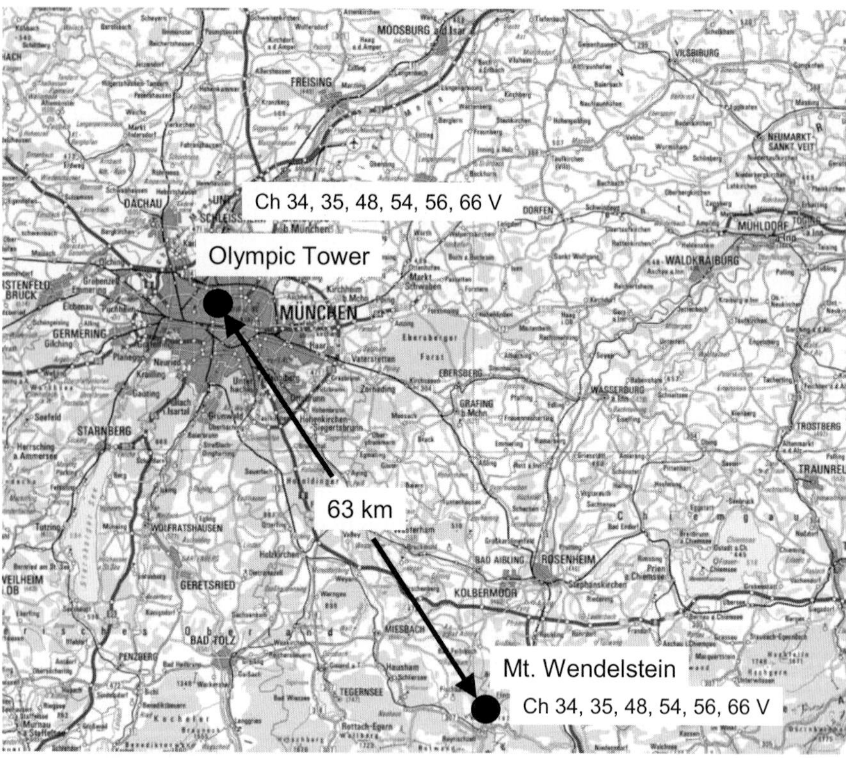

Fig. 31.1. DVB-T single-frequency network Southern Bavaria (DTK500; © Landesamt für Vermessung und Geoinformation Bayern, Nr. 4385/07)

The Mt. Wendelstein transmitter is located at about 1750 m above sea level on the mountain of the same name which has a total height of 1850

m. Although it is not the highest mountain in Bavaria, or Germany, for that matter, it has certainly one of the most beautiful panoramic views in Bavaria. The TV transmitter there is the oldest one in Bavaria, and possibly also the one with the most beautiful location. Anyone who has been able to experience a sunrise or sunset there - which is something not many people are able to do because of the lack of an hotel at the top - will be able to confirm this.

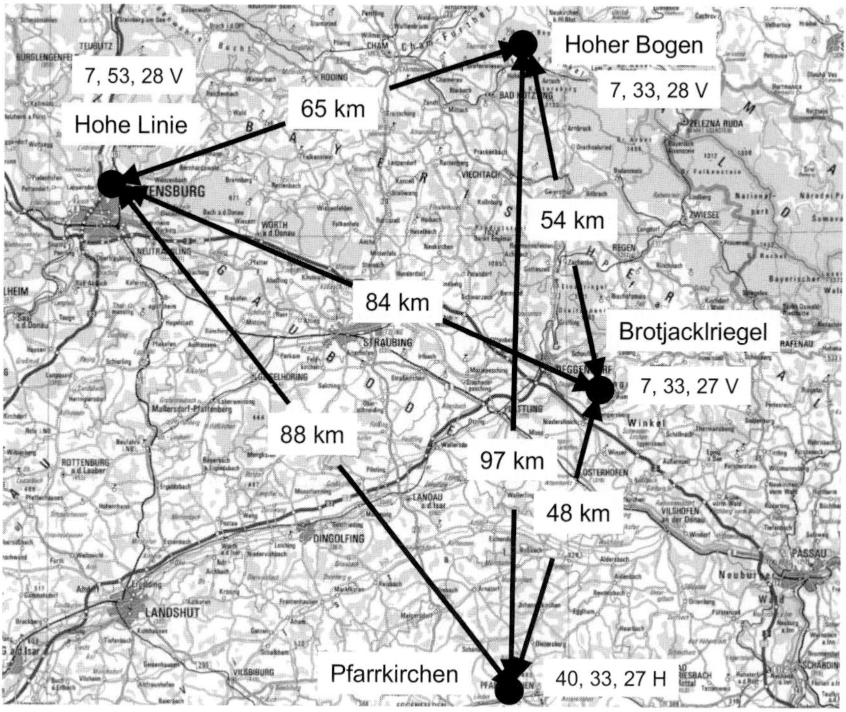

Fig. 31.2. DVB-T single-frequency network Eastern Bavaria (DTK500; © Landesamt für Vermessung und Geoinformation Bayern, Nr. 4385/07)

The Wendelstein transmitter belongs to the Bayerischer Rundfunk. The two Munich Olympic Tower and Wendelstein transmitters form the SFN Southern Bavaria which was taken into operation in the night of the 30th May 2005. At the same time, this was the end of analog terrestrial television in Southern Bavaria. The Munich Olympic Tower and the Wendelstein transmitters broadcast 6 DVB-T channels completely synchronously on the same frequencies as DVB-T single frequency network. The data rates are about 13 Mbit/s each and each carries about 4 TV programs per data stream. Altogether, the viewer is thus provided with about 22 programs over terrestrial digital antenna TV. These TV programs, which are both public service programs and private programs, form a viable alternative to the satellite and cable media. The transmitting frequencies are now only located in the UHF band.

The DVB-T single-frequency network Eastern Bavaria consists of the 4 TV transmitting stations Pfarrkirchen (T-Systems/Deutsche Telekom, now Media Broadcast GmbH), Brotjacklriegel (BR), Hoher Bogen (BR) and Hohe Linie (BR). Two of these transmitters (Brotjacklriegel and Hoher Bogen) are located in the low mountain ranges of the Bavarian Forrest at approx. 1000 m above sea level. The 4 transmitting stations broadcast 3 DVB-T transport streams, some of them at the same frequencies. Only public service programs are distributed. The data rate per data stream is also approx. 13 Mbit/s. Alltogether, 12 programs are distributed. Fig. 31.1. shows the sites of the transmitters in the DVB-T single-frequency network Southern Bavaria and Fig. 31.2. shows the sites of the transmitters in the DVB-T network Eastern Bavaria. With 63 km, the distances between transmitters in the DVB-T single-frequency network Southern Bavaria, consisting of the Olympic tower and Mt. Wendelstein, are right on the limit of what is still allowed. In the DVB-T single-frequency network Eastern Bavaria, the permissible distances between transmitters have been greatly exceeded in some cases and, without inbuilt delay times, would lead to guard interval violations at some locations. Table 31.1. and 31.2. list the technical parameters of both networks.

31.2 Playout Center and Feed Networks

The Playout Center of the Bayerischer Rundfunk is located on the site of the München-Freimann television studio. Here, 2 multiplexed data streams are formed consisting of the ARD stream ("Das Erste, ..."), and the BR stream ("Bayrisches Fernsehen“, "BR-Alpha“, ...). Throughout the country, the ZDF stream comes directly from Mainz and contains (“ZDF, ...). The other transport streams in the Playout Center directly at the Olympic Tower in Munich are formed by Media Broadcast GmbH, partially by reception back via DVB-S. These are 3 further transport streams containing private TV programs. These 3 transport streams are only broadcast in the DVB-T network Southern Bavaria and not in Eastern Bavaria at the moment. The Olympic Tower is linked to the BR Playout Center by fibre optic, the transport streams of the Playout Center from Media Broadcast GmbH being supplied to the broadcast studio via a few meters of cable.

The ZDF transport stream arrives from Mainz by optical fiber. All the transport streams are then beamed up by microwave from the Olympic tower to the Mt. Wendelstein by way of Schnaitsee-Hochries (Media Broadcast GmbH) or via Freimann-Wendelstein (BR). The MIP inserters for synchronizing the transmitters in the single-frequency network are located at the output of the respective Playout Center. The feed links and the components in the Playout Center are all provided redundantly. The transport streams are fed to the DVB-T SFN Eastern Bavaria by microwave (ARD-MUX, BR-MUX) or optical fibre (ZDF-MUX) via an ATM network.

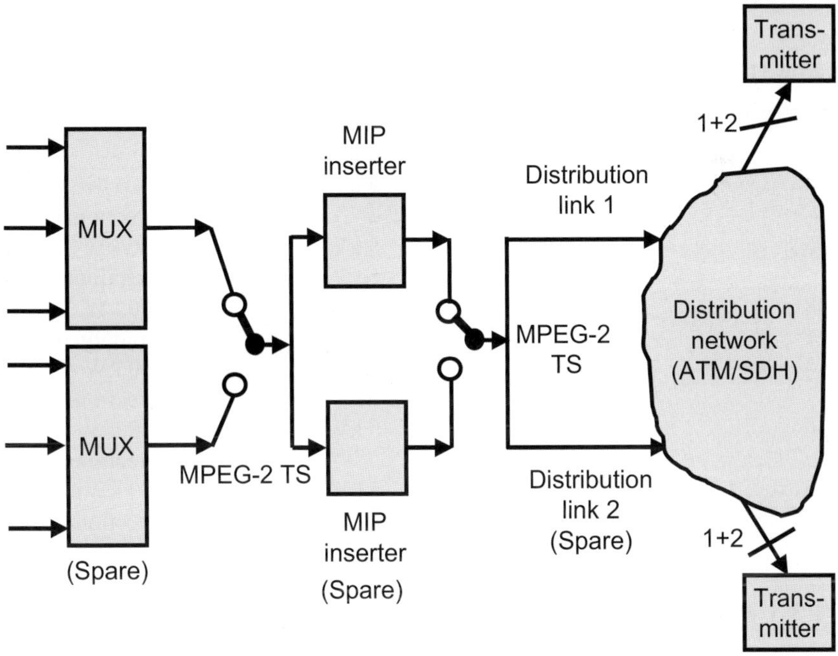

Fig. 31.3. Feed of a transport stream from the playout center to the DVB-T transmitters in the single-frequency network

31.3 Technical Configuration of the Transmitter Sites

In this section, the configuration of a DVB-T site will be explained by means of three examples. These are the Olympic tower Munich (Media Broadcast GmbH), Wendelstein (BR), and Brotjacklriegel (BR) transmitters. As described in the previous chapter, the MPEG-2-transport stream is supplied to the TS-ASI input interface over the most varied networks, but finally via 75 Ohm coaxial cable. There is always a spare link provided. The important factor is that, naturally, the same MIP inserter always feeds the spare link. Each MIP inserter per se operates completely asynchronously and only creates synchronous information in the MPEG-2 transport stream by itself inserting MIP packets. Each MIP inserter inserts these special transport stream packets virtually completely freely into gaps in the MPEG-2 transport stream instead of null packets (PID=0x1FFF).

The DVB-T transmitters at all three sites are liquid cooled solidstate transmitters of the NX7000 and NX8000 series by Rohde&Schwarz, constructed in various power classes and in various spare capacities.

Table 31.1. Channel allocation and power Mt. Wendelstein transmitter

Channel	Frequency	Mean transmitter output power	ERP
34	578 MHz	5 kW	100 kW directional
35	586 MHz	5 kW	100 kW directional
48	690 MHz	5 kW	100 kW directional
54	738 MHz	5 kW	100 kW directional
56	754 MHz	5 kW	100 kW directional
66	834 MHz	5 kW	100 kW directional

31.3.1 Mount Wendelstein Transmitter

The Mt. Wendelstein transmitter is the oldest television transmitter site in Bavaria and may well be one of the oldest in all of Germany. It was taken into operation as television transmitter in 1953, is sited at 1740 m above sea level and its antenna is at approx. 1840m. The Wendelstein mountain itself is very popular as destination for outings by wanderers and skiers. From there, six transport streams are being broadcast by DVB-T since May 30th, 2005. Like from the Olympic Tower, these are channels 34, 35, 48, 54, 56 and 66. In addition to FM transmitters and one DAB transmitter, an analog TV transmitter had been operating there on VHF channel 10 for many decades. At the place where the analog TV transmitter was located, all 6 UHF-transmitters are now installed. From all 6 UHF transmitters, out-of-band components, which are not permissible and could interfere both with the adjacent channels and other channels, are first removed by in each case one mask filter (critical mask). The required shoulder attenuation is generally about 52 dB.

Table 31.2. Program offer and technical parameters, DVB-T Network Southern Bavaria (channel 10 has been moved to channel 54 in July 2009)

Channel	Modulation	Code rate	Guard	Data rate [Mbit/s]	Number of programs	Programs
54	16QAM	2/3	1/4	13.27	4	Das Erste, Phoenix, arte, 1plus
34	16QAM	2/3	1/4	13.27	4	RTL, RTL2, VOX, SuperRTL
35	16QAM	2/3	1/4	13.27	3 + MHP	ZDF, 3sat, Doku/Kika
48	16QAM	2/3	1/4	13.27	4	Sat1, Pro7, Kabel 1, N24
56	16QAM	2/3	1/4	13.27	3 + MHP	Bayr. Fernsehen, BR alpha, SWR
66	16QAM	2/3	1/4	13.27	4	Tele5, Eurosport, HSE24, München TV

The mask filters are made by the company Spinner and are designed as dual mode and cavities filters. The pass-band attenuation of such a filters is about 0.6 dB. The group delay of this filter can be easily pre-equalized, and thus compensated for, by the TV transmitter. Like the transmitters themselves, these mask filters are here located in the same room and are coupled to the transmitters "from the top" via 50 Ohm coaxial copper tubing. I.e., the transmitter output is at the top, which is not always the case. From the mask filters it continues in this case to the combiners which are located underneath and via which in each case one transmitter is connected, decoupled from the other ones, to the antenna line.

The combiner is also made by Spinner. The configuration of the combiners for one channel is shown in Fig. 31.10. and consists of two 3 dB couplers and two band-pass filters tuned to the supplied channel in each case. The channel to be coupled in is thus supplied separately from the others.

Thus, 6 UHF channels are coupled together and conducted up to the transmitting antenna via a coax cable. The length of the line to the antenna is about 280m. Depending on frequency, the cable attenuation can be assumed to be about 0.5 dB per 100 m length of transmitting cable.

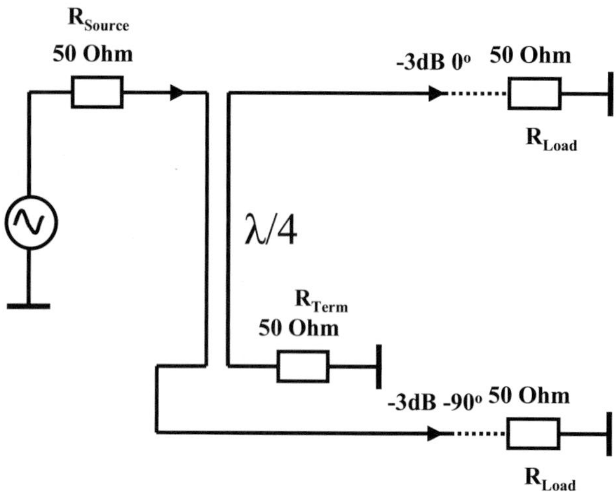

Fig. 31.4. 3-dB splitter

31.3.1.1 Transmitter Technology Used

All of the 6 DVB-T transmitters on the Mt. Wendelstein are liquid-cooled high-power solid-state transmitters of the NX7000 and now also NX8000 series by Rohde&Schwarz. Such a transmitter can be imagined simply as a high-power amplifier driven by an exciter and produced by connecting together a large number of power amplifier modules. Each transistor in these single power amplifier stages generates a average power of about 25 W. The output signals of the amplifiers are added together via 3 dB couplers. Using 3 dB couplers, more and more amplifiers are coupled together one by one so that a total power of about 450 W per amplifier module is obtained in the case of the NX7000 series by Rohde&Schwarz. The output powers of the various amplifier modules are then combined again via couplers to give the total output power of the transmitter of about 5 kW (mean power) in the case of the Mt. Wendelstein transmitter. These couplers can be implemented as Wilkinson coupler or again as 3dB coupler. A Wilkinson coupler is a 0 degree coupler, whereas the 3dB coupler is a 90 degree coupler.

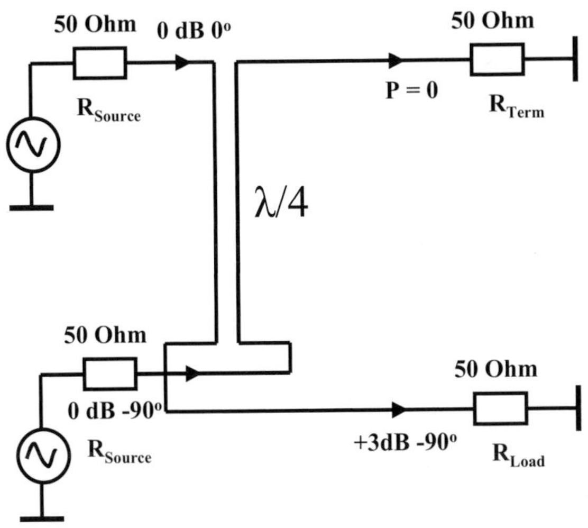

Fig. 31.5. 3-db combiner

31.3.1.1.1 3dB Coupler as Splitter and Combiner

In principle, a directional coupler consists of two closely adjacent parallel lines with a length of λ/4. The spacing between the lines determines the overcoupling attenuation; if this is 3 dB, it is called a 3 dB coupler. If a signal is fed into an input of a 3 dB coupler, 3 dB of it are coupled out with 0 degree phase shift at the output opposite the input, and another 3 dB of it are coupled out with 90 degree phase shift at the output of the λ/4 line section connected electrically to the input.

A 3 dB coupler can be used for adding powers by feeding a signal into one input with 0 degree phase shift and feeding it with a 90 degree phase shift into the input of the parallel λ/4 line section (3 dB combiner). The signals then cancel at one output and the aggregate power is present at the other output with 90 degrees phase rotation. The unused output is terminated with 50 Ohms (load matching resistor). In the case of phase errors or power differences in the signals supplied, power is also absorbed in the load matching resistor.

Fig. 31.6. shows a simplified circuit symbol of a 3 dB coupler which will be used in the following sections.

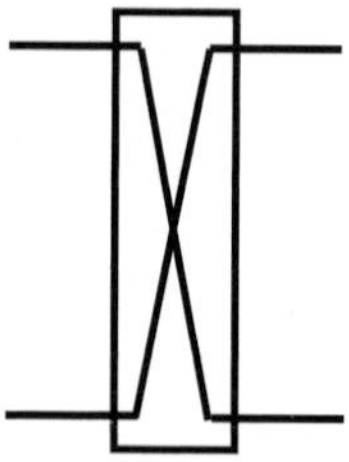

Fig. 31.6. Simplified circuit symbol of a 3-dB coupler

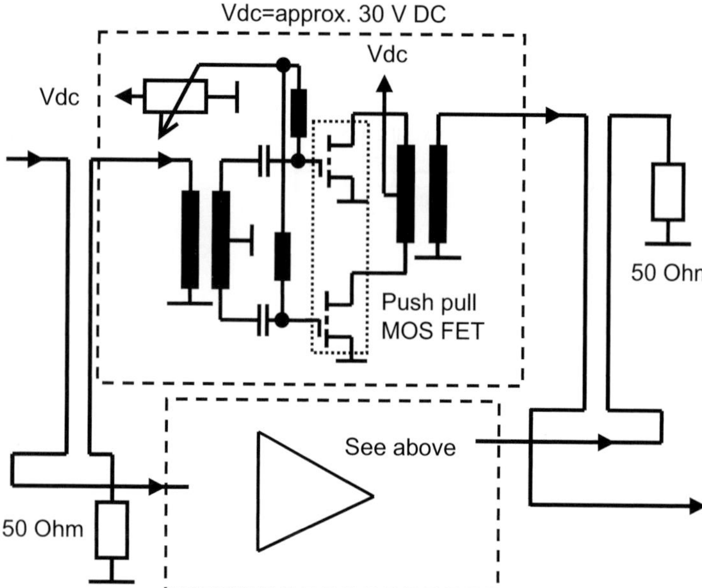

Fig. 31.7. Principle of a single power amplifier stage (50 W), used to build up a power amplifier module

31.3.1.1.2 Single Power Amplifier Stage (appr. 50 W)

Firstly, the basic principle of a single power amplifier stage for DVB-T/ATSC/Analog TV of the 50 W averaged power will be explained. Such power amplifiers also form the high-power transmitter for DVB/ATSC/Analog TV. The input signal is split into two signals of –3dB power each and 90 degrees phase difference by means of a 3 dB coupler. Each amplifier path consists of a class AB amplifier which, in turn, consists of a dual transistor operated in push-pull mode. The operating point is set in such a way that similar conditions are achieved for all transistors in class AB mode and the transfer distortion is minimal. In contrast to FM transmitters, the amplifiers used here are very linear transmitters but must still be pre-corrected. In FM transmitters, class C amplifiers are used which are quite nonlinear but have much greater efficiency.

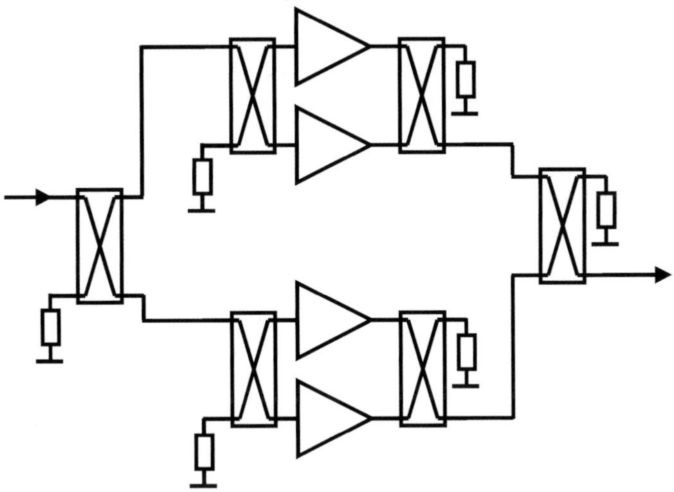

Fig. 31.8. Principle of a power amplifier module for VHF or UHF

In the FM transmitter, the required power can even be chosen by controlling the supply voltage of the class C amplifiers. In television transmitters, where very good linearity is required, power is controlled via the amplifier feed power. This applies especially to analog television, but also to digital television (DVB-T, ATSC, ISDB-T). These amplifiers operate with VMOS transistors (VHF band), or LDMOS transistors (UHF band) and are already "pre-equalized" via the exciter, i.e. the characteristic is simulated in the equalizer and used for comparison.

In principle, a class AB amplifier consists of a push-pull stage which is set by the quiescent current of the transistor in such a way that the transfer distortion is already minimized. The transistor supply voltage is about 30 V.

To build up a power amplifier module, the input signal is first amplified from approx. 0 dBm to a reasonable order of magnitude and then split with the correct power and phase over a number of 3 dB couplers and supplied to the respective individual amplifiers. The output powers of the individual amplifiers are then combined again and again via 3 dB couplers to form the total output signal of an amplifier module. The average total power of an power amplifier module is then approx. 450 W. Then the power amplifier modules are coupled together building a power amplifier unit.

31.3.1.2 Mask Filter

The mask can be implemented as an uncritical mask or as a critical mask, depending on the requirements of the relevant regulating authorities. In the case of the Mt. Wendelstein and Olympic Tower transmitters, they are filters with a critical mask, i.e. the "shoulder" of the DVB-T signal must be lower by more than 51 dB in the adjacent channel. The manufacturer of these filters is the company Spinner in Munich. The filters are so-called dual mode filters. The filters are passive mechanical cavity resonators and are relatively large because of their power rating and weigh more than 100 kg. The mask filters are simply used for lowering or suppressing adjacent channel emissions. The mask filters are tuned to the respective channel and have an attenuation of about 0.3 to 0.6 dB in the passband which is also easily noticed as heat.

31.3.1.2.1 Filter Technology

In the band-pass filters used here, a technological distinction is made between coaxial filters and waveguide filters. In both technologies, band-pass filters with 3 to 10 sections are then constructed as required. They both have their advantages and disadvantages. A coaxial filter has the following characteristics:

- continuous tuning in Band III or IV/V
- greater attenuation (e.g. 0.31 dB in mid-band)
- good temperature stability

The waveguide filter characteristics are:

- less attenuation (e.g. 0.17 dB in mid-band)
- poorer temperature stability
- tunable only over some channels (5 - 6 channels)
- larger dimensions

In combiners, 3- or 4-section filters are used. In mask filters, 6-section filters are needed for the uncritical mask and 8-section filters for the critical mask.

31.3.1.2.1.1 Coaxial Filter

In principle, coaxial filters are nothing else than λ/4-long coaxial lines which are short-circuited at one end and open at the other end. These can be used for constructing resonators. The resonant frequency of the resonators can be varied by varying the length of the line. The impedance of a coaxial line with round inner conductor and round outer conductor is:

$$z_L = 377\Omega \cdot \ln(D/d)(2\pi\sqrt{\varepsilon_r}) = 60\Omega \cdot \ln(D/d)/\sqrt{\varepsilon_r};$$

where

d = diameter of the inner conductor;
D = diameter of the outer conductor;
ε_r = relative dielectric constant.

The impedance for the combination of round inner conductors and rectangular outer conductors is:

$$z_L \approx 60\Omega \cdot \ln(1.07 \cdot D/d)/\sqrt{\varepsilon_r};$$

The following applies to coaxial systems:

Minimum attenuation at $Z_L = 77\Omega/\sqrt{\varepsilon_r}$
maximum dielectric strength at $Z_L = 60\Omega/\sqrt{\varepsilon_r}$
maximum transferable power at $Z_L = 30\Omega/\sqrt{\varepsilon_r}$

The mask filter and the antenna combiner require mainly a low attenuation to keep the losses as low as possible. The dielectric is air, resulting in ε_r = 1. The selected impedance is about 77Ω since the attenuation is lowest there.

The wavelength is calculated as:

$$\lambda = c_0/(f\sqrt{\varepsilon_r});$$

where

c_0 = velocity of light = 3 x 10^8 m/s
f = frequency;
ε_r = relative dielectric constant.

This results in a λ/4 resonator line length of about 36 cm to about 8 cm in the frequency band of between about 200 and 900 MHz with an air dielectric. The cavity diameter depends on the power class, choosing round or square cavities. The line can also be shortened by loading it capacitively at its open end. Coupling into and out of the cavity can be capacitive or inductive. Fig. 31.9. shows the basic configuration of a two-section coaxial filter. The signal is coupled in capacitively by a plunger the engaged depth of which is adjustable. The resonator lengths l can be adjusted by inserting plungers which are short-circuited at one end, to a greater or lesser depth. The resonator length then determines the resonant frequency which can be influenced additionally by trimming screws.

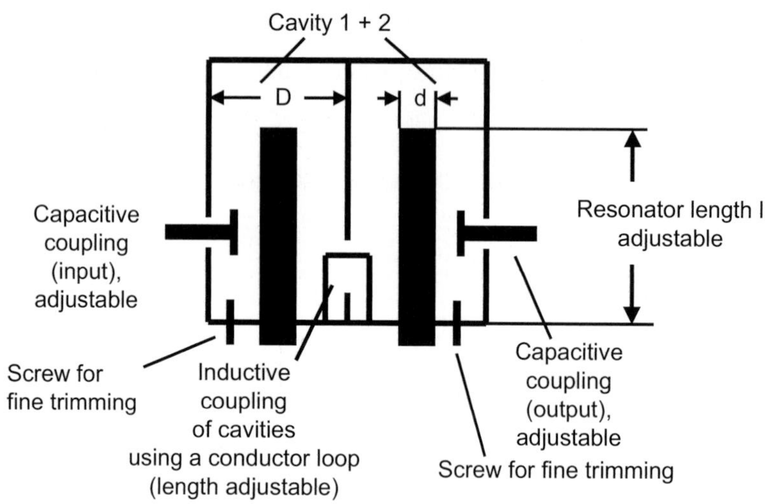

Fig. 31.9. Structure of a coaxial filter

Coupling from cavity 1 to cavity 2 is carried out here by a conductor loop which can be inserted to a greater or lesser depth, making the coupling between the two resonant cavities controllable. The degree of coupling determines the bandwidth of the filter. The output coupling from cavity 2 is also capacitive and is also adjustable. To make the sides of multi-section

filters even steeper, further non-adjacent cavities are also adjustably connected.
The linear expansion of the resonators must be compensated for as far as possible by a suitable choice of materials since otherwise the resonant frequency will become detuned.

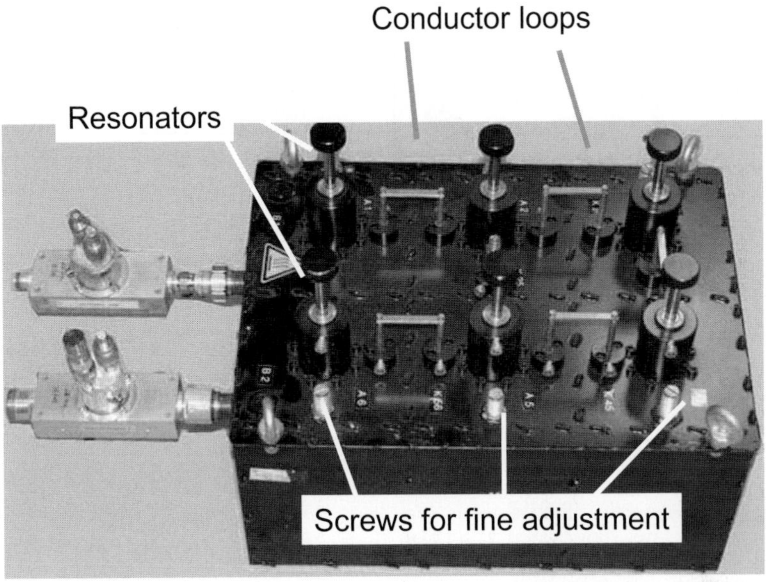

Fig. 31.10. Trimmers on a 6-section coaxial filter (Spinner)

31.3.1.2.1.2 Waveguide Filters

Waveguide filters are hollow chambers, or cavities, which are really hollow without inner conductor and in which real waveguide modes are generated. Cavities can also be used for creating resonators the resonant frequency of which ultimately depends on the volume of the waveguide. The shape of the filters used in mask filters is cylindrical; the two parameters D = diameter and h = height of the cylinder then determine the resonant frequency and also the Q factor. In the cylindrical cavity, two modes which do not interfere with one another can be excited orthogonally to one another (by 90 degrees). This makes it possible to use this cavity resonator virtually twice, thus saving space. It is then called a dual-mode filter, implementing two sections with one cavity. A 6-section filter thus requires 3 cavities and an 8-section filter requires 4. Fig. 31.11. shows the basic configuration of a dual-mode waveguide filter.

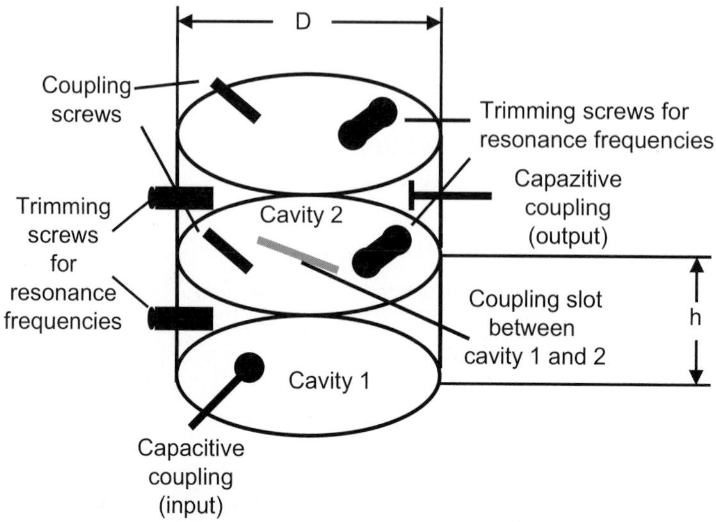

Fig. 31.11. Structure of a dual-mode waveguide filter

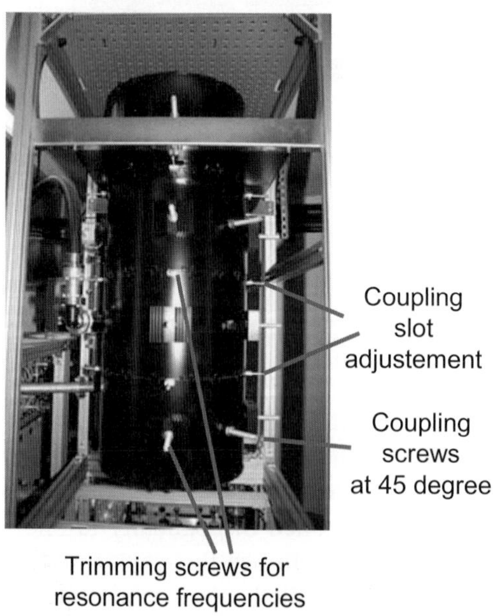

Fig. 31.12. Trimmers of a dual-mode filter (Spinner)

The signal is coupled capacitively into cavity 1 by a plunger which can be inserted to a greater or lesser depth. Opposite to it there are trimming bolts which can be screwed in more or less deeply and can be used for tuning the resonant frequency. A coupling screw which rotates the field and generates an orthogonal waveguide mode at an angle of 90 degrees from the first one is arranged at an angle of 45 degrees to the bolts. Coupling between cavities 1 and 2 is via a variable-length coupling slot. In the second cavity there are also trimming bolts, and a couling screw. The degree of coupling between the orthogonal waveguide modes is determined by the depth of insertion of the coupling screws. When the waveguide filter becomes hot, the cavities expand and the resonant frequency thus becomes detuned towards lower frequencies. This can be prevented by using metal alloys (Invar) which have a particularly low coefficient of expansion. Waveguide filters are larger than coaxial filters; both the cavity height and its diameter must accommodate a half wavelength. The waveguide mode excited is called an H_{111} wave.

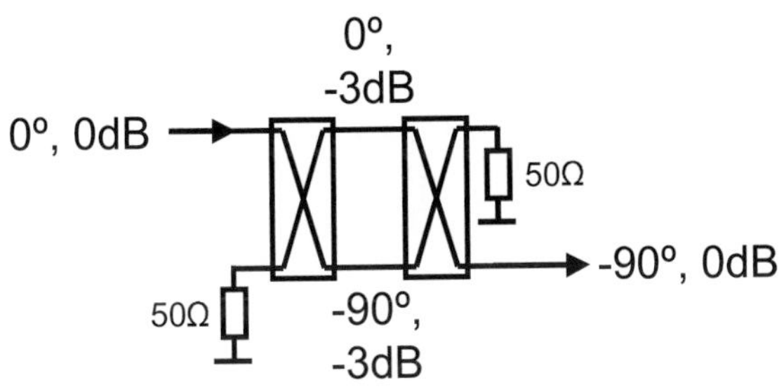

Fig. 31.13. Cascading two 3-dB couplers

31.3.1.3 Combiner

The combiner has the task of combining the various TV channels to form one signal which is then supplied to the respective TV transmitting antenna via one cable in the VHF band and one cable in the UHF band. The transmitters themselves must be well decoupled from one another which is precisely what the filters in the combiner are doing. The combiner has an approximate passband attenuation of 0.3 dB.

Each channel separating filter of the combiner consists of two channel filters which are tuned to the respective TV channel to be supplied. This is preceded and followed by a 3-dB coupler. To understand the operation, two cascaded 3 dB couplers can first be considered (Fig. 31.13.).

If a signal is fed into the first coupler, it divides into 2 signals of 0 degrees and 90 degrees in phase. The second coupler adds these signals again to form a signal which is now shifted by 90 degrees compared with the input signal. In the channel separating filter, band-pass filters are connected between the two couplers.

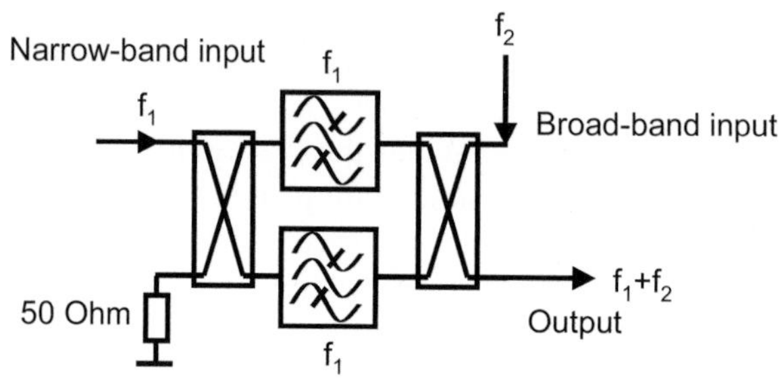

Fig. 31.14. Principle of an antenna combiner

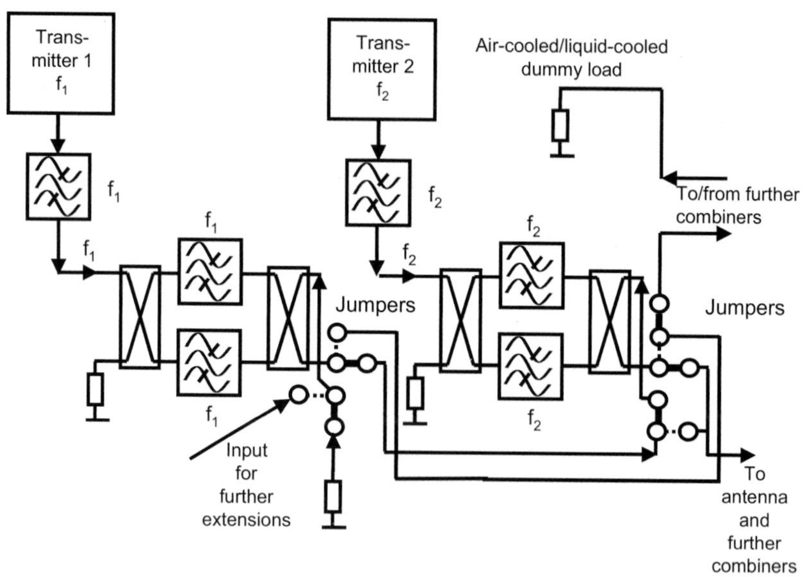

Fig. 31.15. Transmitter, mask filter and antenna combiner

The channel separating filter has a narrow-band input and a wide-band input and is basically constructed in exactly the same way as a vision/sound diplexer in analog television. The channel to be supplied passes through the filter with 0 degrees and 90 degrees phase shift and the signal

of the other transmitters is immediately totally reflected at the filters in the wideband input after the coupler and comes out again in aggregate with the supplied channel at the wideband output. Both filters of the channel separating filter must be tuned at least relatively identically.
Fig. 31.15. shows a combiner with two transmitters, two mask filters, two combiners, and patch panels at which the combiner can be bridged and the transmitter output can be switched to a dummy load, if required, in order to be able to carry out tuning and measuring work without applying the signal to the antenna.

31.3.1.4 Antenna and Feed System

From the transmitter building on the Wendelstein mountain to the transmitting antenna, a total of three lines are run which are "flexible" 50 Ohm coaxial cables. The very first transmitter cable was pulled up unwound on the rail by the rack railway in a "special action" at the beginning of the fifties. The new cables were flown up the mountain by helicopter, among them the last UHF cable which, with a diameter of almost 20 cm, is the thickest cable ever used on the Wendelstein mountain. One cable is used for the VHF band and one is used for the UHF band, and the third cable is a spare for emergency cases.

Such coax cables have approximately the following attenuations over 100 m length in each case:

Table 31.3. Technical parameters of HELIFLEX (©RFS) coax cables

Diameter of the coax cable	Max. avg. power at 500 MHz	Att. [dB] at 200 MHz	Att. [dB] at 500 MHz	Att. [dB] at 800 MHz
4-1/8"	35 kW	0.4 dB	0.7 dB	0.9 dB
5"	55 kW	0.3 dB	0.5 dB	0.7 dB
6-1/8"	75 kW	0.3 dB	0.4 dB	0.6 dB
8"	120 kW	0.2 dB	0.4 dB	0.5 dB

At the top, at the solar observatory and the meteorological observatory of the German Meteorological Service (Deutscher Wetterdienst), there is also the so-called antenna house from which the cables are then run to the actual transmitting antenna. It contains another patch panel via which the upper and lower half antenna of the transmitting antenna can be selectively fed or disconnected both in the VHF band and in the UHF band. If necessary, this is done via 20 cm-large coax-type U-links.

The antenna itself consists of the following components by Kathrein, Rosenheim, which are housed in a FRP (fiber-reinforced plastic) cylinder.

The FRP cylinder has a total height of approx. 24 m and the antenna as a whole is appr. 65 m high; the tip of the antenna being about 1900 m above sea level.

The VHF antenna is composed of 6 levels with 6 vertically polarized Band-III dipole antenna arrays. The bottom 3 levels form the lower half VHF antenna, the upper 3 levels form the upper half VHF antenna. Both half antennas are supplied via their own feed cable each. The UHF antenna consists of 12 levels with 8 Band-V/V antenna arrays each which, just like the VHF antenna, is built up from the lower and upper half antenna (with 6 levels each in this case). Here, too, each half antenna is driven by its own coaxial cable. Above the UHF antenna, there is the mechanical vibration absorber which is intended to prevent oscillations in the case of wind loading.

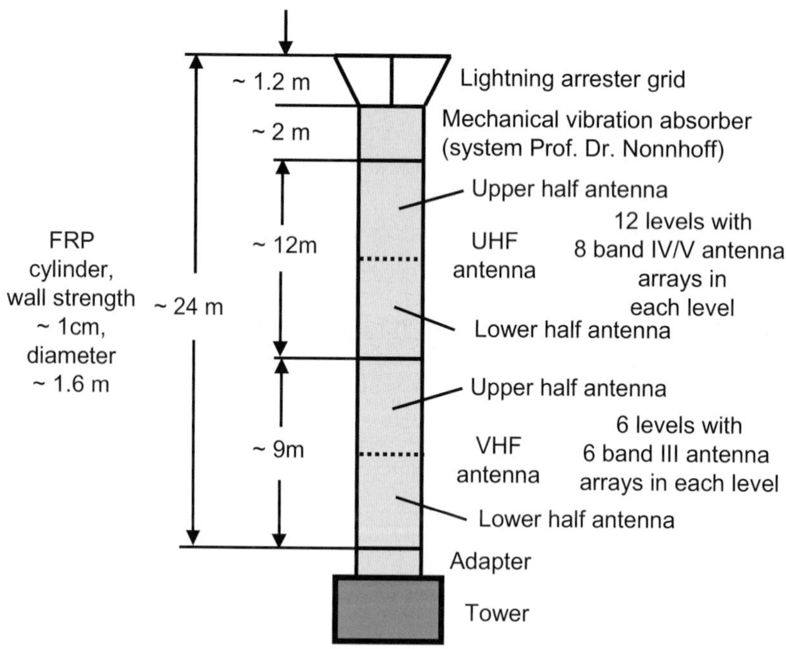

Fig. 31.16. Structure of a VHF/UHF transmitting antenna constructed in FRP

31.3.2 Olympic Tower Transmitter, Munich

The Munich Olympic Tower has originally been built in 1968 as telecommunications tower. Since April 2005, it has at its top the DVB-T transmitting antennas which, together with the Mt. Wendelstein transmitter, form the DVB-T single-frequency network Southern Bavaria since May 30th, 2005. From here, too, a total of 6 channels are broadcast in single frequency completely synchronized with the Mt. Wendelstein transmitter. The output power of the DVB-T-transmitters is approximately twice as high as that of the transmitters on the Wendelstein mountain. It is about 10 kW per channel in the UHF band. In contrast to the Mt. Wendelstein transmitter, where the main pattern of the antenna points to the North, the Olympic Tower has an omni-directional pattern. The ERP is about 100 kW in the UHF band from the Olympic Tower.

Fig. 31.17. Olympic Tower Munich; antenna installation for DVB-T using a helicopter (left) and Wendelstein antenna (right)

The high-power transmitters are also liquid-cooled NX7000/8000 series solid state transmitters by Rohde&Schwarz, but with much higher power capacity. As far as the installation is concerned, the system is set up for heating the Olympic pool close-by with the waste heat of the 6 high power transmitters – which, in spite of their very effective efficiency, is not inconsiderable – instead of dissipating it to the environment via heat exchangers.

Table 31.4. Technical parameters Olympic Tower transmitter

Channel	Frequency	Average transmitter output power	ERP
34	578 MHz	9.3 kW	100 kW omnidir.
35	586 MHz	9.3 kW	100 kW omnidir.
48	690 MHz	9.3 kW	100 kW omnidir.
54	738 MHz	9.3 kW	100 kW omnidir.
56	754 MHz	9.3 kW	100 kW omnidir.
66	834 MHz	9.3 kW	100 kW omnidir.

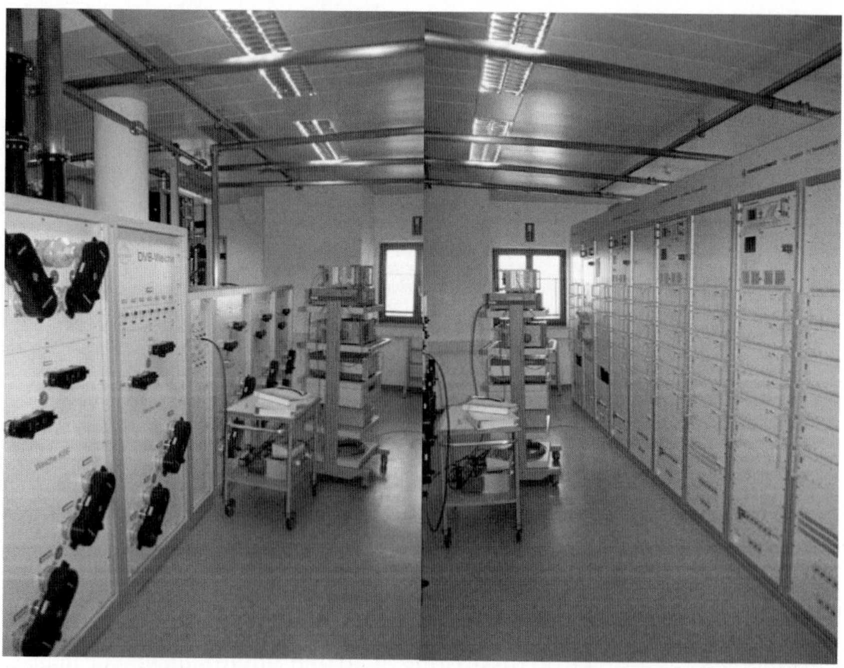

Fig. 31.18. Wendelstein transmitter hall; antenna combiner (left) and UHF transmitters (right)

Mask filter and combiner are made by Spinner, but are also of much larger dimensions. The transmitting antenna is of similar design as that on the Wendelstein mountain and has also been manufactured by Kathrein.

Table 31.5. Technical parameters Brotjacklriegel transmitter

Channel	Frequency	Average transmitter output power	ERP
7	191.5 MHz	4.6 kW	25 kW
33	570 MHz	5 kW	50 kW
28	522 MHz	3.4 kW	100 kW

Table 31.6. Range of programs distributed by the Brotjacklriegel transmitter

Channel	Modulation	Code rate	Guard	Data rate [Mbit/s]	Number of programs	Programs
7	16QAM	3/4	1/4	13.06	4	Das Erste, Phoenix, arte, 1plus
33	16QAM	2/3	1/4	13.27	3 + MHP	Bayr. Fernsehen, BR-Alpha, SWR
28	16QAM	2/3	1/4	13.27	3 + MHP	ZDF, Info/3sat, Doku /Kika

31.3.3 Brotjacklriegel Transmitter

The Brojacklriegel transmitter belongs to the DVB-T network Eastern Bavaria and, like the other transmitter sites in Eastern Bavaria, has 3 DVB-T transmitters. The DVB transmitter is thus currently much smaller than the Wendelstein and Olympic Tower installations but its performance category corresponds to that of the Wendelstein site. Brotjacklriegel currently only broadcasts public service programs which is why 3 data streams or transmitting frequencies are sufficient here. The Brotjacklriegel antenna is also of similar construction to the Wendelstein antenna.

The Hoher Bogen (Bayerischer Wald, Furth im Wald) and Hohe Linie (Regensburg) transmitting stations are comparable sites of the Bayerischer Rundfunk in the Eastern Bavarian DVB-T network.

Channel 7 of the "Brotjacklriegel" transmitter runs synchronously in a single-frequency network with the "Hohe Linie" transmitter at 84 km distance near Regensburg and the "Hoher Bogen" transmitter at 54 km distance. However, the distance to the "Hohe Linie“ would violate the guard interval constraint (77 km in the VHF band) which is why the "Brotjacklriegel" signal in channel 7 is radiated 20 µs earlier which effectively moves the transmitter towards the "Hohe Linie" and also towards the "Hoher Bogen" by 84 km. Because of the mountain chain of the intervening Bavarian Forrest, the "Brotjacklriegel“ and "Hoher Bogen“ transmitters are already well decoupled from one another.

Fig. 31.19. Liquid cooled mask filter (8 kW) for DTV (DVB/ATSC); manufacturer's photo, Spinner.

Fig. 31.20. Antenna combiner Olympic Tower Munich, rear view; manufacturer's photo, Spinner.

Fig. 31.21. DVB-T transmitters Olympic Tower Munich; manufacturer's photo, Rohde&Schwarz.

31.4 Measurements in DVB-T Single-Frequency Networks

Single-frequency networks (SFNs) are "special" networks. They must be

- synchronous in frequency,
- synchronous in time, and
- meet the requirements for the guard interval.

To ensure that these conditions are met in the region of Bavaria, as in the previous examples, they must be measured and monitored during the commissioning and also in later operation. This chapter is the result of numerous inputs from readers and participants in seminars.
An SFN must firstly be planned correctly with knowledge of the topographical and geographical structure. The transmitter distances must not violate the guard interval condition, i.e. they must not exceed a particular maximum distance from one another. Should this be the case, nevertheless, it may help to "shift" one or the other transmitter by advancing or delaying the radiation of the COFDM signal. It is then possible to guarantee reliable reception in areas where otherwise signal paths would have exceeded the guard intervals. However, this gain may be at the cost of problems in other regions. Naturally, the antenna pattern also plays an important role. An SFN can be modelled by "drawing in" the pattern, i.e. reducing the radiated power in a certain direction.
Most of the DVB-T matters discussed in this chapter can be easily applied also to other standards using COFDM as a modulation method, such as, e.g., DAB or ISDB-T.

31.4.1 Test Parameters

Now to answer the question "What is actually measured how in DVB-T single-frequency networks?". Naturally, the matters applicable to measuring SFN coverage are also applicable to the special case of an SFN, the MFN - Multi-Frequency Network, where each transmitter has its own frequency. In comparison with an SFN, the receiver can only expect one signal path from the transmitter, and possibly "correct" echo paths, in this case. But the delay differences are only within a range of about 1- 10 μs instead of up to about 200 μs, by comparison. The parameters to be measured in the field are:

- Level or field strength
- Modulation error ratio

- Bit error ratios
- Channel impulse response
- Constellation diagram (visual assessment).

Naturally, the most important test parameter is firstly the signal level or field strength present on site. The signal level is measured as the output signal of a known test antenna. Its K factor or antenna gain can then be used for calculating the field strength. The formula for this is:

$$E[dB\mu V/m] = U[dB\mu V] + k[dB/m];$$

$$k[dB] = (-29.8 + 20 \cdot \log(f[MHz]) - g[dB]);$$

where

E = electrical field strength
U = antenna output level
k = antenna k factor
F = received frequency
g = antenna gain

The required minimum receiver input level depends on the selected modulation parameters and on the quality of the receiver. As shown in Chapter 20, a noise level of about 10 dBµV can be expected at the receiver input, which leads to a minimum level of about 22 dBµV with 16QAM, code rate 2/3 and of about 28 dBµV with 64 QAM, code rate 2/3. This corresponds quite well to reality in an AWGN (Additive White Gaussian Noise) channel. It won't hurt to add a margin of about 3 dB to this, however. There are implementation losses (antenna, cable) and there are differences in the quality of receivers. However, these minimum receiver input levels only apply to the case of one-way reception, i.e. the pure AWGN channel. In practice, multi-path reception often requires an input level which is higher by 5 to 10 dB. This is due to the actual characteristics of the DVB-T demodulator chips built into the DVB-T receivers. It can be demonstrated, however, that there are distinct differences here and that the latest generations of DVB-T receivers and chips come much closer to expectation. Calculating then firstly the minimum field strengths with an antenna gain of 0 dB, using the theoretical minimum levels without deductions, and then also adding QPSK, with code rate 2/3 (-6 dB compared with 16QAM = 16 dBµV):

Table 31.7. Theoretically required minimum receiver input level (CR=2/3) in an AWGN channel with DVB-T with 0 dB antenna gaain and no implementation losses, receiver noise figure = 7 dB, ambient temperature = 20 °C.

	200 MHz	500 MHz	600 MHz	700 MHz	800 MHz
k factor at 0dB gain	16.4 dB	24.2 dB	25.8 dB	27.1 dB	28.3 dB
Min. level with QPSK	16 dBμV	16 dBμV	16 dBμV	16 dBμV	16 dBμV
Min. received field strength with QPSK	32.4 dBμV/m	40.2 dBμV/m	41.8 dB μV/m	43.1 dBμV/m	44.3 dBμV/m
Min. level with 16QAM	22 dBμV	22 dBμV	22 dBμV	22 dBμV	22 dBμV
Min. received field strength with 16QAM	38.4 dBμV/m	46.2 dBμV/m	47.8 dBμV/m	49.1 dBμV/m	50.3 dBμV/m
Min. level with 64QAM	28 dBμV	28 dBμV	28 dBμV	28 dBμV	28 dBμV
Min. received field strength with 64QAM	44.4 dBμV/m	52.2 dBμV/m	53.8 dBuV/m	55.1 dBμV/m	56.3 dBμV/m

In reality, the implementation losses must be added to this and these, in turn, depend on the chosen receiving situation. There are ultimately four receiving situations which are:

- reception by fixed outdoor antenna,
- reception by portable outdoor antenna,
- reception by portable indoor antenna,
- mobile reception.

With reception by fixed outdoor antenna, the antenna gain of about 6 to 12 dB is added to this and ensures that correspondingly less field strength is required. Field strength is here defined at a corresponding height above ground, in most cases 10 m, and the measurements are therefore also taken under these conditions (mast, 10 m, directional antenna). However, it is advisable also to take into consideration the line losses etc. to the receiver (e.g. 6 dB) and to include these in the calculations. With reception by portable outdoor antenna there is no antenna gain. The reception situation to be considered is then 0 dB antenna gain and e.g. 2 m above ground. With indoor reception, attenuation losses of the walls and windows of up to about 20 dB must be added. Concrete buildings with metallized windows produce especially high attenuation. Polarization losses of 5 to 15 dB are another factor to be considered. Reception with a vertical rod antenna in a horizontally polarized DVB-T network, e.g., leads to a loss of about 15 dB. Portable indoor reception covers a very wide range with respect to minimum field strength. The antenna gain may also exhibit negative values. The most difficult case is mobile reception, DVB-T being a system which was originally not designed for this purpose. The time interleaver value is very short. Although the mobile field strength values measured do no differ from the stationary ones, the Doppler effect plays a very large role and the changing receiving situations play havoc with the receiver. The signal-to noise ratio (SNR) and the modulation error ratio (MER) differ greatly under mobile and stationary conditions and also depend on location. The MER is the aggregate parameter in which all interference effects on the DVB-T reception can be mapped. As explained in Chapter 21, it is the logarithmic ratio of the RMS value of the signal to the RMS value of the error vector in the constellation diagram. If only a noise effect is present, the MER corresponds to the SNR. If the SNR or the MER are measured in mobile operation, the Doppler effect additionally affects a deterioration of the MERs or SNRs in dependence on the speed of travel due to the different types of local reception-related effects and signal paths. This will also be illustrated later in this chapter by providing practical examples from the exemplary DVB-T networks. The MER is thus measured under stationary conditions in accordance with the required nominal conditions of reception, e.g. with a directional antenna at 5 or 10 m height, or with a non-directional antenna. The minimum MER required for reception also depends on the modulation parameters.

Table 31.8. Minimum required MER at code rate = 2/3.

	QPSK	16QAM	64QAM
MER	6 dB	12 dB	18 dB

It is important to know that, as a simple fact of physics, the MER measured under mobile conditions can never correspond even approximately to that measured under stationary conditions. The MER is always, and also in every standard, a function of the speed of travel and of the multi-path reception conditions. The same also applies to the bit error ratios (BER). These are also not only dependent on the received level but can be derived directly from the MER. The minimum required BER before Reed Solomon or after Viterbi with quasi error free DVB-T reception is $2 \cdot 10^{-4}$. There are three bit error ratios in DVB-T:

- BER before Viterbi, or the channel bit error rate ratio,
- BER after Viterbi or before Reed Solomon, and the
- BER after Reed Solomon.

It makes sense to measure all three BERs during the field test. The engineer obtains appropriate information especially from the BER after Viterbi, i.e. before Reed Solomon. Naturally, the measurement of the BERs in mobile operation will also result in quite different values in comparison with stationary measurements. The test results also depend on where in the SFN one is moving, the reason again being simply the effect of Doppler on the multipath reception in the various regions of reception.

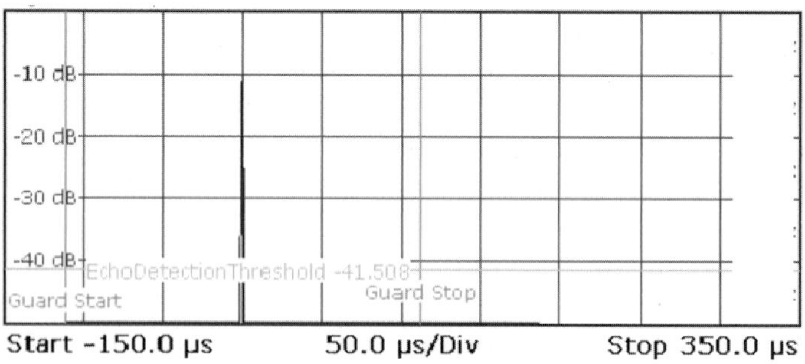

Fig. 31.22. Channel impulse response with one signal path; measured with the TV test receiver ETL

The channel impulse response (Fig. 31.22.) provides reliable information about the multi-path receiving conditions in the various regions in the SFN. It also tells one whether an SFN is running synchronously or not. In addition, the channel impuls response can be used for estimating whether the receiving situation could represent critical states with respect to synchronisation to symbol and guard interval. Critical states are:

- violation of the guard interval,
- the pre-echo,
- the 0-dB echo,
- the quasi mobile receiving situation
- radiation of different TPS bits.

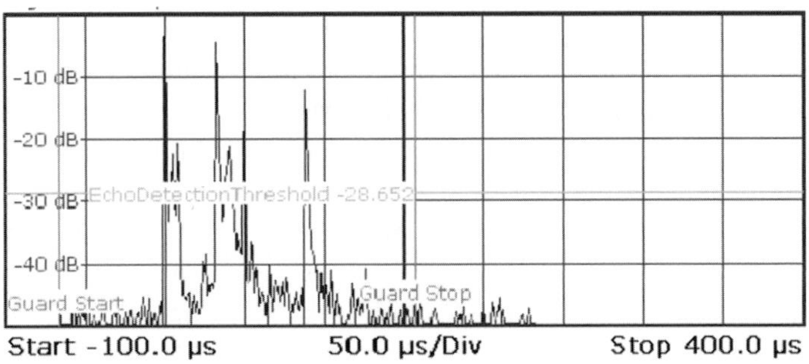

Fig. 31.23. Channel impulse response with multi-path reception with post-echoes, measured here in an SFN with 3 transmitters

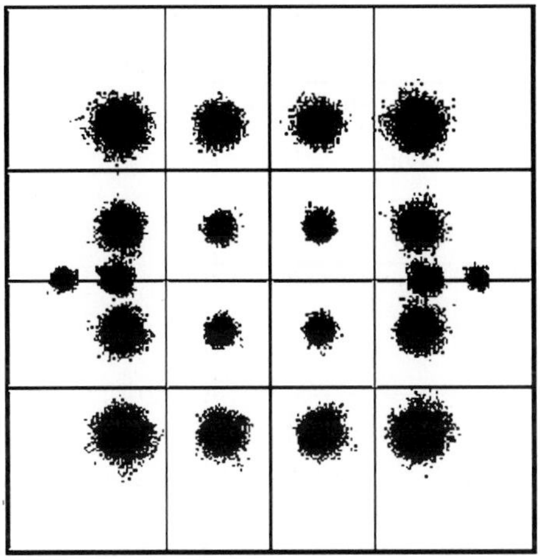

Fig. 31.24. Typical constellation diagram for a guard interval violation (outer points are larger than the inner ones)

31.4.1.1 Guard Interval Violation

The guard interval violation, among the critical receiving states, is the simplest to explain and is considered to constitute an absolute infringement of the SFN conditions. It simply involves the reception of signal paths which are outside the guard interval and still have sufficient energy. Such a problem can arise when transmitters are spaced too far apart and delays have been selected wrongly or unfourably at the transmitter sites, or simply with propagation overshoots. The energy of such a signal path becomes critical if it passes from the attenuation with respect to the main path into the order of magnitude of minimum S/N or minimum MER (fall-off-the-cliff), depending on the selected transmission parameters.

This problem can be solved by a suitable choice of the delay times, i.e. the transmission times, of the COFDM symbols to the transmitter sites, by adapting the antenna patterns and the ERPs of the transmitters. In the case of a guard interval violation, a typical constellation diagram has larger constellation points outside than inside (Fig. 31.41.).

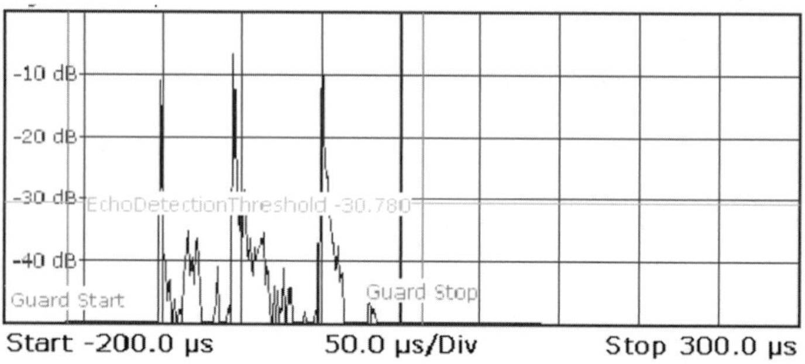

Fig. 31.25. Channel impulse response with pre-echoes

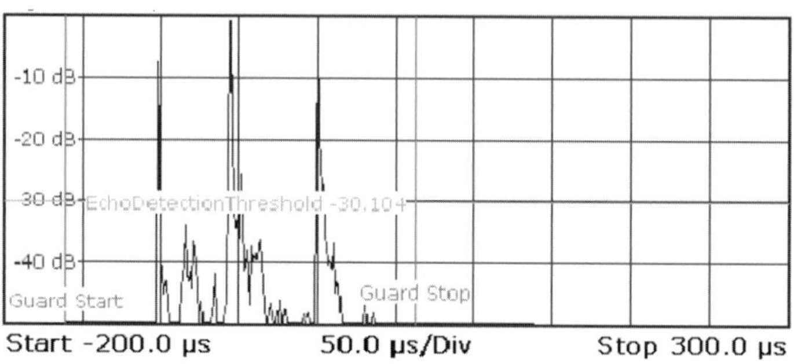

Fig. 31.26. Channel impulse response with pre-echo and 0-dB echo

31.4.1.2 Pre-echoes

Pre-echoes are signal paths in a single-frequency network which appear with a lower level and earlier than the main path (Fig. 31.25.). In theory, it should be possible for all receivers to handle this situation without any problem. In practice, however, it is found that many, mainly older DVB-T receivers cannot manage this. This also depends on the delay time between the 0-dB path, i.e. the main path, and the reduced pre-echo. The problem of the pre-echo can be explained simply by the fact that the receiver simply places the FFT sampling window symmetrically over the main path and thus pushes the pre-echo beyond the guard interval.

31.4.1.3 The 0-dB Echo

It is called a 0-dB echo if 2 or more signal paths having the same level but different delays appear at the receiver (Fig. 31.43.). This can also lead to receiver synchronization problems, mainly in the case of longer delay differences from half the guard interval onward. In theory, a receiver should be able to cope also with this receiving situation without any problem. This problem, too, is explained by the receiver placing the FFT window so badly that one signal path is located outside the guard interval.

31.4.1.4 Quasi Mobile Receiving Situation

It is called a quasi mobile receiving situation if the channel continuously changes due to continuously changing conditions of reflection. The behavior of a receiver in this situation depends on the characteristics of the channel correction of the receiver and, naturally, on the receiving situation. Quasi mobile receiving situations are encountered when e.g. there is no direct line of sight to the transmitters and the reception "lives" mainly from reflections, but these reflections are influenced by cars, trains, trams etc.

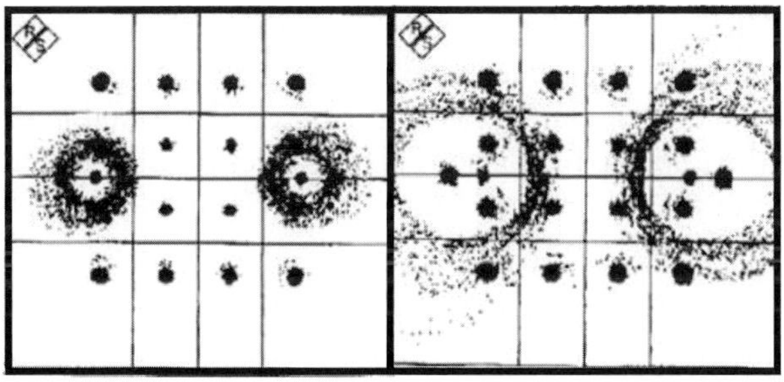

Fig. 31.27. Constellation diagrams in an SFN with differently transmitted TPS bits

31.4.1.5 Transmission of Different TPS Bits

In DVB-T, a total of 67 bits are transmitted as so-called TPS (transmission parameter signalling) bits via 68 symbols. These bits represent a fast information channel from transmitter to receiver for conveying the transmission parameters. These transmission parameters are, among other things, modulation method, error protection etc. Apart from the TPS bits already defined originally in the DVB-T standard, there are the reserved bits, more and more of which are used, e.g., for cell ID and DVB-H.

A length indicator transmitted before the actual TPS payload bits tells how many of the reserved TPS bits are actually currently being used. It is important that all transmitters in an SFN transmit all TPS bits identically and completely synchronously. It has happened a number of times that transmiters in an SFN were configured differently and had transmitted length indicator and reserved bits differently. Depending on their location, the receivers which had then actually evaluated the TPS bits were unable to cope with the receiving situation and could not lock up. The TPS carriers work with DBPSK modulation, i.e. with differential BPSK modulation. The information is transferred in the difference from one symbol to the next. However, this means that from the point in time at which a TPS bit is transmitted differently than at other transmitter sites, the carrier vectors are pointing in the opposite direction and the DBPSK modulation no longer works, causing the circular distortions in the constellation diagram at the TPS points (see Fig. 31.27.) It is strongly recommended always to have one or more test vehicles in the field for all changes being carried out in an SFN and to determine the situation also at the TPS carriers (e.g. carrier No. 50).

31.4.1.6 Frequency Accuracy of the Transmitters

It is important that all transmitters in an SFN transmit at the same frequency, as accurately as possible. The accuracy to be aimed for is $1 \cdot 10^{-9}$ or better. The frequency accuracy can be easily verified with a suitable test receiver by measuring the impulse response. This provides a frequency accuracy of somewhat better than 0.5 Hz, a condition which normally can be easily met.

31.4.2 Practical Examples

31.4.2.1 Pre-echoes

The pre-echoes described occur mainly in regions in which the closer path in terms of distance is attenuated more compared with the 0-dB path due to geographical obstacles (hills, mountains). In the region of the SFN Southern Bavaria described this occurs, e.g, to the North of the Munich airport in the vicinity of the course of the river Isar, where the Olympia Tower is shaded by hills and more distant Wendelstein transmitter thus dominates.

33.4.2.2 The 0-dB Echoes

0-dB echoes occur whenever two or more paths appear with the same power level at the receiver due to the propagation conditions. In the SFN Southern Bavaria, such a situation occurs mainly in the region of the "Erdinger Moos" around Munich airport. It is very flat there and the Wendelstein and Olympia Tower transmitters are received partly with the same level, but with an extreme delay difference of about 140 µs.

31.4.2.3 Quasi Mobile Channel

A quasi mobile channel exists in regions where there is no direct line of sight to the transmitters. This is the case where the transmitters are shielded by obstacles and the reception survives with reflections partly from "variable" obstacles such as cars, trains or trams.

31.4.2.4 TPS Bits

When SFNs are commissioned or re-organized, it may happen that not all transmitters (of one or of different transmitter manufacturers) are identically configured and that the transmitters thus transmit different TPS information. This occurred several times during the commissioning or conversion of the SFN.

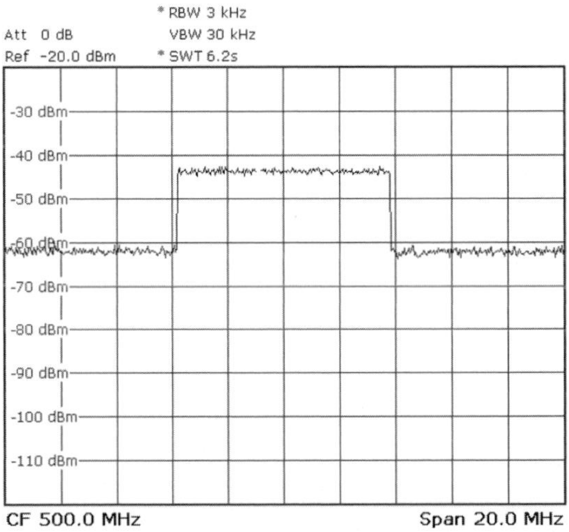

Fig. 31.28. Spectrum of a DVB-T signal in the AWGN channel

31.4.2.5 Mobile DVB-T Reception

A question often asked is "Up to what speed does DVB-T work?", a question which is not easily answered. In principle, it must be said at this point that DVB-T was never intended for mobile reception and, therefore, does not have any characteristics especially provided for this purpose succh as, e.g., a long time interleaver. Mobile reception depends mainly on the multi-path receiving situation. If only one signal path is received, mobile reception does not present a problem. The Doppler effect then only shifts the DVB-T spectrum in the direction of higher or lower frequencies, depending on whether one is moving towards the transmitter or away from it. At the usual travelling speeds, the frequency shift is of the order of 50 to 100 Hz. This frequency shift does not present a problem for DVB-T receivers receiving one signal path. In the case of multi-path reception and Doppler shift, the problem is one of spectrum smearing with all possible intermediate stages which will be presented in examples in the following paragraphs.

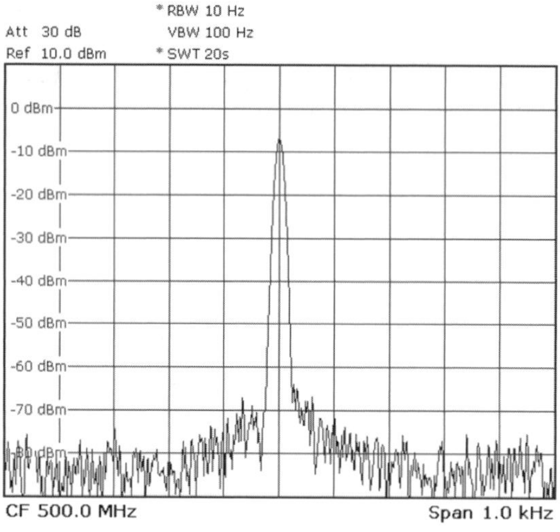

Fig. 31.29. Single COFDM carrier in an AWGN Channel

AWGN channel

In the AWGN channel, the carriers are only affected by noise, as shown in Fig. 31.28. The noise pedestal at about 20 dB below the payload signal can be seen clearly at the shoulder.

The single COFDM carriers are at exactly the right frequency positions in the AWGN channel (Fig. 31.29.). Each carrier is only affected by a greater or lesser "noise fringe".

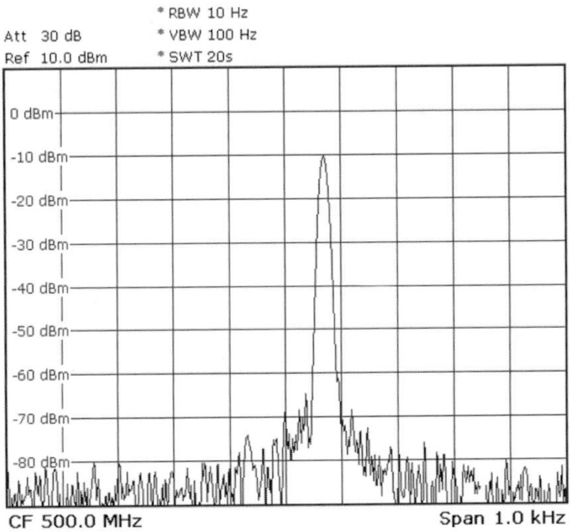

Fig. 31.30. Single frequency-shifted DVB-T carrier in a mobile channel

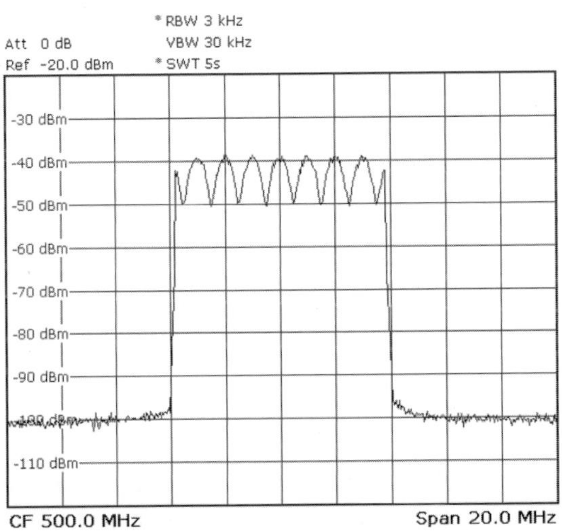

Fig. 31.31. Stationary multi-path reception of two signal paths (0 dB / 0 µs, -5 dB /1 µs)

Doppler Shift

During movement in the mobile channel, the complete DVB-T signal is frequency shifted. All single COFDM carriers are shifted towards higher or lower frequencies depending on whether one is moving towards the transmitter or away from it. Fig. 31.30. shows a single carrier, shifted by 70 Hz, of a DVB-T signal at a speed of 150 km/h moving towards the transmitter.

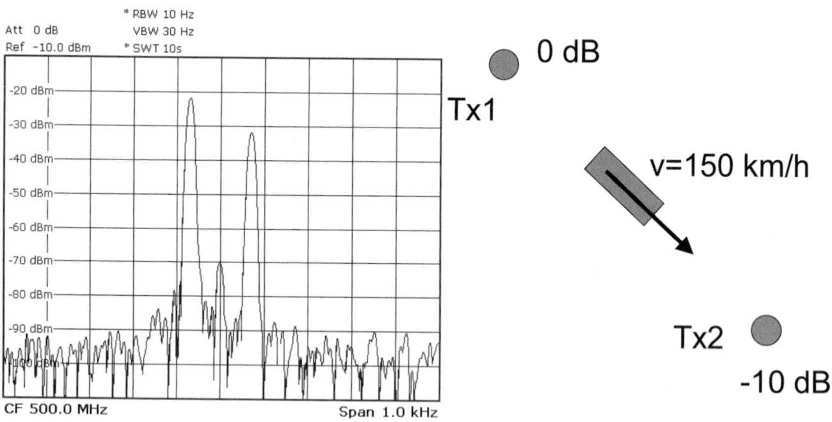

Fig. 31.32. Mobile multi-path reception of two signal paths (0 dB and –10 dB at 150 km/h)

Stationary Multi-path Reception

In stationary multi-path reception the only problem is fading. Depending on the difference in echo delay and echo attenuation, more or less deep dips occur in the signal spectrum as shown in Fig. 31.31. The spacing of the dips corresponds to the inverse of the echo delay difference.

Mobile Multi-path Reception

In mobile multi-path reception, the DVB-T subcarriers are shifted simultaneously upwards and downwards in frequency (Fig. 31.32.) or may not be shifted at all. Depending on the receiving conditions, this frequency smearing results in unwanted amplitude modulation of the DVB-T signal.

Mobile Rice Channel

The model of the Rice channel simulates the case of multiple multi-path reception and dominant main path. Fig. 31.34. shows the spectrum of a single DVB-T carrier in the mobile Rice channel. The dominant main channel can be clearly seen at -10 dB.

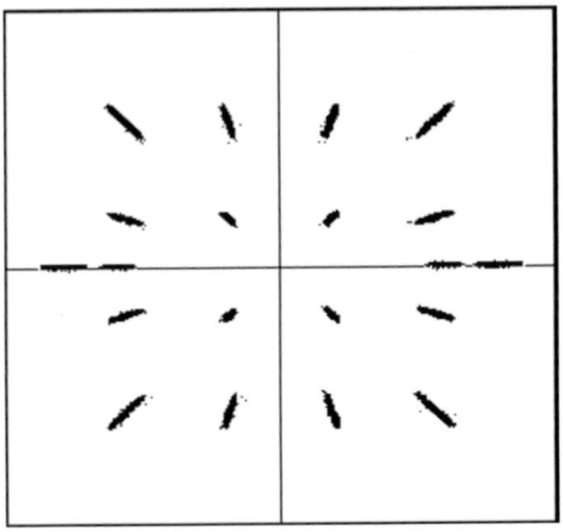

Fig. 31.33. DVB-T constellation diagram with unwanted amplitude modulation caused by mobile multi-path reception (500 MHz, 3 paths, -20 dB/-150 km/h, 0 dB/0 km/h, 150 km/h/-20 dB)

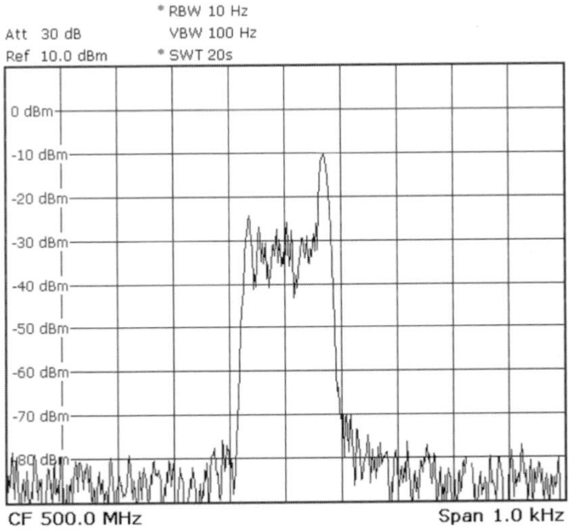

Fig. 31.34. Mobile Rice channel, v=150 km/h, power ratio = 10 dB

Rayleigh Channel

In the Rayleigh channel there is no longer a main path. It corresponds to the Rice channel with a power ratio = 0 dB. Fig. 31.35. shows an example of a single DVB-T carrier in the mobile Rayleigh channel at a speed of 150 km/h.

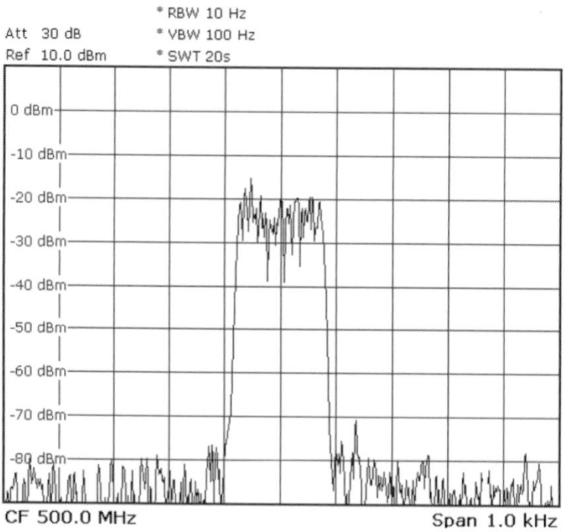

Fig. 31.35. Mobile Rayleigh channel, v=150 km/h, power ratio = 10 dB

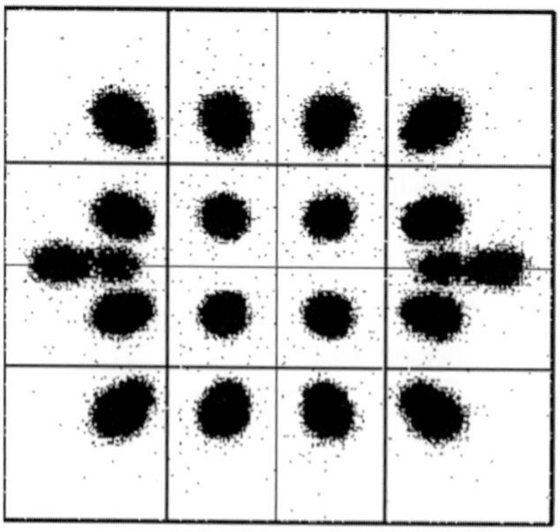

Fig. 31.36. Unwanted amplitude modulation caused by a dried-out electrolytic capacitor in an antenna amplifier

Comparable "mobile situations" can be created in a DVB-T receiver even by dried-out electrolytic capacitors in an antenna amplifier. Superimposed AC hum produced by these can create an unwanted amplitude modulation at 50 or 100 Hz. Fig.31.36. shows a corresponding constellation diagram.

31.4.3 Response of DVB-T Receivers

The response of DVB-T receivers in one or the other receiving situation is greatly dependent on the characteristics of the respective receiver, i.e. mainly on the characteristics of the installed tuner, the DVB-T chip, the MPEG decoder and the firmware of the receiver. In the next section, testing of the receiver will be discussed. The characteristics of the tuner can be differentiated as follows:

- Noise figure
- Phase noise
- RF and IF selectivity characteristics
- Linearity/intermodulation

The characteristics of the tuner essentially determine the minimum received level required and the adjacent-channel compatibility, especially with a high adjacent-channel level.

The DVB-T chip is mainly responsible for how well a receiver can handle different receiving situations such as

- Pre-echo,
- 0-dB echo,
- Multi-path reception in general,
- Mobile reception and quasi mobile reception,
- Adjacent-channel occupancy,
- TPS bits set differently,
- Hierarchical modulation.

The MPEG decoder and the firmware determine how the receiver responds to different transport stream contents. This relates to:

- the channel search (speed and characteristics under critical conditions,
- the PSI/SI tables (e.g. response to dynamic PMT),

- response to network overlap (identical service in different TS)
- decoding of the elementary stream,
- signalling of the source characteristics (4:3/16:9, mono/stereo),
- error concealment,
- switching rate,
- stability,
- receiver configuration such as teletext, VPS, MHP.

31.4.4 Receiver Test and Simulation of Receiving Conditions in Single-Frequency Networks

The characteristics of TV receivers must be tested comprehensively especially in terrestrial broadcasting in order to find out how well they are capable of handling the problem situations described in the previous sections. Receivers are tested in

- the development of receivers,
- production handover (EMI, EMC,...),
- receiver production for final testing,
- comparing receivers in test houses and at network operators.

Experience has shown that DVB-T receivers especially have not been adequately "stress tested". The maximum amount of tests should be performed at least during the receiver development, the production handover and the receiver comparison. These maximum tests are:

- detecting the minimum receiver input level at some frequencies,
- detecting the minimum SNR at some frequencies,
- the response at high adjacent-channel levels (close or more distant),
- the response with co-channel reception of analog TV,
- the response of the receiver during channel search,
- the response of the receiver with network overlap,
- measuring the booting speed,
- measuring the switching speed,
- testing the teletext function,
- testing the VPS function,
- testing the response of dynamic PSI/SI tables,
- testing the firmware configuration and quality,
- EMC tests,

- mechanical construction.

The minimum tests in production must be suitably selected for the respective product by the manufacturer.

33.4.4.1 Minimum Receiver Input Level in the AWGN Channel

The minimum receiver input level should first be determined in the AWGN channel with different transmission parameters (64QAM, 16QAM, QPSK, different code rates) at least 3 frequencies (one VHF and two UHF frequencies). The receiver is supplied with one signal path in this case. Starting with a level of abut 50 dBµV, this level is reduced until the visual and aural assessment of decoded video and audio shows that the receiver is no longer operating correctly. It is important that, when the precise point when the "fall-off-the-cliff" occurs is determined, one always waits for a sufficiently long time (at least 1 minute) to see whether the receiver is really still operating in a stable mode.

31.4.4.2 Minimum SNR

Apart from determining the minimum receiver input level, it is of interest to determine the minimum signal/noise ratio in the AWGN channel. The results should then be compared with the minimum receiver input level measurement and discussed. These tests should also be performed at at least 3 frequencies (the same ones as in paragraph 3.4.4.1, of course), selecting, e.g., a "sensible" DVB-T receiver input level of 50 to 60 dBµV so that the receiver is neither supplied too poorly nor caused to go into attenuating mode. More and more noise is then added progressively until the "fall off the cliff" condition is reached again. This, too, is then determined carefully as in 31.4.4.1.

31.4.4.3 Adjacent-Channel Occupancy

In the adjacent-channel test, the response of a DVB-T receiver with a high adjacent-channel level is determined placing an adjacent DVB-T channel below or above a payload channel. The level of the adjacent channel or channels is then increased more and more until no further reliable reception is possible. This test, too, is performed with different transmission parameters as in 31.4.4.1 and 31.4.4.2. The aim should be to be able to handle an adjacent-channel level which is at least 20 dB above the useful level. Such conditions could easily arise especially with a mixed DVB-

T/DVB-H/DAB scenario. This test should also be performed at 3 frequencies, at the least.

31.4.4.4 Co-channel Reception

Checking the co-channel reception of DVB-T with DVB-T is essentially already done by 31.4.4.2 since a non-synchronous DVB-T interference signal virtually looks like noise. Testing with analog TV in the co-channel, however, is definitely a noteworthy measurement but depends greatly on the ATV image content chosen.

31.4.4.5 Multi-path Reception

In the multi-path reception test, the receiver is presented with a situation which can occur in real life in an MFN or SFN, using a test transmitter with a channel simulator (fading simulator).

31.4.4.6 Channel Search

Testing the channel search function of a receiver mainly tests the search rate and also the search action under different conditions (incl. adjacent-channel occupancy). The test should also involve checking of the performance of a receiver with wrong NIT entries. This also includes its performance in the case of network overlaps, i.e. when receiving the same services from different transport streams. This occurs in regions where the receiver is seeing two or more networks, i.e. at the edges of SFNs.

31.4.4.7 Booting Speed and Action

In the world of computers, "booting" is known to be the initialization of a computer. Since a DVB-T receiver is also nothing else than a computer, it takes a certain time until it is ready for operation and a user will be interested to know how long this will take and how it takes place.

31.4.4.8 Program Change

Users find it particularly bothersome if a program change takes a long time and is "untidy". This test checks the receiver reaction to "zapping".

31.4.4.9 Teletext

DVB provides for the "tunneling" of teletext via private PES packets and this is done mainly by the program providers operating under public law.

In this arrangement, teletext is gated back into the vertical blanking interval of the video signal by the receiver at the analog output interface (SCART or cinch connector). A TV receiver connected there can then decode this teletext. It is also possible for the DVB-T receiver itself to decode the teletext and to output it as a frame signal, storing a number of pages in its buffer. The teletext modes supported by a receiver, either gated into the vertical blanking interval or self-decoded, are a criterion for testing and comparing receivers.

31.4.4.10 VPS Functions

In data line 16 in the vertical blanking interval of the analog TV signal, the VPS information for controlling video recorders has hitherto been transmitted, among other items. This signal, too, can be "tunneled" in DVB in private PES packets in the MPEG-2 data stream and used directly in the receiver (hard disk receiver) and/or gated back into line 16 at the CCVS interface. A video recorder connected there can then respond to this signal and control the recording. These functions, too, must be covered in a receiver comparison test.

31.4.4.11 Dynamic PSI/SI Tables

Dynamic PSI/SI tables means the change of these tables with time. EIT and TOT/TDT are clearly always dynamic but there are also so-called "window programs" which are transmitted only at particular times of the day and are signalled by changing PMTs, so-called dynamic PMTs. A change in the PMT is not recognized by all receivers which is why the response to changes in PAT, PMTs and the SDT should be tested.

31.4.4.12 Firmware Configuration

The way in which a DVB receiver can be operated and how especially the electronic program guide is handled depends greatly on the firmware installed in the receiver. This is another matter to which attention should be paid in a receiver comparison test.

31.4.4.13 Miscellaneous

Naturally, the receiver test also includes adherence to the EMC regulations but this will not be discussed in greater detail at this point. As well, an assessment of the mechanical construction of the receiver case is of importance in a comparison of receivers in test establishments but this again will not be discussed any further here.

31.5 Network Planning

Naturally, a network expansion is preceded by network planning which today is done with the support of software tools. This involves simulation and determination of the network data such as antenna patterns, transmitter powers, error protection. guard interval, delays etc and calculation of the coverage of the regions on the basis of geographical, topographical and morphological data and with knowledge of the possible transmitter sites. Firstly, the frequencies and powers, direction of radiation etc are assigned by the regulatory authorities. In border regions this requires international coordination. Examples of planning tools in the German-speaking area are:

- tools by the Deutsche Telekom,
- tools by the Institut für Rundfunktechnik (IRT),
- tools by the company LStelcom.

The software CHIRplus_BC© by the company LStelcom (Lichtenau near Baden-Baden, Germany), in particular, is encountered throughout the world. By now, the author also has gained the experience that with the appropriate use of planning tools, the problem areas described above for receivers (0-dB echo, pre-echoes) can be unambiguously identified, e.g. by clicking on corresponding buttons in the planning software.

31.6 Filling the Gaps in the Coverage

Even in the days of analog television, gaps in the coverage were normally filled by so-called gap fillers, called translators. In analog television, alternative channels with guard band without adjacent-channels being occupied were selected which supply a signal received from a master transmitter for covering a limited region. In this set-up, TV signals of a master transmitter, received by rebroadcasting reception, were translated into another TV channel and then retransmitted via a transmitter, thus covering a region which was otherwise shaded. In digital terrestrial television it is firstly assumed that many areas are automatically covered by the characteristics of digital television. Nevertheless, depending on the required coverage which, in turn, is dependent on the country concerned, additional gap fillers cannot be avoided. This is because, in analog television, reflections have not only led to reception not being possible at all, but in moderate cases have simply caused unsightly "ghost images". There are no longer any "ghost images" in digital television and, because of COFDM, neither do echoes present problems to the same extent as in analog television. In theory, a COFDM system should be able to handle such a situation quite easily, of course. But if the received field strength is too low because of shading, such regions must still be covered by gap fillers even with digital terrestrial television. In digital television, these gap fillers can be operated both at the same frequency and at other frequencies. In analog television, these transmitters had to be operated at other frequencies. There are:

- transmitters transmitting at the same frequency (gap fillers, SFN), and
- frequency-converting transmitters (transposers, MFN), and
- frequency-converting transmitters with remodulation (retransmitters, MFN).

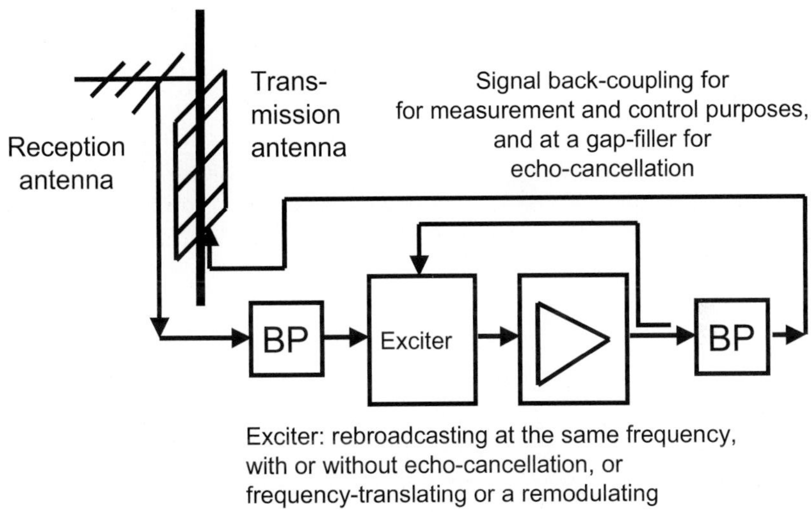

Fig. 31.37. Principle of a gap-filler or translator

To fill gaps in the coverage of SFNs, only gap fillers are used. In these transmiters transmitting at the same frequency, remodulation is impossible since going back to data stream level (demodulating) and remodulating would involve too much delay. This is why in this case the approach of downconverting to a low intermediate frequency and upconverting again to RF and amplifying was adopted. It is important here that receiving and transmitting antenna must be sufficiently well decoupled. Up to a certain extent, an equalizer can be of assistance here in providing echo cancellation. The minimum isolation necessary between receiving and transmitting antenna is (current state of the art):

- + 10 dB gain without echo cancellation,
- - 10 dB gain with echo cancellation.

Fig. 31.38. Practical example of a translator or gap-filler site; the 2 (log periodic) receiving antennas are located in the lower part of the extended pinnacle of the tower, the 3 transmitting antennas (8-element bays) are located in its upper part

Apart from the transmitting antenna and the receiving antenna being sufficiently well decoupled, the correct orientation of the receiving antenna is also of great importance. If possible, only one signal path, and not several as in a multi-path situation, should be forwarded from the SFN. Under no circumstances should pre-echo situations or a 0-dB echo path be radiated since this will lead to problems in many receivers as is well known. Otherwise, receiver problems are created not only over a small area at some locations but over a large area in the gap-filling region and possibly beyond. In the case of a frequency-converting transmitter, the approach via the IF can be selected as in the case of the gap filler, or one can choose a remodulation process. Remodulation is more expensive and means delay, of course. This is also the reason why remodulation is not possible in the gap filler because otherwise the SFN timing would be violated completely. However, remodulation is more stable and, above all, results in a better signal quality. When the transmitter powers are greater, however, the recommended approach is always that of remodulation or of using the retransmitter, respectively.

Fig. 31.39. „Fall off the Cliff" artifacts in digital television

31.7 Fall-off-the-Cliff

Blocking artifacts caused by the compression are too often mistaken for artifacts caused by the transmission link. An image at the output of a receiver which, due to bit errors, has been brought to the limit of decodability, i.e. the "fall-off-the-cliff" state, looks quite different from an image which has been rendered "unsightly" due to too much MPEG compression. In the case of bit errors, entire slices are missing or the entire image freezes or no image can be seen at all. Fig. 31.39. shows an image in which entire groups of blocks, so-called slices, are missing within a line.

31.8 Summary

The empirical values described in this chapter with reference to DVB-T can also be easily applied to other terrestrial transmission standards. The details in error protection and the modulation methods are different in DAB, ISDB-T or other standards, but the principle always remains the same. Although this chapter is thus tailored for DVB-T, due to the author's experience in this field, it is not limited to DVB-T alone. In this section, experiences and problems from real networks were presented. It must be hoped that these experiences will assist in recognizing, and solving, most of the real problems with DVB-T, and in removing the apprehension felt about the new digital television. Digital television differs from analog television but, with the appropriate experience, it is not unfathomable.

Bibliography: [NX7000], [KATHREIN1], [LVGB], [RFS], [VIERACKER], [LSTELCOM], [NX8000]

32 DTMB

32.1 DMB-T, or now DTMB

DTMB - Digital Terrestrial Multimedia Broadcasting - is a Chinese standard which, like DVB-T - has the aim of broadcasting television economically terrestrially by digital means and with modern supplementary services. DMB-T was published in 2006 – at least in excerpts, as "GB20600-2006 – Framing Structure, Channel Coding and Modulation for Digital Terrestrial Broadcasting System". It was renamed DTMB, having combined two proposals to form one standard in 2007. In one proposal, a multi-carrier method was stipulated, in the other one a single-carrier method is suggested. The favoured proposal of the multi-carrier method comes from Tsinghua University in Beijing and was called DMB-T for a long time. The single-carrier method is called ADTB-T and originates from Jiaotong University in Shanghai. DTMB has similarities with DVB-T whilst ADTB-T is derived from the North American ATSC.

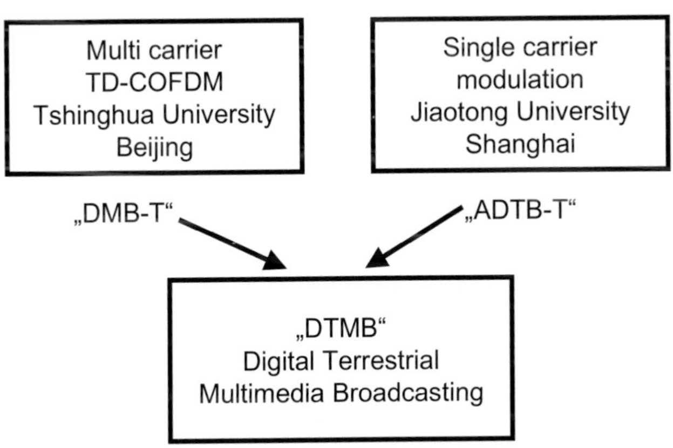

Fig. 32.1. Joining two proposals to form DTMB

W. Fischer, *Digital Video and Audio Broadcasting Technology*, Signals and Communication Technology, 3rd ed., DOI: 10.1007/978-3-642-11612-4_32, © Springer-Verlag Berlin Heidelberg 2010

32.2 Some more Details

It is used single carrier modulation and TD-COFDM (Time Domain Coded Orthogonal Frequency Division Multiplex), among other things. In multicarrier mode the guard interval is here not filled with the end of the next symbol following but with a PRBS. This symbol preamble is called frame header and has a length of 56.6 μs, 78.7 μs or 125 μs with a channel bandwidth of 8 MHz. In multicarrier mode DTMB runs in 4K mode with 3780 used carriers which are spaced apart at 2 kHz in the 8 MHz channel. The symbol period is therefore 500 μs. 3744 of these 3780 carriers are modulated data carriers and 36 are signalling carriers, i.e. virtually TPS carriers. DTMB supports channel bandwidths of 8, 7 and 6 MHz. The useful spectrum is 7.56 MHz wide in the 8 MHz channel. The net data rate is between 4.813 Mbit/s and 32.486 Mbit/s. The spectrum is roll-off filtered with a roll-off factor of r=0.05. The transmission method is intended for SDTV and HDTV transmissions and should work both in stationary and in mobile operation. It is possible to implement both MFN and SFN networks.

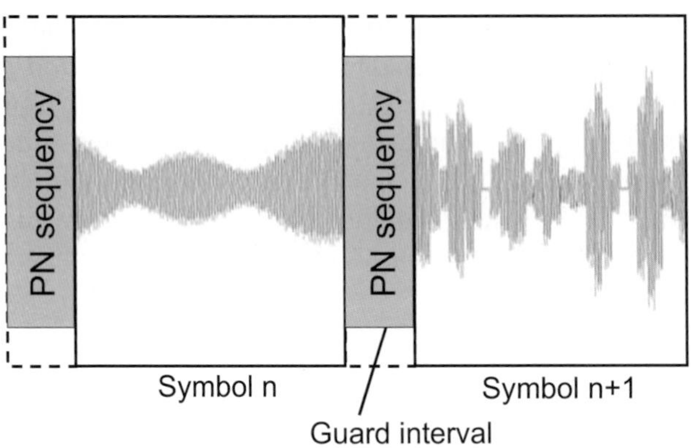

Fig. 32.2. DTMB TD-COFDM

The following can be selected as modulation methods on the 3744 data carriers:

- 64QAM
- 32QAM
- 16QAM
- 4QAM
- 4QAM=NR (Nordstrom Robinson).

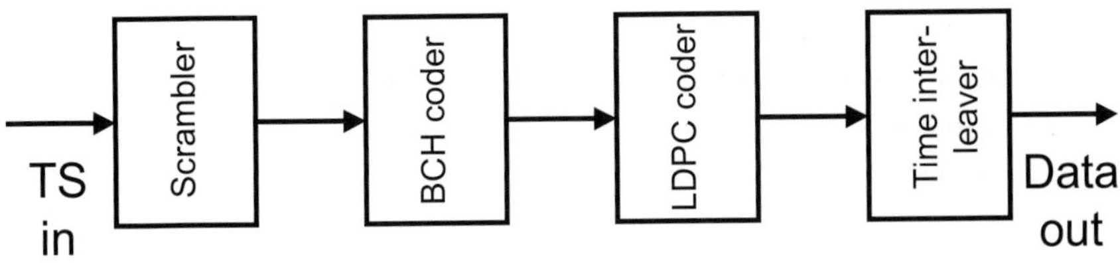

Fig. 32.3. DTMB Forward Error Correction

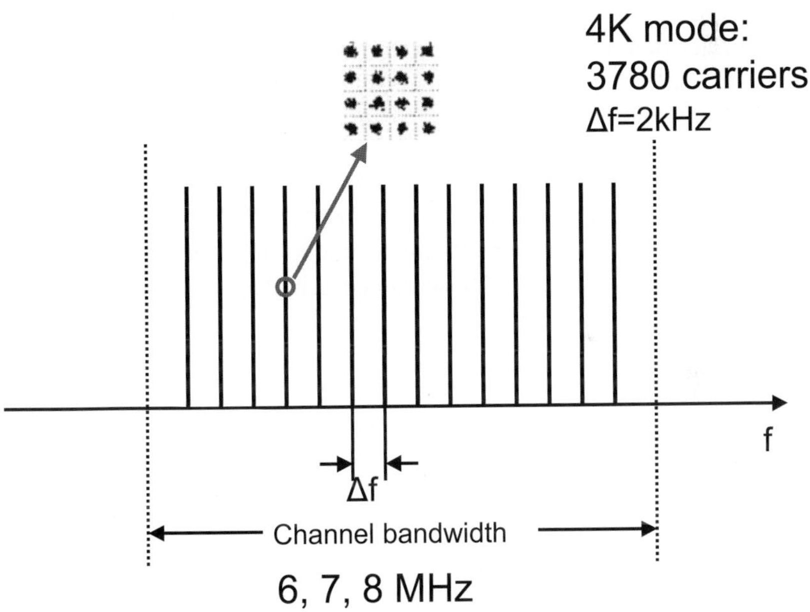

Fig. 32.4. Characteristics of the DTMB multi-carrier mode

The error protection in DTMB (Fig. 32.3.) consists of a

- Scrambler
- BCH coder
- LDPC coder
- Time interleaver

The DMB-T signal is made up out of a

- Signal frame (frame header + frame body = virtually guard + symbol)
- Super frame = N1 • signal frame
- Minute frame = N2 • super frame
- Calendar day frame = N3 • minute frame

As in other transmission methods too, the input signal of a DTMB transmitter is the MPEG-2 transport stream.

Unfortunately, it is difficult to provide more details of DTMB. Not much has been published and not all of the published papers appear to agree with one another, either. At this stage it appears to be prudent to say nothing rather than to provide false information. It is not really clear what advantages are to be gained by a guard interval filled with a PN sequence. The only thing that is clear is that licensing rights have moved towards being less binding with regard to DVB and ATSC and that some details of standards may well have something to do with this. Neither is it clear from where and to what purpose the roll-off characteristic has been introduced. The roll-off characteristic in multi-carrier mode may well come from the guard interval filled with a PN sequence (single-carrier method in the guard interval?).

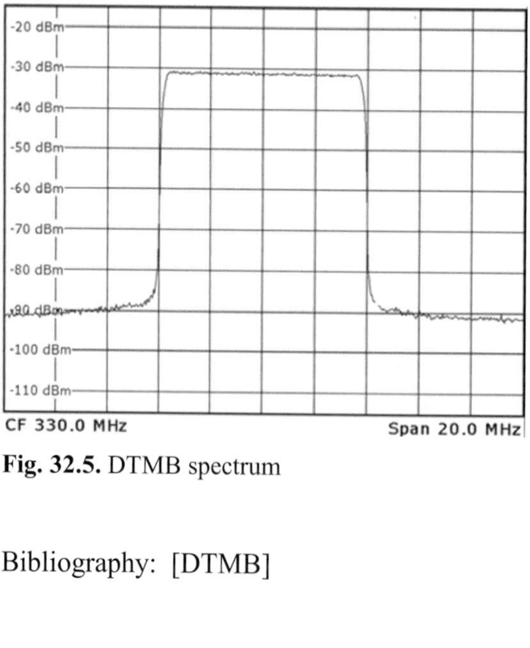

Fig. 32.5. DTMB spectrum

Bibliography: [DTMB]

33 Return Channel Techniques

Return channels have been provided for some time in digital television, having been defined both in DVB-T, in DVB-S and in DVB-C (DVB-RCT, DVB-RCS and DVB-RCC, resp.), but only the cable return channel has gained any real significance. It serves to provide rapid Internet access by broadband cable. There are two standards for return channel arrangements in cable TV: DVB-RCC/DAVIC or DOCSIS/EURO-DOCSIS. All the networks known to the author in Europe use EURO-DOCSIS for the return channel which is located here within the 5 - 65 MHz band. To implement the return channel, this frequency band was completely emptied, i.e. the channels located there were moved into the special-channel band. Today, both the traditional telephone companies and TV cable network operators are promoting "Triple Play", i.e. everything out of one socket - telephone, television and the Internet. This combination can also bring with it a distinct price advantage for the end user.

DOCSIS stands for "Data Over Cable Service Interface Specification" and has its origin in the US. It has also been standardized by ETSI in the [ETSI201488] document and is provided in several versions. Communication with the cable modem takes place both in the downstream and in the upstream direction. Downstream is a continuous MPEG-2 transport stream transmitted physically via ITU-TJ83B (DOCSIS) and ITU-TJ83A (EURO-DOCSIS). The modulation method used there is 64QAM or 256QAM. Downstream, in the frequency band between 5 and 65 MHz, is divided into burst packets (time slots). The modulation method used there is either QPSK or 16QAM. Due to man-made ingress noise, the frequency band between 5 and 65 MHz is not uncritical and the house cabling must be installed with the appropriate care. Thanks to the HFC (hybrid fibre coax) networks provided in most cases, where only the last 1000 meters are still implemented in coax, the distribution network has become quite uncritical. By now, the optical fibre cable can also be run right into one's house. Ingress noise can be measured by means of a spectrum analyzer by comparing the burst peaks in the return channel with the burst intervals (Fig.33.1.).

W. Fischer, *Digital Video and Audio Broadcasting Technology*, Signals and Communication Technology, 3rd ed., DOI: 10.1007/978-3-642-11612-4_33,

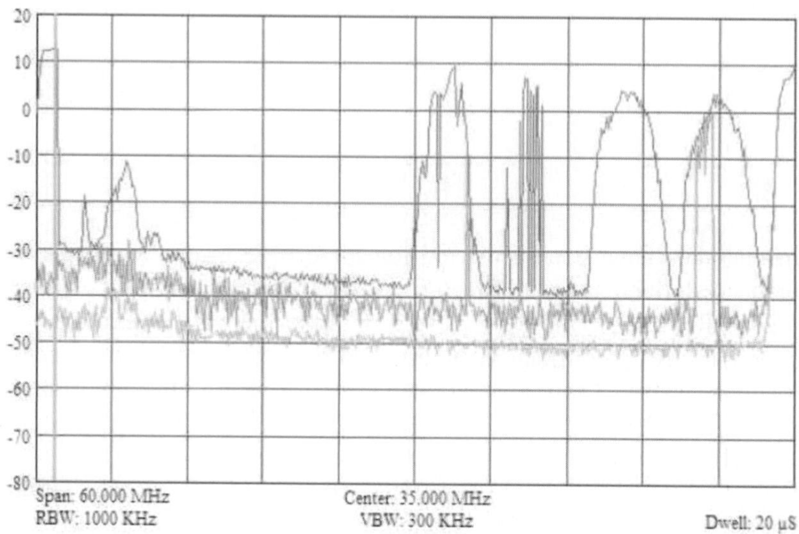

Fig. 33.1. Example of a return channel spectrum (internet and telephony) (picture supplied by UPC Telekabel Klagenfurt, Austria)

Bibliography: [ETSI201488], [ETSI300800], [KELLER]

34 Display Technologies

The cathode ray tube (CRT) has long been dominant as an essential electronic component both on the recording side and on the reproduction side. It forms the basis for many parameters and characteristics of a video baseband signal (composite video signal) such as, e.g., the horizontal and vertical blanking interval and the interlaced scanning method for reducing flicker, none of which would have been necessary with the new technologies where, in fact, they prove to be troublesome.

Before the CRT, however, there had already been attempts, since 1883, in fact, to transmit images electronically from one place to another. Paul Nipkow had invented the rotating Nipkow disk which provided the stimulus for thinking about sending picture information - moving picture information, what's more - from place to place. The Nipkow disk already slices images into transmittable components which are basically lines. Present-day modern display technologies are pixel-oriented, in so-called progressive scanning, i.e. without line interlacing, and exhibit distinctly higher resolutions. Today, we have basically

- partly still the "old" cathode ray tube (CRT),
- the flat screen, and
- projection systems such as beamers and back-projectors.

The underlying technologies for these are

- the cathode ray tube (CRT),
- liquid crystal displays (LCD),
- plasma displays,
- digital micro-mirror chips (DLP=Digital Light Processing chips), and
- organic light-emitting diodes (OLEDs).

Video compression methods such as MPEG-1 and MPEG-2 had still been developed for the cathode ray tube as reproduction medium at the beginning of the 90s. On a CRT, MPEG-2-coded video material still looks

W. Fischer, *Digital Video and Audio Broadcasting Technology*, Signals and Communication Technology, 3rd ed., DOI: 10.1007/978-3-642-11612-4_34, © Springer-Verlag Berlin Heidelberg 2010

quite tolerable even at relatively low data rates (below 3 Mbit/s), whereas it produces clearly visible blocking effects, blurring and so-called mosquito noise (visible DCT structures) on a large modern flat screen. MPEG-4 video, however, still looks clean on these modern displays, even at distinctly lower data rates. Interlaced material must be de-interlaced before being displayed on flat screens if it is not to result in disturbing "line feathering" (Fig. 34.10.). This can be seen most clearly in the case of continuous caption (text) inserts. New technologies result in new problems and artefacts. The new effects in image reproduction, caused by the reproduction system, are:

- burn-in,
- resolution conversion problems,
- motion blur,
- the appearance of compression artefacts,
- phosphor lag,
- dynamic false contours,
- and the rainbow effect.

In this chapter, the operation of the various display technologies and their characteristics and the artefacts which occur, and the respective attempts at compensating for them, will be described.

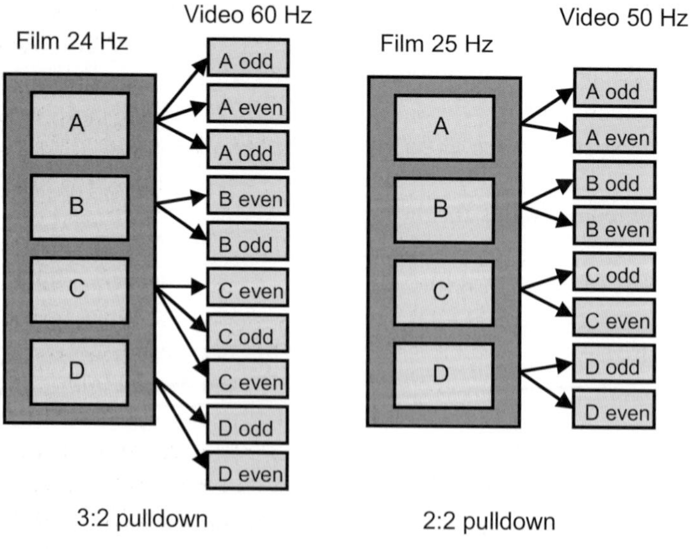

Fig. 34.1. Conversion from cinema-film to video (3:2 and 2:2 pulldown)

The source material for many films transmitted on television was, and still is, original cine-film. This is produced with 24 frames per second, independently of where it is produced. When it is reproduced in television, it must be adapted to the 25 or 30 frames per second used there and the frames must be converted into fields in the line interlace method (Fig. 34.1.). In the 25 frames/second technology, the films are simply run at 25 frames per second instead of at 24 frames per second and each frame is scanned twice by the film scanner, with half a video line offset in each case. The film is thus running imperceptibly slightly faster in television than in the cinema. This type of conversion does not present any problems and does not lead to any visible conversion artefacts. In the case of the 30 (59.94) frames/second standard, a so-called 3:2 pull-down conversion takes place (Fig. 34.1.). Instead of 24 frames per second, the film is here run at 23.976 frames per second, i.e. by 0.1 % more slowly, and 4 frames are then stretched to 5 frames using the line interlace method. For each frame, 2 or 3 fields are then generated alternately. Such converted image material can be juddery and is not easily converted again. A conversion from 30 to 25 frames/s is called an inverse 3:2 pull-down conversion. All this also affects the image processing in the respective TV displays.

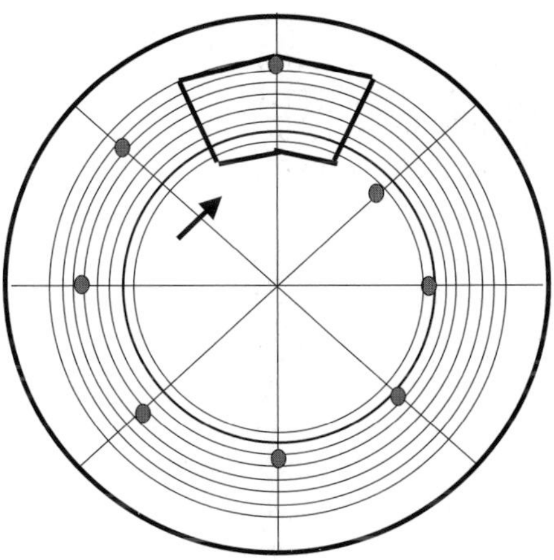

Fig. 34.2. Nipkow disk

34.1 Previous Converter Systems - the Nipkow Disk

Paul Nipkow had the idea of splitting images into "lines" back in 1883, using a rapidly rotating disk both as pickup and for replaying purposes. The

Nipkow disk had holes drilled spirally into the disk so that an original picture was scanned by the holes "line by line". In front of the Nipkow disk (Fig. 34.2.) there was an optical system which projected the original picture onto a selenium cell located behind the disk. This selenium cell was then supplied virtually line by line with the luminance variation of the original picture. At the receiving end, there was a reverse arrangement of a controllable light source and a replay disk, rotating in synchronism with the pickup disk, plus replay optics. Nipkow disks were used until well into the 1940s. The greatest problem was the synchronization of pickup and replay disks. The first trials were conducted with disks mounted on a common axis.

Televisors based on John Logie Baird's principles also used Nipkow disks and were used in so-called television rooms until the 30s. The narrow-band television signal was transmitted by wire or wirelessly by radio. Demonstration models still available today (see, e.g., NBTV - Narrow Band Television Association) show that the signals used then can be handled by narrow-band audio transmission channels. The devices used line numbers of about 30 lines per frame. To simplify synchronization, synchronous motors were used which virtually tied the system to the alternating mains frequency of 25 or 30 Hz. It is interesting to note that the link of the frame rate with the AC frequency of the respective national grid has been maintained to the present day. To synchronize transmitter and receiver, synchronization signals were additionally keyed into the narrow-band video signal which is similar to a current composite color video signal. Modern Nipkow disk demonstration models use black/white line patterns on the disks in conjunction with phase locked loops. The signals derived from these are compared with the synchronization pulses and the error signal is used for correcting the speed of rotation of the disk. Between the sync pulses there is a linearly modulated narrow-band video signal which controls the brightness of the replay element (an LED today, a neon tube in those days) while the respective associated hole in the Nipkow disk scans the visible screen area. The image quality of these simple systems does not require much comment but the principle employed is still fascinating even for the experienced video specialist. It simply illustrates the basics and the history of television technology. Nipkow disk equipment operates in accordance with the principle of stroboscopy, i.e. of a short-time display without image retention, just like its successor, the cathode ray tube.

34.2 The Cathode Ray Tube (CRT)

Cathode ray tubes (CRT) are based on the principle of the Braun tube in which an electron beam which is deflected by two orthogonal magnetic fields and the intensity of which can be controlled by a grid writes an image line by line onto a luminous phosphor coating on the back of the display screen (Fig. 34.3. and 34.4.). The first display tubes, but also the first camera tubes, were monochrome devices. The camera tubes operated inversely, i.e. the electron beam in these read out an optical storage layer line by line. On the pickup side, the cathode ray tubes have been replaced by charge-coupled devices, so-called CCD chips, back in the 1980s. On the display side, they have been dominant until about 2005. Today, there are virtually only the "new screen technologies" in existence. All the characteristics of an analog video baseband signal, the so-called composite colour video signal, are based on the cathode ray tube.

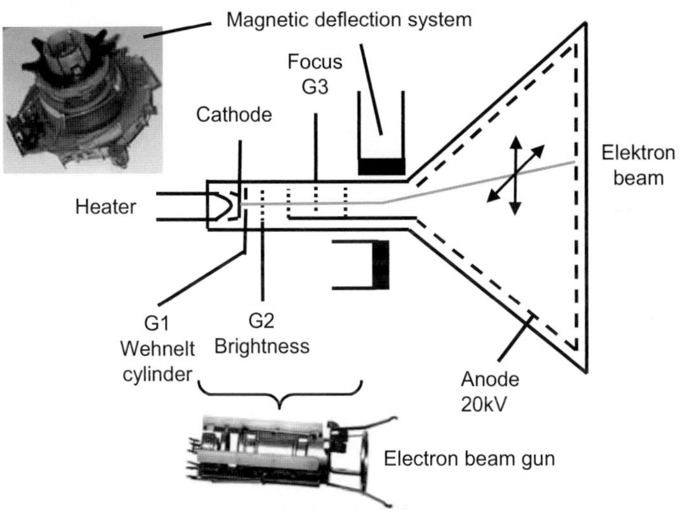

Fig. 34.3. Cathode ray tube with magnetic electron beam deflection

Horizontal and vertical blanking intervals provided for beam retrace blanking and this could only be done in finite time. This is because magnetic fields can only be re-polarized within a finite time with manageable energy consumption. To reduce the flickering effect, the frame was additionally split into fields, i.e. into odd- and even-numbered lines which were then reproduced with an offset. It was thus possible to create virtually a sequence of fields with twice the number of images (50 or 60 fields) from 25 frames or 30 frames per second. Persistence and filtering characteristics of the controllable deflected electron beam were significant components of the display characteristics. The first video compression methods (MPEG-1 and MPEG-2) used in most cases to the present day were developed with these reproduction technologies and for these reproduction technologies. However, the monochrome screens differed from the color CRTs in significant details. In the color CRTs, there is a so-called shadow mask located immediately behind the luminous phosphor coating; at the point where they pass through the slotted mask, the three Red, Green and Blue beams intersect if the convergence is set correctly.

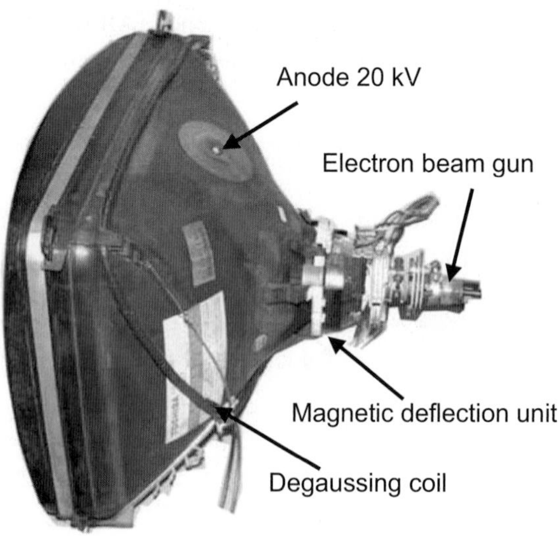

Fig. 34.4. Cathode ray tube (CRT)

I.e., in the case of the color CRT screens, there was already virtually a pixel structure in existence. This pixel structure can also be seen if one approaches a CRT-type monitor very closely (Fig. 34.5.). If it was still very complicated to establish convergence, i.e. the targeting precision of the three electron beams with the delta-type shadow mask, it was much simpler with the slotted-mask or in-line tube. The three electron guns were arranged here in one row (Fig. 34.6. right) so that the beam systems only had to be brought into convergence in the horizontal plane. From more than 30 controls in delta-type shadow mask systems, the convergence adjustments were reduced initially to just a few and later to none in slotted-mask or in-line tubes.

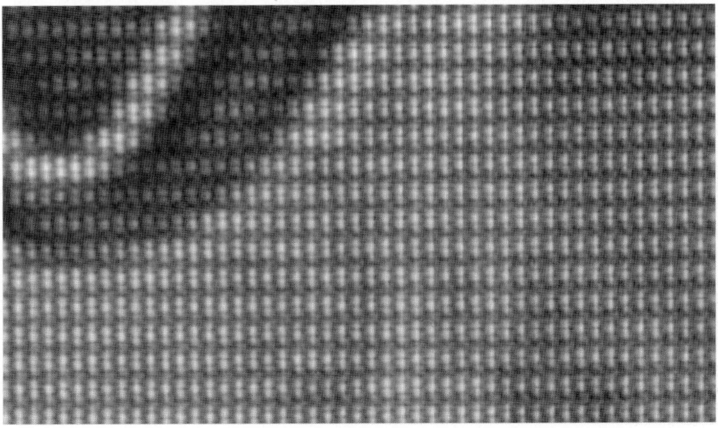

Fig. 34.5. Pixel structure of a slotted, in-line-type shadow mask tube

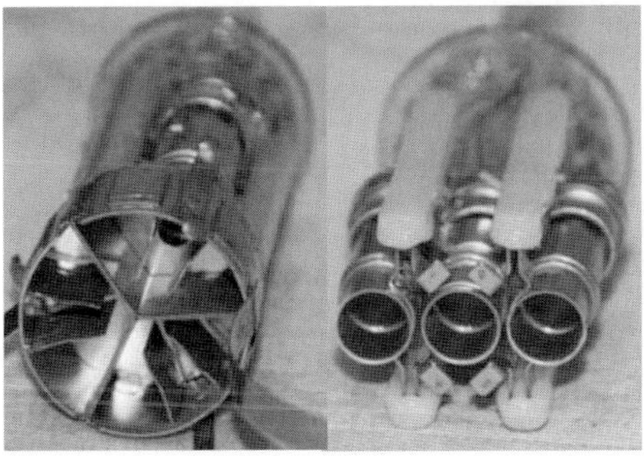

Fig. 34.6. Electron beam gun of a delta-type (left) and in-line-type (right) shadow mask tube

Picture tube monitors have completely different characteristics from modern flat displays. The CRT monitors operate

- with line interlace (fields and frames),
- with low-pass filtering due to their physics,
- stroboscopically, i.e. with short-time reproduction.

For these reasons, they do not show certain artefacts which become visible in connection with compressed image material on modern flat screens. And these artefacts on modern flat displays are noticed not only by the experts; it is mainly the consumer, the viewer, who notices these unsightly image patterns which are now visible whereas they were not on picture tubes.

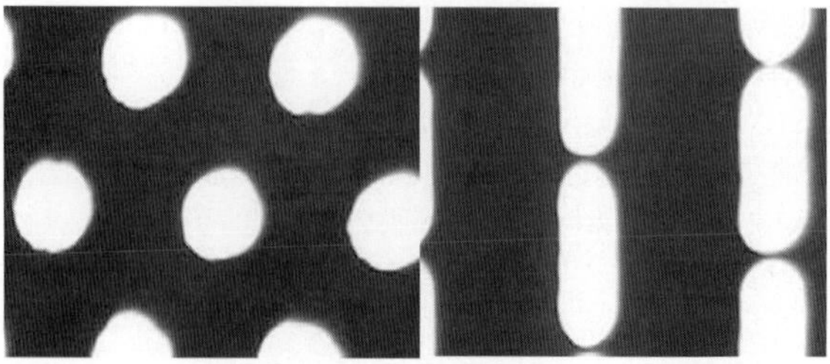

Fig. 34.7. Delta-type (left) and in-line-type (slotted) shadow mask (right)

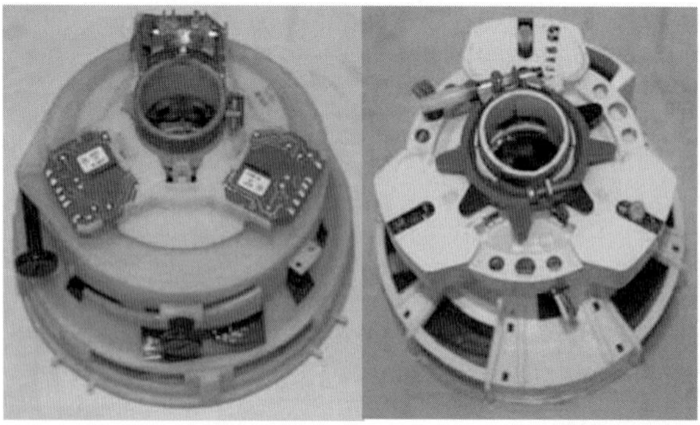

Fig. 34.8. Deflection systems with delta-type (left) and in-line-type (slotted) shadow mask (right)

This will increase the pressure on the program providers to progress more quickly towards HDTV, the high-resolution television which will be broadcast in any case in the new MPEG-4 AVC technology. In summary, it can be said that the picture tube was very tolerant - it did not show things which we can now see. It even filtered out most of the noise. But the picture tube had been developed for television with simple resolution, SDTV - standard definition television. And now it is to be hoped that we will soon really enter the age of high-definition television - HDTV. In the good old "picture tube TV", the power consumption depends on the active image content. The more light there is to be displayed, the more current is consumed.

34.3 The Plasma Screen

A plasma screen operates in accordance with the principles of a gas discharge in a more or less completely evacuated environment. A gas discharge of a so-called plasma - an ionised, rarefied air mixture such as, e.g. in a fluorescent tube, is ignited by a high voltage, producing ultraviolet light. Coloured light can be generated by converting this ultraviolet light of the gas discharge by means of the appropriate phosphors. Each pixel of a plasma screen consists of three chambers of the colours Red, Green and Blue. The three colours can be switched on or off by applying a high voltage and igniting the corresponding cell or can also be switched on or off simultaneously. The deciding factor is only that they can only be switched on or off and not incrementally more or less. To be able to achieve gradations in the respective brightness levels of the colours, a trick has to be used: the respective Red, Green or Blue cell is fired only for a certain length of time, i.e. applying pulse duration modulation. Short firing means darker, long firing means brighter. This principle can also lead to display artefacts (phosphor lag, wrong colours due to different response times of the phosphors). The plasma screen differs from the picture tube in this respect. They are very similar, however, in respect to their energy consumption. More light means more power and the energy consumption is thus dependent on the active picture content both in the picture tube TV and in the plasma screen TV. For a long time, large displays could only be implemented in plasma technology. Current indications are that this technology is fading into the background again in favour of LCD screens. The essential characteristics of plasma screens are:

- power consumption depends on image content,
- less weight than a CRT,
- less overall depth ("flat screen")
- high contrast,
- wide viewing angle,
- tendency to burn-in effects with static images (aging of phosphors),
- progressive scanning, i.e. tendency to line feathering with interlaced material,
- service life used to be a problem, is 100,000 hours today,
- retentive and non-stroboscopic system due to the drive system,
- insensitive to magnetic fields,
- possible short wave radiation with poor shielding,

- possible phosphor lag, wrong colour display due to phosphor characteristics,
- dynamic false contours.

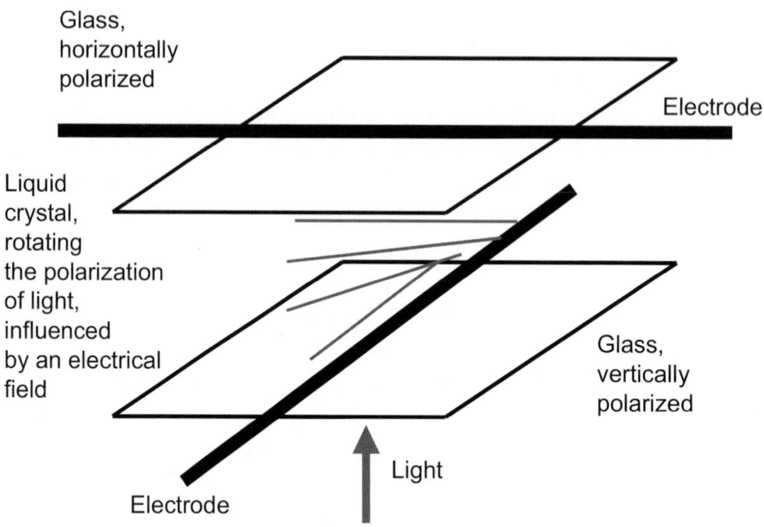

Fig. 34.9. Liquid crystal technology

34.4 The Liquid Crystal Display Screen

LCD displays (Fig. 34.9.) have been on the market for consumers since the 1970s. Today there are scarcely any applications in which LCDs are not found. Simple liquid crystal displays are cheap to produce and very energy-saving. If they are statically controlled they use virtually no energy, which is part of the reason why especially wrist watches have long been equipped with LCDs. Following the monochrome variant, the colour variant has also been in use since about the middle of the 1990s. Since this time, especially TFT - thin film transistor - displays have been increasingly used. Modern computer monitors are employing this technology almost without exception. The CRT monitor has virtually faded away since the beginning of the millenium and it no longer exists in the computer field. And the CRT monitor would not have any advantages there, either, neither in image quality nor in price, weight or size. In the computer field, the main reason for this is the technology used from the beginning - the frame-sequential scanning technique. Fields and line interlace are not used in computers and were never necessary since the images were relatively static and flickering was never a problem there. Computers and computer monitors have always operated with progressive scanning.

In LCD displays, the quantity of light through the display is controlled by more or less rotation of polarized light between two polarizing filters (Fig. 34.9.). The speed of control was a major item for discussion for a long time. As a result, these displays were initially very inert. In LCD displays, the quantity of light through the display can be controlled relatively linearly by the applied cell voltage by more or less rotation of the polarization of the light. If the polarizing filters in front of and behind the liquid crystal are identically polarized, the light will pass through unimpeded when no drive is applied and is attenuated only by applying a control voltage; if the polarizing filters are crossed, light will only pass when a control voltage is applied and the crystal thus rearranges itself and thus rotates the polarisation of the light. The power consumption of LCD displays essentially depends on the background illumination and is constant and thus independent of the active picture content. LCD displays belong to the category of retentive, i.e. non-stroboscopic display systems and may, therefore, exhibit so-called motion blur. The essential characteristics of an LCD screen are

- small overall depth ("flat screen"),
- less weight compared with previous screen technologies,
- less contrast compared with CRT and plasma displays,
- slower response times,
- constant, image-independent power consumption,
- possible motion blur,
- service life of approx. 100,000 hours,
- smaller, but now not inadequate viewing angle compared with CRT and plasma displays,
- retentive and non-stroboscopic system due to the drive system,
- insensitive to magnetic fields,
- progressive scanning, i.e. tendency to line feathering with interlaced material,
- faulty pixels,
- pixel response time depends on the step height (signal change).

34.5 Digital Light Processing Systems

At the beginning of the current millenium, Texas Instruments marketed a new display technology for projection systems called Digital Light Processing (DLP), involving so-called digital micro mirror chips - DMM chips. The concept itself goes back to the year 1987 (Dr. Larry Hornbeck, Texas Instruments). In this technology, the light intensity per pixel is controlled by infinitesimally small movable mirrors. Each pixel is a mirror which can be tilted by about 10 degrees. There is either one chip per colour or only one chip for all colours which is then divided for Red, Blue and Green in time-division multiplex via a rotating color disk. There are systems in use which have three mirror systems for Red, Green and Blue. A mirror can only be switched into or out of the beam path which is why pulse duration modulation is used here, too, for controlling the light intensity. When rotating color disks are used, a so-called rainbow effect occurs, i.e. a color spectrum appears, possibly due to the physiological perception characteristics of the human eye. DLP systems are used in projection systems such as beamers or back-projectors. They cannot be used for constructing real flat screens. They are applied in home cinema systems and may also be used in professional movie theatres because of their good light yield and high resolution and color rendition properties.

34.6 Organic Light-Emitting Diodes

Organic Light-Emitting Diodes (OLEDs) could become the most modern display technology. They are currently being offered only for relatively small displays. Their stability or service life still presents a problem. However, they can be applied in very thin layers even to flexible material, thus making it possible to implement displays which can be rolled up. There are already first small TV monitors with this technology on the market. For each pixel, there are three OLEDs the intensity of which can be controlled linearly.

34.7 Effects on Image Reproduction

For many years, cathode ray tube television was adapted relatively optimally to the characteristics of the human optical system of perception and had been developed further. But now there are new revolutionary technologies which are replacing the picture tube but have quite different characteristics. The decisive factor for image reproduction is the type of display technology in conjunction with the drive technology, i.e. the signal processing. And it must be said that new is not always better. A distinction can be made here between

- stroboscopic reproduction systems with short frame retention time, and
- hold-type (retention) systems with frame retention times within the range of one frame.

However, the perception characteristics of a moving picture depend not only on the physics of the display but also on the anatomy of the human optical perception system (eye, eye tracking and brain).

A distinction must also be made between the type of color reproduction where there are systems

- which share one reproduction element for all colors (e.g. via rotating color filters) or
- which have discrete color reproduction elements for RGB.

And, naturally, the response time of the screen plays an important role. This is greatly dependent on the display technology. LED systems and plasma screens have an inherently very short switch-over time between states of brightness. They also exhibit the largest ratio between light which is switched on and that which is switched off, and thus have the greatest contrast ratio.

In addition, there is a difference between display technologies which operate in

- line interlace scanning mode, and in
- progressive scanning mode.

All modern displays can be categorized as "progressive". When using a display operating with progressive scan it is absolutely necessary to carry out de-interlacing before the scan if unsightly line feathering known from the early 100 Hz TV technology is to be avoided. The reason can be found in the mismatch of fields due to movement between them. Pure cinematographic material is non-critical in this respect since there is no movement between the two fields.

Displays can also be classified as

- linearly driven display elements, and
- switchable display elements.

The display elements which can only be switched on and off are then intensity-controlled by pulse duration modulation.

The display technology thus gives rise to the following problems:

- blurring due to display delay times,
- blurring due to interpolation of unmatched resolution of broadcast material and display system,
- motion blur with non-stroboscopic displays,
- blurring due to the display frame rate,
- line feathering (line tearing) due to poor de-interlacing (Fig. 34.10. left),
- color shift (plasma, phosphor lag),
- different contrast with different steps in brightness (LCD),
- rainbow effect (chromatic separation),
- obvious appearance of compression artefacts due to the high resolution,
- losses in contrast,
- power consumption depending on picture content,
- burn-in effects (plasma).

Fig. 34.10. Different picture quality after de-interlacing (e.g. feathering left)

34.8 Compensation Methods

In modern electronics, there are countermeasures to act as improvement for every shortcoming. Their effectiveness depends on the state of the art. All modern display technologies operate with progressive scanning technology. The simplest remedy for interlace artefacts is not to use interlaced material. This is the aim especially in HDTV - s. a. 720p. But in the SDTV field, interlaced material must be offered because there are still countless television sets with cathode ray tubes. And the archives still contain any amount of interlaced material. De-interlacing is, therefore, an absolute necessity with modern displays if unsightly line feathering effects are to be avoided. In de-interlacing, it is necessary to interpolate between the fields. This is not necessary with material which is inherently progressive such as, e.g., the good old cine-film which only has frames, the fields being produced by repeated scanning in the intermediate lines. In the cine-projector, the film is interrupted twice per frame by a rotating aperture wheel to reduce flicker. However, in material recorded with electronic cameras there is already movement between the two fields. From these fields, virtually motion-compensated interpolated frames must now be generated otherwise the lines will become "frayed". To counter motion blur due to non-stroboscopic displays there are so-called 100 Hz or 200 Hz systems which, in contrast to earlier 100 Hz CRT displays, do into need this for reducing flicker but virtually simulate for the human eye a movement of an object on the display by interpolating fewer or a greater number of intermediate images. Otherwise, the eye and the entire "human optical pickup system" will dwell on an intermediate state of the image display which, time and again, is virtually static which then leads to smearing or, in other words, blurring.

34.9 Test Methods

Video signal generators [DVSG] contain test sequences for detecting the problems caused by the display technology. These sequences deliberately generate stress signals for the displays in order to reveal , e.g., resolution, interpolation characteristics, motion blur, rainbow effect and line feathering. It is also of major interest to see how a display system behaves when it adjusts itself to a source resolution not corresponding to the system (e.g. a laptop with unmatched screen resolution), requiring interpolation between pixels (scaling up or down). It is also of interest to see how the display devices handle compression artefacts such as

- blurring,
- blocking, or
- mosquito noise.

The good old picture tube (CRT) was relatively tolerant in this respect as can be seen impressively in demonstrations. But we are not concerned here with hanging on to a proven old system but are only comparing new systems with it and are correspondingly astonished or disappointed. Certain reproduction systems simply do not show some of the things because these are concealed by them. The normal Gaussian noise evident in analog TV channels, in particular, produces different effects in different display systems which causes viewers of analog TV with flat screens in broadband cable networks increasingly to protest even though the comfort of not requiring a special DVB-C receiver is appreciated so much in these quarters. But there are also screens which allow noise to be suppressed but this can only be done at the cost of image sharpness. Noise can only be eliminated by averaging between frames which again leads to blurring. New interfaces such as HDMI also exchange data between display and receiver, e.g. for determining the correct resolution from the possible features provided by the display. These things must also be tested. Another interesting test is how the various displays behave in a comparison of

- 25/50 Hz interlaced material,
- 30/60 Hz material (3:2 pull-down) (Fig. 34.1.), and
- inverse 3:2 pull-down (material reconverted from 30 to 25 frames/s) (Fig. 34.1.).

34.10 Current State of the Technology

Conclusion: The quality of reproduction on displays is a matter not only of the installed display technology but mainly also of the type of compensation measures applied. Some consume a great amount of memory space and computer power. It is expected that LCD and LED technologies will displace the plasma systems. Cathode ray tube systems have provided good service for many decades but their end has come. It cannot be predicted what significance will be accorded to e.g. DLP systems or other, mainly new technologies. They may well be the active element in cine-projectors of the future. And it can also be said that for most of the picture material available at present (status of 2009 in Europe), the cathode ray tu-

tube still offers the best solution in most cases. However, this will change as SDTV becomes HDTV.

Bibliography: [ERVER], [NBTV], [DVSG]

35 The New Generation of DVB Standards

In 2003, DVB-S2 appeared as the first new transmission standard for Digital Video Broadcasting - DVB. Because the transmission of High Definition Television (HDTV) now requires a much higher data rate, the transmission capacity, i.e. the net data rate per satellite transponder, had to be increased by at least 30% compared with DVB-S. On the other hand, the hardware now available is much more powerful in comparison with that of the early 90s. Both the memory capacity and the computing speed of the chips have increased significantly which made it possible to use Forward Error Correction (FEC) which, although it swallowed up an enormous amount of resources on the receiver side, also yielded a significantly higher net data rate (approx. 30%). This error protection has been known since 1963 and is based on Robert Gallager's work. Today, all the new transmission standards, either for mobile or for broadcasting applications, are using either turbo codes (1993) or LDPC (low density parity check) codes. DVB-S2 now enables 48 to 49 Mbit/s to be transmitted over a satellite transponder which previously provided a data rate of 38 Mbit/s via DVB-S. Using MPEG-4 Part 10 AVC (H.264) video coding, 4 HDTV programs will then fit into one satellite transponder. But with DVB-S2, the direct tie with the MPEG-2 transport stream as baseband input signal was also relinquished for the first time. The catchphrase is now GS - Generic Stream. I.e., apart from the MPEG-2 transport stream, quite general data streams such as, e.g. IP, are now also provided as the input signal for a DVB-S2 modulator, making it possible to feed either one or even several streams into the modulator. At present, this capability is not being exploited for DVB-S2, however, where only pure, standard MPEG-2 transport streams are currently being transmitted. As well, only parts of the capabilities provided in the Standard are currently utilized. The DVB-S2 Standard has already been discussed in Chapter 14 "Transmitting Digital TV Signals by Satellite - DVB-S/S2". In 2008, the new, very powerful standard DVB-T2 - "Second Generation Digital Video Broadcasting Terrestrial" was then published which also contains this new error protection, with Generic Streams - GS - and Multiple Inputs also being a subject. The possibility of having multiple input streams, either transport streams or generic streams, is here called Physical Layer Pipes (PLP) and will probably

W. Fischer, *Digital Video and Audio Broadcasting Technology*, Signals and Communication Technology, 3rd ed., DOI: 10.1007/978-3-642-11612-4_35,

also be made use of. The ability to chose between different modulation parameters (error protection and type of modulation) for different contents will probably be applied mainly with DVB-T2, the keywords being Variable Coding and Modulation (VCM). It will thus be possible to broadcast HDTV programs, e.g. less robustly but equipped with a higher net data rate than, e.g., SDTV programs. DVB-T2 was being promoted by the BBC; the first field trials in the so-called BBC mode have been running since 2008 as Single PLP, i.e. with one (transport stream) input. In 2009, the new, also very powerful DVB-C2 cable standard then appeared as draft standard. DVB-C2 is based on DVB-T2 and also has the capability of GSE and Multiple PLP with VCM. The LDPC coding is also used as FEC. The novel feature in DVB-C2 is the possibility of combining several 8- or 6-MHz-wide channels to form channel groups. This is intended to increase the effectiveness by avoiding gaps between the channels. Like DVB-T2, DVB-C2 uses COFDM, i.e. a multicarrier method, whereas DVB-S2 only uses a single-carrier method.

35.1 Overview of the DVB Standards

The following DVB standards exist at present:

- DVB-S - first generation satellite transmission standard
- DVB-C - first generation cable transmission standard
- DVB-T - first generation terrestrial transmission standard
- DVB-SI - Service Information
- DVB-SNG - satellite transmission standard for professional applications
- DVB-H - extension for handheld mobiles in DVB-T
- DVB-IP - transport stream transmission via IP networks, e.g. VDSL,
- DVB-SH - hybrid method for handheld mobiles via terrestrial and satellite
- DVB-S2 - second generation satellite transmission standard
- DVB-C2 - second generation cable transmission standard
- DVB-T2 - second generation terrestrial transmission standard.

Naturally, there are numerous other DVB documents but those listed above are the ones which are most meaningful in the present context.

35.2 Characteristics of the Old and the New Standards

In this section, the essential characteristics of the old and new DVB standards are compared with one another. The essential factor is that the old DVB standards were firmly tied to the MPEG-2 transport stream as an input signal and a combination of Reed Solomon coding and convolutional coding was used as error protection. In the new DVB standards, the tie with the transport stream is relinquished and the error protection has been modernized.

Table 35.1. DVB standards

Standard	Applicatio n	Input signal	Forward Error Correction (FEC)	Modulation
DVB-S	Satellite	MPEG-2 TS	Reed-Solomon and convolutional coding	Single Carrier, QPSK
DVB-C	Cable	MPEG-2 TS	Reed-Solomon	Single Carrier, 64QAM, 256QAM, additional QAM orders below 256 are possible
DVB-T	Terrestrial	MPEG-2 TS	Reed-Solomon and convolutional coding	CODFM, QPSK, 16QAM, 64QAM
DVB-SNG	Satellite	MPEG-2 TS	Reed-Solomon and Trellis-coding	Single Carrier, QPSK, 8PSK, 16QAM
DVB-H	Terrestrial	MPEG-2 TS with MPE	Reed-Solomon and convolutional coding and additional Reed-Solomon over IP	like DVB-T
DVB-IP	Twisted pair lines or Ethernet	MPEG-2 TS	--	--
DVB-SH	Terrestrial and	MPEG-2 TS with MPE	Turbo coding	largely like DVB-T and

	satellite			DVB-S2
DVB-S2	Satellite	Single or multiple MPEG-2-TS or GS	BCH and LDPC	Single Carrier, QPSK, 8PSK, 16APSK, 32APSK
DVB-T2	Terrestrial	Single or Multiple MPEG-2-TS or GS	BCH and LDPC	COFDM, QPSK, 16QAM, 64QAM, 256QAM
DVB-C2	Cable	Single oder Multiple MPEG-2-TS oder GS	BCH and LDPC	COFDM, QPSK, 16QAM, 64QAM, 256QAM, 1024QAM, 4096QAM

35.3 Capabilities and Aims of the New DVB Standards

It is especially the new error protection which provides about 30% increase in data rate from DVB-S2 onward. The Shannon limit is thus coming closer. DVB-S2 also enables several different input data streams to be radiated via one satellite. And it is intended to break the tie with the MPEG-2 transport stream. DVB-T2 additionally exploits other possibilities for gaining even more data rate compared with DVB-T. The symbols used are longer, thus reducing, e.g., the overhead in the guard interval; widening of the useful spectrum more in the direction of the adjacent channels is being offered etc.. In addition, variable coding and modulation can be activated in the multiple PLP mode, i.e. contents can be transmitted with different degrees of robustness similar to DAB and ISDB-T. DVB-T2 provides for much more such as, e.g., the application of multi-antenna systems at the transmitting end, the active reduction of the crest factor and much else besides. The details are described in the "DVB-T2" chapter. DVB-C2 is based on DVB-T2 and uses similar "tricks" for raising the data rate. It provides multiple PLP as well as VCM. DVB-T2 and DVB-C2 provide for up to about 50% more data rate compared with the comparable old DVB standards. DVB/T2 and DVB/C2 also provide higher modulation types such as, e.g. 256QAM with DVB-T2 and up to 4096QAM with DVB-C2.

The essential features and capabilities of the new DVB standards are thus:

- 30 - 50% increase in net data rate,
- multiple inputs (TS or GS),
- with T2 and C2, the possibility of variable coding and modulation.

With the change from SDTV to HDTV looming ahead, increased amounts of net data rate will become necessary. DVB-S2 will eventually assert itself over DVB-S, but not as yet. There are only a few transponders with DVB-S2 on air and the future will tell how quickly DVB-T2 and DVB-C2 will be employed.

In the chapters following, the baseband signals of DVB-x2 and then the DVB-T2 and DVB-C2 standards will be described. DVB-S2 has already been discussed in the chapter on DVB-S/S2 and the associated DVB-S2 test methods are already found in the joint chapter "DVB-S/S2 Measurements".

Bibliography: [EN302307], [TR102376]. [EN302755], [A138], [TS102773], [TS102606], [TS102771], [A133]

36 Baseband Signals for DVB-x2

In the first generation DVB standards (DVB-S, DVB-C and DVB-T), the format of the input data was confined only precisely to MPEG-2 transport streams where all the modulation and demodulation steps are firmly synchronously linked to the 188-bytes-long transport stream packet structure. An MPEG-2 transport stream packet begins with a 4-bytes-long header which, in turn, begins with a sync byte having the value 0x47. This limitation to the transport stream structure no longer exists in the new DVB-x2 standards. In the first generation DVB standards, it was also only possible to feed precisely one transport stream into the modulator, the only exception being DVB-T in its "hierarchical modulation" mode of operation where the modulator could be supplied with up to two transport streams. In the new DVB-x2 standards, up to 255 transport streams or generic streams, or both, can be fed into the modulator and transmitted. The present chapter deals with the input signals for the new DVB-x2 standards and how they are processed and conditioned in the input interfaces of the DVB-x2 modulators.

36.1 Input Signal Formats

The new DVB-x2 modulators can accept four different input signal formats, which are:

- MPEG-2 transport streams (TS),
- Generic Fixed Packetized Streams (GFPS),
- Generic Continuous Streams (GCS),
- and Generic Encapsulated Streams (GSE).

At the time of publication of the DVB-S2 standards, the DVB-GSE standard did not yet exist. DVB-S2 only provided for GFPS streams and GCE streams as generic input streams, supporting both a single input stream (TS or generic) and multiple input streams. Multiple input streams

W. Fischer, *Digital Video and Audio Broadcasting Technology*, Signals and Communication Technology, 3rd ed., DOI: 10.1007/978-3-642-11612-4_36, © Springer-Verlag Berlin Heidelberg 2010

can also have different formats. The multiple input streams are called PLPs - Physical Layer Pipes - in DVB-T2 and DVB-C2.

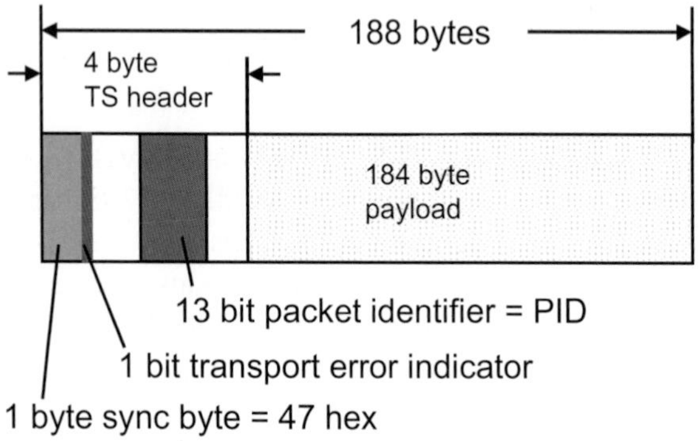

Fig. 36.1. MPEG-2 Transport Stream (TS)

36.1.1 MPEG-2 Transport Streams - TS

An MPEG-2 transport stream consists of packets having a constant length of 188 bytes (Fig. 36.1.). The packet itself is divided into a header component of 4 bytes length and a payload component of 184 bytes. The first byte of the header is the sync byte which has a constant value of 0x47. The next three bytes in the header are used for signalling important transport parameters. The rest is described in detail in Chapter 3.

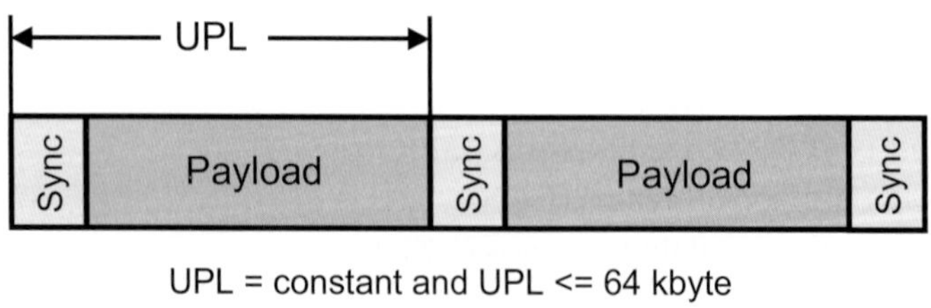

Fig. 36.2. Generic Fixed Packetized Stream (GFPS)

36.1.2 Generic Fixed Packetized Streams - GFPS

Generic Fixed Packetized Streams (GFPS) are data streams which have a packet structure and the packet length of which is known and is constant (Fig. 36.2.). The beginning of a packet is marked with a special sync byte. An example of a relevant application would be ATM data signals with a constant length of 53 bytes. The length of a GFPS must not exceed 64 kbytes for else it would be a Generic Continuous Stream - a GCE.

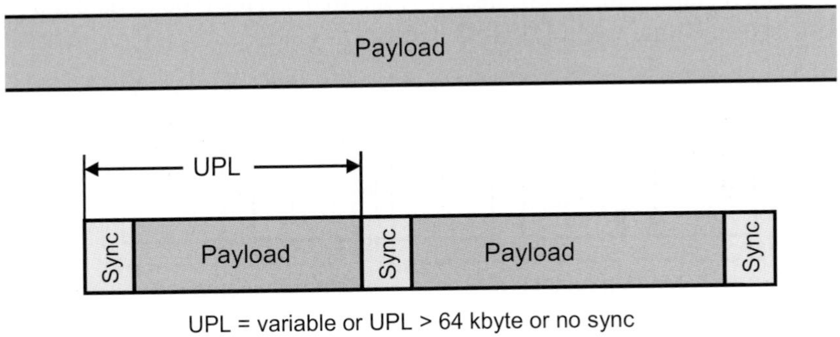

Fig. 36.3. Generic Continuous Stream (GCS)

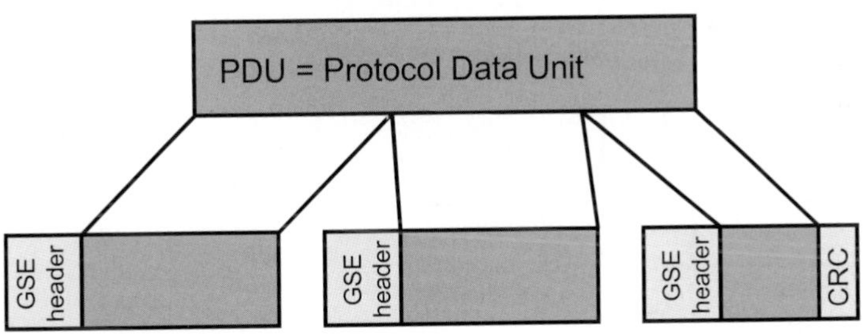

Fig. 36.4. Generic Stream Encapsulation (GSE)

36.1.3 Generic Continuous Streams - GCS

Generic Continuous Streams (GCS) do not have any packet structure (Fig. 36.3.). Thus, the modulator interface does not recognize any boundaries in the data stream. Generic continuous streams are the most generalized form of data streams. Generic continuous streams are also data streams which have a packet structure which don't differs from that of GFPS but the packet length of which varies or is longer than 64 kbytes.

36.1.4 Generic Encapsulated Streams - GSE

Generic Encapsulated Streams (GSE) have a packet structure the packet length of which varies. The beginning of the packet is provided with a special GSE header as defined in the DVB GSE standard [TS102606]. The user data packet is prefixed by a GSE header in the GSE encapsulator and a CRC is formed over the entire data packet, which CRC is then appended. The user data packets can also be divided into a number of packets where each packet starts with a GSE header (Fig. 36.4.). This type of data did not yet exist at the time when DVB-S2 became a standard.

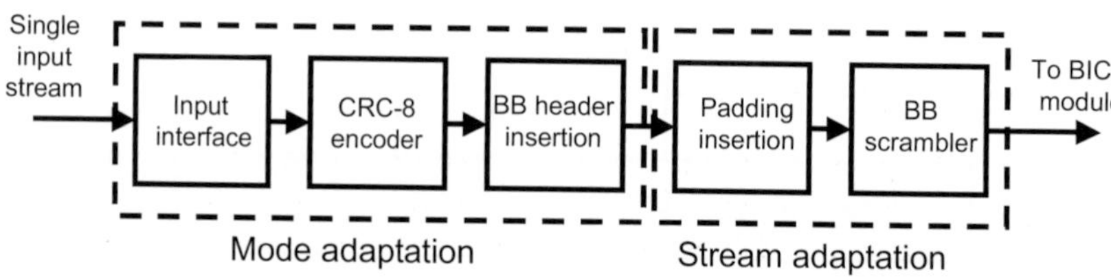

Fig. 36.5. Input Processing with a single input stream

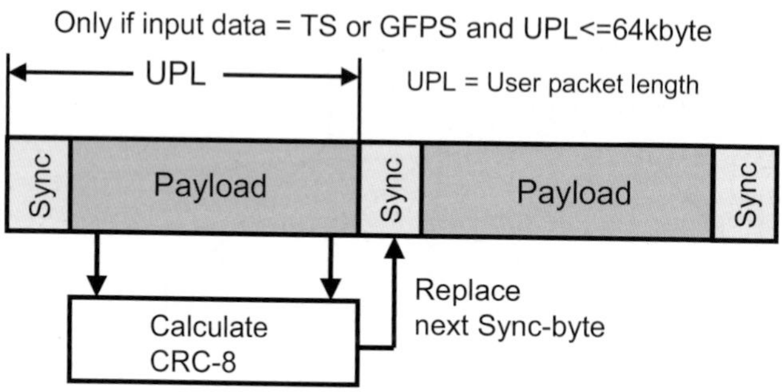

Fig. 36.6. CRC-8 encoding

36.2 Signal Processing and Conditioning in the Modulator Input Section

The following paragraphs describe how one or more DVB-x2 input streams are conditioned in the input section of the modulator. The difference lies in the conditioning of a single input stream in comparison with multiple input streams. Conditioning a single data stream, either a transport stream or a generic stream is clearly easier than conditioning a number of input streams.

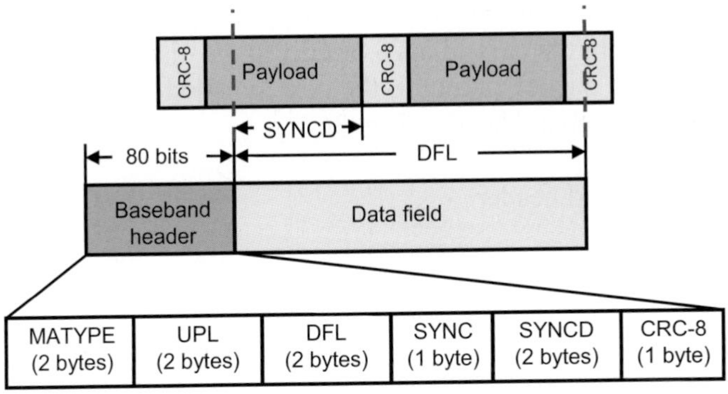

Fig. 36.7. Baseband header insertion with TS or GFPS

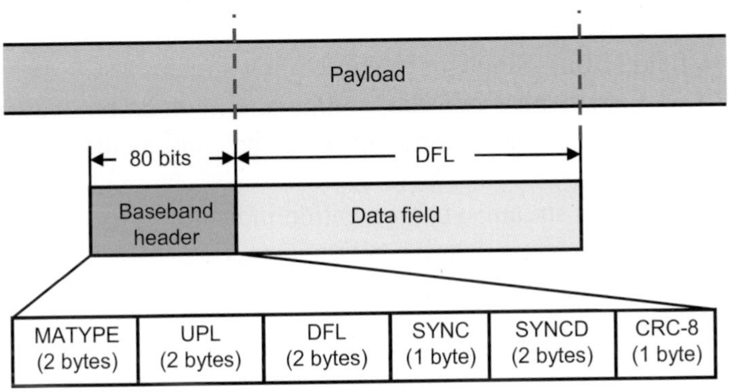

Fig. 36.8. Baseband header insertion with GCS

36.2.1 Single Input Stream

If only one data stream is fed into the DVB-x2 modulator (called Mode A in DVB-T2), the modulator first synchronizes itself in the input interface to the data stream supplied (Fig. 36.5.). This is followed by the CRC-8 encoder (Fig. 36.6.) which inserts a checksum into the data stream at a particular point unless this is a continuous data stream. In the case of a transport stream, a CRC-8 is formed over the 187 bytes preceding the next sync byte and the subsequent sync byte is then replaced by this checksum. If it is a GFPS, the CRC-8 is formed over all data except the sync byte and the sync byte following the packet is also replaced by a CRC-8.

Following this, a piece of corresponding length is cut out of the data stream and a ten-byte-long, i.e. 80-bit-long, baseband header is placed in front (Fig. 36.7.). This is done continuously piece by piece and occurs completely asynchronously with the data steam supplied, whether this is a TS, GFPS, GCS or GSE. In the case of a TS or GFPS, the distance in bytes from the beginning of the packet cut out to the next sync byte is entered in the SYNCD part of the baseband header (Fig. 36.7.).

Components of the baseband header:

MATYPE (2 bytes) – Mode Adaptation Type

MATYPE-1

- TS/GS field (2 bits), input stream format: Generic Fixed Packetized Stream, Transport Stream, Generic Continuous Stream, Generic Encapsulated Stream
- SIS/MIS field (1 bit): Single or multiple input streams
- CCM/ACM field (1 bit): Constant coding and modulation or variable coding and modulation (or adaptive coding and modulation in DVB-S2)
- ISSY (1 bit): Input stream synchronization indicator
- NPD (1 bit): Null-packet deletion active/not active
- EXT (2 bits): media specific, reserved for future use

MATYPE-2

- If multiple input streams, MATYPE-2 = ISI = Input stream identifier (up to 255 input streams).

Other components of the baseband header:

- UPL (2 bytes): User packet length in bits (0…65535)
- DFL (2 bytes): Data field length in bits (0 …53760)
- SYNC (1 byte): A copy of the user packet sync-byte
- SYNCD (2 bytes): Distance between beginning of data field to the beginning of the first user packet which starts in the data field
- CRC-8 mode (1 byte): The XOR of the CRC-8 field with the MODE field
- CRC-8 is the CRC over the first 9 bytes of the baseband header

- MODE: 0 = Normal Mode (NM), 1 = High Effiency Mode (HEM), other
- values reserved for future use

The first 3 blocks of input processing, namely input interface, CRC-8-encoding und baseband header insertion are called Mode Adaptation (Fig. 36.5.). This is followed by the stream adaptation (Fig. 36.5.) consisting of padding und baseband scrambler.

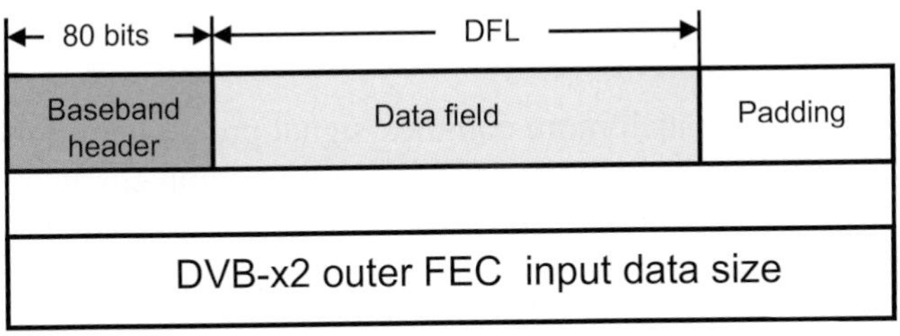

Fig. 36.9. Padding

Padding (Fig. 36.9.) means filling up or stuffing, i.e. if there are not sufficient user data available, a DVB-x2 FEC frame is filled up with stuffing or padding bytes until it is completely filled with data. This corresponds to an adaptation to the FEC frame structure of the outer DVB-x2 (BCH) error protection and is dependent on the code rate set for the error protection.

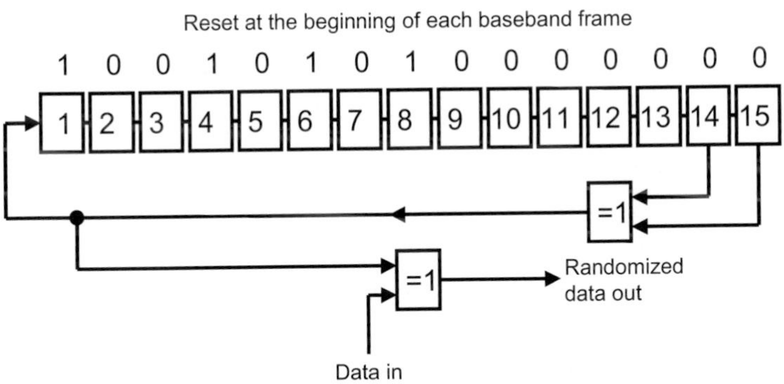

Fig. 36.10. Baseband scrambler

This is followed by the baseband scrambler (Fig. 36.10.) which has the task of randomizing the data as much as possible, i.e. adjacent long sequences of zeroes and ones are broken up and a pseudo-random data stream is generated. For this purpose, the data are Exclusive-ORed with a pseudo-random sequence. The PRBS generator has exactly the same structure as the energy dispersal stage of the first generation DVB standard and is always reset at the beginning of a baseband frame.

The DVB-x2 input data are now ready for the next signal processing steps which depend on the respective DVB-x2 standard.

36.2.2 Multiple Input Streams

In the sections following, the much more complex signal processing in the input stage of a DVB-x2 modulator in the case of multiple input streams is described. In DVB-T2, the operating mode with multiple input streams, whether they are transport streams or generic streams, is called Mode B or Multiple PLP.

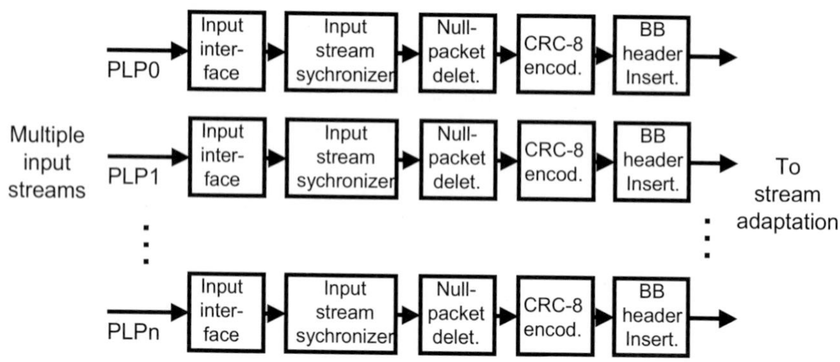

Fig. 36.11. Mode adaptation with multiple input streams

The multiple input streams (TS, GFPS, GCS or GSE) are present at the respective inputs of the input interface where synchronization to the input streams takes place. The input streams are then supplied to the further signal processing blocks of the mode adaptation section (Fig. 36.11.). The first one is the optional input stream synchronizer (ISSY) (Fig. 36.12.). The ISSY can be used for removing jitter from the input streams in the receiver; which is done via an internal clock similar to the STC/PCR in the case of the MPEG-2 transport stream. However, this clock is controlled via the modulator clock. The clock consists of a 22-bit counter the current value of which is continuously repeatedly written into optional ISSY fields which are appended to the end of the user packet (UP) and have a length of 2 or 3 bytes.

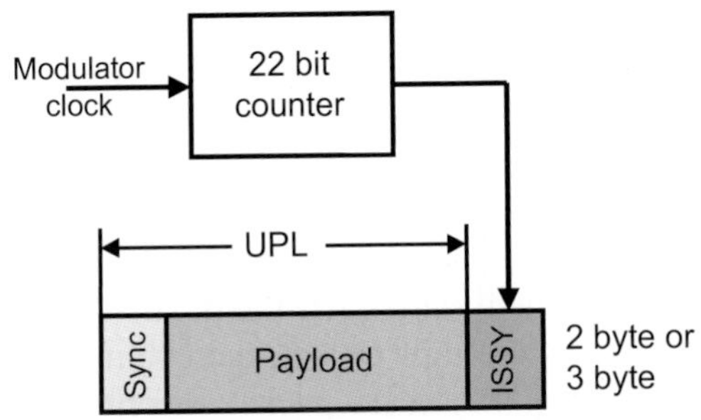

Fig. 36.12. Input stream synchronization (ISSY)

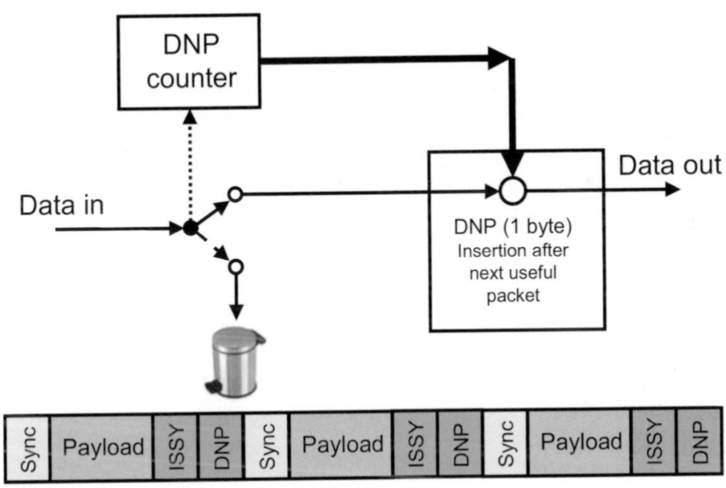

Fig. 36.13. Null Packet Deletion

The use of ISSY is signalled via the baseband header, followed by the "Null Packet Deletion" block (Fig. 36.13.). This block is optional and is only used if an MPEG-2 transport stream is present as input stream. An MPEG-2 transport stream contains null packets which run on PID=0x1FFF, i.e. on the highest PID. These packets do not carry any payload data but only pad the transport stream to a constant total data rate. The "Null Packet Deletion" processing block has the aim of removing these null packets from the transport stream. Null packets are unnecessary ballast which does not have to be transmitted. However, the null packets must be removed in such a way that the receiver is able to add them again to the transport stream in the correct number and at the correct position. For this purpose, a DNP byte (Fig. 36.14.) is inserted after each transport stream packet (after the ISSY field or, if there is none, directly after the transport stream packet). In the DNP (Delete Null Packet) field, the number of null packets removed from in front of this packet is entered. If the value in the DNP field is zero, no null packets have been removed from in front of this packet. To this end, the removal of the null packets is tracked in parallel by a null packet counter which is incremented with each removal. After the next transport stream packet which is not a null packet, the count is then entered in the DNP field of this transport stream packet carrying "real data" and the counter is reset. If the counter reaches its maximum of 255, this value is entered in the DNP field of the next transport stream packet and this is also transmitted even if it is a null packet. Following this, the DNP counter is also reset.

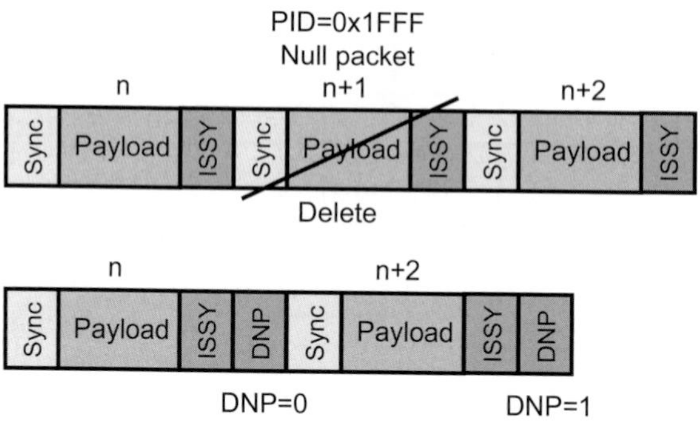

Fig. 36.14. Deletion of Null Packets

As in the case of the single input stream, too, the CRC-8 checksum and the baseband header are then inserted. The operation of these two blocks has already been described in the "Single Input Streams" section (Fig. 36.6.).

In the next block "Stream Adaptation" (Fig. 36.15.), all the streams prepared in the Mode Adaptation block are then combined. The block which combines the streams is called either merger or scheduler, depending on the standard, DVB-S2, DVB-T2 or DVB-C2. The "Stream Adaptation" block also differs slightly in detail with these standards. In the present section, only the common features of all standards will be discussed. The details and differences will be described later.

As in the case of the single input stream, this is followed by the padding and the baseband scrambler. These two blocks have also been described already in the "Single Input Stream" chapter,

The data streams are now ready for the "Bit Interleaved Coding and Modulation" signal processing block following, which, however, differs greatly in the different standards and will be explained in the respective chapter on DVB-S2, DVB-T2 and DVB-C2.

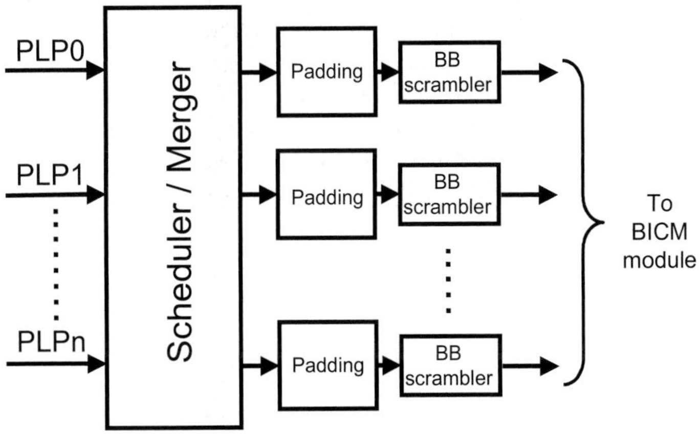

Fig. 36.15. Stream Adaptation in multiple PLP

36.3 Standard-related Special Features

We will now briefly discuss standard-related special features in the input signal processing of DVB-S2, DVB-T2 and DVB-C2.

36.3.1 DVB-S2

The signal processing in DVB-S2 corresponds very much to that already described in this chapter. At the time when the DVB-S2 standard was being fixed, the "Generic Encapsulated Stream" (GSE) data format had not yet been defined. Neither is there any mention of the term PLP (Physical Layer Pipe) in the DVB-S2 standard where simply multiple input streams are referred to. The signal processing section in the "Stream Adaptation" block is called "merger/slicer" in DVB-S2 and "Multiple Input Streams" has not as yet appeared as an operating mode in any DVB-S2 application.

36.3.2 DVB-T2

The term PLP (Physical Layer Pipe) was used for the first time in DVB-T2. And in DVB-T2, the Multiple Input Mode is called Mode B and will also be used as such. It is especially the possibility of being able to use this standard for transmitting contents with different robustness and with different data rates which will be made use of and is here called VCM - Variable Coding and Modulation.

The Stream Adaptation block (Fig. 36.16.) contains further processing steps such as

- dynamic scheduling information,
- frame delay and
- in-band signalling.

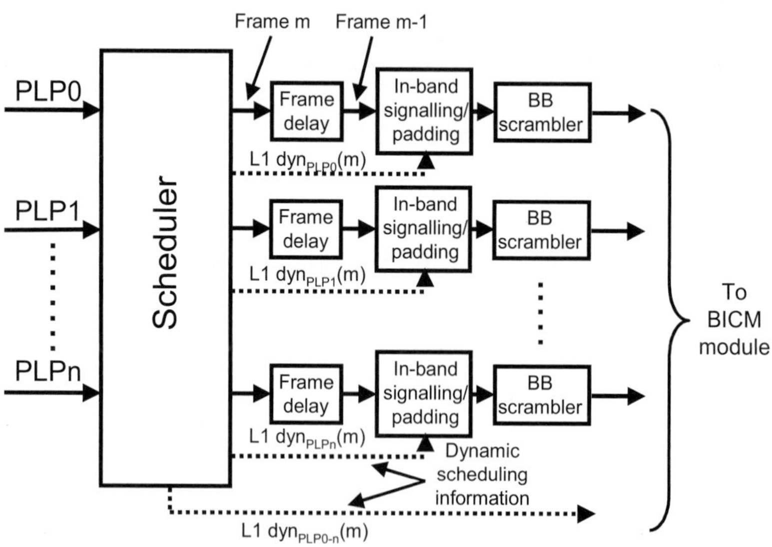

Fig. 36.16. Stream Adaptation Block in DVB-T2

Instead of padding data, the padding field can also contain in-band signalling data. This can be used for dynamic Layer-1 (L1) signalling for subsequent frames. I.e. it can be used for dynamically signalling and altering e.g. modulation parameters and error protection. Since the signalling information relates to subsequent frames, each PLP path will require a frame delay block.

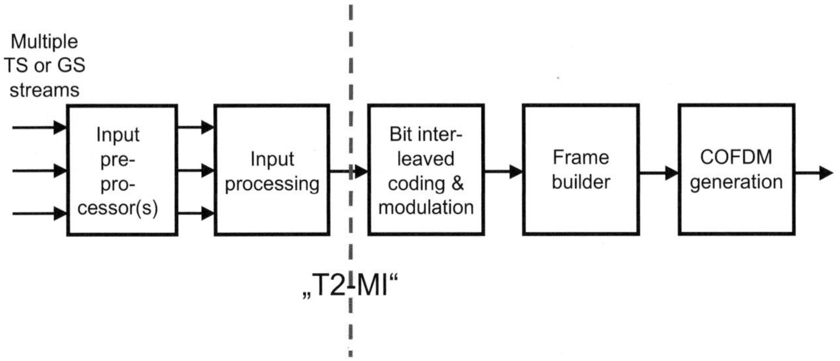

Fig. 36.17. Interface between DVB-T2 multiplexer and modulator

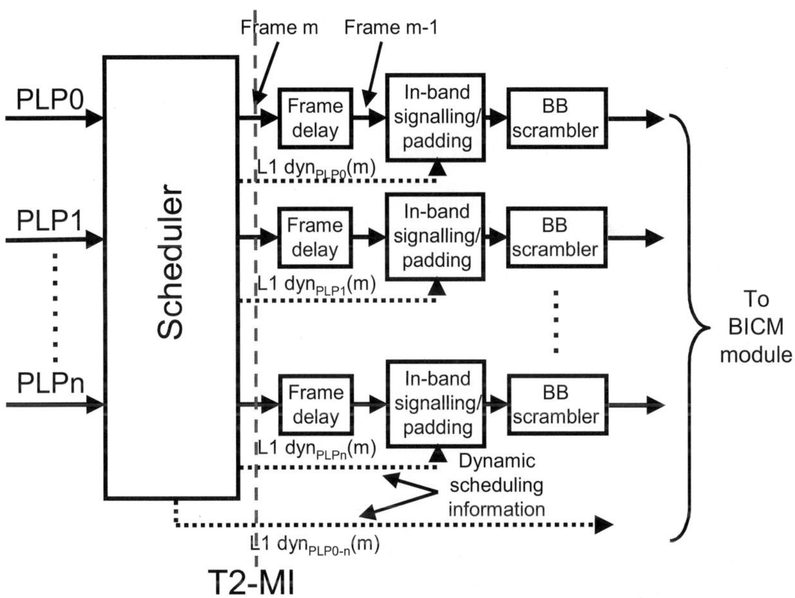

Fig. 36.18. Precise interface between DVB-T2 gateway and modulator (T2-MI)

Since DVB-T2 is also intended for forming single-frequency networks, these multiple input streams must be supplied completely synchronously to all modulators. This would never be possible over n feed lines which is why the PLPs are combined in the DVB-T2 multiplexer/DVB-T2 gateway outside the modulator, a separate interface, the DVB-T2 modulator interface, T2-MI in brief, being defined for this purpose (Fig 36.17.).

In principle, the interface between DVB-T2 multiplexer or DVB-T2 gateway and modulator is located between the input processing block and Bit Interleaved Coding and Modulation (Fig. 36.17.). The T2-MI signal contains all PLPs. The DVB-T2 modulator is only supplied with a single special input signal. In precise terms, the T2-MI interface is located after the scheduler (Fig. 36.18.). However, the padding is still carried out in the DVB-T2 gateway.

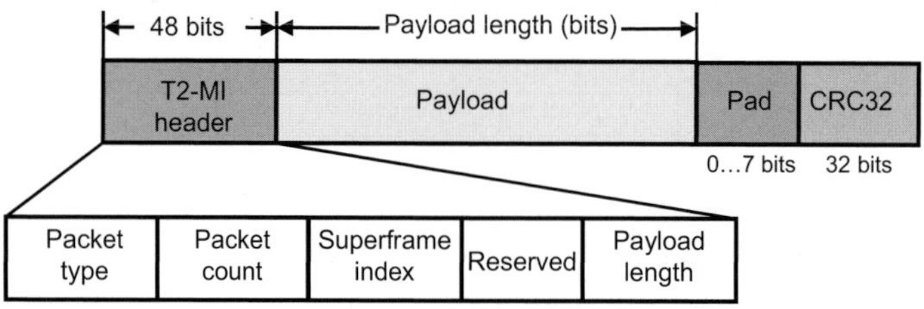

Fig. 36.19. T2-MI packet structure

Table 36.1. Packet Type in the T2-MI header

T2-MI Packet Type	Description
0x00	Baseband frame
0x01	Auxiliary stream I/Q data
0x10	L1 current
0x11	L1 future
0x20	DVB-T2 timestamp
0x21	Individual addressing
0x30	FEF part: Null
0x31	FEF part: I/Q data
All other values	Reserved for future use

For the T2-MI DVB-T2 modulator interface, a separate packet structure was defined, namely T2-MI packets (Fig. 36.19.) with header and payload. After the payload field, the frame is padded with bits to provide an integral number of bytes overall. This is followed at the end with a CRC-32 checksum. The T2-MI header contains the following components:

- Packet Type (8 bits)
- Packet Count (8 bits)
- Superframe Index (4 bits)
- Reserved for future use (12 bits)

- Payload Length (16 bits)

The Packet Type is used for signalling which data are currently being transmitted in the paylaod-field.

Packet count is a counter which is incremented by one continuously and independently of payload always from T2-MI packet to T2-MI packet. The counter runs from 00 to FF and then starts again at 00. Packet Count can be used to determine at the modulator input, e.g., if packets have been lost. The superframe index is constant for all T2-MI packets belonging to the same superframe. The "reserved for future use" bits are currently meaningless. "Payload length" signals the length of the payload component in bits.

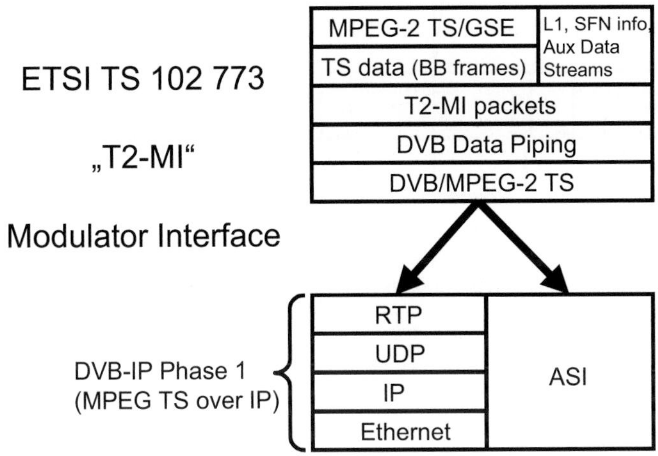

Fig. 36.20. Physical T2-MI interface

The T2-MI packets are physically packaged into MPEG-2 transport stream packets via DVB Data Piping (Fig. 36.21.). i.e. the T2-MI packets are cut into 184-byte-long pieces and then packaged into the payload component of the MPEG-2 transport stream packets. To utilize the transport stream interface particularly effectively, it is intended to work with a pointer field immediately after the transport stream header similar to MPEG-2 sections. If the transport error indicator in the TS header is set to One, it marks the beginning of a new T2-MI packet in the payload component of the transport stream packet when the T2-MI packet is embedded. However, it is also possible to transmit the rest of the preceding T2-MI packet at the beginning in the TS packet. The pointer then points to the beginning of the next T2-MI packet in the payload proportion of the TS packet. This saves stuffing in the transport stream packet containing the end part of the last T2-MI packet and thus gains transmission capacity on the feed link.

The interface type provided is then either TS-ASI or DVB-IP (Fig. 36.20.). The Gigabit Ethernet interface is becoming more and more popular as a TS interface and it makes sense, therefore, to provide both currently very popular TS interfaces TS-ASI and DVB-IP as the physical interface for T2-MI, making it possible to use the existing TS infrastructure also for the distribution of T2-MI signals.

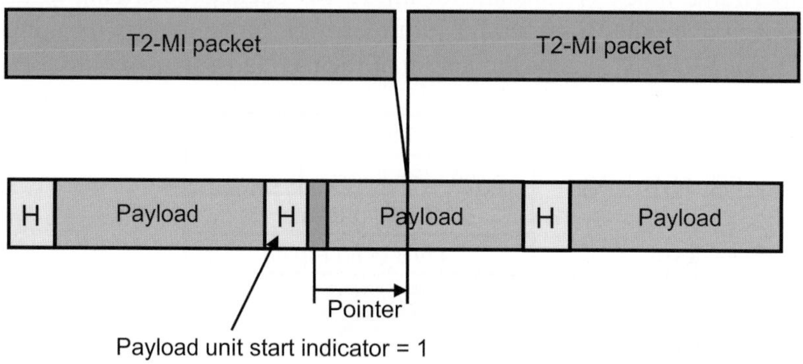

Fig. 36.21. Data piping – T2-MI packets are transmitted in MPEG-2 transport stream packets

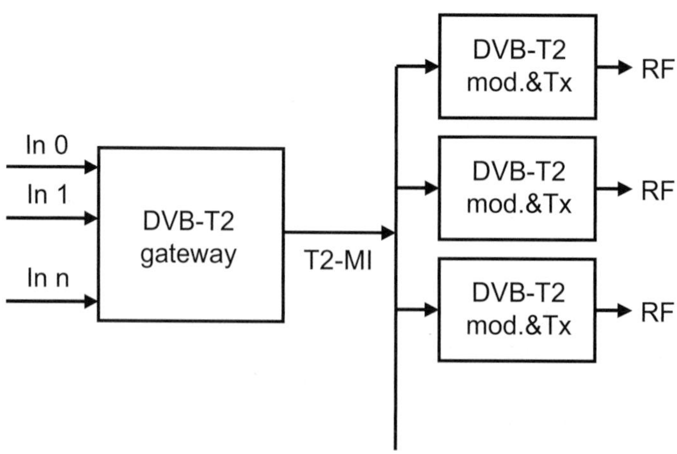

Fig. 36.22. DVB-T2 network with DVB-T2 gateway and DVB-T2 modulators and transmitters

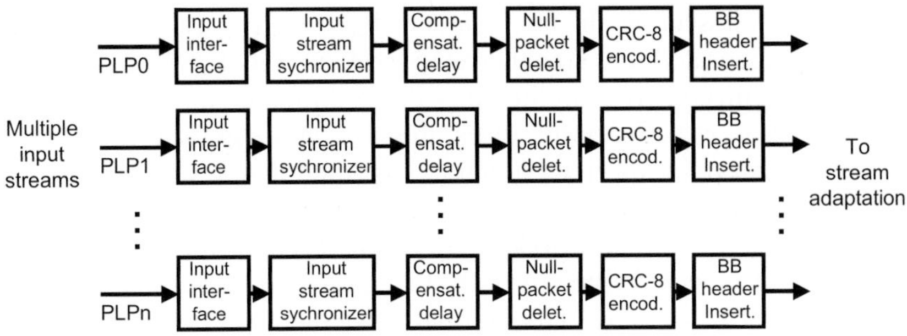

Fig. 36.23. Extended Mode Adaptation Block in DVB-T2

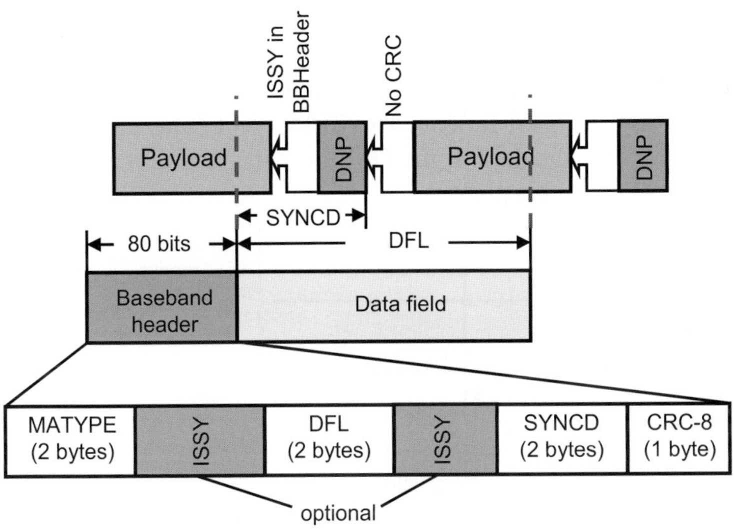

Fig. 36.24. High Efficiency Mode in DVB-T2 for TS

T2-MIP packets for synchronizing DVB-T2 single-frequency networks (SFNs) are also transmitted over the MPEG-2 transport stream interface serving as a T2-MI interface, where MIP stands for Modulator Information Packet. This is a normal MPEG-2 transport stream packet and not a T2-MI packet. The DVB-T2 modulators are also configured (Layer-1 signalling) via special T2-MI packets over the T2-MI interface. Further details are described in the DVB-T2 chapter since this requires more prior DVB-T2 knowledge.

DVB-T2 contains more special features. Firstly, there is also the concept of a Common PLP. This is a physical layer pipe which carries information for several PLPs. The mode adaptation block also contains the compensating delay circuit section. In this section, delay differences between the PLPs are compensated for. Together with the ISSY block, a Common PLP can be synchronized here with the other PLPs. Although this also provides for synchronisation between the PLPs, this does not appear to be relevant from the current point of view. It can be expected that a receiver will demodulate exactly one PLP plus Common PLP at the same time.

Furthermore, a Normal Mode (NM) and a High Efficiency Mode (HEM) have also been defined in DVB-T2. These two modes only relate to the input signal processing. The Normal Mode actually does not require any further comment since it corresponds precisely to the subject matter discussed before. The NM is also the mode which is compatible with DVB-S2. It applies to all four input signal formats. The High Efficiency Mode is restricted to only TS and GSE. In this mode, no CRC-8 is formed and transmitted. In addition, the ISSY field is transported in the baseband header in the UPL and SYNC fields, which are now free. UPL and SYNC are both known in the signal formats TS (UPL = 188 bytes and SYNC = 0x47) and GSE (UPL signalled in the GSE header). It is thus possible to save a few more bytes at this point.

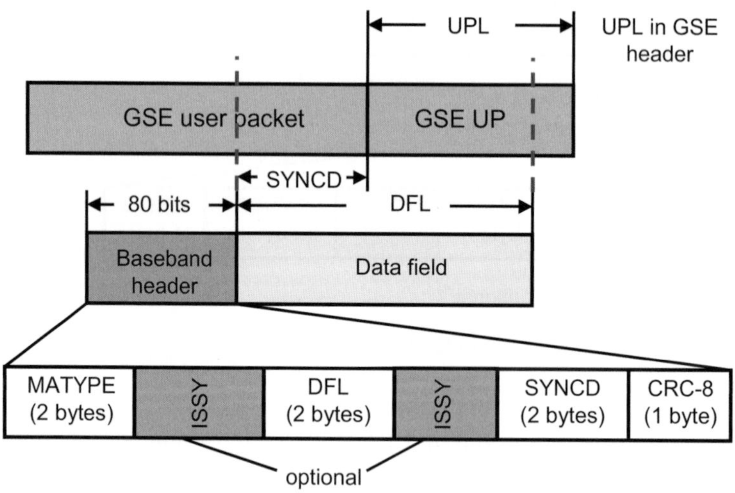

Fig. 36.25. High Efficiency Mode in DVB-T2 for GSE

36.3.3 DVB-C2

As far as DVB-C2 is concerned, no additional features can be currently reported. The signal processing is almost exactly the same as that discussed in the present chapter. DVB-C2 also mentions PLPs. At present, no modulator interface is defined. DVB-C2 also contains the Normal Mode (NM) and the High Efficiency Mode (HEM).

Bibliography: [EN303307], [TR102376], [EN302755], [A138], [TS102773], [TS102606], [TS102771], [A133]

37 DVB-T2

DVB-T2 – „Second Generation Digital Terrestrial Video Broadcasting“ [DVB A122r1], [ETSI EN 302755] is a completely new DVB-T-Standard which no longer has anything in common with the conventional DVB-T-standard. Just like DVB-T, DVB-T2 was mainly promoted by the BBC. In the UK, it was intended to use it to push the terrestrial HDTV-coverage in conjunction with MPEG-4-source coding. DVB-T2 is intended to achieve an approximately at least 30% to 50% higher net data rate compared with DVB-T, as well as better suitability for mobile use. There are applications which, on the one hand, require either a higher data rate, but also applications which must be very rugged in the mobile environment and must be able to manage with very narrow channels in some cases. It was not possible to cover both with DVB-T itself. It wasn't without reason that the bandwidth defined for DVB-T2 was also 1.7 MHz which is why it will be interesting to see what this will mean for DAB.

37.1 Introduction

Just as in DVB-T, the modulation method used is also COFDM, but with altered and extended constellation diagrams. The error protection used is the FEC defined in DVB-S2, i.e. BCH coding in the outer error protection and LDPC coding in the inner error protection, followed by bit interleaving.

The FEC frame structure as a whole corresponds to the frame structure of DVB-S2. The LDPC-coding (Low Density Parity Check Codes) has already been known since the 1960s but requires very much more computing power in the receiver and could only be implemented recently due to the chip technologies now available. Since the Spring of 2006 until March 2008, there have been seven DVB-T2-meetings lasting several days each. In the March 2008 meeting, a preliminary paper was adopted which was published by ETSI as a Draft in May 2008. In the autumn of 2008, the Implementation Guidelines and the T2-MI (T2-Modulator Interface) then appeared. DVB-T2 is probably coming too early or too late for many coun-

W. Fischer, *Digital Video and Audio Broadcasting Technology*, Signals and Communication Technology, 3rd ed., DOI: 10.1007/978-3-642-11612-4_37, © Springer-Verlag Berlin Heidelberg 2010

tries – depending on your point of view -, since DVB-T has already been introduced in these countries and a change-over to DVB-T2 would not be appropriate at the present time or would not find any acceptance. The DVB-T networks in the UK are the very first ones and have been running since 1998. A jump in technology from SDTV to HDTV would probably justify a parallel introduction of DVB-T2 in this country, especially since terrestrial propagation is dominant here. In countries like Germany where terrestrial TV coverage has no longer been the main path for TV propagation for years, HDTV by terrestrial means, and thus DVB-T2, does not seem to make any sense at the moment. In principle, the scenarios for the introduction of DVB-T2 can be subdivided as follows:

- into countries which are already operating DVB-T,
- countries which are still operating analog TV country-wide, and
- countries wishing to introduce additional new applications which cannot be implemented using DVB-T.

New applications are, e.g. applications in frequency bands which are now free and which were or are intended for DAB.

37.2 Theoretical Maximum Channel Capacity

However, before discussing the DVB-T2-standard in detail, the theoretical limits of the terrestrial transmission channel will first be considered on the basis of an 8-MHz-wide channel, looking at various receiving conditions from portable indoor antenna to fixed outdoor antenna, with characteristics known from DVB-T. The maximum possible data rate in theory is expressed in approximation by the Shannon-limit via the following formula if the signal-to-noise ratio is about or more than 10 dB:

$C = 1/3 \cdot B \cdot SNR$;
where C = channel capacity (in bits/s);
B = bandwidth (in Hz);
SNR = signal/noise ratio (in dB);

An 8-MHz-wide terrestrial TV channel will then provide the following theoretical maximum channel capacity:

Table 37.1. Theoretical maximum channel capacity of an 8-MHz-wide TV channel

SNR[dB]	Theor. max. channel capacity [Mbit/s]	Comment
10	26.7	
12	32	Poor portable indoor reception
15	40	Portable indoor reception
18	48	Good portable indoor reception
20	53.3	Poor reception with outdoor antenna
25	66.7	Good roof antenna reception
30	80	Very good roof antenna reception

In DVB-T, the data rates in an 8-MHz channel in DVB-T networks designed for portable indoor-reception are frequently about

13.27 Mbit/s (16QAM, CR = 2/3, g= 1/4, SFN, limit SNR = 12 dB)

and in DVB-T networks designed for roof antenna reception they are in most cases about

22.39 Mbit/s (64QAM, CR = 3/4, g=1/4, SFN, limit SNR = 18 dB).

The aim in DVB-T2 is to achieve data rates which are higher by at least 30 to 50%. Without familiarity with the DVB-T2 Standard, it can thus be expected that, given comparable conditions, the following data rates can be achieved:

Portable indoor-reception (SFN, long guard interval):
17.3 to 19.9 Mbit/s

Roof antenna reception (SFN, long guard interval):
29.1 to 33.6 Mbit/s.

The error protection alone will bring 30 %. Additional features such as

- the 16K- and 32K-mode,
- the extended carrier mode,
- 256QAM modulation,
- the rotated Q-delayed constellation diagrams

will bring further improvements in the data rate.

37.3 DVB-T2 - Overview

The essential core parameters of DVB-T2 are:

- Several MPEG-2 transport stream inputs or possibly generic-streams as baseband signals (up to 255)
- approx. at least 30% higher net data rate mainly due to the improved BCH+LDPC error protection already used in DVB-S2
- compatibility with the Geneva 2007 Frequency Plan (8, 7, 6 MHz bandwidths)
- additional bandwidths 1.7 MHz and 10 MHz
- stationary, but also mobile applications
- COFDM
- 1K, 2K, 4K, 8K, 16K and 32K-Mode
- guard interval 1/4, 1/8, 1/16, 1/32, 19/256 and 1/128
- modulation scheme QPSK, 16QAM, 64QAM and 256QAM
- Q-delayed „rotated" constellation diagrams
- RF frame-structure with P1 and P2-symbol at the beginning of the frame
- flexible pilot structures with fixed and distributed pilots
- PAPR reduction (Peak to Average Power Ratio) reduction, i.e. reduction of the crest-factor (2 different methods)
- variable coding and modulation (the transmission parameters can be changed in operation)
- time interleaving
- time slicing
- optional MISO-principle (Multiple Input, Single Output)
- inbuilt FEFs (Future Extension Frames) for later extensions
- auxiliary data streams as an option
- time frequency slicing (TFS) mentioned in the Appendix of the Standard.

The details of DVB-T2 will now be discussed in the following sections.

37.4 Baseband Interface

The DVB-T2 baseband interface provides one or more data inputs. DVB-T2 is no longer orientated only towards MPEG-2 transport streams but also provides for generic-streams as possible input streams. Up to 255 input streams are possible. Initially it was left open where these streams will be multiplexed. The answer came with the "T2-MI“ standard, the modulator-interface for DVB-T2. The streams are combined in the playout center and the DVB-T2 modulator is supplied with only one data stream via DVB-T2-MI. Similar to the ETI stream in DAB, this data stream is provided with all necessary information for the modulator. It also contains the time stamp for synchronizing single-frequency networks. The baseband interface has already been described in Chapter 36. There are two modes in DVB-T2, namely Mode A = Single PLP (Physical Layer Pipe) and Mode B = Multiple PLPs. It is only in the Mode A case that all processing steps take place in the modulator itself, whereas in the case of Mode B, the T2-MI interface follows immediately after the scheduler.

There is one special feature of DVB-T2 which also influences the input signal processing. In Mode B, it is possible to work with variable coding and modulation and this can also be done dynamically. I.e., in the next DVB-T2 frame, the transmission parameters could change and this may have to be signalled dynamically. This is done in the padding field of the baseband header. And there is an optional Common PLP which contains information for all or some PLPs. In Mode B, an additional distinction is made between the

- HEM = High Efficiency Mode (for MPEG-2 transport streams and GSE) and the
- NM = Normal Mode (compatible with DVB-S2).

Further details of the signal processing in the baseband interface can be found in Chapter 36 “Baseband Signals für DVB-x2”.

37.5 Forward Error Correction

Just as in DVB-S2, the modified error protection (Fig. 37.1.) leads to a - gain in S/N, thus coming closer to the Shannon-Limit overall. The net data rate is increased by 30% by this measure alone. As in DVB-S2, the error protection in DVB-T2 consists of a baseband scrambler, a BCH-block encoder, and an LDPC-block encoder followed by the bit-interleaver. In the DVB-T2 modulator, the baseband-frame including baseband header and padding-block are first scrambled (baseband scrambler) and then supplied to the FEC-block where the BCH code is first added. After that, a further error protection, the length of which depends on a selectable code rate, is appended in the LDPC encoder, the possible code rates being:

- 1/2,
- 3/5,
- 2/3,
- 3/4,
- 4/5 or
- 5/6.

Code rate 1/2 means maximum error protection and minimum net data rate and code rate 5/6 means minimum error protection and maximum net data rate.

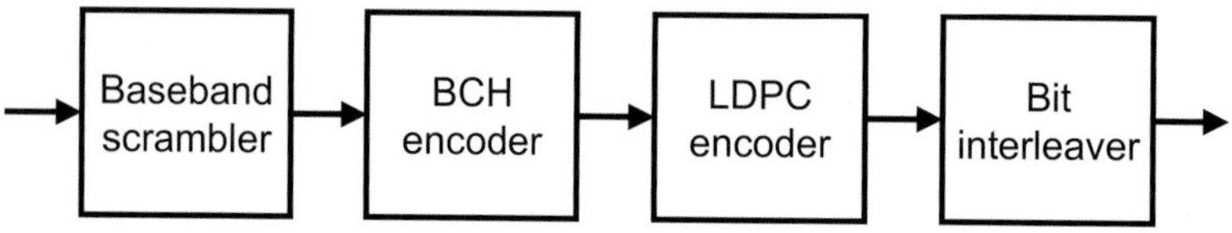

Fig. 37.1. DVB-T2-Error protection (BCH = Bose-Chaudhuri-Hoquenghem, LDPC = Low Density Parity Check Code)

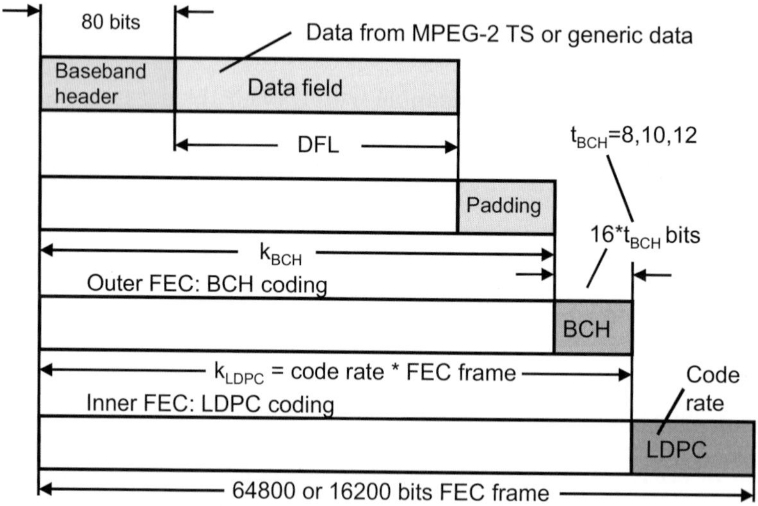

Fig. 37.2. DVB-T2 FEC frame

Just as in DVB-S2, it is possible to use a short (16K) or a long FEC frame (64K) in DVB-T2 (Fig. 37.2.). The differences in performance with respect to the required signal/noise ratio are minimum and are within a range of a few tenths dB. The short FEC frame is possibly more advantageous for low-rate data streams and the long frame is better for higher-rate data streams. The data rates now possible in DVB-T2 lie between 7.49 Mbit/s (QPSK, CR = 1/2) and 50.32 Mbit/s (256QAM, CR = 5/6). The lowest required signal/noise ratios are between 0.4 and 25.9 dB (Table 37.2. and 37.3.). In comparison with DVB-T, the fall-off-the-cliff effect is much steeper in DVB-T2. The transition from Go- to No-Go takes place within only a few steps of a hundredth dB each. The reason for this lies in the concatenation of two block-codes. Examples of data rates are listed in Table 37.4. for the case of an 8-MHz-channel, 32K mode, g = 1/128, PP7. Other transmission parameters will lead to numerous other data rates. There are many possible combinations which would fill many pages.

Table 37.2. C/N limits for a BER of $1 \cdot 10^{-4}$ after LDPC, long 64K FEC frame

Modulation	Code rate	C/N Gaussian channel [dB]	C/N Rice channel [dB]	C/N Rayleigh channel [dB]	C/N 0dB echo channel @ 90% GI [dB]
QPSK	1/2	0.8	1.0	1.8	1.5
	3/5	2.1	2.4	3.4	3.0
	2/3	2.9	3.3	4.6	4.2
	3/4	3.9	4.3	5.9	5.5
	4/5	4.5	5.0	6.8	6.4
	5/6	5.0	5.6	7.2	7.2
16QAM	1/2	5.7	6.1	7.3	7.0
	3/5	7.4	7.7	9.1	8.8
	2/3	8.6	8.9	10.5	10.2
	3/4	9.8	10.3	12.2	11.9
	4/5	10.6	11.1	13.4	13.2
	5/6	11.2	11.8	14.4	14.2
64QAM	1/2	9.6	10.0	11.7	11.5
	3/5	11.7	12.1	13.8	13.6
	2/3	13.2	13.6	15.4	15.1
	3/4	14.9	15.3	17.5	17.3
	4/5	15.9	16.4	19.0	18.9
	5/6	16.6	17.2	19.9	20.1
256QAM	1/2	12.8	13.3	15.4	15.3
	3/5	15.6	16.0	18.1	18.2
	2/3	17.5	17.8	20.0	20.0
	3/4	19.7	20.2	22.5	22.5
	4/5	21.1	21.5	24.2	24.4
	5/6	21.8	22.3	25.3	25.7

Table 37.3. C/N limits for a BER of $1 \cdot 10^{-4}$ after LDPC, short 16K FEC frame

Modulation	Code rate	C/N Gaussian channel [dB]	C/N Rice channel [dB]	C/N Rayleigh channel [dB]	C/N 0dB echo channel @ 90% GI [dB]
QPSK	1/2	0.4	0.7	1.5	1.2
	3/5	2.2	2.4	3.5	3.2
	2/3	3.1	3.4	4.7	4.4
	3/4	4.0	4.5	6.0	5.7
	4/5	4.6	5.1	6.9	6.5
	5/6	5.1	5.7	7.8	7.4
16QAM	1/2	5.2	5.5	6.6	6.3
	3/5	7.5	7.9	9.3	9.0
	2/3	8.8	9.1	10.7	10.4
	3/4	10.0	10.5	12.4	12.1
	4/5	10.8	11.3	13.6	13.3
	5/6	11.4	12.0	14.6	14.4
64QAM	1/2	8.7	9.1	10.7	10.5
	3/5	12.0	12.4	14.2	14.0
	2/3	13.4	13.8	15.7	15.5
	3/4	15.2	15.6	17.8	17.6
	4/5	16.1	16.6	19.1	18.9
	5/6	16.8	17.4	20.3	20.3
256QAM	1/2	12.1	12.4	14.4	14.3
	3/5	16.5	16.9	18.8	18.8
	2/3	17.7	18.1	20.3	20.3
	3/4	19.9	20.4	22.6	22.7
	4/5	21.2	21.7	24.2	24.3
	5/6	22.0	22.5	25.6	25.9

Table 37.4. DVB-T2 channel capacity in the 8-MHz channel, 32K mode, g=1/128, PP7 (Source: DVB-T2 Implementation Guidelines, February 2009)

Modulation	Code rate	Bit rate [Mbit/s]	Frame length [symbols]	FEC blocks per frame
QPSK	1/2	7.44	60	50
	3/5	8.94	60	50
	2/3	9.95	60	50
	3/4	11.20	60	50
	4/5	11.95	60	50
	5/6	12.46	60	50
16QAM	1/2	15.04	60	101
	3/5	18.07	60	101
	2/3	20.11	60	101

	3/4	22.62	60	101
	4/5	24.14	60	101
	5/6	25.16	60	101
64QAM	1/2	22.48	60	151
	3/5	27.02	60	151
	2/3	30.06	60	151
	3/4	33.82	60	151
	4/5	36.09	60	151
	5/6	37.62	60	151
256QAM	1/2	30.08	60	202
	3/5	36.14	60	202
	2/3	40.21	60	202
	3/4	45.24	60	202
	4/5	48.27	60	202
	5/6	50.32	60	202

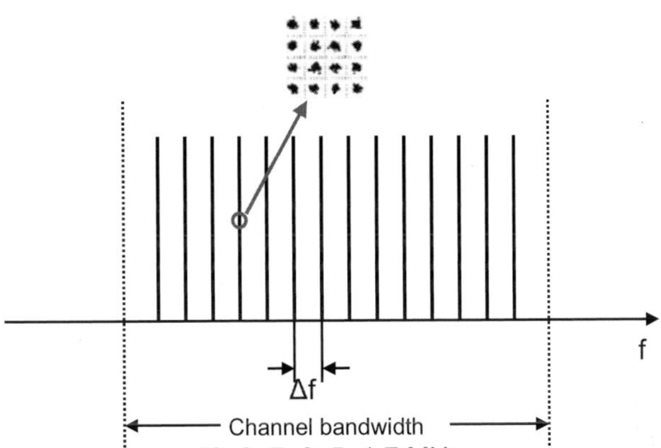

Fig. 37.3. DVB-T2 channel

37.6 COFDM Parameters

DVB-T2 supports channel bandwidths of 1.7, 5, 6, 7, 8 and 10 MHz (Fig. 37.3.). The actual signal bandwidth is slightly narrower because of the guard band at the upper and lower end of the DVB-T2 channel (Table 37.5.). Table 37.3. shows the CODFM parameters in the 8-MHz channel possible in DVB-T2. In the case of the 7- and 6-MHz and 1.7- and 10-

MHz channel, respectively, (Fig. 37.3.), the parameters must be adapted correspondingly by a factor of 7/8 and 6/8 etc.. As can be seen in Table 37.6., not every guard interval is possible in every COFDM mode. Apart from one exception (P1 symbol), the guard interval (Fig. 37.4.) is also a cyclic prefix (CP) in DVB-T2, i.e. a repetition of the symbol end in the corresponding length.

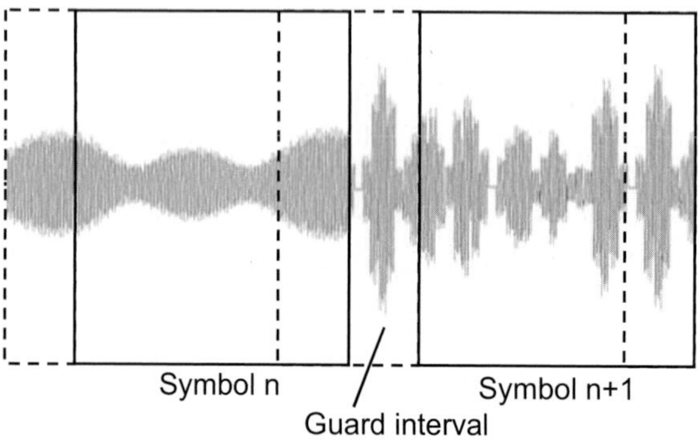

Fig. 37.4. Symbol and guard interval as a cyclic prefix (CP)

Table 37.5. DVB-T2 channel- and signal bandwidths

Bandwidth [MHz]	1.7	5	6	7	8	10
Elementary period [µs]	71/131	7/40	7/48	1/8	7/64	7/80
Signal bandwidth [MHz]	1.54	4.76	5.71	6.66	7.61	9.51

Table 37.6. DVB-T2 COFDM parameters in the 8-MHz channel

FFT	Symbol period [ms]	Carrier spacing [kHz] Δf	g = 1/128	g = 1/32	g = 1/16	g = 19/256	g = 1/8	g = 19/128	g = 1/4
32K	3.584	0.279	X	X	X	X	X	X	--
16K	1.792	0.558	X	X	X	X	X	X	X
8K	0.896	1.116	X	X	X	X	X	X	X
4K	0.448	2.232	--	X	X	--	X	--	X
2K	0.224	4.464	--	X	X	--	X	--	X
1K	0.112	8.929	--	--	X	--	X	--	X

A 16K and a 32K mode was provided in order to achieve less time overhead and thus a higher net data rate (6% overhead in the 32K mode, 25% in the 8K mode) with the same absolute guard interval length. The longest guard interval is now more than twice as long as the longest guard interval in DVB-T (0.224 ms); 0.532 ms in DVB-T2, 32K, g=19/128, correspond to a maximum transmitter distance of almost 160 km (Table 37.17.). Narrower signal bandwidths (7, 6, 5, 1.7 MHz) lead to even longer symbols and thus also lead to longer guard intervals. With these guard interval-lengths nationwide single-frequency networks can be implemented. The 32K mode is the mode providing the longest symbols and thus has the least overhead in the guard interval; at the same time it serves to implement the largest single frequency networks. However, the 32K mode, due to its much narrower subcarrier spacing, is the mode least suitable for mobile use. The mode most suitable for mobiles is the 1K mode which has the largest subcarrier spacing. But it thus also has the shortest symbols and is, therefore, the one least suitable for forming large SFNs.

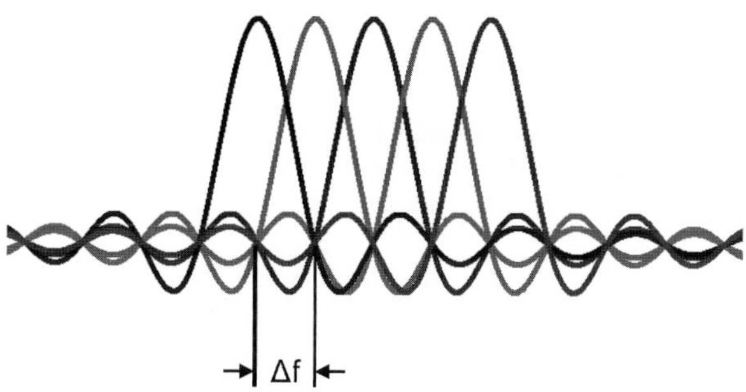

Fig. 37.5. sin(x)/x-shaped spectra of the OFDM carriers

37.6.1 Normal Carrier Mode

In the DVB-T2 Normal Carrier Mode, the bandwidths of the useful signal approximately correspond to the bandwidths of DVB-T. Between the useful signal spectrum and the beginning of the adjacent channel there is at the lower and the upper end of the DVB-T2 channel the so-called guard band which has a width of up to about 200 kHz. The guard-band has several tasks, the most important of which is to protect the adjacent channels (Fig. 37.7.). The shoulders of the OFDM signal must decay within the guard band. One cause of the shoulders is the superimposition of the tails of the sin(x)/x-functions of each modulated single carrier (Fig. 37.5.). It can be demonstrated, however, that the more carriers are used, the more the resultant shoulders are suppressed; i.e. the 1K mode inherently has higher shoulders than the 32K mode. These shoulders are lowered as much as possible by digital filtering measures in the modulator. Nevertheless, it can also be demonstrated with a very good test transmitter that in the short range, the shoulders are much lower in the 32K mode than in the 1K-Mode (Fig. 37.6.). There are simple mathematical reasons for this. The more carriers are used, the better the sin(x)/x tails will cancel.

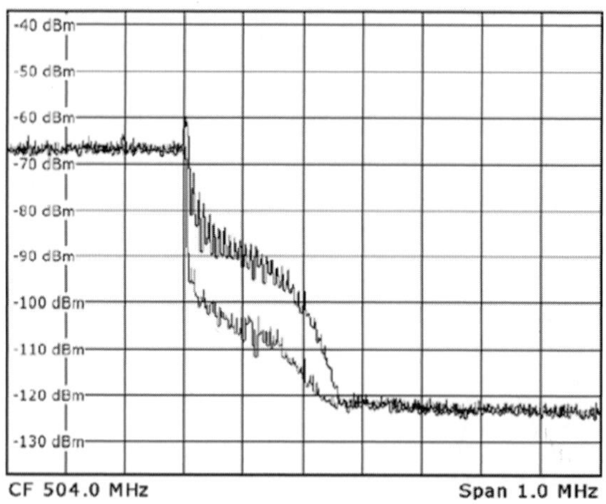

Fig. 37.6. Shoulders of the DVB-T2 signal in 32K and 1K mode in comparison

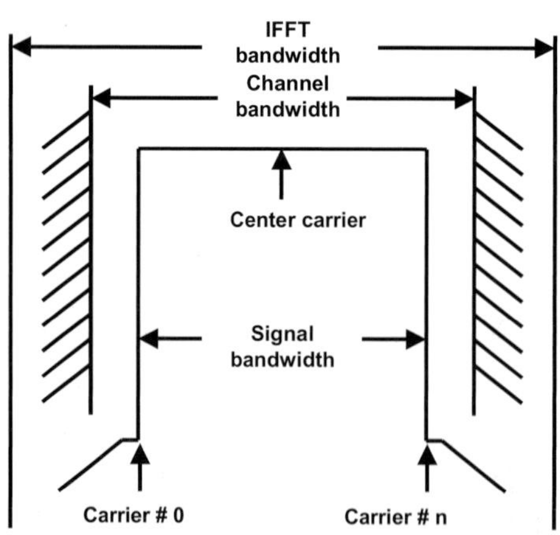

Fig. 37.7. DVB-T2 spectrum with top and bottom guard bands

37.6.2 Extended Carrier Mode

Since the sin(x)/x tails, and thus the shoulders, drop more towards the adjacent channels in the modes having more carriers, it was provided in DVB-T2 to support a wider spectrum above the 8K mode in the Extended Carrier Mode. The advantage is that this increases the data rate of the DVB-T2 signal.

Fig. 37.8. and 37.9. clearly show the wider DVB-T2 spectrum of the 32K mode in the Extended Carrier Mode compared with the Normal Carrier Mode. In the selected example, the difference in the net data rate is about 1 Mbit/s.

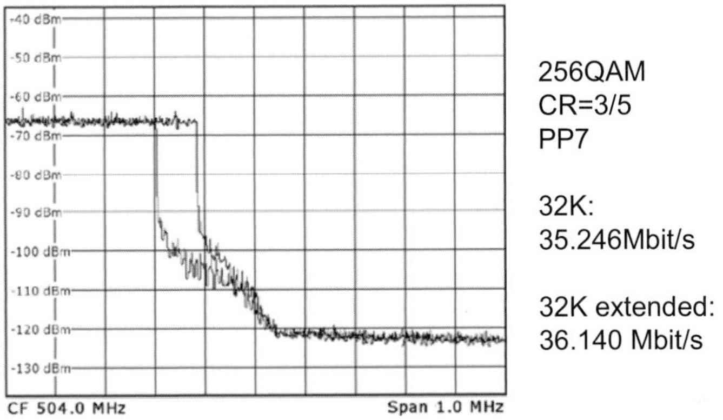

Fig. 37.8. DVB-T2 spectrum in Normal and Extended Carrier Mode (32K)

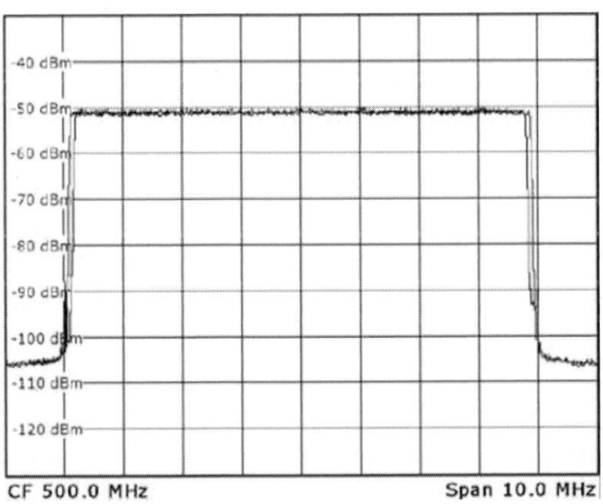

Fig. 37.9. Overall DVB-T2 spectrum in Normal and Extended Carrier Mode (32K)

Table 37.7. DVB-T2 OFDM parameters in the 8 MHz channel

Parameter	1K Mode	2K Mode	4K Mode	8K Mode	16K Mode	32K Mode
Number of carriers in Normal Mode	853	1705	3409	6817	13633	27265
Number of carriers in Extended Carrier Mode	--	--	--	6913	13921	27265
Additional carriers in Extended Carrier Mode	0	0	0	96	288	596
IFFT	1024	2048	4096	8192	16384	32768
Symbol period [µs]	112	224	448	896	1792	3584
Carrier spacing Δf [kHz]	8.929	4.464	2.232	1.116	0.558	0.279
Signal bandwidth in Normal Mode [MHz]	7.61	7.61	7.61	7.61	7.61	7.61
Signal bandwidth in Extended Carrier Mode [MHz]	--	--	--	7.71	7.77	7.77

37.7 Modulation Patterns

The modulation patterns (Fig. 37.10.) used in DVB-T2 are coherent Gray-coded QAM orders. The constellations possible in DVB-T2 are:

- QPSK,
- 16QAM,
- 64QAM and
- 256QAM.

Differential modulation is not supported in DVB-T2. The first three QAM orders fully correspond to the mapping used in DVB-T. It is a special feature that the constellation diagrams can be either reversed or rotated ("flipped") to the left by a certain angle; so-called rotated, Q-delayed constellation diagrams (Fig. 37.12.).

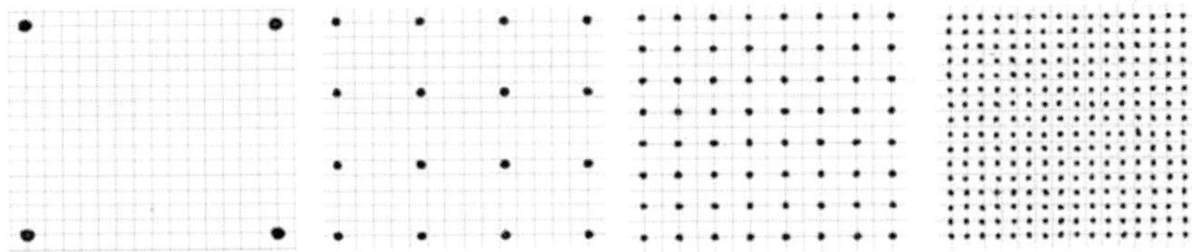

Fig. 37.10. Non-rotated "normal“ constellation diagrams in DVB-T2 (QPSK, 16QAM, 64QAM and 256QAM)

37.7.1 Normal Constellation Diagrams

In the case of QPSK, 16QAM and 64QAM, the normal non-rotated constellation diagrams exactly correspond to the constellation diagrams of DVB-T. In addition, 256QAM was also defined as a possible constellation in DVB-T2. 256QAM makes sense because of the improved error protection. Fig. 37.10. shows the non-rotated variants of the constellation diagrams possible in DVB-T2.

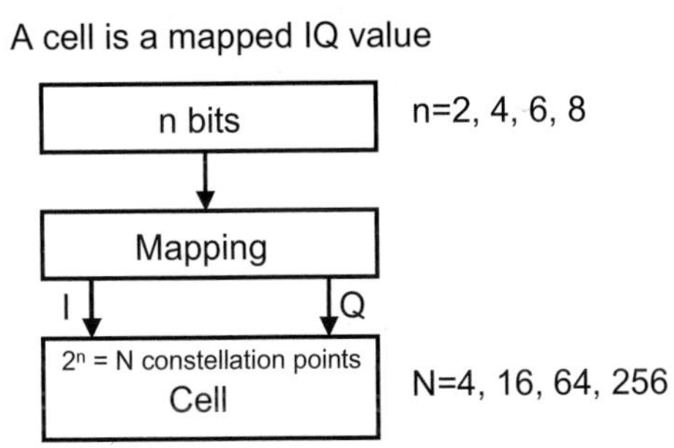

Fig. 37.11. Definition of 'cell'

37.7.2 Definition of 'Cell'

The term "cell“ (Fig. 37.11.) will now have to be defined. This term is mentioned time and again in the DVB-T2 Standard. Thus, there is also, e.g. a so-called cell interleaver. A cell is simply the result of mapping a later carrier. In contrast to DVB-T, mapping is not carried out after all interleaving processes but relatively early, after the error protection and after the bit interleaver. However, this is still followed by the cell-interleaver, the time interleaver and the frequency interleaver which is why the mapping result cannot yet be allocated to a carrier and why the term "cell" was introduced. A cell is, therefore, a complex number consisting of an I-component and a Q-component, i.e. a real part and an imaginary part (Fig. 37.11.).

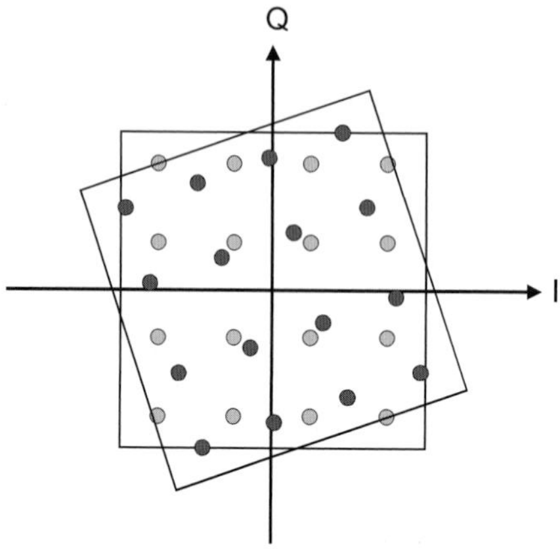

Fig. 37.12. Rotated Q-delayed constellation diagram

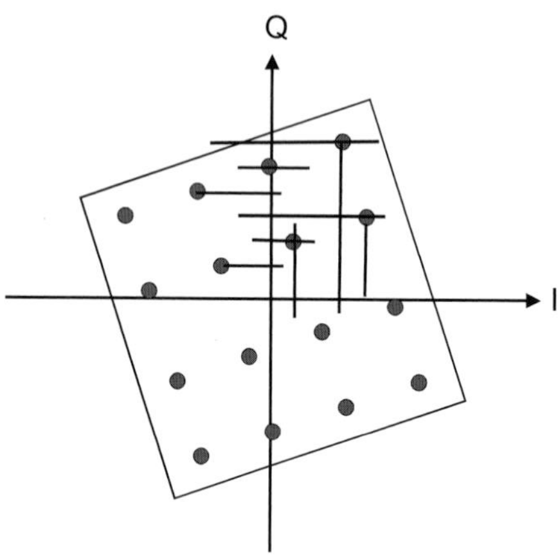

Fig. 37.13. Discrete mapping of constellation points on the I and Q axis with a rotated constellation diagram

37.7.3 Rotated Q-delayed Constellation Diagrams

If rotated constellation diagrams (Fig. 37.18.) are used, the information about the position of a constellation point is contained both in the I component and in the Q component of the signal (Fig. 37.13.). In a case of disturbance, this can be used for providing more reliable information about the position of the constellation point, in contrast to a non-rotated diagram (Fig. 37.14.), contributing to better decodability. In contrast to a non-rotated constellation diagram, the IQ information, which is now discrete, can be used for soft-decisions if necessary. Practice will show how much actual benefit can be derived from this. Table 37.5. lists the angles of rotation of the various DVB-T2-constellations as a function of the QAM mode.

Table 37.8. Angles of rotation of the constellation diagrams

Mod.	QPSK	16QAM	64QAM	256QAM
φ [degrees]	29.0	16.8	8.6	atan(1/16)

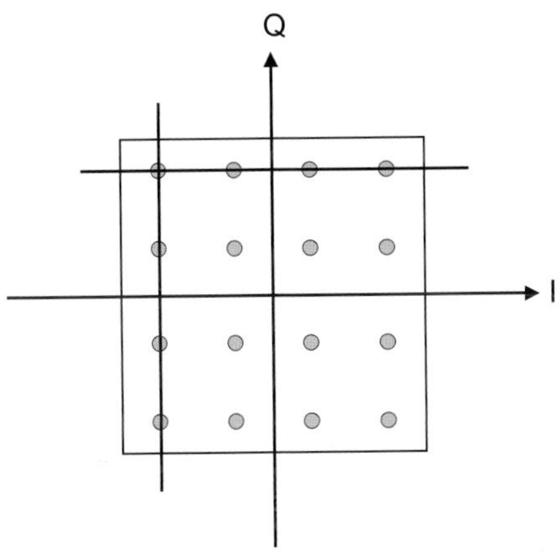

Fig. 37.14. The constellation points on the I and Q axis in a non-rotated constellation diagram

In reality, however, the whole process is slightly more complex. With a rotated diagram, the Q component is not transmitted on the same carrier, or more precisely in the same "cell“, but with delay on another carrier (Fig. 37.15. and 37.16.) or better in another cell. From one QAM, virtually two ASKs (Amplitude Shift Keying modulations) in the I and Q-direction are then produced which are then transmitted on independent carriers – "cells“ which are disturbed differently in practice and are thus intended to contribute to the reliability of demodulation.

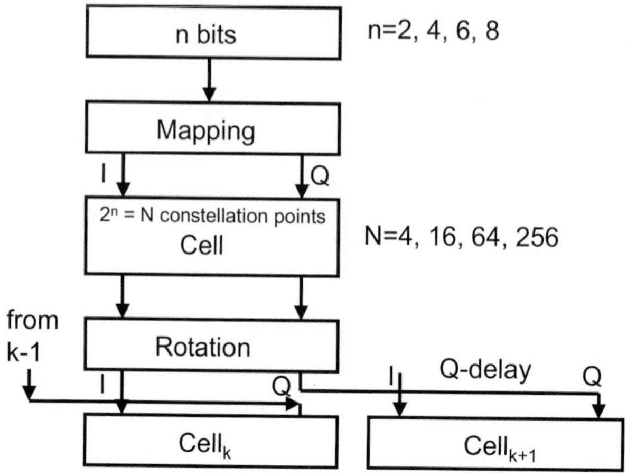

Fig. 37.15. Mapping, rotation and Q delay

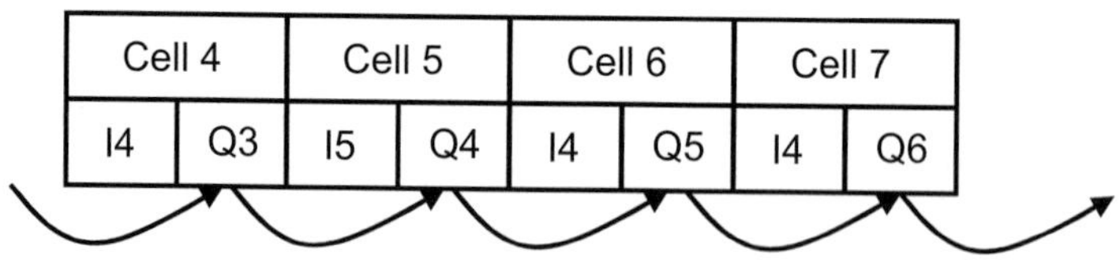

Fig. 37.16. Cyclic Q delay between adjacent cells

37.8 Frame Structure

A Physical Layer Frame (Fig. 37.17.) in DVB-T2 begins with a P1 symbol used for synchronization and frame-finding, followed by one or more P2-symbols containing Layer-1 (L1)-signalling data for the receiver. This is followed by symbols which carry the actual payload data. The theoretically up to 255 input data streams are transmitted in so-called Physical Layer Pipes (PLPs) in which the different contents can be transmitted with higher or lower data rate and more or less robustly (error protection and modulation). This is called Variable Coding and Modulation (VCM). In addition, the transmission parameters of the PLPs can also be changed dynamically from T2 frame to T2 frame. The current transmission parameters of all PLPs are signalled in the P2 symbols; dynamic L1 signalling for the receiver takes place in the padding-field of the baseband frame.

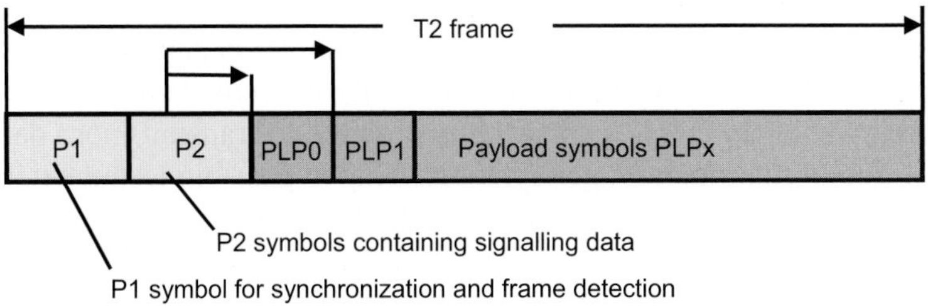

Fig. 37.17. Structure of a DVB-T2 frame

Table 37.9. Maximum length of a DVB-T2 frame in numbers of OFDM symbols

FFT	Symbol duration in a 8 MHz channel [µs]	g= 1/128	g= 1/32	g= 1/16	g= 19/256	g= 1/8	g= 19/128	g= 1/4
32K	3584	68	66	64	64	60	60	--
16K	1792	138	135	131	129	123	121	111
8K	896	276	270	262	259	247	242	223
4K	448	--	540	524	--	495	--	446
2K	224	--	1081	1049	--	991	--	892
1K	112	--	--	2098	--	1982	--	1784

Table 37.10. Number of P2 symbols per DVB-T2 frame as a function of the DVB-T2 FFT mode

FFT mode	Number of P2-symbols per DVB-T2 frame
1K	16
2K	8
4K	4
8K	2
16K	1
32K	1

Apart from the FFT mode, almost all transmission parameters can be changed from Physical Layer Pipe (PLP) to Physical Layer Pipe. As already mentioned, their signalling and the addressing of the PLP (Start, Length etc.) is handled via the P2 symbols; called L1-signalling. The number of P2 symbols depends on the FFT mode (Table 37.10.); the reason being simply the different data capacity of the symbols, depending on the FFT mode. The 1K mode has the shortest symbols and thus the lowest data capacity per symbol. In the 32K mode, it is possible to transmit more data per symbol due to the much longer symbols. This is possible to the extent that the data transmission can even begin in the P2 symbols due to unused capacity. Table 37.10. lists the number of P2-Symbols per DVB-T2 frame. A DVB-T2 frame thus consists of

- a P1 symbol
- 1 … 16 P2 symbols (depending on FFT mode)
- N data symbols (PLP data, FEF, auxiliary data, dummy cells)

A frame can have a maximum length of 250 ms, resulting in a maximum number of data symbols which, in turn, is dependent on the FFT mode and the guard interval. The net data rate per PLP can fluctuate due to different transmission parameters. A data stream carrying HDTV services, e.g., requires a higher data rate than a data stream transporting SDTV services or a data stream transporting pure audio broadcasting services (Fig. 37.18.).

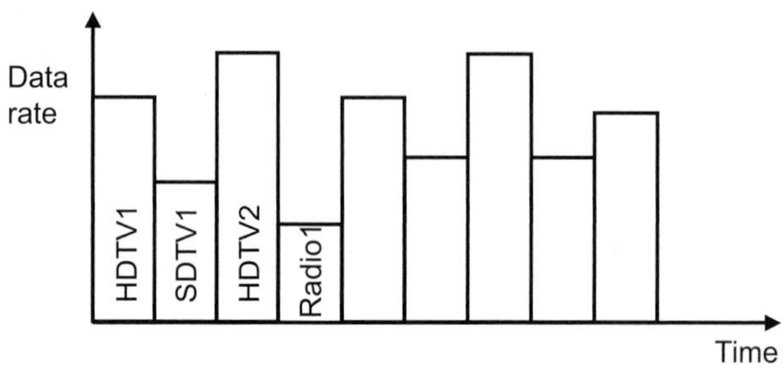

Fig. 37.18. Variable Coding and Modulation (VCM)

37.8.1 P1 Symbol

A P1 symbol (preamble symbol 1) marks the beginning of a frame, similar to the null symbol in DAB. Overall, the P1 symbol is used for

- marking the beginning of the DVB-T2 frame,
- time and frequency synchronization,

- signalling the basic transmission parameters (FFT mode, SISO/MISO),

The P1 symbol has the following characteristics:

- FFT mode = 1K,
- 1/2 guard interval with frequency offset before and after the P1 symbol,
- carrier DBPSK modulated,
- 7-bit signalling data (SISO/MISO/Future Use (3 bits), use of FEF (1 bit), FFT (3 bits)).

So that the P1 symbol (Fig. 37.19.) could be identified easily and reliably, two guard intervals were appended – one in front and one behind. This results virtually in a double correlation during the autocorrelation. In addition, the carriers are all shifted upward in frequency in the cyclical pre- and post-fix. Pre- and post-fix are not exactly of the same length.

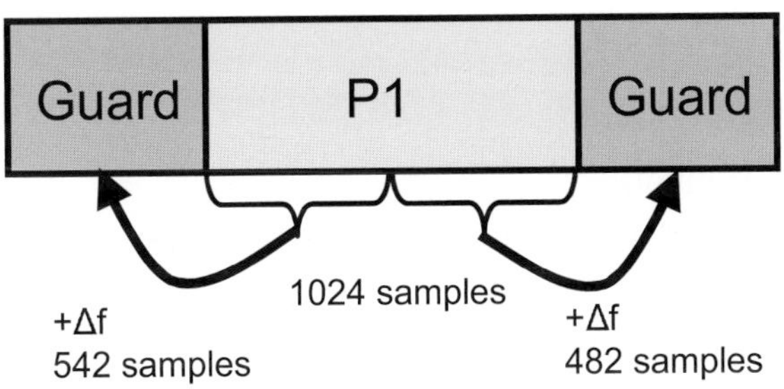

Fig. 37.19. P1 symbol

37.8.2 P2 Symbols

The Layer 1 signalling (L1 signalling) is transmitted from the modulator to the receiver in 1 ... 16 P2 symbols (preamble symbols 2) per DVB-T2 frame. Physically, a preamble symbol 2 has almost the same structure as later data symbols. The FFT mode corresponds to that of the data symbols and is already signalled in the P1 symbol. However, the pilot density is greater.

A P2 symbol (Fig. 37.20.) consists of a pre- and post-signalling component. Both components are differently modulated and error protected. The pre-signalling-component is permanently BPSK-modulated and protected with a constant error protection known to the receiver. The transmission parameters of the P2-pre-signalling component are:

- BSPK modulation,
- FEC = BCH+16K LDPC,
- LDPC code rate =1/2.

The transmission parameters of the P2 post-signalling component are:

- BPSK, QPSK, 16QAM or 64QAM,
- FEC = BCH+16K LDPC,
- LDPC code rate =1/2 or 1/4 with BPSK,
- LDPC code rate =1/2 with QPSK, 16QAM or 64QAM.

P2 data in part 1 (constant length, L1 pre-signalling):

- Guard interval
- Pilot pattern
- Cell ID
- Network ID
- PAPR use
- Number of data symbols
- L1 post-signalling parameters (FEC and mod. of L1 post)

P2 data in part 2 (variable length, L1 post-signalling):

- Number of PLPs
- RF frequency
- PLP IDs
- PLP signalling parameters (FEC and mod. of PLP)

In the higher FFT modes, not all the carriers are needed for the L1 signalling. This free capacity can then be used already for the actual data transmission. I.e., the transmission of the PLPs can start already in the P2 symbols. Although this sounds somewhat adventurous to someone with years of experience and probably doesn't simplify its implementation, either, it does bring additional capacity.

37.8.3 Symbol, Frame, Superframe

A DVB-T2 frame is composed of a P1 symbol, 1 to 16 P2 symbols and N data symbols which can contain PLP data, Future Extension Frames and Auxiliary Data , as well as dummy cells. Several frames, in turn, become one superframe.

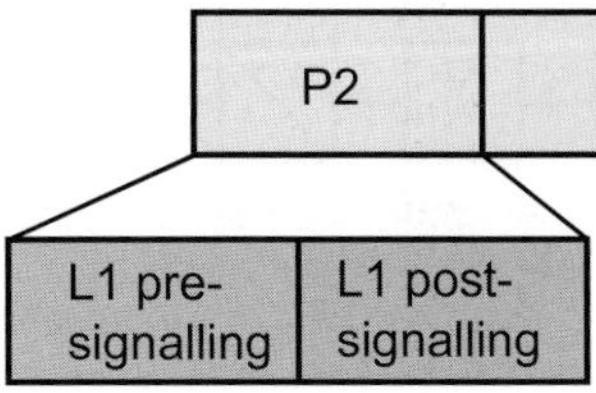

Fig. 37.20. P2-Symbols

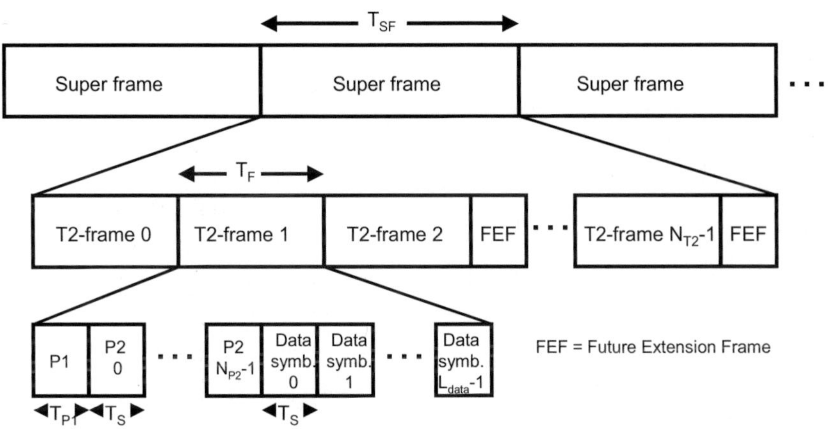

Fig. 37.21. Symbol, frame and superframe

37.9 Block Diagram

The time has come to turn to the complete block diagram of the DVB-T2 modulator. A comparison with the DVB-T modulator will demonstrate the colossal scale of the DVB-T2 Standard (Fig. 37.22.).

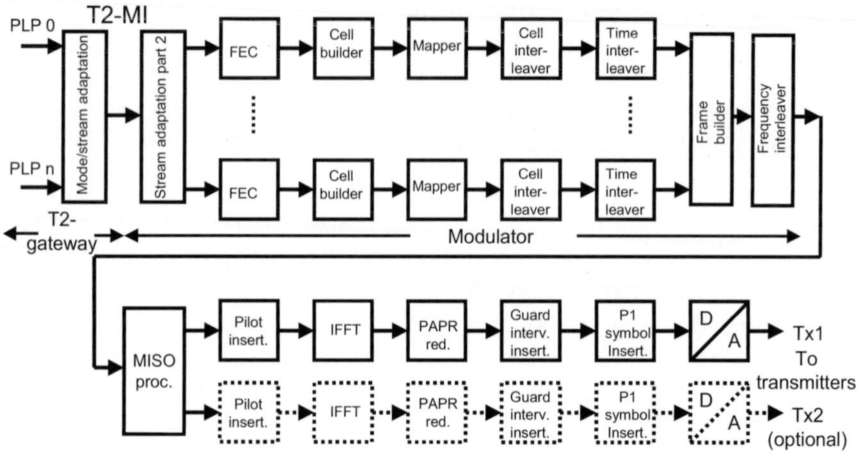

Fig. 37.22. Block diagram of the DVB-T2 modulator

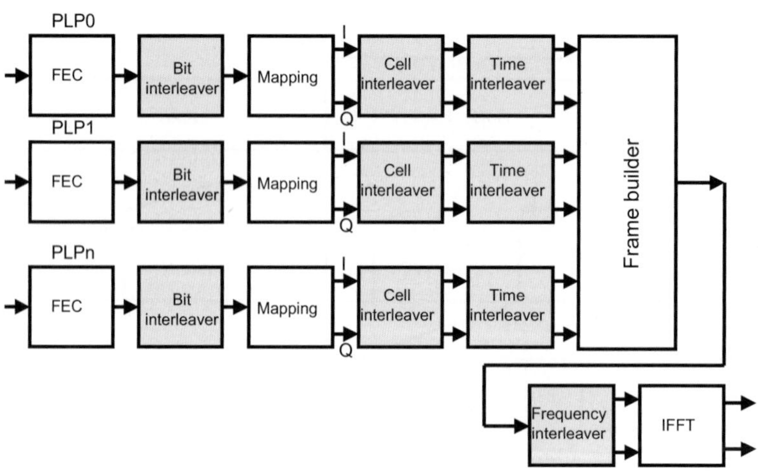

Fig. 37.23. Interleavers in DVB-T2

37.10 Interleavers

In DVB-T2, interleaving (Fig. 37.23.) is separated into

- bit interleaving,
- cell interleaving,

- time interleaving, and
- frequency interleaving.

The bit interleaver is a very short interleaver. It corresponds to the DVB-S2 bit interleaver and like that operates at the FEC frame level. The bit interleaver has the task of optimizing the characteristics of the error protection for immunity against burst errors, independently of the other interleavers. The cell interleaver also operates at FEC frame level and interleaves the cells already mapped, i.e. the IQ values. It improves the performance mainly in conjunction with the rotated and Q-delayed constellations. The time interleaver then distributes the information over a time which can be adjusted within wide ranges. The time interleaver is of assistance mainly with mobile reception, with long burst errors and with impulsive noise when the frequency interleaver distributes the information as randomly as possible to the various DVB-T2 OFDM carriers. Notches due to multipath reception will then lead to reduced losses.

37.10.1 Types of Interleaver

Back in the 1960s, David Forney had the idea of improving error protection with regard to susceptibility to burst errors by applying time interleaving. Interleavers are intended to distribute a data stream in time or frequency during transmission as randomly as possible, but so as to be recoverable by receivers. There are

- block interleavers and
- PRBS interleavers.

The interleavers used in DVB-T2 are either block interleavers or PRBS interleavers.

Table 37.11. Interleaver types in DVB-T2

Interleaver	Interleaver type
Bit interleaver	Block interleaver
Cell interleaver	PRBS interleaver
Time interleaver	Block interleaver
Frequency interleaver	PRBS interleaver

A block interleaver will first read the data, e.g. line by line, into a block and then read them out again, e.g. column by column, or in zig-zag form.

A PRBS interleaver is controlled by a pseudo random sequence and distributes the data even more randomly.

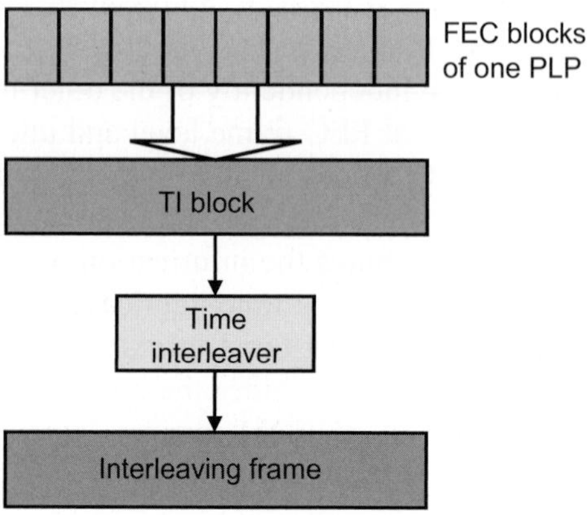

Fig. 37.24. DVB-T2 time interleaving

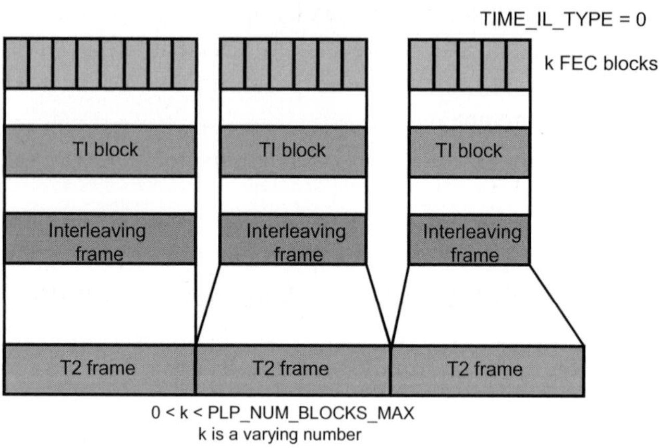

Fig. 37.25. Time interleaver, type 1

37.10.2 DVB-T2 Time Interleaver Configuration

The time interleaver (Fig. 37.24.) has the task of distributing the data of a PLP over a very long period if possible (several hundred milliseconds).

This increases robustness against burst errors. Burst errors can occur mainly in mobile reception and with impulsive noise. In the DVB-T2 time interleaver, several FEC blocks are combined to form one or more time interleaving blocks and these are then interleaved and result in an interleaving frame. The time interleaver in DVB-T2 can be set by the following configuration parameters:

- TIME_IL_TYPE (1 Bit) 0 or 1
- TIME_IL_LENGTH (8 Bit) in blocks per interleaving frame
- FRAME_INTERVAL (8 Bit) in frames
- PLP_NUM_BLOCKS_MAX (10 Bit) in blocks 0 … 1023

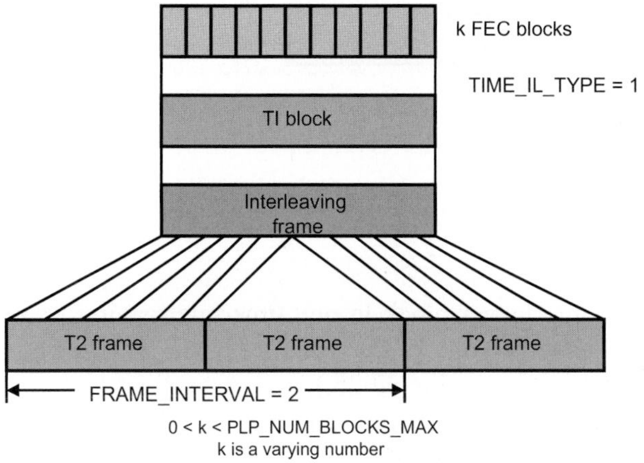

Fig. 37.26. Time interleaver, type 2

These enable the time interleaver to be set within wide ranges. The TI configuration parameters of each PLP are signalled via the L1 post-signalling part in the P2 symbols.

The TIME_IL_TYPE can be used for deciding whether the time interleaving is to take place within one T2 frame (TIME_IL_TYPE = 0) or distributed over several T2 frames (TIME_IL_TYPE = 1).

TIME_IL_LENGTH specifies the number of interleaving blocks per time interleaving frame.

FRAME_INTERVAL defines the interval between two frames carrying the time interleaving data of a PLP. I.e., gaps or T2 frames can now be inserted between interleaving data of PLPs and thus greater interleaving intervals can be achieved.

PLP_NUM_BLOCKS_MAX specifies how many FEC frames may be maximally combined to form one time interleaving block.

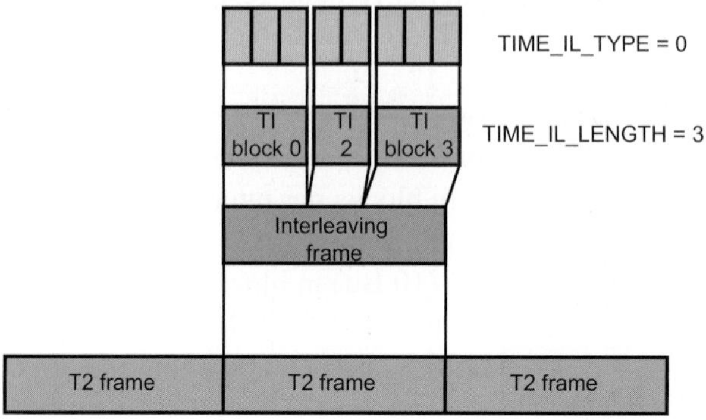

Fig. 37.27. Time Interleaver, Typ 3

Having selected these TI configuration parameters, 3 time interleaver types can now be implemented, namely

- Typ 1: a time interleaver block in one time interleaving frame, mapped in exactly one T2 frame (TIME_IL_TYPE=0, TIME_IL_LENGTH=1) (Fig. 37.25.),
- Typ 2: a time interleaver block in one time interleaving frame, mapped in several (n) T2 frames with a definable frame interval between them (TIME_IL_TYPE=1, FRAME_INTERVAL=n) (Fig. 37.26.) and
- Typ 3: a definable number (m) in one time interleaving frame mapped in one T2 frame (TIME_IL_TYPE=0, TIME_IL_LENGTH=m) (Fig. 37.27.).

Each of these 3 time interleaver-types has different TI characteristics which are listed in Table 37.12.

Table 37.12. Characteristics of the TI types in DVB-T2

TI type	Characteristics/Application
1	Better interleaving at medium data rates than with type 3
2	Better interleaving at low data rates
3	Application at higher data rates, but less interleaving than with Type 1

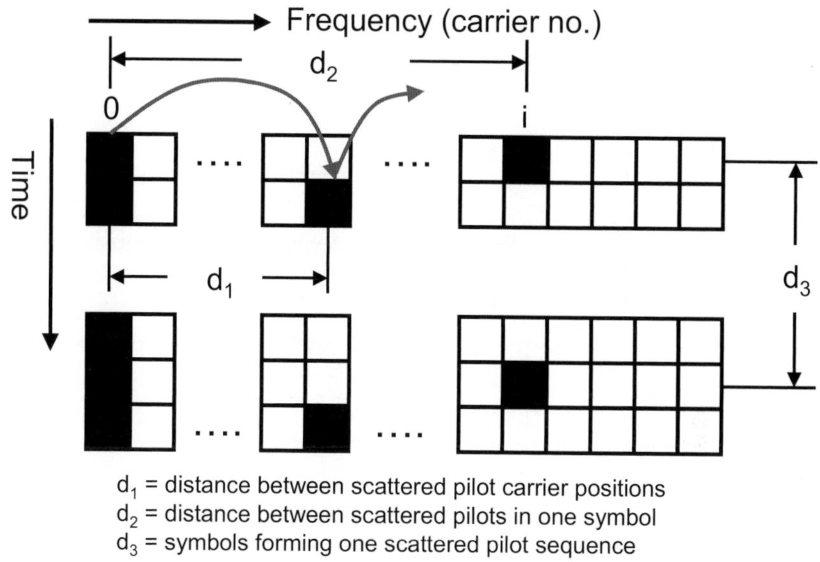

Fig. 37.28. Parameters for pilot pattern in DVB-T2

37.11 Pilots

In COFDM systems, the following tasks must basically always be implemented by special-carriers:

- Frequency lock (AFC = Automatic Frequency Control),
- Channel estimation – and channel correction, and the
- Signalling of the transmission parameters.

For this purpose, DVB-T has the following pilot signals, namely

- the Continual Pilots for the AFC,
- the Scattered Pilots for the channel estimation and
- the TPS carriers for the signalling.

DVB-T2 has the following pilots, namely

- Edge pilots at the beginning and the end of the channel,
- Continual pilots,

- Scattered pilots,
- P2 pilots at every 3rd carrier position,
- Frame-closing pilots for cleanly closing a frame.

Table 37.13. Pilot pattern in DVB-T2

Pilot pattern	Distance d_1 between pilot carrier positions (and distance d_2 of the pilots within a symbol)	Number of symbols d_3 forming a pilot sequence
PP1	3 (12)	4
PP2	6 (12)	2
PP3	6 (24)	4
PP4	12 (24)	2
PP5	12 (48)	4
PP6	24 (48)	2
PP7	24 (96)	4
PP8	6 (96)	16

Table 37.14. Scattered pilot pattern in SISO mode

FFT Mode	g=1/128	g=1/32	g=1/16	g=19/256	g=1/8	g=19/128	g=1/4
32K	PP7	PP4, PP6	PP2, PP8, PP4	PP2, PP8, PP4	PP2, PP8	PP2, PP8	--
16K	PP7	PP7, PP4, PP6	PP2, PP8, PP4, PP5	PP2, PP8, PP4, PP5	PP2, PP3, PP8	PP2, PP3, PP8	PP1, PP8
8K	PP7	PP7, PP4	PP8, PP4, PP5	PP8, PP4, PP5	PP2, PP3, PP8	PP2, PP3, PP8	PP1, PP8
4K, 2K	--	PP7, PP4	PP4, PP5	--	PP2, PP3	--	PP1
1K	--	--	PP4, PP5	--	PP2, PP3	--	PP1

In DVB-T2, the information of the previous TPS carriers is contained in the P2 symbols. There are no longer any TPS carriers, but there are continual pilots and scattered pilots. In addition, the term Edge Pilot was introduced; it is not actually a special feature and was already provided in DVB-T where it fell under the category of Continual Pilot. The edge pilots are simply the first and last pilot in the spectrum. The scattered pilots have several selectable, more or less dense, pilot patterns. Less dense pilot patterns means that there are more payload carriers, resulting in a higher net data rate. Denser pilot patterns (Fig. 37.28. and Tab. 37.13.), however, allow for a better channel estimation especially in the presence of difficult reception conditions such as multipath reception and mobile reception. When planning the network, the corresponding pilot pattern can then be selected in dependence on the planned coverage. Not all pilot patterns (called PP1 to PP8) can be used in all mode and guard interval configurations.

Table 37.15. Scattered pilot pattern in MISO mode

FFT Mode	g=1/128	g=1/32	g=1/16	g=19/256	g=1/8	g=19/128	g=1/4
32K	PP8, PP4, PP6	PP8, PP4	PP2, PP8	PP2, PP8	--	--	--
16K	PP8, PP4, PP5	PP8, PP4, PP5	PP3, PP8	PP3, PP8	PP1, PP8	PP1, PP8	--
8K	PP8, PP4, PP5	PP8, PP4, PP5	PP3, PP8	PP3, PP8	PP1, PP8	PP1, PP8	--
4K, 2K	--	PP4, PP5	PP3	--	PP1	--	--
1K	--	--	PP3	--	PP1	--	--

Table 37.16. Amplitudes of the scattered pilots

Scattered pilot pattern	Amplitude	Equivalent boost [dB]
PP1, PP2	4/3	2.5
PP3, PP4	7/4	4.9
PP5, PP6, PP7, PP8	7/3	7.4

37.12 Sub-Slicing

Without sub-slicing (Fig. 37.29.), a PLP will arrive in the receiver in one piece in a timeslot in DVB-T2. I.e. the peak data rate may be relatively high for a PLP and then there may be no further reception of these data for a relatively long period. Sub-slicing divides the PLPs into smaller "morsels“ which are then transmitted synchronously from PLP to PLP in the DVB-T2 frame. A PLP can be divided into 2 to 6480 subslices in sub-slicing. The subslices of the various PLPs then follow one another synchronously within a T2 frame. More subslices means more time diversity and less buffer memory demand, and fewer subslices means less time diversity, but offers the possibility of saving more energy in the receiver.

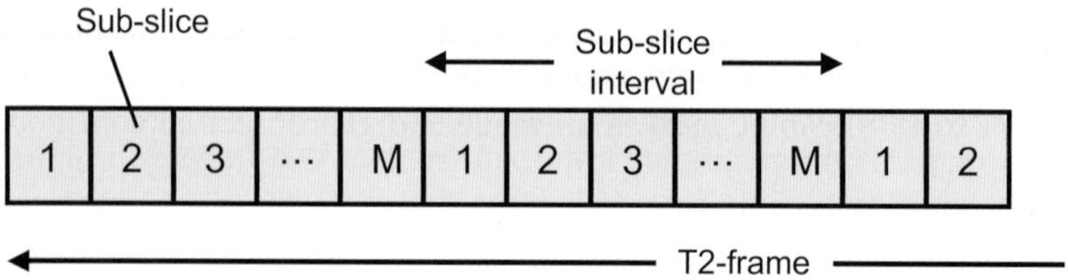

Fig. 37.29. Sub-slicing

37.13 Time-Frequency-Slicing (TFS)

Time-Frequency-Slicing (TFS) is mentioned as an option in the Appendix to the DVB-T2 Standard. This would make it possible to radiate the PLPs or their subslices in up to 8 different RF channels. The complexity would be very high at the transmitting end since up to 8 transmitting trains would have to be implemented and the receiver would then require at least two tuners. It is questionable whether TFS will become reality, although this is already contained as a requirement in the first version of the NorDig-Spec for DVB-T2 receivers.

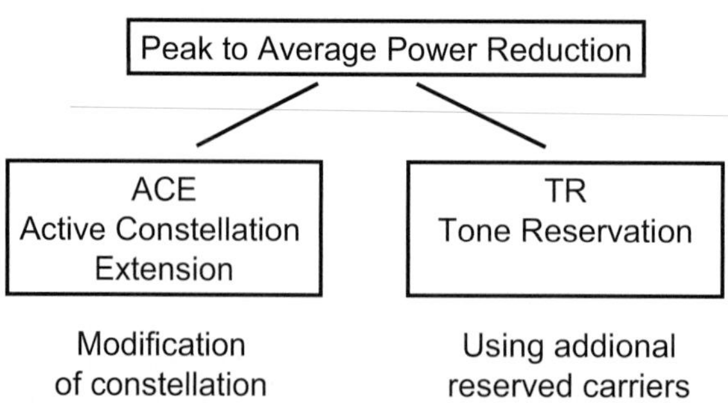

Fig. 37.30. PAPR (Peak to Average Power Ratio Reduction)

37.14 PAPR Reduction

PAPR reduction stands for Peak to Average Power Ratio Reduction (Fig. 37.30.) and means nothing else than the reduction of the crest factors. The crest factor is the ratio of the maximum peak voltage to the RMS value. In theory, the crest factor can assume very high values in COFDM systems. In practice it is maximally about 12 to 15 dB, clipped at about 12 to 13 dB in the power transmitter. There have been relevant discussions and contributions since the beginning of the applications of COFDM. In order to be able to limit this crest factor in DVB-T2, two methods of PAPR are provided, namely Active Constellation Extension (ACE) (Fig. 37.31.) and Tone Reservation (TR).

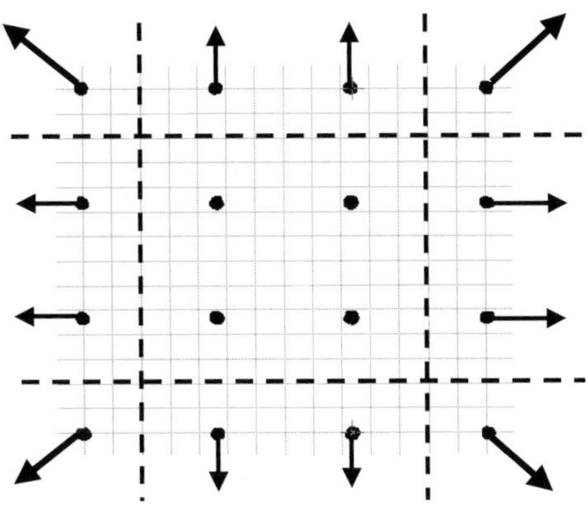

Fig. 37.31. PAPR – ACE – Active Constellation Extension

In the case of the Active Constellation Extension, the fact is used that the outermost constellation points could be shifted further out within certain limits, without restriction in the demodulation, in order to reduce the current crest factor by the summation of all carriers and by suitably adapting certain carrier amplitudes. However, the ACE is not possible with rotated constellation diagrams which is why this method will probably not be used in DVB-T2. In the case of the Tone Reservation, certain carrier bands are not intended for payload data transmission and also not for pilot tones. If necessary, these carriers, which are normally not switched on, can be activated so that they reduce the crest-factor. It is then up to the respective DVB-T2 modulator in its characteristics determined by the respective manufacturer to set these carriers in amplitude and phase in such a way that they really decisively reduce the crest-factor at the transmitter site.

This is of minor importance in the interplay with neighboring sites in an SFN.

PAPR can be used mainly for increasing the efficiency of the transmitter output stages which would be a real cost saving for the operation. It remains to be seen whether the gain is so great with regard to the dielectric strength of the downstream passive transmitter components. However, due to the crest factor, there has always been a great need for discussion with respect to the correct dimensioning especially in the case of the mask filters, the antenna combiner, the antenna cable and the antenna itself.

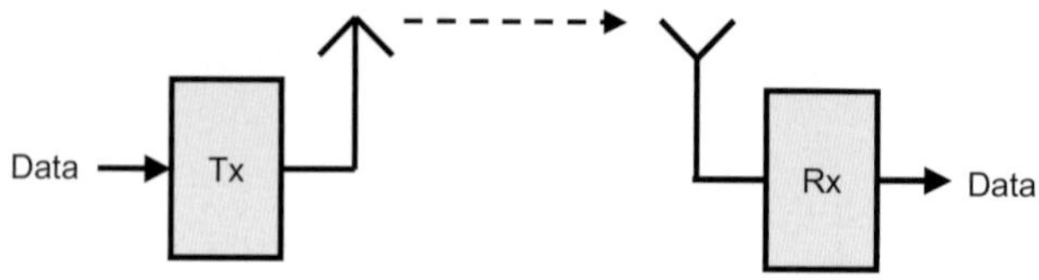

Fig. 37.32. SISO = Single Input – Single Output

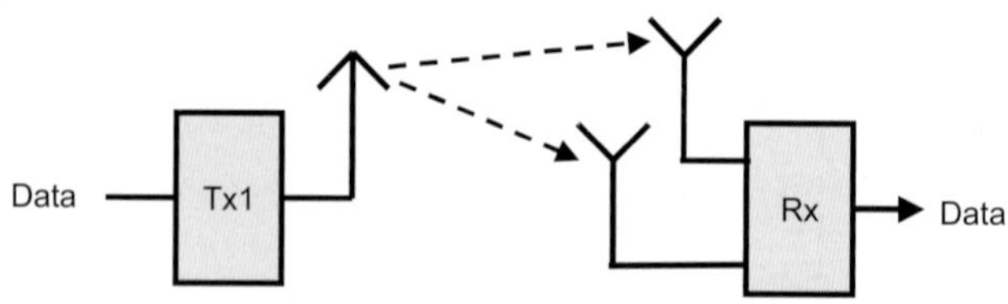

Fig. 37.33. SIMO = Single Input – Multiple Output

37.15 SISO/MISO Multi-Antenna Systems

DVB-T2 contains MISO (Multiple Input/Single Output) as an option. This means that possibly two transmitting antennas may be used which, however, do not radiate the same transmitted signal as in an SFN. Instead, adjacent symbols are transmitted repeatedly once by one and once by the other transmitting antenna in accordance with the modified Alamouti principle. This is an attempt to come closer to the Shannon limit, especially in the mobile channel. SISO (Single Input/Single Output) is the traditional case of a terrestrial transmission link (Fig. 32.32.).

This arrangement uses exactly one transmitting and receiving antenna (Fig. 32.32.). SIMO (Single Input/Multiple Output) corresponds to diversity-reception with one transmitting antenna and several receiving antennas (Fig. 32.33.) In motor vehicles, in some cases 2 – 4 receiving antennas bonded to the vehicle's windows are employed for mobile DVB-T reception.

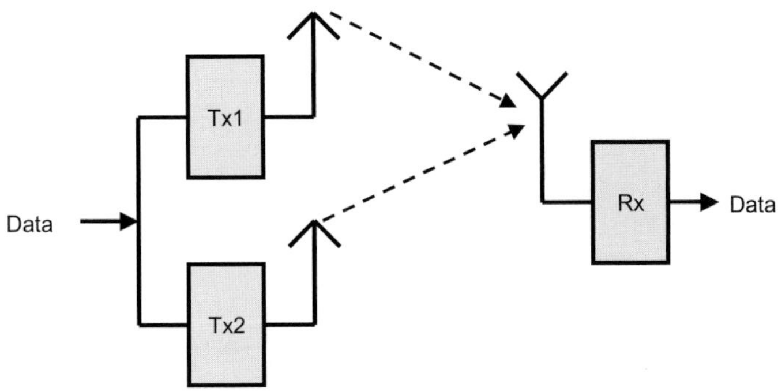

Fig. 37.34. MISO = Multiple Input – Single Output

37.15.1 MISO according to Alamouti

In the case of MISO according to Alamouti, however, two transmitting antennas and one receiving antenna are used. The aim is to save on antenna expenditure in the receiver by changing to transmit diversity (Fig. 37.34.). This is also called space/time diversity according to [ALAMOUTI]. A further possibility would also be MIMO (Multiple Input/ ultiple Output) with several transmitting and receiving antennas (Fig. 37.35.). The idea of the MISO principle goes back to [ALAMOUTI], 1998. This principle is already being used in mobile radio (WiMAX) where adjacent symbols (COFDM symbols) are repeated at the two transmitting antennas. At the receiving antenna, a superimposed grouping of adjacent symbols always arrives which, without modification, would result in mutual interference and could thus no longer be separated in the receiver. In the case of the Alamouti principle, on the other hand. the adjacent symbols are not radiated unmodified at the various transmitting antennas, but in accordance with the Alamouti code (Fig. 37.36.). According to [ALAMOUTI], the two adjacent symbols s_n are first present at antenna 1 and s_{n+1} at antenna 2. Then symbol s_{n+1} is applied in negative conjugate complex form to antenna 1 and at the same time symbol s_n is radiated in conjugate complex form at transmitting antenna 2.

This will enable the receiver (Fig.32.37.) to separate two adjacent symbols again by means of suitable complex mathematical operations on these symbols (Fig. 37.38.).

In addition, the channel transfer function from transmitting antenna 1 to the receiver and transmitting antenna 2 to the receiver must be known. I.e., it is necessary to perform a channel estimation over all transmitting and receiving paths.

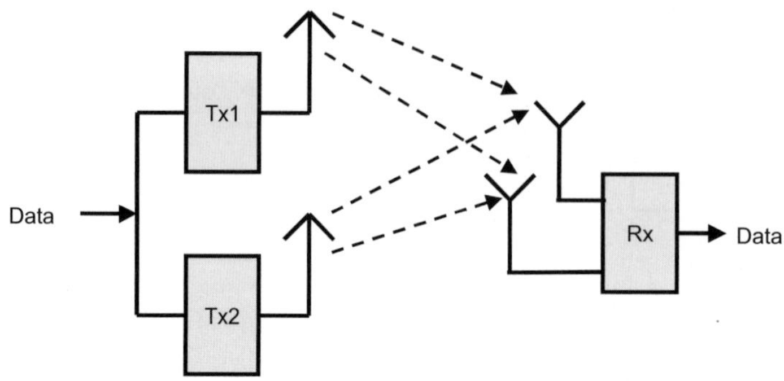

Fig. 37.35. MIMO - Multiple Input – Multiple Output

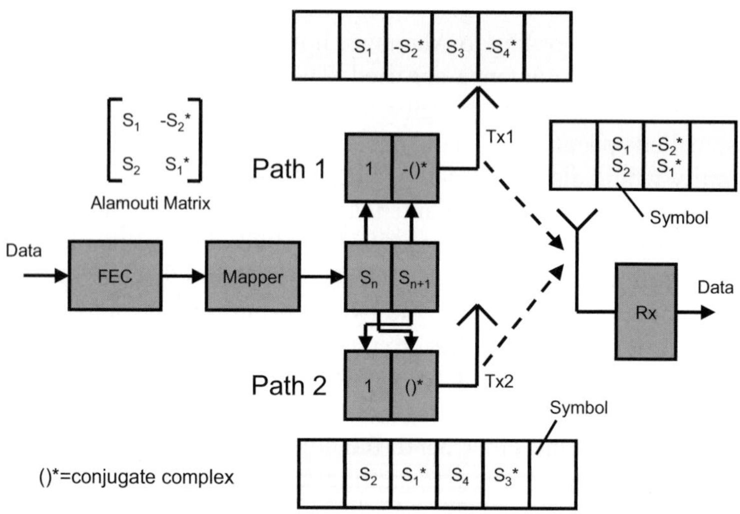

Fig. 37.36. MISO principle according to Alamouti

37.15.2 Modified Alamouti in DVB-T2

It is intended to employ the MISO principle both at only one site and distributed over SFN sites. Used at one site, this can be done by horizontal and vertical polarization. However, DVB-T2 uses a modified Alamouti coding (Fig. 32.39.). At antenna 1, the cells c_1, c_2, c_3, c_4, ... are present unchanged. It is only at antenna 2 that correspondingly changed cells are radiated. This has the advantage that the DVB-T2 system can be easily reduced to SISO by simply omitting the second transmitting path.

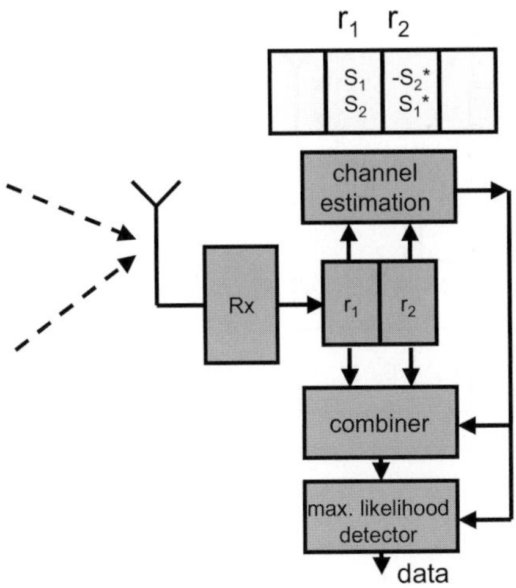

Fig. 37.37. MISO signal reception in the receiver

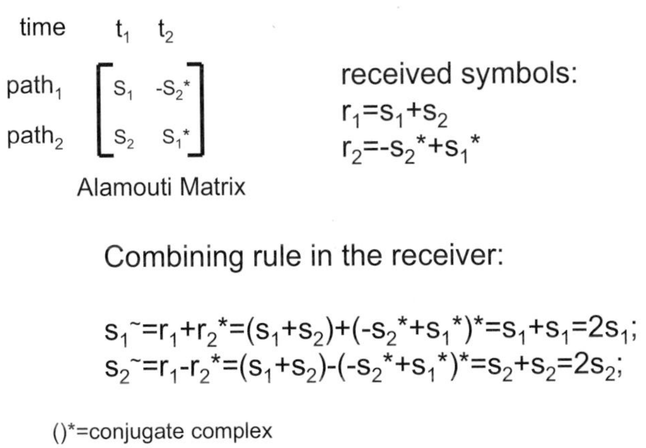

time t1 t2
path1 s1 -s2*
path2 s2 s1*
Alamouti Matrix
received symbols:
$r_1=s_1+s_2$
$r_2=-s_2^*+s_1^*$
Combining rule in the receiver:
$s_1^\sim=r_1+r_2^*=(s_1+s_2)+(-s_2^*+s_1^*)^*=s_1+s_1=2s_1;$
$s_2^\sim=r_1-r_2^*=(s_1+s_2)-(-s_2^*+s_1^*)^*=s_2+s_2=2s_2;$
()*=conjugate complex

Fig. 37.38. MISO signal processing in the receiver

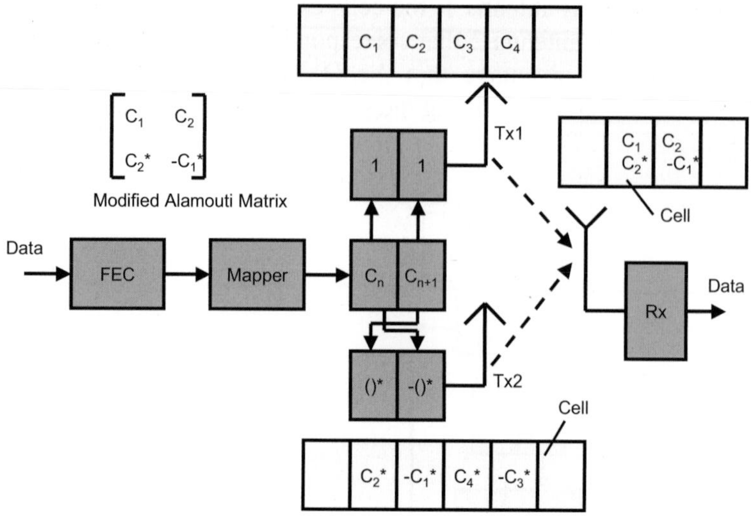

Fig. 37.39. Modified Alamouti in DVB-T2

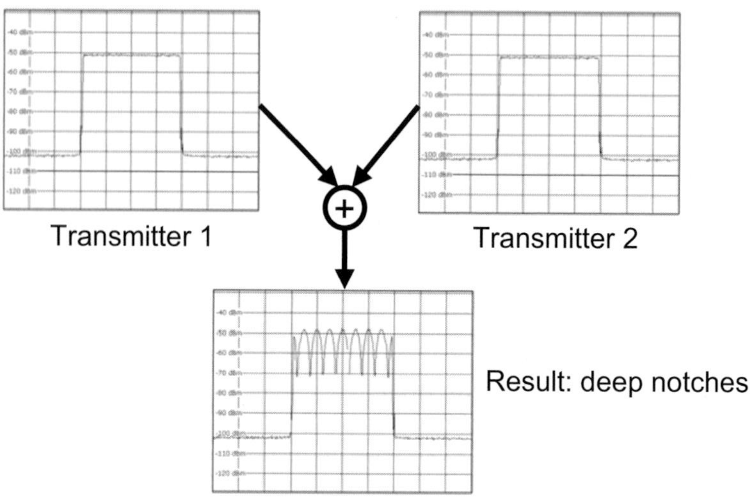

Fig. 37.40. Reception of two signal paths in an SFN without MISO, with fading notches

This alone is not all, however. In DVB-T2, MISO is not applied via space/time diversity but via space/frequency diversity (based on adjacent cells in the spectrum). I.e. at transmitting antenna 2, adjacent pairs of carriers are radiated interchanged compared with those radiated at transmitting antenna 1. One great advantage of this modified Alamouti principle in DVB-T2 is that the signals from transmitting antenna 1 and 2 are no longer correlated with one another. This makes it possible to avoid the notches prevalent in DVB-T and DAB, especially when using "distributed MISO" in an SFN (Fig. 37.40. and 37.41.). The principle of MISO by H- and V-polarization is called co-located MISO in DVB-T2.

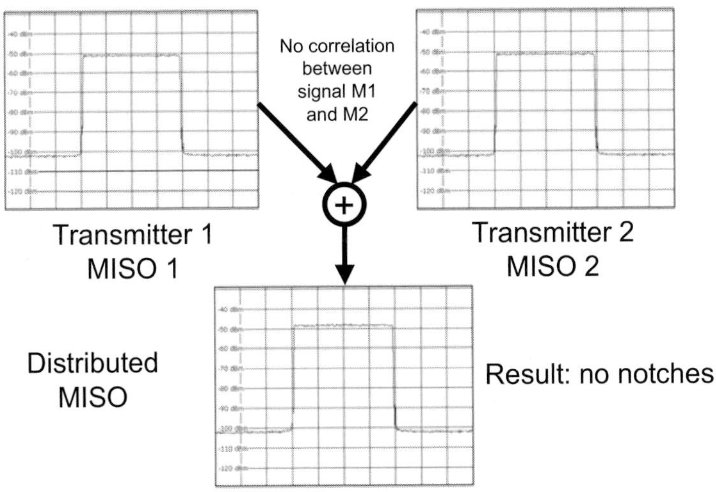

Fig. 37.41. Reception of two signal paths in a distributed DVB-T2-MISO, without fading notches

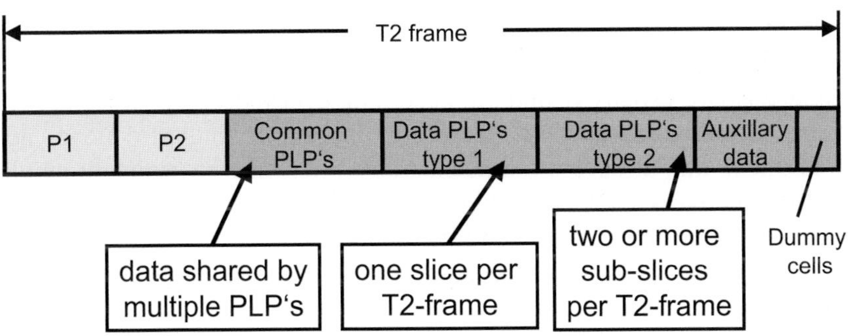

Fig. 37.42. DVB-T2 frame with auxiliary stream data

37.16 Future Extension Frames

DVB-T2 inherently already provides for possible expansion in so-called Future Extension Frames. These are special frames with as yet undefined transmission parameters which can be tied into the DVB-T2-frame structure. They are signalled via the P2 symbols.

37.17 Auxiliary Data Streams

At the end of a DVB-T2 frame (Fig. 37.42.), auxiliary stream data can still be appended. These are customer-designed error-protected and mapped IQ values. A normal DVB-T2 receiver does not need to be able to evaluate these data.

37.18 DVB-T2-MI

In order to be able to conduct several (up to 255) data streams synchronously to the DVB-T2 modulators and transmitters in Mode B (Multiple Physical Layer Pipes), the T2-MI modulator interface was defined. Apart from supplying the data streams, it also handles the control and signalling. The DVB-T2-MI interfaces are described in Chapter 36.

37.19 SFNs in DVB-T2

Naturally, it should be possible to implement single frequency networks also in DVB-T2, the main reason being the economic use of frequencies. Frequencies are expensive and are becoming ever more scarce. Being able to reuse the same frequency is, therefore, important. Single frequency networks allow this at several adjacent transmitter sites in isolated, single-frequency networks. DVB-T2 additionally allows larger interference-free single frequency networks to be formed.

Single frequency networks must meet the following conditions:

- Frequency synchronism,
- Time synchronism,
- Data synchronism,

- Guard interval condition, i.e. maximum transmitter spacing must not be exceeded (see table 37.17 for a 8 MHz wide DVB-T2 channel).

Table 37.17. Guard interval sizes in DVB-T2 (8 MHz-channel)

Mode	Symbol duration [ms]	g= 1/128 t [ms] d [km]	g= 1/32 t [ms] d [km]	g= 1/16 t [ms] d [km]	g= 19/256 t [ms] d [km]	g= 1/8 t [ms] d [km]	g= 19/128 t [ms] d [km]	g= 1/4 t [ms] d [km]
32K	3.584	0.028 8.4	0.112 33.6	0.224 67.2	0.266 79.7	0.448 134.3	0.532 159.5	--
16K	1.792	0.014 4.2	0.056 16.8	0.112 33.6	0.133 39.9	0.224 67.2	0.266 79.75	0.448 134.3
8K	0.896	0.007 2.1	0.028 8.4	0.061 16.8	0.067 19.8	0.112 33.6	0.133 39.89	0.224 67.2
4K	0.448	--	0.014 4.2	0.031 8.4	--	0.056 16.8	--	0.112 33.6
2K	0.224	--	0.07 2.1	0.016	--	0.028 8.4	--	0.056 16.8
1K	0.112	--	--	4.2	--	0.014 4.2	--	0.028 8.4

With d = t • 299792458 m/s;
correction factor for
10 MHz = 8/10, 8 MHz = 1, 7 MHz = 8/7, 6 MHz = 8/6,
5 MHz = 8/5 and 1.7 MHz = 8/1.7.

Frequency synchronism is achieved by frequency standards at the transmitter site, generally a professional GPS receiver providing a 10-MHz reference. Time and data synchronism is achieved by time stamps in the baseband feed signal. This is the T2-MI signal in DVB-T2. The DVB-T2 modulator synchronizes its frame structure to these time stamps. The guard interval condition is met by suitable network planning with planning software [LStelcom]. The new factor in DVB-T2 is the possibility of distributed MISO. There are transmitter sites which radiate either MISO Mode 1 or 2. The advantage of this method is that destructive fading no longer occurs between two adjacent transmitter sites. However, the appropriate choice and simulation of the MISO modes is also a new challenge for the network planning.

37.20 Transmitter Identification Information in DVB-T2

A the time the present edition went to press, an ETSI Draft was being prepared which is intended to enable transmitters in an SFN to be identified by using measurement techniques as is done in DAB. DAB has the TII signal in the Null symbol for this purpose. Transmitter identification was not possible in DVB-T since this would have violated synchronism and would have led to severe disturbances in an SFN. The cell ID in DVB-T only allowed an SFN cell to be identified. The cell ID is here built into reserved TPS bits.

DVB-T2 has the capability of inserting Auxiliary Streams and FEFs (Future Extension Frames) into a T2-frame. Normal DVB-T2 receivers will ignore the Auxiliary Streams and FEFs. DVB-T2 provides for such transmitter identifiers or transmitter signatures to be inserted into the T2 frames either via the Auxiliary Streams or via the FEFs. Similar to DAB, these would be certain active carriers which are activated only at individual carriers within the Auxiliary Streams and would be switched off at all other transmitters in an SFN. By this means, the signals of the channel impulse response in an SFN can then be correlated with the individual transmitters. The signalling of the transmitter identification via the FEFs provides for broadcasting certain different signatures or sequences over the transmitters which can be identified by correlation and thus also enable the individual transmitters in an SFN to be recognized. The additional signalling of the transmitter identification is an overhead which, however, should occupy only very little payload data rate. In DVB-T, transmitter identification has always been wished for by the maintenance technicians who had to perform the coverage tests, but it had been a wish which could not be fulfilled. In DVB-T2, as with DAB, however, this provides both for more comfortable calibration and balancing of delays in a single-frequency network, and for monitoring these.

37.21 Capacity

The aim was to achieve 30% better capacity compared with DVB-T. In some cases, the data rate will be up to 50% higher under comparable conditions. And that is already enormous. To achieve this, this Standard is also more complex than DVB-T by about 150%. In combination with the new video and audio compression standards such as MPEG-4 H.264/AVC and MPEG-4 AAC this results in a huge increase in effectiveness in respect to program variety and quality. It may well be that this Standard is so far ahead of its time that there will be peace and quiet in this corner for some years to come.

37.22 Outlook

In areas where DVB-T has just been successfully introduced and analog television belongs to the past it will be initially difficult to apply new models of coverage without harming the existing one. And that applies to many countries in Central Europe, but also worldwide. In a lecture at the IRT in 2008, it was quoted, e.g. "For Germany, DVB-T2 is either 5 years too early or 5 years too late" [IRT2008_KUNERT]. But there may also be applications which do not have the aim of initially replacing traditional DVB-T but to find a replacement for applications which were planned for DAB. There are standards which have been on air longer than DVB-T and have not really been able to gain the attention of the consumer. DVB-T2 can also be used for broadcasting pure audio.

Bibliography: [DVB_A122r1], [DVB-T2], [ALAMOUTI], [FKTG2008_GUNKEL], [IRT2008_KUNERT]

38 DVB-C2 – the New DVB Broadband Cable Standard

It has been apparent for some time that after DVB-T2, there would also be a new DVB broadband cable standard, the "Call for Papers" having been issued at the end of 2007. At the IBC 2008, rough outlines were then published on a DIN A4 sheet. But it was not quite clear then whether this would only amount to a multi-carrier method or whether single-carrier modulation would also be a component of DVB-C2. DVB-C2 working documents based on DVB-T2 were then available at the end of 2008 and in the spring of 2009 the time had come: DVB-C2 was published as a Draft.

38.1 Introduction

There are many formulations in DVB-C2 which are derived straight from DVB-T2. And it is also true that it is relatively easy to find one's way around DVB-C2 if one knows DVB-T2. Like DVB-T2, DVB-C2 uses COFDM as a modulation method, the only difference being that the guard intervals are very short and there is only a 4k mode. And the constellation diagrams extend from QPSK up to 4096QAM. At first glance, it appears to be surprising that up to 4096QAM is provided but rough estimates of parameters known from DVB-T2 show that 4096QAM is possible with signal-to-noise ratios which are easily achieved in modern broadband cables. Such high-level types of modulation are possible mainly because of the error protection used in DVB-C2, which corresponds to the error protection used in DVB-S2 and DVB-T2. DVB-C2 thus also uses BCH and LDPC coding. And modern broadband cable networks are continuously improving with respect to the signal-to-noise ratio. Fiber optics are coming closer and closer to the end user terminal, i.e. only the last few meters up to perhaps 1000 meters are still run in coaxial cable technology. Even in purely coaxial cable networks, signal-to-noise ratios within a range of more than 30 dB are easily possible in most cases and modern broadband cable networks provide signal-to-noise ratios within a range of up to 40 dB. The "old" DVB-C standard operates with a single-carrier modulation method

W. Fischer, *Digital Video and Audio Broadcasting Technology*, Signals and Communication Technology, 3rd ed., DOI: 10.1007/978-3-642-11612-4_38, © Springer-Verlag Berlin Heidelberg 2010

and normally uses either 64QAM or 256QAM. However, modulation schemes down to QPSK would also be possible but are not normally used. In DVB-C, the "fall-off-the-cliff" occurs at pre-Reed Solomon bit error ratios of more than $2 \cdot 10^{-4}$, corresponding to a signal-to-noise-ratio of about

- 26dB with 64QAM or
- 32dB with 256QAM.

Before considering a new standard which replaces another one, it is always of interest to know the limitations of the previous standard. Broadband cable networks operate within a frequency range of about 30 to 860 MHz. However, the range below 65 MHz is in most cases used for the return channel (Internet, telephony) today. The upper end of the broadband-channel frequency band is followed closely by the GSM900 band. To protect other radio services, certain frequency bands in the cable must be kept free due to possible leaks in the cable. There are currently attempts in the terrestrial domain to utilize 9 previous TV channels for mobile radio applications (WiMAX, LTE) at the upper end of the TB band ("Digital Dividend"). This will correspond to a bandwidth of about 80 MHz. However, investigations have shown that it may then no longer be possible to use this frequency band in the cable since even fractions of the transmitting powers provided for these mobile radio applications will have such severe effects, especially on the receivers, thus making reception impossible. But there have previously been frequency bands in the cable which could be disturbed by other services. The reason could be especially radiation-induced interference on the final meters at the end user but the broadband cable also produces its own interference. Multi-channel allocation and non-linear amplifiers generate intermodulation products. And the end user's wiring can also be quite eccentric at times. Wrong terminations and simple T-junctions are no rarities. Amplifiers which are overdriven or set at the wrong level can be frequently found at the end user's. The cable network operator does not always have a compulsory influence on the correct state of the network at level 4, i.e. the end user, and this especially is the main cause of the limitations of DVB-C. The reason for this is the relatively simple error protection (only Reed-Solomon block coding) and the single-carrier modulation method used. Fitted with a channel equalizer, DVB-C receivers can only compensate for frequency response errors to a limited extent. It is easier to detect and compensate for frequency response errors with the aid of a multi-carrier modulation method and the assistance of pilot carriers which is why COFDM is also the correct approach in cable applications. At the moment it appears in any case that the majority of modern transmission standards back COFDM and will continue to do so. This also applies more and more to the mobile radio field. And it is clear, therefore, that COFDM will also be used in DVB-C2 where a single-carrier modulation method is not provided for. It can be said that the DVB-T2 standard was taken as the basis for DVB-C2, modes and features irrelevant for DVB-C2 were initially deleted from it and then new applications were implemented for broadband cable applications and the modulation schemes were also adapted to the world of cable.

38.2 Theoretical Maximum Channel Capacity

Before considering the DVB-C2 standard in greater detail, the theoretical limits of broadband cable will first be discussed. The Shannon Limit specifies that with signal-to-noise ratios from 10 dB, approximately the following formula for the theoretical maximum channel capacity applies:

$C = 1/3 \cdot B \cdot SNR$;
where
C = channel capacity (in bit/s);
B = bandwidth (in Hz);
SNR = signal/noise ratio (in dB);

A 8-MHz-wide broadband channel then provides the following theoretical maximum channel capacity:

Table 38.1. Shannon-Limits in a broadband cable

SNR[dB]	Theoretical max. channel capacity [Mbit/s]	Remarks
26	69.3	Fall off the cliff with DVB-C at 64QAM
30	80	Poor coax cable system
32	85.3	Fall-off-the-cliff with DVB-C at 256QAM
35	93.3	
40	106.7	Good HFC
45	120	In the headend

Assuming that a symbol rate of 6.9 MS/s is used, the data rates in an 8-MHz channel in DVB-C were in most cases:

- 38.15 Mbit/s with 64QAM, and
- 50.87 Mbit/s with 256QAM.

In DVB-C2, the minimum aim was to achieve a 30% higher data rate. Without having any knowledge of the DVB-C2 standard, it can be expected, therefore that, instead of the data rates shown above, either at least about 50 Mbit/s should be achieved with signal/noise ratios of around 26 dB, or 66 Mbit/s with a signal/noise ratio of around 32 dB, in an 8-MHz-wide channel. The aims were even exceeded. The reasons for the increase in channel capacity in DVB-C2 are:

- Better forward error correction,
- higher-order types of modulation,
- channel bundling and, as a result, no use of guard bands.

38.3 DVB-C2 – An Overview

In the following section, an overview of the most important DVB-C2 details will be given briefly.

- Based on DVB-T2
- COFDM (4K Mode, 2 short guard intervals)
- FEC based on LDPC (… DVB-S2, DVB-T2, …)
- multiple TS (transport streams) and GS (generic streams)
- single and multiple PLPs (physical layer pipes)
- data slices (fixed mapping of certain PLPs onto certain fixed carrier groups)
- QPSK … 4096QAM (16K QAM under discussion)
- Variable Coding and Modulation
- channel-raster bandwidth of 6 or 8 MHz
- channel bundling up to approx. 450 MHz bandwidth
- supports notches (notching out disturbed or interfering frequency bands)
- reserved carriers for PAPR reduction (Peak to Average Power Ratio reduction)

38.4 Baseband Interface

The DVB-C2 input interface allows use of a simple MPEG-2 transport stream, to feed in several transport streams, a generic stream and several generic streams. The input signal formats fully correspond to those of DVB-T2 and have already been described in Chapter 36 and 37. Like DVB-T2, DVB-C2 supports the Normal Mode and High Efficiency Mode (HEM). The term Physical Layer Pipe (PLP) is also used in DVB-C2 and corresponds to the physical transport of one of these input signals. There is provision for common PLPs for a group of PLPs in order to save on signalling. A group of up to 255 PLPs can be combined to form data slices.

38.5 Forward Error Correction

The forward error correction in DVB-C2 corresponds to that of DVB-S2 and DVB-T2. It consists of a baseband scrambler, a BCH-block coder, an LDPC block coder followed by a bit interleaver. The error protection in DVB-C2 can be adjusted by selecting one of 5 code rates, listed in Table 38.2.

Table 38.2. Code rates in DVB-C2

Long frame (64800 bits)	Short frame (16200 bits)
Code rate	Code rate
2/3	2/3
3/4	3/4
4/5	4/5
5/6	5/6
8/9	8/10

38.6 COFDM Parameters

In DVB-C, only the 4k mode is possible, corresponding to 3408 carriers used (Fig. 38.1.). However, the IFFT operates with carriers to a power of two, i.e. 4096 carriers overall. The channel raster bandwidths are 6 or 8 MHz. The subcarrier spacings and the actual signal bandwidths are then:

Table 38.3. COFDM parameters in DVB-C2

Channel raster bandwidth [MHz]	Subcarrier spacing [kHz]	Signal bandwidth [MHz]
6	1.674	5.71
8	2.232	7.61

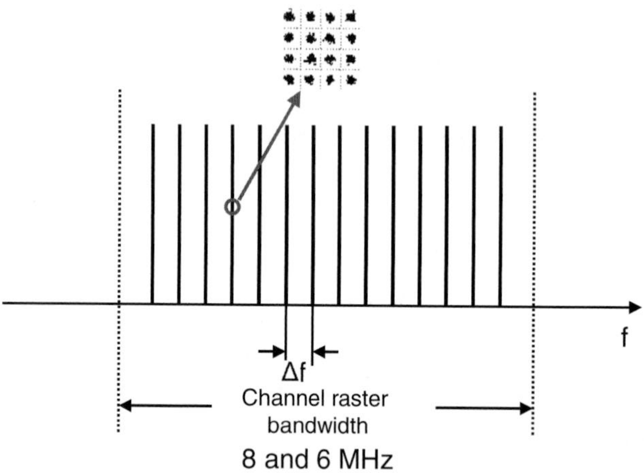

Fig. 38.1. COFDM parameters

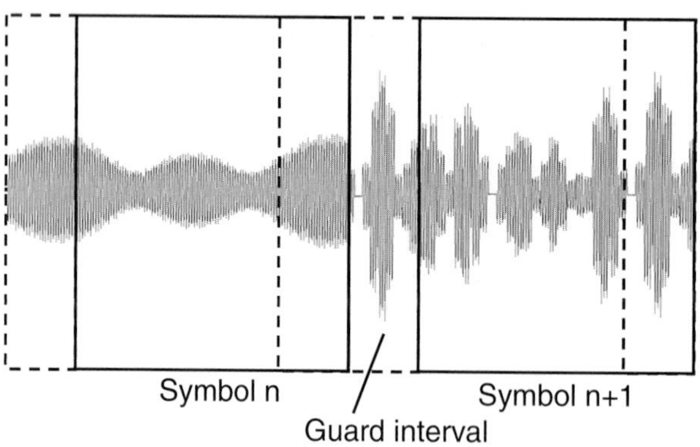

Fig. 38.2. COFDM symbol and guard interval

Only very short guard intervals (Fig. 38.2.) will be used since only very short reflections can be expected in a broadband cable. The selectable guard intervals are 1/64 or 1/128 of the symbol period. The guard interval is a cyclic prefix, i.e. a prefixed copy of the symbol end.

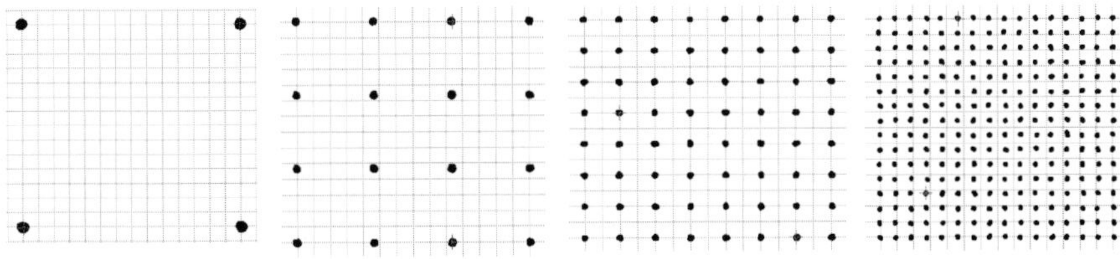

Fig. 38.3. QAM orders QPSK, 16QAM, 64QAM and 256QAM in DVB-C2

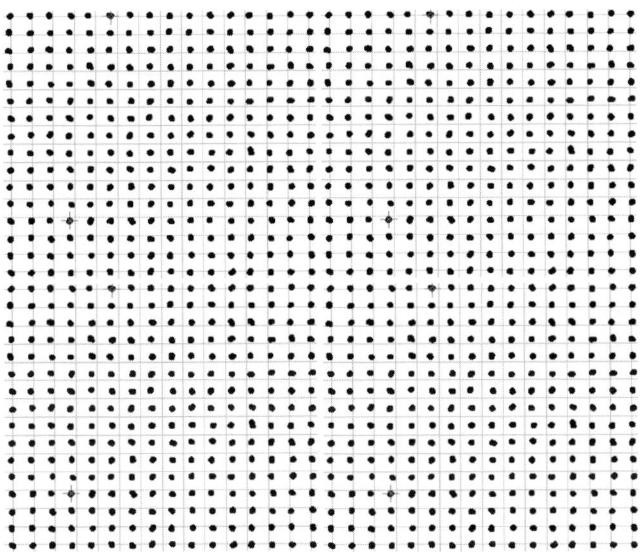

Fig. 38.4. QAM order 1024 in DVB-C2 (simulated)

38.7 Modulation Pattern

DVB-C2 operates with coherent Gray-coded modulation. The following QAM orders (Fig. 38.3., 38.4. and 38.5.) are supported:

- QPSK
- 16QAM

- 64QAM
- 256QAM
- 1024QAM
- 4096QAM.

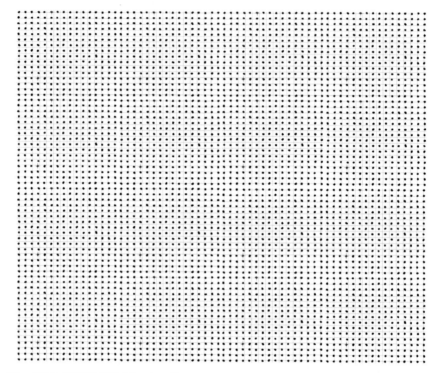

Fig. 38.5. 4096QAM (simulated)

Although 4096QAM (Fig. 38.5.) initially appears to be an interesting or eccentric choice, it is quite practicable when considered more closely and compared with the minimum signal/noise ratios necessary known from DVB-T2. Although no tables about the minimum signal/noise ratios necessary with DVB-C2 have been publishes as yet, a minimum SNR of about 30 dB can be expected with 4096QAM on the basis of estimates from parameters of the DVB-T2 standard. Such QAM orders are made possible by the signal/noise ratios of 30 dB up to almost 40 dB in cables which have become quite good in the meantime. DVB-C2 mentions even modulation methods of up to 16k QAM.

38.8 Definition of a Cell

Similar to DVB-T2, the term "cell" (Fig. 38.6.) was introduced in DVB-C2. The reason is that bit groups are combined relatively early to form later carriers which are mapped, i.e. modulated, but some interleaving processes still have to be carried out. In DVB-T2, a cell is an IQ value whereas in DVB-C2 a cell corresponds to a 2-, 4-, 8-, 10- or 12-bit-wide bit group depending on the selected QAM order transported later on a carrier.

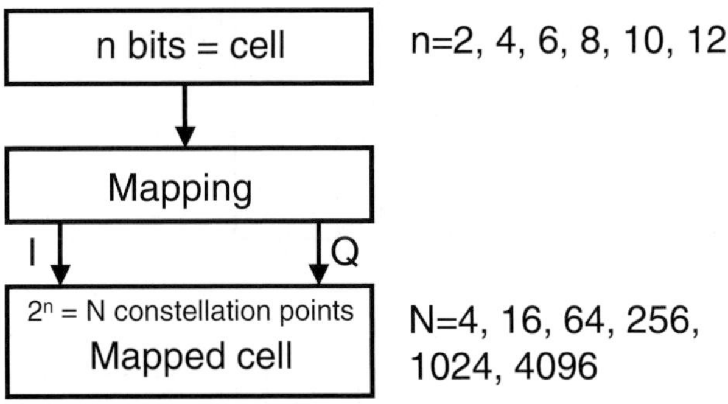

Fig. 38.6. Definition of a cell in DVB-C2

38.9 Interleavers

There are altogether 3 interleavers (Fig. 38.7.) in DVB-C2, which are:

- a bit interleaver which belongs directly to the FEC,
- a time interleaver, and
- a frequency interleaver.

There is no cell interleaver here as there is in DVB-T2.

The bit interleaver operates at FEC level and optimizes the characteristics of the error protection. Exactly as in DVB-S2 and DVB-T2, however, it is a short interleaver. The time interleaver makes DVB-C2 more robust against relatively long burst errors. The time interleaver is adjustable in its interleaving depth deep time interleaving as in DVB-T2 is thus not necessary here. And the frequency interleaver distributes the data as randomly as possible over many COFDM carriers. Time and frequency interleaving takes place in DVB-C2 within a data slice to which a certain number of PLPs are allocated.

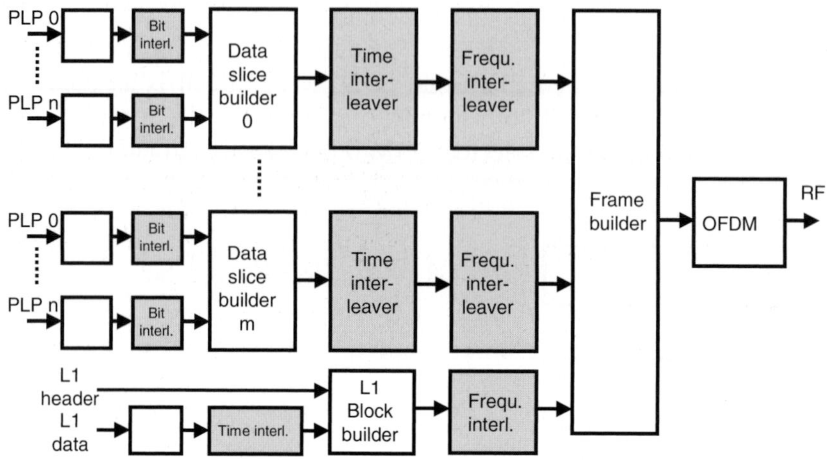

Fig. 38.7. Interleavers in DVB-C2

38.10 Variable Coding and Modulation (VCM)

In DVB-C2, each input stream is transmitted in its own Physical Layer Pipe (PLP). All interleaving-processes are restricted not to a PLP but to a group of PLPs which are combined to form a data slice. Within each PLP, however, various transmission parameters such as error protection and modulation method can be selected. This is called Variable Coding and Modulation (VCM), a term which is also known from DVB-T2. However, due to the uniformity of the broadband cable, it must be assumed that VCM, i.e. the selection of different transmission parameters in different frequency bands or PLPs in the cable will not be used so intensively.

38.11 Frame Structure

Just like other standards, DVB-C2, too, has the concept of a frame. A DVB-C2 frame begins with preamble symbols which are repeated every 7.61 MHz and have a width of 3408 carriers each. These are followed by the data symbols, a total of 448 data symbols. The preamble symbols are used both for time and frequency synchronization and for signalling of the

Layer-1 (L1) parameters. The preamble symbols are arranged with respect to frequency in such a way that a receiver with a receiver bandwidth of 7.61 MHz will get all the data necessary for finding the Layer-1 parameters.

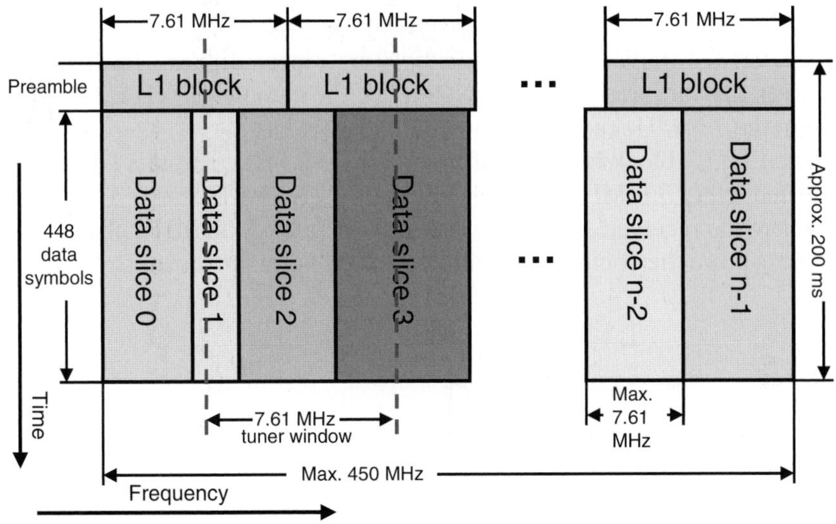

Fig. 38.8. Framing and channel bundling

38.12 Channel Bundling and Slice Building

There are actually no longer any channels in DVB-C2 but only two channel rasters of either 6 or 8 MHz. Channels can be bundled together to form a channel with a total width of approx. 450 MHz (Fig. 38.8.). There are then no longer any gaps between the original channels. The lack of any more gaps (guard bands) then enables the frequency spectrum to be used more effectively and allows a higher data rate overall. However, there are very frequently also disturbed frequency bands in the cable or frequency bands which could interfere with other radio services (e.g. aircraft radio). These can be notched out in DVB-C2 by simply switching off certain OFDM carriers (Fig. 38.10.). This is called nothing which produces gaps in the frequency spectrum. The channel bundling leads to increased demands on the modulator but not on the receiver. The receiver bandwidth is limited to 7.6 MHz in DVB-C2, i.e. the receiver only needs to be capable of demodulating channel rasters with a maximum width of 7.6 MHz. In DVB-C2, frequency slices with a maximum width of up to 7.6 MHz are formed for this purpose in which a whole number of Physical Layer Pipes (PLPs) are mapped. During the demodulating, the receiver selects the frequency slice containing the data stream to be demodulated, i.e. the relevant PLP. Every 7.6 MHz there is a Layer-1 signalling so that the receiver will find the slices and know the transmission parameters in the slices. I.e., every 7.6 MHz, special signalling symbols are inserted at every beginning of a DVB-C2 frame. Since the symbols are repeated every 7.6 MHz, the receiver can arbitrarily locate itself over the bundled DVB-C2 channel within the 7.6-MHz receiver-bandwidth.

The first L1 block begins mathematically at 0 MHz; and this L1 block is repeated every 7.6 MHz at the beginning of a DVB-C2 frame. Regardless of where the receiver, having a bandwidth of at least 7.6 MHz, places itself it will capture the complete signalling of the Layer-1 parameters.

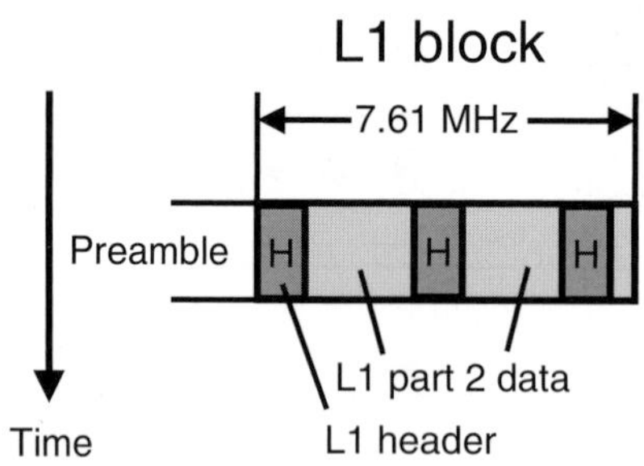

Fig. 38.9. Preamble symbols

38.13 Preamble Symbols

The preamble symbols (Fig. 38.9.) are used for

- time synchronization,
- frequency synchronization and for
- signalling the Layer-1 parameters (guard -interval, modulation, error protection etc.).

They are located at the beginning of a DVB-C2-frame in the L1 block and also mark the beginning of the frame. The preamble symbols consist of a header and an L1 time interleaving block. The header contains over 32 bits, consisting of 16 data bits and error protection, the basic signalling of the most important basic L1 parameters.

The preamble header is 32 OFDM cells wide. The following are transmitted in the header:

- L1_INFO_SIZE (14 bits), and
- L1_TI_MODE (2 bits).

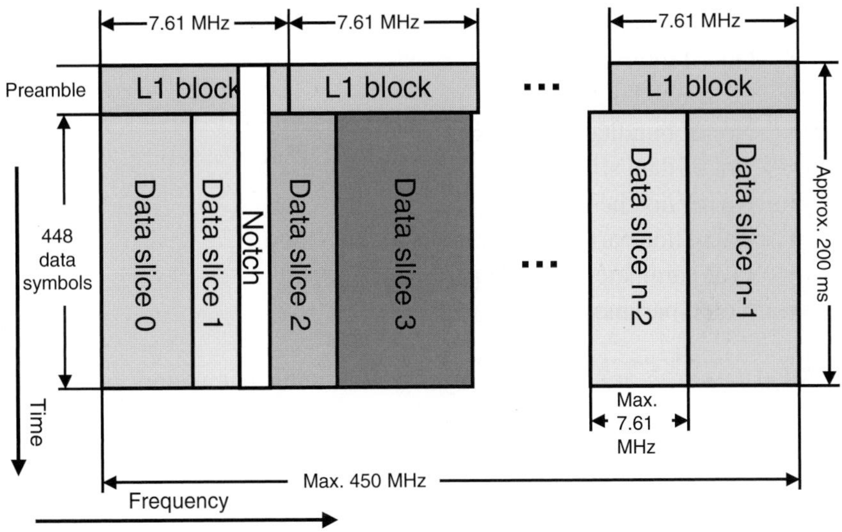

Fig. 38.10. Forming of notches (carrier bands switched off) in DVB-C2

Altogether, the preamble header consists of 32 bits which are composed of the 16 payload data bits and additional 32 FEC bits (Reed-Mueller-coded). The 32 OFDM carriers of the preamble header are QPSK-modulated.

L1_INFO_SIZE signals in 14 bits half the width of L1-Part 2 (L1 time interleaving block, consisting of data and stuffing bits).

L1_TI_MODE provides information about the Time Interleaving Mode of L1 Part 2 (Layer 1 Time Interleaving Block) used. These are 2 bits which convey the following:

- 00 = no time interleaving
- 01 = best fit
- 10 = 4 OFDM symbols
- 11 = 8 OFDM symbols.

The OFDM data carriers in the Layer-1 Part 2 block are 16QAM-modulated; the error protection consists of BCH and 16K-LDPC coding with a code rate of ½.

In Part 2, the following L1 parameters are then signalled:

- Network ID,
- C2 system ID,
- Start frequency,
- Guard interval,
- C2 frame length,
- No. of bundled channels,
- No. of data slices,
- No. of notches,
- Data slice parameters in the data slice loop,
- PLP parameters in PLP parameter loops,
- Notch parameters.

All data slices are described by the following parameters in the data slice loop:

- Data slice ID,
- Data slice tune position,
- Data slice offset left,
- Data slice offset right,
- Data slice time interleaver depth,
- Data slice type,
- No. of PLPs per data slice,
- PLP loop with all PLP descriptions per data slice.

All PLPs of a data slice are described in PLP loops. These loops contain the following parameters:

- PLP ID,
- PLP bundled,
- PLP type,
- PLP payload type,
- PLP start,
- PLP FEC type,
- PLP modulation.

L1 signalling header and L1 time interleaving block can be repeated several times within 7.6 MHz in the direction of frequency depending on the length of the L1 time interleaving blocks.

38.14 Pilots in DVB-C2

Like DVB-T2, DVB-C2 has

- edge pilots,
- continual pilots and
- scattered pilots.

Edge-pilots and continual pilots are used for frequency synchronization (AFC). Additionally, denser pilot structures are formed by inserting jumping scattered pilots (Fig. 38.11.) for channel estimation and channel correction. In DVB-C2, the pilot structures do not need to be as dense as in DVB-T2 due to the cable channel being physically simpler for the receiver. DVB-C2 supports a pilot structure for each of the two guard interval lengths. All DVB-C2 pilots are boosted by 7/3 compared with the data carriers.

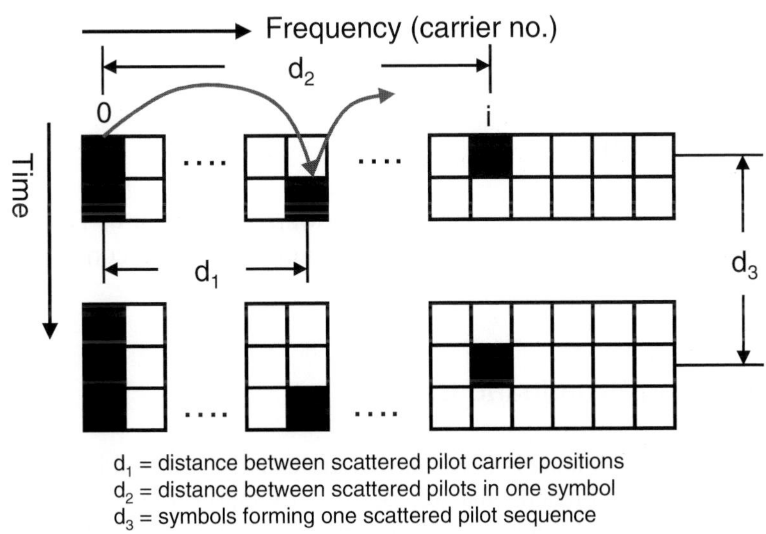

Fig. 38.11. Pilot structures in DVB-C2

Table 38.4. Scattered pilot structures in DVB-C2

Guard interval	Distance between carriers carrying pilot carriers (and distance between the pilot carriers within a COFDM symbol	Number of symbols forming one pilot sequence
1/64	12 (48)	4
1/128	24 (96)	4

Table 38.5. Continual pilot positions in DVB-C2

Continual Pilot No.
96 216 306 390 450 486 780 804
924 1026 1224 1422 1554 1620 1680 1902
1956 2016 2142 2220 2310 2424 2466 2736
3048 3126 3156 3228 3294 3366

38.15 PAPR

Like DVB-T2, DVB-C2 also provides for the reservation of carriers in order to reduce the crest factor. If needed, these carriers can then be switched on by the modulator and set in such a way that they reduce the current crest-factor of the signal. This method corresponds to the tone reservation in DVB-T2.

38.16 Block Diagram

The DVB-C2-block diagram is much more powerful than that of DVB-C but still less complex than that in DVB-T2 although some parts correspond to those of DVB-T2 (Fig. 38.12.).

38.17 Levels in Broadband Cables

The levels in DVB-C channels (Fig. 38.13.) are usually adjusted in such a way that 64QAM-modulated channels are 12 dB below the analog TV reference level (vision carrier - sync peak power) and 256QAM-modulated channels are 6 dB below that (RMS), in order to obtain sufficient distance from interfering noise in the broadband cable. DVB-C2 channels with 1024QAM-modulation will probably be 4 ... 6 dB below the

1024QAM-modulation will probably be 4 ... 6 dB below the ATV reference level and with 4096QAM, the level is then approx. 0 ... +2 dB above the analog-TV-reference level.

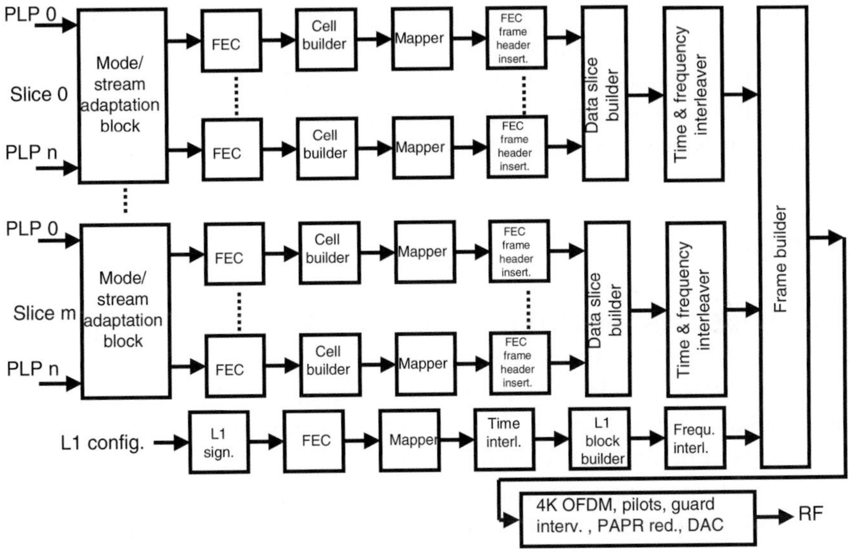

Fig. 38.12. DVB-C2 block diagram

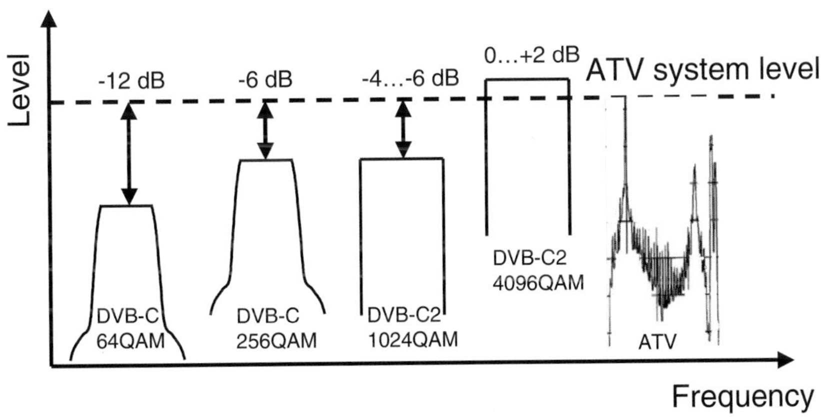

Fig. 38.13. Levels in broadband cables

38.18 Capacity

DVB-C2 can now handle up to more than 70 Mbit/s in an former 8-MHz channel which could previously handle only 51 Mbit/s. The reason for this is mainly the modern LDPC forward error correction which can now be applied, in conjunction with channel bundling, i.e. the avoidance of gaps between the channels. Using COFDM modulation, the broadband cable can now be used even more effectively, since frequency-selective problems can be eliminated more easily. It is intended to bundle up to approx. ca. 450-MHz-wide channels. This will result in about 380000 individually usable carriers in 862 MHz bandwidth.

38.19 Outlook

With DVB-C2, as with DVB-T2 and DVB-S2, a very modern, high-capacity new DVB transmission standard has been created. Its applications will lie in the field of HDTV and fast downstreams for broadband Internet via broadband cable. It is not yet known which countries or cable network operators will be the first to employ this standard. DVB-C is still fighting for customer acceptance in many broadband cable networks. Analog television (ATV) by cable is still very comfortable since no special receivers are needed. Every normal television set or flat panel screen still has a conventional analog TV tuner. On the other hand, these will not provide HDTV and many customers now have a flat panel screen which, in any case, reproduces ATV only with modest quality and displays artefacts even with SDTV material.

Bibliography: [DVB_A138]

39 DVB-x2 Measuring Techniques

In the DVB-x2-Standards, some revolutionary new approaches were implemented. It was clear from the beginning that this will also require new and enhanced DVB-x2 measuring techniques. The changes in the new Standards, which will have to be considered as having an influence on the new measuring techniques, are:

- The high degree of complexity of DVB-x2,
- a new optimized but computationally highly complex error protection,
- multiple input streams,
- generic streams,
- higher-level modulation methods,
- rotated, Q-delayed QAM,
- variable coding and modulation,
- MISO,
- TFS (Time Frequency Slicing),
- grouping of channels.

The points of high complexity and new error protection apply similarly to all new DVB-x2 Standards. This requires simply more computing power in the test instruments, i.e. hardware. of even higher power. But this can be achieved simply by "work" and has been applied in practice in DVB-S2 for some time. A huge innovation compared with the old DVB Standards is the provision for multiple input streams and also the departure from the MPEG-2-transport stream format. The details will now be discussed in the standard-specific chapters.

39.1 DVB-S2

DVB-S2 has been established firmly since 2004 and there are now also numerous satellite transponders which radiate DVB-S2-modulated HDTV programs. A chapter on the subject of DVB-S2 measuring techniques is al-

W. Fischer, *Digital Video and Audio Broadcasting Technology*, Signals and Communication Technology, 3rd ed., DOI: 10.1007/978-3-642-11612-4_39, © Springer-Verlag Berlin Heidelberg 2010

ready contained in this book under the heading of DVB-S/S2 Measuring Technology and is referred to here. Multiple input streams and data formats deviating from MPEG-2-TS are not being used up to the present day. Measuring the bit error ratios differs from DVB-S and there are now

- the pre-LDPC bit error ratio,
- the pre-BCH bit error ratio
- and the post-BCH bit error ratio or better the post-BCH packet error rate.

In comparison with DVB-S, the MER must now be measured on higher-level modulation methods but it does not require any further rethinking. There are test transmitters (e.g. Rohde&Schwarz SFU, SFE) and also instruments with integrated DVB-S2 test receivers (e.g. Rohde&Schwarz DVM400).

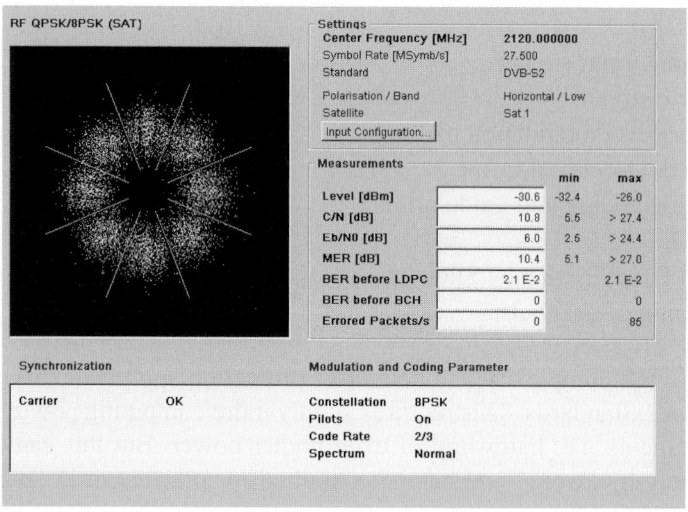

Fig. 39.1. DVB-S2 Constellation Diagram [DVM] and DVB-S2 Measurement Parameters

39.2 DVB-T2

DVB-T2 represents a new and greater challenge for measuring techniques. This is where almost all the new approaches of the Standard are transformed into reality. In the signal generation, the new input stream formats must now be supported. However, using only precompiled stream libraries makes testing of the output units too inflexible, i.e. the requirement is for tools by means of which meaningful scenarios for input stream combinations can be edited. In the case of multiple PLPs, the transmission parameters of the modulators are also controlled via the T2-MI modulator input stream. I.e., the modulator can then no longer be controlled in such a simple way directly in its modulation parameters such as mode, error protection and modulation method as in DAB. If TFS (Time Frequency Slicing) is used at some time, a test transmitter must also be capable of simulating several channels. MISO, too, requires at least one second RF path with its own fading simulator. And MISO will be used because, above all, distributed MISO costs nothing extra in the SFN but avoids the destructive fading notches or at least renders them minimisable in a large SFN. Incidentally, the planning of such MISO-SFNs also represents a new challenge for the planning tools. The transmitter sites can then operate in two different modes, i.e. these two types of coverage must be simulated. All new test receivers must now be capable of handling the new error protection which requires very much additional computing power, and thus hardware.. The BER definitions are the same as in DVB-S2. QAM orders of up to 256QAM are nothing special and have been known already since DVB-C times at the latest, i.e. since 1995.

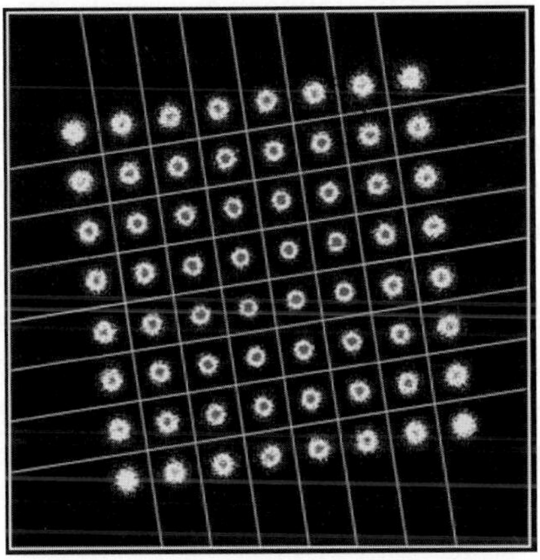

Fig. 39.2. Rotated DVB-T2 Constellation Diagram [ETL]

However, rotated and especially Q-delayed constellation diagrams will change the way in which the MER is calculated. Since the I and Q value of a cell are then no longer transported by the same cell but are transmitted on quite different pairs of carriers by multiple interleaving, new calculation methods, definitions and forms of representation must be considered, e.g. MER of I and Q separately as a function of frequency or number of carriers. If TFS is used (which is currently improbable), two receiving sections will be needed. In MISO operation, each signal path must be channel-estimated, i.e. the signal path from MISO Mode 1 and MISO Mode 2 must be evaluated. The results must also be presented correspondingly to the test technician. This may also have an influence on the MER calculation and its definition. All this will only become evident through practical experience. In the case of multiple PLPs, the test receiver is confronted by a different modulation method and a different FEC code rate in every PLP. And the transmission parameters can also change dynamically. There are already test transmitters available which support especially the Single PLP mode and will soon support other features (Rohde&Schwarz SFU).

39.3 DVB-C2

In DVB-C2, too, multiple PLPs will be used. The requirements for the test transmitter and test receiver will here be comparable to DVB-T2. The special feature of DVB-C2 is the provision for channel grouping in a total bandwidth of 450 MHz. However, the receiver bandwidth is limited to a maximum of 8 MHz but the channel tuning is no longer necessarily fixed on channels; there are narrow data slices and also notches which are masked out. The extended requirements arise here on the modulator side or on the test transmitter side, respectively. Both the modulator and a test transmitter must be capable of generating not only one 8-MHz-wide channel but two and more continuous channels with a maximum width of 450 MHz. The new requirement on the receiver side is the new error protection and especially the modulation method with up to 4096QAM. It is the bit resolutions in the receiver and also in the test receiver which must be designed for this. Another challenge is increasingly the representation of constellation diagrams of greater than 256QAM. Without having a zoom-function, artifacts in the constellation diagram will no longer be recognizable at all. The BER definitions correspond to those of DVB-S2 and DVB-T2; The MER definition will not need to be changed compared with DVB-C but is only performed on higher-level modulation methods. At the time when the third edition of this book goes to press, DVB-C2 is still too "young" for measuring techniques to be available for this technology. It is a safe bet, however, that here, too, test transmitters will be the first to provide signals.

39.4 Summary

DVB-S2 has been operating for years; the associated measuring techniques are in place. The additional demands made on the measuring techniques have previously been approximately comparable to those of DVB-S. DVB-T2 and DVB-C2, on the other hand, necessitate increased demands also on the measuring techniques. At the time of this book going to press, the range of measuring techniques required is available only in parts or not at all.

Bibliography: [SFU], [DVM]

40 CMMB – Chinese Multimedia Mobile Broadcasting

CMMB – Chinese Multimedia Mobile Broadcasting is also a mobile TV standard. CMMB is comparable to DVB-SH; it is a hybrid system and supports the terrestrial service route and the service route via satellite. Just as in DVB-SH, gap-fillers are provided which are fed via satellite. CMMB has nothing to do with DTMB, and the baseband signal is not the MPEG-2 transport stream, either, in this case. The radiated data signal can consist of up to 39 service channels and a logical control channel. These are broadcast in up to 40 time slots. The modulation method used is OFDM, the types of modulation are BPSK, QPSK and 16QAM. The error protection used is a combination of Reed-Solomon coding and LDPC coding and bit interleaving. The Standard is published only in excerpts and only under the seal of „Confidential". The channel bandwidths supported are 8 MHz and 2 MHz, the signal bandwidths are 7.512 MHz and 1.536 MHz. In the 8-MHz channel, the 4k mode is used with 3077 carriers actually being used. The 2-MHz channel uses the 1k mode with 629 carriers actually being set. The subcarrier spacing is 2.441 kHz, similar to the other OFDM-based systems, there are Continual Pilots, Scattered Pilots and data carriers. The contents broadcast are certainly MPEG-4 AVC and AAC signals. Unfortunately it is not possible to report further details at present.

Bibliography:

W. Fischer, *Digital Video and Audio Broadcasting Technology*, Signals and Communication Technology, 3rd ed., DOI: 10.1007/978-3-642-11612-4_40, © Springer-Verlag Berlin Heidelberg 2010

41 Other Transmission Standards

Apart from the standards for Digital Video and Audio Broadcasting and Mobile TV, there are by now other methods which are only touched upon to some extent in this chapter but are partly also discussed in greater detail as far as possible. The situation is frequently that it is impossible to discuss details since they have not been published, or are only known by their status. Naturally, it is not intended to violate any rights.

The standards discussed in this chapter are:

- MediaFLO
- IBOC/HD Radio
- FMextra

41.1 MediaFLO

MediaFLO stands for Media Forward Link Only and is a proprietary US standard for mobile TV developed by Qualcomm.

The technical parameters of MediaFLO are:

- COFDM
- 4k mode, QPSK or 16QAM
- Guard interval = 1/8
- Channel bandwidth = 5, 6, 7 or 8 MHz
- Concatenated Reed-Solomon(16, k = 12, 14 or 16) and turbo code forward error correction (code rate = 1/3, 2/3 1/2)
- Net data rate (6 MHz, CR_{Inner}=2/3, CR_{Outer}=12/16, 16QAM) = 8.4 Mbit/s
- Data rates of up to 11.2 Mbit/s
- Source encoding: MPEG-4 Part 10 AVC Viddeo, MPEG-4 AAC+ Audio or IP

W. Fischer, *Digital Video and Audio Broadcasting Technology*, Signals and Communication Technology, 3rd ed., DOI: 10.1007/978-3-642-11612-4_41, © Springer-Verlag Berlin Heidelberg 2010

41.2 IBOC - HD Radio

IBOC (In Band on Channel) or HD Radio (Hybrid Radio) (iBiquity Digital Coporation, US) are synonymous terms for digital radio in combination with analog FM radio COFDM-modulated bands (Fig. 41.1 and 41.2.), in which digital audio signals are transmitted, are here added above and below an FM carrier. IBOC has its origin in the US and is currently also being tested in some regions in Europe. It represents a current possibility of operating FM and digital radio adjacently to one another.

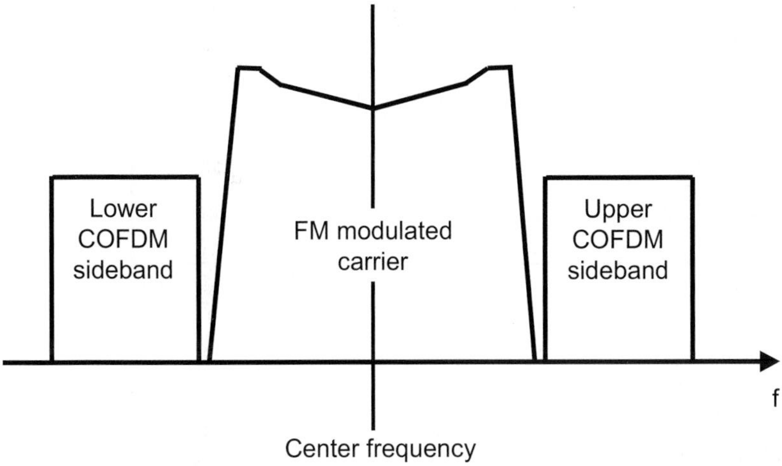

Fig. 41.1. IBOC spectrum

In VHF sound broadcasting, the channel spacing is 300 kHz, or 200 kHz (US), and the frequency deviation is typically 75 kHz. The VHF multiplex signal consists of the L+R signal with a width of 15 kHz, the pilot with 19 kHz and the L-R component in two sidebands around a suppressed AM carrier at 38 MHz, and a modulated auxiliary carrier at 57 kHz (RDS, previously ARI); the baseband bandwidth is thus 57 kHz. According to Carson's formula, the required RF bandwidth is a little over 250 kHz with a deviation of 75 kHz; the gaps towards the adjacent channels can be filled up with COFDM-modulated signals. It is also possible here to play with the deviation of the FM carrier, i.e. it may be possibly reduced in order to be able to widen the COFDM spectra. In this context, channel spacing does not necessarily imply available RF bandwidth. The baseband bandwidth actually available in VHF sound broadcasting, too, is up to 100 kHz and is being utilized in the FMextra standard described later.

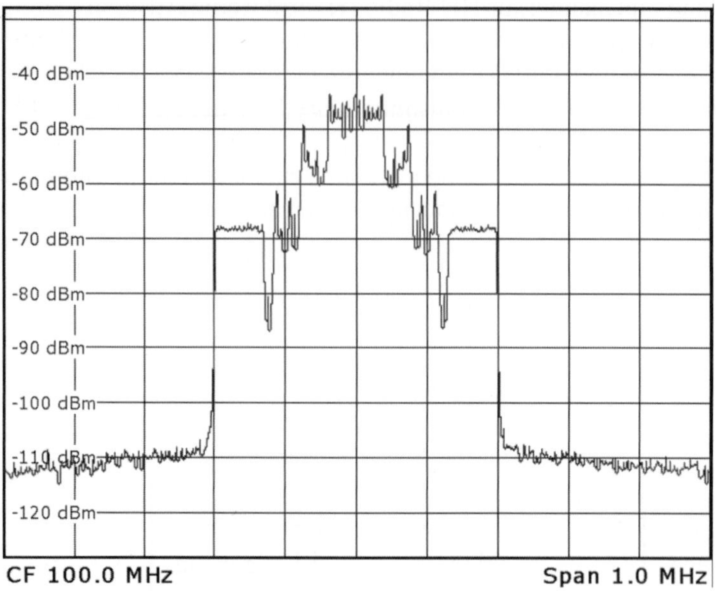

Fig. 41.2. Real HD-Radio spectrum

41.3 FMextra

The baseband bandwidth of a VHF FM channel is 100 kHz, of which about 60 kHz are currently being utilized. In the US, the range between 60 and 100 kHz is partly being used for additional information. VHF transmitters have SCA inputs which allow for a spectrum occupancy between 60 and 100 kHz. This is where FMextra comes into play and enables this range to be used in the baseband. The channel spacing in the RF band remains unchanged.

41.4 Effects of the Digital Dividend on Cable and Terrestrial TV Networks

Due to the switchover from analog terrestrial television to digital terrestrial television (DVB-T) in Europe, much fewer frequencies are needed for propagating the same number of programs since DVB-T can accommodate at least 4 SDTV programs per channel. At the last World Radiocommunication Conference (WRC07, Geneva, October/November 2007), the upper TV channels 61 – 69 (790 – 862 MHz) were therefore reserved for mobile radio applications (UMTS, LTE, WiMAX) under the heading of "Digital Dividend". Naturally, this has effects on applications operating in the same frequency band like DVB-T-broadcasts in these channels, but also broadband cable networks using the full frequency band from 5 to 862 MHz. The effects differ and have now also been verified in practice by trials. The author was actively involved in some of the first investigations. In the following sections, the influences of the use by mobile radio of channels 61 - 69 on DVB-T networks and on broadband cable networks will be discussed.

41.4.1 Anatomy of the Mobile Radio Signals

One might think at first that the anatomy of the mobile radio signals, i.e. the modulation methods used, which with the digital dividend are now creating interference signals for DVB-T and DVB-C networks, would play a role, but this is not the case. All mobile radio standards use digital modulation methods, just like digital television. And these are not much different from one another. Regardless of whether it is a single-carrier modulation method, WCDMA (Wide-Band Code Division Multiple Access) or OFDM which is used, the signals all look like band-limited noise signals with a more or less large crest factor. GSM uses a single-carrier modulation method, UMTS uses WCDMA, and LTE and WiMAX operate with OFDM methods. And the general trend in all areas is in any case towards multi-carrier methods. Anatomically, these signals can be distinguished by the following parameters:

- bandwidth,
- envelope shape (rectangular, or roll-off or Gaussian filtering),
- crest factor.

The main influence on applications in the same frequency band is produced by the bandwidth of these mobile radio signals. And only the uplink, i.e. the return channel from the mobile radio to the base station, can have an interfering effect. As far as the power budget is concerned, the downlink is too weak to have any interfering effect at the location where terrestrial and cable receivers are used. Neither does a dense cable network exert any interfering effect on uplink or downlink. DVB-T transmitters and mobile radio applications cannot coexist in the same channel; a DVB-T transmitter would displace a mobile radio downlink and a mobile radio uplink would radiate into a DVB-T antenna and make it impossible to receive DVB-T channels in the same frequency band. There are, therefore, effects which do not even need to be investigated since their interactions with applications in the same frequency band can be explained or estimated simply by physics, mathematics and experience.

The permissible transmitting power in the uplink of mobile radios is up to 24 dBm (250 mW) in this frequency band. This needs to be taken into consideration.

41.4.2 Terrestrial TV Networks and Mobile Radio

Co-channel operation of mobile radio and DVB-T is not possible. DVB-T would interfere with the mobile radio downlink since the DVB-T field strength is much higher than the field strength of the mobile radio channel at the receiving location. This applies especially to DVB-T networks which are designed for portable indoor reception. The mobile radio uplink would interfere with the DVB-T reception since the uplink channel would radiate either directly into the DVB-T receiving antenna; and the DVB-T receivers are also certainly leaky and would admit stray radiation. Neither is there any need to confirm this by investigations since the characteristics of DVB-T receivers are well known. The extent to which a mobile radio uplink is noticed in the adjacent channel greatly depends on the characteristics of the DVB-T receiver. Virtually all DVB-T receivers can handle adjacent channels which are up to 20 dB higher, but many can also handle adjacent channels which are up to 40 dB and more higher. The greater the separation from the frequency of an adjacent channel the easier it becomes for the DVB-T receiver. An adjacent mobile radio channel is virtually like an adjacent DVB-T channel for a DVB-T receiver. Analyses and investigations of the interference effects between mobile radio network and analog terrestrial television network are no longer required since there will soon be no more analog terrestrial television (ATV) networks in Europe.

41.4.3 Broadband Cable TV Networks and Mobile Radio

The situation is quite different in the case of broadband cable networks. Broadband cable networks currently handle the following four applications:

- analog television,

- digital television (now DVB-C, later DVB-C2),
- fast Internet access (Euro-DOCSIS with DVB-C signals in the downlink), and telephony.

It is not clear for how long analog television will still be supplied. Many broadband customers love especially this simple type of reception without additional equipment apart from the actual screen with integrated analog TV tuner. A broadband provider will have to think long and hard before he switches off the last analog TV channels even if it would be more economical for him to do so.

It was found in trials that up to network level 4 (= subscriber network level), a broadband cable network using modern multi-shielded cables, modern amplifiers and termination boxes is leakproof to such an extent that only slight or no influences were detectable with an external irradiation of up to approx. 24 dBm (250 mW) power. It was also found that most of the terminals were not leakproof enough with co-channel irradiation. And this applies to all types of broadband reception. In the adjacent channel, up to 20 dB more interfering radiation can be accepted in some cases which was shown in tests of modern DVB-C receivers. I.e., the problem is mainly the terminals themselves and naturally also the cables from the socket to the terminal.

41.4.3.1 Influence of Co-channel Mobile Radio on DVB-C Reception

Tests have shown that with a modern network level 4, the terminals are the only problem. Virtually all the DVB-C receivers showed symptoms of interference from about 8 to 15 dBm irradiated power from 1 to 2 meters distance and displayed slicing, freezing and picture loss. The plane of polarization (horizontal or vertical) played a great part in this and amounts to between 3 and 10 dB. Naturally, the quality of the transmitting antenna in the mobile unit also plays a great role and easily accounts for another 5 dB difference. From about 5 to 10 m distance and with properly attenuating walls, no further interference effects can be expected for the co-channel mobile radio downlink. However, the variance of immunity from interference of the terminals is very great and lies within a range of around 10 dB. The quality of the connecting cables used is always decisive and does not need to be investigated. The immunity of the plugs is also a problem. The starting point of the disturbances depends on the type of QAM (64QAM or 256QAM) and on the RF level present. The worst-case constellation is 256QAM with a level of 54 dBµV (minimum level) and the best is

64QAM and a level of 74 dBµV (maximum level). The standard should be assumed to be 256QAM with a level of 64 dBµV (usual average level).

41.4.3.2 Influence of Co-channel Mobile Radio on Analog TV Reception

Analog TV reception in the broadband cable network also showed that the only leakage is to be expected at the terminal or in the terminal wiring. Terminals show symptoms of interference more or less from about 1 to 2 m distance and depending on polarization The interference is visible as noise or moire in the picture. At more than 5 m distance or with sufficiently leakproof walls, there will be no further detectable interference but this, too, depends on the terminal.

41.4.3.3 Influence of Co-channel Mobile Radio on Other Digital Broadband Services

The influence of co-channel interferences of mobile radio leakages in the broadband cable network on the other digital broadband cable services can be seen as being particularly dramatic. Depending on polarization, transmitting powers of approx. 10 to 15 dBm here lead to shorter dropouts or lower data rates, respectively, and, in addition, to resynchronization pauses of the cable modem lasting minutes. The end user will not especially enjoy the resynchronization pauses, in particular. The major leakage in co-channel reception corresponds to that experienced in pure DVB-C reception. It is the downlink which is affected and this is also pure DVB-C with Euro-DOCSIS. A similar effect will be experienced in telephony via broadband cable.

41.4.4 Electromagnetic Field Immunity Standard for Sound and Television Broadcast Receivers

The electromagnetic field immunity of sound and television broadcast receivers and related terminals is regulated in standards EN 55020 and CISPR 20, resp. [CISPR20]. Such receiving units must not display any symptoms of interference at noise field intensities of up to 3 V/m, apart from the operating frequency itself ("tuned frequency excluded“). Naturally, this is appropriate for terrestrial receivers but this standard is also applied to broadband cable receivers.

41.4.5 Summary

The broadband cable networks are leakproof – both with respect to the power radiated and to the incident power - as long as they are of the correct and modern design. Naturally, this will not apply to all networks, especially the older ones of network level 4. Only new networks have been tested. New DVB-C receivers and cable modems have also been identified as particular leakage points. And this also applies to analog TV receivers. DVB terminals already show symptoms of interference at a radiated interference of 10 dB less than what a mobile receiver is allowed to radiate in the uplink. This should not be understood to be a wholesale opposition against the "Digital Dividend“ but simply represents the current problems and facts. The electromagnetic field immunity standard for broadcast receivers explicitly allows susceptibility to interference in the operating channel set. The co-existence of broadband cable networks or DVB-T networks with mobile radio networks in the same frequency band was never intended or planned. If the broadband cable terminals are made leakproof, this will not be a problem for broadband cable networks of the more modern type. To achieve this, however, a corresponding electromagnetic field immunity in the operating channel should be backed up with a corresponding electromagnetic field immunity standard. If only adjacent channels are occupied in DVB-T and mobile radio networks, this will not be a problem, either.

Bibliography: [SFU], [LESNIK], [CISPR20]

42 Digital Television throughout the World - an Overview

The numerous technical details of the various digital television standards have now been discussed. The only thing that is still missing is a report about the current development and spread of these technologies, and a look at the future. Digital satellite television (DVB-S) is available in Europe over numerous transponders of the ASTRA and Eutelsat satellites. Many streams can be received unencrypted. Complete receiving systems for DVB-S are available at low cost in many department stores. DVB-C, too, has become well established in the meantime. Digital terrestrial television has also become widely used in numerous countries and above all in Great Britain, where DVB-T started in 1998. DVB-T first spread in Scandinavia where Sweden is covered completely by DVB-T. Australia, too, was one of the first countries to have introduced DVB-T. In Australia, DVB-T is available mainly in the population centers along the Eastern and Southern coast. DVB-T is also being built up in South Africa and in India. In Europe, the current status is as follows: Autumn 2002 saw the start of DVB-T in Berlin and in August 2003, 7 data streams with more than 20 programs were on the air and analog television was being operated in parallel for only a brief period in simulcast mode and then switched off completely in August 2003, which certainly represented a minor revolution! DVB-T was designed to implement portable indoor reception. Reception is possible using simple indoor antennas from the heart of Berlin out to the outer suburbs in some cases. Naturally, there are restrictions in indoor reception due to the attenuation of buildings and other shadowing. In the years of 2003, 2004 and 2005, this type of reception known as "Anywhere Television" then also spread to the North-Rhine-Westphalia, Hamburg, Bremen, Hanover and Frankfurt regions and since May 30th 2005 also to the Munich and Nuremberg conurbation areas in Germany. The data rates per DVB-T channel are 13.27 Mbit/s, providing space for about 4 programs per channel. In most cases there are about 4 - 6 frequencies in the air at any time. Mecklenburg followed in autumn 2005, and Stuttgart in 2006. The networks implemented in Germany are all designed as small isolated SFN regions with few transmitters. In the meantime, the upgrad-

W. Fischer, *Digital Video and Audio Broadcasting Technology*, Signals and Communication Technology, 3rd ed., DOI: 10.1007/978-3-642-11612-4_42, © Springer-Verlag Berlin Heidelberg 2010

ing to DVB-T has been completed in Germany since November 2008, apart from a few additional low-power transmitters. Analog terrestrial television no longer exists in Germany.

In Italy, DVB-T was expanded greatly in 2004. MHP is not only "in the air" but is also accepted there. Switzerland followed in 2004/2005 and in Austria, DVB-T started in 2006. In Greenland, for example, DVB-T is a very inexpensive alternative for offering TV to the population in the small towns, each of which is an isolated self-contained community. Satellite reception is very difficult in Greenland because of its location and requires very large antennas, making it a logical choice to rebroadcast the received channels inexpensively terrestrially by means of DVB-T, using transmitting powers of around 100 W. DVB-T is also transmitted in Belgium and Holland and is being expanded greatly, especially in Holland.

In the US, Canada, Mexico and South Korea, ATSC is used and analog terrestrial television was switched off in June 2009 in the US. ATSC will probably remain restricted to these countries.

Japan has its own ISDB-T standard which has also been adopted by Brazil.

On the other hand, it appears that the initial attempts to introduce mobile TV by means of DVB-H and T-DMB in Europe seem to have failed in most countries.

Europe is currently also in the initial stages of introducing HDTV (High Definition Television). This is based on new technologies such as MPEG-4 AVC / H.264, and the new satellite standard DVB-S2. The German public-law broadcasters will commence regular DVB operations in February 2010.

It is only digital audio broadcasting which is still having problems, the main reason being the very good quality and acceptance of FM audio broadcasting. Australia is currently engaged in starting up DAB+ and some transmitter sites have already been set up. In the UK, DAB has been operating successfully for some years.

There are other TV or mobile TV standards and extensions to standards which were or are being developed since the first (English and German) editions and information about these has been included in this book as far as possible. But somewhere there must be a cut-off point. It already includes reports about the new terrestrial digital DVB-T2 TV standard and about the new DVB-C2 cable TV standard. At some stage, DVB-T2 and DVB-C2 will replace the DVB-T and DVB-C standards developed in the mid-nineties. The same applies to DVB-S2. The new DVB-x2 standards provide a distinctly higher net data ate.

Following suggestions by many participants of seminars on digital television all over the world, this book was created to provide the man in the field, be he a transmitter network planner, a service technician at a transmitter site, a technician responsible for MPEG-2 encoders and multiplexers in a studio or in a playout center, an engineer working in a development laboratory or even a student, with an insight into the technology and measuring techniques of digital television. It concentrates deliberately on the practical things of importance and attempts to include as little mathematical "ballast" as possible.

The author of this book was able to participate personally in the introduction of digital television both in his work in the TV test instrumentation development department of Rohde & Schwarz and especially also later in his many seminar trips including eleven journeys to Australia alone. Direct participation in the installation of the DVB-T network of Southern Bavaria with the Olympic Tower transmitter in Munich and the Mt. Wendelstein transmitter right from the start up to switch-on at 1.00 am on May 30th, 2005 at the Mt. Wendelstein transmitter also left a deep impression. Many subsequent field measurements and experiences from trips ranging from Australia to Greenland have thus found their way into this book.

The author's assistance with smaller or greater technical problems or problems involving test technology has always been readily welcomed and the resultant insights can also be found in this book. The continuing very close contact with the development departments of TV test instrumentation and transmitter technology of Rohde & Schwarz, is still of especially great value to me. Just copying a TV standard would never result in a useful book.

Greetings and many heartfelt thanks to the many participants in these courses throughout the world, for the discussions and suggestions in the seminars and for the lively interest shown in this work. Lots of feedback has shown that this book is really being used in practice. Is is now considered to be one of the standard textbooks for television and radio engineering which is by far the most rewarding compensation for the many hours of work spent on producing it.

Bibliography

[ALAMOUTI] Alamouti, S.: A Simple Transmit Diversity Technique for Wireless Communications, IEEE Journal, October 1998
[ARIB] Association of Radio Business and Industries, Arib Std.-B10 Version 3.2, Service Information for Digital Broadcasting System, 2001
[ATSC-MH] ATSC-M/H Standard, Part 1, A/153, 2009
[A53] ATSC Doc. A53, ATSC Digital Television Standard, September 1995
[A65] ATSC Doc. A65, Program and System Information Protocol for Terrestrial Broadcast and Cable, December 1997
[A110B] ATSC Synchronization Standard for Distributed Transmission, A/110B, Revision B, 2007
[A133] Implementation Guidelines for a Second Generation Digital Terrestrial Television Broadcasting System (DVB-T2), DVB Document A133, February 2009
[A138] Digital Video Broadcasting (DVB), Frame Structure Channel Coding and Modulation for a Second Generation Digital Transmission System for Cable Systems (DVB-C2), April 2009
[A153] Candidate Standard: ATSC-MH, Part 1 – 5, Mobile Handheld Television System, A/153, 2009
[BEST] Best, R.: Handbuch der analogen und digitalen Filterungstechnik. AT Verlag, Aarau, 1982
[BOSSERT] Bossert, M.: Kanalcodierung. Teubner, Stuttgart, 1998
[BRIGHAM] Brigham, E. O.: FFT, Schnelle Fouriertransformation. Oldenbourg, München 1987
[BRINKLEY] Brinkley, J.: Defining Vision - The Battle for the Future of Television. Harcourt Brace, New York, 1997
[BRONSTEIN] Bronstein, Semendjajew: Teubner-Taschenbuch der Mathematik, Stuttgart, Leipzig, 1996
[BUROW] Burow, R., Mühlbauer, O., Progrzeba, P.: Feld- und Labormessungen zum Mobilempfang von DVB-T. Fernseh- und Kinotechnik 54, Jahrgang Nr. 3/2000
[CHANG] Robert W. Chang, Orthogonal Frequency Data Transmission System, United States Patent Office, 1970

W. Fischer, *Digital Video and Audio Broadcasting Technology*, Signals and Communication Technology, 3rd ed., DOI: 10.1007/978-3-642-11612-4,

[CISPR20] International Electrotechnical Commission, Sound and Television Broadcast Receivers and Associated Equipment – Immunity Characterstics, 2006

[COOLEY] Cooley, J. W., and Tukey J. W.: An Algorithm for Machine Calculation of Complex Fourier Series, Math. Computation, Vol. 19, pp. 297-301, April 1965

[DAMBACHER] Dambacher, P.: Digitale Technik für Hörfunk und Fernsehen. R. v. Decker, 1995

[DAVIDSON] Davidson, G., Fielder, L., Antill, M.: High-Quality Audio Transform Coding at 128 kBit/s, IEEE, 1990

[DIBCOM] Kabelnetzhandbuch, Dibcom, 4. Auflage, Mai 2007

[DREAM] DREAM; Version 1.5 cvs, Open-Source Software Implementation of a DRM Receiver, Darmstadt University of Technology, 2005

[DTMB] see [GB20600]

[DVB_A122r1] Digital Video Broadcasting (DVB), Frame Structure Channel Coding and Modulation for a Second Generation Digital Terrestrial Television Broadcasting System (DVB-T2), DVB, June 2008

[DVB_A138] Digital Video Broadcasting (DVB); Frame Structure Channel Coding and Modulation for a second Generation Digital Transmission System for Cable Systems, 2009

[DVB-T2] Digital Video Broadcasting (DVB); Frame Structure Channel Coding and Modulation for a Second Generation Digital Terrestrial Television Broadcasting System (DVB-T2), January 2008

[DVB-C2] Digital Video Broadcasting (DVB); Frame Structure Channel Coding and Modulation for a Second Generation Digital Transmission System for Cable Systems (DVB-C2), DVB Document A138, April 2009

[DVB-T2-MI] ETSI TS 102773, Digital Video Broadcasting (DVB); Modulator Interface (T2-MI) for a Second Generation Digital Terrestrial Television Broadcasting System (DVB-T2), 2009

[DVM] MPEG-2 Analyzer DVM, Rohde&Schwarz, Munich, 2005

[DVMD] Digital Measurement Decoder DVMD, Rohde&Schwarz, Munich, 2001

[DVSG] Digital Video Signal Generator, Rohde&Schwarz, Munich, 2009

[DVQ] Digital Picture Quality Analyzer DVQ, Rohde&Scharz, Munich, 2001

[DVG] Digital Video Generator DVG, Rohde&Schwarz, Munich, 2001

[EFA] TV Test Receiver EFA, Rohde&Schwarz, Munich, 2001

[EIA608A] EIA-608-A, Recommended Practice for Line 21 Data Service, December 1999

[EIA708B] EIA-708-B, Digital Television Closed Captioning, December 1999

[ERVER] Erver, M.: Mess-Sequenzen für TV-Displays, Diplomarbeit TU-Dortmund, 2008

[EN302583] Framing Structure, Channel Coding and Modulation for Satellite Services to Handheld Devices below 3 GHz, ETSI 2007

[ETL] TV Analyzer ETL, Rohde&Schwarz, Munich, 2009

[ETS101812] Digital Video Broadcasting (DVB); Multimedia Home Platform (MHP) Specification 1.1.1., ETSI 2003

[ETS101980] Digital Radio Mondiale (DRM); System Specification, ETSI, 2001

[ETS102006] ETSI TS 102 006, Digital Video Broadcasting (DVB); Specifications for System Software Update in DVB Systems, ETSI, 2004

[ETS102034] ETS 102034, Digital Video Broadcasting (DVB); Transport of MPEG-2 TS based DVB Services over IP based Networks, 2007

[ETSI201488] Data over Cable Service Information Specification, ETSI, 2000

[ETR290][ETS100290][TR100290] ETSI Technical Report ETR 290, Digital Video Broadcasting (DVB), Measurement Guidelines for DVB systems, 2000

[ETS300401] Radio Broadcasting Systems ; Digital Audio Broadcasting (DAB) to mobile, portable and fixed receivers, ETSI May 2001

[ETS300472] ETS 300472 Digital broadcasting systems for television, sound and data services; Specification for conveying ITU-R System B Teletext in Digital Video Broadcasting (DVB) Bitstreams, ETSI, 1995

[ETS300421] ETS 300421, Digital broadcasting systems for television, sound and data services; Framing structure, channel coding and modulation for 11/12 GHz satellite systems, ETSI, 1994

[ETS300429] ETS 300429, Digital Video Broadcasting; Framing structure, channel coding and modulation for cable systems, ETSI, 1998

[ETS300468] ETS 300468 Specification for Service Information (SI) in Digital Video Broadcasting (DVB) Systems, ETSI, March 1997

[ETS300744] ETS 300744, Digital Video Broadcasting; Framing structure, channel coding and modulation for digital terrestrial television (DVB-T), ETSI, 1997

[ETSI300800] ETSI 300800, Digital Video Broadcasting (DVB), Interaction Channel for Cable TV Distribution Systems, ETIS 1998

[ETS300797] Digital Audio Broadcasting (DAB); Distribution Interfaces ; Service Transport Interface (STI), ETSI 1999

[ETS300799] ETS300799, Digital Audio Broadcasting (DAB); Distribution Interfaces ; Ensemble Transport Interface (ETI), ETSI, Sept. 1997

[ETS301192] [EN301192]ETSI EN 301 192, Digital Video Broadcasting (DVB); DVB specification for data broadcasting

[ETS301234] ETS301234, Digital Audio Broadcasting (DAB) ; Multimedia Object Transfer (MOT) protocol, ETSI 1999

[ETS301775] Digital Video Broadcasting (DVB); Specification of Vertical Blanking Information (VBI) Data in DVB Bit Streams

[EN301210] ETSI EN 301 210 ; Digital Video Broadcasting (DVB), Framing Structure, Channel Coding and Modulation for Digital Satellite News Gathering (DSNG) and other Contribution Applications by Satellite, ETSI, 1999

[ETS302304] Digital Video Broadcasting (DVB); Transmission Systems for Handheld Terminals (DVB-H), ETSI 2004

[ETS302307] ETSI EN 301 307; Digital Video Broadcasting (DVB), Second Generation of Framing Structure, Channel Coding and Modulation for Broadcasting, Interactive Services, News Gathering and other Broadcast Satellite Operations, ETSI, 2006

[EN302583] Digital Video Broadcasting (DVB); Framing Structure, Channel Coding and Modulation for Satellite Services to Handheld Divices (SH) below 3 GHz, 2008

[EN302755] EN 302755, Digital Video Broadcasting (DVB), Frame Structure, Channel Coding and Modulation for a second Generation Digital Terrestrial Video Broadcasting System, ETSI, 2009

[FASTL] Fastl, H., Zwicker, E.: Psychoacoustics, Springer, Berlin, Heidelberg, 2006

[FISCHER1] Fischer, W.: Die Fast Fourier Transformation - für die Videomesstechnik wiederentdeckt. Vortrag und Aufsatz Fernseh- und Kinotechnische Gesellschaft, FKTG, Mai 1988

[FISCHER2] Fischer, W.: Digital Terrestrial Television: DVB-T in Theory and Practice, Seminar Documentation, Rohde&Schwarz, Munich, 2001

[FISCHER3] Fischer, W.: DVB Measurement Guidelines in Theory and Practice, Seminar Documentation, Rohde&Schwarz, Munich, 2001

[FISCHER4] Fischer, W.: MPEG-2 Transport Stream Syntax and Elementary Stream Encoding, Seminar Documentation, Rohde&Schwarz, Munich, 2001

[FISCHER5] Fischer, W.: Picture Quality Analysis on Digital Video Signals, Seminar Documentation, Rohde&Schwarz, Munich, 2001

[FISCHER6] Fischer, W.: Digital Television, A Practical Guide for Engineers, Springer, Berlin, Heidelberg, 2004

[FISCHER7] Fischer, W.: Einführung in DAB, Seminardokumentation, Rohde&Schwarz München, 2004

[FITZEK] Fitzek, F., Katz, M., Cooperation in Wireless Networks: Principles and Applications, Springer, 2006

[FKTG_GUNKEL] Gunkel, G.: Presentation DVB-T2, Fernseh- und Kinotechnische Gesellschaft, Jahrestagung Mai 2008, München

[GB20600] GB20600-2006, Framing Structure, Channel Coding and Modulation for Digital Television Terrestrial Broadcasting System, China, 2006

[GIROD] Girod, B., Rabenstein, R., Stenger, A.: Einführung in die Systemtheorie. Teubner, Stuttgart, 1997

[GRUNWALD] Grunwald, S.: DVB, Seminar Documentation, Rohde&Schwarz, Munich, 2001

[HARRIS] Harris, Fredrik J.: On the Use of Windows for Harmonic Analysis with the Discrete Fourier Transform, Proceedings of the IEEE, Vol. 66, January 1978

[HOEG_LAUTERBACH] Hoeg W., Lauterbach T.: Digital Audio Broadcasting, Principles and Applications of Digital Radio, Wiley, Chichester, UK, 2003

[HOFMEISTER] Hofmeister, M.: Messung der Übertragungsparameter bei DVB-T, Diplomarbeit Fachhochschule München, 1999

[H.264] ITU-T H.264 Advanced video coding for generic audiovisual services, International Telecommunication Union, 2005

[IRT2008_KUNERT] Kunert, C.: Presentation DVB-T2, April 2008, IRT, Institut für Rundfunktechnik, München

[ISO/IEC13522.5] Information technology – Coding of multimedia and hypermedia information – part 5: Support of base-level interactive applications. ISO/IEC 13522-5, 1997

[ISO13818-1] ISO/IEC 13818-1 Generic Coding of Moving Pictures and Associated Audio: Systems, ISO/IEC, November 1994

[ISO13818-2] ISO/IEC 13818-2 MPEG-2 Video Coding

[ISO13181-3] ISO/IEC 13818-3 MPEG-2 Audio Coding

[ISO/IEC13818-6] Digital Storage Media Command and Control (DSM-CC), ISO/IEC 13818-6, 1996

[ISO14496-10] ISO/IEC 14496-10 AVC

[ITU205] ITU-R 205/11, Channel Coding, Frame Structure and Modulation Scheme for Terrestrial Integrated Services Digital Broadcasting (ISDB-T), ITU, March 1999

[ITU500] ITU-R BT.500 Methology for Subjective Assessment of Quality of Television Signals

[ITU601] ITU-R BT.601

[ITU709] ITU-R BT.709-5, Parameter values for the HDTV standards for production and interactive programme exchange, ITU, 2002

[ITU1120] ITU-R BT.1120-3, Digital interfaces for HDTV studio signals, ITU, 2000

[ITU-T G.993] ITU-T G.993 Very High Speed Digital Subsrciber Line

[ITUJ83] ITU-T J83: Transmission of Television, Sound Programme and other Multimedia Signals; Digital Transmission of Television Signals, April 1997

[JACK] Jack, K.: Video Demystified, A Handbook for the Digital Engineer, Elsevier, Oxford, UK, 2005

[JAEGER] Jäger, D.: Übertragung von hochratigen Datensignalen in Breitbandkommunikationsnetzen, Dissertation. Selbstverlag, Braunschweig, 1998

[KAMMEYER] Kammeyer, K.D.: Nachrichtenübertragung. Teubner, Stuttgart, 1996

[KATHREIN1] Kathrein, Broadcast Antennas for FM, TV, MMDS, DAB and DVB, Cathalogue, Rosenheim, 2005

[KELLER] Keller, A.: Datenübertragung im Kabelnetz, Springer, Berlin, Heidelberg, 2005

[KORIES] Kories, R.: Schmidt-Walter, H.: Taschenbuch der Elektrotechnik. Verlag Harri Deutsch, Frankfurt, 2000

[KUEPF] Küpfmueller, K.: Einführung in die theoretische Elektrotechnik. Springer, Berlin, 1973

[LASS] Lassalle, R., Alard, M.: Principles of Modulation and Channel Coding for Digital Broadcasting for Mobile Receivers. EBU Review no. 224, August 1987

[LEFLOCH] Le Floch, B.: Halbert-Lassalle, R., Castelain, D.: Digital Sound Broadcasting to Mobile Receivers. IEEE Transactions on Consumer Electronics, Vol. 35, No. 3, August 1989

[LESNIK] Lesnik, F.: Messtechnische Untersuchung über Einfluss der Belegung in den UHF-Kanälen K61 – 69 mit UMTS/3G FDD oder WIMAX-Signalen und Auswirkungen auf die Übertragung in Kabel-TV-Netzen, Februar 2009

[LOCHMANN] Lochmann, D.: Digitale Nachrichtentechnik, Signale, Codierung, Übertragungssysteme, Netze. Verlag Technik, Berlin, 1997

[LStelcom] LStelcom, CHIRplus BC, Lichtenau, 2009

[LVGB] Landesamt für Vermessung und Geoinformation Bayern, TOP50 and DTK500 Maps, 2006

[MAEUSL1] Mäusl, R.: Digitale Modulationsverfahren. Hüthig, Heidelberg, 1985

[MAEUSL2] Mäusl, R.: Modulationsverfahren in der Nachrichtentechnik. Huethig, Heidelberg, 1981

[MAEUSL3] Mäusl, R.: Refresher Topics - Television Technology, Rohde&Schwarz, Munich, 2000

[MAEUSL4] Mäusl, R.: Von analogen Videoquellensignal zum digitalen DVB-Sendesignal, Seminar Dokumentation, München, 2001

[MAEUSL5] Mäusl, R.: Fernsehtechnik, Von der Kamera bis zum Bildschirm. Pflaum Verlag, München, 1981
[MAEUSL6] Mäusl R.: Fernsehtechnik, Vom Studiosignal zum DVB-Sendesignal, Hüthig, Heidelberg, 2003
[MPEG-4] see [ISO/IEC14496-10]
[NBTV] Narrow-Bandwidth Television Association
[NELSON] Nelson, M.: Datenkomprimierung, Effiziente Algorithmen in C. Heise, Hannover, 1993
[NEUMANN] Neumann, J.: Lärmmesspraxis. Expert Verlag, Grafenau, 1980
[NX7000] Rohde&Schwarz, High Power Transmitters, Munich, 2007
[NX8000] Rohde&Schwarz, High Power Transmitters, Munich, 2009
[PRESS] Press, H. W.: Teukolsky, S. A., Vetterling, W. T., Flannery, B. P.: Numerical Recipes in C. Cambridge University Press, Cambridge, 1992
[RAUSCHER] Rauscher, C.: Grundlagen der Spektrumanalyse. Rohde&Schwarz, München, 2000
[REIMERS] Reimers, U.: Digitale Fernsehtechnik, Datenkompression und Übertragung für DVB. Springer, Berlin, 1997
[REIMERS1] Reimers U.: DVB, The Family of International Standards for Digital Video Broadcasting, Springer, Berlin, Heidelberg, New York, 2004
[REIMERS2] Reimers U.: DVB-Digitale Fernsehtechnik, Springer, Berlin, Heidelberg, 3. Auflage, 2007
[R&S_APPL_1MA91] Test of DVB-H Capable Mobile Phones in Development and Production, Application Note, Rohde&Schwarz, April 2005
[RFS] RFS, Documentation, Heliflex Cable, www.rfsworld.com, 2007
[RICHARDS] Richards, J.: Radio Wave Propagation, Springer, Berlin, Heidelberg, 2008
[SIGMUND] Sigmund, G.: ATM - Die Technik. Hüthig, Heidelberg, 1997
[STEINBUCH] Steinbuch K., Rupprecht, W.: Nachrichtentechnik. Springer, Berlin, 1982
[SFN1] System Monitoring and Measurement for DVB-T Single Frequency Networks with DVMD, DVRM and Stream Explorer®, Rohde&Schwarz, April 2000
[SFQ] TV Test Transmitter SFQ, Rohde&Schwarz, Munich, 2001
[SFU] TV Test Transmitter SFU, Rohde&Schwarz, Munich, 2009
[TAYLOR] Taylor, J.: DVD Demystified, McGraw-Hill, 2001
[T-DMB] Levi, S.: DMB-S/DMB-T Receiver Solutions on TI DM342, May 2004
[TEICHNER] Teichner, D.: Digitale Videocodierung. Seminar, 1994
[THIELE] Thiele, A.N.: Digital Audio for Digital Video, Journal of Electrical and Electronics Engineering, Australia, September 1993
[TM2939] ETSI TM 2939, DVB-H Overview, 2003

[TODD] Todd, C. C.: AC-3 The Multi-Channel Digital Audio Coding Technology. NCTA Technical Papers, 1994

[TOZER] Tozer E.P.J.: Broadcast Engineer's Reference Book, Elsevier, Oxford, UK, 2004

[TR101190] TR101190, Implementation Guidelines for DVB Terrestrial Services, ETSI, 1997

[TR101496] Digital Audio Broadcasting (DAB) ; Guidelines and rules for implementation and operation, ETSI 2000

[TR102376] TR102376, DVB, User Guidelines for the second Generation System for Broadcasting, Interactive Services , News Gathering and other Broadband Satellite Applications (DVB-S2)

[TS102006] DVB; Specification for System Software Update in DVB Systems, 2004

[TS102606] TS102606, Digital Video Broadcasting (DVB), Generic Stream Encapsulation Protocol, ETSI, 2007

[TS102773] TS102773, DVB, Modulator Interface for a second Generation Digital Terrestrial Video Broadcasting System (DVB-T2), ETSI, 2009

[TS102771] TS102771, Digital Video Broadcasting (DVB), Generic Stream Encapsulation (GSE), Implementation Guidelines, ETSI, 2009

[VIERACKER] Vieracker, T.: Analyse von DVB-T-Empfänger-Synchronisationsproblemen in DVB-T-Gleichwellennetzen, Diplomarbeit Berufsakademie Ravensburg, Sept. 2007

[VSA] Video Analyzer VSA, Rohde&Schwarz, Documentation, 1995

[WATKINSON] Watkinson, J.: The MPEG Handbook, Elsevier, Oxford, UK, 2004

[WEINSTEIN] Weinstein, S. B., Ebert, P. M.: Data Transmission by Frequency-Division Multiplexing Using the Discrete Fourier Transform. IEEE Transactions and Communication Technology, Vol. Com. 19, No. 5, October 1971

[WOOTTON] Wootton C.: A Practical Guide to Video and Audio Compression, From Sprockets and Rasters to Macro Blocks, Elsevier, Oxford, UK, 2005

[ZEIDLER] Zeidler, E., Bronstein, I.N., Semendjajew K. A.: Teubner-Taschenbuch der Mathematik, Teubner, Stuttgart, 1996

[ZIEMER] Ziemer, A.: Digitales Fernsehen, Eine neue Dimension der Medienvielfalt. R. v. Decker, Heidelberg, 1994

[ZWICKER] Zwicker, E.: Psychoakustik. Springer, Berlin 1982

[7EB01_APP] Application Note, ATSC Mobile DTV, Rohde&Schwarz, 2009

Definition of Terms

AAL0	ATM Adaptation Layer 0
AAL1	ATM Adaptation Layer 1
AAL5	ATM Adaptation Layer 5
ASI	Asynchronous Serial Interface
ATM	Asynchronous Transfer Mode
ATSC	Advanced Television Systems Committee
BCH	Bose-Chaudhuri-Hocquenghem Code
BAT	Bouquet Association Table
CA	Conditional Access
CAT	Conditional Access Table
CI	Common Interface
COFDM	Coded Orthogonal Frequency Division Multiplex
CRC	Cyclic Redundancy Check
CVCT	Cable Virtual Channel Table
DAB	Digital Audio Broadcasting
DDB	Data Download Block
DII	Download Information Identification
DMB-T	Digital Multimedia Broadcasting Terrestrial
DRM	Digital Radio Mondiale
DSI	Download Server Initializing
DSM-CC	Digital Storage Media Command and Control
DTS	Decoding Time Stamp
DVB	Digital Video Broadcasting
ECM	Entitlement Control Messages
EIT	Event Information Table
EMM	Entitlement Management Messages
ES	Elementary Stream
ETT	Extended Text Table
FEC	Forward Error Correction
IRD	Integrated Receiver Decoder
ISDB-T	Integrated Services Digital Broadcasting Terrestrial
J83	ITU-T J83
LDPC	Low Density Parity Check Code
LVDS	Low Voltage Differential Signalling

MGT	Master Guide Table
MHEG	Multimedia and Hypermedia Information Coding Experts Group
MHP	Multimedia Home Platform
MIP	Megaframe Initialization Packet
MP@ML	Main Profile at Main Level
MPEG	Moving Picture Experts Group
NIT	Network Information Table
OFDM	Orthogonal Frequency Division Multiplex
PAT	Program Association Table
PCR	Program Clock Reference
PCMCIA	PCMCIA
PDH	Plesiochronous Digital Hierarchy
PES	Packetized Elementary Stream
PID	Packet Identity
PMT	Program Map Table
Profile	MP@ML
PS	Program Stream
PSI	Program-Specific Information
PSIP	Program and System Information Protocol
PTS	Presentation Time Stamp
QAM	Quadrature Amplitude Modulation
QPSK	Quadrature Phase Shift Keying
RRT	Rating Region Table
SDH	Synchronous Digital Hierarchy
SDT	Service Description Table
SI	Service Information
SONET	Synchronous Optical Network
SSU	System Software Update
ST	Stuffing Table
STD	System Target Decoder
STT	System Time Table
T-DMB	Terrestrial Digital Multimedia Broadcasting
TDT	Time and Date Table
TOT	Time Offset Table
TS	Transport Stream
TVCT	Terrestrial Virtual Channel Table
VSB	Vestigial Sideband Modulation

Adaptation Field
The adaptation field is an extension of the TS header and contains ancillary data for a program. The program clock reference (PCR) is of special importance. The adaptation field must never be scrambled when it is to be transmitted (see Conditional Access).

Advanced Television Systems Committee (ATSC)
The North American Standards Committee which determined the standard of the same name for the digital transmission of TV signals. Like DVB, ATSC is also based on MPEG-2 systems as far as transport stream multiplexing is concerned and on MPEG-2 video for video compression. However, instead of MPEG-2, standard AC-3 is used for audio compression. ATSC specifies terrestrial transmission and transmission via cable while transmission via satellite is not taken into account.

Asynchronous Serial Interface (ASI)
The ASI is an interface for the transport stream. Each byte of the transport stream is expanded to 10 bits (energy dispersal) and is transmitted with a fixed bit clock of 270 MHz (asynchronous) irrespective of the data rate of the transport stream. The fixed data rate is obtained by adding dummy data without information content. Useful data is integrated into the serial data stream either as individual bytes or as whole TS packets. This is necessary to avoid PCR jitter. A variable buffer memory at the transmitter end is therefore not permissible.

Asynchronous Transfer Mode (ATM)
Connection-oriented wideband transmission method with fixed-length 53-byte cells. Both payload and signalling information is transmitted.

ATM Adaptation Layer 0 (AAL0)
The ATM AAL0 Layer is a transparent ATM interface. The ATM cells are forwarded here directly without having been processed by the ATM Adaptation Layer.

ATM Adaptation Layer 1 (AAL1)
The ATM Adaption Layer AAL1 is used for MPEG-2 with and without FEC. The payload is 47 bytes, the remaining 8 bytes are used for the header with the forward error correction and the sequence number. This makes it possible to check the order of incoming data units and the transmission. The FEC allows transmission errors to be corrected.

ATM Adaptation Layer 5 (AAL5)
The ATM Adaption Layer AAL5 is basically used for MPEG-2 without FEC. The payload is 48 bytes, the remaining 7 bytes are used for the header. Data with transmission errors cannot be corrected on reception.

Bose-Chaudhuri-Hocquenghem Code (BCH)
Cyclic block code used in the FEC of the DVB-S2 satellite transmission standard.

Bouquet Association Table (BAT)

The BAT is an SI table (DVB). It contains information about the different programs (bouquet) of a broadcaster. It is transmitted in TS packets with PID 0x11 and indicated by table_ID 0x4A.

Cable Virtual Channel Table (CVCT)

CVCT is a PSIP table (ATSC) which comprises the characteristic data (eg channel number, frequency, modulation type) of a program (= virtual channel) in the cable (terrestrial transmission → TVCT). TVCT is transmitted with the PID 0x1FFB in TS packets and indicated by the table_id 0xC9.

Channel Coding

The channel coding is performed prior to the modulation and transmission of a transport stream. The channel coding is mainly used for forward error correction (FEC), allowing bit errors occurring during transmission to be corrected in the receiver.

Coded Orthogonal Frequency Division Multiplex (CODFM)

COFDM is basically OFDM with error protection (coding - C), which always precedes OFDM.

Common Interface (CI)

The CI is an interface at the receiver end for a broadcaster-specific, exchangeable CA plug-in card. This interface allows scrambled programs from different broadcasters to be de-scrambled with the same hardware despite differences in CA systems.

Conditional Access (CA)

The CA is a system allowing programs to be scrambled and for providing access to these programs at the receiver end only to authorized users. Broadcasters can thus charge fees for programs or individual broadcasts. Scrambling can be performed at one of the two levels provided by an MPEG-2 multiplex stream, e.g. the transport stream or the packetized elementary stream level. The relevant headers remain unscrambled. The PSI and SI tables also remain unscrambled except for the EIT.

Conditional Access Table (CAT)

The CAT is a PSI table (MPEG-2) and comprises information required for descrambling programs. It is transmitted in TS packets with PID 0x0002 and indicated by table_ID 0x01.

Continuity Counter

A continuity counter for each elementary stream (ES) is provided as a four-bit counter in the fourth and last byte of each TS header. It counts the TS packets of a PES, determines the correct order and checks whether the packets of a PES are complete. The counter (fifteen is followed by zero) is incremented with each new packet of the PES. Exceptions are permissible under certain circumstances.

Cyclic Redundancy Check (CRC)

The CRC serves to verify whether data transmission was error-free. To this effect, a bit pattern is calculated in the transmitter based on the data to be monitored. This bit pattern is added to the data in such a way that an equivalent computation in the receiver yields a fixed bit pattern in case of error-free transmission after processing of the data. Every transport stream contains a CRC for the PSI tables (PAT, PMT, CAT, NIT) as well as for some SI tables (EIT, BAT, SDT, TOT).

Decoding Time Stamp (DTS)

The DTS is a 33-bit value in the PES header and represents the decoding time of the associated PES packet. The value refers to the 33 most significant bits of the associated program clock reference. A DTS is only available if it differs from the presentation time stamp (PTS). For video streams this is the case if delta frames are transmitted and if the order of decoding does not correspond to that of output.

Digital Audio Broadcasting

A standard for digital radio in VHF band III, and the L band, defined as part of the EUREKA Project 147. The audio is coded to MPEG-1 or MPEG-2 Layer II. The modulation method used is COFDM with DSPSK modulation.

Digital Multimedia Broadcasting Terrestrial (DMB-T)

Chinese standard for digital terrestrial television.

Digital Radio Mondiale

Digital standard for audio broadcasting in the medium- and short-wave bands. The audio signals are MPEG-4 AAC coded. The modulation method used is COFDM.

Digital Storage Media Command and Control (DSM-CC)

Private sections according to MPEG-2 which are used for the transmission of data services in object carousels or for datagrams such as IP packets in the MPEG-2 transport stream.

Digital Video Broadcasting (DVB)

The European DVB project stipulates methods and regulations for the digital transmission of TV signals. Abbreviations such as DVB-C (for transmission via cable), DVB-S (for transmission via satellite) and DVB-T (for terrestrial transmission) are frequently used as well.

Data Download Block (DDB)

Data transmission blocks of an object carousel, logically organized into modules.

Download Information Identification (DII)

Logical entry point into modules of an object carousel.

Download Server Initializing (DSI)

Logical entry point into an object carousel.

Elementary Stream (ES)

The elementary stream is a 'continuous' data stream for video, audio or user-specific data. The data originating from video and audio digitization are compressed by means of methods defined in MPEG-2 Video and MPEG-2 Audio.

Entitlement Control Messages (ECM)

ECM comprise information for the descrambler in the receiver of a CA system providing further details about the descrambling method.

Entitlement Management Messages (EMM)

EMM comprise information for the descrambler in the receiver of a CA system providing further details about the access rights of the customer to specific scrambled programs or broadcasts.

Event Information Table (EIT)

EIT is defined both as an SI table (DVB) and a PSIP table (ATSC). It provides information about program contents like a TV guide.

In DVB the EIT is transmitted in TS packets with PID 0x0012 and indicated by a table_ID from 0x4E to 0x6F. Depending on the table_ID, it contains different information:

Table_ID 0x4E	actual TS / present+following
Table_ID 0x4E	actual TS / present+following
Table_ID 0x4F	other TS / present+following
Table_ID 0x50...0x5F	actual TS / schedule
Table_ID 0x60...0x6F	other TS / schedule

EIT-0 to EIT-127 are defined in ATSC. Each of the EIT-k comprises information on program contents of a three-hour section where EIT-0 is the current time window. EIT-4 to EIT-127 are optional. Each EIT can be transmitted in a PID defined by the MGT with table_id 0xCB.

Extended Text Table (ETT)

ETT is a PSIP table (ATSC) and comprises information on a program (channel ETT) or on individual transmissions (ETT-0 to ETT-127) in the form of text. ETT-0 to ETT-127 are assigned to ATSC tables EIT-0 to EIT-127 and provide information on the program contents of a three-hour section. ETT-0 is with reference to the current time window, the other ETTs refer to later time sections. All ETTs are optional. Each ETT can be transmitted in a PID defined by MGT with table_id 0xCC.

Forward Error Correction (FEC)

Error protection in data transmission, channel coding.

Integrated Receiver Decoder (IRD)

The IRD is a receiver with integrated MPEG-2 decoder. A more colloquial expression would be set-top box.

Integrated Services Digital Broadcasting Terrestrial (ISDB-T)

Japanese standard for digital terrestrial television. The modulation method used is OFDM. The baseband signal is an MPEG-2 transport stream.

ITU-T J83

Collection of various standards for digital television over broadband cable.

J83A = DVB-C

J83B = North American standard for digital television over broadband cable (64QAM, 256QAM).

J83C = Japanese standard for digital television over broadband cable (6-MHz variant of DVB-C)

J83D = ATSC variant for digital television over broadband cable (16VSB); not used.

Low Density Parity Check Code

Block code used in the FEC of the DVB-S2 satellite transmission standard.

Low Voltage Differential Signalling (LVDS)

LVDS is used for the parallel interface of the transport stream. It is a positive differential logic. The difference voltage is 330 mV into 100 Ω.

Master Guide Table (MGT)

MGT is a reference table for all other PSIP tables (ATSC). It lists the version number, the table length and the PID for each PSIP table with the exception of the STT. MGT is always transmitted with a
Section in the PID 0x1FFB and indicated by the table_ID=0xC7.

Main Profile at Main Level (MP@ML)

MP@ML is a type of source coding for video signals. The profile determines the source coding methods that may be used while the level defines the picture resolution.

Megaframe Initialization Packet (MIP)

The MIP is transmitted with the PID of 0x15 in transport streams of terrestrial single frequency networks (SFNs) and is defined by DVB. The MIP contains timing information for GPS (Global Positioning System) and modulation parameters. Each megaframe contains exactly one MIP. A megaframe consists of n TS packets, n being dependent on the modulation parameters. The transmission period of a megaframe is about 0.5 seconds.

Moving Picture Experts Group (MPEG)

MPEG is an international standardization committee working on the coding, transmission and recording of (moving) pictures and sound.

MPEG-2

MPEG-2 is a standard consisting of three main parts and written by the Moving Picture Experts Group (ISO/IEC 13818). It describes the coding and compression of video (Part 2) and audio (Part 3) to obtain an elementary stream, as well as the multiplexing of elementary streams to form a transport stream (Part 1).

Multimedia Home Platform (MHP)

Program-associated DVB data service. HTML files and JAVA applications are broadcast via object carousels for MHP-enabled receivers and can then be started in the receiver.

Multimedia and Hypermedia Information Group (MHEG)

Program-associated data service in MPEG-2 transport streams, based on object carousels and HTML applications Broadcast in the UK as part of DVB-T.

Network Information Table (NIT)

The NIT is a PSI table (MPEG-2/DVB). It comprises technical data about the transmission network (eg orbit positions of satellites and transponder numbers). The NIT is transmitted in TS packets with PID 0x0010 and indicated by table_ID 0x40 or 0x41.

Null Packet

Null packets are TS packets with which the transport stream is filled to obtain a specific data rate. Null packets do not contain any payload and have the packet identity 0x1FFF. The continuity counter is undefined.

Orthogonal Frequency Division Multiplex (OFDM)

The modulation method is used in DVB systems for broadcasting transport streams with terrestrial transmitters. It is a multicarrier method and is suitable for the operation of single-frequency networks.

Packet Identity (PID)

The PID is a 13 bit value in the TS header. It shows that a TS packet belongs to a substream of the transport stream. A substream may contain a packetized elementary stream (PES), user-specific data, program specific information (PSI) or service information (SI). For some PSI and SI tables the associated PID values are predefined (see 1.3.6). All other PID values are defined in the PSI tables of the transport stream.

Packetized Elementary Stream (PES)

For transmission, the "continuous" elementary stream is subdivided into packets. In the case of video streams one frame constitutes the PES, whereas with audio streams, the PES is an audio frame which may represent an audio signal between 16 ms and 72 ms. Each PES packet is preceded by a PES header.

Payload

Payload signifies useful data in general. With reference to the transport stream all data except for the TS header and the adaptation field are payload. With reference to an elementary stream (ES) only the useful data of the ES without the PES header are payload.

Payload Unit Start Indicator

The payload unit start indicator is a 1 bit flag in the second byte of a TS header. It indicates the beginning of a PES packet or of a section of PSI or SI tables in the corresponding TS packet.

PCMCIA (PC Card)

PCMCIA is a physical interface standardized by the Personal Computer Memory Card International Association for the data exchange between computers and peripherals. A model of this interface is used for the common interface.

PCR Jitter

The value of a PCR refers to the exact beginning of a TS packet in which it is located. The reference to the 27 MHz system clock yields an accuracy of approx. ±20 ns. If the difference of the transferred values deviates from the actual difference of the beginning of the packets concerned, this is called PCR jitter. It can be caused, for example, by an inaccurate PCR calculation during transport stream multiplexing or by the subsequent integration of null packets on the transmission path without PCR correction.

PES Header

Each PES packet in the transport stream starts with a PES header. The PES header contains information for decoding the elementary stream. The presentation time stamp (PTS) and decoding time stamp (DTS) are of vital importance. The beginning of a PES header and thus also the beginning of a PES packet is indicated in the associated TS packet by means of the set payload unit start indicator. If the PES header is to be scrambled, it is scrambled at the transport stream level. It is not affected by scrambling at the elementary stream level.

PES Packet

The PES packet (not to be confused with TS packet) contains a transmission unit of a packetized elementary stream (PES). In a video stream, for example, this is a source-coded image. The length of a PES packet is normally limited to 64 kbytes. It may exceed this length only if a video image requires more capacity. Each PES packet is preceded by a PES header.

Plesiochronous Digital Hierarchy (PDH)

The Plesiochronous Digital Hierarchy was originally developed for the transmission of digitized voice calls. In this method, high-bit-rate transmission systems are generated by time-interleaving the digital signals of low-bit-rate subsystems. In PDH, the clock rates of the individual subsystems are allowed to fluctuate and these fluctuations are compensated for by appropriate stuffing methods. The PDH includes E3 and DS3, among others.

Presentation Time Stamp (PTS)

The PTS is a 33 bit value in the PES header and represents the output time of the contents of a PES packet. The value refers to the 33 most significant bits of the associated program clock reference. If the order of output does not correspond to the order of decoding, a decoding time stamp (DTS) is additionally transmitted. This is the case for video streams containing delta frames.

Program and System Information Protocol (PSIP)

PSIP is the summary of tables defined by ATSC for sending transmission parameters, program descriptions etc. They contain the structure defined by MPEG-2 systems for 'private' sections. The following tables exist:

Master Guide Table (MGT),
Terrestrial Virtual Channel Table (TVCT),
Cable Virtual Channel Table (CVCT),
Rating Region Table (RRT),
Event Information Table (EIT),
Extended Text Table (ETT),
System Time Table (STT).

Program Association Table (PAT)

The PAT is a PSI Table (MPEG-2). It lists all the programs contained in a transport stream and refers to the associated PMTs containing further information about the programs. The PAT is transmitted in TS packets with PID 0x0000 and indicated by table_ID 0x00.

Program Clock Reference (PCR)

The PCR is a 42-bit value contained in an adaptation field and helps the decoder to synchronize its system clock (27 MHz) to the clock of the encoder or TS multiplexer by means of PLL. In this case, the 33 most significant bits refer to a 90 kHz clock while the 9 least significant bits count from 0 to 299 and thus represent a clock of 300 x 90 kHz (= 27 MHz). Each program of a transport stream relates to a PCR which is transmitted in the adaptation field by TS packets with a specific PID. The presentation time stamps (PTS) and decoding time stamps (DTS) of all the elementary streams of a program refer to the 33 most significant bits of the PCR. PCRs have to be transmitted at intervals of max. 100 ms according to MPEG-2 and at intervals of max. 40 ms according to the DVB regulations.

Program Map Table (PMT)

The PMT is a PSI table (MPEG-2). The elementary streams (video, audio, data) belonging to the individual programs are described in a PMT. A PMT consists of one or several sections each containing information about a program. The PMT is transmitted in TS packets with a PID from 0x0020 to 0x1FFE (referenced in the PAT) and indicated in table_ID 0x02.

Program Stream (PS)

Like the transport stream, the program stream is a multiplex stream but only contains elementary streams for a program and is only suitable for the transmission in 'undisturbed' channels (e.g. recording in storage media).

Program Specific Information (PSI)

The four tables below defined by MPEG-2 are summed up as program specific information:

Program Association Table (PAT),
Program Map Table (PMT),
Conditional Access Table (CAT),
Network Information Table (NIT).

Quadrature Amplitude Modulation (QAM)

QAM is the modulation method used for transmitting a transport stream via cable. The channel coding is performed prior to QAM.

Quadrature Phase Shift Keying (QPSK)

QPSK is the modulation method used for transmitting a transport stream via satellite. The channel coding is performed prior to QPSK.

Rating Region Table (RRT)

The RRT is a PSIP table (ATSC). It comprises reference values for different geographical regions for the classification of transmissions (e.g. 'suitable for children older than X years'). RRT is transmitted with a section in the PID 0x1FFB and indicated by the table_ID=0xCA.

Running Status Table (RST)

The RST is an SI table (DVB) and contains status information about the individual broadcasts. It is transmitted in TS packets with PID 0x0013 and indicated by table_ID=0x71.

Section

Each table (PSI and SI) may comprise one or a number of sections. A section may have a length of up to 1 kbyte (for EIT and ST up to 4 Kbytes). Most of the tables have 4 bytes at the end of each section for the CRC.

Service Description Table (SDT)

The SDT is an SI table (DVB) and contains the names of programs and broadcasters. It is transmitted in TS packets with PID 0x0011 and indicated by table_ID 0x42 or 0x46.

Service Information (SI)

The following tables defined by DVB are called service information. They have the structure for 'private' sections defined by MPEG-2 systems:

Bouquet Association Table (BAT),
Service Description Table (SDT),
Event Information Table (EIT),
Running Status Table (RST),

Time and Date Table (TDT),
Time Offset Table (TOT).
Sometimes, the Program Specific Information (PSI) is also included.

Source Coding

The aim of source coding is data reduction by eliminating redundancy to the greatest possible extent whilst affecting the relevance in a video or audio signal as little as possible. The methods to be applied are defined in MPEG-2. They are the precondition for the bandwidth required for the transmission of digital signals being narrower than that for the transmission of analog signals.

Stuffing Table (ST)

The ST is an SI table (DVB). It has no relevant content and is obtained by overwriting tables that are no longer valid on the transmission path (eg at cable headends). It is transmitted in TS packets with a PID of 0x0010 to 0x0014 and indicated by table_ID 0x72.

Sync byte

The sync byte is the first byte in the TS header and thus also the first byte of each TS packet. Its value is 0x47.

Synchronous Digital Hierarchy (SDH)

The Synchronous Digital Hierarchy (SDH) is an international standard for the digital transmission of data in a uniform frame structure (containers). All bit rates of the PDH can be transmitted, like ATM, by means of SDH. Although SDH differs due to the pointer management, it is compatible with the American PDH and SONET standards.

Synchronous Optical NETwork (SONET)

The Synchronous Optical NETwork (SONET) is an American standard for the digital transmission of data in a uniform frame structure (containers). All bit rates of the PDH can be transmitted, like ATM, by means off SONET. SONET differs due to the pointer management and is thus not compatibel with the European SDH standard.

System Software Update (SSU)

Standardized system software update for DVB receivers according to ETSI TS102006.

System Target Decoder (STD)

The system target decoder describes the (theoretical) model for a decoder of MPEG-2 transport streams. A 'real' decoder has to fulfil all the conditions based on STD if it is to be guaranteed that the contents of all transport streams created to MPEG-2 are decoded error-free.

System Time Table (STT)

STT is a PSIP table (ATSC). It comprises date and time (UTC) as well as the local time difference. STT is transmitted in TS packets with the PID 0x1FFB and indicated by the table_ID 0xCD.

Table_ID

The table_identity defines the type of table (eg PAT, NIT, SDT, etc) and is always located at the beginning of a section of the table. The table_ID is necessary especially because different tables can be transmitted with one PID in one substream (eg BAT and SDT with PID 0x11, see Table 1-3).

Terrestrial Digital Multimedia Broadcasting (T-DMB)

South Korean standard for digital TV reception for mobile receivers, based on DAB and MPEG-4 AVC and AAC

Terrestrial Virtual Channel Table (TVCT)

TVCT is a PSIP table (ATSC) comprising the characteristic data of a program (eg channel number, frequency, modulation method) for terrestrial emission (transmission in cable → CVCT). TVCT is transmitted in TS packets with the PID of 0x1FFB and indicated by the table_id 0xC8.

Time and Date Table (TDT)

The TDT is an SI table (DVB) and contains date and time (UTC). It is transmitted in TS packets with PID 0x0014 and indicated by table_ID 0x70.

Time Offset Table (TOT)

The TOT is an SI table (DVB) and contains information about the local time offset in addition to date and time (UTC). It is transmitted in TS packets with PID 0x0014 and indicated by table_ID 0x73.

Transport Error Indicator

The transport error indicator is contained in the TS header and is the first bit after the sync byte (MSB of the second byte). It is set during channel decoding if channel decoding could not correct all the bit errors generated in the corresponding TS packet on the transmission path. As it is basically not possible to find the incorrect bits (e.g. the PID could also be affected), the errored packet must not be processed any further. The frequency of occurrence of a set transport error indicator is not a measure of the bit error rate on the transmission path. A set transport error indicator shows that the quality of the transmission path is not sufficient for an error-free transmission despite error control coding. A slight drop in transmission quality will quickly increase the frequency of occurrence of a set transport error indicator and finally transmission will cease.

Transport Stream (TS)

The transport stream is a multiplex data stream defined by MPEG-2 which may contain several programs that may consist of a number of elementary streams. A program clock reference (PCR) is carried along for each program. Multiplexing is done by forming TS packets for each elementary stream and by stringing together these TS packets originating from different elementary streams.

TS Header

The TS header is provided at the beginning of each TS packet and has a length of four bytes. The TS header always begins with the sync byte 0x47. Further important elements are the PID and the continuity counter. The TS header must never be scrambled when it is to be transmitted (see Conditional Access).

TS Packet

The transport stream is transmitted in packets of 188 bytes (204 bytes after channel coding). The first four bytes form the TS header which is followed by the 184 payload bytes.

Vestigial Sideband Modulation (VSB)

The vestigial sideband amplitude modulation method is used in ATSC systems. For terrestrial transmission, 8VSB with 8 amplitude levels is used while 16VSB is mainly for cable transmission.

TV Channel Tables

The channels listed in the following tables are possible examples for analog television and for DVB-C, DVB-T, ATSC and J83B.

Analog TV:
Vision carrier at 7 MHz bandwidth 2.25 MHz below center frequency,
vision carrier at 8 MHz bandwidth 2.75 MHz below center frequency,
vision carrier at 6 MHz bandwidth 1.75 MHz below center frequency.

ATSC:
Pilot carrier at ATSC (6 MHz bandwidth) 2.69 MHz below center frequency.

Europe, Terrestrial and Cable

Table 45.1. TV channel occupancy, Europe

Channel	Band	Center frequency [MHz]	Bandwidth [MHz]	Remarks
2	VHF I	50.5	7	
3	VHF I	57.5	7	
4	VHF I	64.5	7	
	VHF II			FM 87.5...108.0 MHz
5	VHF III	177.5	7	
6	VHF III	184.5	7	
7	VHF III	191.5	7	
8	VHF III	198.5	7	
9	VHF III	205.5	7	
10	VHF III	212.5	7	
11	VHF III	219.5	7	
12	VHF III	226.5	7	
S 1	special channel	107.5	7	not in use (FM)

S 2	special channel	114.5	7	cable, midband
S 3	special channel	121.5	7	cable, midband
S 4	special channel	128.5	7	cable, midband
S 5	special channel	135.5	7	cable, midband
S 6	special channel	142.5	7	cable, midband
S 7	special channel	149.5	7	cable, midband
S 8	special channel	156.5	7	cable, midband
S 9	special channel	163.5	7	cable, midband
S 10	special channel	170.5	7	cable, midband
S 11	special channel	233.5	7	cable, superband
S 12	special channel	240.5	7	cable, superband
S 13	special channel	247.5	7	cable, superband
S 14	special channel	254.5	7	cable, superband
S 15	special channel	261.5	7	cable, superband
S 16	special channel	268.5	7	cable, superband
S 17	special channel	275.5	7	cable, superband
S 18	special channel	282.5	7	cable, superband
S 19	special channel	289.5	7	cable, superband
S 20	special channel	296.5	7	cable, superband
S 21	special channel	306	8	cable, hyperband
S 22	special channel	314	8	cable, hyperband
S 23	special channel	322	8	cable, hyperband
S 24	special channel	330	8	cable, hyperband
S 25	special channel	338	8	cable, hyperband
S 26	special channel	346	8	cable, hyperband
S 27	special channel	354	8	cable, hyperband
S 28	special channel	362	8	cable, hyperband
S 29	special channel	370	8	cable, hyperband
S 30	special channel	378	8	cable, hyperband
S 31	special channel	386	8	cable, hyperband
S 32	special channel	394	8	cable, hyperband
S 33	special channel	402	8	cable, hyperband
S 34	special channel	410	8	cable, hyperband
S 35	special channel	418	8	cable, hyperband
S 36	special channel	426	8	cable, hyperband
S 37	special channel	434	8	cable, hyperband
S 38	special channel	442	8	cable, hyperband
S 39	special channel	450	8	cable, hyperband
S40	special channel	458	8	cable, hyperband
S41	special channel	466	8	cable, hyperband
21	UHF IV	474	8	
22	UHF IV	482	8	
23	UHF IV	490	8	
24	UHF IV	498	8	
25	UHF IV	506	8	
26	UHF IV	514	8	

27	UHF IV	522	8
28	UHF IV	530	8
29	UHF IV	538	8
30	UHF IV	546	8
31	UHF IV	554	8
32	UHF IV	562	8
33	UHF IV	570	8
34	UHF IV	578	8
35	UHF IV	586	8
36	UHF IV	594	8
37	UHF IV	602	8
38	UHF V	610	8
39	UHF V	618	8
40	UHF V	626	8
41	UHF V	634	8
42	UHF V	642	8
43	UHF V	650	8
44	UHF V	658	8
45	UHF V	666	8
46	UHF V	674	8
47	UHF V	682	8
48	UHF V	690	8
49	UHF V	698	8
50	UHF V	706	8
51	UHF V	714	8
52	UHF V	722	8
53	UHF V	730	8
54	UHF V	738	8
55	UHF V	746	8
56	UHF V	754	8
57	UHF V	762	8
58	UHF V	770	8
59	UHF V	778	8
60	UHF V	786	8
61	UHF V	794	8
62	UHF V	802	8
63	UHF V	810	8
64	UHF V	818	8
65	UHF V	826	8
66	UHF V	834	8
67	UHF V	842	8
68	UHF V	850	8
69	UHF V	858	8

Australia, Terrestrial

Table 45.2. TV terrestrial channel occupancy, Australia (terrestrial)

Channel	Band	Center frequency [MHz]	Bandwidth [MHz]	Remarks
0	VHF I	48.5	7	
1	VHF I	59.5	7	
2	VHF I	66.5	7	"ABC Analog" Sydney
3	VHF II	88.5	7	
4	VHF II	97.5	7	
5	VHF II	104.5	7	
5A	VHF II	140.5	7	
6	VHF III	177.5	7	s.t. "Seven Digital"
7	VHF III	184.5	7	s.t. "Seven Analog"
8	VHF III	191.5	7	s.t. "Nine Digital"
9	VHF III	198.5	7	s.t. "Nine Analog"
9A	VHF III	205.5	7	
10	VHF III	211.5	7	s.t. "Ten Analog"
11	VHF III	219.5	7	s.t. "Ten Digital"
12	VHF III	226.5	7	s.t. "ABC Digital"
28	UHF IV	529.5	7	"SBS Analog" Sydney
29	UHF IV	536.5	7	
30	UHF IV	543.5	7	
31	UHF IV	550.5	7	
32	UHF IV	557.5	7	
33	UHF IV	564.5	7	
34	UHF IV	571.5	7	"SBS Digital" Sydney
35	UHF IV	578.5	7	
36	UHF V	585.5	7	
37	UHF V	592.5	7	
38	UHF V	599.5	7	
39	UHF V	606.5	7	
40	UHF V	613.5	7	
41	UHF V	620.5	7	
42	UHF V	627.5	7	
43	UHF V	634.5	7	
44	UHF V	641.5	7	
45	UHF V	648.5	7	

46	UHF V	655.5	7	
47	UHF V	662.5	7	
48	UHF V	669.5	7	
49	UHF V	676.5	7	
50	UHF V	683.5	7	
51	UHF V	690.5	7	
52	UHF V	697.5	7	
53	UHF V	704.5	7	
54	UHF V	711.5	7	
55	UHF V	718.4	7	
56	UHF V	725.5	7	
57	UHF V	732.5	7	
58	UHF V	739.5	7	
59	UHF V	746.5	7	
60	UHF V	753.5	7	
61	UHF V	760.5	7	
62	UHF V	767.5	7	
63	UHF V	774.5	7	
64	UHF V	781.5	7	
65	UHF V	788.5	7	
66	UHF V	795.5	7	
67	UHF V	802.5	7	
68	UHF V	809.5	7	
69	UHF V	816.5	7	

North America, Terrestrial

Table 45.3. TV terrestrial channel occupancy, North America

Channel	Band	Center frequency [MHz]	Bandwidth [MHz]	Remarks
2	VHF	57	6	
3	VHF	63	6	
4	VHF	69	6	
5	VHF	79	6	
6	VHF	85	6	
7	VHF	177	6	
8	VHF	183	6	
9	VHF	189	6	
10	VHF	195	6	
11	VHF	201	6	
12	VHF	207	6	

13	VHF	213	6
14	UHF	473	6
15	UHF	479	6
16	UHF	485	6
17	UHF	491	6
18	UHF	497	6
19	UHF	503	6
20	UHF	509	6
21	UHF	515	6
22	UHF	521	6
23	UHF	527	6
24	UHF	533	6
25	UHF	539	6
26	UHF	545	6
27	UHF	551	6
28	UHF	557	6
29	UHF	563	6
30	UHF	569	6
31	UHF	575	6
32	UHF	581	6
33	UHF	587	6
34	UHF	593	6
35	UHF	599	6
36	UHF	605	6
37	UHF	611	6
38	UHF	617	6
39	UHF	623	6
40	UHF	629	6
41	UHF	635	6
42	UHF	641	6
43	UHF	647	6
44	UHF	653	6
45	UHF	659	6
46	UHF	665	6
47	UHF	671	6
48	UHF	677	6
49	UHF	683	6
50	UHF	689	6
51	UHF	695	6
52	UHF	701	6
53	UHF	707	6
54	UHF	713	6
55	UHF	719	6
56	UHF	725	6
57	UHF	731	6
58	UHF	737	6

59	UHF	743	6	
60	UHF	749	6	
61	UHF	755	6	
62	UHF	761	6	
63	UHF	767	6	
64	UHF	773	6	
65	UHF	779	6	
66	UHF	785	6	
67	UHF	791	6	
68	UHF	797	6	
69	UHF	803	6	
70	UHF	809	6	
71	UHF	815	6	
72	UHF	821	6	
73	UHF	827	6	
74	UHF	833	6	
75	UHF	839	6	
76	UHF	845	6	
77	UHF	851	6	
78	UHF	857	6	
79	UHF	863	6	
80	UHF	869	6	
81	UHF	875	6	
82	UHF	881	6	
83	UHF	887	6	

North America, Cable

Especially in the cable the current occupancy can not be guaranteed.

Table 45.4. Channel occupancy North America, Cable

Channel	Band	Center frequency [MHz]	Bandwidth	Remarks
2		57	6	
3		63	6	
4		69	6	
5		79	6	
6		85	6	
7		177	6	
8		183	6	
9		189	6	

10	195	6
11	201	6
12	207	6
13	213	6
14	123	6
15	129	6
16	135	6
17	141	6
18	147	6
19	153	6
20	159	6
21	165	6
22	171	6
23	219	6
24	225	6
25	231	6
26	237	6
27	243	6
28	249	6
29	255	6
30	261	6
31	267	6
32	273	6
33	279	6
34	285	6
35	291	6
36	297	6
37	303	6
38	309	6
39	315	6
40	321	6
41	327	6
42	333	6
43	339	6
44	345	6
45	351	6
46	357	6
47	363	6
48	369	6
49	375	6
50	381	6
51	387	6
52	393	6
53	399	6
54	405	6
55	411	6

56	417	6
57	423	6
58	429	6
59	435	6
60	441	6
61	447	6
62	453	6
63	459	6
64	465	6
65	471	6
66	477	6
67	483	6
68	489	6
69	495	6
70	501	6
71	507	6
72	513	6
73	519	6
74	525	6
75	531	6
76	537	6
77	543	6
78	549	6
79	555	6
80	561	6
81	567	6
82	573	6
83	579	6
84	585	6
85	591	6
86	597	6
87	603	6
88	609	6
89	615	6
90	621	6
91	627	6
92	633	6
93	639	6
94	645	6
95	93	6
96	99	6
97	105	6
98	111	6
99	117	6
100	651	6
101	657	6

102	663	6
103	669	6
104	675	6
105	681	6
106	687	6
107	693	6
108	699	6
109	705	6
110	711	6
111	717	6
112	723	6
113	729	6
114	735	6
115	741	6
116	747	6
117	753	6
118	759	6
119	765	6
120	771	6
121	777	6
122	783	6
123	789	6
124	795	6
125	801	6
126	807	6
127	813	6
128	819	6
129	825	6
130	831	6
131	837	6
132	843	6
133	849	6
134	855	6
135	861	6
136	867	6
137	873	6
138	879	6
139	885	6
140	891	6
141	897	6
142	903	6
143	909	6
144	915	6
145	921	6
146	927	6
147	933	6

148	939	6
149	945	6
150	951	6
151	957	6
152	963	6
153	969	6
154	975	6
155	981	6
156	987	6
157	993	6
158	999	6

Europe, Satellite

Fig. 45.1. shows the occupancy of the Ku band for direct broadcasting TV satellite–reception.

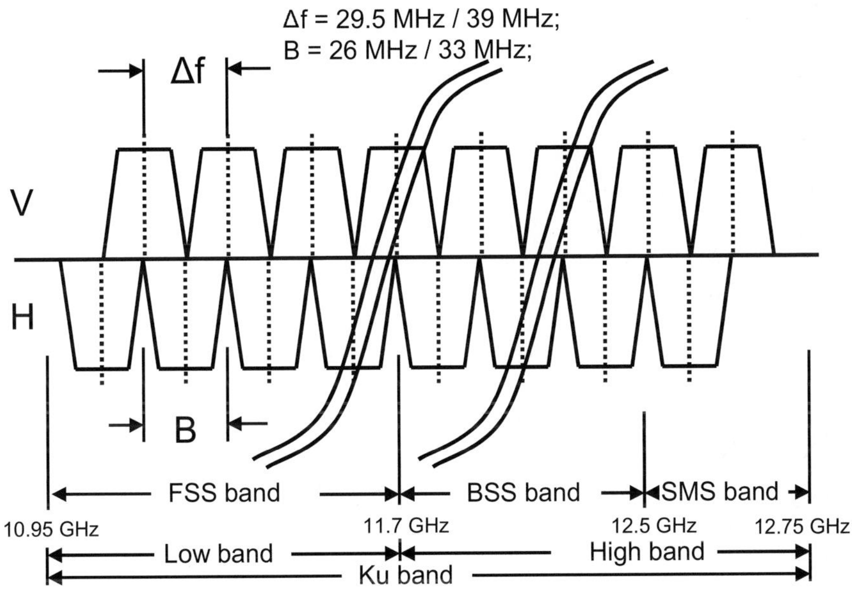

Fig. 35.1. Ku band for direct broadcasting TV satellite-reception

Bibliography:

Index

Typical Test Instruments and Broadcasting Systems for TV Signals

R&S®SFU, R&S®SFE and R&S®SFE100 Broadcast Test Transmitters

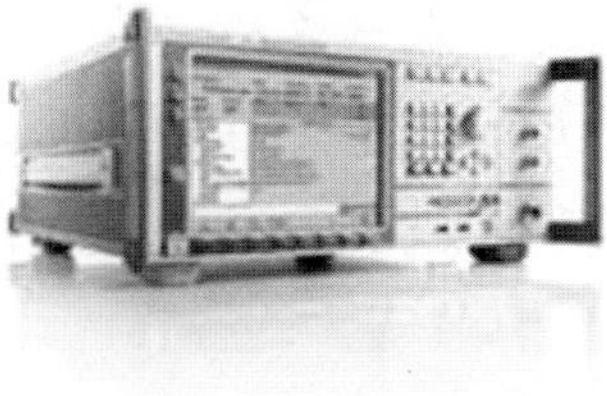

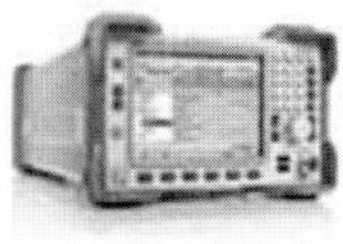

Test transmitters provide reference RF signals for testing digital receivers. The intentional degradation of the ideal signal by superimposed noise as well as the simulation of mobile reception scenarios help to make the receivers operational under any conditions and to ensure the interference-free reception of TV programs.

R&S®DVSG Digital Video Signal Generator

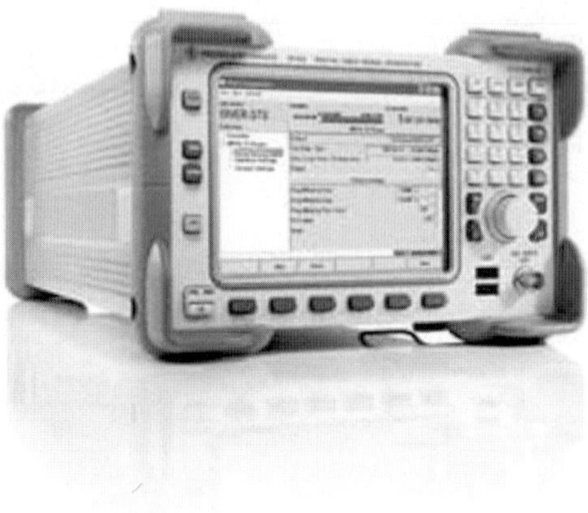

Comprehensive tests are required to assess the functioning of the picture processing unit of a TV set or the quality of built in display panels: All resolution, timing and color depth modes must be verified. To make sure that the quality of the display unit under test is assessed, the test signal source must satisfy extremely high quality requirements. Digital video signal generators are reference signal sources for development and quality assurance applications of latest-generation TV sets and projectors.

R&S®ETL TV Analyzer and R&S®ETH Handheld TV

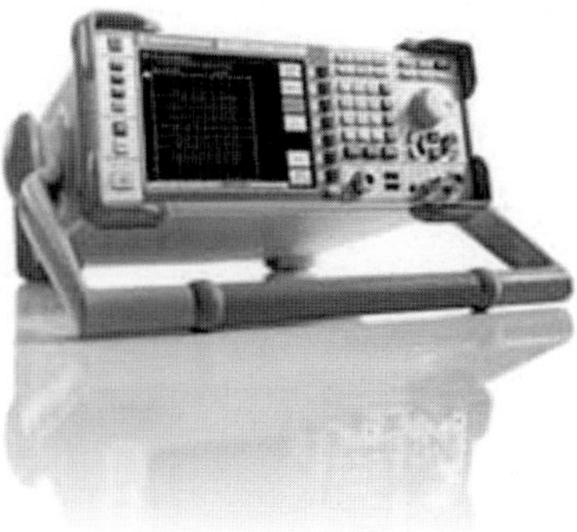

Analyzers

TV analyzers measure high-precision RF signals of both analog and digital TV systems. Measurements performed directly on an antenna, a TV transmitter, or cable headend make it possible to clearly assess the quality of digital transmission. Error sources can be identified and specifically eliminated.

Monitoring, Analysis, Recording, and Generation of MPEG Transport Streams with the R&S®DVM Family

An MPEG transport stream is modified at many places in the transmission chain. For example, after the satellite signal has arrived at a cable headend, various programs are taken from the transport stream multiplex and replaced by local programs. This is a deep intervention in the transport stream structure. MPEG analyzers check the entire MPEG protocol syntax and indicate any errors and discrepancies, thus ensuring the secure transmission of signals.

Transmitters, Transposers and Gap Fillers for ATV, DTV and Mobile TV from 1 W to 60 kW

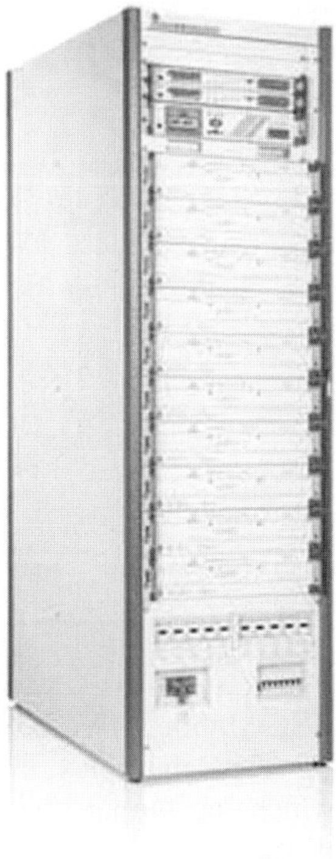

TV transmitters are one of the most important components of terrestrial transmitter networks. They convert the MPEG transport stream signal into high-quality RF signals of different power classes from low power to high power. These signals are then distributed via antennas. Reliability, small footprint, and high efficiency are key parameters for securely providing viewers with digital TV signals via antenna.

R&S AEM100 Multiplexer for ATSC Mobile DTV

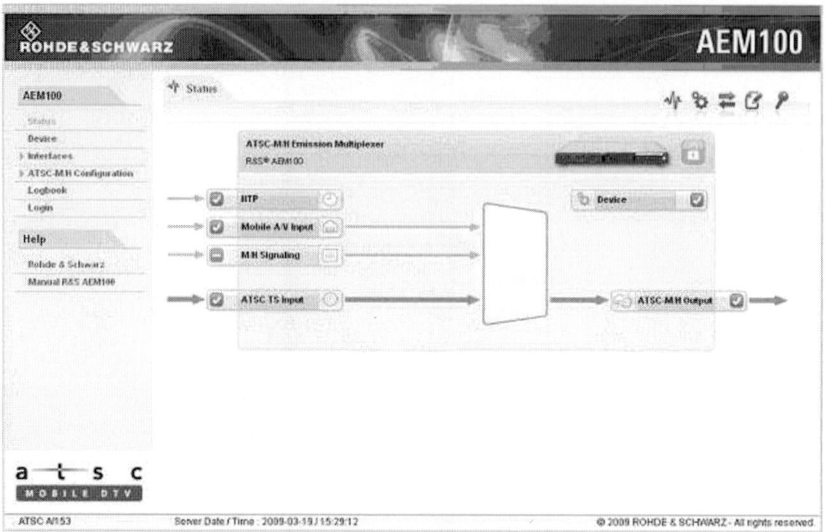

As a key component for ATSC Mobile DTV, the R&S AEM100 inserts the mobile data into the existing ATSC multiplex. It generates the required robust transport stream, which is output via ASI or Ethernet. The R&S AEM100 includes all of the required functions for ATSC Mobile DTV. The complete functionality can be configured, controlled and monitored using a web browser locally or remotely. All commands for automatic monitoring and for the instrument settings are available over an SNMP interface.

Printing and Binding: Stürtz GmbH, Würzburg